Lecture Notes in Intelligent Transportation and Infrastructure

The series "Lecture Notes in Intelligent Transportation and Infrastructure" (LNITI) publishes new developments and advances in the various areas of intelligent transportation and infrastructure. The intent is to cover the theory, applications, and perspectives on the state-of-the-art and future developments relevant to topics such as intelligent transportation systems, smart mobility, urban logistics, smart grids, critical infrastructure, smart architecture, smart citizens, intelligent governance, smart architecture and construction design, as well as green and sustainable urban structures. The series contains monographs, conference proceedings, edited volumes, lecture notes and textbooks. Of particular value to both the contributors and the readership are the short publication timeframe and the world-wide distribution, which enable wide and rapid dissemination of high-quality research output.

More information about this series at https://link.springer.com/bookseries/15991

Olegas Prentkovskis · Irina Yatskiv (Jackiva) ·
Paulius Skačkauskas · Raimundas Junevičius ·
Pavlo Maruschak
Editors

TRANSBALTICA XII: Transportation Science and Technology

Proceedings of the 12th International Conference TRANSBALTICA, September 16–17, 2021, Vilnius, Lithuania

Springer

Editors
Olegas Prentkovskis
Vilnius Gediminas Technical University
Vilnius, Lithuania

Irina Yatskiv (Jackiva)
Transport and Telecommunication Institute
Riga, Latvia

Paulius Skačkauskas
Vilnius Gediminas Technical University
Vilnius, Lithuania

Raimundas Junevičius
Vilnius Gediminas Technical University
Vilnius, Lithuania

Pavlo Maruschak
Ternopil Ivan Puluj National
Technical University
Ternopil, Ukraine

ISSN 2523-3440 ISSN 2523-3459 (electronic)
Lecture Notes in Intelligent Transportation and Infrastructure
ISBN 978-3-030-94773-6 ISBN 978-3-030-94774-3 (eBook)
https://doi.org/10.1007/978-3-030-94774-3

This Springer imprint is published by the registered company Springer Nature Switzerland AG
The registered company address is: Gewerbestrasse 11, 6330 Cham, Switzerland

Preface

The international scientific conference "Transbaltica: Transportation Science and Technology" is a traditional, bi-annual event of the Faculty of Transport Engineering in Vilnius Gediminas Technical University (VILNIUS TECH), organized in cooperation with partners since 2001. The twelfth edition of "Transbaltica XII: Transportation Science and Technology" was held on September 16–17, 2021, remotely from Lithuania. Authors from 20 countries presented their research, covering a number of scientific problems within the research field of transport engineering, transportation and logistics, as well as other disciplines and interdisciplinary areas related to transport system.

The proceedings of Transbaltica 2021 has been organized in five parts, featuring its main areas of interest:

1) **Vehicle Engineering and Dynamics** discusses timely issues in vehicle modeling and simulations, vehicle safety systems and railway transport, and relating technologies and applications
2) **Engine Combustion, Energy Management and Emissions** provides useful information regarding alternative fuels, fuel mixtures, combustion process control methods, and efficient use of energy.
3) **Logistics and Transportation** describes most recent research trends regarding green logistics, supply chain connectivity, computational logistics, as well as new circumstances affecting the carriage of goods and passengers, and possible solutions to them.
4) **Intelligent Vehicles and Infrastructure** covers highly automated and autonomous driving and infrastructure for autonomous and connected vehicles.
5) **Human–Machine Interaction, Behavioural Sciences** covers human factors in transportation and research problems relating to driver monitoring, driver behavior, as well as other road users behavior.

The program committee of the international scientific conference "Transbaltica XII: Transportation Science and Technology", the organizers, and the editors of these proceedings would like to acknowledge all reviewers who helped evaluate conference submissions and refine contents of this volume. Moreover, our thanks go also to all program committee members:

- Prof. Olegas Prentkovskis, Vilnius Gediminas Technical University, Lithuania—Chairman
- Prof. Raimundas Junevičius, Vilnius Gediminas Technical University, Lithuania—Chairman
- Prof. Vidas Žuraulis, Vilnius Gediminas Technical University, Lithuania—Co-Chairmen
- Dr. Paulius Skačkauskas, Vilnius Gediminas Technical University, Lithuania—Secretary
- Darius Aleksonis, Vilnius Gediminas Technical University, Lithuania—Secretary
- Prof. Darius Bazaras, Vilnius Gediminas Technical University, Lithuania
- Prof. Edgar Sokolovskij, Vilnius Gediminas Technical University, Lithuania
- Prof. Zdenek Dvorak, University of Žilina, Slovakia
- Prof. Marianna Jacyna, Warsaw University of Technology, Poland
- Prof. Jolanta Janutėnienė, Klaipeda University, Lithuania
- Prof. Artūras Keršys, Kaunas University of Technology, Lithuania
- Prof. Pavlo Maruschak, Ternopil Ivan Puluj National Technical University, Ukraine
- Prof. Irina Yatskiv (Jackiva), Transport and Telecommunication Institute, Latvia
- Dr. Giedrius Garbinčius, Vilnius Gediminas Technical University, Lithuania
- Dr. Saugirdas Pukalskas, Vilnius Gediminas Technical University, Lithuania
- Dr. Metin Mutlu Aydin, Ondokuz Mayıs University, Turkey
- Dr. Olja Čokorilo, University of Belgrade, Serbia
- Dr. Dalibor Barta, University of Žilina, Slovakia
- Dr. Rafal Burdzik, Silesian University of Technology, Poland
- Dr. Viktoriia Ivannikova, National Aviation University, Ukraine
- Dr. Ilya Jackson, Transport and Telecommunication Institute, Latvia
- Dr. Rolandas Makaras, Kaunas University of Technology, Lithuania
- Dr. Josep Maria Salanova Grau, Center for Research and Technology Hellas (CERTH)—Hellenic Institute of Transport, Greece
- Dr. Yong-Gang Wang, Chang'an University, China
- Dr. Paulius Yamin Slotkus, Vilnius Gedimintas Technical University, Lithuania
- Dr. Tadeusz Szymczak, Motor Transport Institute, Poland

We acknowledge all the authors who have chosen "Transbaltica XII: Transportation Science and Technology" as the publication platform for their research and would like to express our hope that their papers will foster further developments in the design and analysis of complex transport systems, offering a valuable and timely resource for scientists, researchers, practitioners, and students who work in all the areas mentioned above.

Olegas Prentkovskis
Irina Yatskiv (Jackiva)
Paulius Skačkauskas
Raimundas Junevičius
Pavlo Maruschak

Contents

Vehicle Engineering and Dynamics

Study of Air Flow Around a Moving Vehicle as a Source of Energy

Dalibor Barta(✉), Vladimir Pavelcik, and Milos Brezani

University of Žilina, Univerzitna, 8215/1, 010 26 Žilina, Slovakia
{dalibor.barta, vladimir.pavelcik, milos.brezani}@fstroj.uniza.sk

Abstract. The growing number of vehicles is related to the overall growth in the consumption of fuel from fossil fuels or electricity, so new ways are being sought to increase the efficiency of propulsion or the recovery of energy from driving. This article deals with the possibility of recovering kinetic energy from the movement of vehicles, specifically from the air flowing around a passing car. The subject of measurements and simulations were parameters influencing the amount of recovered energy such as the location of the wind turbine near the road, its distance from the side of the moving vehicle, turbine height from the road, speed and size of the vehicle front (car, bus, truck). The measured values were compared with the values from the simulations performed in the program Fluent, and it was shown that in passenger cars the highest air flow is 0.5–1 m above the road and the closest possible to vehicles. With regard to feasibility, the smallest distance was considered to be 0.5 m from the side of the vehicle. Measurements in real traffic showed delay of the air flow behind the vehicle by 1.5–3 s depending on the size of the vehicle and the need for sufficient traffic density. The high intensity of vehicles in urban and suburban areas predestines them to deploy this system. Conversely, in rail transport, given the low frequency of vehicles, the use of this system is unjustified.

Keywords: Vehicle · Airflow · Energy recuperation · Vehicle distance

1 Introduction

Modern society is very energy intensive. It makes extensive use of fossil fuels in transport, and therefore considerable pressure is being exerted to increase the efficiency of propulsion and, if possible, to reuse propulsion energy. In order to reduce the amount of energy required, research into new, lighter materials is underway [1] and the driving characteristics of vehicles are being optimized [2]. In internal combustion engines, where only about one third of the energy stored in the fuel is converted into efficient engine power, the rest goes mainly to the cooling and exhaust system [5], conventional fossil fuels, whose combustion products have a serious impact on the environment, are replaced by alternative [3, 4]. In order to meet the ever-tightening emission standards, several strategies have been developed over time. One of them is the change of the type of propulsion unit from an internal combustion engine to an electric motor, the efficiency of which is much higher than the efficiency of the internal

O. Prentkovskis et al. (Eds.): TRANSBALTICA 2021, LNITI, pp. 3–15, 2022.
https://doi.org/10.1007/978-3-030-94774-3_1

combustion engine and in which no gaseous emissions are produced locally. The advantages of both types of drives, sufficient range, the ability to achieve local zero emissions, or the ability to recover energy from braking, are provided by their combination in hybrid vehicles [6–8]. According to Jinglai et al. [9], in the hybrid drives the electric motors with internal combustion engines can be connected to wheels in different ways, some of which are more efficient than others.

There are many ways to recover energy in vehicles. Some of them are described below. Closer attention will be paid to the possibilities of obtaining energy from the air flowing around the oncoming vehicle and its conversion into electrical energy. Not much attention has been paid to this topic so far, so CFD simulations and real measurements in operation were performed, which showed what has a significant impact on the amount of energy obtained.

2 Energy Recovery

2.1 Braking Energy Recovery

Various energy recovery systems have been designed to make more efficient use of the energy supplied in the fuel and thus to further reduce the total energy wasted by the vehicle. These systems are based on different physical principles (Fig. 1), which aim to convert as much of the kinetic energy of the vehicle as possible to some other usable form instead of converting it to heat in the disc brake.

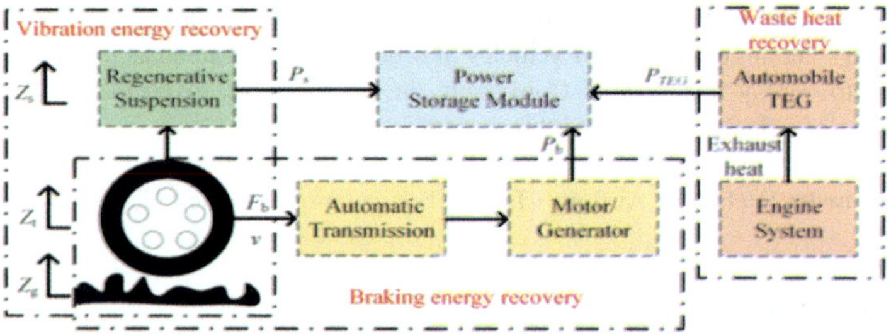

Fig. 1. General architecture of the comprehensive vehicle energy recovery system [10].

Electrical system of recuperation uses the drive motor itself switched to the generator mode. The generator then feeds the battery with the resulting current. This braking effect can be used only on driving wheels. Guo et al. [11] discussed the importance and effect of strategy of braking force distribution in such case. With the progress in optimization theories, more braking energy changing algorithms (intelligent neuro-fuzzy systems, such as proposed by Medvedev and Rudakov [12]) have been developed in order to maximize recuperated energy. On top of algorithms, whole strategies have been developed, based on the vehicle mass and road parameters (Boisvert et al. [13]) or based on the number of driven wheels (four-wheel independent drive strategy proposed by Li et al. [14]). Energy recovery strategy based on pedals positions of electric vehicle suggested by JI et al. [15] extended driving range of electric vehicle by more than 20% when compared with the mode of no energy recovery. A genetic-fuzzy control method proposed by Liu et al. [16] was proven to increase the recoverable energy ratio by 2.7% compared to model before optimization.

The electric recuperation system is the most common recuperation system used in vehicles. It uses the drive motor itself switched to generator mode for deceleration. The generator generates braking torque in the production of electricity. The obtained electrical energy then powers the battery. This braking effect can only be used on the drive wheels. Guo et al. [11] discussed the importance and effect of a brake force distribution strategy in such a case. With advances in optimization theories, several algorithms have been developed to change the braking energy (intelligent neurofuzzy systems as proposed by Medvedev and Rudakov [12]) in order to maximize the energy recovered. In addition to algorithms, whole strategies have been developed based on mass and traction parameters (Boisvert et al. [13]) or on the number of driven wheels (an independent all-wheel drive strategy proposed by Li et al. [14]). An energy recovery strategy based on the pedal positions of an electric vehicle proposed by JI et al. [15] can extend the range of an electric vehicle by more than 20% compared to the mode without energy recovery. The genetic fuzzy control method proposed by Liu et al. [16] demonstrably increased the usable energy ratio by 2.7% compared to the model before optimization.

Mechanical system of recuperation uses flywheels in two configurations. First of them is purely mechanical, where the kinetic energy is transferred from wheel into flywheel during braking and is returned to wheel during acceleration. Second one is electro-mechanical, where energy from wheels is converted in generator/motor into electric energy, but instead of battery, electricity is stored via motor/generator in flywheel. Flywheel can be used in vehicle without electric engine as well as in hybrid vehicles, where energy accumulated in flywheel can be converted into electricity and stored in battery. Summary of energy recovery technology was made in detail by Fahlbeck and Kindahl [17]. Minimizing losses in flywheel system is a matter of decreasing friction and storing the flywheel in vacuum.

Hydraulic recuperation is based on accumulating energy in pressurized gas. The gas is pressurized by hydraulic pump pumping hydraulic liquid into bladder accumulator during braking. Energy is then used to propel the car with hydraulic engine powered by expanding gas. There are more concepts, however, mostly operating on the same principle. The main disadvantage of this system is its weight and dimensions, thus its useability in casual cars is limited. On the contrary, according to Schwalbe et al. [18], strongly influenced by the velocity profile, the energy consumption savings can reach up to 74% for harmonic sinusoidal profile.

2.2 Waste Heat Recovery

In internal combustion engine, about 30% of energy stored in fuel is exhausted through exhaust pipe. Some part of this energy can be recuperated using thermoelectric materials. Thermoelectric generator is made of couples of thermoelectric legs thermally connected in parallel and electrically in series. The generated electrical power is given by the amount of heat and effectivity of the generator. Durand et al. [19] have shown in the study that the amount of heat is directly affected by position of the generator (Fig. 2) (the energy generated after the exhaust of after-treatment system (ATS) is 5 times lower when compared to the position before turbocharger). At higher engine loads, about 150 W can be expected to be generated this way.

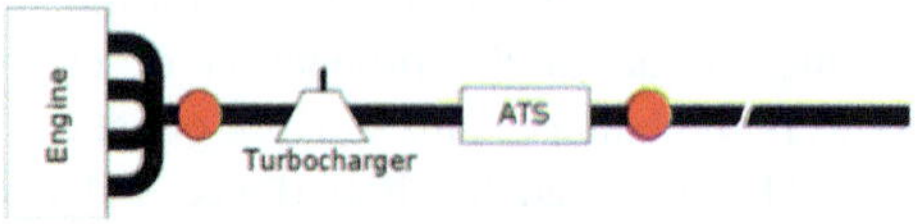

Fig. 2. Investigated locations of thermoelectric generators [19].

2.3 Vibration Energy Recovery

Because damping is essential part of any vehicle, there have been some experiments with collecting as much of the energy wasted in dampers as possible. At present, active shock absorbers are already in use, which make it possible to obtain energy when the vehicle passes over uneven ground. Caban et al. [20], for example, dealt with the issue of possibilities of energy harvesting from the engine mount damping system of the vehicle. For instance, rack-pinion damper aims to convert damped motion into rotation of motor/generator via gear rack and gear box, however, this system is subject to greater losses in efficiency. Average power output of this system was 19 W per damper when tested on relatively smooth road [21].

There are other recuperation concepts using damping, such as damping of linear movements, hydraulic electromagnetic damping, damping of ball screws, etc. [17], but a detailed overview of them is beyond the scope of this article.

2.4 Energy Recovery on the Road

Zahibi and Saafi [22] reviewed current possibilities of collecting energy from road infrastructure. In the article, piezoelectric, thermoelectric, photovoltaic, and electromagnetic energy harvesters are compared using method proposed by Clausing and Holmes [23]. According to their results, electromagnetic technology tops all other in power output, however, it influences the traffic flow. Photovoltaic systems power output is highly dependent on weather conditions and its initial costs as well as maintenance are relatively high. Both thermoelectric and piezoelectric systems are easy to install, but they provide very low power output at expensive initial costs.

3 Wind Energy Recovery

As has already been mentioned, energy is consumed to accelerate and move the vehicle. However, the vehicle acts on its surroundings and, for example, sets the ambient air in motion. The vehicle transmits kinetic energy into the air, which can later be captured by wind turbines and used to generate electricity. In order to design a wind turbine to convert energy from traffic and to evaluate the velocity of the air that drives the turbine, a CFD simulation was performed.

In order to know the real air velocities at different locations near the oncoming vehicle, experiments were performed in real traffic. The experiments also made it possible to determine the approximate distance between successive vehicles, i.e. the frequency of vehicles at which the rotation of the turbine would not be stopped. The measurements were made in two locations in and around Žilina city, with the speed limits of 70 and 90 km•h^{-1} respectively, generally well representing ratio of number of smaller cars to number of SUVs and of trucks and busses. Both mechanical and electrical anemometers intended for meteorological purposes were used to measure the speed of the air and its direction. Horizontal distance was measured between anemometer and vehicle's right wheels.

The results in Table 1 show that smaller vehicles tend to drive more from the roadside than larger vehicles. However, the difference is roughly correlated with the difference in the average widths of these vehicle types. The values of the distance of the vehicle from the measuring point were influenced by the fact that drivers have a natural tendency to avoid a place on the road where a person is located (measuring point). However, it can be assumed that the wind turbine itself located near the track will not affect the behavior of the drivers and thus it will be possible to place it closer to moving vehicles.

Table 1. Comparison of distance between various vehicle types and the anemometer.

Vehicle type	Average measured distance [m]
Smaller cars	1.99
SUV	1.77
Trucks, buses	1.65
Total average	1.8

A CFD simulation was performed in the ANSYS Fluent program, the aim of which was to reveal the direction of air flow around the vehicle after passing the measuring point, so that the turbine could be positioned as best as possible in terms of direction and speed of air flow. A coordinate system was used where the positive values on the x-axis are in the road plane and against the direction of travel, the y-axis is perpendicular to the x-axis in the road plane and the z-axis is vertical. The starting point of the coordinate system is at the rear of the vehicle. According to Hamat et al. [24], although the situation is a complex transition problem, considerable simplification can be made to reduce computational time. The simulation settings are summarized in Table 2.

The model of the vehicle was derived from a CAD model of Opel Zafira. Because rear view mirrors contribute to aerodynamic drag of the vehicle in considerable way, (causing turbulences), the velocity magnitudes are expected to be higher in reality.

Table 2. Simulation settings.

Geometry	Simplified CAD model of Opel Zafira (without wheels and rear-view mirrors)
Fluid domain dimensions	Distance before the vehicle - lengths of 4 vehicles (VL), above the vehicle - 3 VL, next to the vehicle - 3 VL, distance behind the vehicle - 16 VL
Mesh	Overset mesh applied near the vehicle Polyhedral elements Symmetry applied
Mesh quality	$1.2.10^6$ cells, minimum orthogonal quality 0.558, maximum aspect ratio 53.7; y^+ according to turbulence model requirements
Turbulence model	k-ε realizable
Solver	Coupled
Steady-state simulation	Yes, inlet velocity 25 $m{\cdot}s^{-1}$

Steady-state simulation could have been made because of interest only in velocity fields around the vehicle (Fig. 3). To compensate the simplification of constant inlet velocity, the initial velocity was subtracted from the velocity v_x (in the vehicle movement direction). The velocity v_x, the velocity v_y (in the direction perpendicular to the vehicle movement and in the horizontal plane) and their combination magnitude in the distance of 180 cm away from the vehicle (selected according to the measurement) in different height levels above the ground are displayed in the Fig. 4 (v_x), Fig. 5 (v_y) and in the Fig. 6 (comparison of them).

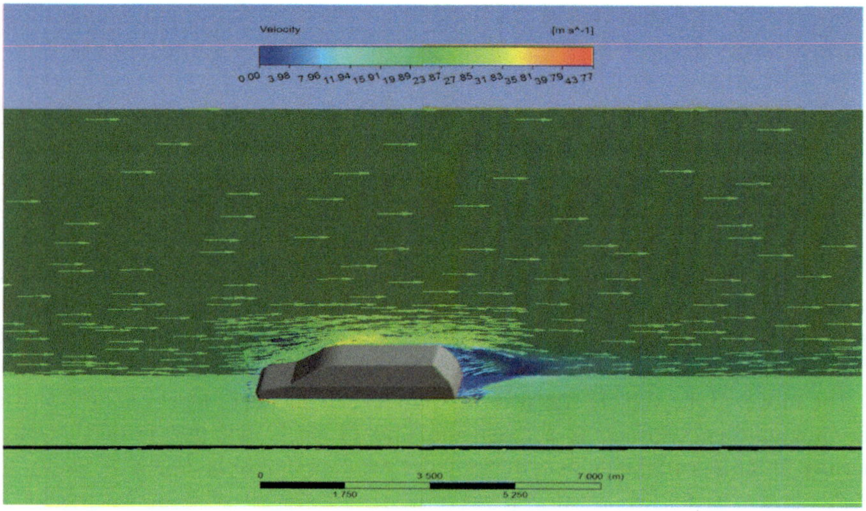

Fig. 3. Relative velocity vectors (black line is 1.8 m away from the vehicle).

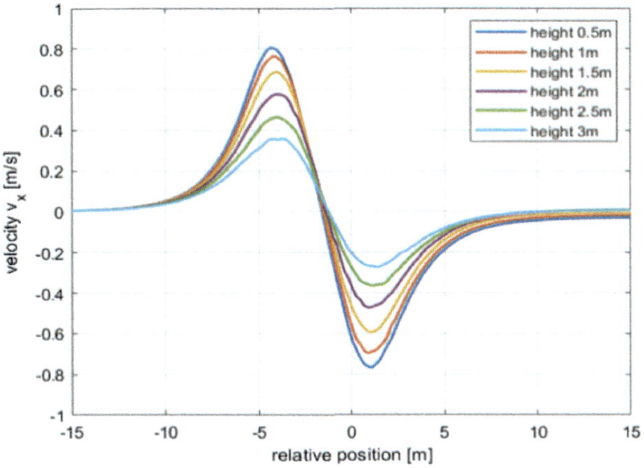

Fig. 4. Comparison of velocities v_x in different heights.

The curves in the Fig. 4 show the size of components x of the air velocity vectors (horizontal axis is parallel to the 'x' direction). The first peak values are achieved approx. 4 m before the rear end of the vehicle (which means this peak is located at the front end/near front wheels), the second peak ca. 1 m behind the vehicle. The change from positive to negative values indicate the change of direction of this vector component. Similarly, in the Fig. 5, the size of components y of the air velocity vectors is presented (horizontal axis represents the same line as in the Fig. 4). The peak value is now in the location corresponding to the front door of the vehicle. In both cases, the biggest peak values are observed in the height 0.5 m to 1 m over the road.

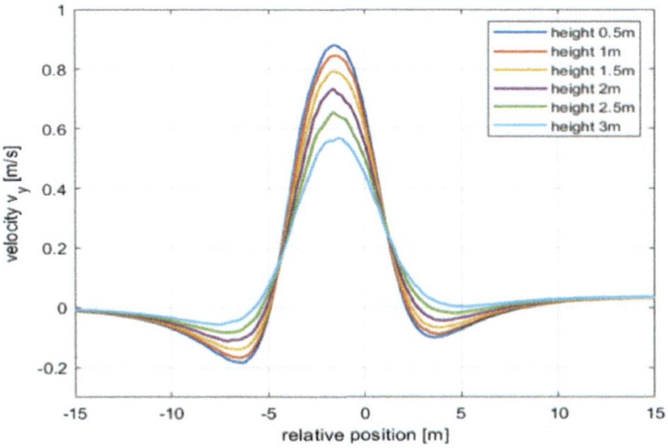

Fig. 5. Comparison of velocities v_y in different heights.

The resulting velocity magnitude v_{mag} presented in the Fig. 6 is obtained from velocities v_x and v_y via following equation:

$$v_{mag} = \sqrt{v_x^2 + v_y^2}. \tag{1}$$

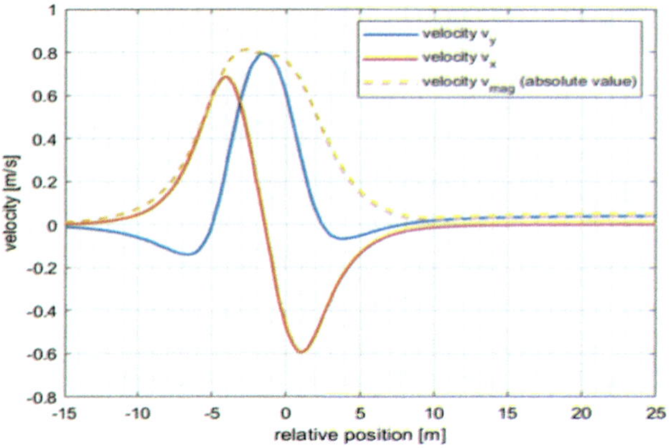

Fig. 6. Comparison of velocity profiles at height 1.5 m above the ground.

Resulting velocity magnitudes in different height levels are compared in the Fig. 7. Vertical part of velocity vector is omitted, because this part of the vector is not able to propel the vertical axis wind turbine without curved blades.

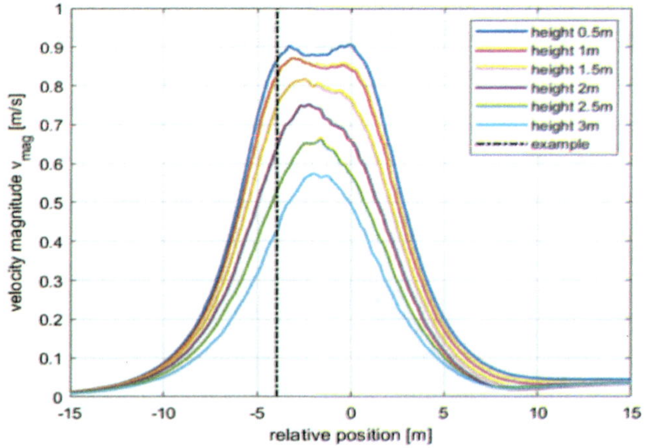

Fig. 7. Comparison of velocity magnitudes at different height levels.

With increasing height, the velocity magnitude changes. The decrease of the velocity magnitude is showed in the Fig. 8. Hereby a conclusion can be made, that by decreasing the distance between the turbine and the ground from 1.5 m to 1 m can increase velocity magnitude by 6.6% (in reality, this could be higher, because of the simplifications applied to the simulation model).

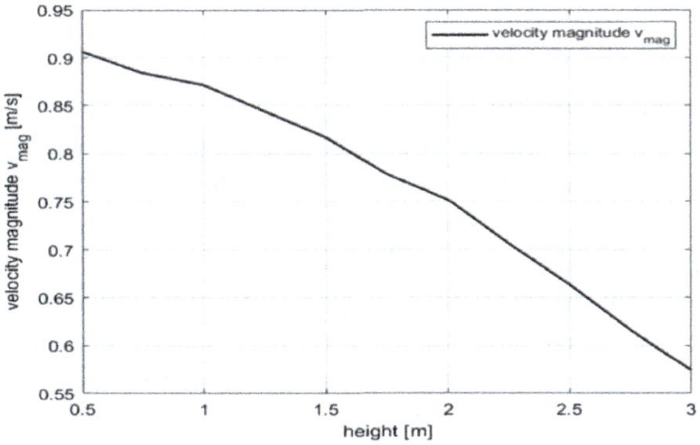

Fig. 8. Effect of height on the velocity magnitude xy.

The effect of changing the distance between the vehicle and the turbine is demonstrated in following figures by one example. The position of the measuring point for which the calculation was made is shown in Fig. 7 by a black dashed line. It is the position where the velocity v_x reaches its maximum and the velocity magnitude is close to its local maximum.

Figure 9 shows the course of the speed as a function of the changing distance of the vehicle from the turbine. As can be seen, moving the turbine 20 cm closer (from the distance 1.8 m) to the vehicle (or the vehicle closer to the turbine) results in a 13% increase in air velocity and thus an increase in turbine power.

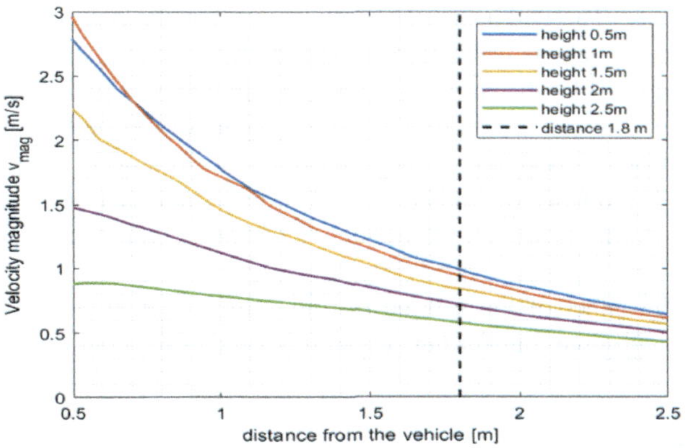

Fig. 9. Effect of increasing distance from the vehicle on velocity magnitude.

The power P_w, which can be passed by wind to an object, is given by the formula [26]:

$$P_w = \frac{1}{2} \cdot A \cdot \rho \cdot v^3, \quad (2)$$

where A is the projected area (of a blade), ρ is density of air and v is velocity of air. This formula shows the importance of correct placement of the wind turbine, because even small addition to the velocity is powered by 3. The simulation was verified by the above-mentioned measurements. Air velocity magnitude induced by smaller vehicles reached up to 1–2 m•s^{-1}, SUVs and smaller vans up to 4.5 m•s^{-1} and buses and trucks up to 7 m•s^{-1}. These values are average, as their size can be influenced by the shape of the vehicle body, the speed limit on the road and also by the wind from the surroundings, even if the measurements were made in windless or light breeze. Savonius turbine or Banki rotor are considered the best choices for such a case, because of only low velocity needed to start their rotation. The difference between them and their comparison for similar case is summarized by Tian et al. [25]. However, these turbines have only small efficiency (about 15%), so the power output P_t of the turbine can be estimated as [25]:

$$P_t = 0.15 \cdot P_w. \quad (3)$$

The power output for whole day can then be calculated by statistically counting vehicles and their types (taking into account different wind velocity magnitudes of different types of vehicles). The whole methodology is proposed by Penne et al. [26].

Because of the effect shown in previous figures, the easiest way to maximize the power output of a wind turbine is to move it closer to the vehicle. This can be hardly achieved on a highway, where the dimensions and no-obstacle-spaces are strictly given by local rules. However, there is a possibility to move the turbine closer to vehicle on the roads in the city. The width of such roads is smaller, the roadside is nearer to the vehicle. Even though vehicle drive at lower speed, there are road sections in the cities, where speed limit is 70 km•h^{-1} or even higher, where potential of this system could be used successfully. Another possibility for this system would be to collect the wind energy generated by rail vehicles. A single train, due to its length, can provide longer-lasting "impulse", probably of higher velocity (the turbine can be placed in smaller distance from the train), however, the frequency of trains could be too low to keep this system efficient.

In general, the effect of the turbine on the vehicle and its aerodynamic drag is estimated to be negligible, but further research is needed for confirmation.

4 Conclusion

There are many options for recovering energy from driving. Measurements in real traffic as well as simulations performed in the Fluent program have shown that it also has the potential to capture wind energy generated by driving the vehicle around wind turbines located near the road. As it is shown in simulation results, side effect of overcoming air resistance is arising of airflow around the vehicle. For example, near the front part of the simulated car, in the distance of 0.5 m, velocity magnitude can reach up to 3 $m{\cdot}s^{-1}$, decreasing with increasing distance from the vehicle to 1 $m{\cdot}s^{-1}$ at 1.8 m away from the vehicle. Velocity magnitude is also influenced by height between the measured place and the road level, and by aerodynamic parameters as well. As shown the measurements in real traffic, the increment of velocity magnitude provided by vans, trucks and buses was significant.

Interesting values have been measured behind the vehicles. Average velocity of all types of vehicles was ca. 4 $m{\cdot}s^{-1}$ and was measured 1 s after the vehicle passed by the measurement setup. For trucks with semi-trailers, this could reach up to 7 $m{\cdot}s^{-1}$ and 3 s behind the vehicle. Taking into account the complexity and time-consumption of transient simulation of such case in Fluent, it will be interesting to further focus on this problematic and to implement the influence of climatic conditions and surrounding environment on the velocity magnitude. Placement, type and dimensions are the next part of solution of effective usage of this source of wind energy. The study has demonstrated the importance of suitable placement of wind turbines from the power-maximization point of view. To keep the rotation of the turbine – generator system, sufficiently dense traffic is needed (around 1 vehicle per second) with speed above 50 $km{\cdot}h^{-1}$. Given requirement denies the possibility of using this principle on railways. Even though a higher amount of energy can be collected from a single train, the distance between trains (and in a conclusion low frequency) does not make this solution effective. Due to significant decrease of air velocity with increasing distance from vehicle, the usability of this system could be constrained on the highways with wide emergency lane. For further research, placement of turbines between fastest lines could be considered, where although the frequency of vehicle is lower, they tend to drive faster, and the turbine could be propelled by vehicles on both sides. Analysis results show estimation of effective usage of this system in city traffic ideally on four-lane roads with high density of vehicles moving at velocity higher than 50 $km{\cdot}h^{-1}$ with wind turbines installed in between lanes of opposing direction as well as near the roadsides. Of course, with the use of turbines in the vicinity of moving vehicles, the question arises as to what extent their presence may affect traffic safety. However, it can be assumed that their influence should not be higher than the influence of barriers. And since their use proves to be advantageous only in certain places, these can be labeled.

Acknowledgement. The paper was supported by the Scientific Grant Agency of the Ministry of Education of the Slovak Republic and the Slovak Academy of Sciences in project KEGA, no. KEGA 023ŽU-4/2020: Development of advanced virtual models for studying and investigation of transport means operation characteristics.

"This publication was realized with support of Operational Program Integrated Infrastructure 2014 – 2020 of the project: Innovative Solutions for Propulsion, Power and Safety Components of Transport Vehicles, code ITMS 313011V334, co-financed by the European Regional Development Fund".

References

1. Blatnicky, M., Saga, M., Dizo, J., Bruna, M.: Application of light metal alloy EN AW 6063 to vehicle frame construction with an innovated steering mechanism. Materials **13**(4), 817 (2020)
2. Dizo, J., Blatnicky, M.: Investigation of ride properties of a three-wheeled electric vehicle in terms of driving safety. Transp. Res. Procedia **40**, 663–670 (2019)
3. Lebedevas, S., Pukalskas, S., Žaglinskis, J., Matijošius, J.: Comparative investigations into energetic and ecological parameters of camelina-based biofuel used in The 1Z diesel engine. Transport **27**(2), 171–177 (2012)
4. Korsakas, V., Melaika, M., Pukalskas, S., Stravinskas, P.: Hydrogen addition influence for the efficient and ecological parameters of heavy-duty natural gas si engine. Procedia Eng. **187**, 395–401 (2017)
5. Caban, J., Drozdziel, P., Ignaciuk, P., Kordos, P.: The impact of changing the fuel dose on chosen parameters of the diesel engine start-up proces. Transp. Prob. **14**(4), 51–62 (2019)
6. Pukalskas, S., et al.: Comparison of conventional and hybrid cars exploitation costs. Adv. Sci. Technol. Res. J. **12**(1), 221–227 (2018)
7. Małek, A., Caban, J., Wojciechowski, L.: Charging electric cars as a way to increase the use of energy produced from RES. Open Eng. **10**(1), 98–104 (2020)
8. Brumercik, F., Lukac, M., Caban, J.: Unconventional powertrain simulation. Commun. Sci. Lett. Univ. Zilina **18**(2), 30–33 (2016)
9. Jinglai, W., et al.: Efficiency comparison of electric vehicles powertrains with dual motor and single motor input. In Mechanism and Machine Theory. **128**, 569–585 (2018)
10. Wei, Y., Ruochen, W., Runze, Z.: A Comparative research on the energy recovery potential of different vehicle energy regeneration technologies. Energy Procedia **158**, 2543–2548 (2019)
11. Guo, J., Wang, J., Cao, B.: Study on braking force distribution of electric vehicles. In: Proceedings of Asia-Pacific Power and Energy Engineering Conference, Wuhan, China, pp. 1–4 (2012)
12. Medvedev, V., Rudakov, I.: Self-adaptive algorithm for changing braking energy recuperation of an electric vehicle based on neuro-fuzzy inference system. In: Proceedings of IASF - Technologies and Components of Land Intelligent Transport Systems, Moscow, Russia (2019)
13. Boisvert, M., et al.: Comparison of two strategies for optimal regenerative braking, with their sensitivity to variations in mass, slope and road condition. IFAC Proc. **46**(21), 626–630 (2013)
14. Li, L., et al.: Energy recovery strategy for regenerative braking system of intelligent four-wheel independent drive electric vehicles. IET Intell. Transp. Syst. **15**, 119–131 (2021)
15. JI, F.Z., et al.: Energy recovery based on pedal situation for regenerative braking system of electric vehicle. Veh. Syst. Dyn. **58**(1), 144–173 (2020)
16. Liu, Z.Q., Lu, S., Du, R.H.: A genetic-fuzzy control method for regenerative braking in electric vehicle. Int. J. Comput. Sci. Math. **11**(3), 263–277 (2020)
17. Fahlbeck, J., Kindahl, J.: Energy recovery systems in cars and detail study of flywheel regenerative braking systems. Bachelor thesis, Chalmers University of technology, Gothenburg, Sweden (2016)
18. Schwalbe, K., et al.: Recuperation gain for a hydraulic energy storage in automotive applications. Appl. Thermal Eng. **175**, 115275 (2020)
19. Durand, T., et al.: Potential of energy recuperation in the exhaust gas of state of the art light duty vehicles with thermoelectric elements. Fuel **224**, 271–279 (2018)

20. Caban, J., Litak, G., Ambrożkiewicz, B., Gardyński, L., Stączek, P., Wolszczak, P.: Possibilities of energy harvesting from the suspension system of the internal combustion engine in a vehicle. Commun. Sci. Lett. Univ. Zilina **23**(2), 106–116 (2021)
21. Li, Z., et al.: electromagnetic energy-harvesting shock absorbers: design, modeling, and road tests. IEEE Trans. Veh. Technol. **62**, 1065–1074 (2013)
22. Zahibi, N., Saafi, M.: Recent developments in the energy harvesting systems from road infrastructures. Sustainability **12**(17), 6738 (2020)
23. Clausing, D., Holmes, M.: Technology readiness. Res. Control Manag. **53**, 52–59 (2010)
24. Hamat, A.M.A., Kahar, O., Md, N.M.: Renewable energy from induced airflow generated by cruising ground vehicles in tandem using RANS. Energy Procedia **14**, 1877–1882 (2012)
25. Tian, W., Song, B., Mao, Z.: Numerical investigation of wind turbines and turbine arrays on highways. Renew. Energy **147**(1), 384–398 (2020)
26. Penne, M., Qiao, W., Hudgins, J.: Converting waste vehicle aerodynamic energy into electricity. In: IEEE Energy Conversion Congress and Exposition (ECCE), pp. 416–422 (2018)

Methodology for Determination Coefficients Values of the Proposed Rheological Model for the Tire Tread

Marijonas Bogdevičius(✉), Mykola Karpenko, and Daiva Rožytė

Vilnius Gediminas Technical University, 10223 Vilnius, Lithuania
{marijonas.bogdevicius,mykola.karpenko,
daiva.rozyte}@vilniustech.lt

Abstract. The article presents theoretical and experimental researches on investigation properties of the vehicle wheel tire tread. The approach presented in the research of the pre-experimental measurement, based on spectrum analysis, and combined with theoretical investigation is able to describe a viscoselastic behaviour of the tire tread material. Proposed mathematical model of the vehicle wheel tire tread material represents five parameters rheological model of the Maxwell and Kelvin-Voigt units' hybrid. The research included optimization task based on minimization objective function for determination a stiffness and damping coefficients. In the proposed rheological model not only the displacements are unknown, but also forces that are described by second–order differential equations. Validation between experimental measurement and theoretical investigation based on the frequency spectrum analysis was made.

Keywords: Tire tread · Rubber · Frequency · Rheological model · Optimization

1 Introduction

Rubber materials, as one kind of soft matter, are used for sealing, transmission processes, damping shock absorption, heat insulation, corrosion protection, etc., and are widely used in transport, aerospace, industrial and other fields [1]. In the vehicles structure usually are used different components based on the rubber, for example – pneumatic tires, sealing, tread elements etc. The pneumatic tire is usually composed of a dozen of structural components (see Fig. 1), joined together and even slightly overlapped during the vulcanization process, which are made either of rubber or rubber-based composites [2]. One of the main element of the pneumatic tire is tread cover which has a protective function and is responsible for adhesion to the road surface. The main dynamic processes between the road and the vehicle wheel occur directly through the contact of tire tread and road surface.

Nevertheless, development of the pneumatic tire structure is associated with carrying out a series of experimental studies for determining stability and reliability of its implementation. A numerical modelling using the Finite Element Method or similar

O. Prentkovskis et al. (Eds.): TRANSBALTICA 2021, LNITI, pp. 16–27, 2022.
https://doi.org/10.1007/978-3-030-94774-3_2

methods is an alternative to experimental testing, since, its time and economical values are more effective compared to experimental test [3]. Thus, for effective and correct analysis, a numerical model should be developed with particular attention and care. Therefore, an accurate assessment of mechanical properties of the tire rubber is essential in the tire numerical modelling [4].

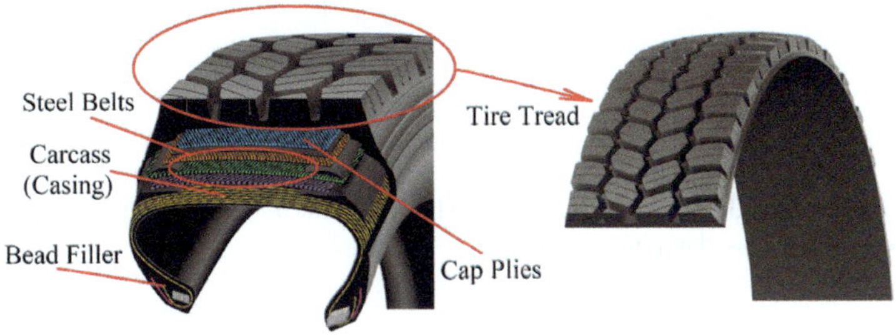

Fig. 1. Vehicle part – composite pneumatic tire construction.

To develop tires in a better way to be used efficiently in different conditions (off-road, winter, city etc. conditions of using), the factors that influencing on the friction and the contact mechanics of the tire with the road surface should be well studied. To investigate the friction mechanisms, a behaviour of the tire tread rubber material should be investigated as well [5]. The viscoelastic material properties of tire tread rubber cause a complicate friction behaviour that depends on the contact area, sliding velocity, normal pressure, ambient temperature and material behaviour under loads.

Already started theoretical and experimental researching for the determination of elastic and dynamic properties of the vehicle tire composite materials based on rubber, but the damping properties are still difficult to assess and difficult to include at the analysis and design stage according to [6]. According to [7] the most used methods for experimentally determining the dynamic and damping properties of material – are various combination of the non-destructive measurement methods. For proper investigation of vehicle wheel tire materials behaviours should be taken in an account the loads under operation, according to [8]. According to [9] the investigation of tire composite materials under the loadings can be held for improving blast resistance of objects and according to [10] can be optimized material structures by combining a finite element method approach with mathematical modelling and different algorithms. Therefore, it is recommended, according to [11], to take a great care when comparing damping values measured by different methods and in different environments with taking in account frequency response of material. That`s why experimental researches with combination of computational as well as advanced constitutive modelling is used in a wide range for investigation and improving vehicle tire materials.

For mathematical reproduction of the real behaviour of tire material, it is necessary to introduce a viscous-elastic model. Constitutive relation of linear viscous-elasticity is built up, considering the material as a sum of elements such as linear springs (elastic – k) and linear dashpots (viscous – c), with taking into account the time-dependent relaxation of the material wall. The simplest model is able to reproduce properly the stress-strain behaviour of the rubber composite material that is the 3-parameter model known as Standard Linear Solid Model (SLSM), which can be structured with a Maxwell or a Kelvin–Voigt units according to [12] (shown in Fig. 2).

Fig. 2. Schemes of the 3-parameter SLSM: a – Maxwell unit; b – Kelvin-Voigt unit: k_1 and k_2 – stiffness coefficients; c_1 – damping coefficient.

Theoretically, the more elements in a connection, the more accurate model will be in describing the real response of the material. By review of all previous researches, it can be observed that none of them does not take in account a detail investigation of coefficients of viscos-elastic models and doesn`t take in account frequency response of vehicle wheel tire materials. In this case, it`s not possible to describe completely the deformation and damping behaviour of the tire tread rubber material.

The approach presented in this research of the experimental measurement and combined with theoretical investigation sequence enables the engineer to determine efficiently dynamic properties of vehicle wheel tire tread rubber material and frequency dependence at the analysis and design stage. The present research is based on investigation of the tire tread material characteristics and in future will be held in simulation of vehicle wheel tire contacts with the road surfaces under loads conditions. The present research is directed on investigated values of damping and stiffness coefficients purposed to rheological model of the tire tread material with taking in account frequency response of material.

2 Pre-experimental Investigation

In the pre-experiment investigation is used a piece of the tire tread element according to standard [13, 14]. The dimension of cut outed piece – 15 × 15 × 76 (mm) and the investigation material – manufactured synthetic rubber (NBR). The top mass, made from steel with dimension Ø55 × 25 (mm) and mass of load – 0.5 kg. The values of physical and geometrical parameters of the tire tread material and top mass is presented in the Table 1. An experimental bench for the pre-experimental investigation on spectrum analysis of tire tread material is shown in Fig. 3. The pre-experimental test was performed via Single Measurement Design and is based on One-Sample Statistical Method with Estimating Uncertainty in Repeated Measurements [15].

Table 1. Physical and geometrical parameters of the tire tread material and top mass.

Object	Dimension	Tensile strength	Mass	Density
A piece of tire tread	15 × 15 × 7 mm	24.1 MPA	$2.49 \cdot 10^{-3}$ kg	1,522 kg/m^3
Top mass	Ø55 × 25 mm	420 MPa	0.5 kg	8,050 kg/m^3

The experiment measurement includes several measuring's of displacement, speed and acceleration on a top mass of investigation object. For decrease, errors and not equal force of impact with an IEPE (integrated electronics force transducers) impact hammer in each measurement, for experimental investigation it was accepted and presented average results of several measuring's.

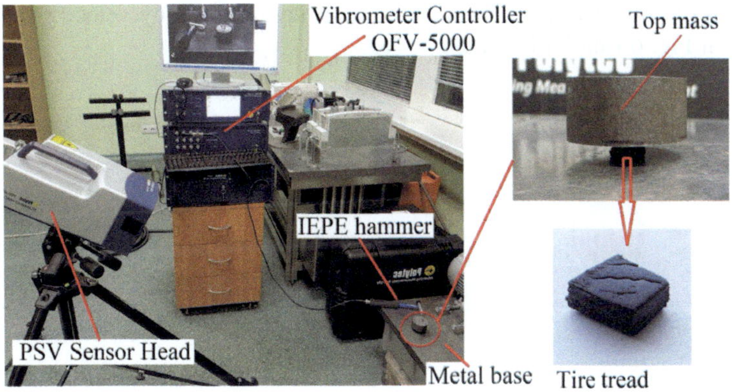

Fig. 3. The pre-experimental bench for research on spectrum analysis of the rubber material of the tire tread.

The time of one measuring by PSV Sensor Head is about 4 s, during this time, between 2 and 3 s, IEPE impact hammer generates an impulse on metal base with a force shown in Fig. 4a, and velocity measuring by PSV Sensor Head on a metal base, shown on Fig. 4b. Velocity measuring by PSV Sensor Head on top mass of research object, during test, shown on Fig. 5.

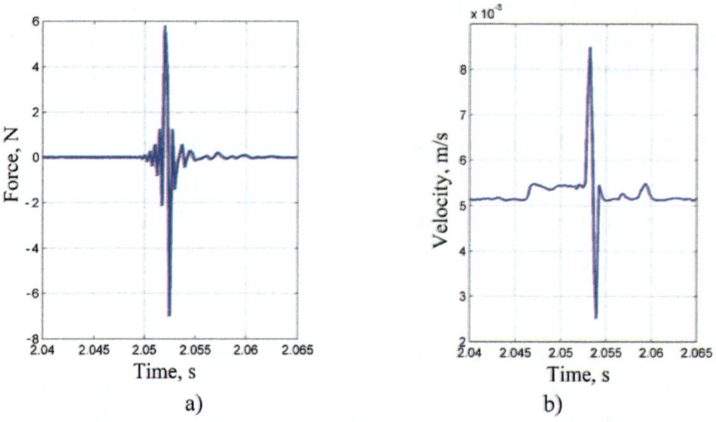

Fig. 4. Pre-experimental measuring (1/2): a – signal from IEPE impact hammer; b – velocity measurement from PSV sensor head on metal base.

In the result by spectrum analysis the frequency response by velocity variation is presented in Fig. 6 (until 1,000 Hz).

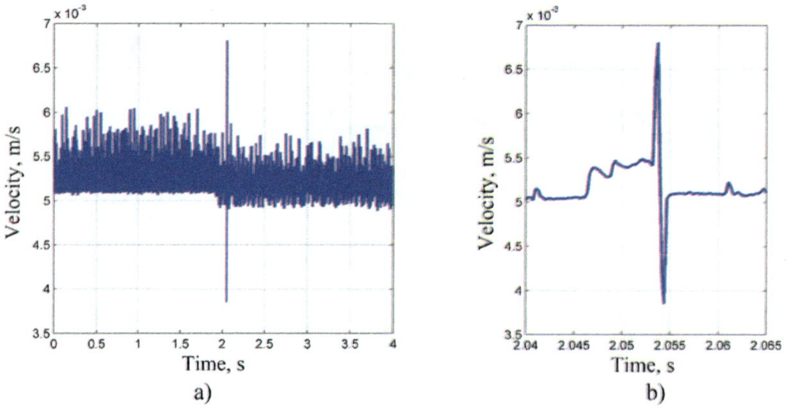

Fig. 5. Pre-experimental measuring (2/2): a – velocity measurement from PSV Sensor Head on top mass during all test time; b – velocity measurement from PSV Sensor Head on top mass during time from 2.04 to 2.065 s. (when generated impulse by IEPE impact hammer).

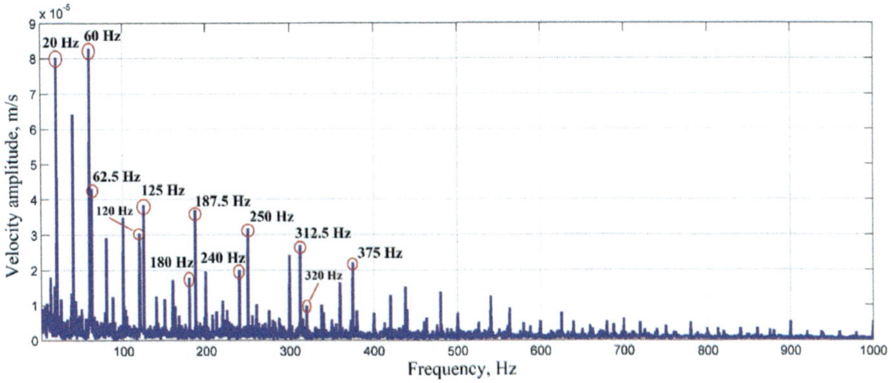

Fig. 6. Frequency response by velocity changings on top mass.

From the pre-experiment results the peak of velocity impulse, from IEPE impact hammer on metal base is $8.62 \cdot 10^{-3}$ m/s (Fig. 4b) and a peak of velocity impulse on top mass is $6.79 \cdot 10^{-3}$ m/s (Fig. 5b). That confirms the damping properties of tire tread material. The difference in time representation on graphs can be explained that impulse from IEPE impact hammer during some amount of time is going from metal base to the top mass through the tire tread material. From frequency analysis (Fig. 6) it seems that main resonance frequency of tire tread material are in the low frequency 0 ÷ 200 Hz range and in the beginning of middle frequency 200 ÷ 500 Hz range. It should be noted that the main interest for researching is the frequency range up to 500 Hz, since, in this frequency range we can observe the main resonant modes. By frequency analysis the main and first resonance frequency of investigated material is equal to 20 Hz with harmonic by each 20 Hz (20/40…80/100…etc. Hz). Also it should be pointed that frequency's 62.5 Hz with harmonic by each 62.5 Hz (62.5/125… 250/312.5…etc. Hz). The existing of second resonances frequencies is explained by manufacturing processes of creating a tire tread material and by complex structure of tire tread material.

3 Mathematical Model of the Tire Tread Investigation

The mathematical model of the tire tread is based on model from [16]. The tire tread material is derived by *n* masses (from m_1 to m_n) with a top mass (m_{n+1}) on a top (Fig. 7a) and is established one system. The thickness of each elements (Δh) of the system is individual and depend from the thickness of the material. The scheme of forces distribution and displacement between masses and elements position of proposed mathematical model is shown in Fig. 7b. According to pre-experimental investigation, the impulse is generated from bottom of the model and is transferred through elements to top mass. In first force equation of mathematical model it was used the values of the displacement (z), velocity, and acceleration of impulse from IEPE impact hammer. The each element from e_1 to e_{n+1} of mathematical model represents the rheological model of hybrid Maxwell unit and Kelvin-Voigt unit in a connection (Fig. 7c).

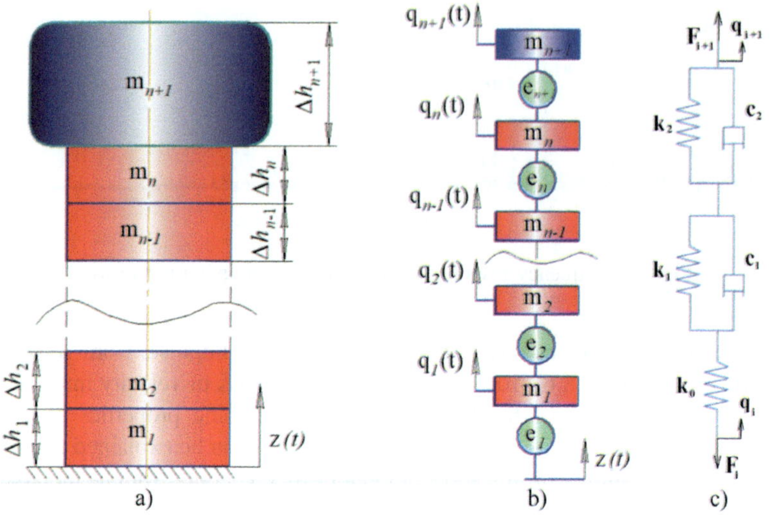

Fig. 7. Schemes of calculation mathematical model: a – principal scheme of derived material on masses with a load mass; b – scheme of forces and displacement between elements; c – rheological model of elements: k_i and c_i stiffness and damping coefficients, respectively; q_i – displacement between rheological model units.

In the proposed five parameters rheological model (Fig. 7c) the displacements and forces, described by second–order differential equations, are unknown. The range of min and max value of coefficients, for optimization task is taken from [16]. According to mathematical model, a system of equations for each mass is:

$$\begin{aligned} m_1\ddot{q}_1 &= -F_{0,1} + F_{1,2} - m_1 g, \\ m_2\ddot{q}_2 &= -F_{1,2} + F_{2,3} - m_2 g, \\ &\dots \\ m_n\ddot{q}_n &= -F_{n-1,n} + F_{n,n+1} - m_n g, \\ m_{n+1}\ddot{q}_{n+1} &= -F_{n,n+1} - m_{n+1} g. \end{aligned} \tag{1}$$

where $F_{i,i+1}$ – force between neighbour elements in mathematical model, g – the standard gravity defined by standard as 9.80665 m/s^2.

The each force between neighbour elements in mathematical model can be describe by formulas:

$$\begin{aligned}
&\ddot{F}_{0,1}+a_{10,1}\cdot\dot{F}_{0,1}+a_{10,1}\cdot F_{0,1}=b_{10,0}(q_1-z)+b_{10,1}(\dot{q}_1-\dot{z})+b_{10,2}(\ddot{q}_1-\ddot{z}),\\
&\ddot{F}_{1,2}+a_{12,1}\cdot\dot{F}_{1,2}+a_{12,0}\cdot F_{1,2}-b_{12,0}(q_2-q_1)-b_{12,1}(\dot{q}_2-\dot{q}_1)-b_{12,2}(\ddot{q}_2-\ddot{q}_1)=0,\\
&\ldots\\
&\ddot{F}_{n-1,n}+a_{n,1}\cdot\dot{F}_{n-1,n}+a_{n,0}\cdot F_{n-1,n}-b_{n,0}(q_n-q_{n-1})-b_{n,1}(\dot{q}_n-\dot{q}_{n-1})-b_{n,2}(\ddot{q}_n-\ddot{q}_{n-1})=0,\\
&\ddot{F}_{n,n+1}+a_{n+1,1}\cdot\dot{F}_{n,n+1}+a_{n+1,0}\cdot F_{n,n+1}-b_{n+1,0}(q_{n+1}-q_n)\\
&\quad-b_{n+1,1}(\dot{q}_{n+1}-\dot{q}_n)-b_{n+1,2}(\ddot{q}_{n+1}-\ddot{q}_n)=0.
\end{aligned}\tag{2}$$

Coefficients of variables in force equation are described by relationships between elements coefficients of rheological models:

$$\begin{aligned}
a_{i,0}&=\frac{k_0\cdot k_1+k_0\cdot k_2+k_1\cdot k_2}{c_1\cdot c_2},\ a_{i,1}=\frac{c_1\cdot k_0+c_1\cdot k_2+c_2\cdot k_0+c_2\cdot k_1}{c_1\cdot c_2},\\
b_{i,0}&=\frac{k_0\cdot k_1\cdot k_2}{c_1\cdot c_2},\ b_{i,1}=\frac{c_1\cdot k_0\cdot k_2+c_2\cdot k_0\cdot k_1}{c_1\cdot c_2},\ b_{i,2}=k_0.
\end{aligned}\tag{3}$$

The systems of Eqs. (1) and (2) is combined in one general equation:

$$[M]\{\ddot{X}\}+[C]\{\dot{X}\}+[K]\{X\}=\{F(t)\},\tag{4}$$

where [M]; [C] and [K] are matrices of variables vector. The matrices [M], [C] and [K] – are non-symmetrical and have a view according to [16]. The variables vectors $\{X\}$ and forces vector $\{F(t)\}$ is presented as:

$$\begin{aligned}
\{X\}^T&=\left[q_1,F_{0,1},q_2,F_{1,2}\ldots q_n,F_{n-1,n},q_{n+1},F_{n,n+1}\right],\\
\{F(t)\}^T&=\left[-m_1g,-b_{10,0}z-b_{10,1}\dot{z}-b_{10,2}\ddot{z},-m_2g,0\ldots-m_ng,0,-m_{n+1}g,0\right].
\end{aligned}\tag{5}$$

For determination closer values of rheological elements coefficients, it was generated a task of optimization and algorithm of optimization task based on minimization of object function according to [16]. After optimization, by using Fourier analysis, is obtained the theoretical frequencies and spectrum of theoretical oscillations. From the obtained spectrum, it was determined the frequencies at which normalized the function and refined the coefficients that stand for the entire function.

4 Results

A values of object function in each iteration shown in Fig. 8 and optimization parameters values depended from iteration shown in Fig. 9. The obtained result showed that at initial iteration an object function is in high range. The solution of optimization continues until the object function reaches tolerance (*Toler* = 10^{-4}). After 636 iterations, object function reaches tolerance. Solution time according to experimental part is equal to 4 s. The parameters $\{k_0, k_1, c_1, k_2, c_2\}$ were optimized until the object function values reached tolerance. After the object function reaches the tolerance, the optimization of parameters is stopped. By graph results in Fig. 8 it seems that at interval 636…710 iterations the optimization parameters are in stable forms and can be stopped.

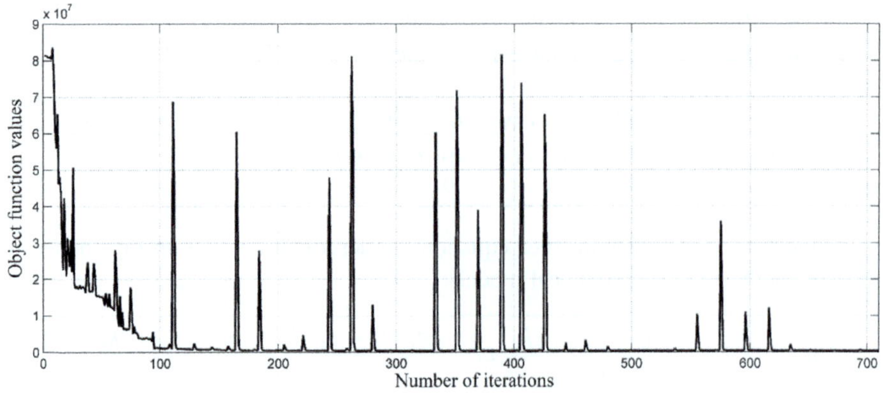

Fig. 8. Graph of relation of object function from iterations.

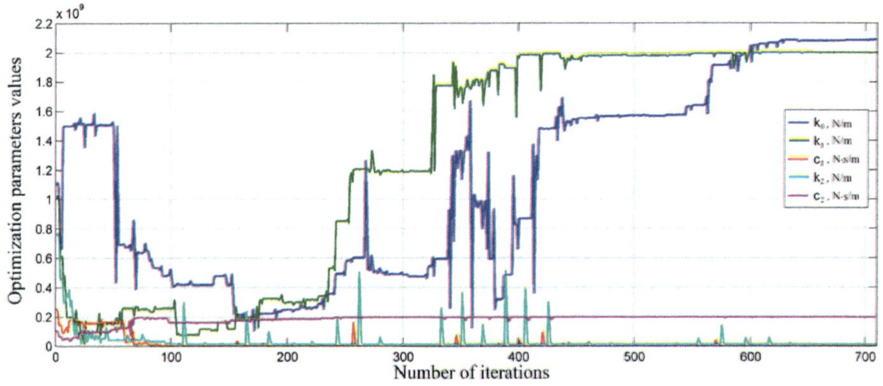

Fig. 9. Graph of relation of optimization parameters from iterations.

The results obtained comparing the velocity amplitudes depend on frequencies between experimental measurements and theoretical modelling shown in Fig. 10.

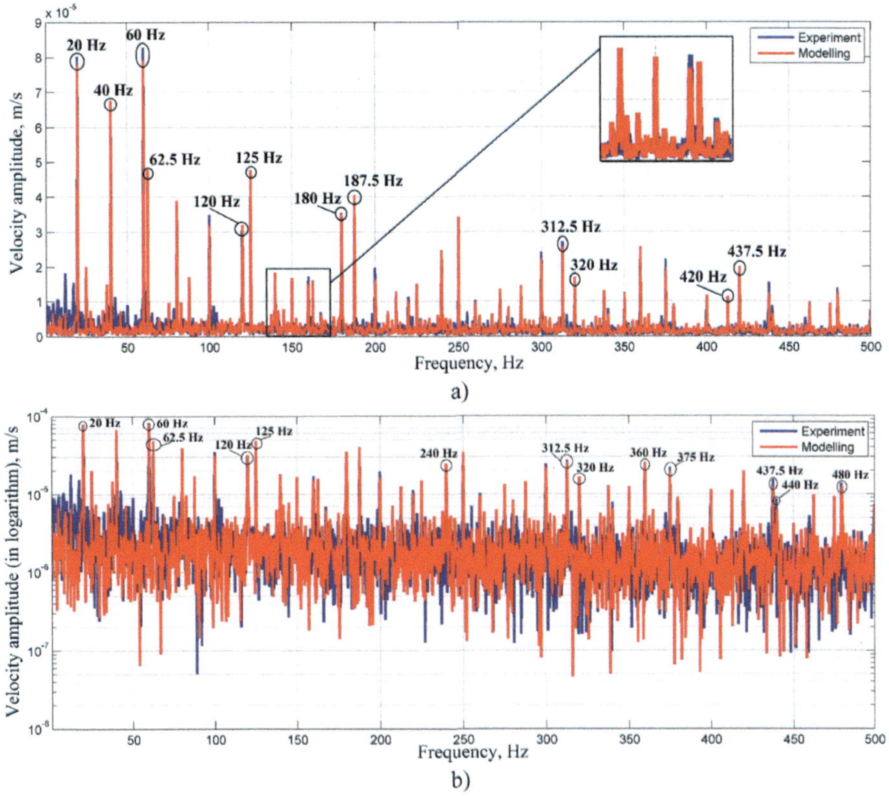

Fig. 10. Comparisons of frequency response between experimental and theoretical investigations: a) spectrum of velocity amplitude; b) velocity amplitude (in logarithm).

The difference of frequency response between experimental measurements and theoretical modelling is less than <5% at more frequencies (see Fig. 10). Also it should be pointed that a difference on resonances points is no more than 3%. The conducted research disclosed that coefficients values for a rheological model of the tire tread material for future modelling can be easily determinate by spectrum analyses and optimization of present mathematical model relying on frequency response.

5 Conclusions

With reference to theoretical research on determination coefficients values of proposed rheological elements (five parameters rheological model – hybrid of Maxwell and Kelvin-Voigt units in a connection) of tire tread material, the obtained results showed that, proposed methodology is able to describe the viscous-elastic properties of the material based on rubber. The results of experimental measurements is showed that by spectrum analyses the main resonance frequency are in the low frequency (0 ÷ 200 Hz) range and beginning of middle frequency ranges (200 ÷ 500 Hz). The main resonance frequency of investigated material is equal 20 Hz with harmonic by each 20 Hz (20/40…80/100…etc. Hz). Also should be pointed frequency's 62.5 Hz with harmonic by each 62.5 Hz (62.5/125…250/312.5…etc. Hz). The existing of second resonance frequencies is explained by manufacturing processes of creating a tire tread material and by complex structure of tire tread material. The present mathematical model with optimization in pair was able to determinate the coefficient values of proposed rheological elements for obtained closer results to experimental measurement relying on frequency response. The obtained results showed that it took around 636 iterations for the objective function to reach the specified tolerance. A verification of obtained value carried out on comparing frequency responses from experimental and theoretical investigations. A difference of amplitudes between experimental measurements and theoretical modelling is less than <5% at more frequencies. The present research is held for future researches in area of composite tire investigation, with taking in account viscos-elastic properties of rubber material, for obtaining a realistic model and simulation dynamic processes of vehicle wheel tire contacts with a road surfaces under loads.

References

1. Zhao, J., Dong, J., Liu, Z., Xie, H.: Characterization method of mechanical properties of rubber materials based on in-situ stereo finite-element-model updating. Polym. Testing **79**, 1–6 (2019)
2. Korunović, N., Fragassab, C., Marinkovićc, D., Vitkovića, N., Trajanović, M.: Performance evaluation of cord material models applied to structural analysis of tires. Compos. Struct. **224**, 1–13 (2019)
3. Reid, J., Boesch, D., Bielenberg, R.: Detailed tire modelling for crash applications. Int. J. Crashworthiness **12**(5), 521–529 (2007)
4. Baranowski, P., Janiszewski, J., Małachowski, J.: Tire rubber testing procedure over a wide range of strain rates. J. Theor. Appl. Mech. **55**(2), 727–739 (2017)
5. Bogdevičius, M., et al.: Investigation of tire force transmission on interaction with slush. Problemy Transportu – Transp. Prob. **14**(1), 13–21 (2019)
6. Treviso, A., van Genechten, B., Mundo, D., Tournour, M.: Damping in composite materials: properties and models. Compos. B Eng. **78**, 144–152 (2015)
7. Wesolowski, M., Barkanov, E.: Air damping influence on dynamic parameters of laminated composite plates. Measurement **85**, 239–248 (2016)
8. Wang, W., Yan, S., Zhao, S.: Experimental verification and finite element modelling of radial truck tire under static loading. J. Reinf. Plast. Compos. **32**(7), 490–498 (2013)
9. Baranowski, P., Małachowski, J., Mazurkiewicz, L.: Local blast wave interaction with tire structure. Defence Technol. **16**(3), 520–529 (2020)
10. Abe, A., Kamegawa, T., Nakajima, Y.: Optimization of construction of tire reinforcement by genetic algorithm. Optim. Eng. **5**, 77–92 (2004)
11. Avril, S., Bonnet, M., Bretelle, A.: Overview of identification methods of mechanical parameters based on full-field measurements. Exp. Mech. **48**(4), 381–402 (2008)
12. Bertaglia, G., Ioriatti, M., Valiani, A., Dumbser, M., Caleffi, V.: Numerical methods for hydraulic transients in visco-elastic pipes. J. Fluids Struct. **81**, 230–254 (2018)

13. ISO 10499–1:2019. Industrial tyres and rims – Rubber solid tyres (metric series) for pneumatic tyre rims – Part 1: Designation, dimensions and marking. International Organization for Standardization, p. 6 (2019)
14. ISO 10499–2:1998. Industrial tyres and rims—Rubber solid tyres (metric series) for pneumatic tyre rims—Part 2: Load ratings. International Organization for Standardization, p. 5 (1998)
15. Vardeman, S., Jobe, M.: Statistical Methods for Quality Assurance: Basics, Measurement Control, Capability, and Improvement, p. 437. Springer-Verlag, New York (2016)
16. Bogdevičius, M., Karpenko, M., Bogdevičius, P.: Determination of rheological model coefficients of pipeline composite material layers based on spectrum analysis and optimization. J. Theor. Appl. Mech. **59**(2), 265–278 (2021)

Development of an Innovative Technical Solution to Improve the Efficiency of Rolling Stock Friction Brake Elements Operation

Juraj Gerlici, Kateryna Kravchenko, and Yuliia Fomina(✉)

University of Zilina, Univerzitná, 8215/1, 01026 Žilina, Slovakia
{juraj.gerlici,kateryna.kravchenko,
yuliia.fomina}@fstroj.uniza.sk

Abstract. Braking system is the most important component of a vehicle. Traffic safety directly depends on its correct function and proper operation. When braking, a large amount of heat is generated, which can lead to overheating of the friction elements and subsequent damage. The authors describe a method for stabilizing the temperature of the brake pad and wheel contact in the article. In this context, it is proposed to use inserts made of expanding agents in the brake pad design. The presence of holes in the pad helps to eliminate wear particles, that reduces the intensity of the pad wear. The design of such a pad and the operation of the proposed brake system are described in the article. The diameter and depth of the brake pad holes are substantiated. The described solution will allow to cool the contact area of the brake pad and the wheel, and, thus, increase the braking efficiency.

Keywords: Brake pad · Friction wear · Expanding agent

1 Introduction

One of the priority development fields of railways is the creation or the improvement of the new design elements of vehicle which will meet the modern needs [1–4]. Traffic safety, the possibility of increase in the speed, and, accordingly, increasing the carrying and traffic capacity of the railway line largely depend on the efficiency of the friction elements of the rolling stock brake system. When braking, the train kinetic energy is converted into other forms of energy, most of which is a heat one. This process is accompanied by an increase in the temperature of the friction elements [5]. In cases of thermal overloads, which mainly occur as a result of long-term braking on down-grade railroads, the generation of stresses and deformations occurs, whose consequences are the appearance of cracks on the rim of a wheel and the final total wheel defect [6, 7]. Effective train braking depends on the coefficient of friction, which is influenced by temperature changes in tribocontact [8–10]. Hence a topical issue in the operation of rolling stock is the development of innovative technical solutions to stabilize temperature conditions in the interaction zone of brake elements.

O. Prentkovskis et al. (Eds.): TRANSBALTICA 2021, LNITI, pp. 28–38, 2022.
https://doi.org/10.1007/978-3-030-94774-3_3

2 Analysis of the Problem

When braking, the place of contact between the wheel and the brake pads heats up due to friction forces. The temperature determines thermo-mechanical behavior of the structure. In the braking surface, high temperatures and thermal gradients are produced. This generates stress and deformations in which the consequences are manifested by the appearance of cracks [11, 12].

The authors of papers [13, 14] revealed serious thermal effects on the rolling surface of a wheel during braking (heating up to 700–800 °C). It was found that thermal loads are in inverse relation to the contact area of the brake pad with the wheel. Temperature fields with significant gradients are an unstoppable physical phenomenon of the transformation of mechanical energy into thermal energy. This will lead to a change for the worse in both frictional and strength properties due to structural changes in the materials. With prolonged exposure to high temperatures, thermal crack initiation with their exit to the outer face is possible [5, 6, 15]. Under such conditions of friction, there is a rapid thermal wear of the surfaces of the wheel and brake pad [5, 6]. In this case, the wear rate is determined by the occurrence and development of high temperatures in the friction zone. For brake discs, an increase in the disc temperature leads to axial skewing (umbrella) of the disc, and thermal deformation in the rib area necessarily causes waviness or roughness of the disc friction surface, thereby increasing the pressure on the friction linings and creating local temperature spots on the disc friction surface [16].

The problem of resource saving of shoe brakes with which the rolling stock is equipped is multi-scale. It needs a solution of technical and economic, technological, metallurgical and tribological knowledge associated with the choice of wear-resistant materials and methods of cooling the working surfaces of the brake elements.

There are various ways of implementing thermoregulatory and energy-efficient functions in brake systems. They can be considered within the framework of the following classification.

1. Thermal control based on the heat absorption or release by metallic or non-metallic friction elements of braking devices. In this case, the following items can be used:
   1.1. Chemical reactions to friction lining materials with the heat emission or absorption.
   1.2. The effects of energy emission and absorption when changing the aggregative state of friction elements (melting, evaporation, sublimation, crystallization, etc.).
   1.3. Physical properties of friction elements materials that provide high heat transfer.
2. Thermal regulation based on the heat elimination from the friction pairs of friction elements into the environment.
   2.1. Ventilation and self-ventilation.
   2.2. Heat elimination by special cooling elements.

The staff of the open joint-stock company «Heat-resistant products and engineering development» proposes to improve the railway brake pad design [17] to reduce temperature differences by placing a flat element made of material having lower coefficient of linear thermal expansion than polymer friction material in the surface layer of the back of the pad. This solution makes it possible to reduce the temperature stresses in the surface layer of the pad during periodic heating and cooling due to unequal expansion of its individual areas. This reduces the possibility of the surface layer cracking and maintains the integrity of the back of the pad, which is especially important for pads reinforced with only a metal frame or a shortened metal rear part. In the latter case, the flat element is placed in the unreinforced back of the pad. Better interaction of the pad with the brake block holder and better distribution of pressure on the pad during frictional contact provides maintaining the integrity of the pad back during operation, as a result of which the pad operational factors are improved.

In the patent RU2319871 "Brake pad of a railway vehicle (options)", a structure that contains a metal frame and a composite friction element fixed on it is proposed. The composite friction element is made of two longitudinal layers differing in thermal conductivity. The less heat-conducting layer is made of a composite friction material that has greater adhesion to metal and strength compared to the layer located on the pad working surface. The thickness of the heat-conducting layer is less than the minimum pad thickness allowed for operation, but not more than the thickness from the back surface of the pad to the protruding parts of the metal frame (Fig. 1, a).

a)

b)

c)

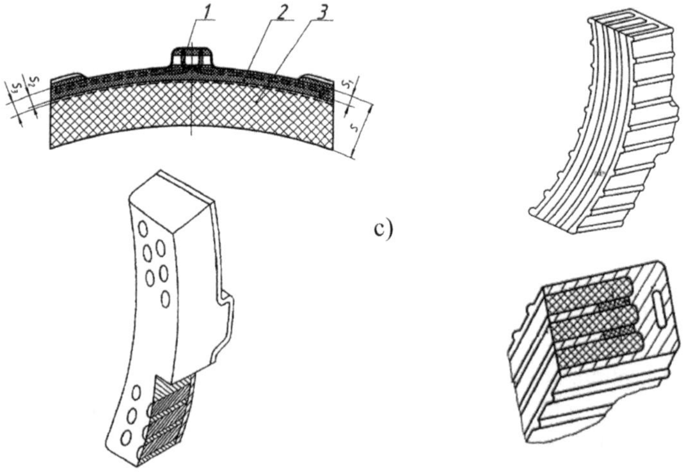

Fig. 1. Analysis of the brake pad designs with improved operational and frictional properties: a) a brake pad of a railway vehicle with a metal frame of reduced thermal conductivity: 1 – is a metal mesh-wire frame, 2 – is a longitudinal less heat-conducting layer of a composite friction element located on the back surface of the pad, 3 – is a longitudinal more heat-conducting layer of the composite friction element, located on the working surface of the pad (working layer); S – is a pad thickness S1 – is a thickness from the back surface of the pad to the protruding parts of the metal frame; S2 – is a thickness of the less heat-conducting layer of the composite friction element; S3 – is the minimum pad thickness allowed for operation; b) bimetallic brake pad; c) binary brake pad.

There is a bimetallic brake pad (Fig. 1, b) [18], which contains an arcuate curved metal bar made by cast iron casting and a steel plate from which steel cylindrical filling pieces extend into the cast iron bar. The filling pieces are located radially, in two parallel rows, unfolded relative to the longitudinal axis of the pad in one direction. The filling pieces are made of a material with a higher abrasiveness than the material of the bar.

The authors of the patent of Ukraine 75974 "Brake pad" proposed a brake pad made of two materials of different thermal conductivity. The body of the pad is made of a material with high thermal conductivity, for example, of cast iron in the form of a box with back and side walls and longitudinal partitions in the middle, and the cavities between them are filled with a composite material. In addition, transverse heat sinking fins of the cooling system are made on the outer surfaces of the body side walls, and fastening holes filled with a composite material are made near the back wall perpendicular to the side walls and partitions (Fig. 1, c). That is, such a pad, due to its design features, with the same resistance against wear, will be able to remove the remaining heat from the contact zone of the pad and the wheel rolling surface. Thus it will ensure the elimination of undesirable effects arising from the use of composite brake pads, and the creation of effective conditions for the operation of railway brake systems while maintaining high resistance against wear of the brake pads used in them.

A wide variety of brake element designs does not solve the problem of significant heating at the contact of the brake elements. An integrated approach is needed to reduce the heating of the brake elements. It is necessary not only to cool the contact, but also to remove heat from the contact through the use of heat-absorbing elements. In the studies of the article authors, the task of stabilizing the temperature in contact of tribological elements through the use of several structural elements is posed. Depending on the type of operation and the type of pads, these solutions can be used either together or separately.

3 Problem Solution Elaboration

To solve the problem of temperature stabilization in the tribological elements of the brake system, several technical solutions have been developed. These technical solutions can be used both together and separately. The maximum temperature stabilization effect will be achieved using all three technical solutions.

1. Let us consider the issue of stabilizing the temperature of the frictional contact of the brake elements. To cool the tribopairs, it is proposed to use inserts made of expanding agents which form an active gaseous medium in the pads when the temperature rises.

In this case, when the railway rolling stock is braking, the pads are pressed against the wheel or disc. The temperature in the tribopair rises. Under the influence of temperature, the process of thermal decomposition of the pad expanding agents begins in the pad, that leads to the release of a significant amount of gas products at a high rate [19]. One of the main decomposition products is nitrogen gas, which interacts with thin surface layers of friction units. This has a positive effect on the frictional properties of the friction pair, increases the adhesive coefficient and sharply strengthens and stabilizes the surface of materials, thereby increasing their durability and wear resistance.

2. Cast iron microparticles formed in the process of wear in the brake pad and the wheel friction zone are relatively large. This also leads to an increase in the intensity of wear in the contact zone of the brake elements, and their movement in the friction zone during motion reduces the coefficient of the brake pad friction on the wheel rolling surface. This affects a decrease in the effectiveness of mechanical braking and an increase in the braking distance, and reduces traffic safety.

Elimination of solid microparticles in contact and increasing resource saving when using the brake system is proposed by introducing new elements into brake system design. These elements ensure the accumulation of compressed air and cleaning the contact with it.

The proposed braking system works in the following way.

Before the train departs, the brake is charged. For this purpose the handle of the brake valve 4 is put into the release position (Fig. 2). In this case, the brake cylinder 1 through the brake control valve 2 and the one-way valve 3 is connected to the bellows 4, in which compressed air is accumulated. The one-way valve 3 allows the compressed air to move in only one direction from the brake cylinder into the bellows and prevents the air from moving in the opposite direction.

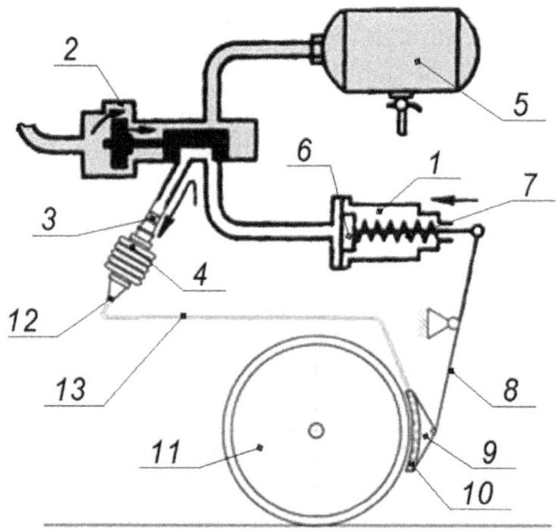

Fig. 2. The braking system of the locomotive with air storage and contact cooling elements.

To brake the train the brake control valve 6 comes into action and disconnects the brake cylinder 1 from the bellows 4 and combines it with a control reservoir 5 filled with compressed air. When compressed air is supplied to the brake cylinder 1, the piston 6 with the push rod 7 moves to the right, and the brake beams 8 at the same time presses the brake shoe 9 with the brake pad 10 to the wheel 11. Then the control valve 12 of the bellows 4 is activated, and the compressed and accumulated air is fed under pressure into the holes 13 (Fig. 3) of the brake pad 10 through the rubber supply line

14. This helps to cool the "brake pad – wheel" contact and to carry the grit along the grooved channels 15 into the environment (Fig. 3).

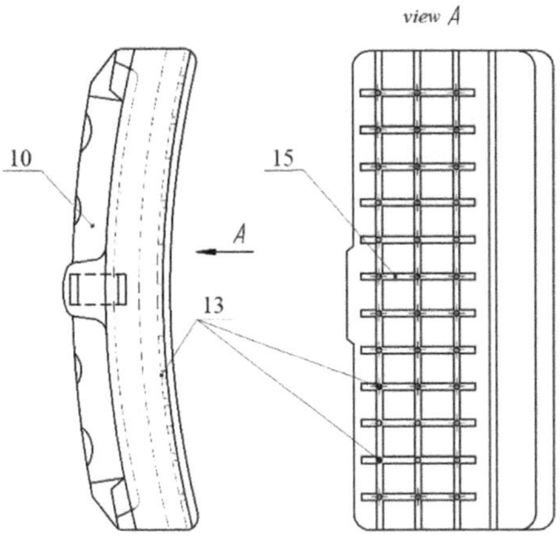

Fig. 3. Brake pad with cooling holes.

To release the brake, the brake valve handle is put in the release position. The supply line is combined with the brake line, as a result of what the pressure rises in it and the brake control valve 2 connects the brake cylinder 1 to the bellows 4 through the one-way valve 3, and connects the line to the control reservoir 5. In this case, compressed air accumulates in the bellows 4. If the emergency braking valve (conductor valve) is opened on the wagon, brakes are automatically activated.

3. It is proposed to use compressed air from the locomotive sanding system to supply cooled air into the contact of the brake pad with the wheel. For this, the nozzle of the sanding system must be equipped with a vortex tube, the design of which is based on the Ranque effect [20]. In this case, the nozzle of the sanding system has three holes: the first one is for cooled air, the second one is for hot air, and the third one is with abrasive material. The outlet for the cooled air is combined with the brake pad holes through the pipeline, while ensuring both the transfer of solid microparticles from the contact and its cooling. Hot air should be used when hauling on dirty rails.

4 Result of Research

During braking, the temperature of the brake pad changes from T_1 to T_2 depending on the time t (Fig. 4). In the absence of contact cooling, the dependence of temperature T on time t has a linear increasing character. When compressed air is supplied to the holes of the brake pad, the curve of the dependence of the brake pad temperature T on the of brake pad pressing time t changes (Fig. 4). An increase in the diameters d of the holes to a certain value leads to a decrease in the temperature T in the contact. It is advisable to make holes from 2 mm to 8 mm, a further increase in the diameter of holes will not affect the temperature characteristics of the brake pad.

When compressed air is supplied, a backpressure force Q arises between the brake pad and the wheel. It is necessary that the backpressure force Q does not impair the braking performance of the rolling stock. To do this, it is necessary to control the diameters d and depth h of holes (Fig. 5). The studies (Fig. 5) show that at a hole depth of $h > 0.5$ mm, regardless of the diameter d, the value of Q is much less than the pressing force of the pads and, accordingly, does not affect the braking characteristics of the rolling stock.

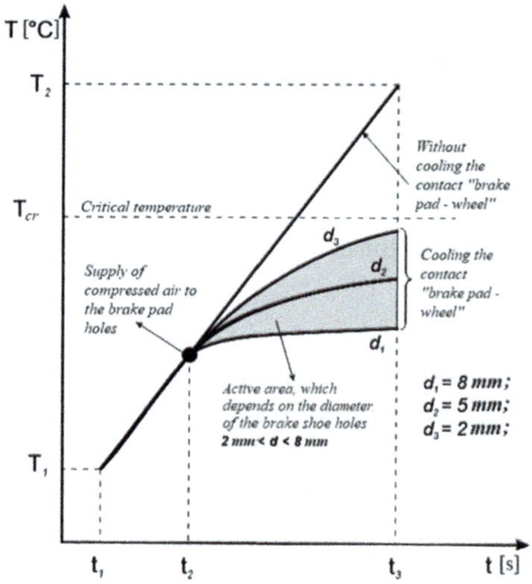

Fig. 4. Brake pad temperature plot against the time of pressing the pads.

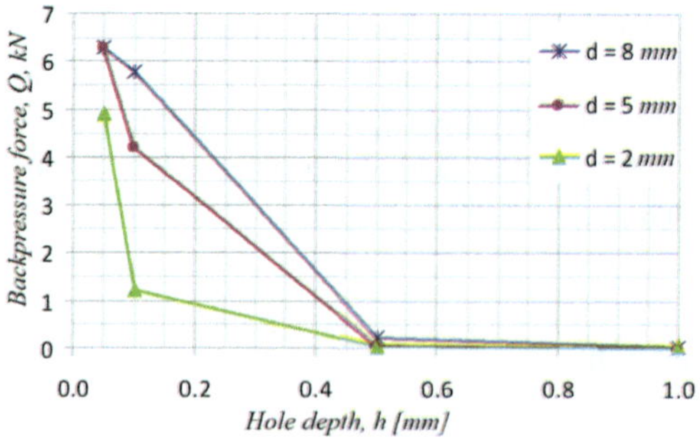

Fig. 5. The backpressure force Q plot against the depth h of the brake pad holes.

The amount of the brake pad wear ΔH in thickness for the case of emergency braking for the time t_s until a complete stop is determined by the formula [8]:

$$\Delta H = \frac{4}{3}\frac{\alpha_p^2 B_b^2 \Delta h v_0^2 t_s}{A^2(\chi F_p)^2(\Delta\tau_{\max} - \Delta\tau_{av})^2 \pi\lambda\gamma c}, \tag{1}$$

where a_p- is a heat flow distribution coefficient, which shows how much of the heat generated by friction has entered the pad; B_b- is a braking force; $\Delta h-$ is an average breaking height of actual contact protrusions; υ_o- is an initial braking speed; $A-$ is a mechanical equivalent of thermal work; $\chi-$ is the coefficient of the actual contact area; F_p- is a geometric area of the pad and wheel contact; $\Delta\tau_{\max}-$ is an increase in temperature and the value at which the destruction of the roughness protrusions occurs (it depends on the pad material and is constant); $\Delta\tau_{av}-$ is an average increasing temperature in the working surface of the pad friction; $\gamma-$ is the specific gravity of the heating wheel; $c-$ is the specific heat.

It can be seen from the formula (1) that the presence of holes and grooves contributes to the effective removal of the third body (wear cast iron microparticles) from the friction zone, that makes it possible to reduce the average destructible height of the protrusions of the actual contact during braking, thereby reducing the wear rate of the pad and increasing its friction coefficient, that improves braking efficiency.

The average increase in the temperature of the working surface of the pad friction is determined by the formula:

$$\Delta\tau = \frac{1}{t_b}\int_0^{t_b} \Delta\tau_s dt, \tag{2}$$

where $\Delta\tau_s-$ is the excess temperature of a surface on which heat is generated.

The change in temperature $\partial(\Delta\tau_s)$ up to time t is caused by the amount of heat generated per surface unit:

$$\partial(\Delta\tau_s) = \frac{q(t)}{\sqrt{\pi\lambda\gamma c}\sqrt{t_1 - t}}dt. \tag{3}$$

We have while taking into account the transfer of heat from the friction surface into the environment:

$$q(t) = q_a\left(1 - \frac{t}{t_r}\right) - \alpha \cdot \Delta\tau_s, \tag{4}$$

where q_a- is the amount of heat released per surface unit per time unit (heat flux density) at the beginning of the temperature rise process; $\alpha-$ is the coefficient of heat transfer from the friction surface into the environment.

The heat transfer coefficient from a heated surface into the environment is defined as the sum of the convection α_c and radiation α_r heat transfer coefficients:

$$\alpha = \alpha_c + \alpha_r. \quad (5)$$

The radiation heat transfer coefficient α_r depends on the given measure of the colour scheme tone (dark or light) of the emitting pad surface, the absolute pad temperature and the environment temperature. In the modified pad design, the radiation heat transfer coefficient α_r is not considered, since it does not significantly affect the wear of the pad, since there is little heat removal from the friction zone of the rolling surface of the wheel and pad into the environment.

The convection heat transfer coefficient α_c is calculated depending on the Reynolds and Prandtl numbers [8].

The presence of grooves in the proposed pad design reduces the intensity of their wear ΔH, especially at high motion speeds, by stabilizing the temperature regime by increasing the efficiency of heat removal from the friction zone into the environment. During braking, the cold air blown under the pad heats up and is efficiently released into the environment. As a result, the coefficient α_c increases, that reduces the average friction temperature of the pad working surface $\Delta\tau_{av}$. So the time to reach the maximum temperature $\Delta\tau_{\max}$, at which the destruction of the protrusions of the actual contact occurs, increases according to the formula (1). In a typical pad, the heating intensity increases when increasing the initial braking speed.

5 Discussion of Results

When braking, the train kinetic energy is converted into other forms of energy, most of which is a heat one. This process is accompanied by an increase in the friction elements' temperature, which affects the braking efficiency and the wear resistance of the brake elements. The problem of resource saving of shoe brakes with which the rolling stock is equipped is multi-scale. It needs a solution of technical and economic, technological, metallurgical and tribological knowledge associated with the choice of wear-resistant materials and methods of cooling the working surfaces of the brake elements. In research, innovative technical solutions to stabilize temperature conditions in the interaction zone of brake elements developed.

To cool the tribopairs, it is proposed to use inserts made of expanding agents which form an active gaseous medium in the pads when the temperature rises. One of the main decomposition products is nitrogen gas, which interacts with thin surface layers of friction units. This has a positive effect on the frictional properties of the friction pair, increases the adhesive coefficient and sharply strengthens and stabilizes the surface of materials, thereby increasing their durability and wear resistance.

Elimination of solid microparticles in contact and increasing resource saving when using the brake system is proposed by introducing new elements into brake system design. These elements ensure the accumulation of compressed air and cleaning the contact with it.

When compressed air is supplied to the holes of the brake pad, the curve of the dependence of the brake pad temperature on the of brake pad pressing time changes. An increase in the diameters of the holes to a certain value leads to a decrease in the temperature in the contact. It is advisable to make holes from 2 mm to 8 mm, a further increase in the diameter of holes will not affect the temperature characteristics of the brake pad.

The presence of grooves in the proposed pad design reduces the intensity of their wear, especially at high motion speeds, by stabilizing the temperature regime by increasing the efficiency of heat removal from the friction zone into the environment. During braking, the cold air blown under the pad heats up and is efficiently released into the environment. It is proposed to use compressed air from the locomotive sanding system to supply cooled air into the contact of the brake pad with the wheel. For this, the nozzle of the sanding system must be equipped with a vortex tube.

6 Conclusion

In the process of research, innovative technical solutions have been developed to improve the conditions for the interaction of tribological elements of the brake system.

These solutions can be effectively used both together and separately. These studies were carried out at an inventive level. According to the calculated data, the presented technical solutions will stabilize the temperature at the contact of the brake elements, remove wear elements. The most efficient brake pad air hole is 2–8 mm.

Acknowledgements. This research was supported by the Slovak Research and Development Agency of the Ministry of Education, Science, Research and Sport of the Slovak Republic in Educational Grant Agency of the Ministry of Education of the Slovak Republic in the project No. VEGA 1/0558/18: Research of the interaction of a braked railway wheelset and track in simulated operational conditions of a vehicle running in a track on the test bench.

References

1. Fomin, O.: Modern requirements to carrying systems of railway general-purpose gondola cars. Sci. Tech. J. Metall. Mining Ind. **5**, 31–43 (2014)
2. Dižo, J., Blatnický, M., Sága, M., Harušinec, J., Gerlici, J., Legutko, S.: Development of a new system for attaching the wheels of the front axle in the cross-country vehicle. Symmetry **12**(7), 1156 (2020)
3. Fomin, O.: Experimental confirmation of the theory of implementation of the coupled design of center girder of the hopper wagons for iron ore pellets. East.-Eur. J. Enterp. Technol. **5**(1 (89)), 11–18 (2017). https://doi.org/10.15587/1729-4061.2017.109588
4. Štastniak, P., Kurčík, P., Pavlík, A.: Design of a new railway wagon for intermodal transport with the adaptable loading platform. In: MATEC Web of Conferences, vol. 235, p. 00030 (2018)
5. Wasilewski, P.: Frictional heating in railway brakes: a review of numerical models. Arch. Comput. Methods Eng. **27**(1), 45–58 (2018). https://doi.org/10.1007/s11831-018-9302-3
6. Milošević, M.S., Stamenković, D.S., Milojević, A.P., Tomić, M.M.: Modeling thermal effects in braking systems of railway vehicles. Therm. Sci. **16**(2), S515–S526 (2012)
7. Suchánek, A., Harušinec, J., Loulová, M., Strážovec, P.: Analysis of the distribution of temperature fields in the braked railway wheel. In: MATEC Web of Conferences, vol. 157, p. 02048 (2018)

8. Kazarinov, V.M., Inozemtsev, V.G., Yasentsev, V.F.: The theoretical basis for the design and operation of automatic brakes. Transport, Moscow, p. 400 (1968). (in Russian)
9. Vodiannikov, Yu.Ya., Svistun, S.M., Makeyeva, Ye.G.: Computational and experimental method for determining the heating temperature of a freight wagon wheel during braking. Bull. Volodymyr Dahl East Ukr. Natl. Univ. Part 1 **5**(176), 50–55 (2012). (in Russian)
10. Barta, D., Dižo, Ján., Blatnický, M., Vaňko, J.: Experimental investigation of the motorcycle braking properties when riding on different road surfaces. IOP Conf. Ser. Mater. Sci. Eng. **1002**(1), 012030 (2020)
11. Belhocine, A., Bouchetara, M.: Thermal analysis of a solid brake disc. Appl. Thermal Eng. **32**, 59–67 (2012)
12. Ghadimi, B., Kowsary, F., Khorami, M.: Thermal analysis of locomotive wheel-mounted brake disc. Appl. Therm. Eng. **51**(1–2), 948–952 (2013)
13. Inozemtsev, V.G., Vinogradov, B.M.: Improving the reliability of freight wagon mechanical brake equipment. In: Proceedings of the Central Research Institute of the Ministry of Railways "Development and Improvement of Automatic Brakes, vol. 607, pp. 15–24 (1974). (in Russian)
14. Vukolov, L.A,. Fomina, A.G.: Methods of accelerated assessment of the operational reliability of composite brake pads. In: Proceedings of the Central Research Institute of the Ministry of Railways "Improvement of Automatic Brakes", no. 413, pp. 47–66 (1970). (in Russian)
15. Smolyaninov, A.V., Smolyaninov, V.S., Smolyaninov, P.V.: Improving the efficiency of freight wagon brakes by improving the repair technology: technological support for repair and improving the dynamic qualities of railway rolling stock. In: Proceedings of the All-Russian Scientific and Technical Conference with International Participation, pp. 115–122. Omsk State Transport University, Omsk (2011). (in Russian)
16. Damdyn, S. I.: Increasing the energy intensity of wheel brakes by introducing a reverse friction pair with liquid-cooled disks. Ph.D. thesis, Moscow (1994). (in Russian)
17. Ignatiev, D.M., et al.: Patent for invention No. 2115044. Brake pad for railway rolling stock. Accseed 07 Oct 1998. (in Russian)
18. Bychkov, V.N.: Patent No. 2413643. Sectional bimetallic brake pad. Accessed 03 Oct 2011. (in Russian)
19. Mitrofanov, A.P.: Increasing the efficiency of grinding by impregnating an abrasive tool with substances from the class of expanding agents, Volgograd, p. 16 (2012). (in Russian)
20. Martynov, A.V., Brodyansky, V.M.: What is a vortex tube? Energia, Moscow, p. 152 (1976). (in Russian)

3D Simulation and Analysis of the Course of a Bus/Train Accident at a Railway Crossing

Michal Ballay[1(✉)], Bohuš Leitner[1], and Ľudmila Macurová[2]

[1] Faculty of Security Engineering, University of Zilina, 1. mája 32, 010 26 Zilina, Slovakia
{michal.ballay,bohus.leitner}@fbi.uniza.sk

[2] Institute of Forensic Research and Education, University of Zilina, 1. mája 32, 010 26 Zilina, Slovakia
ludmila.macurova@uzvv.uniza.sk

Abstract. Rail crossings are very important elements in the field of land transport infrastructure, as they allow the combination of two different transport systems, whose mode of operation and management is subject to different regulations, transport performance and design solutions. Traffic accidents at railway crossings have become a relatively common negative phenomenon (meeting of road users with rail vehicles) with fatal consequences and extensive material damage. For this reason, level crossings are high-risk locations, with a high potential for an accident. The aim of the article is to present the results of the simulation and subsequent analysis of the accident of category A3 - collision of a train with a bus at a railway crossing and selected results from the realized computer simulation of the course of the pre-accident event and the predicted impacts of the modeled accident.

Keywords: Railway crossing · Safety · Traffic accident · Simulation

1 Introduction

Crossings are unique in the world of transport, as they represent the only possibility of a collision between two different infrastructures, the management system of which is subject to different regulations and, last but not least, vehicles with dramatically different performance and design. As a result, level crossings are high-risk locations where the potential for an accident is high, due to the fact that that the railways control only half of the level crossing environment. Unfortunately, traffic accidents have happened in recent decades a recurring phenomenon resulting in human deaths and major economic losses. Although there are legal norms, traffic regulations, standards in the design of roads, ultimately the movements of participants, they are not organized and monitored by one specific entity, as are railway movements.

Statistics show that road accidents account for more than a quarter of all rail accidents in the EU (Fig. 1) [1].

O. Prentkovskis et al. (Eds.): TRANSBALTICA 2021, LNITI, pp. 39–47, 2022.
https://doi.org/10.1007/978-3-030-94774-3_4

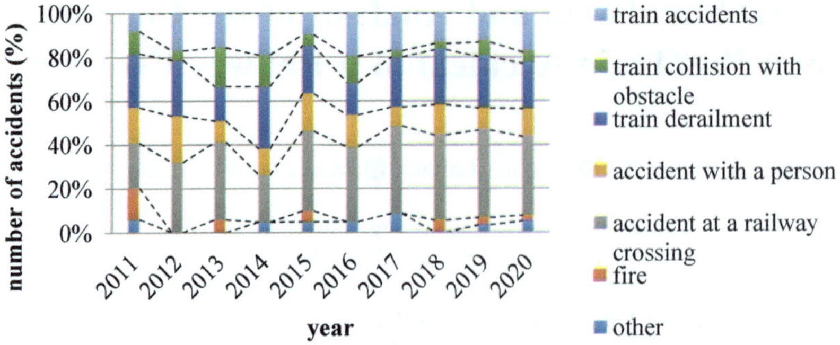

Fig. 1. Accident in rail transport in the EU [1].

In addition to statistical indicators of accidents at level crossings, the authors of the post had after the expiry of the statutory period. Details of the 2009 tragic accident, where 12 people were killed, 6 suffered severe injuries and 19 minor injuries. Based on a computer simulation of an accident [2–4], an analysis of the causes and course of an accident, it has been shown that the use of state-of-the-art safety features is necessary at level crossings where level crossing of road and railway infrastructure is not possible (overpass over the railway).

2 Analysis of Real Accident Event in Railway Crossing

To perform an accident analysis to identify the possible impacts the emergence and course of an accident in this article, we will deal with an accident of category A3 - a train collision with a bus on a rail crossing. From the point of view of accident events within the Slovak Republic it's the most tragic accident which killed 12 people, 6 suffered severe injuries and 19 were slightly injured. Accident event became an unsecured railway crossing in the Hel'pa - Polomka intersection section. The overall situation of the traffic accident site is shown in Fig. 2. The traffic situation and conditions are listed in Table 1 [1, 5].

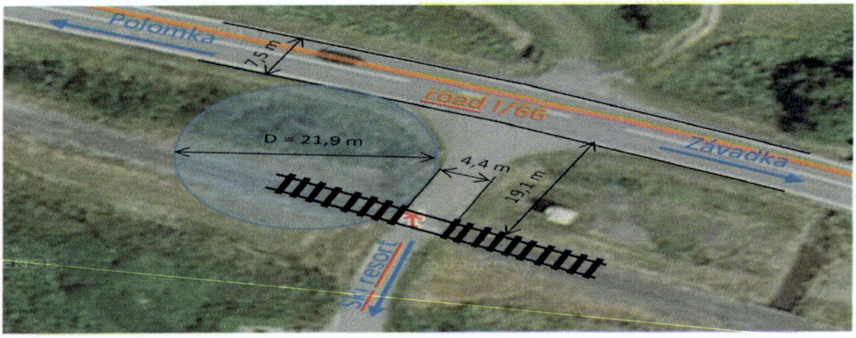

Fig. 2. Situation of a city accident [2].

Table 1. Traffic situation, road surface condition and weather conditions.

Directional conditions:	Straight section
Road surface:	Resin
Surface quality:	Good
Surface condition:	Dry surface, not dirty
Edge condition:	Curb is missing
Road fault:	No fault
Road division:	Two-lane
Special objects:	Railway crossing not secured by barriers or light warning devices
Speed limit:	60 km/h
Weather:	Unimproved weather conditions
Visibility:	During the day, visibility not reduced due to weather conditions
Viewing conditions:	Good

The course of the accident was known on the basis of the testimony of witnesses to the accident and the subsequent analysis of the accident. The accident happened so that the bus turned right off the main road and without stopping he entered the railroad crossing which is located about 19 m from the right edge of the main road. On the rail crossing, the bus entered when crossing the railway crossing from a bus driver's perspective, a train approached composed of a motor car and a wagon. The bus was thus pushed 26 m until the train stops. Figure 3 shows a 3-dimensional impact calculation [1, 6].

Fig. 3. Three-dimensional representation of the shock calculation [2].

When analysing an accident event - a train collision with a bus, it is necessary to find out how did the bus move before the accident what was the time-spatial relationship with the incoming train, what was the technical condition of the vehicle and its possible impact on emergence and the course of a traffic accident [8]. These questions can be answered mathematical-graphical analysis of vehicle movement using computer simulation and detailed diagnosis of the brake system, steering, transmission system and chassis parts. Impact analysis was performed using the PC CRASH computer program. The actual situation at the scene of the accident with the technical parameters was modelled in the program [7–9].

The analysis of the view conditions at the accident site was processed into a precise plan, which was processed using 3-dimensional graphics. Technical parameters for modelling include: specific vehicles, bus operating weight, number of passengers, average weight with luggage, total weight (luggage + persons), total weight of the vehicle, weight of the locomotive, weight of the wagon. With the aid of the above graphics,

Figure 4 and Fig. 5 shows exactly the viewing conditions at the accident site [2, 6, 10].

Fig. 4. Viewing distance in the PC CRASH 8.3. [2].

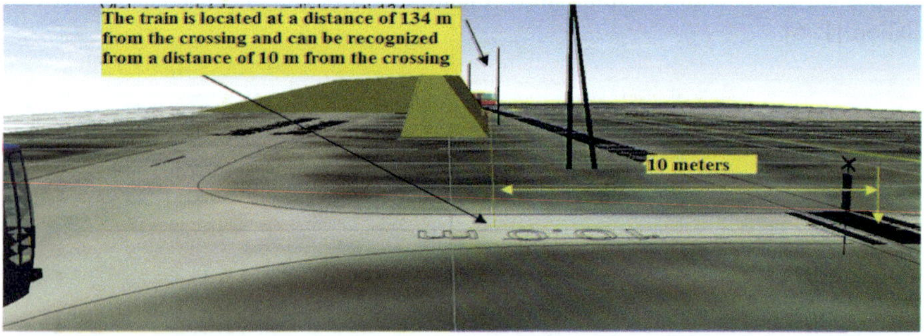

Fig. 5. Three-dimensional representation of viewing conditions for the distance of 10 m from the crossing [2].

The impact calculation showed that the train had a speed at the moment of impact corresponding to the data according to the results of the investigation at the level of 54 $km.h^{-1}$. According to the calculation of the impact, at the moment of impact, the bus was moving at a speed of 8 $km.h^{-1}$. The calculated speed data needs to be corrected due to the fact that it was not possible to model the guiding properties of the railway wheels and rails with sufficient accuracy during the calculation. The profile of the wheels puts resistance in the lateral direction to a greater extent than could be modeled using the PC CRASH 8.2 software. Therefore, it can be admitted that the speed of the bus was slightly higher than calculated at the moment of impact. Based on the above, it can be argued that the speed of the bus at the moment of impact was about 10 km.h^{-1} [1, 3, 4, 6].

When analyzing the mutual movement of the bus with train before impact it is important to note. It was found in the impact calculation that the train had a speed of 54 km.h^{-1} at the moment of impact and the bus was moving at between 8 and 10 km.h^{-1}. In the following sections article is an analysis of the overall situation before the collision [2, 6].

- **Time 11.0 s before collision** - Train is about 200 m from the crossing, bus starts turning off the main road. The bus driver is not yet in direct sight of the train because the train is located in the obstructed space for the bus driver. The train is already recognizable from the crossing. Plan view of train and bus position 11 s before impact is presented in Fig. 6 [2, 6].

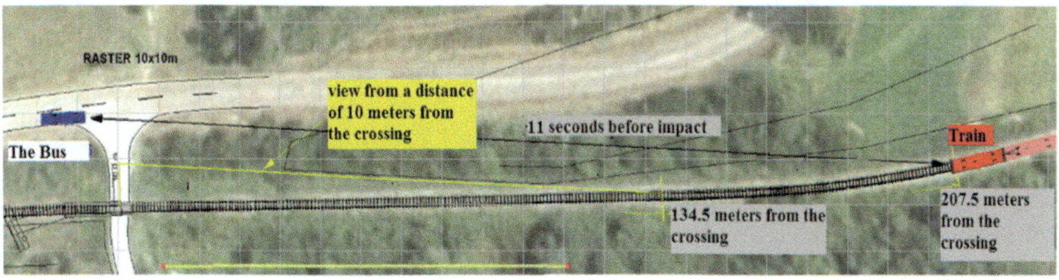

Fig. 6. 3 – dimensional view of the train and bus position 11 s before the impact [1].

- **Time 7.2 s before impact** (Fig. 6) – Train is located 132 m from the crossing. The train is already recognizable 10 m away from crossing, but the bus driver hasn't seen the train yet because the bus was more than 10 m away [2, 6].

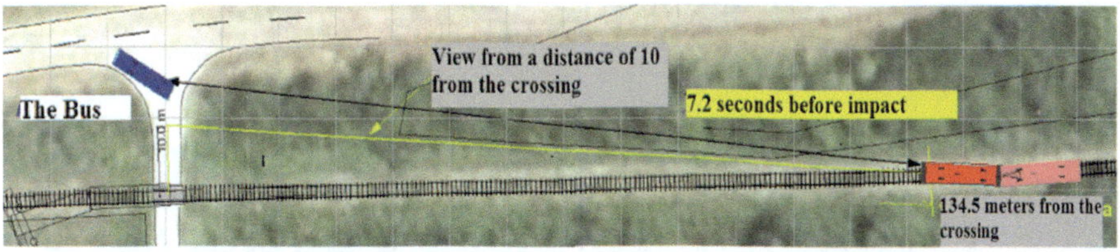

Fig. 7. 3 – dimensional view of the train and bus position 7.2 s before the impact [2].

- **4.8 s before impact** (Fig. 7) - The train is about 85 m from the crash site. The bus is located 6.8 m from the rail axis. The bus is just a short distance away should the driver start to react to the speed of 10 km.h^{-1} he could stop with the bus at a safe distance from the track that is, 4.0 m from the track axis. The track required to stop the vehicle calculation is done as follows:

$$S_Z = \frac{v^2}{2.a_s} + v.t_r, \tag{1}$$

where a_s is medium vehicle deceleration (m.s^{-2}), v is vehicle speed (m.s^{-1}), t_r is driver reaction time: $t_r = 0.7$ s, S_z is track required to stop the vehicle 2.74 m.

Distance from rail for the latest reaction 4 m + 2.74 m = 6.74 m. With a bus speed of 10 km.h^{-1}, the driver had 1.0 s to recognize the train to be able to react and stop the bus at a safe distance from the track after recognition [2, 5, 6, 10].

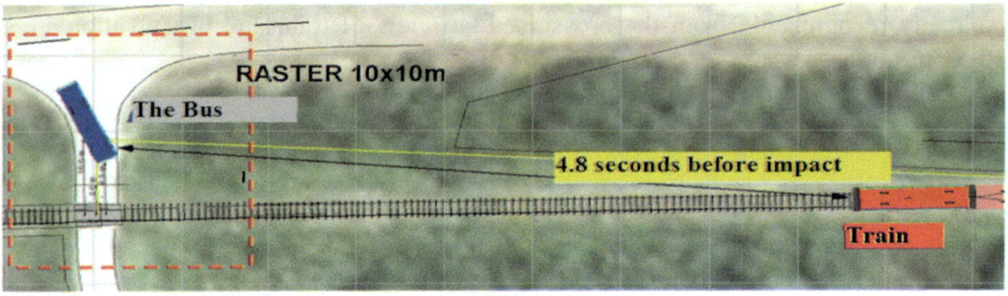

Fig. 8. Bus movement 4.8 s before impact [2].

- **Time of 2.0 s before impact** (Fig. 8) - At this point, a quick-acting brake starts to operate. The train has no way to stop. Despite the use of a high-speed brake the train will slow down from 70 km/h to a crash speed of 54 km/h [2, 6] (Fig. 9).

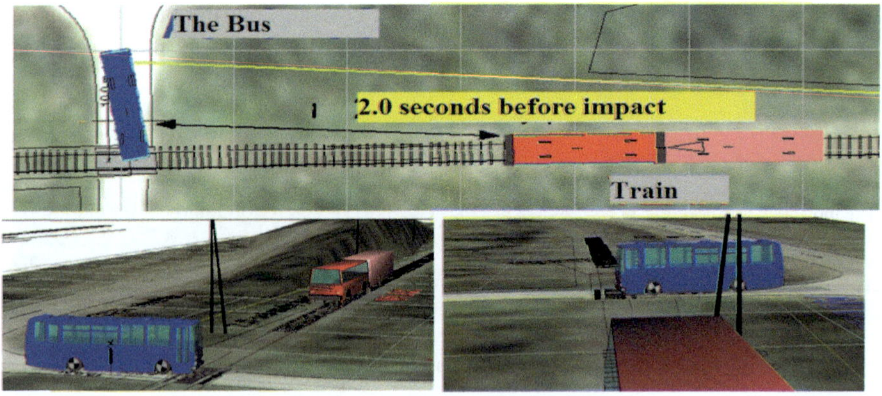

Fig. 9. Bus movement 2 s before impact [2].

It follows, that the bus driver at 10 km.h^{-1} was able to observe the incoming train at least 1.0 s to 3.2 m long stretch in front of where he should start responding at the latest and then brake, to stop at a safe distance from the railway. We consider the time and section as sufficient, while nothing prevents the bus driver from driving slower thereby making the necessary time to detect the incoming train. According to the detected bus movement the bus driver did not respond to the incoming train at all [2, 6].

Analysis found that the driver reacted when he had the opportunity to first recognize that bus approaching can't stop to the track at a safe distance from the track (Fig. 10). According to the analysis, the driver reacted in time as soon as the risk of collision could be recognized with the approaching bus [2, 6].

Fig. 10. Dimensional representation of the shock calculation [2].

We didn't find any circumstances by analyzing the accident that would keep the driver away to see an incoming train from the crossing and respond properly. The train driver responded in time when he had the opportunity to first recognize that the bus approaching the rail crossing will not stop. After the reaction, the driver of the train exploited in a maximum way technical possibilities to stop the train to avert and mitigate the effects of the impact the train to the bus. In the present case traffic accident participants only bus driver options to prevent a traffic accident. In the overall accident analysis and the technical condition of the bus, we found no circumstances to the driver before entering the crossing in time recognize an incoming train respond properly to the situation and stop the bus at a safe distance from the level crossing. However, in the present case several conflicting circumstances have been identified, namely:

- The level of railway crossing (as well as the condition of the road ahead of the rail crossing) he was like that that it contained inequalities and the allegations that made it logical that drivers are increasingly paying attention to directional guidance in front of the crossing, as well as driving through the railroad crossing (in order to drive the vehicle away from bulges and bumps). As a result of this factor, there was a phenomenon that drivers have less attention to drivers the situation in terms of incoming train.
- For car drivers coming in the direction of driving the bus arose on the left side of the covered view area as a result of a bush stand next to the track. For such conditions (with reduced viewing ratios) there is a phenomenon where time goes down available to the driver for proper evaluation whether the train is approaching the railway crossing.
- For drivers of longer vehicles (including buses) is for crossing the railroad crossing (under conditions of reduced vision - as happened in the present case). You need to drive through the rail crossing faster. However, this factor is contrary to the state of

rail crossing as a result of which drivers tend to move at a slower speed. However, the bus speed increases via the railroad crossing thereby reducing the driver's time to evaluate the situation in the presence of any incoming train, since the driver's view of the direction from where he came from is revealed train just before entering the railway crossing.

These factors combined with reality that bus passengers (including driver) after a relatively long journey, they were just before their destination (and that was the ski slope - to which they have seen the outlook just before the accident occurred, when the driver's attention to some extent this phenomenon ultimately led to that the bus driver responded late on an incoming train. It was clearly appropriate in the circumstances the railway crossing in question was fitted with gates with light signalling device possibly reduced maximum permitted train speed [2, 5, 6, 8, 10].

3 Conclusions

The railway crossing is a place of crossing the road with the railway network. In the event of various accidents at level crossings, road users are most at risk. From the point of view of the analysis of a traffic accident at a railway crossing, which is analysed in the article in question is clear that that the only cause of the accident from a procedural point of view was the misconduct of the bus driver. Due to the accident in question, which was analysed in the article, there was a long discussion from forensic experts and various experts. It lasted about 3 years. Alpha and omega in this accident were the conclusions drawn by the Institute of Forensic Engineering in Žilina. The bus driver was not convinced before crossing, whether it is possible to cross the level crossing safely and entered the crossing in time, when it was already possible to observe the incoming train set. In this case, there was a failure of the human factor, however, the conditions around the railway line also contributed to it, in particular the high growth around the line and the directional conditions of the line before crossing. The advantage of using computer modelling and traffic accident simulations and analysis of the conditions of any accident is in particular the possibility of experimenting with the model, changes in conditions before and during the accident, determining the dependence between the monitored quantities, etc.

One of the possibilities is the use of modelling of accident events, creating scenarios of the occurrence, the course of the accident, its expected consequences, but also the adequate response of intervening rescue services and their preparedness for common but specific traffic accidents.

References

1. European Railway Agency. Railway safety performance in the European union 2016–2020, France (2021). Accessed 10 May 2021, https://www.era.europa.eu/sites/default/files/library/docs/safety_interoperability_progress_reports/report_on_railway_safety_and_interoperability_in_the_eu_2020_en.pdf
2. Kohút, P.: Expert opinion. Institute of Forensic Engineering, University of Zilina (2009)

3. Kim, I., Ferreira, L., Tey, L.S., Wallis, G.: Integration of driving simulator and traffic simulation to analyse behavior at railway crossings. Proc. Inst. Mech. Eng. Part F J. Rail Rapid Transit **227**(5), 427–438 (2013)
4. Tey, L.-S., Ferreira, L., Wallace, A.: Measuring driver responses at railway level crossings. Accid. Anal. Prevent. **43**(6), 2134–2141 (2011)
5. Macurova, L., Kohut, P., Copiak, M., Imrich, L., Redl, M.: Determinig the energy equivalent speed by using software based on the finite element method. In: International Scientific Conference on LOGI - Horizons of Autonomous Mobility in Europe. Ceske Budejovice, Czech Republic (2019)
6. Kohút, P., Kasanický, G.: Evaluation of crash tests of the institute of forensic engineering, University of Zilina, aimed at EES. Commun. Sci. Lett. Univ. Zilina **15**(2), 124–130 (2013)
7. Sventekova, E., Urbancova, Z.: Increasing of transport safety by using the intelligent signalling device RMS. In: Transport Means 2018. Part II: Proceedings of the International Scientific Conference, pp. 807–810. Kaunas University of Technology, Kaunas (2018). ISSN 1822–296X
8. Lovecek, T., Sventekova, E., Maris, L., Rehak, D.: Determining the resilience of transport critical infrastructure element: use case. In: Transport Means 2017. Part I-III: Proceedings of the International Scientific Conference. Kaunas University of Technology, Kaunas (2017)
9. Mikhailov, E., Semenov, S., Dižo, J., Kravchenko, K.: Research of possibilities of reducing the driving resistance of a railway vehicle by means of the wheel construction improvement. In: Trancom 2019 [Electronic]: Conference Proceedings, pp. 831–838. Elsevier Science, Amsterdam (2019). ISSN 2352–1465
10. Kasanický, G., Kohút, P.: Analysis of single-track vehicle accidents.In: EDIS, p. 488. University of Žilina, Žilina (2003). ISBN 80-7100-599-1
11. Dižo, Ján., Blatnický, M., Steišūnas, S.: Multibody simulation of rail vehicle running considering track flexibility. In: Gopalakrishnan, K., Prentkovskis, O., Jackiva, I., Junevičius, R. (eds.) TRANSBALTICA 2021. LNITI, pp. 309–319. Springer, Cham (2020). https://doi.org/10.1007/978-3-030-38666-5_33

Optimization of a Trestle Weight of an Operating Hydraulic Jack Used During Wagons Repairing

Miroslav Blatnický[1], Ján Dižo[1(✉)], Alexander Kravchenko[2], and Stasys Steišūnas[3]

[1] University of Žilina, Univerzitná, 8215/1, 010 26 Žilina, Slovak Republic
{miroslav.blatnicky,jan.dizo}@fstroj.uniza.sk
[2] Zhytomyr State Technological University, 103, Chudnivska Street, Zhytomyr 10005, Ukraine
avtoap@ukr.net
[3] Vilnius Gediminas Technical University, Saulėtekio al. 11, 10223 Vilnius, Lithuania
Stasys.steisunas@vilniustech.lt

Abstract. Safety of a technical solution is the first and very important factor for the production of any product. It is reached by a suitable type of a structure, and also by a suitable solution of a relation between a person and a product itself. A proper ergonomic rules of a design require to prevent any injury in a case, which a human body occurs in a gap between a trestle and the ground. In term of physics, it is necessary to reduce the control forces as much as possible. This article is focused on optimization of a trestle weight of an operating hydraulic jack. The research is aimed at finding a compromise between a net weight and the safety of the design. For this purpose, numerical calculations were performed. Reached results have shown, that it is possible to reduce the weight of an original trestle in compliance of design safety in term of strength and stability. An optimization process of the trestle structure results to an optimal design in term of safety as well as the weight.

Keywords: Optimization · Steel structure · FEM analysis · Trestle · Manipulation

1 Introduction

Transportation and manipulation of material are activities, which move of required of amount of materials to a certain location in the required moment, in the certain sequence, to the proper position and for as low price as possible. Generally, there are hundred thousand tons of materials processed every day, which requires using of large amount of labor [1]. An elemental condition, which has to be met for the proper choice of material of structures, is the ability to transmit static and dynamic loads. For manipulation machines, steel is the most important material for production for structures [2]. Common quality steels offer sufficient mechanical properties, at which it is able to withstand high loads thanks to the well-designed geometry [3].

O. Prentkovskis et al. (Eds.): TRANSBALTICA 2021, LNITI, pp. 48–57, 2022.
https://doi.org/10.1007/978-3-030-94774-3_5

All components of a structure interact with each other. Therefore, it is necessary to transmit loads safely. Every joint has its specifications, which relate with load capacity, stresses, assembly rules etc. A designer must take into account these factors [3, 4].

Fig. 1. A configuration of the hydraulic jack and a steel structure of the original trestle in operation – repairing of a wagon bogie, RailClininic Company.

This article is focused on optimization of weight of a steel structure of a hydraulic jack trestle, which is marked as TH 600/300–450 (Fig. 1). It is used during a repairing processes of railway wagons [4, 5]. Generally, jacks of this type are made of high-strength alloys of light metals [6]. They work on the principle of a hydraulic over-pressure system with the operational pressure of 30 MPa.

2 Materials and Methods

The used hydraulic jack (Fig. 2) includes an adjustable check valve as an overload protection. This prevents the load from slipping in case of the hose breaks and also it protect the lifting elements against overload. Safe and smooth operation is also ensured under the full load. It is possible to extend stroke of the jack by means of supporting components of a piston and cylinder.

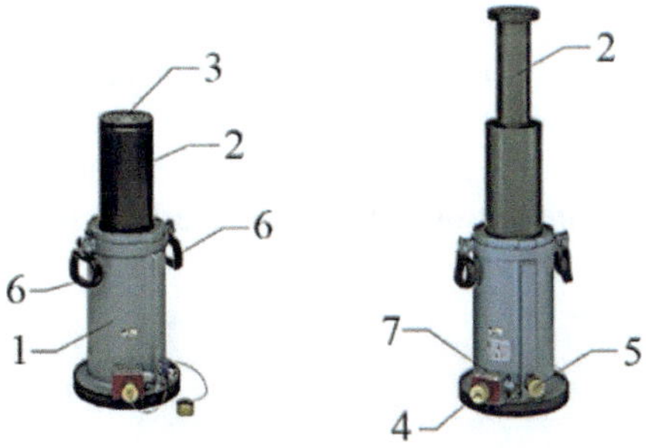

Fig. 2. Basic parameters of the hydraulic jack TH 600/300–450: 1 – a cylindrical body, 2 – a piston, 3 – a push part, 4 – hydraulic connection with a protection (for burden lifting), 5 – hydraulic connection with a protection (for burden releasing), 6 – handles for manipulation, 7 – an adjustable overload valve.

Operational parameters of the hydraulic jack TH 600/300–450 are listed in Table 1. These values will be important for definition of boundary conditions within of numerical analysis of a trestle. A smooth operation of a jack means, that the load is static without jerky running, which would lead to the dynamic load of a structure. The maximal load value as well as the trestle diameter are the most important parameters for its engineering design. At the same time, these are the boundary conditions of a calculation.

Table 1. Operational parameters of the hydraulic jack TH 600/300–450.

Parameter	Value
Height	450 mm
Number of pistons	2
Total stroke	566 mm
Stroke on the first degree	278 mm
Stroke of the second degree	288 mm
The lifting force on the first degree	681 kN
The lifting force on the second degree	285 kN
Volume of hydraulic oil	9.1 l
Diameter of a pushing part	140 mm
Diameter of a bottom flange	295 mm
Operational weight	58 kg
Operational pressure	30 MPa

The operator of the trestle, the RailClinic Company, requires to optimize this component from several points of view. Ergonomics is the first point of view. An operator must not to be in contact with sharp edges, corners or other parts, which prevent to capture human body parts or the clothes. Open profiles must be modified, sharp edges of sheet metal parts must be bent or modified by a similar way.

The other important requirement of the customer is trouble-free maintenance of the structure. Therefore, it is necessary to take this fact into account already in the design in terms of sufficient accessibility for maintenance with regard to the body dimensions of the worker, together with the tools used and any protective equipment. Furthermore, ease of maintenance and elimination of a number of special aids. In the view of the different body predispositions, the structure must be designed in such a way that it can be easily maneuvered by different operators with the least possible force.

Fig. 3. An original trestle under a hydraulic jack with the weight of 140 kg.

The operational conditions is the other point of view. The trestle has to be used on a gravel ballast of a railway track out of sleepers. Therefore, it is necessary to fulfil the requirement of the longitudinal and lateral stability, thus, the trestle has to be wider and shorter. The required load capacity is of 80 tons, i.e. 800 kN. The original trestle is shown in Fig. 3.

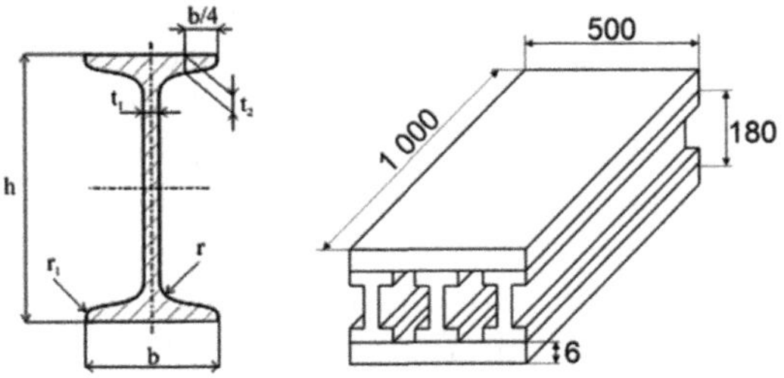

Fig. 4. Dimensions of an original trestle under a hydraulic jack with the weight of 140 kg.

Because of its high weight, the manipulation by workers is problematic. The design itself has consisted in choice of material. The structural steel S235JR was a preselected material for the trestle due to its availability and the price. The "I" profile (Fig. 4 left) is proposed as a main part of a trestle beam. Its geometrical parameters are listed in Table 2. This profile corresponds to the primary customer's requirements, which comes from the given design (Fig. 4 right).

Table 2. Geometrical parameters of the normalized "I 160" profile.

Parameter	Value
b	74 mm
h	160 mm
t_1	6.3 mm
t_2	9.5 mm
r	6.3 mm
r_1	3.8 kN
Cross-section	2,280 mm^2
Theoretical weight	17.9 km·m^{-1}

Thus, this profile was used for calculation of reduced stresses under the maximal load of the jack structure. The mass m of steel plates is given their dimensions a, b, and c and is calculated using the well-known formulation:

$$m = \rho \cdot V. \tag{1}$$

The density ρ of the used steel is supposed of 7,860 kg·m^{-3}. The volume V of a plate is calculated as following:

$$V = a \cdot b \cdot c. \tag{2}$$

By substituting of corresponding parameters, the plate volume is of $3 \cdot 10^{-3}$ m^3. Then, the plate mass is of 24 kg. This value is up-rounded, because we have to also consider increasing of material due to welding. The unit mass of the "*I* 180" profile is of 22 kg·m^{-1}. As the designed technical solution consists of three "*I*" profiles and two plates, the total weight of the trestle is of 114 kg. It is 14% deviation of the required weight. It would be optimal, the weight will not exceed the weight of 100 kg. In such a case, manipulation with the trestle by two workers would be significantly easier. This solution would represent reducing of the total weight of 30%.

The trestle will be manufactured by means of the welding technology. The critical cross-section of the welding joint *S* is considered as the thinner thickness of the welded parts. A cant of a weld face is not considered due to safety. However, we have to take into account residual stresses [7–9]. The welding process causes internal heating changes in the material, which lead to plastic deformation of the welded parts because of these residual stresses [10, 11].

2.1 Strength Analysis of the Original Trestle Structure

Numerical analyses are carried out in a commercial simulation software Adina [12]. This software works based on the finite element method [13, 14]. The trestle model has been created as a shell structure because of advantages resulting from its properties, it has been set-up as it can be seen in Fig. 4. Material properties were prescribed for all variant solution (Table 3).

Table 3. Material properties prescribed in the FE model.

Young's modulus [MPa]	Poisson's ratio [–]	Density [kg·m^{-3}]
$2.1 \cdot 10^5$	0.33	7,850

As a next step, the boundary conditions definition was performed. They are depicted in Fig. 5 by the coupling *C*, *y* direction. In one corner, all degrees of freedom are restrained, Fig. 5, coupling *D*, in the opposite corner, the rotation of the body in the *B* point is restrained, Fig. 5 right, *z* direction.

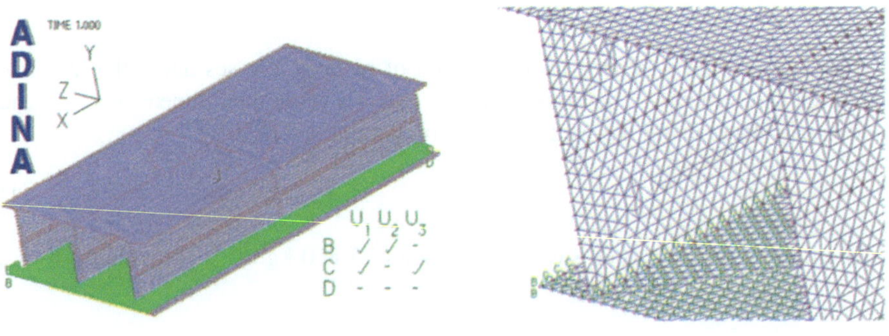

Fig. 5. Boundary conditions (left) and the FE mesh of the model (right).

Parameters of the finite element mesh are listed in Table 4.

Table 4. The FE mesh parameters.

Element size	Mesh algorithm in Adina software	Element type
6 mm	Advanced front	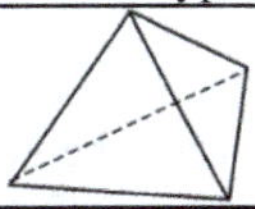

The load was applied to the symmetry center of the structure in the form of the continuous load. This load has been calculated to the jack pressure value through its base plate with the diameter of d = 295 mm. The load of F = 800 kN (the gravitational force of a burden) was applied to the structure (Fig. 6).

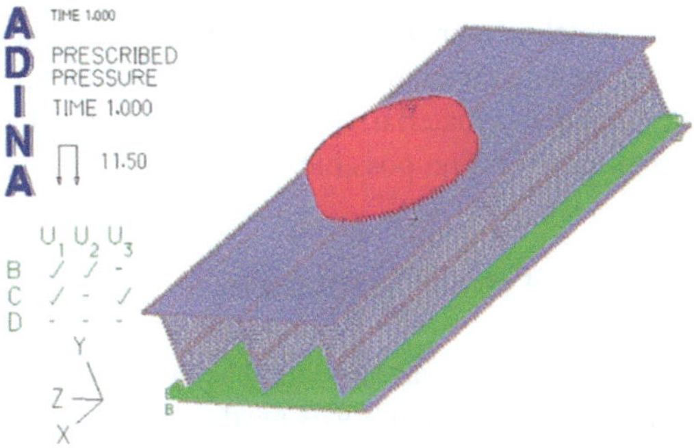

Fig. 6. The load of the structure according to the RailClininc requirements.

As it can be seen in Fig. 7, the calculated stresses in the structure do not meet requirements of safe use in the practice.

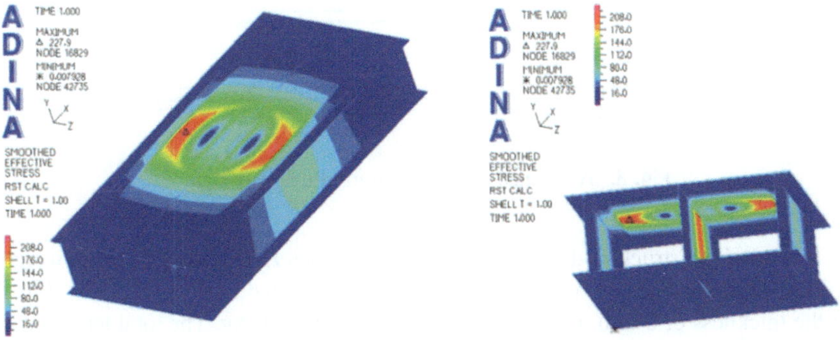

Fig. 7. Distribution of the equivalent stresses in the trestle structure under the load of 800 kN implemented to the axis of symmetry on the jack bottom with the diameter of 295 mm (left), the view of the middle profile web (right).

The used material, S235 steel, has the yield of strength of R_e = 235 MPa. The design stress is chosen regarding to the static loads of the welded structure with the safety coefficient of k = 1.6. The permissible stress σ_{redper} of the welded structure material is given by:

$$\sigma_{redper} = R_e/k'. \tag{3}$$

Substituting the input values, we get the value of σ_{redper} = 150 MPa. The maximal critical value of 228 MPa calculated by the software is lower by 3% than the yield of strength, however, we have to take into account the calculation accuracy as well as the fact, that the structure is affected by the welding. In compliance with the theoretical recommendation for the manual welding process (likely, it will be used), it is necessary to reduce the yield of strength approx. by 30%. Therefore, the permissible value is of 165 MPa. The jack loads the structure in a static way, by the disappearing load. It will be used just sporadically, the fatigue analysis is not needed. Within the safety, we propose to increase the safe stress to the value of 150 MPa for the S235 steel (under the described vertical load), which is more of 119 kN then the maximal lifting force of the TH600/300–450. It is possible to conclude, that this design of the trestle is not possible to use safely with the considered material. The material can be changed by the S355 steel. In such a case, the use could be possible.

3 Optimization of the Trestle Structure

The original design of the structure was not suitable in term of stresses distribution. Therefore, the optimization of the structure was needed. There were several choices of the structure modification. This optimization process included various combinations of vertical and horizontal profiles layouts, which were subsequently analyzed in the Adina software. The optimization resulted to the final design of the structure, which is shown in Fig. 8.

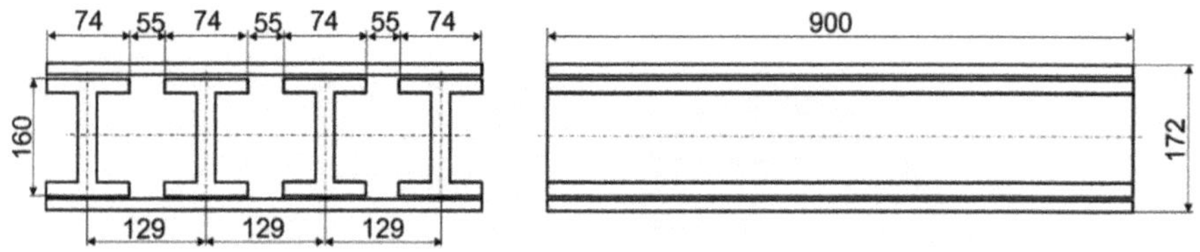

Fig. 8. A geometry of the optimized trestle design.

The optimized structure design includes four profiles I160. Every I profile weights of 16 kg. These profiles are placed and welded between two horizontal plates (Fig. 8) with the thickness of 6 mm. Every plate has the weight of 20 kg. The total length of the trestle is of 900 mm, which allowed to reduce the unit weight (the original design has the length of 1,000 mm, Fig. 4). The gaps between vertical I profiles is proposed of

55 mm. Thus, the total width of the trestle is of 461 mm (originally 500 mm), what also contributes to reduce the total weight of the trestle. The achieved final total weight of the trestle is of 104 kg, which is acceptable value in term of requirements.

3.1 Strength Analysis of the Optimized Trestle Structure

The optimized trestle structure was analyzed in the Adina software. Boundary conditions were prescribed at the same manner as in the previous case (Sect. 2.1). They included definition of degrees of freedom and the acting force. Results of analyses are shown in Fig. 9. The distribution of stresses in the upper plate is shown in Fig. 9 left.

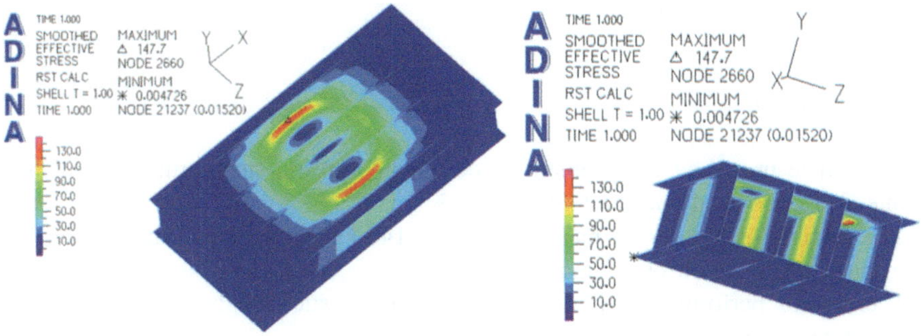

Fig. 9. Distribution of the equivalent stresses in the optimized trestle design (left), the view of the stresses distribution inside the structure (right).

As we can see, the maximal stresses in the structure is in locations, where the upper plate is supported by I profiles. The value of the maximal stress is of 147 MPa. Therefore, we can conclude, the safety condition is met. Maximal values of stresses in I profiles are even lower, namely of 110 MPa (Fig. 9 right).

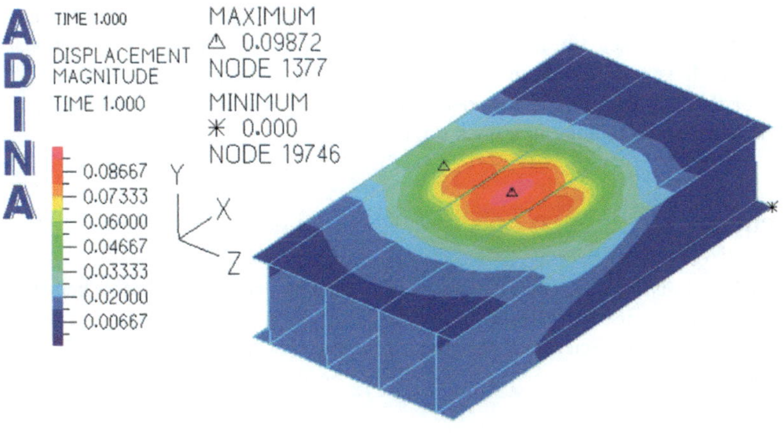

Fig. 10. Total displacement of the final optimized structure.

We also observed the deformation of the optimized trestle structure, which is shown in Fig. 10. The maximal displacement of the structure is lower than 0.1 mm under the maximal load. This value is acceptable and the structure has sufficient stiffness.

The optimization process of the trestle structure resulted to the modified design of the trestle, which meets the requirements in term of safety as well as of its weight.

To summarize, the original trestle structure did not meet the requirements of the RailClinic Company. After taking into account all inputs, the trestle structure was modified and optimized. The optimized design of the trestle has even slightly smaller dimensions, thus the trestle is more compact. At the same time, however, the requirements of the safety and weight are fully met. The final trestle is only of 4 kg, which is an acceptable result.

4 Conclusion

- The article presented the problem of design of a trestle steel structure, which is intended to be used under a hydraulic jack.
- The original trestle design had a weight of 140 kg, which was unacceptable for manual manipulation by persons.
- There were performed activities to reduce the trestle weight, at which, safety requirements have to be met.
- The optimized trestle design has the final weight of 104 kg, which is higher only by 4%. It is acceptable for using in the operational conditions.
- The stress analyses have proven, that the trestle structure meets safety requirements within the determined permissible values.

Acknowledgement. "This research was supported by the Cultural and Educational Grant Agency of the Ministry of Education of the Slovak Republic in the project No. KEGA 023ŽU-4/2020: Development of advanced virtual models for studying and investigation of transport means operation characteristics."

"This research was also supported by the Slovak Research and Development Agency of the Ministry of Education, Science, Research and Sport of the Slovak Republic in Educational Grant Agency of the Ministry of Education of the Slovak Republic in the project No. VEGA 1/0558/18: Research of the interaction of a braked railway wheelset and track in simulated operational conditions of a vehicle running in a track on the test bench."

References

1. Blatnický, M., Dižo, J., Kurčík, P.: Design of a handling machine for disable persons, Part I. Technológ **10**(1), 36–41 (2018). (in Slovak)
2. Skočovský, P., et al.: Science of Material for Engineers. University of Žilina, Žilina (2001). (In Slovak)

3. Leitner, B., Dvorak, Z.: Special railway crane PKP 25–20i-dynamic loads and a fatigue life prediction of its load-bearing structure. In: 17th International Conference on Transport Means, Kaunas, Lithuania, pp. 262–265 (2013)
4. Leitner, B., Mocova, L., Hromada, M.: A new approach to identification of critical elements in railway infrastructure. In: 10th Transbaltica International Scientific Conference – Transportation Science and Technology, Vilnius, Lithuania, pp. 143–149 (2017)
5. Wachnik, R., Drzewiecki, A.: Maintenance management system of railway vehicles (MMS). Sci. J. Silesian Univ. Technol. Ser. Transp. **73**, 115–123 (2011)
6. Xia, C., Sun, Y.: Research on vehicle lifting jack structure failure trend on basis of static strength analysis. Appl. Mech. Mater. **275–277**, 751–755 (2013)
7. Feng, Z.: Processes and Mechanisms of Welding Residual Stress and Distortion. Woodhead Publishing Limited and CRC Press LLC, Sawston Cambridge (2005)
8. Baniari, V., Blatnicka, M., Sajgalik, M., Vasko, M., Saga, M.: Measuremrnt and numerical analyses of residual stress distribution near weld joint. In: 12th International Scientific Conference of Young Scientists on Sustainable, Modern and Safe Transport, High-Tatras, Slovakia, vol. 192, pp. 22–27 (2017)
9. Muravev, V.V., Muraveva, O.V., Volkova, L.V., Saga, M., Sagova, Z.: Measurement of residual stresses of locomotive wheel treads during the manufacturing technological cycle. Manag. Syst. Prod. Eng. **27**(4), 236–241 (2019)
10. Strazovec, P., Suchanek, A., Sťastniak, P., Harusinec, J.: Detection of residual stress in a railway wheel. In: 13th International Scientific Conference on Sustainable, Modern and Safe Transport, vol. 40, pp. 898–905 (2019)
11. Cheng, X., Fisher, J.W., Prask, H.J., Gnäupel-Herold, T., Yen, B.T., Roy, S.: Residual stress modification by post-weld treatment and its beneficial effect on fatigue strength of welded structures. Int. J. Fatigue **25**, 1259–1269 (2003)
12. Adina. Theory and Modelling Guide – Helpmanual. 71 Elton Avenue, MA 02472, Watertown, USA (2010)
13. Fomin, O., Gerlici, J., Lovskaya, A., Kravchenko, K., Burlutski, O., Hauser, V.: Pecularities of the mathematical modelling of dynamic loading on containers in flat wagons. In: 23rd Polish-Slovak Conference on Machine Modelling and Simlations, Rydzyna, Poland, vol. 254 (2019)
14. Stastniak, P., Moravcik, M., Smetanka, L.: Investigation of strength conditions of the new wagon prototype type Zans. In: 23rd Polish-Slovak Conference on Machine Modelling and Simlations, Rydzyna, Poland, vol. 254 (2019)

Investigation of Dynamic Behaviour of an Overhead Crane Using Multivariant Calculations

Viačeslav Petrenko(✉) and Pavel Ževžikov

Vilnius Gediminas Technical University, Plytinės Street 27,
10105 Vilnius, Lithuania
viaceslav.petrenko@vilniustech.lt,
pavel.zevzikov@stud.vilniustech.lt

Abstract. In this study, the problems of increasing the speed of movement of the crane along the rail track and, accordingly, changes in safety rules in the design and operation of cranes are considered. The process of creating models of a crane and crane railway tracks is also described. The paper presents the results of modelling the dynamic behaviour of the overhead crane at various speeds. The obtained results of modelling allow to optimize the speed of the overhead crane depending on the irregularities of the crane tracks.

Keywords: Overhead crane · Simulation · Dynamic behaviour

1 Introduction

One of the main efficiency problems with overhead cranes application is the long length crane tracks. In the case when the length of the working area is relatively small (up to 200 m), the problem is not so urgent, but when the lengths reach kilometres, it seriously limits the efficiency of the crane. The length of the crane tracks in some cases often exceeds 1,000 m and, in that case, an important efficiency factor is the travel time of the crane between work positions, which depends on the maximum speed of the crane. Increasing the speed of the crane could help significantly increase the productivity of these companies with the long length crane tracks, but the maximum speed of the crane is regulated by the certain national safety rules and the state standards.

According to the established long-term practice of designing cranes, the influence of crane dynamics is sought to be compensated as follows:

- strengthening the crane railway;
- reducing the speed of the crane.

It is obvious, that the simplest way to increase the crane speed creates an increased dynamic load. An overhead crane is a multiply statically indeterminate system with a variable number of many degrees of freedom. The dynamics of the crane is very complex [1–7]. Movement is influenced by many factors, the main ones are:

O. Prentkovskis et al. (Eds.): TRANSBALTICA 2021, LNITI, pp. 58–67, 2022.
https://doi.org/10.1007/978-3-030-94774-3_6

- technical characteristics of electric engines and transmissions;
- the geometric parameters of the wheels;
- weight, dimensions and shape of metal structures;
- cargo weight, size, shape and the method of fastening;
- the position of the crane bogie;
- rail tracks parameters.

The purpose of this study is to create computer models of the overhead crane and crane railway tracks and to study the dynamic behaviour of the crane at different speeds.

2 Overhead Crane Computer Simulation

The Universal Mechanism (UM) program package was used for simulation of kinematics and dynamics of overhead crane. That software allows to simulate the dynamics of fully parameterized crane models. The simulation is performed in time domain by means of numeric integration of differential or differential-algebraic equations of motion [8]. Also, for testing flexible elastic crane elements, the ANSYS engineering simulation software package was used, the geometry of the crane parts was preliminarily created in the KOMPAS 3D graphics package.

2.1 Overhead Crane Model

The Fig. 1 (left) shows the coordinate system that is relevant for all the models presented in the work, in which the X-axis is red, the Y-axis is green, and the Z-axis is blue (the positive directions of the axes are shown, the default text refers to the global coordinate system of the model). Also, the Fig. 1 (right) shows the numbering (1, 2, 3 and 4) of the crane wheels adopted in this article. The crane model was created in several stages. At the first stage, a rigid model of a gantry crane was created, with a span of 10 m, a lifting height of 4 m and without load. The prototype is a gantry crane of the KPB type, widely used in Lithuania. Crane constructions weight is 10 t (Fig. 1 (left)).

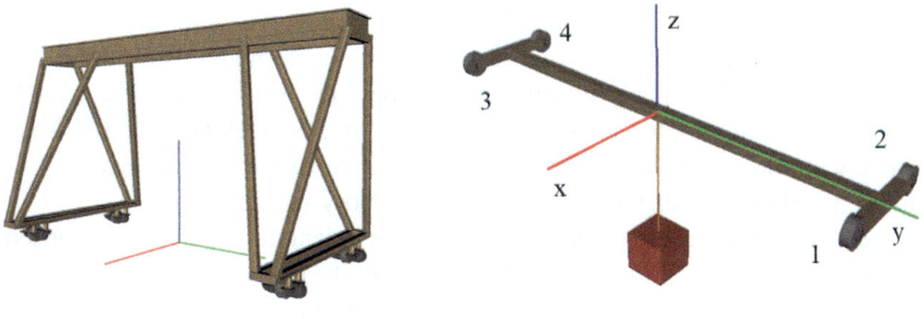

Fig. 1. First (left) and final (right) versions of the crane's models.

At the next stage, several simplified crane models were developed. Gantry crane model without vertical supports transformed into a hybrid overhead crane (Fig. 1 (right)). A load weighing 7.8 t is attached to the model, the total weight of the model is 17.8 t.

Wheels and axle boxes were integrated from the Universal Mechanism software Railway module. The wheel profile is created according to GOST 3569–74 (Fig. 2), wheel diameter 0.5 m. It is also important to note that wheel geometry deviations were not considered in this study. The wheel system is detailed and represents wheels of the same diameter, ideally symmetrically installed in the crane structure.

A steel rod 3 m long and 40 mm in diameter imitates the suspension of the load (steel rope). We believe that the elasticity of the bar can be neglected, that is, in the model we set it as rigid. One end of the rod is attached with a joint adopted in the UM software to the central point of the main beam of the crane, and the other to the upper point of the load. The load is presented in the form of a steel cube with an edge of 1 m, the weight of the load is 7.8 t. Joints have three degrees of freedom of rotation (each along all three axes), friction in the joints is not considered in this model.

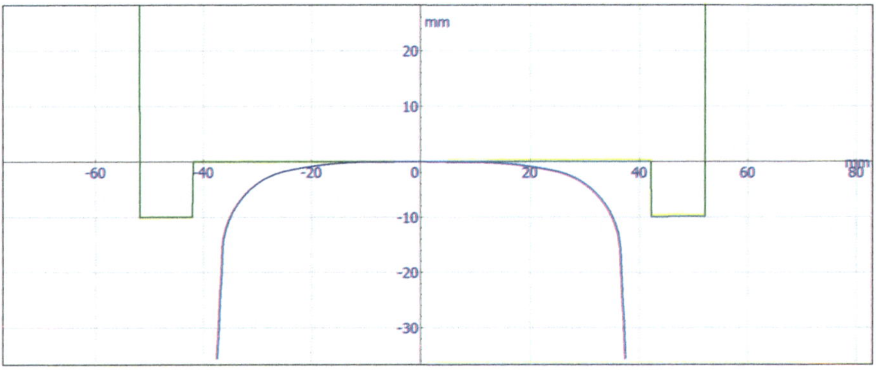

Fig. 2. Crane wheel and R65 rails profiles.

2.2 Crane Tracks Model

Crane track model presented as a massless rail track model, where rails, sleepers and other track parts interact as massless force elements. Track model uses new R65 type (GOST R 51685–2013) rails, with dissipative properties: vertical stiffness 66 MN/m, lateral stiffness 24 MN/m, torsional stiffness 6.6 MN/m, vertical damping 0.4 MN s/m and lateral damping 0.1 MN s/m. For wheel and rail contact forces estimation Fastsim algorithm, based on Kalker's creep forces linear theory was used.

To create irregularities in the crane track (100 m in length), an algorithm for setting irregularities by spectra developed by the Federal Railways Agency (FRA) was used. Power spectral density (PSD) functions corresponded to railway tracks of half-sum of horizontal track irregularities:

$$\Phi(\Omega) = \frac{A_a\Omega_c^2}{\Omega^2\left(\Omega^2 + \Omega_c^2\right)} \cdot \Omega > 0. \tag{1}$$

PSD of half-sum of vertical track irregularities:

$$\Phi(\Omega) = \frac{A_v\Omega_c^2}{\Omega^2\left(\Omega^2 + \Omega_c^2\right)} \cdot \Omega > 0. \tag{2}$$

PSD of half-difference of horizontal track irregularities:

$$\Phi(\Omega) = \frac{4A_a\Omega_c^2}{\left(\Omega^2 + \Omega_c^2\right)\left(\Omega^2 + \Omega_c^2\right)} \cdot \Omega > 0. \tag{3}$$

PSD of half-difference of vertical track irregularities:

$$\Phi(\Omega) = \frac{4A_v\Omega_c^2}{\left(\Omega^2 + \Omega_c^2\right)\left(\Omega^2 + \Omega_c^2\right)} \cdot \Omega > 0, \tag{4}$$

where A_a, A_v – track quality parameter, cm^{2}rad/m; Ω – circular frequency, rad/m.

In the track model, a spectrum of irregularities of the sixth quality class of railways was used: $A_a = 0.0339$ cm^2 rad/min, $A_v = 0.0339$ cm^2 rad/min, $\Omega_s = 0.4380$ cm^2, $\Omega_c = 0.8245$ cm^2. The requirements for the geometry of crane tracks irregularities differ from the requirements for the railway track [9, 10], so the spectra of vertical and horizontal irregularities were multiplied by a factor 0.1. Irregularities of the crane railway track (Fig. 3) were checked using a special module for assessing the condition of the track. The generated track widening, narrowing and skewing (Fig. 4) meet the requirements [10] for the crane track geometry and the track can be used for further modelling.

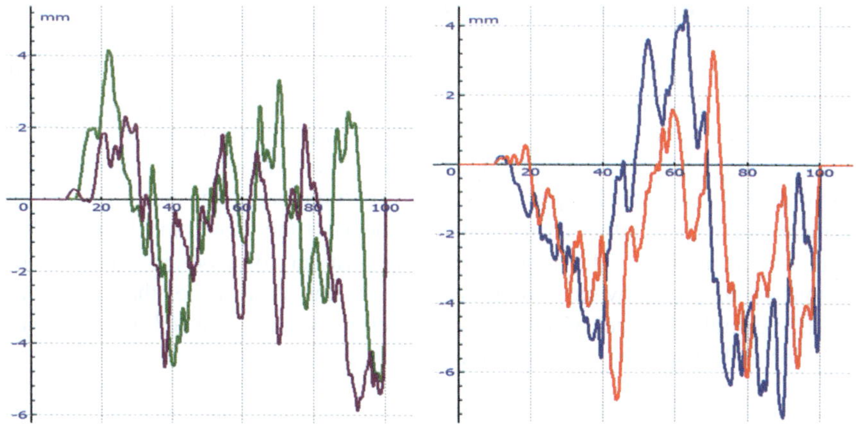

Fig. 3. Left and right crane rails vertical (left) and horizontal (right) irregularities.

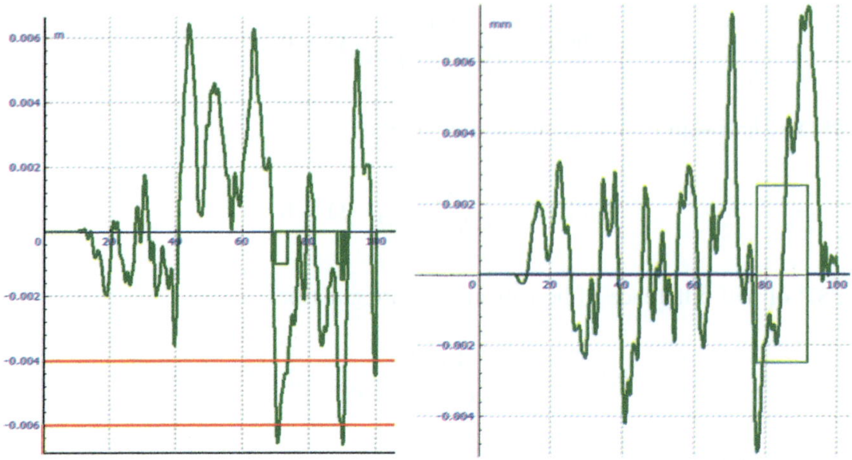

Fig. 4. Crane railway track widening (left) and narrowing (right).

To analyse the dynamic behaviour of the crane model, the following criteria were determined when creating simulation results:

- normal reactions N along Z-axes, acting on crane wheels;
- total lateral force acting on the wheel along the Y-axes of the local track coordinate system;
- lateral displacement of the wheel relative to the rail, along the Y-axes of the local wheel coordinate system.

These criteria allow analysing the dynamic processes of movement of the overhead crane.

3 Crane Software Simulation Process

Simulation experimental part was divided into three steps. During the first phase of the simulation, the various crane models created were tested and optimized using numerical simulation methods. As a result of the simulation, a model was selected that most realistically simulated the dynamic behaviour of the crane. At the second stage, a series of different numerical experiments was carried out at certain speeds of the crane movement and with a further change (improvement) of the crane model, crane tracks and moving conditions. The most appropriate criteria for assessing the dynamic behaviour of the crane were also determined. At the third stage, the created and adequately operating model of a numerical experiment is investigated using a special module for multi-parameter optimization UM Experiments.

3.1 Testing and Optimization of Various Crane Models

At the first stage of numerical simulation, three main stages of this simulation were developed:

1. Created and improved models of cranes of various designs.
2. Models and parameters of crane tracks were selected.
3. The conditions for the numerical experiment were developed.

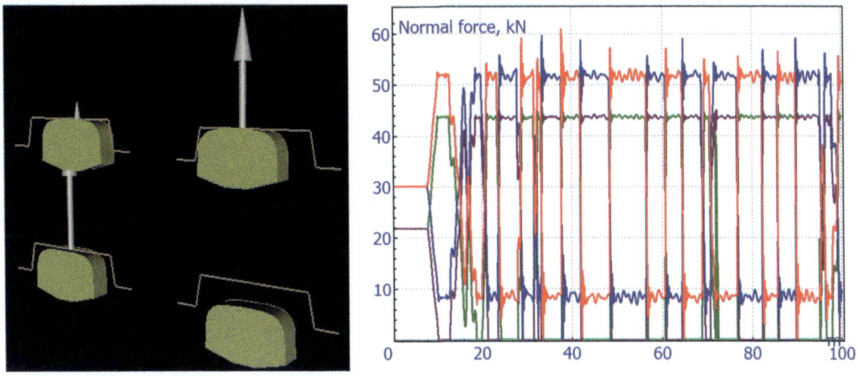

Fig. 5. Wheels and rails contact image (left) and normal force distribution between wheel and rail (right).

At this stage, on the first step, a rigid model of a gantry crane was created, similar to the crane of the KPB model. The primary crane track model was developed based on a massless model of a classic railway track. The irregularities of the crane rail track met the requirements for irregularities of the railway track. As can be seen in Fig. 5, the dynamic behaviour of this model does not correspond to reality, since there is a constant lifting of the crane wheels from the rail and the overhead crane practically moves on three wheels, the normal force between the wheel and the rail is unevenly distributed (Fig. 5).

3.2 Simulation of the Dynamic Behaviour of a Crane

At this stage, the created models of the overhead crane and crane rail tracks were used for a series of experiments. The purpose of the experiments was to create certain conditions for the experiment: the crane movement mode, the range of movement speeds, the distribution of the traction force on the crane wheels, the parameters of the load and load suspension and others.

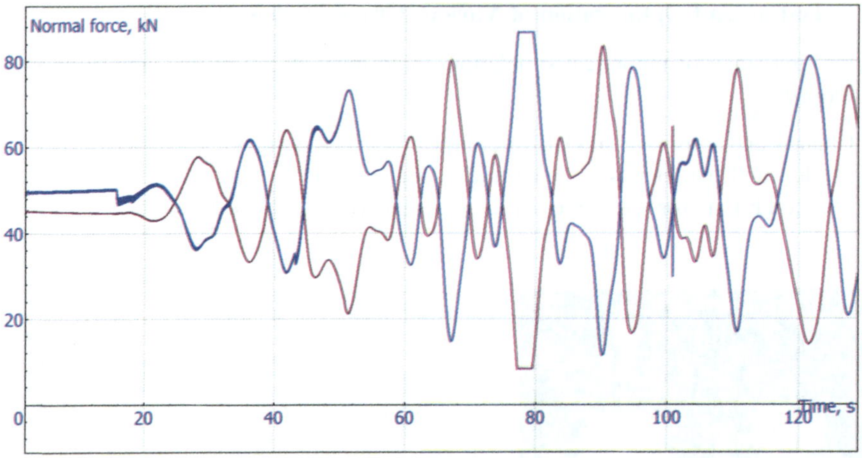

Fig. 6. Distribution of normal force of 2 (blue) and 4 (purple) crane wheels at a speed of 0.5 m/s.

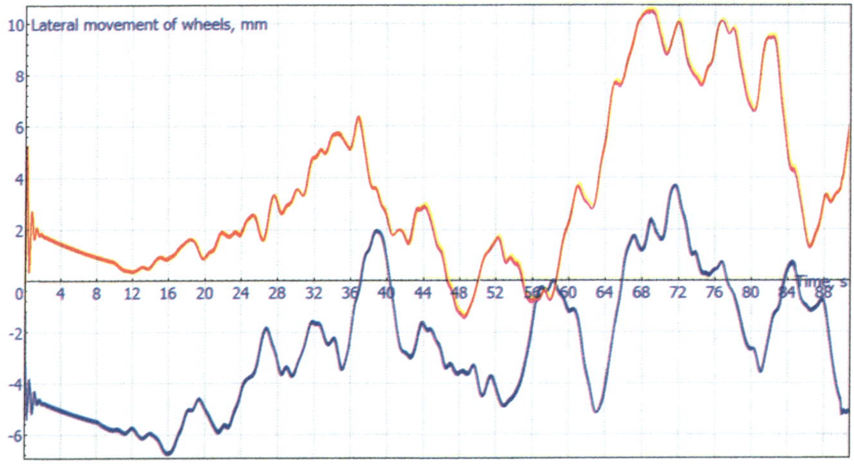

Fig. 7. Lateral movement of 2 (blue) and 4 (red) crane wheel in the rail gauge.

At the same time, the evaluation criteria for the dynamic behaviour of the girder crane were selected and determined: distribution of normal force of crane wheels (Fig. 6), lateral movement of crane wheel in the rail gauge (Fig. 7). At the same time, the models of the rail tracks model and the overhead crane model were corrected. At the final stage, a series of experiments was carried out, determining the effect of increasing the speed of the crane on the change in dynamic parameters.

3.3 Multi-parameter Optimization of the Crane Model

At the last stage, a series of numerical experiments, to analyse dynamical behaviour of overhead crane was carried out.

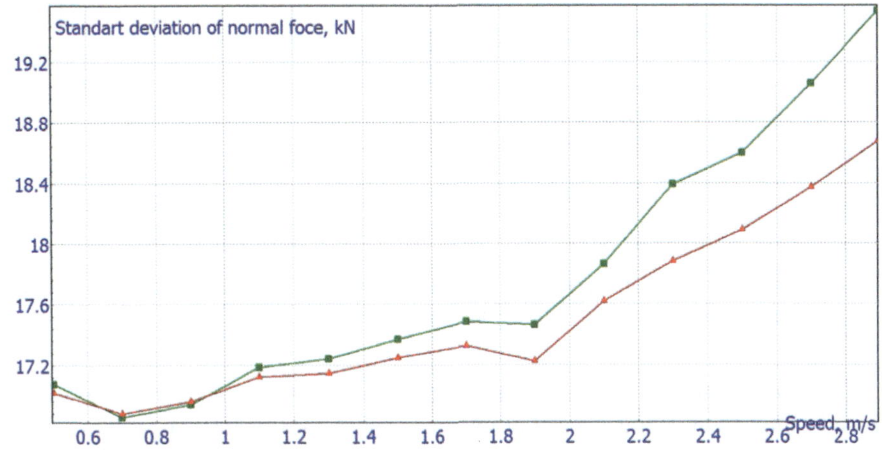

Fig. 8. Standard deviation of normal force of 2 (green) and 4 (red) crane wheels for speed range from 0.5 till 2.9 m/s.

The built-in Universal Mechanism software Experiments module a scanning tools for advanced analysis of dynamics of overhead crane was used.

In a series on numerical experiments the overhead crane speed changed from 0.5 m/s till 2.9 m/s in 0.2 m/s steps.

To analyse the dynamic behaviour of the crane model, the several variables were determined when creating simulation results: normal force, standard deviation of normal force, total lateral force acting on the wheel, lateral movement of wheels and wheel sets in the rail gauge step and others. Standard deviation of normal force of overhead crane wheels for speed range from 0.5 till 2.9 m/s shown in Fig. 8. Obviously, with increasing speed, the contact forces between the wheel and the rail increase. For example, for 0.7 m/s speed standard deviation of the normal forces for first wheel did not exceed 16.8 kN and for 2.9 m/s speed standard deviation of the normal forces did not exceed 19.6 kN.

Lateral force acting on the wheel also increasing (Fig. 9). For second wheel for 0.7 m/s speed standard deviation of lateral force acting on the wheel does not exceed 1.4 kN and for 2.9 m/s speed standard deviation of lateral force did not exceed 2.6 kN.

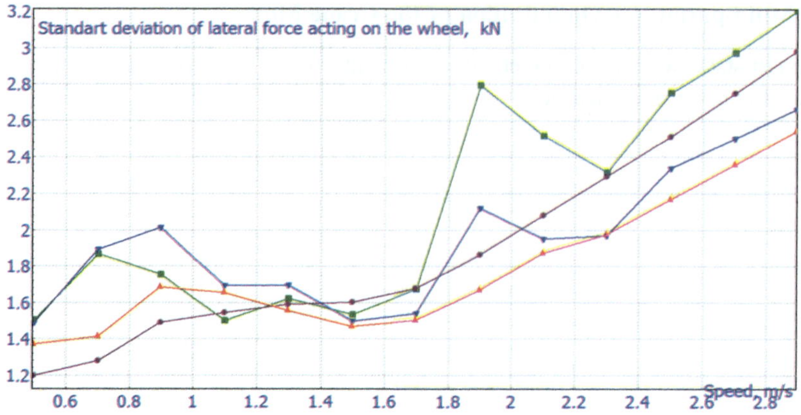

Fig. 9. Standard deviation of lateral force acting on the wheel of 1 (blue), 2 (green), 3 (purple) and 4 (red) crane wheels for speed range from 0.5 till 2.9 m/s.

Analyzing numerical experiments results, we can create certain recommendations for evaliuation of the critical speed of overhead cranes. The maximum speed (1 m/s) for KPB type cranes is set by the manufacturer in strict accordance with the Lithuanian national safety rules, but the numerical experiments results prove that the contact forces between the wheel and the rail increase sharply from 1.6 − 1.8 m/s. This means that an increase in the critical speed of cranes is possible, which will allow to improove efficency of the overhead cranes.

4 Conclusions

1. In this study, an overhead crane model, crane tracks and numerical experiment conditions were developed to investigate the dynamic behaviour of cranes.
2. At the first research stage a rigid model of a gantry crane and crane rail track massless model was created. The results show that a rigid crane model cannot be used to investigate dynamic processes.
3. To study the dynamic processes of crane movement, it is necessary to use a hybrid crane model with elastic elements.
4. The simulation results prove that in some cases it is possible to safely increase the efficiency of the overhead crane by increasing the critical speed of the crane.

References

1. Eglynas, T., et al.: Modelling and experimental research of quay crane cargo lowering processes. Adv. Mech. Eng. **11**(12), 1–9 (2019)
2. Jaafara, H.I., Jamianb, Z.M., Abidinb, A.F.Z., Kassima, A.M., Ghani, Z.A.: Dynamic behaviour of a nonlinear gantry crane system. In: The 4-th International Conference on Electrical Engineering and Informatics, pp. 25−32 (2013)

3. Sowa, L., Saternus, Z., Kubiak, M.: Numerical modelling of mechanical phenomena in the gantry crane beam. In: XXI International Polish-Slovak Conference Machine Modelling and Simulations, pp. 24−36 (2016)
4. Spicina, D.N., Jurin, A.N.: Vlijanie netocnosti ukladki putej na urovenž nagruzok, dejstvijuscie na mostovije krani. Izvestija visshich ucebnych zavedenij Nr. **8**, 150–157 (2013)
5. Kulka, J., Mantic, M., Fedorko, G., Molnar, V.: Analysis of crane track degradation due to operation. Eng. Failure Anal. **59**, 384–395 (2016)
6. Zrnić, N.Đ, Gašić, V.M., Bošnjak, S.M.: Dynamic responses of a gantry crane system due to a moving body considered as moving oscillator. Arch. Civil Mech. Eng. **15**(1), 243–250 (2014). https://doi.org/10.1016/j.acme.2014.02.002
7. Ghasemi, P., Aslani, M., Ashlock, J.C., Williams, R.C., Schaefer, V.R.: Evaluating gantry crane-way pavement performance: an inverse approach. Transp. Geotech., 250−261 (2021)
8. Laboratory of Computational Mechanics. Simulation of Rail Vehicle Dynamics Homepage. http://www.universalmechanism.com/download/80/eng/08_um_loco.pdf, Accessed 05 May 2021
9. LST EN 15011:2021. Kranai. Tiltiniai ir ožiniai kranai
10. GOST R 56944–2016 Krany gruzopodjemnyje. Puti relsovije nadzemnyje. Obschie technicheskie uslovija

Application of Pass-Band Step Filtering Method for Identification the Vibration-Acoustic Signature of a Moving Train

Rafał Burdzik(✉) and Paweł Słowiński

Faculty of Transport and Aviation Engineering, Silasian University of Technology, Krasińskiego 8, 40-019 Katowice, Poland
{rafal.burdzik, pawel.slowinski}@polsl.pl

Abstract. The paper presents research on application of pass-band step filtering method for identification the vibration-acoustic signature of a moving train. Authors present results of applying new methods of filtering vibration-acoustic signatures. The signal comb filtering was applied and the RMS in the time course and the sum of the FFT amplitudes in the spectra were used as a general measure after filtering the signal. Modern rail vehicles, as well as those modernized due to reaching high speed (HSR), arouse interest in terms of vibration and acoustic signals that they generate, especially in the context of the changing market of rail transport services in Central Europe. Rail transport and railway infrastructure managers perceive the problem of acoustic comfort in the vicinity of railway routes, therefore this subject becomes an important aspect from the point of view of further analyzes. The issue of the use of various filtering methods to identify the vibration and acoustic signature of a moving train is of key importance for further analyzes of research on environmental comfort as well as other research on vibroacoustic signals on the railway.

Keywords: Vibration · Noise · Train · Vibroacoustic signature

1 Introduction

1.1 Vibrations and Noise of Railway Trains - Classification of Random Signals

The classification of vibrations coming from all means of transport is classified as random-non-stationary paraseismic vibrations. Signals corresponding to random physical phenomena cannot be described with exact mathematical relationships, because the result of each observation of a signal is the only one (non-reproducible). A single time function describing a random phenomenon is called a random function or an execution, and at a finite time interval - an observed signal [5].

Four statistical functions are used to describe the main properties of random signals:

- Mean square value.
- Probability density function.

O. Prentkovskis et al. (Eds.): TRANSBALTICA 2021, LNITI, pp. 68–78, 2022.
https://doi.org/10.1007/978-3-030-94774-3_7

- Autocorrelation function.
- Power spectral density function.

The higher vibrations occur when the excitation frequency is close to the natural frequency of the vehicle rail-track system [1, 2, 15]. This pseudo-resonance phenomena is important for vibration analysis. Especially when vibroacoustic signature is analysed and other parameters of the signal related with train dynamics and types are investigated [11]. The dominant source of noise from the low velocity moving train set is the rotating wheel of the trolley on the rail, and more precisely the wheel-rail contact mechanism, which depends on the speed and unevenness of the running surfaces or damage [3, 4, 6, 16–18].

Many of papers focus on the problem of noise generated by moving rail vehicles [9]. The sources of external noise emitted by rolling stock are listed below [10, 14]:

a) primary noise - noise during rolling stock movement,
b) basic - caused by the wheel rolling on the rail,
c) other factors - noise from traction and auxiliary machinery, air turbulence, noise generated by pantographs and vibrations of individual vehicle parts,
d) secondary - signals from the locomotive, noise from manoeuvres.

There are three basic categories of noise sources in moving trains: traction noise, rolling noise and aerodynamic. At a train speed of up to 30 km/h, the noise emitted by the vehicle itself, e.g. a working diesel engine, is the most important. Above 30 km/h to 200 km/h, the main source of noise is the noise caused by rolling of wheel on the rail causing all rail-wheel contact phenomenon. However, aerodynamic noise is the main source of noise only above 200 km/h, i.e. for HSR passenger trains. Below approx. 70 km/h the aerodynamic noise is below 70 dB (A).

In the case of rail vibrations, the problems are different than in the case of wheel vibrations [7, 8, 12]. The rail has a different characteristic in which the wave is supported and spreads along its entire length.

To sum up, from the point of view of the operation of means of railway transport, the most important sources of vibrations include:

- contact phenomena in the wheel-rail system;
- operation of the drive unit;
- air vibrations when the vehicle is in motion;
- effects of unbalance of mechanical parts in vehicles.

1.2 Research Problem and Research Method

The authors formulated new research goals as the analysis of synchronized vibroacoustic images in the aspect of assessing the information capacity of the recorded signals. This forced an increase in the scope of research and continuation of the currently conducted research and experiments. The research method was developed, the assumption of which was the synchronous registration of acoustic signals in the immediate vicinity of the track and vibrations in three directions recorded directly on the rail. The signals were recorded by the LabVIEW software.

The measurement system used for this research included:

- Three Endevco accelerometers with the following parameters: Sensitivity: 100 mV/g | 10.2 mV/m/s^2, maximum measuring range: 50 g pk | 491 m/s^2 pk. Sampling rate 42,000 Hz;
- Data acquisition module: 24 bits max 50 kHz;
- Integrating sound level meter SON-50 type with accuracy class 1 and measuring range in dB in 4 sub-ranges.

The accelerometers were attached to the special metal cube and by the magnets to the rail (on the web). However, tests are also possible on the foot of the rail [13]. The signal was recorded in three orthogonal axes X, Y, Z, where the X axis (along the track axis), the Y axis (transversely to the track axis), and the Z axis (vertical to the track axis). The ambient temperature was 6 degrees Celsius. An example of the mounting method (mounting through a magnet attached to the rail) is shown in the graphic below.

Fig. 1. The method of mounting the sensor to the rail. Own source.

2 Methods of Analysis

The authors used a band-pass filter - an algorithm that passes the spectral components of the signal in a specific frequency range, called the passband. The bandwidth of a bandpass filter is defined as the range between the low f_L and upper f_H cut-off frequency, or as a range around a specific center frequency of this filter. The relationship between the cutoff frequencies and the center frequency and bandwidth B is expressed by the formulas:

$$f_0 = \sqrt{f_L f_H},$$
$$B = f_H - f_L.$$

The filtering range was from 100 Hz to 2,400 Hz in 100 Hz steps. The signal processing contain pass-band filtering 100–200 Hz, waveform analysis (RMS), FFT analysis and the spectrogram. The scheme of the procedure is presented below:

- Performing vibration measurements in three axes.
- Performing noise measurements.

- Selecting from the course of vibrations and noise the fragments related to the passage over the sensor.
- Prior to filtering, categorized into four categories of trains.
- Analysis of unfiltered signals and determination of the "boundary" for the performance of the bandpass filter.
- Running a script in a MATLAB environment containing a band-pass filter, e.g. 0 to 2,400 Hz in 100 Hz intervals.
- The saved files after filtering are subjected to further analysis, e.g. RMS, FFT determination.
- After performing the FFT analysis in a given frequency range, the maximum values | dft (f) | are determined and rewrites to the table.
- The written values from the table are ready to be presented in the form of graphs and compared to each other.

The correct operation of the filter with a slight transmission within its operating limits can be noticed, which is confirmed by the FFT analysis as well as the spectrogram.

The recording of vibrations and noise was performed synchronously Also the correct operation of the filter with slight transmission within its operating limits can be confirmed, which is confirmed by the FFT analysis as well as the spectrogram. Higher values related to the filter pass-through were noticed at frequencies above 5,000 Hz.

3 Results

3.1 Vibration Measurements

The results were recorded on the railway route located in Poland, on the main railway line on the Wroclaw-Katowice section. Due to the variety of measurement points in which the authors made the measurements, the focus was only on those carried out at one point, on the same day, with the same weather and environmental conditions. The characteristics of the trains are:

- Electric multiple unit (EMU) - 120 km/h - Quanity of car numbers: 3 – Railcar.
- High-speed rail (HSR) – 160 km/h - Quanity of car numbers: 8.
- Electric locomotive with passenger railroad cars (PO) – 120 km/h - Quanity of car numbers: 4.
- Electric locomotive with hopper wagons (TO) – 60 km//h - Quanity of car numbers 41.

The vibration signals of vehicles differs significantly in various respects, but the most important of them include, first of all, the length of the train (and the time of its registration depending on the speed of the train - in this case the speed of the EMU, HSR and PO trains was higher) than cut on the trail due to the measurements on the main bus the number of amplification of the signal (indicating the wheel passing over the sensor). The amplitude in the course of the above-mentioned trains has the greatest gain in the Electric locomotive with hopper wagons and Electric locomotive with passenger railroad cars (TO, PO) compositions, while the ED250 (HSR) composition has the lowest (Table 1).

Table 1. Calculated RMS, Peak 2 Peak and Max values for the waveforms before filtering.

Type	RMS	PEAK 2 PEAK	MAX value
Electric multiple unit (EMU)	8,326	112,8	56,39
High-speed rail (HSR)	2,476	65,20	33,50
Electric locomotive with passenger railroad cars (PO)	11,201	112,8	56,39
Electric locomotive with hopper wagons (TO)	3,821	108,1	51,66

Also in the case of the FFT analysis for the X axis of unfiltered signals, differences in gain levels for given frequencies can be noticed. A characteristic feature is the fact that the gains are up to 2,500 Hz, which also takes place in further analysis using a band-pass filter at frequencies from 100 Hz to 2,500 Hz every 100 Hz.

Summary of the results after filtering, taking into account the highest amplitude value for a given analysed frequency with the bandpass filter 100–2,400 Hz (Fig. 2).

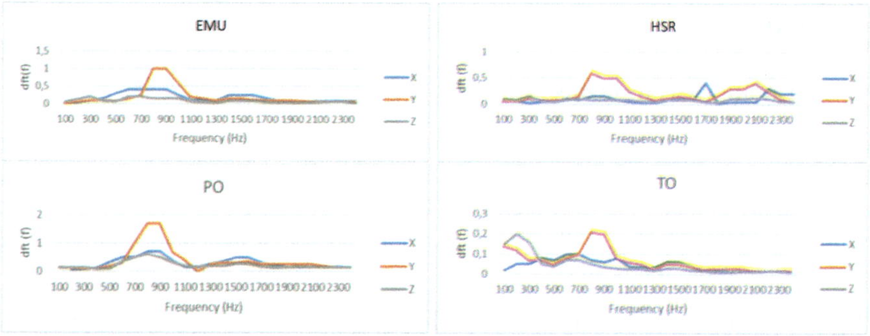

Fig. 2. Results after filtering with the band-pass step method in the range from 100 to 2,400 Hz in 100 Hz windows - the highest amplitude. Y axis - gain, X axis - frequency (Hz). The axes are designated X Y Z (see at Fig. 1). Results for EMU. Values from the signal spectra.

For EMU the highest value of the spectrum can be noticed for frequencies from 700 to 1,100 Hz, especially around 900 Hz where this value reaches 1. However, the X and Z axes are characterized by a lower gain, while for the X axis this gain of 0.4 and 0.3 is found in the ranges of 600–900 Hz and 1,400–1,600 Hz. In the Z axis, no significant reinforcements are found (the highest value 0.2). For High Speed Railways, the highest value of the spectrum can be noticed for frequencies from 700 to 1,150 Hz, especially around 800 Hz where this value reaches 0.6 and in frequencies between 1,900–2,300 Hz. On the other hand, the X and Z axes are characterized by a lower gain, while for the X axis the gain of 0.4 and 0.3 is found in the 1,600–1,800 Hz and 2,200 Hz ranges. In the Z axis, no significant reinforcements are found (the highest value of 0.15). For Electric locomotive with passenger railroad cars, the highest value of the spectrum can be noticed for frequencies from 600 to 1,150 Hz, especially around 800–900 Hz, where this value reaches 1.7. On the other hand, the X and Z axes are characterized by a lower gain, while for the X axis the gain of 0.7 and 0.5 is found in the 600–1,000 Hz and 1,500–1,600 Hz ranges. In the Z axis, the highest value was found between 700–900 Hz for a value of approx. 0.5. For a freight train, the highest value of the spectrum can be noticed for frequencies up to 300 Hz and from 750 to 1,000 Hz, where this value is 0.15 and 0.2 respectively. On the other hand, the X and Z axes are characterized by a lower gain, while for the X axis the gain of 0.05 to 0.1 is found in the 400–800 Hz ranges. In the Z axis, the highest value was found for the frequency of about 250 Hz for the value of 0.2.

The following statement of results after filtering takes into account the RMS values for a given analyzed frequency using the Bandpass 100–2,400 Hz filter (Fig. 3).

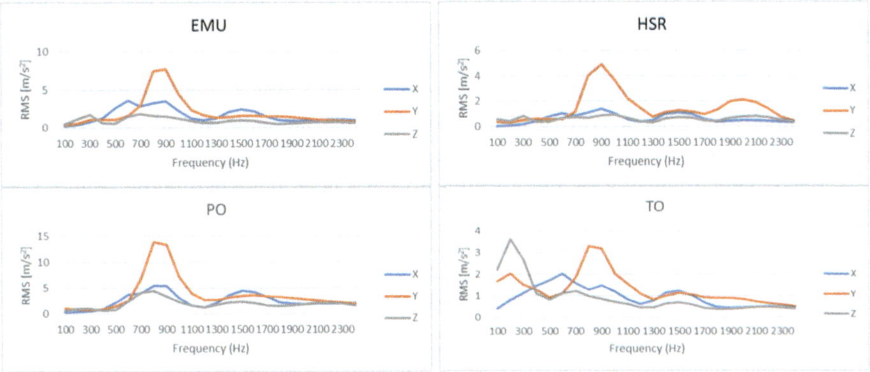

Fig. 3. Results after band-pass step method filtering in the range from 100 to 2,400 Hz in 100 Hz windows. The axes are designated X Y Z (see at Fig. 1). Results for EMU. Determined RMS values for given frequencies.

For EMU, the highest RMS value can be noticed for frequencies from 700 to 1,100 Hz in the Y axis, especially around 800–900 Hz where this value reaches approx. 7. The X and Z axes are characterized by lower RMS values, but for the X axis these values are not more than 3.5 for 600 Hz. In the Z axis, the highest value was found for 700 Hz and it amounts to 2. For HSR, the highest RMS value can be noticed for frequencies from 700 to 1,100 Hz in the Y axis, especially for 900 Hz where this value reaches 5. However, the X and Z axes are characterized by smaller RMS values, but for the X axis these values are no more than 1, 2 for 900 Hz. In the Z axis, the highest value was found for 1,000 Hz and it amounts to 1. For the PO, the highest RMS value can be noticed for the frequencies from 700 to 1,000 Hz in the Y axis, especially around 800–900 Hz where this value reaches approx. 14. The X and Z axes are characterized by smaller RMS values, but for the X axis these values are not more than 6 in the range 800–900 Hz. In the Z axis, the highest value was found for 800 Hz and it amounts to 4.

For TO, the highest RMS value can be noticed for frequencies from 800 to 900 Hz in the Y axis, where this value reaches approx. 3. A significant difference is the Z axis characterized by values exceeding the Y axis in the range of approx. 250 Hz in which the RMS value is over 3, 5. The X axis is characterized by max. 2 values for 600 Hz.

As would be expected, these values are largely analogous to the analysis of the maximum values in the FFT spectrum in the same bandpass configuration. Each of the passenger train signals is characterized by the maximum gain in the Y axis, but it is not dominant for the freight train, where the Z axis gives us a greater role in the information capacity. A correlation between all trains in the Y axis has been noticed, i.e. higher RMS values for frequency 800 Hz + - 200 which gives the basis for further analyses. Further analysis of the commodity composition is also necessary, where an increase in RMS was also noticed at lower frequencies, especially around 200–300 Hz.

3.2 Acoustics Measurements

In order to analyze the vibroacoustic signature, analogous processing of the recorded acoustic signals was carried out. Synchronously recorded vibration and acoustic signals are the basic source and carrier of the vibroacoustic image of a passing train and the source of its vibroacoustic signature. In the initial phase of the analysis, the similarities in the phenomena and the information transmitted by vibration and acoustic signals were assumed. In order to verify these assumptions, the same signal analyzes were carried out in order to compare the separated information components.

Acoustic signal waveforms before filtering have been depicted below (Fig. 4).

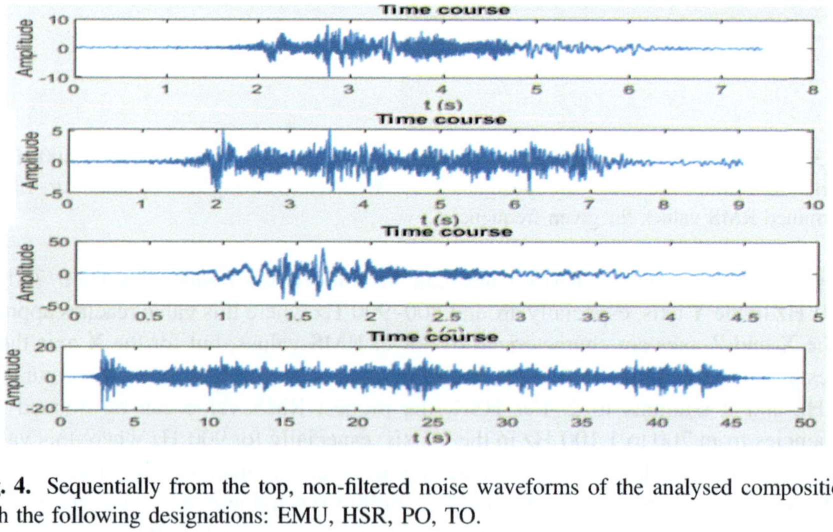

Fig. 4. Sequentially from the top, non-filtered noise waveforms of the analysed compositions with the following designations: EMU, HSR, PO, TO.

As it can be seen in the above graphic, the acoustics signal of vehicles differ significantly in various respects, similarly to the vibration analysis. This is important information for the further handling and analysis of files.

The table below shows the basic characteristics of the noise waveforms for the above-mentioned compositions (Table 2).

Table 2. Calculated RMS, Peak 2 Peak and Max values for noise waveforms before filtering.

Type	RMS	PEAK 2 PEAK	MAX value
Electric railcar (EMU)	1,051	18,49	8,044
High-speed rail (HSR)	0,765	10,19	5,290
Electric locomotive with passenger railroad cars (PO)	5,443	79,92	39,51
Electric locomotive with hopper wagons (TO)	2,337	39,97	17,90

Summary of the results after filtering, including the highest dft (f) value for a given analyzed frequency with the Bandpass filter 100–20,000 Hz every 1,000 Hz.

All compositions were characterized by low frequency in the ranges up to 4,000 Hz. This gives rise to interest and further analysis of data below this value. It is also information that the equipment used for recording does not have to meet the sampling requirements above 20,000 Hz. Further analysis of the sound in the range below 1,000 Hz is also necessary. A visible distinction is made only by PO and EMU - the gain is greater than in the other two (Figs. 5 and 6).

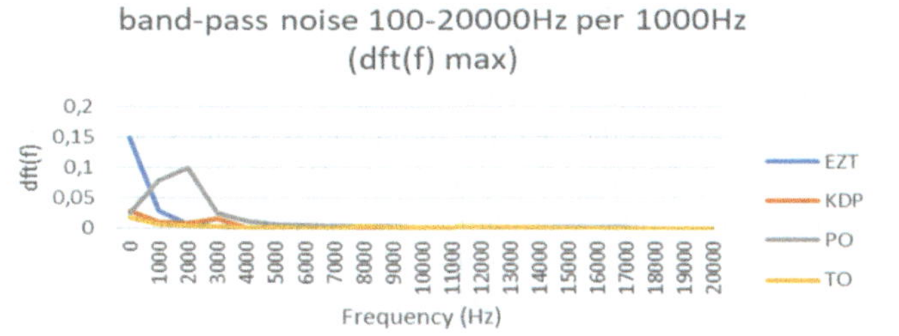

Fig. 5. Results after band-pass step method filtering in the range from 10 to 20,000 Hz in windows 1,000 Hz - the highest amplitude. Y axis - FFT gain, X axis - frequency (Hz). Diagram markings as in the figure.

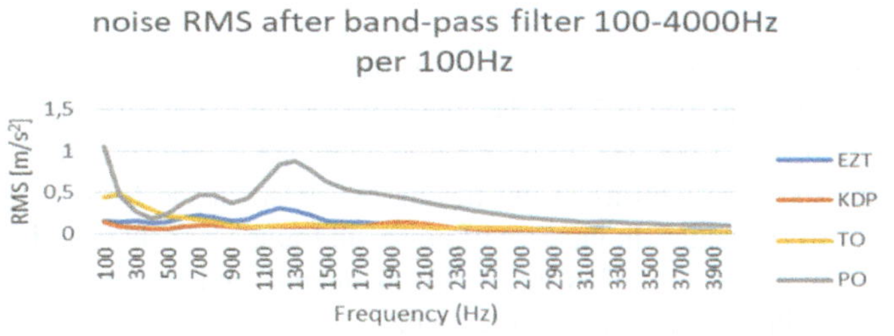

Fig. 6. Results after band-pass filtering in the range from 100 to 4,000 Hz in windows of 100 Hz - calculation of the RMS value. Diagram markings as in the figure.

The results from the RMS calculation after filtering the signals gave significant differences compared to the max dft (f) calculation. All compositions were characterized by a low frequency in the ranges up to 4,000 Hz, which was used here with a focus on lower frequencies. Due to this relationship, a band-pass step method filter was used, filtering in the 100 Hz windows than before 1,000 Hz. A significant difference can be noticed in the case of the PO composition, which, in contrast to HSR and EMU which are similar structures, i.e. EMU and HSR, is characterized by higher RMS values, in two frequency ranges, i.e. up to 300 Hz and between 1,100 and 2,500 Hz. Higher values can also be noticed in the case of the TO composition which is characterized by frequencies up to 500 Hz. From the aerodynamic point of view, the EMU and HSR trainsets are better adapted to higher speeds, therefore their values are lower than those of other structures.

4 Summary and Conclusion

The paper summarizes the extended analysis of the influence of a passing railway train on the spectrum and the course of vibrations in three axes as well as the FFT analysis. The article presents the results of tests of rail vibrations forced by the passage of a rail vehicle. For comprehensive analysis, vibrations were recorded in three axes: longitudinally, transverse and vertically. The analysis was carried out in the time domain as a distribution of vibration accelerations in successive recording periods, which allows for the identification of the torn time window. Additionally, in order to analyze the signal dynamics, frequency spectra have been defined, which allow the evaluation of the frequency components. Due to the fact that the recorded signals belong to the group of non-stationary signals, they were also transformed into the t-f representation. This allows you to observe the time distribution for specific frequency bands. Comprehensive comparison of the obtained results allows the observation of vibroacoustic images that can be treated as a trace and signature of a passing train. As a consequence, a set of data is obtained that can be analyzed according to various criteria, including assessment of the impact of vibrations and noise on means of transport. An additional assumption of the authors is the possibility of identifying railway trains only with the help of vibroacoustic signals generated by them, which according to the existing premises it is possible. An additional correlation with sound gives attitudes to the premises for the validity of such a thesis. The analyzed noise generated by passing railway trains was recorded synchronously with the vibration waveforms listed above. The analysis was performed with the use of a band-pass step method filter for vibrations at frequencies from 100 to 2,500 Hz every 100 Hz and noise for frequencies from 10 to 20,000 Hz every 1,000 Hz. In both cases, there are significant differences in the amplitude gain levels as well as in the results presented in the form of graphs with calculated max dft (f) values. In the case of noise, all trains were characterized by a low frequency (below 4,000 Hz). This gives rise to interest and further analysis of data below this value. Further analysis of the sound in the range below 1,000 Hz is also necessary. On the other hand, the analysis of vibrations may require further analyzes in the future, such as a comparison of several dozen compositions of one type and recalculation of the max dft (f) and RMS for each of them. On the other hand, the data presented in the article reflect a well-selected research sample (the four most popular types of trains traveling on Polish tracks). The role of aerodynamic train structures (EMU and HSR) was also noticed in the case of noise analysis for RMS values.

Acknowledgement. Work presented in this study was supported by the European Union through the European Social Fund as a part of a Silesian University of Technology as a Centre of Modern Education based on research and innovation project, number of grant agreement: POWR.03.05.00 00.z098/17-00.

References

1. Burdzik, R., Słowiński, P., Juzek, M., Nowak, B., Rozmus, J.: Dependence of damage to the running surface of the railway rail on the vibroacoustic signal of a passing passenger train. Vibroeng. Procedia **19**, 226–229 (2018)
2. Chromański, W.: Simulation and Optimization in the Dynamics of Rail Vehicles, p. 141. The Publishing House of the Warsaw University of Technology, Warsaw (1999)
3. Thompson, D.: Railway noise and vibration: the use of appropriate models to solve practical problems. In: The 21st International Congress on Sound and Vibration, Beijing, China, 13–17 July 2014 (2014)
4. Wszolek, T., Majchrowicz, J.: Analysis of the usefulness of distinctive noise features from rail and wheel in assessing their impact on the overall railway noise level. Diagnostyka **20** (2019)
5. Durka, P.J.: Between time and frequency: elements of modern signal analysis. Faculty of Physics, University of Warsaw (2004)
6. Cisielski, R., Maciąg, E.: Road Vibrations and their Influence on Buildings, p. 248. Publishing House of Communication and Communications, Warsaw (1990)
7. Jeong, D., Choi, H., Choi, Y., Jeong, W.: Measuring acoustic roughness of a longitudinal railhead profile using a multi-sensor integration technique. Sensors **19**(7), 1610 (2019). https://doi.org/10.3390/s19071610
8. Nader, M.: Vibration and Noise in Transport - Selected Issues. Publishing House of the Warsaw University of Technology, Warsaw (2016)
9. Railway noise. Gijsjan van Blokland M+P Ard Kuijpers M+P sources: Müller-BBM (D), D. Thompson (GB), M.Dittrich (TNO). topics. Relevance Sources Rolling noise Propulsion noise Aero dynamic noise Model of generation process of rolling noise Force generation in wheel/rail contact
10. Zvolenský, P., Grenčík, J., Pultznerová, A., Kašiar, L.: Research of noise emission sources in railway transport and effective ways of their reduction. In: MATEC Web of Conferences, vol. 107, p. 00073 (2017)
11. Burdzik, R., Nowak, B.: Application of vibration signals in the rail vehicle identification system. In: Diagnostyka Maszyn. XLVII Ogólnopolskie Sympozjum (2020)
12. Burdzik, R., Nowak, B., Konieczny, Ł., Mańka, A.: Analysis of the propagation of vibration waves in the railroad rails caused by the vehicle's passage. In: WibroTech 2019, Zawiercie (2019)
13. Bhore, C., Andhare, A., Padole, P., Korde, M.: Analysis of track vibration for metro rail. In: Kalamkar, V.R., Monkova, K. (eds.) Advances in Mechanical Engineering: Select Proceedings of ICAME 2020, pp. 67–73. Springer Singapore, Singapore (2021). https://doi.org/10.1007/978-981-15-3639-7_9

14. Trimpe, F., Salander, C.: Wheel–rail adhesion during torsional vibration of driven railway wheelsets: vehicle system dynamics. Int. J. Veh. Mech. Mobility **59**(5), 785–799 (2020)
15. Chen, C., Fang, C., Qu, G., He, Z.: Vertical vibration modelling and vibration response analysis of Chinese high-speed train passengers at different locations of a high-speed train. Proc. Inst. Mech. Engineers Part F J. Rail Rapid Transit **235**, 35–46 (2020)
16. He, X., Shi, K., Wu, T.: An efficient analysis framework for high-speed train-bridge coupled vibration under non-stationary winds. Struct. Infrastruct. Eng. Maintenance Manag. Life-Cycle Des. Perf. **16**(9), 1326–1346 (2019)
17. Khan, M.R., Dasaka, S.M.: Characterisation of high-speed train vibrations in ground supporting ballasted railway tracks. Transp. Infrastruct. Geotechnol. **7**(1), 69–84 (2019). https://doi.org/10.1007/s40515-019-00091-w
18. Liratzakis, A., Tsompanakis, Y., Psarropoulos, P.N.: Assessment of high-speed train-induced vibrations using efficient numerical models. In: Sapountzakis, E.J., Banerjee, M., Biswas, P., Inan, E. (eds.) Proceedings of the 14th International Conference on Vibration Problems. Lecture Notes in Mechanical Engineering. Springer, Singapore (2021)

Definition of the Decisive Criteria for Estimation of Measurement System of Rail Vehicle Wheel Positioning

Gintautas Bureika(✉), Gediminas Vaičiūnas, and Viktor Skrickij

Vilnius Gediminas Technical University, Saulėtekio al. 11,
10223 Vilnius, Lithuania
{gintautas.bureika,gediminas.vaiciunas,
viktor.skrickij}@vilniustech.lt

Abstract. The safety and security of railway transport strongly depends on the technical state of rail track elements. Railway infrastructure and rolling-stock operators faces the need to constantly monitor the technical condition of their assets. As the diversity of vehicle wheel tracking and positioning systems raises the demand to decide of which measurement system for technical state monitoring to choose is actual as well. This article provides an example of how to identify the derive criteria for estimation of measurement system based on expert survey data.

Keywords: Railway transport · Rail vehicle wheel · Measurement system · Sensors · Wheel positioning · Multi-criteria assessment

1 Introduction

With the development of rolling stock operating systems, rail vehicle parameter monitoring systems are becoming increasingly relevant. The temperature dynamics of the wheel bearings and the dimensions of the wearing wheel surfaces can be monitored [10]. The rail vehicle wheel impact on rail and locomotive driven wheel adhesion with the rail are investigated intrusively too [1, 3].

The development of various vehicle wheel tracking and positioning (measurement) systems raises the question of which type of sensors to choose for this monitoring. The following two aspects need to be considered in the selection. In particular, these are the technical parameters specified in the equipment documentation [13]. Another issue is the opinion of the professionals who are used this type of equipment. Authors of this article focus more on this aspect.

Multi-criteria assessment methods (thereafter – MCAMs) are widely used and numerous, but choosing the right ones to solve a particular problem is not so simple. Although a wide variety of MCAMs have been proposed in the scientific literature. They differ from each other in their depth and complexity. However, it is not clear (definitively) which method is best suited for which case (s). The application of MCAM to a specific case is examined on the basis of a number of features: the number of technology-specific challenges addressed by this approach; the maximum possible

O. Prentkovskis et al. (Eds.): TRANSBALTICA 2021, LNITI, pp. 79–86, 2022.
https://doi.org/10.1007/978-3-030-94774-3_8

number of objects (alternatives) to be assessed; the maximum possible number of criteria describing the object (alternative); the level of qualification of the decision-makers that is relevant to the application of this method; possibility to evaluate (validate) the reliability of the obtained evaluation results; the time of the employee's mastering the use of this method when he has no knowledge of it; the time required for a multi-criteria assessment of an employee capable of applying this method; the resources needed to solve a multi-criteria problem in a specific way. In most cases, the decision maker is concerned with the content of his MCAM calculation algorithm. Thus, in addition to the features listed above, the comprehensibility of the method is very important. It can be simple, medium or complex [7].

MCAMs can be classified in different ways. For example, they can be divided into classical and technological [11], compensatory and compensatory [4].

Some researchers classify methods according to the nature of the available information on assessment objects (alternatives) [5]. When choosing a method, it is important to consider where it will be applied. In this study, the method will be used to select the sensors to be attached to the vehicle running gear. An example of mounting the sensor to the vehicle running gear is shown in the Fig. 1 [12].

Fig. 1. Example of mounting of a sensor on a vehicle running gear.

Various mathematical methods are available to summarize the respondents' survey data, the authors chose the AHP method and used it to summarize the data. The hierarchical method was chosen because the criteria in question are not hierarchically homogeneous.

2 Applying of the Analytic Hierarchy Process Method

(AHP) method was suggested in [8]. This method allows us to determine the weights (significances) of hierarchically non-structured or particular hierarchical level criteria in respect to those belonging to a higher level.

The method is based on the pairwise comparison matrix $P = \|p_{ij}\|, (i,j = 1,2,\ldots,m)$. Experts compare all the evaluation criteria R_i and R_j ($i, j = 1, 2, \ldots, m$), where m is the number of criteria compared. In the ideal case, the elements of the matrix present the relationships between the unknown criteria weights:

$$p_{ij} = \frac{\omega_i}{\omega_j}. \tag{1}$$

The comparison is qualitative and easy to perform. It indicates if one criterion is more significant than the other and to what level the priority it belongs. This technique allows the qualitative estimates elicited from experts to be converted to quantitative ones.

The pairwise comparison matrix is an inverse symmetrical (reciprocal) matrix [6, 9]:

$$\mathbf{P} = \begin{vmatrix} 1 & p_{12} & \cdots & p_{1m} \\ \frac{1}{p_{12}} & 1 & \cdots & p_{2m} \\ \cdots & \cdots & \cdots & \cdots \\ \frac{1}{p_{1m}} & \frac{1}{p_{2m}} & \cdots & 1 \end{vmatrix}. \tag{2}$$

The main principle of filling in the matrix is simple, as an expert should indicate how much more important a particular criterion is than another one. Saaty suggested a widely known 5-point scale (1-3-5-7-9) be used for evaluation. The evaluation of the criteria ranges from $p_{ij} = 1$, when R_i and R_j are equally significant, to $p_{ij} = 9$, when the criterion R_i is much more significant than the criterion R_j concerning the research aim [8]. Also the parameters 2-4-6-8 can be used. The necessary condition of the comparison matrix' consistency is the transitivity of the matrix elements.

The normalised subjective weight or relative importance can be calculated:

$$\omega_i = \frac{\sqrt[m]{\prod_{j=1}^{m} e_{ij}}}{\sum_{i=1}^{m} \sqrt[m]{\prod_{j=1}^{m} e_{ij}}}. \tag{3}$$

The eigenvalue should be calculated by multiplying pairwise the comparison matrix with the weight vector (3):

$$Pw = \lambda_{MCDM} w; \tag{4}$$

where λ_{MCDM} is an eigenvalue.

The weights in Saaty's approach, the vector ω are normalized components of the eigenvector corresponding to the largest eigenvalue $\lambda_{MCDM,max}$ that can be calculated:

$$det A = P - \lambda_{MCDM} E = 0; \tag{5}$$

It is known that the largest eigenvalue of the inverse symmetrical m-order matrix is $\lambda_{MCDM,max} \geq m$. The AHP method assesses the consistency of each expert's estimates using the consistency in [2]:

$$C.I. = \frac{\lambda_{MCDM,max} - m}{m - 1}. \tag{6}$$

The relationship between the calculated consistency index *C.I.* of a particular matrix and the random consistency index *R.I.* (presented in Table 1) is referred to as consistency relationship. It determines the ratio of matrix consistency:

$$C.R. = \frac{C.I.}{R.I.}. \tag{7}$$

If the value of the ratio matrix consistency *C.R.* is smaller than or equal to 0.1 is acceptable, implying that the matrix is consistent.

Table 1. The values of a random consistency index *R.I.*

Matrix order	3	4	5	6	7	8	9	10	11	12	13	14	15
R.I.	0.58	0.90	1.12	1.24	1.32	1.41	1.45	1.49	1.51	1.48	1.56	1.57	1.59

3 Criteria Setting and Evaluation

It was found that the most important factors of the technology used for measuring the rail vehicle wheel transversal position are:

1. *Dimensions and Weight* including geometrical parameters of the system, its weight, mounting options etc.
2. *Energy Consumption* also including energy efficiency, input voltage, etc.
3. *Life Cycle Cost* (LCC) including system price, maintenance expenses, long-life, durability, reliability, etc.
4. *System Robustness* including vibration and shock resistance; temperature and dust resistance, disruptive from railway signalling/ automation, ambient conditions (rain, snow, slob, sleet, fog, etc.), etc.
5. *Measurement Accuracy* including sampling rate, system dependency on train speed.
6. *Technical Compatibility (Interoperability)* including the possibility of system installation under different running gears, with different gauges, signalling/automation system, interfaces, etc.
7. *Output Data* including the complexity of final parameter estimation from output data, data format, amount of data, additional data such as location, time, date, visual data, etc.
8. *Measurement Repeatability* including the reliability of results.

In order to determine the most important criteria, a questionnaire was created, experts made rankings and identified weights of the drivers (Table 2).

These procedures allowed to identify the weights in three different ways: Rank correlation method, AHP method and Direct Score Method. A total of twelve questionnaires were filled out. Summarizing of ranked drives according Rank correlation method is presented in Table 2.

Afterwards, the AHP method, was applied on the expert rankings and calculations were performed. The concordance coefficient was calculated; the estimated numerical value is W = 0.805. The value of the parameter χ^2 is 67.61, which is greater than the critical value χ^2_{kr} = 14.067. The degrees of freedom ν = 8–1 = 7, importance level α = 0.05.

Table 2. Ranking of drivers according Rank correlation method from Questionnaires.

Criteria		Expert												Weight
		1st	2nd	3rd	4th	5th	6th	7th	8th	9th	10th	11th	12th	Avr.
1	Dimensions and weight	2	6	7	8	8	3	6	3	8	7	6	8	5.83
2	Energy consumption	6	5	8	2	4	8	7	8	7	1	7	7	3.75
3	Life cycle cost	5	3	6	4	1	5	3	2	1	3	8	4	1.92
4	System robustness	3	1	1	1	3	2	1	1	4	2	1	3	3.67
5	Measurement accuracy	4	7	2	3	7	4	2	4	2	4	4	1	4.17
6	Technical compatibility	1	2	4	7	5	1	5	6	5	6	3	5	6.42
7	Output data	8	8	5	5	6	7	8	5	6	8	5	6	4.25
8	Measurement repeatability	7	4	3	6	2	6	4	7	3	5	2	2	5.83

As a result, it can be stated that the assessment of experts is in agreement. During the next step, the AHP matrix was developed using the ranking of one expert (whose competence is beyond doubt) as reference (Table 3).

Table 3. AHP ranking matrix from the 1st questionnaire.

Criteria	1	2	3	4	5	6	7	8	ω_i
1	**1**	1/2	1/7	1/9	1/8	1/6	2	1/5	0.025
2	2	**1**	1/5	1/9	1/7	1/3	3	1/2	0.040
3	7	5	**1**	1/5	1/3	2	5	3	0.139
4	9	9	5	**1**	2	3	7	5	0.339
5	8	7	3	1/2	**1**	3	7	5	0.255
6	6	3	1/2	1/3	1/3	**1**	5	2	0.109
7	1/2	1/3	1/5	1/7	1/7	1/5	**1**	1/5	0.022
8	5	2	1/3	1/5	1/5	1/2	5	**1**	0.071
-									Σ = 1.0

The eigenvector "ω_i" (the last column of Table 3) is calculated according Formula (3) – by normalizing the values of the geometric means of the rows. The consistency index is calculated as *C.I.* = 0.070, while the degree of matrix consistency is *C.R.* = 0.050. The value of consistency index *C.R.* is acceptable as it is smaller than 0.1.

In parallel, the calculations were made, using the weights according Direct Score Method provided directly by 12 experts (the evaluation form 0 before 30 score). The results are presented in Table 4.

Table 4. Weighting of drivers according Direct Score Method.

Criteria		Expert												Weight
		1st	2nd	3rd	4th	5th	6th	7th	8th	9th	10th	11th	12th	Avr.
1	Dimensions and weight	20	10	5	5	5	15	9	16	2	5	10	5	8.9
2	Energy consumption	10	10	5	17	15	5	7	4	5	15	5	5	8.6
3	Life cycle cost	9	15	10	14	20	10	16	18	30	12	5	10	14.1
4	System robustness	12	30	20	20	15	15	20	20	10	20	25	15	18.5
5	Measurement accuracy	9	5	18	15	5	15	18	15	20	10	15	25	14.2
6	Technical compatibility	20	15	15	7	15	20	12	10	8	18	15	10	13.8
7	Output data	10	5	10	12	5	10	3	12	10	5	10	10	8.5
8	Measurement repeatability	10	10	17	10	20	10	15	5	15	15	15	20	13.5

The numerical values of weights evaluated using different methods are presented in Table 5.

The row "Weights of Rank correlation" contains the normalized data values of the right (last) column of Table 2. The row "Weights AHP" was obtained from the eigenvector "ω_i" – the last column of Table 3. The row "Weights of direct Score" contains the normalized data values of the right (last) column of Table 4.

Table 5. The comparison of evaluation results.

Criteria	1	2	3	4	5	6	7	8
Weights of rank correlation	**0.167**	0.162	0.104	0.053	0.102	0.116	**0.178**	0.118
Rank	**2**	3	6	**8**	**7**	5	**1**	4
Weights AHP	0.025	0.040	0.139	**0.339**	**0.255**	0.109	0.022	0.071
Rank	7	6	3	**1**	**2**	4	8	5
Weights of direct score	0.089	0.086	0.141	**0.185**	**0.142**	0.138	0.085	0.135
Rank	6	7	3	**1**	**2**	4	8	5
Average of ranks	5.00	5.33	4.00	**3.33**	**3.67**	4.33	5.67	4.67
Final rank	6	7	3	**1**	**2**	4	8	5

Summarizing the results of the three methods, it can be concluded that the most important criteria are System Robustness; Measurement accuracy; Life Cycle Cost and Technical Compatibility.

4 Conclusions

The example presented in this article confirms that it is possible to name the decisive criteria for measurement system based on the data of the Questionnaire of respondents. Various mathematical methods are available to summarize the respondents' survey data.

This research was performed by Rank correlation method, AHP method and Direct Score Methods.

The results vary when considered by different methods. The comparison of the results of measurement system assessment by the three methods showed that the most important criteria are System Robustness, Measurement accuracy, Life Cycle Cost and Technical Compatibility.

Acknowledgment. The research is funded by the EU Shift2Rail project Assets4Rail (Grand number: 826250) under Horizon 2020/ Shift2Rail Framework Programme.

References

1. Bureika, G., Levinzon, M., Dailydka, S., Steišunas, S., Žygiene, R.: Evaluation criteria of wheel/rail interaction measurement results by trackside control equipment. Int. J. Heavy Veh. Syst. **26**(6), 747–764 (2019). https://doi.org/10.1504/IJHVS.2019.102682
2. Ginevičius, R., Podvezko, V., Ginevičienė, V.B.: A complex evaluation of the work productivity of university teaching staff. In: seminar proceedings of 8th Baltic Region Seminar on Engineering Education, pp. 44–46. Kaunas University of Technology, Kaunas (2004)
3. Gorbunov, M., et al.: Estimation of sand electrification influence on locomotive wheel / rail adhesion processes. Eksploatacja i Niezawodnosc – Maintenance Reliab. **21**(3), 460–467 (2019). http://dx.doi.org/10.17531/ein.2019.3.12
4. Hwang, C.L., Lin, M.J.: Group Decision Making under Multiple Criteria: Methods Applications. Springer, Berlin (1987). https://doi.org/10.1007/978-3-642-61580-1
5. Hwang, C.L., Yoon, K.: Multiple Attribute Decision Making – Methods and Applications. A State of the Art Survey. Springer, Heidelberg (1981). https://doi.org/10.1007/978-3-642-48318-9
6. Maskeliūnaitė, L., Sivilevičius, H.: Expert evaluation of criteria describing the quality of travelling by international passenger train: technological, economic and safety perspectives. Technol. Econ. Dev. Econ. **18**(3), 544–566 (2012)
7. Ozernoy, V.: A framework for choosing the most appropriate discrete alternative multiple criteria decision-making method in decision support systems and expert systems. In: Sawaragi, Y., Inoue, K., Nakayama, H. (eds.) Toward interactive and intelligent decision support systems, pp. 56–64. Springer, Heidelberg (1987). https://doi.org/10.1007/978-3-642-46609-0_6
8. Saaty, T.L.: The Analytic Hierarchy Process. McGraw-Hill, New York (1980)
9. Sivilevičius, H., Maskeliūnaitė, L.: The model assessing the impact of price and provided services on the quality of the trip by train: MCDM approach. Econ. Manag. **22**(2), 51–67 (2019)

10. Vaiciunas, G., Bureika, G., Steišunas, S.: Rail vehicle axle-box bearing damage detection considering the intensity of heating alteration. Eksploatacja i Niezawodnosc – Maintenance Reliab. **22**(4), 724–729 (2020). http://dx.doi.org/10.17531/ein.2020.4.16
11. Zimmermann, H.-J., Gutsche, L.: Multi-Criteria Analyse. Springer, Heidelberg (1991). https://doi.org/10.1007/978-3-642-58198-4
12. Equipmentimes (Dalian) E-Commerce Co., Ltd. Homepage. https://www.equipmentimes.com/product/details, Accessed 01 May 2021
13. Skrickij, V., Šabanovic, E., Shi, D., Ricci, S., Rizzetto, L., Bureika, G.: Visual measurement system for wheel–rail lateral position evaluation. Sensors **21**(4), 1–16 (2020). https://doi.org/10.3390/s21041297

The Second Generation of Electric Vehicles: Integrated Corner Solutions

Paulius Kojis[1(✉)], Algimantas Danilevičius[2], Eldar Šabanovič[1], and Viktor Skrickij[1]

[1] Transport and Logistics Competence Centre, Transport Engineering Faculty, Vilnius Gediminas Technical University, Saulėtekio al. 11, 10223 Vilnius, Lithuania
{paulius.kojis,eldar.sabanovic, viktor.skrickij}@vilniustech.lt

[2] Department of Mobile Machinery and Railway Transport, Transport Engineering Faculty, Vilnius Gediminas Technical University, Plytinės Street 27, 08101 Vilnius, Lithuania
algimantas.dalinevicius@vilniustech.lt

Abstract. The impact of the transportation sector on pollution and emissions is very high. To solve this issue, manufacturers are shifting from conventional combustion engines to new electric solutions. As a result, the opportunity to redesign vehicle architecture appears. This paper provides a review of driveline topologies of the new generation vehicles, targeting wheel corner modules with in-wheel solutions. To understand the impact of the new solution on the vehicle dynamic properties, we analysed the literature and performed a SWOT analysis.

Keywords: Vehicle corner · Electric vehicle · Vehicle dynamics

1 Introduction

Average daily travelling distance is growing significantly for the last decades and harms ecology. Fuel consumption is one of the main parts of vehicle maintenance costs. The impact of the transportation sector on pollution and emissions is very high. It generates 23% of global CO_2 emissions [1], and fleet change from conventional to zero-emission is not as fast as predicted. Hence, a solution is needed.

The European Union's regulation of passenger car emissions has failed when it comes to nitrogen oxide emissions (NO_x) by diesel engines [2]. The main problem of diesel engines is NO_x emissions that have the highest proportion of diesel pollutant emissions with a rate of more than 50% [3]. After NO_x emissions, particulate matter (PM) has the second-highest proportion in pollutant emissions. Previous studies have shown a strong relationship between exposure to PM 2.5–10 and mortality [4, 5]. There are three possible solutions for the minimisation of emission problem [6]: *i)* alternative fuels; *ii)* technological advancements of transportation components; *iii)* optimised vehicle design.

This paper comprehensively reviews electric vehicle designs targeting integrated wheel corner solutions to replace conventional internal combustion engines that

O. Prentkovskis et al. (Eds.): TRANSBALTICA 2021, LNITI, pp. 87–100, 2022.
https://doi.org/10.1007/978-3-030-94774-3_9

decrease CO_2 emissions. Widely used hybrids can potentially decrease local CO_2 emissions by 25%; plug-in hybrids by 50%. Vehicles equipped with Range Extended Electric drive (RXBEV) by 90%, 100% reduction can be reached using the battery-electric drive and fuel cell drive (FCBEV) technologies [7].

Several technological aspects guide the full potential of electric vehicles (EV). Currently, batteries are the main obstacle to EV wider adoption. The battery pack is the most expensive component of any EV. Nevertheless, battery capacity has increased significantly and now reached a margin of 200 kWh. However, fully charging a vehicle may require extended periods, between 1 and 3 h, depending on whether it is charged at the standard electrical outlet or a charging station. In addition, there are a variety of EV charging connectors; architecture and charging modes vary according to different standards. Therefore, compatibility issues may arise [8]. Out of the existing battery types, lithium-ion batteries (Li-ion) shows the best performance in terms of capacity, cell voltage and operating temperature range. Considering cycle durability, sodium-sulfur batteries (Na-S) perform better than Li-ion ones, but that is the only advantage [9]. The outcome of this section is that battery technology must advance, charging connector types become unified, charging station infrastructure must expand to access various areas.

Another consideration is the propulsion system. The industry is moving away from induction machines toward the use of permanent magnet machines for traction applications. The shift is caused by increasing requirements for higher efficiency, specific power, power density and higher constant power speed range. Several other types of motors used in EVs have switched reluctance machines and synchronous reluctance machines. All of described motors are direct current machines [10]. For in-wheel motor (IWM) applications, a permanent magnet synchronous motor (PMSM) is used. Efforts in research and development must be placed to perfect and reduce manufacturing costs of propulsion systems [11].

This paper review technological solutions in the field of EVs drivetrains, with a focus on corner designs that can be used for the second generation of EVs. The platform can be redesigned for energy efficiency and maximum comfort instead of repeating architectural similarities concerning combustion engine vehicle (e.g., centralised motor). In Sect. 2, drivetrain topologies of electric vehicles are reviewed. Implementation of IWMs for the vehicle corners is analysed. In Sect. 3, vehicle corner design impact on vehicle dynamics is reviewed. In Sect. 4, the discussion is performed.

2 Electric Vehicle's Driveline Topologies

Fully EVs are designed to be propelled by only electric power. The battery pack group powers the electric motor through an inverter driven by a control module that receives driver commands. Based on the mounting of the electric motor, several drivetrain topologies are applied in fully EVs. Depending on the electric motor count, location and power distribution scheme, it is configured as central drives, near the wheel drives and in-wheel drives [12, 13]. If the reduction gear is used, direct and indirect drive expands the classification. Figure 1 details the schematic representation of the possible configurations for EV architecture.

The central drive (CD) variant is the most common way to retrofit the internal combustion vehicle to fully electric. Traditional driveline is simply replaced by an electric motor and an appropriate gear. The first EVs were implemented in this way. As described before, the central drive configuration can be categorised as direct and indirect drive. In the direct drive variant, fewer components are defining the driveline Fig. 1(a, b). In the topologies using a near-wheel drive (NWD) unit, the drivetrain is further simplified. Using a single electric motor per wheel eliminates the necessity to connect both axles, and both axles can be driven independently. The electric motors are operating at different speeds; therefore, no differential is required. Figure 1(c, d) shows that the near-wheel drive unit could be direct and indirect.

The IWMs is the latest topology of EVs. In the following section, it will be discussed in more detail.

2.1 In-Wheel Motors

IWMs are positioned in the rims of EV's wheels. This enables independent control of individual wheels. With the separately controllable electric motors, torque vectoring improved electronic stability control (ESC), driving force distribution and slip angle control are available [15]. Combinations between the use of reduction gear and direct drive are possible (Fig. 1 e, f).

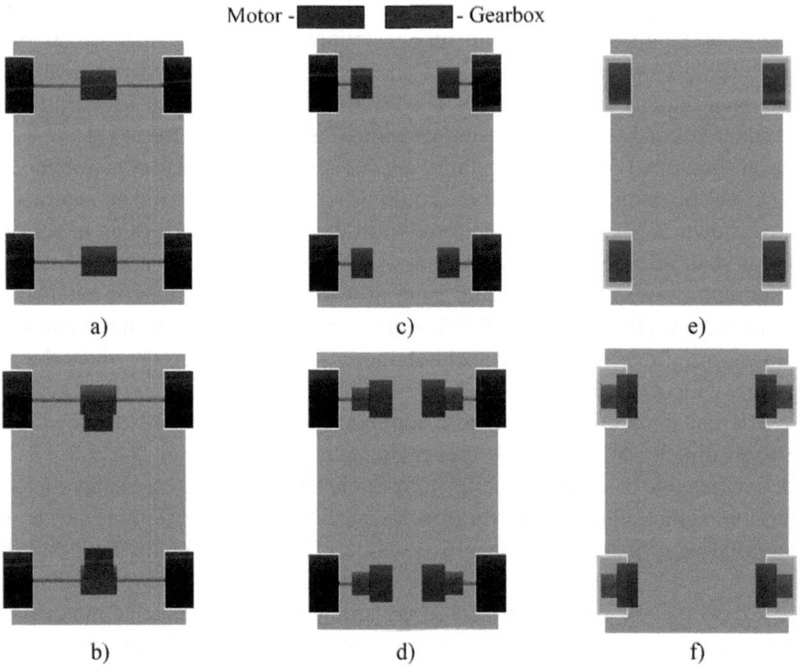

Fig. 1. EVs topologies [14]: a) Direct (CD); b) Indirect (CD); c) Direct (NWD); d) Indirect (NWD); e) Direct (IWMs); f) Indirect (IWMs).

By fully abandoning any mechanical gearing, power delivery is done without mechanical losses or additional interconnections (Cardan or axle shafts). It is also possible to use two or four IWMs to design vehicles with two or four driven wheels. In this way, identical motors can be used for different vehicle classes and applications [7].

The application of IWMs highlights fundamental advantages for EV's. As stated above, IWMs offers a deficiency of complex and heavy transfer mechanisms. For this reason, power is used more efficiently, the overall mass of the vehicle is reduced, and space is freed up that can be used for more battery packs. Energy recuperation is more easily implemented and provided by regenerative braking [12]. The IWM provide fast and dynamic response. A greater level of redundancy is reached as torque load can be transferred to any individual wheel if problems occur in any other motor. According to a scientific study by [16], positive or negative torques are applied within milliseconds using IWMs. As a result, the improved performance of ESC, anti-lock braking system (ABS) and traction control systems is achieved. This benefits safety, stability and decreases braking distance.

Despite the advantages over traditional propulsion systems, there are disadvantages and several technical barriers. The increase in unsprung mass leads to a reduction of the comfort and handling properties of the vehicle [24]. Added weight decreases unsprung mass resonance frequency. This results in resonance that may occur in the 4–8 Hz range, to which humans are most sensitive [12]. Due to the electric motor being mounted inside the wheel, cooling, sealing and ingress problems occur. Also, power output is limited because wheel dimensions limit weight and dimensions. Since regenerative braking strongly depends on constraining parameters such as battery state of charge, vehicle velocity, motor speed and motor temperature, it needs assistance from frictional braking in emergencies. [13]. Maintaining the air gap between the rotor and the stator is complex due to material stiffness and dynamic loads [11].

Several different IWM classifications are made based on their construction. First considers rotor position; therefore, there could be an internal rotor and an external rotor design, as shown in Fig. 1(d, c). The rotation in the internal rotor design is performed by a motor shaft attached to the wheel hub. In the external rotor variant, the rotating part is the motor housing. In addition, there are two derivatives for the external rotor configuration: partially integrated IWM with external rotor, where the rotor is connected to the rim; fully integrated wheel hub motor with external rotor, where the motor housing is a part of the rim [18].

The external rotor solution is a more common application to IWMs since the rotor can be fixed directly to the rim. A bigger diameter between rotor and stator in such construction results in higher specific torque; this makes them a better choice for application in vehicles [19]. However, for the machines with an external rotor cooling, the rotor windings, where the highest losses appear, are located in the inner part of the machine.

The second type of IWMs is based on the arrangement of the active motor parts in relation to the axle. The axial or radial orientation is possible, as shown in Fig. 1(b, a). An axial layout of active motor parts is often called a bell-shaped rotor, and a radial is called a disc rotor [16].

The third type of IWMs depends on the torque transmission. This could be done directly or using additional gear. Such indirect configuration expands the torque and rotational speed interval, has a smaller size, lighter weight and lower costs. Direct drive layout has a bigger size and weight, but construction is simpler, the overall reliability and efficiency are higher [17].

As depicted in the DHL report [20], it is widely estimated that around 55% of all new car sales in Europe will be electric vehicles by 2030. Therefore, IWM manufacturers are presenting new products and concepts. Currently, there are two most promising manufacturers: Elaphe and Protean Electric. Elaphe's signature IWM L1500 produces 77 kW continuous power, 650 Nm of continuous torque and weights 34.8 kg [21]. On the other camp, Protean Electric offers a Pd18 motor, which generates 60 kW and 650 Nm continuous power and torque, respectively [22].

2.2 Wheel Corner Modules

The Wheel Corner Module (WCM) offers a high degree of modularity since it is held within the boundaries of a conventional wheel. The module is mounted to a vehicle body using bolted connections and an electrical connector port. Generally, the module consists of an in-wheel propulsion system combined with suspension elements and electro-mechanical components. The described combination allows complete vehicular motion control, comprising longitudinal, lateral and vertical motions [13].

Most of the functions of this module are electrically controlled based on input from an operational device in connection with control units. Therefore, it represents a pure x-by-wire based vehicle manoeuvrings system. In conjunction with various sensors mounted around the vehicle, control units can gather information about the vehicle position and state [14].

WCM technology offers versatility since motion control systems can all be configured individually [24]. This drastically enhances handling properties which can be set according to driver characteristics and road conditions. The level of ride comfort is reached using active suspension systems, which effectively counteract angular and vertical motions [14].

One of the significant drawbacks of WCMs is increased unsprung mass, which brings adverse effects. The rise in the vertical accelerations and the dynamic wheel load fluctuations deteriorates the ride comfort and vehicle's road-holding ability [25]. In the article by Anderson and Harty [26], the impact of unsprung mass increase is perceptible but small and unlikely to be apparent to an average driver. Therefore, such differences in research data require clear solutions and proper evaluations.

Below known examples of WCMs are presented with a focus on operating principle and applied technologies.

Michelin Active Wheel concept comprises an indirect drive IWM that provides the tractive force. Pinion and gearwheel are used to transfer the torque from the motor to the hub. This results in a more compact design. The wheel design is presented in Figure 3(a). Motor rated power is 30 kW. Each active wheel module weighs 42 kg [27]. This corner module has an integrated active suspension system, including a

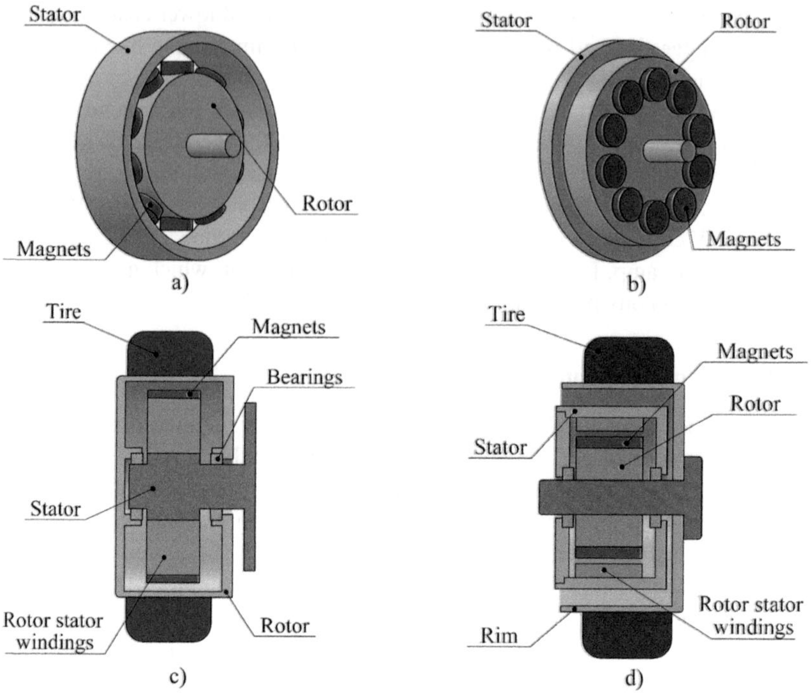

Fig. 2. In-wheel motor classification [18, 23]: a) radial arrangement; b) axial arrangement; c) external rotor IWM; d) internal rotor IWM.

passive coil spring and an active electric actuator. Combined elements isolate the chassis from vibrations induced by the road. A conventional brake calliper coupled with a brake disk is used to assist the regenerative braking system [28].

An active camber control system for this module is created and patented by Michelin. It can provide high camber angle variations up to $\pm 15°$ in each direction. The passenger compartment is tilted synchronous with the camber angle. However, the proposed control system takes up a large space and contradicts the compact vehicle design [29].

In Bridgestone Dynamic-Damping In-Wheel Motor Drive, the IWM is mounted separately to decrease the unsprung mass, as shown in Fig. 3(e). The stator of the IWM is mounted to a knuckle on the axle via the springs and dampers. Therefore, it is an internal rotor design. The flexible coupling is used to transmit the torque from the motor to the wheel hub. This coupling allows vertical movement of up to 50 mm between the rotor and wheel. No active suspension components are incorporated [30].

The effectiveness of this proposed configuration of the Wheel Corner Module was investigated during experimental research [31]. In the investigation, the module was compared to a conventional EV with an unsprung mass of 40 kg per quarter and with an IWM EV with an unsprung mass of 70 kg per wheel corner. The results indicate that such a solution decreases the load fluctuation of the tire. According to this article, Bridgestone's solution decreased vertical accelerations in the vehicle body. However, accelerations were only reduced in the short frequency band from 10 to 15 Hz.

In the Siemens VDO eCorner direct-drive, the IWM with an external rotor design is used. In addition, an active suspension system and an electro-mechanical disc brake are fitted. The motor is constructed using the radial placement of magnets. In this way, higher specific torque is achieved. Regenerative braking is working in parallel with frictional brake. The module can change the steering angles of each corner [32]. The corner module is presented in Fig. 3(b).

Due to high frictional force demand, an Electronic Wedge Brake (EWB) system is implemented. Linear actuator force is applied onto a wedge-shaped element pushed against another wedge-formed element, forcing the brake pad towards the brake disc. The wedge brake is based on a self-amplifying process. Therefore a lower force actuator can be used [33].

An indirect drive IWM propels Volvo's WCM. The power output is in the range of 10–15 kW per wheel. Active suspension components are combined with passive ones. The active part maintains the specific ride height, handles anti pitch and roll systems. A combined rotary damper and rubber torsion spring unit are integrated into the module. The suspension features a pretensioning function. Therefore, the system can be utilised to vary the suspension's stiffness, thereby obtaining better handling and comfort properties. Besides, an active steering, toe and camber angle variation is possible with this setup. The described module is shown in Fig. 3(d).

Advanced control algorithms are used in order to achieve optimal steering, toe and camber angles. All the input data is adopted from driving style, tire characteristics and road surface conditions [34].

Protean 360+ WCM features 360° steering capability and IWM, an active suspension and pneumatic ride height control. Due to the compact design of this WCM, a flat-floor vehicle platform could be adopted. An innovative multi-link suspension system with an additional lower wishbone pivot keeps the suspension system within the rigid packaging envelope. The high degree of steering angle allows to reach tight spaces and improves overall mobility. The selected direct-drive IWM provides 60 kW of continuous power but weighs 36 kg. The motor features integrated electronics, enabling real-time monitoring of wear and tear on areas such as bearings and suspension, thus improving predictive maintenance [35]. Protean's module is shown in Fig. 3(c).

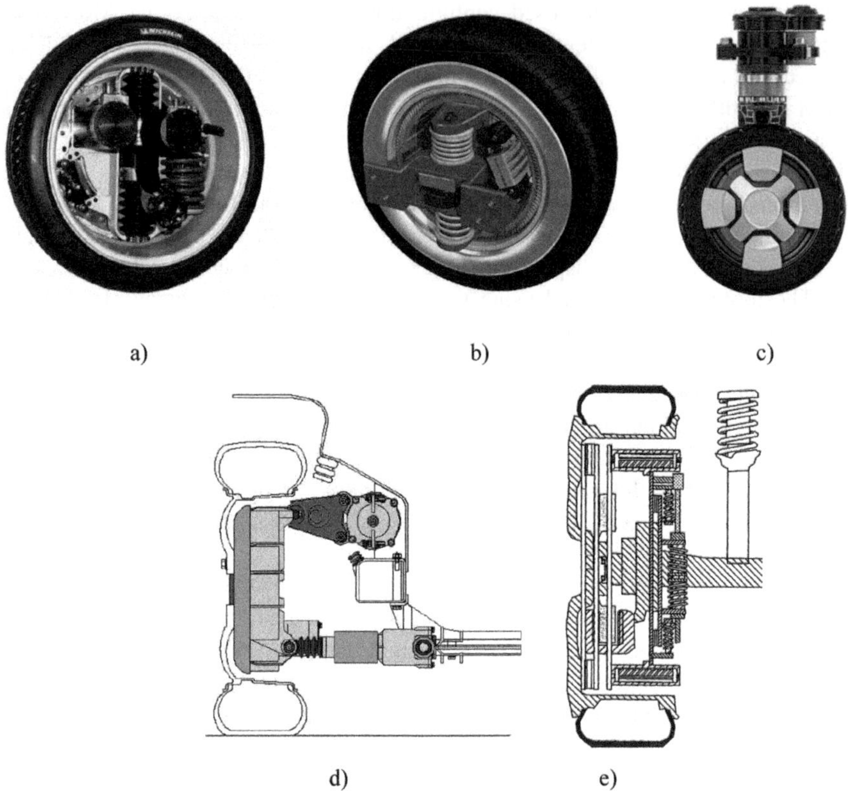

Fig. 3. Wheel corner modules: a) Michelin active wheel [27]; b) Siemens VDO eCorner [32]; c) Protean 360+ [35]; d) Volvo's WCM [19]; e) Bridgestone's WCM [30].

3 Wheel Corner Module Impact on Vehicle Dynamics

The higher degree of integration between powertrain and chassis and the shift toward corner solution promises improved performances regarding new vehicle architecture design, energy efficiency, vehicle stability and comfort [36, 37]. The remarkable merit of IWM is the ability to precisely generate the driving/braking torques with a fast response at individual wheels using drive-by-wire or other innovations. X-by-wire solutions remove mechanical link related problems, allows integration of torque management with cruise control, traction control and stability control, reacts more quickly, noiseless operation and eliminates vibration, reduces the weight of the overall system [38]. Technology fits well for WCM equipped with IWMs, as all motors can be controlled individually.

One of the best-known examples of drive-by-wire technologies is electronic throttle control (ETC), where a mechanical connection between the accelerator pedal and the throttle is eliminated. The system realised using an accelerator pedal position sensor and an electronically operated throttle. Another example is brake–by–wire, an automotive technology that eliminates traditional mechanical and hydraulic components and replaces them with electronic sensors and actuators to control the brakes in vehicles [13]. However, additional problems may also occur. Under heavy braking or due to road surface unevenness, airgap variation between the stator and rotor of the IWM is induced. Such oscillations result in unbalanced electric magnetic force, decrease the braking force, and increase stopping distance, therefore diminishing braking performance [39].

Longitudinal motion control schemes could be realised using WCM with IWM as a propulsion system. Control algorithms can be classified into the following groups: (i) slip ratio control, (ii) anti-slip control and (iii) driving force control [15]. The slip ratio control strategy improves traction and reduces energy losses while a vehicle accelerates on a slippery road. This is possible due to the ease of measurements of motor torques acting on each tire. Therefore, driving forces, which are generated by tire-road interaction, could be estimated. Out of this information, detection of wheel slip occurrence is available [40]. The anti-slip algorithm controls torque output. If excessive torque is detected, the system constrains it to follow the maximum possible while maintaining the friction between the tire and the road surface. This provides high adaptability to different tire-road conditions [41]. Driving force control enables force distribution between the individual wheels, the reduction of total driving force and generation of yaw moment prevention. As a result, EVs can be driven on low friction coefficient surfaces with less effort [42].

Lateral motion control can be realised via direct-yaw-moment control. By differentiating the torques distributed to the left-side and right-side wheels, a yaw-moment is generated about the vertical axis of the vehicle body. Many studies have demonstrated that direct yaw moment control is an effective method to improve vehicle manoeuvrability and stability. The steering efforts are reduced by approximately 25% compared to the case without the control at high-speed cornering [43]. It is also possible to generate the roll moment about the longitudinal axis by allocating the motor torques suitably, thus improving comfort [15]. Considering four IWM configuration in EV, independent braking, drive and steering combinations are possible. In the article [44], such combinations are compared by the impact on vehicle stability. The best vehicle stability results are obtained by combining four-wheel independent drive and steering combinations [45]. Since lateral dynamics are mainly defined by steering, Wang et al. [46] introduced assisted steering system. The working principle applies a varying torque between the right and left front wheels to minimise the difference between the actual steering torque and the mapped steering torque. Therefore, reduced steering efforts, driving safety and comfort is achieved. However, for comfort, pitch, as well as vertical dynamics are fundamental.

Vertical dynamics are influenced mainly by WCM related increase of unsprung mass. In addition, the chassis supports the drive torque of the IWM. This will reduce the pitch angle during acceleration and braking. Due to higher unsprung mass, wheel eigenfrequencies tend to be lower. Therefore, resonance may occur in the range where humans are more sensitive [37].

Large unsprung mass is often connected with the accompanying increase in the sprung mass. According to an experiment conducted by [25], an increase of 15 kg on front wheels results in a decrease in ride comfort index within a range from 13.4% to

25.5%, depending on the road surface. Respectively increase in rear unsprung mass results in a decrease which ranges from 1.1% to 31.3%. This overall deteriorates the ride comfort index from 5 to 10%. Considering related increase to sprung mass, the comfort of the heavier EV with four IWMs is still higher by 1.5 to 6.3% than the baseline vehicle. Newton's second law can explain this: increased system mass causes an acceleration decrease. In contrast, the dynamic wheel load and suspension travel are increased by 40 and 16%, respectively [25].

Several solutions are presented, which helps to cope with those problems, considering negative aspects of increased unsprung mass to vertical dynamics. In the article by [47], the authors propose several solutions to diminish unsprung mass adverse effects to ride comfort and road holding ability. Three approaches are presented. *i)* Close to wheel drive system (CWDS) is introduced. In this system, the motor and reducer is part of the suspension system. This reduces the equivalent inertia to joints and reduces the unsprung mass. However, the design contradicts the IWM approach. *ii)* Dynamic vibration absorber that involves additional damper and spring assembly to suspend the motor and reducer. This architecture reduces suspension vibrations, but relative motion between the motor and the gearbox requires flexible coupling. *iii)* Active and regenerative energy suspension (AERS) comprises an active suspension system with an actuator that supplies active force and enhances suspension performance. At the same time, an electromagnetic actuator is utilised to harvest energy from vibration. Proposed solutions are compared numerically based on the root mean square of body acceleration (BA), dynamic tyre load (DTL) and suspension working stroke (SWS). AERS showed the lowest BA and SWS values, while DVA produced the smallest DTL variations. Nie et al. [48] studied a combined structure to eliminate the unsprung adverse effect of in-wheel electric vehicles, constituting a controlled suspension and a controlled tuned mass damper. The well-tuned tuned mass damper counters the random vibrations of the unsprung mass by making the same movement in the opposite direction. This can improve ride comfort by 31.2% and road-holding by 2.2%.

4 Discussion

In this section, SWOT analysis is performed to evaluate strengths, weaknesses, opportunities, and technology threats.

WCM strengths are mainly associated with better manoeuvrability and stability capabilities. Due to the fast response of individual wheels and the ability to distribute torques from the left-side to the right-side wheels, enhanced dynamical behaviour of the vehicle is achieved. Electronically controlled IWMs are able to generate the driving/braking torques precisely. This feature also provides an opportunity to control the slip ratio improving traction and reducing energy losses. Therefore, it is easier to drive the vehicle on slippery roads. In addition, active control actuators are able to configure suspension settings based on the road surface and driving manner. Values like toe, caster, camber and steering angles can be changed following the required performance. As a result, better handling and control of a vehicle is achieved. Vibrations are constantly generated while driving; it is possible to use electromagnetic actuators to gather energy from these vibrations.

The weakness of the technology is braking force fluctuations that occur during sudden braking or severe road bumps because of air gap variation between the stator and rotor of the IWM. This can lead to a longer stopping distance. However, this can also be solved with the help of controllers and brake blending implementation. The negative impact on comfort is mainly due to the large unsprung mass, one of the most significant weaknesses of this technology. However, there is an opportunity to eliminate the unsprung mass adverse effect. Active WCM with ride dynamics control or active WCM with integrated wheel positioning control can be developed for this task (Table 1).

Table 1. WCM SWOT analysis.

Strengths	Opportunities	Weaknesses and Threads
• Ability to precisely generate the driving/braking torques • Better vehicle manoeuvrability and stability • Possible to generate the roll moment about the longitudinal axis by allocating the motor torques suitably • Reduced the pitch angle during acceleration and braking • Ability to configure suspension settings	• The slip ratio control strategy improves traction and reduces energy losses • Implementing torque control systems, EVs can be driven on low friction coefficient surfaces with less effort • Harvest energy from vibration • Increased comfort level using controlled active suspension components and wheel positioning control	• Decrease the braking force and increases stopping distance because of air gap variation inside the stator and rotor of IWM • Lower ride comfort using passive suspension components because of larger unsprung mass • Increased vehicle inertia about vertical axis impairs vehicle stability and control, when passive suspension components are used • Larger unsprung mass induces higher loads on suspension components, this requires modifications and increases module price

5 Conclusion

Vehicle electrification allows to redesign its architecture, and an integrated corner solution may be an essential part of future car design. The majority of research focuses on lateral and longitudinal dynamics. However, issues related to vertical dynamics and driving comfort requires increased attention. The possibility of implementing active suspension components and integrated wheel positioning control will lead to higher ride comfort and may be used for automated driving.

Acknowledgement. This research was funded from the European Union Horizon 2020 Framework Program, Marie Skłodowska-Curie actions, under grant agreement no. 872907.

References

1. Liu, Z., Li, L., Zhang, Y.J.: Investigating the CO_2 emission differences among China's transport sectors and their influencing factors. Nat. Hazards **77**, 1323–1343 (2015)
2. Hooftman, N., Messagie, M., Mierlo, J.V., Coosemans, T.: A review of the European passenger car regulations - real driving emissions vs local air quality. Renew. Sustain. Energy Rev. **86**, 1–21 (2018)
3. Reşitoğlu, İA., Altinişik, K., Keskin, A.: The pollutant emissions from diesel-engine vehicles and exhaust aftertreatment systems. Clean Technol. Environ. Policy **17**(1), 15–27 (2014). https://doi.org/10.1007/s10098-014-0793-9
4. Sosa, B.S., Porta, A., Lerner, J.E.C., Noriega, R.B., Massolo, L.: Human health risk due to variations in PM10-PM2.5 and associated PAHs levels. Atmos. Environ. **160**, 27e35 (2017)
5. Zanobetti, A., Schwartz, J.: The effect of fine and coarse particulate air pollution on mortality: a national analysis. Environ. Health Perspect. **117**(6), 898e903 (2009)
6. Watts, A., Vallance, A., Whitehead, A., Hilton, C., et al.: The technology and economics of in-wheel motors. SAE Int. J. Passeng. Cars Electron. Electr. Syst. **3**(2), 37–57 (2010)
7. Basics of Electric Vehicles - Design and Function. Self Study Program 820233, Volkswagen Group of America, Inc. 62 (2013). https://www.academia.edu/35166844/Basics_of_Electric_Vehicles_Design_and_Function, Accessed 15 May 2021
8. Lilly, C.: EV connector types (2020). https://www.zap-map.com/charge-points/connectors-speeds/, Accessed 9 June 2021
9. Sanguesa, J.A., Torres-Sanz, V., Garrido, P., Martinez, F.J., Marquez-Barja, J.M.: A review on electric vehicles: technologies and challenges. Smart Cities **4**(1), 372–404 (2021)
10. Agamloh, E., von Jouanne, A., Yokochi, A.: An overview of electric machine trends in modern electric vehicles. Machines **8**(2), 20 (2020)
11. Biček, M., Pepelnjak, T., Pušavec. F.: Production aspect of direct drive in-wheel motors. Procedia CIRP. **81**, 1278–1283 (2019)
12. Perekopskiy, S.: On the analysis and design of in-wheel motor for vehicle application (2020) Dissertation: https://opendata.uni-halle.de/bitstream/1981185920/35004/1/Perekopskiy_Sergey_Dissertation_2020.pdf
13. Jneid, M.S., Harth, P., Ficzere, P.: In-wheel-motor electric vehicles and their associated drivetrains. Int. J. Traffic Transp. Eng. **10**(4), 415–431 (2020)
14. Hag, J.: Wheel Corner Modules Technology and Concept Analysis, Master Thesis (2011). https://www.diva-portal.org/smash/get/diva2:503940/FULLTEXT01.pdf, Accessed 25 May 2021
15. Nguyen, B.M., Nguyen, H.V., Ta-Cao, M., Kawanishi, M.: Longitudinal modelling and control of in-wheel-motor electric vehicles as multi-agent systems. Energies **13**(20), 5437 (2020)
16. Murata, S.: Innovation by in-wheel-motor drive unit. Int. J. Veh. Mech. Mob. **50**, 807–830 (2012)
17. Liu, M., Gu, F., Huang, J., Wang, C., Cao, M.: Integration design and optimization control of a dynamic vibration absorber for electric wheels with in-wheel motor. Energies **10**(12), 2069 (2017)
18. Christensen, L.: Designing In-Hub Brushless Motors (2014). https://www.machinedesign.com/motors-drives/article/21833665/designing-inhub-brushless-motors, Accessed 20 May 2021
19. Biček, M., Miljavec, D., Gotovac, G., Zupan, S.: Mechanical failure mode causes of in-wheel motors. Strojniški vestnik – J. Mech. Eng. **61**(1), 74–85 (2015)

20. DHL report. Automotive trend report (2021). https://www.dhl.com/content/dam/dhl/global/dhl-supplychain/documents/pdf/SC_Automotive_Trend_Report_2021_EN_.pdf, Accessed 8 May 2021
21. Elaphe's Site. Direct-drive in-wheel motors. https://in-wheel.com/en/solutions-2/direct-drive-in-wheel-motors/, Accessed 10 May 2021
22. Protean Electric Site. ProteanDrive. https://www.proteanelectric.com/technology/#overview, Accessed 15 May 2021
23. Moreels, D., Leijnen, P.: This Inside-Out Motor for EVs Is Power Dense and (Finally) Practical (2019). https://spectrum.ieee.org/transportation/alternative-transportation/this-insideout-motor-for-evs-is-power-dense-and-finally-practical, Accessed 23 May 2021
24. Shyrokau, B., Wang, D.: Control allocation with dynamic weight scheduling for two-task integrated vehicle control. In: Proceedings 11th International Symposium on Advanced Vehicle Control, pp. 1–6 (2012)
25. Vos, R., Besselink, I., Nijmeijer, H.: Influence of in-wheel motors on the ride comfort of electric vehicles. In: International Symposium on Advanced Vehicle Control, Loughborough, pp. 835–840 (2010)
26. Anderson, M., Harty, D.: Unsprung mass with in-wheel motors - myths and realities. AVEC **10**, 261–266 (2010)
27. Evans, P.: Michelin active wheel system to hit roads in 2010. https://newatlas.com/michelin-active-wheel-production-electric-car-by-2010/10489/, Accessed 3 June 2021
28. Laurent, D., Sebe, M., Walser, D.: Unit comprising a wheel and a suspension integrated in the wheel hub. Europ Patent Publication Numer: EP0878332 B1, Application Number: EP19980107687 (2002). https://worldwide.espacenet.com/patent/search/family/009507098/publication/EP0878332B1?q=EP0878332%20B1, Accessed 2 June 2021
29. Rajaie, M.: Design and Fabrication of a Novel Corner Wheel Module for Urban Vehicles. Master thesis (2016)
30. Nagaya, G.: In-wheel motor system. Patent US0 247 496A1 (2005). Bridgestone https://patentimages.storage.googleapis.com/15/01/33/c1dd288400d001/EP1547844B1.pdf, Accessed 30 May 2021
31. Nagaya, G., Wakao, Y., Abe, A.: Development of an in-wheel drive with advanced dynamic-damper mechanism. JSAE Rev. **24**(4), 477–481 (2003). ISSN 0389–4304
32. Hanlon, M.: Siemens VDO visualises the electric wheel hub motor (2006). https://newatlas.com/go/5996/, Accessed 25 May 2021
33. Herr, M.: Braking device having a wedge mechanism. Patent US0 230 330A1 (2008). Siemens: https://patentimages.storage.googleapis.com/4a/99/64/ef8366f3042ed8/US20080230330A1.pdf, Accessed 29 May 2021
34. Zetterstrom, S.: Wheel suspension arrangement in a vehicle. Patent EP1144212B1 (2004). Volvo: https://worldwide.espacenet.com/patent/search/family/020413903/publication/EP1144212B1?q=EP1144212B1, Accessed 2 June 2021
35. Brich, S.: Protean Electric aims for pirouetting autonomous EVs (2019). SAE International: https://www.sae.org/news/2019/09/protean-electric-360-degree-steering, Accessed 2 June 2021
36. Shyrokau, B., Wang, D., Lienkamp, M.: Integrated vehicle dynamics control based on control allocation with subsystem coordination. In: Proceedings of 23rd IAVSD International Symposium Dynamic Vehicles Roads Tracks, pp. 1–10 (2013)
37. Armengaud, E., Eitzinger, S., Pirker, H., Buh, J., et al.: E-Mobility-opportunities and challenges of integrated corner solutions. SAE Technical Paper (2021)
38. Parsania, P., Saradava, K.: Drive-by-wire systems in automobiles. In: Conference: Journal of systematic computing at: VVP Engineering College, vol. 6 (2012)

39. Qin, Y., Zhao, Z., Wang, Z., Li, G.: Study of longitudinal-vertical dynamics for in-wheel motor-driven electric vehicles. Autom. Innov. **4**, 227–237 (2021)
40. Nam, K., Hori, Y., Lee, C.: Wheel slip control for improving traction-ability and energy efficiency of a personal electric vehicle. Energies **8**(7), 6820–6840 (2015)
41. Yin, D., Oh, S., Hori, Y.: A novel traction control for EV based on maximum transmissible torque estimation. IEEE Trans. Ind. Electron. **56**(6), 2086–2094 (2009)
42. Maeda, K., Fujimoto, H., Hori, Y.: Four-wheel driving-force distribution method for instantaneous or split slippery roads for electric vehicle with in-wheel motors. In: 2012 12th IEEE International Workshop on Advanced Motion Control (AMC), pp. 1–6 (2012)
43. Kim, H.S., Hyun, Y.J., Nam, K.H.: Disturbance observer-based sideslip angle control for improving cornering characteristics of in-wheel motor electric vehicles. Int. J. Autom. Technol. **19**(6), 1071–1080 (2018). https://doi.org/10.1007/s12239-018-0105-1
44. Nah, J., Yim, S.: Vehicle stability control with four-wheel independent braking, drive and steering on in-wheel motor-driven electric vehicles. Electronics **9**(11), 1934 (2020)
45. Shyrokau, B., Wang, D., Savitski, D., Hoepping, K., Ivanov, V.: Vehicle motion control with subsystem prioritization. Mechatronics **30**, 297–315 (2015)
46. Wang, J., Wang, Q., Jin, L., Song, C.: Independent wheel torque control of 4WD electric vehicle for differential drive assisted steering. Mechatronics **21**(1), 63–76 (2011)
47. Chen, X., Yin, J., Wang, W., Wu, L., Tang, F.: Approaches to diminish large unsprung mass negative effects of wheel side drive electric vehicles. J. Adv. Mech. Des. Syst. Manuf. **10**(4), JAMDSM0064 (2016)
48. Nie, S., Zhuang, Y., Chen, F., Wang, Y., Liu, S.: A method to eliminate unsprung adverse effect of in-wheel motor-driven vehicles. J. Low Freq. Noise Vib. Active Control **37**(4), 955–976 (2018)

Study on the Vertical Dynamics of a Passenger Car with Independently Rotating Wheelsets

Stasys Steišūnas and Gediminas Vaičiūnas(✉)

Vilnius Gediminas Technical University,
Saulėtekio al. 11, 10223 Vilnius, Lithuania
{stasys.steisunas,gediminas.vaiciunas}@vilniustech.lt

Abstract. In this paper adjustment of the passenger car suspension parameters and adapting them to independently rotating wheels were studied. The study examines impact of changing primary and secondary suspension parameters on passenger car vertical dynamics. The most appropriate suspension parameters were selected and regularities of the vertical accelerations of the car body were analyzed at different running speeds in curve and in straight track.

Keywords: Passenger car · Vertical dynamics · Vertical accelerations · Suspension parameters

1 Introduction

Stability and comfort technologies for rolling stock are being developed around the world. They are being tested to increase passenger safety and comfort. To increase passenger comfort and safety are not enough only old technological solutions [6, 8]. Improvement of all mechanical parts requires the installation of active systems with hydraulic and electronic elements. With the improvement of rolling stock running speed increases with safety requirements and requires more focus on dynamic processes of rolling stock [13, 19]. Studies on the effect of wheel defects on rolling stock dynamics must be mentioned as a specific area of research [16]. Natural and forced vibrations occur when a vehicle moves along a railway track at an appropriate speed. Forced vibrations occur due to track irregularities and/or damaged wheels of vehicles [7]. These vibrations are transferred to vehicle bogies via the initial suspension and the bogie vibrations are transferred to the vehicle carbody [12]. Repetition of these vibrations is called vehicle vibration and can adversely affect the comfort of passengers. The developed mathematical model allows the determination of the impacts of track roughness and wheel damages on the vertical and angular accelerations of the body in various seasons [1, 14]. The most important thing is to reduce those vibrations that are most detrimental to human well-being and health. Research has shown that human health and well-being are not affected by a single parameter, but by a set of them. This can be read in the relevant literature [2, 3]. Various technical solutions have been taken to improve passenger comfort [5]. The proposed power steering bogie consists of independently rotating power steering wheels and allows us to eliminate steering vibration while still providing perfect low power steering when cornering

O. Prentkovskis et al. (Eds.): TRANSBALTICA 2021, LNITI, pp. 101–109, 2022.
https://doi.org/10.1007/978-3-030-94774-3_10

[4, 10]. There is proposes the use wheelsets with independently rotating wheels (IRW) with inverse tread conicity to get self-steering ability without any complex bogie structure. Both the experiment and simulation results show that the proposed wheelset with IRW with inverse tread conicity have good performance [15]. The benefits of implementing active steering systems in railway vehicles mounting bogies with wheelsets with IRW and outlines a design methodology for such systems is presented [11]. It is very important to study low-radius track curves (180–300 m), there is a big difference in the rolling radius of the wheel sets and it greatly determines the rolling laws [9].

It is possible to compare the vertical dynamic characteristics of passenger cars with one-piece (OP) wheelsets and wheelsets with IRW by comparing root mean square (RMS) carbody vertical accelerations values. One of the easiest methods to compare the dynamic characteristics of cars with OP wheel sets and wheel sets with an independently rotating wheel is to compare Nadal criterion values of the car running through various railway curves and straight sections. Research performed by the authors has shown that this method is operative. After having calculated Nadal criterion values, it was found that as the speed of the car increases in the straight section, the values increase with a regularity close to the linear, however, Nadal criterion values for wheel sets with independently rotating wheel are approximately twice a s large as for the car with OP wheel sets [17, 18].

The dynamic parameters of vehicles with IRW were found to be worse than that of vehicles with OP wheelsets, according to the values of the Sperling's comfort index and the Nadal criterion [2]. Therefore, it was decided to examine in more detail the peculiarities of the vertical dynamics of a passenger car with IRW.

2 Study

Computer simulation of the Universal Mechanism (UM) software package was chosen for the study for several reasons. In particular, testing with a computer model does not waste railway resources. Tests with this software package can perform much more tests in a relatively short time than tests with real rolling stock. In order to evaluate the influence of the suspension structure on comfort calculations with passenger car models were made. The influence of the suspension parameters on the root mean square (RMS) accelerations of the carbody was evaluated according to the formula (1):

$$a_{RMS} = \sqrt{\frac{1}{n}\left(a_1^2 + a_2^2 + \cdots + a_n^2\right)}, \tag{1}$$

where a_{RMS} – root mean square accelerations of the carbody.

During the assessment of carbody acceleration level and taking into account the passenger comfort, Sperling's comfort index can be used [5]:

$$W_Z = \left(\sum_{i=1}^{n_f} W_{Z_i}^{10} \right)^{\frac{1}{10}}, \quad \text{here}: W_{Z_i} = \left[a_i^2 B(f_i)^2 \right]^{\frac{1}{6.67}}, \tag{2}$$

where a – acceleration of body, m/s^2, f – frequency of vibration, Hz, o B(f) – coefficient estimating the frequency and direction of vibration:

$$B(f) = k \sqrt{\frac{1.911f^2 + (0.25f^2)^2}{(1 - 0.277f^2)^2 + (1.563f - 0.0368f^3)^2}}, \tag{3}$$

where k = 0.737, if oscillations are horizontal, and 0,588 if oscillations are vertical. The relevant indicators are calculated according to formulas (2)–(3). From the formula (2), we can see that one of the main factors influencing passenger comfort is carbody acceleration. Therefore, the RMS of carbody acceleration change was investigated in computer simulations.

Figures 1, 2 and 3 show graphs with the RMS vertical carbody accelerations in different track sections.

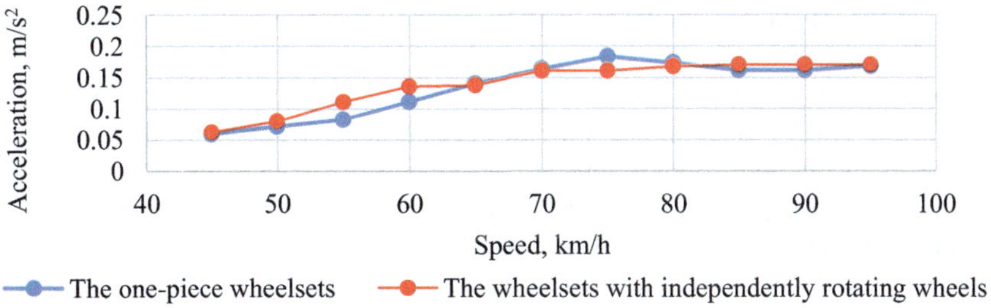

Fig. 1. RMS of carbody acceleration in the vertical direction in a 200 m radius, 1,000 m long "S" type curve.

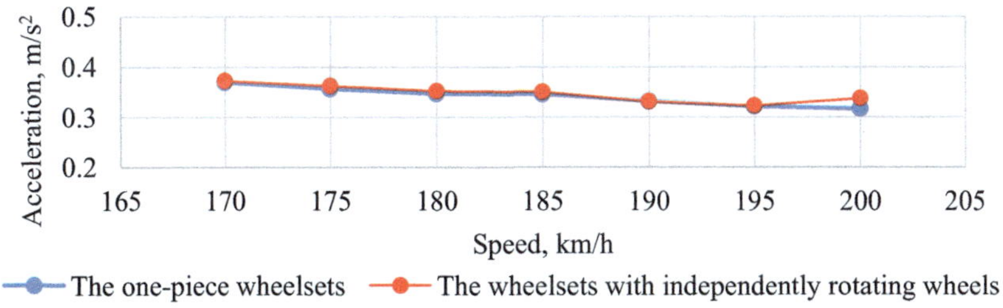

Fig. 2. RMS of carbody acceleration in the vertical direction in a 2,900 m radius, 1,000 m long "S" type curve.

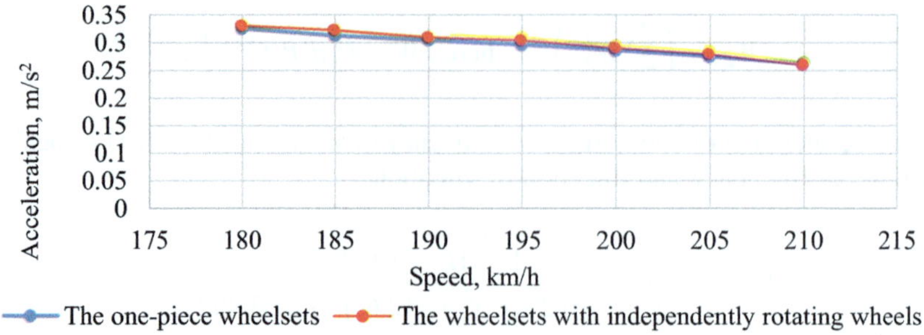

Fig. 3. RMS of carbody acceleration in the vertical direction in a straight track

Figure 1 shows that the RMS values of the vertical accelerations of the model with a IRW on a 200 m radius curve at speeds between 45 km/h and 65 km/h are higher than values of the model with OP wheelset. At speeds between 65 km/h and 80 km/h, the RMS values of the vertical accelerations of the model with OP are higher than those of the models with IRW. In the 200 m radius curve at speeds from 45 km/h to 95 km/h, the model with OP wheelset reached the maximum value of RMS accelerations at 75 km/h (0.185 m/s^2), the model with IRW reached the maximum value at 95 km/h (0.1714 m/s^2).

Comparing the graphs in Fig. 2, it can be seen that at speeds between 170 km/h and 195 km/h, the RMS values of the vertical acceleration on the curve of 2,900 m are similar or equal in both models, but at 200 km/h IRW values are 0.33 m/s^2 when moving at the same speed in the model with OP values are 0.27 m/s^2.

Comparing the graphs in Fig. 3, it can be seen that at speeds from 180 km/h to 210 km/h the RMS values of the vertical acceleration in a straight section of road are similar or equal and only decrease from ~ 0.33 m/s^2 to ~ 0.26 m/s^2.

3 Adjustment of Passenger Car Suspension Parameters

The following figures provide graphs of the root mean square (RMS) body accelerations as a function of the stiffness and damping parameters of the primary and secondary suspension.

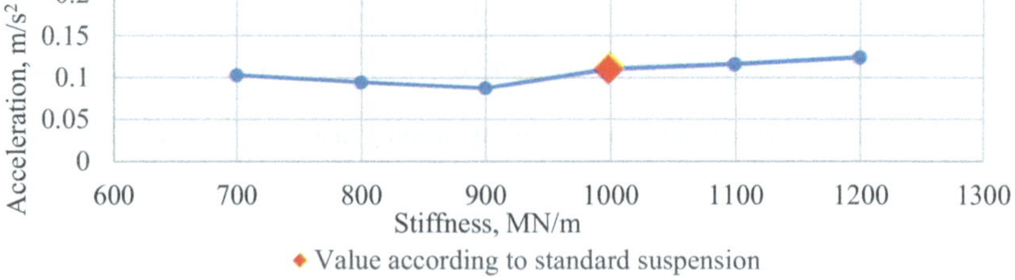

Fig. 4. Dependence of RMS body accelerations on primary suspension vertical stiffness parameters at 210 km/h.

Based on the data in Fig. 4, it is appropriate to select a vertical stiffness value of 900 MN/m for the primary suspension.

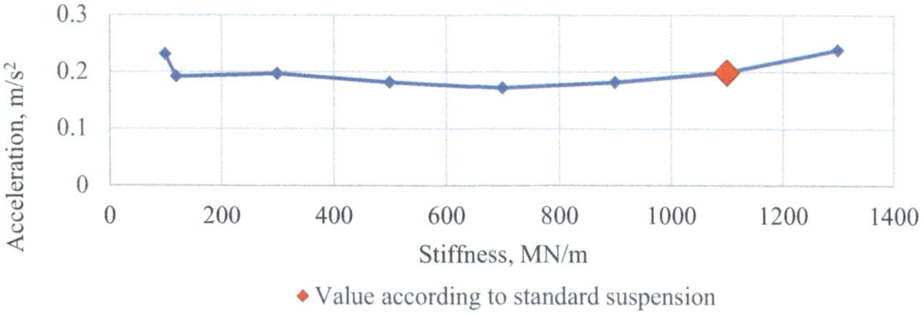

Fig. 5. Dependence of RMS body accelerations on primary suspension horizontal stiffness parameters at 210 km/h.

Based on the data in Fig. 5, it is appropriate to select a horizontal stiffness value of 600–650 MN/m for the primary suspension.

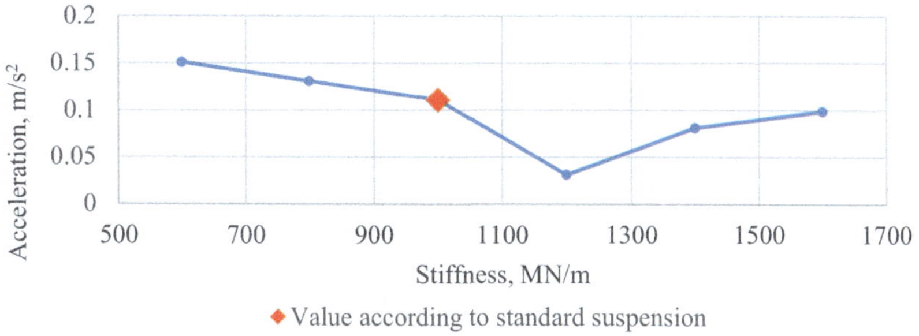

Fig. 6. Dependence of RMS body accelerations on secondary suspension vertical stiffness parameters at 210 km/h.

Based on the data in Fig. 6, it is appropriate to select a vertical stiffness value of 1,200 MN/m for the secondary suspension.

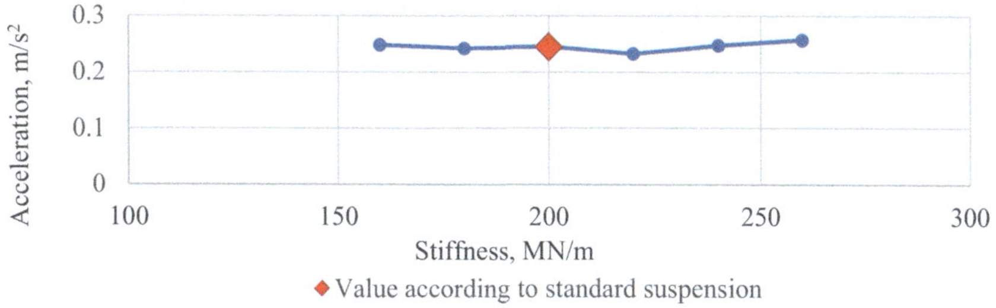

Fig. 7. Dependence of RMS body acceleration on secondary suspension horizontal stiffness parameters at 210 km/h.

Based on the data in Fig. 7, it is appropriate to select a value for the horizontal stiffness of the secondary suspension of 220–225 MN/m.

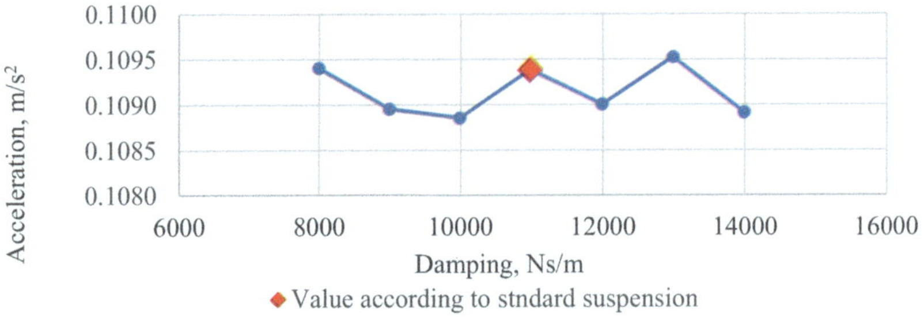

Fig. 8. Dependence of RMS body accelerations on suspension damping parameters at 210 km/h.

Based on the data in Fig. 8, it is appropriate to select a value of 10,000 Ns/m for the primary and secondary suspension damping parameters. According to the determined minimum values of RMS accelerations, the values of the selected parameters of the primary and secondary suspension of a passenger car are presented in Table 1.

Table 1. Selected values of suspension parameters.

Parameters	Direction of oscillations	Meaning
Primary suspension stiffness coefficient, N/m	Vertical	$9{\cdot}10^5$
	Horizontal	$6{\cdot}10^5$
Secondary suspension stiffness coefficient, N m	Vertical	$1.2{\cdot}10^6$
	Horizontal	$2.2{\cdot}10^6$
Total damping factor of primary and secondary suspension, Ns/m		10^4

The dynamic parameters of the passenger car with independently rotating wheels are further modeled using the suspension parameters given in the table. The results of the study are presented below.

According to the suspension parameters presented in Table 1, a new model of a passenger car with IRW was developed and new simulated were performed. The results of the simulations are shown in the Figs. 9, 10 and 11 below.

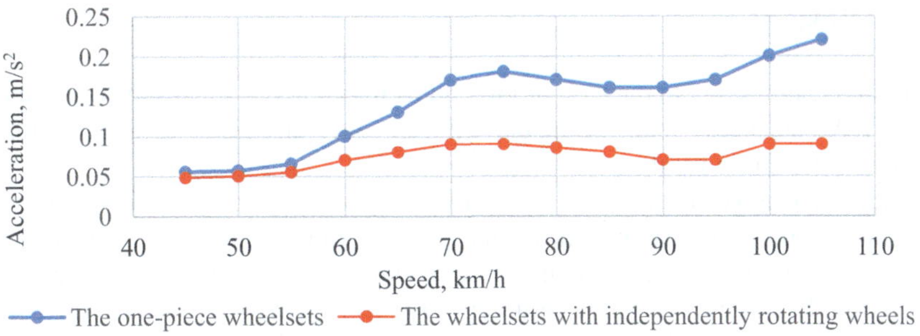

Fig. 9. RMS values of the vertical accelerations of the carbody in a 200 m S-type curve.

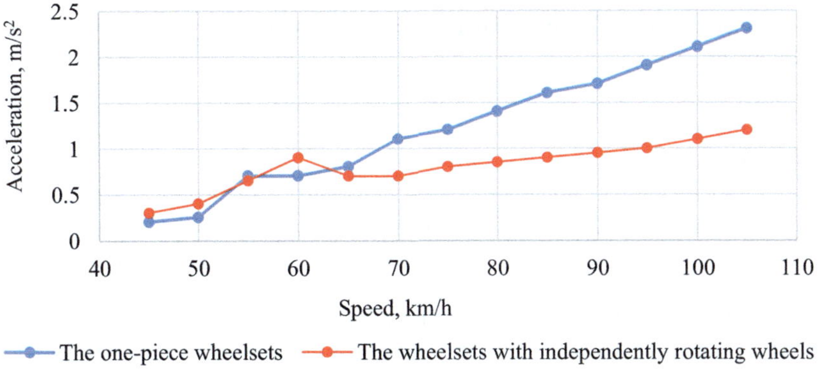

Fig. 10. RMS values of the vertical accelerations of the carbody in a 2,900 m S-type curve.

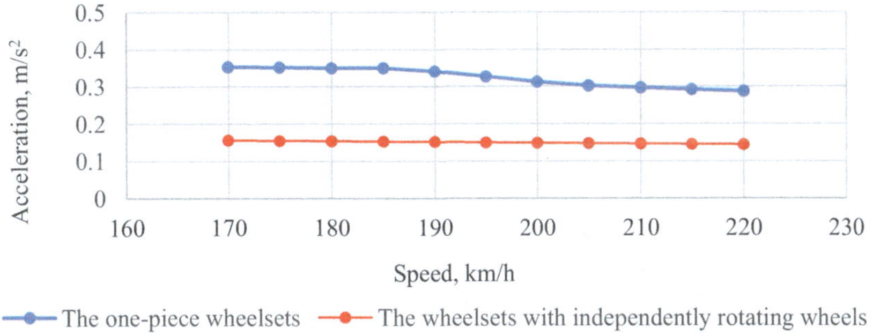

Fig. 11. RMS values of the vertical accelerations of the carbody in a straight track.

It can be seen from the graphs in Figs. 9, 10 and 11 that the RMS acceleration values of the modified model with IRW are lower than those of the OP model at all speeds up to critical, on all track sections (except the 200 m S-type curve at speeds from 55 km/h up to 60 km/h).

4 Conclusions

1. The study shows that in the case of low-radius track curves, the results of the model with IRW are better than the model with OP only at speeds up to 80 km/h. At higher speed these results are worse.
2. In order to achieve better results with NBR at higher speeds, it is necessary to select the most appropriate parameters for the primary and secondary suspension of the passenger car. Most of the parameters are adjusted very slightly, the changes are 10–20%, the largest correction (primary suspension horizontal stiffness) was 45%.
3. After adjusting the stiffness and damping parameters of the primary and secondary suspension of the passenger car, the RMS acceleration values of the modified model with IRW are lower than in the model with OP wheelset moving at different speeds and track sections. In the straight section, in the speed range from 170 to 220 km/h the body acceleration values (with adjusted suspension parameters) are more than 2 times lower.

References

1. Alexandrou, G., Kouroussis, G., Verlinder, O.: A comperhensive prediction model for vehicle/track/soil dynamic response due to wheel flats. Rail Rapid Transit. **230**(4), 1088–1104 (2016)
2. Dizo, J., Harusinec, J., Blatnicky, M.: Modal analysis of modified freight wagon bogie frame. In: 23rd International Conference on Current Problems in Rail Vehicles, pp. 29–36 (2017)
3. Dižo, J., Steišunas, S., Blatnický, M.: Vibration analysis of a coach with the wheel-flat due to suspension parameters changes. Procedia Eng. **192**, 107–112 (2017)
4. Gorbunov, M., Gerlici, J., Kravchenko, K., Dizo, J., Lack, T., Blatnicky, M.: Method of the detection of a defective damper. In: 24th International Conference on Current Problems in Rail Vehicles, pp. 153–160 (2019)
5. Graa, M., Nejlaoui, M., Houidi, A., Affi, Z., Romdhane, L.: Modeling and simulation for vertical rail vehicle dynamic vibration with comfort evaluation. Multiphys. Model. Simul. Syst. Des. Monit. Appl. Cond. Monit. **2**, 47–57 (2015)
6. Huang, C., Zeng, J., Liang, S.: Carbody hunting investigation of a high speed passenger car. J. Mech. Sci. Technol. **27**(8), 2283–2292 (2013). https://doi.org/10.1007/s12206-013-0611-z
7. Jiang, Y., Bernard, K., Chen, B.K., Thompson, C.: A comparison study of ride comfort indices between Sperling's method and EN 12299. Int. J. Rail Transp. **7**(4), 1–18 (2019)
8. Kalker, J.J.: A fast algorithm for the simplified theory of rolling contact. Veh. Syst. Dyn. **11** (1), 1–13 (1982)
9. Matsumoto, A.: Improvement of bogie curving performance by using friction modifier to rail/wheel interface. Wear **258**(7), 1201–1208 (2005)
10. Michitsuji, Y., Suda, Y.: Running performance of power-steering railway bogie with independently rotating wheels. Veh. Syst. Dyn. **44**(1), 71–82 (2006)
11. Perez, J., Mauer, M., Busturia, J.M.: Design of active steering systems for bogie-based railway vehicles with independently rotating wheels. Veh. Syst. Dyn. **37**(1), 209–220 (2002)
12. Soukup, J., Skocilas, J., Skocilasova, B., Dizo, J.: Vertical vibration of two axle railway vehicle. Procedia Eng. **177**, 25–32 (2017)

13. Steišūnas, S.: Research on dynamic processes of wagon wheelsets with flat impact on rail. PhD thesis. VGTU. Vilnius (2017)
14. Steišūnas, S., Bureika, G., Gorbunov, M.: Effects of rail-wheel parameters on vertical vibrations of vehicles using a vehicle-track-coupled model. Transp. Prob. **14**(3), 27–39 (2019)
15. Suda, Y., Wang, W., Nishina, M., Lin, S., Michitsuji, Y.: Self-steering ability of the proposed new concept of independently rotating wheels using inverse tread conicity. Veh. Syst. Dyn. **50**(1), 291–302 (2012)
16. Taletavičius, R.: Rolling stock driving stability study. Master thesis. VGTU. Vilnius (2019)
17. Vaičiūnas, G., Steišūnas, S., Dižo J.: The Nadal criterion study of a passenger car with independently rotating wheels. In: Proceedings of the 24th International Scientific Conference "Transport means 2020", pp. 878–883. KTU, Kaunas (2020)
18. Vaičiūnas, G., Steišūnas, S.: Shperling criterion study in a passenger car with independently rotating wheels. In: XII International Scientific Conference "Transport Problems 2020", pp. 853–861. Silesian University of Technology, Katowice (2020)
19. Yu, Y., Zhao, L., Zhou, C.: A new vertical dynamic model for railwayvehicle with passenger-train-trackcoupling vibration. Inst. Mech. Eng. **234**(1), 134–146 (2020)

The Research of Running Resistance of a Railway Wagon with Various Wheel Designs

Stanislav Semenov[1], Evgeny Mikhailov[1], Ján Dižo[2](✉), and Miroslav Blatnický[2]

[1] Volodymyr Dahl East Ukrainian National University, Central Avenue 59A/403, Severodonetsk 93400, Ukraine
{semenov,mihajlov}@snu.edu.ua

[2] University of Žilina, Univerzitná, 8215/1, 010 26 Žilina, Slovak Republic
{jan.dizo,miroslav.blatnicky}@fstroj.uniza.sk

Abstract. The article presents the research aimed at the description of a mathematical model of the movement of a railway wagon on a railway track. With the help of mathematical modelling, studies of running resistance of a wagon with wheels of various design schemes have been carried out. Two wheel designs are compared, namely the design of a traditional wheel design (TWD) with a monolithic tread surface and the flange and the perspective wheel design (PWD), which allows an independent rotation of the wheel tread surface (a tread surface) and its guiding surface, i.e. its flange. Potential advantages of using the wheels of a promising design scheme in a bogie of a railway wagon are revealed, which consist in reducing running resistance.

Keywords: Railway wagon · Flange · Mathematical modelling · Running resistance

1 Introduction

One of the ways of reduction of the running resistance of rail vehicles can be using of wheels with a perspective wheel design (PWD). This design allows to rotate a flange and a tread surface of a railway wheel independently to each other about their common axis of rotation. The results of studies [1–3] have shown certain peculiarities of the movement this wheel regarding to a rail and the possibility of reduction of the slip speed in a flange contact. This, in turn, leads to reducing the running resistance of a wheel and also wear in contact surfaces [4–6]. In this regard, it seems to be appropriate to further study the possibilities of reducing the running resistance of rail vehicles by means of using the PWD wheels.

A review of the scientific literature in this research area has shown that at present, a significant number of models, which are created within various degrees of a system idealization and the description of details have been proposed including the mathematical description of movement of rail vehicles. These approaches have been derived for various given tasks [7–10]. This article presents a choice of railway wagon design

O. Prentkovskis et al. (Eds.): TRANSBALTICA 2021, LNITI, pp. 110–119, 2022.
https://doi.org/10.1007/978-3-030-94774-3_11

and mathematical modelling of its movement, which has been carried out based on the analysis of works [2, 11–13], and taking into account the peculiarities of the interaction of the PWD wheel with a rail [1–3].

The main purpose of this work is a mathematical description of the movement of a four-axle railway wagon with wheels of various design schemes to study their influence on the running resistance.

2 A Method of the Problem Solution

The research comes from a mathematical model of railway wagon movements, which is presented in the works of scientists from the Dnepropetrovsk National University of Railway Transport named by V. A. Lazaryan [14–18].

A vehicle model was considered as a system including 19 rigid bodies, namely a body, two bolsters, four side frames of bogies, four wheelsets and eight flanges (having the possibility of independent rotation relative to wheels tread surfaces around their mutual axis) with rigid and visco-elastic connections between them. A scheme of the considered dynamic model is shown in Fig. 1.

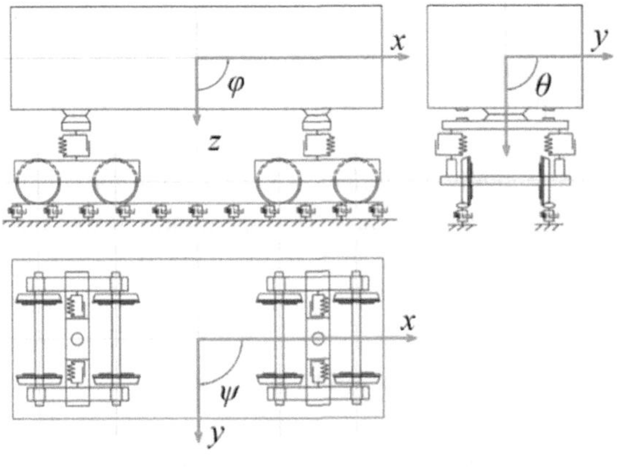

Fig. 1. A dynamic scheme of a railway wagon.

The description of the wagon movements is considered within a coordinate system, which x axis is for a longitudinal movement, y axis is the lateral axis and z axis heading downwards [19, 20]. It was assumed, that the wagon moves at the constant speed v and the track consists of a straight sections and curves. The total number of degrees of freedom of all bodies in the system is of 84.

2.1 A Mathematical Description

The derived mathematical model of the railway wagon includes quite large number of equations of motion, which can be found in [21]. As it is known, a general form of the derived equations of motion is as following:

$$\boldsymbol{M} \cdot \ddot{\boldsymbol{q}} + \boldsymbol{K} \cdot \dot{\boldsymbol{q}} + \boldsymbol{C} \cdot \boldsymbol{q} = \boldsymbol{F}, \tag{1}$$

where $\boldsymbol{M}$ is the mass matrix and it includes masses and moments of inertia of individual bodies, $\boldsymbol{K}$ is the damping matrix, in which, damping coefficients of considered damping elements are written and $\boldsymbol{C}$ is the stiffness matrix with stiffness coefficients of springs mounted in suspensions systems. Further, the right side of the system (1) includes accelerations, velocities and deviations of individual bodies marked as $\ddot{\boldsymbol{q}}$, $\dot{\boldsymbol{q}}$ and $\boldsymbol{q}$, respectively. They includes translations movements as well as angular movements of individual bodies. More detailed description of this model is introduced in [21].

The mathematical model provides for the possibility of using the PWD wheels (with independent rotation of the guide discs of the wheels relative to the corresponding supporting parts of the wheels around their common axis). Each guide surface of the PWD wheel (i.e. the flange) has one degree of freedom. From the side of the rails, a moment $\Phi_{imj}^{r,g}$ acts on it and from the side of the corresponding supporting part of the wheel, the moment $\Phi_{imj}^{w,g}$ acts on it.

Equations of motion of the guiding surface of the PWD wheel is:

$$I_y^g \cdot \ddot{\varphi}_{imj}^g = \Phi_{imj}^{w,g} + \Phi_{imj}^{r,g}, \tag{2}$$

where I_y^g is a moment of inertia of the guiding surface of the PWD wheel relative to its axis of rotation and $\Phi_{imj}^{r,g}$ represents the moment of creep forces acting on the j-th guiding surface of the i-th wheelset of the m-th wagon bogie in the longitudinal vertical plane.

The moment $\Phi_{imj}^{w,g}$ is actually a moment of resistance in the junction of the support part of the wheel and the guiding surface. The magnitude of this moment depends on the chosen design of the mutual node and, in the general case, it is determined by a following expression:

$$\Phi_{imj}^{w,g} = M_0 \cdot sign\left(\dot{\varphi}_{imj}^w - \dot{\varphi}_{imj}^g\right), \tag{3}$$

where M_0 is the maximum value of the moment of resistance at the interface of the supporting part of the wheel and the guiding surface.

The comparison of indicators of the running resistance of a railway wagon with different wheel designs was performed based on a comparison of several indicators. The total specific running resistance of a railway wagon w_u'' [N/kN] is the main of them.

The value of this indicator was determined comes from data sources [18, 19]. According to them, the total resistance of the railway wagon is:

$$w_u'' = w_{um}'' + w_{ua}'' \tag{4}$$

where w_{um}'' is the main resistance of the railway wagon and w_{ua}'' is the additional resistance of the railway wagon.

Among known works, Astahov [22] has examined, that the main components of this resistance are the resistance due to friction in the bearings w_b'', the resistance due to movement of wheels on rails (rolling, friction) marked as w''_{rf} and w''_{sf}, the aerodynamic drag w''_{ad}, the resistance due to the energy dissipation in the track w''_{rt} and the resistance due to the energy dissipation in the environment w''_{e}. The values w''_{b}, w''_{rf}, w''_{ad} and w''_{rt} are calculated according to the recommendations introduced in [22, 23]. Hence, the value of the resistance due to the friction in bearings is determined as:

$$w''_b = \frac{150 \cdot (Q_0 - 1.2) \cdot \mu}{Q_0}, \tag{5}$$

where Q_0 is the axle load of the wagon and μ is a value of the friction coefficient of rolling bearings. The value of the rolling resistance of a wheel on a rail w''_{rf} is difficult to quantify, because microprocesses of sliding friction and wear in contact surfaces arise [24, 25] and it is influenced by the large loads and plastic deformations. Therefore, according to the recommendations defined in the work [22], this value was taken from the range:

$$w''_{rf} = 0.3 \div 0.4\,[\text{N/kN}]. \tag{6}$$

The specific aerodynamic drag w''_{ad} of a railway wagon movement depends on the factors related with the speed [26, 27], the axle load and ambient temperature:

$$w''_{ad} = \frac{0.69 \cdot v^2}{Q_0 \cdot T}, \tag{7}$$

where v is a wagon speed, q_0 is the axle load of the railway wagon and T is air temperature [K]. It is known, that the conditions of a track significantly affects the total running resistance, especially with increasing speeds and the axle load. The expression for determining the running resistance of the railway wagon is determined as [18]:

$$w''_{rt} = 0.000035 \cdot v^2 + 0.0036 \cdot Q_0. \tag{8}$$

The value of the resistance due to the energy dissipation into the environment w''_e is also difficult to quantify using generalized empirical formulas due to the presence of a large number of factors that affect its magnitude. In comparative calculations, the values w''_e were taken according to [19].

The value of the resistance w''_{sf} was determined by converting the specific values [N/kN] of the absolute values [N] of the forces of resistance of the wheels of a railway wagon w_{sf}, obtained by calculating the formula:

$$w''_{sf} = \frac{w_{sf}}{Q_0}. \tag{9}$$

For calculation of the specific additional resistance, only the specific running resistance in a curve w''_{uar} was taken into account. The calculation of the value of this resistance was also determined based on the formula [18]:

$$w''_{uar} = \frac{100}{R} + 1.5 \cdot |\tau|, \tag{10}$$

where R is a radius of a curve [m] and $|\tau|$ the absolute value of the outstanding transverse acceleration in the curve [m/s^2].

3 Results of the Research and Discussion

Calculation of equations of motion was performed by means of the method of the two-step Runge-Kutta of the second order [28]. This method is used due to the presence of a large number of high stiffness springs, one-sided limiters with gaps and dry friction elements in the simulated system that makes smooth a part of the equations system. The integration step was selected experimentally.

Figure 2 shows the comparison of the results of some calculated dynamic indicators with the experimental data for an empty railway wagon (Fig. 2) and for a loaded railway wagon (Fig. 2) in the case of its running in a straight section of a railway track. Dash lines indicate the maximum values at which the vehicle movement is still safe.

The indicators of the vertical $\ddot{z}$ and horizontal $\ddot{y}$ acceleration of the mass centre of the body (in fractions of g), the coefficients of the vertical body dynamics K_{vd} and maximum ratios of horizontal forces in flange contact to the static load Y/G were selected.

The circles show the values of the corresponding indicators obtained as a result of field tests, solid lines - according to the results of calculations on a mathematical model. The horizontal dashed lines show the permissible values of the corresponding dynamic indicators.

The values of the dynamic indicators during field tests on the 1,520 mm rails gauge of the Ukraine and the Russian Federation available in literature sources [29–31], were taken as experimental.

The quantitative agreement between the calculated and experimental values of the quantities under consideration is satisfactory. The maximum discrepancy between the results does not exceed 9% for a loaded railway wagon and 16% for an empty railway wagon. This confirmed the suitability of the presented mathematical model for solving the problem.

The analysis of the calculation results showed that the change of the design of a railway wheel had practically no effect on the dynamic performance of the railway wagon movement. Some results of calculations concerning to the running resistance railway wagons are presented below.

Figures 3, 4, 5 and 6 show the calculated dependences of the total running resistance of an empty railway wagon and a loaded railway wagon with wheels with the different design schemes on the wagon speed, where TWD is the traditional wheel design, PWD is the perspective wheel design and TCR is the traction calculation rules. These are theoretical and experimental studies of railway wagons on a track gauge of 1,520 mm.

Analyzing the obtained results, a rather close coincidence of the nature of the calculated curves with the empirical curves is identified. These curves were constructed base on the formulas described in the normative document - the Rules of traction calculations for train operation [32]. The quantitative agreement between the calculated and experimental values of the considered quantities is also satisfactory. The maximum discrepancy between the results does not exceed 11% for a loaded car and 17% for an empty car.

Based on the analysis of the plotted graphs, it was found out, that the values of the total running resistance of a railway wagon in the empty as well as in the loaded state with the PWD wheel in various modes of the movement are lower than the values of these indicators for railway wagon with the TWD wheel under the same conditions.

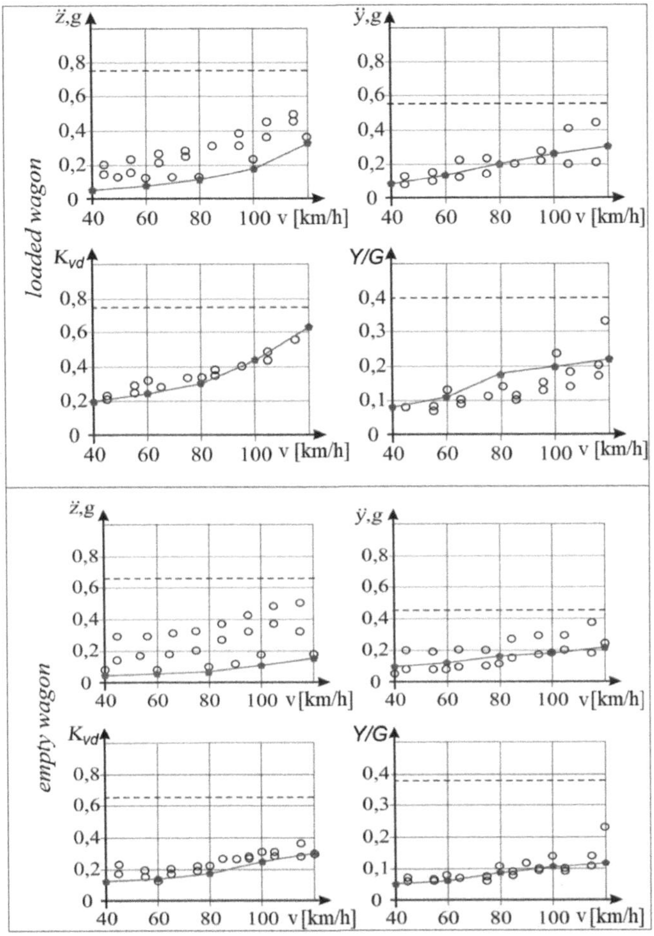

Fig. 2. Dynamic performance comparison results for a loaded railway wagon (above) and for an empty railway wagon (below): O the experimental data, ● the theoretical data.

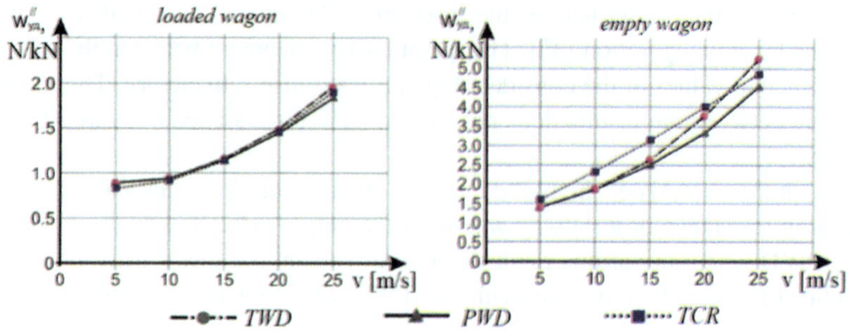

Fig. 3. The total running resistance of a railway wagon in a straight railway track section.

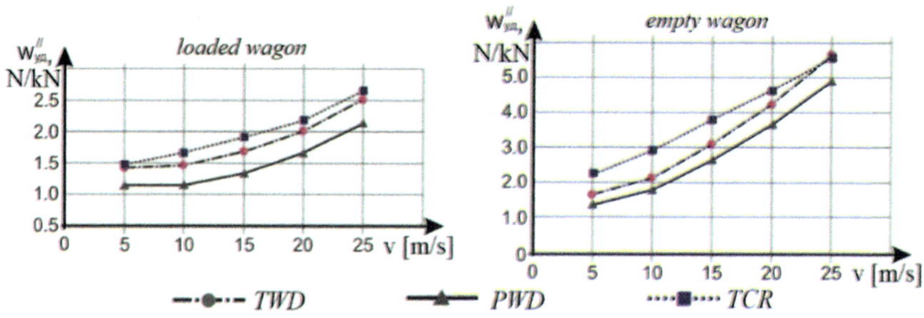

Fig. 4. The total running resistance of a railway wagon in a curve R = 1,200 m.

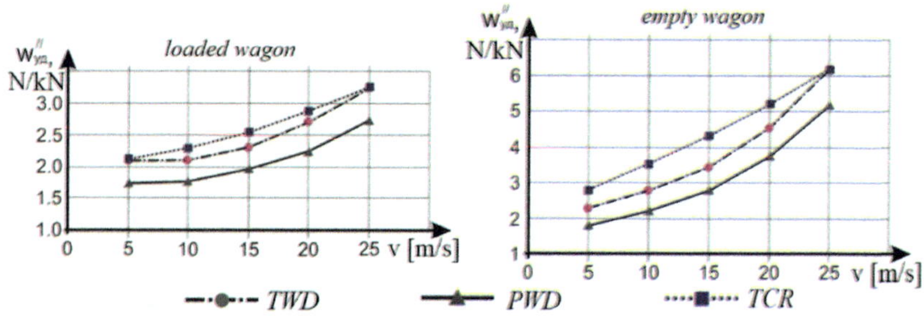

Fig. 5. The total running resistance of a railway wagon in a curve R = 750 m.

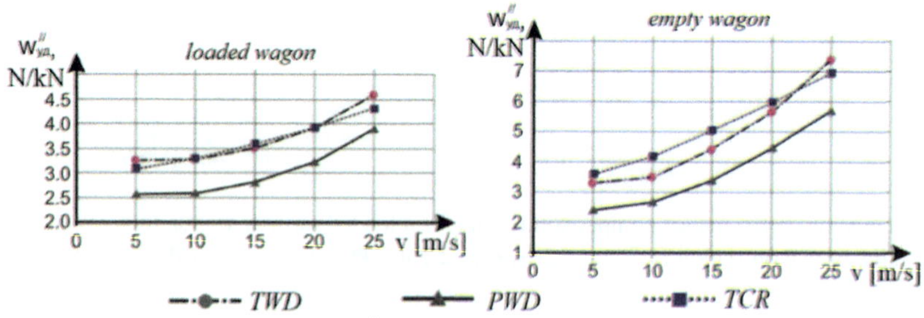

Fig. 6. The total running resistance of a railway wagon in a curve R = 350 m.

In the straight section of a railway track, the reduction of these values can be, depending on the wagon running speed, of 10 to 12% for a loaded railway wagon and 13 to 15% for an empty railway wagon. Reduction of values of these indicators is also reached for wagon running in curves and that is as following: in curves with the radius R = 350 m 20 to 22% for a loaded wagon and 17 to 19% for an empty wagon, in curves with the radius R = 750 m 23 to 25% for a loaded wagon and 13 to 15% for an empty wagon and in curves with the radius R = 1,200 16 to 18% for a loaded wagon and 14 to 16% for an empty wagon.

4 Conclusion

Analysis of the results of mathematical modelling of the movement of a four-axle railway wagon with wheels of different design schemes along a rail track indicates that:

- The use of wheels with the PWD can reduce the resistance to the movement of rail rolling stock by minimizing the kinematic slipping of the wheel flanges on the rails.
- It confirms the expediency of using such wheels in the carriage parts of cars to reduce the cost of traction for trains.
- It reduces wear of the contact surfaces of the wheel flanges and side faces of the rail heads.

Acknowledgement. "This research was supported by the Cultural and Educational Grant Agency of the Ministry of Education of the Slovak Republic in the project No. KEGA 023ŽU-4/2020: Development of advanced virtual models for studying and investigation of transport means operation characteristics."

"This research was also supported by the Slovak Research and Development Agency of the Ministry of Education, Science, Research and Sport of the Slovak Republic in Educational Grant Agency of the Ministry of Education of the Slovak Republic in the project No. VEGA 1/0558/18: Research of the interaction of a braked railway wheelset and track in simulated operational conditions of a vehicle running in a track on the test bench."

References

1. Mikhailov, E., Semenov, S., Tkachenko, V., Sapronova, S.: Reduction of kinematic resistance to movement of the railway vehicles. In: MATEC Web of Conferences, vol. 235 (2018)
2. Mikhailov, E., et al.: Clarification of Features of the Wheel Movement with a Perspective Constructive Scheme on a Rail. Appl. Sci. **10** (2020)
3. Mikhajlov, E.V., et al.: The wheel of the rail vehicle (In Ukrainian). Patent na korisnu model' UA, no. 87418 (2014)
4. Lack, T., Gerlici, J., Stastniak, P.: Wheelset/rail geometric characteristics and contact forces assessment with regard to angle of attack. In: MATEC Web of Conferences 254 (2018)
5. Craciun, C., Cruceanu, C.: Influence of resistance to motion of railway vehicles on the longitudinal trains dynamics. In: MATEC Web of Conferences 178 (2018)

6. Leitner, B., Mocova, L., Hromada, M.: A new approach to identification of critical elements in railway infrastructure. Procedia Eng. **187**, 143–149 (2017)
7. Leitner, B., Figuli, L.: Fatigue life prediction of mechanical structures under stochastic loading. In: MATEC Web of Conferences 157 (2018)
8. Strazovec, P., Suchanek, A., Sťastniak, P., Harusinec, J.: Detection of residual stress in a railway wheel. Transp. Res. Procedia **40**, 898–905 (2019)
9. Gerlici, J., Nozhenko, O., Cherniak, G., Gorbunov, M., Domin, R., Lack, T.: The development of diagnostics methodological principles of the railway rolling stock on the basis of the analysis of dynamic vibration processes of the rail. In: MATEC Web of Conferences 157 (2017)
10. Kurcik, P., Gerlici, J., Lack, T., Suchanek, A., Harusinec, J.: Innovative solution for test equipment for the experimental investigation of friction properties of brake components of brake systems. Transp. Res. Procedia **40**, 759–766 (2019)
11. Golubenko, A.L.: The adhesion of wheel and rail. VUGU Publ, In Russian), Lugansk (1999)
12. Tkachenko, V., Sapronova, S., Zub, E., Morneva, M.: Closed power loops in the guidance of vehicles by railway track system. In: Transport Means - Proceedings of the International Conference 2020, pp. 554–559 (2020)
13. Vershinskij, S.V., Danilov, V.N., Husidov, V.D.: Wagon Dynamics, Hanbook: 360 pages (1991). (in Russian)
14. Danovich, V.D.: Spatial oscillations of wagons on the inertial track. Dnepropetrovsk, 44 pages (1982). (in Russian)
15. Danovich, V.D., Rejdemejster, A.G.: Mathematical model of wheel and rail interaction. In: Transport –Proceedings of DIITu, Nauka i osvita, Dnipropetrovs'k, pp. 17–22 (1999). (in Russian)
16. Ushkalov, V.F.: Mathematical modelling oscillations of rail vehicles, Naukova dumka, 240 pages (1989). (in Russian)
17. Mjamlin, S.V.: Modelling the dynamics of rail vehicles, Dnepropetrovsk: Novaja ideologija, 240 pages (2002). (in Russian)
18. Rejdemejster, O.G.: Influence of parameters the running parts and the shape profile the rolling surface of wheels on dynamic characteristics of freight cars and wear in a pair of "wheel-rail", Dnipropetrovs'k, 185 pages (2000). (in Russian)
19. Gerlici, J., Sakhno, V., Yefymenko, A., Verbitskii, V., Kravchenko, A., Kravchenko, K.: The stability analysis of two-wheeled vehicle model. In: MATEC Web of Conferences 157 (2018)
20. Lack, T., Gerlici, J.: Integration methods for rail vehicle ride dynamics solution assessment. In: XXIII konference s mezinarodni ucasti: soucasne problemy v kolejovych vozidlech/23rd Conference: Current Problems in Rail Vehicles 2017, Ceska Trebova, Czech Republic, 20–22 September, pp. 217–234 (2017)
21. Semenov, S., Mikhailov, E., Dižo, J., Blatnický, M.: Research of resistance of movement of a four-axle wagon with various wheels designs. In: XXV. Konference s mezinarodni ucasti/Cuurent Problems in Rail Vehilces 2021, 15–17 September 2021, Ceska Trebova, Cyech Republic (2021). (to be published)
22. Astahov, P. N.: Resistance to movement of railway rolling stock. Transport, 178 pages (1966). (in Russian)
23. Komarova, A.N.: Influence of bogie characteristics on energy efficiency of freight cars. dis. kand. tehn. nauk: PGUPS, Sankt-Peterburg, 88 pages (2015). (in Russian)
24. Kabzinski, J.: Adaptive, compensating control of wheel slip in railway vehicles. Bull. Polish Acad. Sci. Tech. Sci. **63**(4), 955–963 (2015)
25. Ignesti, M., Innocenti, A., Marini, L., Meli, E., Rindi, A.: Development of a wear model for the wheel profile optimisation on railway vehicles. Veh. Syst. Dyn. **51**(9), 1363–1402 (2013)

26. Arsene, S.: Analysis of the action of the aerodynamic resistances on an electric multiple unit train type. In: IOP Conference Series – Materials Science and Engineering 564 (2019)
27. Uyulan, C., Gokasan, M., Bogosyan, S.: Modeling, simulation and slip control of a railway vehicle integrated with traction power supply. Cogent Eng. **4**(1) (2017)
28. Isakov, V.B.: Elements of numerical methods, Akademija, 192 pages (2003). (in Russian)
29. Bubnov, V.M., Mankevich, N.B.: Dynamics of freight cars on trolleys of model 18–1711 (In Russian). In: Problems and prospects development of railway transport DNUZT, tezi dop. 73 Mizhnar. naukovo-prakt. konf., 23–24 May, pp. 74–75 (2013)
30. Mankevich, N.B.: Improvement the design of cast parts two-axle bogies of freight wagons. Diss. kand.tehn. nauk, Dnepropetrovsk, 263 pages (2015). (in Russian)
31. Ushkalov, V. F., Lapina, L. G., Mashhenko, I. A.: Estimated disturbances for the evaluation of dynamic qualities of freight vagons. In: Bulletin of Dnipropetrovsk National University of Railway Transport named after Academician V. Lazaryan, Vol. 4, No. 46, pp. 135–144 (2013). (in Russian)
32. Traction calculation rules for train operation. Transport, 287 pages (1985). (in Russian)

Modelling and Analysis of a Rolling Wheel with "Inconsistent Abrasion" in Contact with the Rail

Vladas Kukėnas[1,2], Boris Kharitonov[3], Mikhail Levinzon[3], Raimondas Jasvičius[1,4], and Viačeslav Petrenko[1(✉)]

[1] Vilnius Gediminas Technical University, Saulėtekio al. 11, 10223 Vilnius, Lithuania
{vladas.kukenas,raimondas.jasvičius, viaceslav.petrenko}@vilniustech.lt

[2] JSC "Railway Products Conformity Assessment Centre", Panerių g. 14, Vilnius, Lithuania
v.kukenas@gpavc.lt

[3] JSC LLC "Railway Vehicle and Track Interaction Testing Center", 6/3 Slobodskoy pereulok, Office 215, 111033 Moscow, Russia

[4] Institute of Mechanical Science of Vilnius, Gediminas Technical University, Saulėtekio al. 11, 10223 Vilnius, Lithuania

Abstract. When the track interacts with the rolling stock, forces arise that determine all the dynamic loads transmitted to the railway structural elements and rolling stock. In determining the reliability, stability of the operation of rolling stock and infrastructure, it is necessary to refer to the values of the interaction forces between the rail and the wheelsets. At present, these force measurement systems have not been developed during operation, so it is necessary to control the derived (indirect) indicators that determine the interaction. These are wheel defects and track geometry irregularities. Wheel defects are currently detected by their size, measured at wagon inspection points, and track geometry irregularities are detected by track gauges. In order to assess the criteria ensuring traffic safety, special experiments are performed, during which the parameters of the interaction are measured. A mathematical model of the interaction of geometric parameters with force with uneven wear is presented.

Keywords: Detection · Flat spot · Inconsistent abrasion · Interaction · Railways · Rolling wheel

1 Introduction

Based on the detailed analysis of the violation of safe traffic conditions on the railway tracks, including the derailing, it is obvious that the violations of safe traffic sometimes occur even when all indirect safety parameters are adhered to. If in this case the rolling stock and track interaction forces could be controlled, accidents may be avoided.

Usually, measurements are taken to establish the interaction between the geometric damage to the track and the wheel and the force acting on the track structure and the

O. Prentkovskis et al. (Eds.): TRANSBALTICA 2021, LNITI, pp. 120–130, 2022.
https://doi.org/10.1007/978-3-030-94774-3_12

wheel. Identified dependencies do not always work. And systems that determine traffic conditions, track and rolling stock repairs are based precisely on the measurement of geometric indirect parameters.

In order to ensure traffic safety and reduce the number of track residual defects, the rolling stock that permits for operation in the railway infrastructure must be evaluated by their impact for the railway track. This conclusion forces the engineers to develop systems which would evaluate not only the auxiliary, but also the direct parameters which determine the traffic safety and the reduction of the occurrence and accumulation of residual deformation of the track and rolling stock structural elements.

A significant achievement in this area is the development and implementation of the automated detection of wheels with rolling surface defects by the values of the interaction forces. Automated systems for detecting wheels with rolling surface defects are developed and used in numerous countries around the world [1–14]. There are valid prototypes and even industrial specimen which is essentially different in terms of methods for determining the increase of wheel load on the rail.

Experience in operating such systems shows that factors that influence the results registered by the system are rolling-stock speed, rolling-stock type and load, features of the track structure where the system is installed, climatic conditions and etc. Analysis of those systems also shows that there is a good correlation [1] between the results registered by those systems and the results registered by wagon maintenance personnel, although occasionally there are essential deviations from this correlation. Therefore, when creating new criteria, it is also needed to consider the results of operating the automated systems that register the wheel rolling surface defects.

More research is being done on this topic and various mathematical models are being used to determine these criteria. There are many sophisticated mathematical models for determining wheel-to-rail contact forces, considering even the smallest possible effects on determining this force. For example, for calculating the air gap between sleeper and rail for force calculation [15]. Other scholars have developed mathematical models to determine the effect of irregularities through vibrations [16].

In study [17], an adaptive wheel-rail con-tact model with radial spring is developed for prediction of wheel-rail normal contact force and the values of the maximum strength emerging in wheel-rail interaction are compared to the Hertz theory. Based on this mathematical model, the results obtained were compared with other well-known results of numerical and physical experiments in scientific papers [18–20]. On the basis of which the Atlas–LG model and the damage detection criteria were developed.

Proposed [21] multicriteria additive model, which evaluates the normalized weight coefficients of individual criteria. 16 criteria linked to the elements of the train and the technical condition of the track are formulated and justified: rail level, train speed, external condition of the wagon, noise reduction measures, etc. Using the multi-criteria additive model [22], it was determined that the most important parameter of the technical condition of a track is the speed of the wagon.

Forces of interaction between the track and the rolling stock controlled by ATLAS-LG system [23] are determined not only by the wheel rolling surface defects, but also numerous external factors, because the wheel and rail system itself has a feedback. Geometric parameters, weight and rigidity of the system elements can vary due to external impact of the forces and modify the forces themselves. The main external factors are the following:

1. Movement speed.
2. Wheel (axle) load (axial load).
3. Rolling stock structure parameters (rigidity and suspension suppression, mass of structural elements and etc.).
4. Track structure parameters (mass, rigidity, geometry).

Theoretical investigation has been performed in the course of analysis of the impact of the irregularities of the wheel rolling surface on the forces which act on the track in case of different types of rolling stock considering the external factors.

The purpose of this work is to investigate the effect of wheel rolling surface irregularities on the forces acting on the track, with different types of rolling stock considering the following factors: motion speed, wheel (axle) load size (axial load); track construction parameters (mass, stiffness, geometry).

Results and patterns of the investigation were used to create a reference load development methodology for ATLAS-LG system, and equivalent normatives were developed. During the investigation, it was determined how to evaluate the impact of wheel rolling surface damage type and dimensions, the rolling stock movement speed, rolling stock type and their load on the measurement results.

2 Types of Surface Defects and Methods

Wheel rolling surface defects by their impact on the rail are split into two types:

- defects that cause an impact effect of losing the contact (flat spots, compression, abrasion). The defect is characterized by the depth of the flat spot and the compression, and the length in case of abrasion. Such defects will be referred to as "Flat spot" type defects (Fig. 1), the mathematical model can be found [24];
- isolated defects, which amplify the wheel effect on the rail without losing contact (inconsistent abrasion). The defects are characterized by the amplitude and length. Such defects will be referred to as "inconsistent abrasion" type defects (Fig. 2). This paper presents a mathematical model of this type of defect.

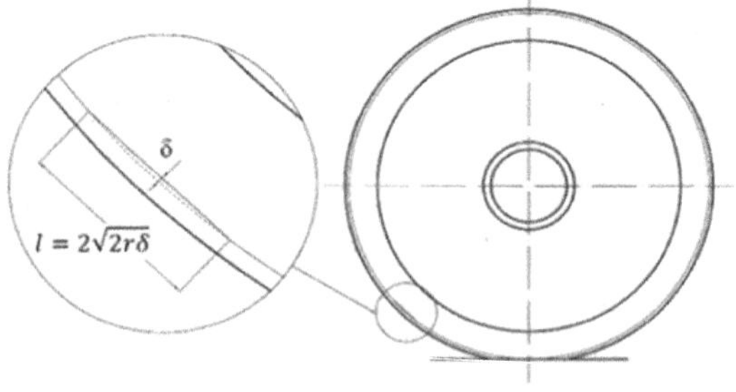

Fig. 1. "Flat spot" type defect scheme. l – length of "flat spot" type defect, mm; δ – flat spot depth of compression spot height, mm; r – wheel radius, m.

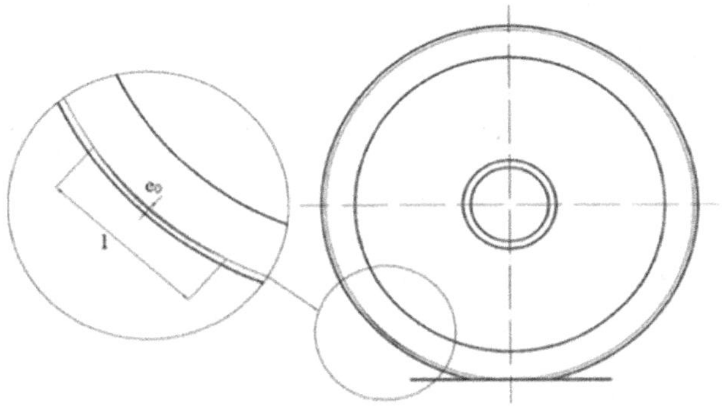

Fig. 2. "Inconsistent abrasion" defect scheme. l – length of "flat spot" type defect, mm; e_0 –. depth of the isolated defect, m.

Special methods are needed to investigate the vertical mutual interactions of the contact between wheel and rail, when it acquires the nature of impact, because the processes under investigation have a high frequency. This is caused by the damage to the rolling stock wheel rolling surface. Respectively, the maximum duration of the force, de-pending on the train travel speed, is several milliseconds or even less.

For creating a mathematical model, we used a methodology to apply reverse Laplace transformation, used in works [6–8, 25–27].

3 Mathematical Model of the Impact of "Inconsistent Abrasion" Type Defect

Frequency characteristics that describe the dynamic system using the input at the uneven location (which occurs as the mass of the wheel center moves) and the output as the dynamic force at the wheel and rail contact location, is expressed using the following equation:

$$W_p = \frac{1}{1/C_k + W_y(0, iw) - W^0(iw)}, \tag{1}$$

where W_p – frequency characteristic, N/m; C_k – stiffness in wheel-rail contact, N/m; $W_y(0, iw)$ – frequency characteristic of the dynamic force transformations into the rail indentation, first argument indicates the distance where the value is determined, m/N; $W^0(iw)$ – frequency characteristic, which related the wheel displacement (under weight located above the spring) with the dynamic force at the point of contact, m/N.

These frequency characteristics are determined as follows:

$$W^0(iw) = \frac{M_1 w^2 - if_1 w - C_1}{M_1 M_0 w^4 - if_1(M_1 - M_0)w^3 \quad C_1(M_1 + M_0)w^2}, \tag{2}$$

where M_1 – weight above the spring distributed for one wheel, 10 N (1,0 kg≈10 N); M_0 – wheel weight, 10 N (1.0 kg≈10 N); f_1 – damping coefficient, N·s/m; C_1 – spring set elasticity distributed for one wheel, N/m.

$$W_y(0, iw) = \frac{1}{EI(r_1 - r_2)(r_1 - r_3)(r_1 - r_4)} + \frac{1}{EI(r_2 - r_1)(r_2 - r_3)(r_2 - r_4)}, \quad (3)$$

where r_j (j = 1, 2, 3, 4) – roots of the algebraic equation with complex indexes:

$$r^4 + \frac{mV^2}{EI}r^2 - \frac{fV + 2imwV}{EI}r + \frac{u - mw^2 + ifw}{EI} = 0, \quad (4)$$

where E – rail steel elasticity module, N/m^2; I – rail moment of inertia (with respect to the neutral axis), m^4; m – given track weight, kg/m; f – distributed track inhibition index, N·s/m^2; u – rail base elasticity module, Pa; V – travel speed, m/s.

Root numbers are selected by the following method (presence of two roots with negative real parts and two with positive real parts is proved by using the Raus criterion Re):

$$Ref_1\langle 0, Ref_2\langle 0, Ref_3\rangle 0, Ref_4\rangle 0 \quad (5)$$

where f_1, f_2, f_3, f_4 – fluctuation suppression index of the weight above the springs, N·s/m.

If we know the wheel centre movement trajectory ξ(Vt), after the inverse Laplace transformation, the dynamic force P_d will be calculated as follows:

$$P_d^{\cdot} = W_p(s)\xi^{\cdot}. \quad (6)$$

Using this calculated expression for the frequency characteristics, which relates the system to the input by a defect and output in the form of dynamic force $W_p(iw)$ it is also possible to determine the dynamic increment when the wheel surface has "inconsistent abrasion" type defect. In this case the trajectory of wheel centre movement can be described as follows:

$$\xi(Vt) = \begin{cases} 0, & t \le 0, \\ \frac{e_0}{2}\left(cos\left(\frac{V_t}{l}\right) - 1\right), & 0 \le t \le \frac{l}{V}; \\ 0, & t \ge \frac{l}{V}. \end{cases} \quad (7)$$

where l – length of the isolated defect, m; e_0 – depth of the isolated defect, m.

In this way, using the $P_D^{\cdot} = W_p(S)\xi^{\cdot}$ ratio to determine the dynamic force, the required load can be determined as reverse Laplace transformation:

$$P_D(t) = \frac{1}{2\pi}\int_{-\infty}^{+\infty} W_p(iw)F(\xi)e^{iwt}dw. \quad (8)$$

Laplace transform of the defect task function:

$$F(\xi) = L[\xi(\mathrm{Vt})] = L\left[\frac{e_0}{2}(\cos\left(\frac{Vt}{l}\right) - 1)\delta\left(t - \frac{e_0}{2}(\cos\left(\frac{Vt}{l}\right) - 1\right)\delta\left(t - \frac{l}{V}\right)\right]$$
$$= -\frac{V^2 e_0}{2l^2 s\left(s^2 + \frac{V^2}{l^2}\right)} + \frac{V^2 e_0}{2l^2 s\left(s^2 + \frac{V^2}{l^2}\right)} e^{-\frac{l}{V}s} \tag{9}$$

$$h_2(t) == \frac{1}{2\pi}\int_{-\infty}^{+\infty} W_p(iw)\frac{V^2}{l^2 iw\left(-w^2 + \frac{V^2}{l^2}\right)} e^{iwt} dw; \tag{10.1}$$

$$h_2(t) = \frac{1}{2\pi}\int_{-\infty}^{+\infty} \frac{ReW_p(iw) + ilmW_p(iw)}{i} \frac{V^2}{l^2 w\left(-w^2 + \frac{V^2}{l^2}\right)}(coswt + isinwt)dw; \tag{10.2}$$

$$h_2(t) = \frac{V^2}{l^2\pi}\int_{0}^{+\infty} \frac{lmW_p(iw)coswt + ReW_p(iw)sinwt}{w\left(-w^2 + \frac{V^2}{l^2}\right)} dw. \tag{10.3}$$

Using the acquired relations, we can get the finite expression to determine the dynamic force:

$$P_D(t) = \frac{e_0}{2}h_2(t) - \frac{e_0}{2}h_2\left(t - \frac{l}{V}\right). \tag{11}$$

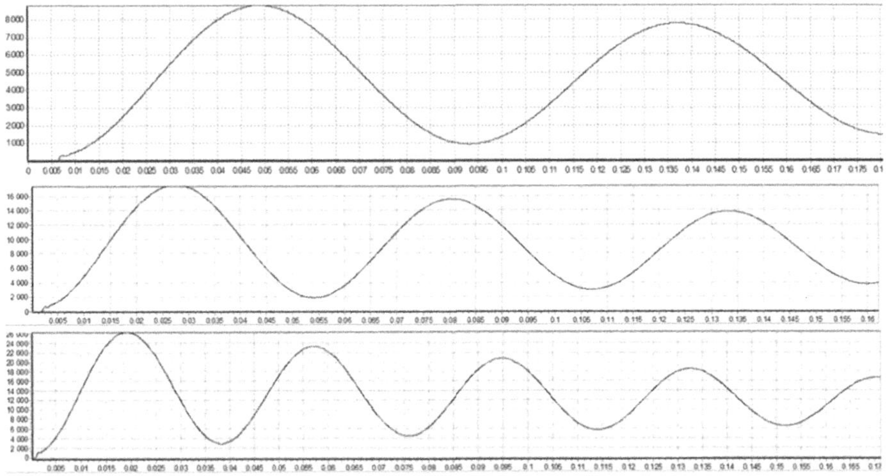

Fig. 3. The additional dynamic force caused by the inconsistent abrasion for the front wagon (wear of the flat spot is 2 mm, track type – on wooden sleepers, season – summer). The horizontal axis represents time, s; the vertical axis represents force, 10 N (1.0 kg*10 N) for 30, 60 and 90 km/h.

The source data for the calculation is the track characteristics (we use data of R65-ŽB-1840 track structure) in winter and summer, for track with different elasticity and composition rolling-stock parameters. Calculations have been performed for the first wagon with different load and dampening in spring block, for passenger wagon and diesel locomotive M62. Calculations are performed for train travel speeds of 30 km/h, 60 km/h and 90 km/h, with 1 mm, 2 mm and 3 mm "inconsistent abrasion" depth and 30 mm, 90 mm, 150 mm and 300 mm length. Example of calculations of additional vertical forces for 2 mm deep inconsistent abrasion in the track with wooden sleepers in summer is given in the Fig. 3.

4 Results of "Inconsistent Abrasion" Type Defects Calculations

"Inconsistent abrasion" type defect calculation results for different types of rolling stock, on different track structure, at different rolling stock travel speeds and amplitudes and lengths of defects are given in Table 1 and Table 2. Applying mathematical models of "Flat spot" and "Inconsistent abrasion" wheel, comparing the results of numerical experiment with known numerical and physical experiments [18–20], that similar results are used to determine maximum force.

Table 1. Results of calculation of wheel vertical force impact when front wagon wheels have "irregular abrasion" type defects (wooden sleepers (wood), reinforced concrete sleepers (r. conc.). During the winter, the difference in stiffness between wooden and reinforced concrete sleepers were not considered).

Speed, km/h	Depth of defects, mm	Length of wear, cm	Front open wagon with varying axial load, tf (tonnes per axle)								
			23.5			15.5			5.5		
			Summer		Winter	Summer		Winter	Summer		Winter
			Wood	r. conc		Wood	r. conc		Wood	r. conc	
	1	30	16 142	16 148	16 125	10 452	10 461	10 437	3 917	3 918	3 896
	2		20 534	20 547	20 499	13 405	13 422	13 375	5 084	5 086	5 041
	3		24 927	24 945	24 874	16 357	16 383	16 312	6 251	6 255	6 187
	1	90	13 474	13 477	13 468	8 659	8 662	8 652	3 208	3 208	3 199
	2		15 198	15 202	15 184	9 817	9 824	9 806	3 667	3 667	3 650
	3		16 922	16 929	16 901	10 976	10 987	10 959	4 124	4 125	4 099
	1	150	12 516	12 518	12 513	8 015	8 017	8 012	2 954	2 954	2 949
	2		13 282	13 284	13 276	8 530	8 533	8 525	3 157	3 158	3 150
	3		14 048	14 052	14 039	9 045	9 050	9 037	3 361	3 361	3 350
30	1	30	20 527	20 544	20 501	13 648	13 668	13 621	5 078	5 082	5 038
60	2		29 304	29 338	29 252	19 546	19 586	19 492	7 407	7 414	7 326
	3		38 081	38 131	38 003	25 444	25 504	25 363	9 735	9 745	9 614
	1	90	15 196	15 203	15 187	10 067	10 073	10 056	3 664	3 666	3 648
	2		18 643	18 657	18 624	12 382	12 398	12 360	4 578	4 582	4 547
	3		22 091	22 109	22 059	14 699	14 721	14 666	5 493	5 497	5 445

(continued)

Table 1. (*continued*)

Speed, km/h	Depth of defects, mm	Length of wear, cm	Front open wagon with varying axial load, tf (tonnes per axle)								
			23.5			15.5			5.5		
			Summer		Winter	Summer		Winter	Summer		Winter
			Wood	r. conc		Wood	r. conc		Wood	r. conc	
	1	150	13 282	13 285	13 277	8 779	8 784	8 774	3 157	3 157	3 149
	2		14 813	14 819	14 805	9 809	9 816	9 800	3 562	3 564	3 548
	3		16 346	16 354	16 332	10 838	10 849	10 824	3 969	3 971	3 949
90	1	30	24 908	24 942	24 883	16 591	16 623	16 551	6 238	6 244	6 178
	2		38 066	38 133	38 016	25 431	25 497	25 353	9 726	9 738	9 607
	3		51 224	51 325	51 148	34 272	34 370	34 154	13 214	13 232	13 035
	1	90	16 916	16 930	16 907	11 221	11 234	11 206	4 120	4 121	4 096
	2		22 084	22 111	22 064	14 693	14 720	14 663	5 489	5 494	5 442
	3		27 251	27 291	27 222	18 166	18 204	18 119	6 859	6 865	6 789
	1	150	14 047	14 053	14 042	9 293	9 299	9 286	3 359	3 359	3 348
	2		16 343	16 354	16 334	10 837	10 848	10 822	3 967	3 969	3 947
	3		18 640	18 657	18 627	12 379	12 396	12 359	4 577	4 580	4 545

However, prior known models have not found that in case of "Flat spot" defect the measurement results are mainly (up to 30–40%) influenced by track parameters (elasticity, weight and dampening), which are in turn influenced by the ambient temperature, whereas the impact of such parameters on the "inconsistent abrasion" type defects is very low. In this case the important thing is the frequency of specific impact.

In case of "Inconsistent abrasion" – as isolated, irregular sine curve (Fig. 3) with varying amplitude and length. Elements such as rolling stock type, axle load, passage speed in the measurement area, impact force amplitude and pulse length are also missing.

Table 2. Results of calculation of the vertical impact force 10 N (1,0 kg*10 N) of the wheel when the rolling surface of the wheel of the passenger wagon and diesel locomotive M62 has a flat spot (wooden sleepers (wood), reinforced concrete sleepers (r. conc.). During the winter, the difference in stiffness between wooden and reinforced concrete sleepers was not considered).

Speed, km/h	Depth of defects, mm	Length of wear, cm	Passenger wagon			Diesel locomotive M62		
			15.0 tf (tonnes per axle)			19.7 tf (tonnes per axle)		
			summer		Winter	summer		Winter
			Wood	r. conc		Wood	r. conc	
30	1	30	9 932	9 894	9 886	14 335	14 324	14 313
	2		12 363	12 288	12 272	18 819	18 798	18 775
	3		14 795	14 682	14 658	23 304	23 272	23 238
	1	90	8 455	8 440	8 437	11 610	11 607	11 602
	2		9 409	9 379	9 373	13 370	13 362	13 353
	3		10 364	10 320	10 310	15 130	15 119	15 105

(*continued*)

Table 2. (*continued*)

Speed, km/h	Depth of defects, mm	Length of wear, cm	Passenger wagon 15.0 tf (tonnes per axle)			Diesel locomotive M62 19.7 tf (tonnes per axle)		
			summer		Winter	summer		Winter
			Wood	r. conc		Wood	r. conc	
	1	150	7 924	7 918	7 916	10 632	10 630	10 629
	2		8 348	8 335	8 332	11 415	11 411	11 407
	3		8 773	8 753	8 748	12 197	12 191	12 186
60	1	30	12 356	12 283	12 265	18 815	18 788	18 782
	2		17 213	17 066	17 029	27 779	27 725	27 715
	3		22 069	21 849	21 794	36 744	36 663	36 647
	1	90	9 407	9 379	9 371	13 370	13 359	13 358
	2		11 314	11 256	11 242	16 890	16 870	16 865
	3		13 221	13 134	13 114	20 410	20 379	20 373
	1	150	8 348	8 334	8 331	11 415	11 410	11 408
	2		9 195	9 170	9 163	12 979	12 970	12 968
	3		10 043	10 004	9 994	14 544	14 529	14 526
90	1	30	14 785	14 672	14 644	23 297	23 233	23 231
	2		22 071	21 844	21 788	36 744	36 615	36 611
	3		29 356	29 015	28 932	50 191	49 998	49 992
	1	90	10 360	10 316	10 305	15 131	15 106	15 104
	2		13 222	13 133	13 111	20 410	20 360	20 359
	3		16 083	15 949	15 916	25 691	25 616	25 615
	1	150	8 771	8 752	8 747	12 197	12 186	12 186
	2		10 043	10 004	9 994	14 544	14 522	14 520
	3		11 314	11 256	11 240	16 890	16 857	16 856

5 Conclusions

In order to determine the impact on the track of the rolling stock with wheels with rolling surface defects a mathematical model of this process was developed. The mathematical modelling showed the following results:

1. The effect of different wheel irregularities "inconsistent abrasion" (1, 2, 3 mm deep and 30, 90, 150 mm long) on the forces acting on the track for different types of rolling stock (freight wagon, passenger wagon, locomotive) with regard to factors such as speed of movement (30, 60, 90 km/h), wheel/axle load (axle load – 23.5; 19.7; 15.5; 15.0; 5.0 tf (tonnes per axle)), track construction parameters (reinforced concrete rails, wooden sleepers) and environmental parameters (winter, summer). Their regularities were determined.
2. Analysing the impact on the track, essential differences between "Flat spot" and "Irregular abrasion" types of defects have been determined. In case of "Flat spot" defect, the measurement results are mainly (up to 30–40%) influenced by track

parameters (elasticity, weight and dampening), which are in turn influenced by the ambient temperature [24], whereas the impact of such parameters on the "Irregular abrasion" type defects is very low. In this case the important thing is the frequency of specific impact. In case of "Flat spot" this impact can be described as a del-ta-function, and in case of "Irregular abrasion" − as isolated, irregular sine curve with varying amplitude and length.

3. To bring the results acquired by the ATLAS-LG system to the reference value in case of "irregular abrasion" it is needed to consider the following factors which influence the result: train travel speed and type of rolling-stock. In the resolution of the formulated task, when using the ATLAS-LG system, it is needed to collect the following additional in-formation: type of rolling stock, axial load, speed of passing the measurement area, impact force effect amplitude and pulse length.

References

1. Stone, D.H., Kalay, S.F., Tajaddini, A.: Statistical behavior of wheel impact detectors to various wheel defects, pp. 9–13. International Wheelset Congress, Sydney (Australia) (1992)
2. Zarembski, A.M.: Wheel/Rail: P1 and P2 Forces, Railway Track and Structures, 11 p. (1995)
3. Wheel load detection system, Railways of the World, 11 (1995)
4. Research on the interaction of track and rolling stock in the USA, Railways of the World, 12 (1996)
5. System for monitoring rail loads and technical condition of wagons, Railways of the World, 12 (2000)
6. Kogan, A.Ya.: Vertical dynamic forces acting on a path, Works of the JSC VNIIZHT, 402 208 p. (1969)
7. Kogan, A.Ya.: Track dynamics and its interaction with rolling stock, Transport 328 p. (1997)
8. Kogan, A.Ya., Verkhotin, A.A.: Calculation of the impact on the path of a pair of wheels with a slider, Studies on the possibility of increasing train speeds, JSC VNIIZHT, pp. 31–37 (1984)
9. Weed, D., Lonsdale, C.: Review and analysis of wheel impact load detector (wild) and wheel removal data. Chicago Freight Car Leasing Co. (2004)
10. Contact fatigue damage to freight car wheels. Automatic wheel diagnostics system (detector of defective wheels DKC), VNIIZhT Proceedings, M.: InText (2004)
11. Guidelines to Best Practices for Heavy Haul Railway Operations: Wheel and Rail Interface Issues, International Heavy Haul Association, 198 p. (2001)
12. Vertical dynamic load monitoring system "AGUILA", based on fiber optic pressure sensors, ITSS 14 p. (2014)
13. Mobile monitoring system for train, vehicle and infrastructure under real operating conditions, Innotec System Gmbh (2020). http://www.innotec-systems.de/index.php/en/products/lasca.html
14. WILD Wheel Impact Load Detector: Scalex Wild Wheel Impact load detector and control scale. Salient Systems, Inc, Tampton Finland (2020)
15. Bogdevičius, M., Žygienė, R., Bureika, G., Dailydka, S.: An analytical mathematical method for calculation of the dynamic wheel–rail impact force caused by wheel flat. Vehicle Syst. Dyn. Int. J. Veh. Mech. Mob. **54**, 689–705 (2016). https://doi.org/10.1080/00423114.2016.1153114

16. Steišūnas, S., Bureika, G., Gorbunov, M.: Effects of rail-wheel parameters on vertical vibrations of vehicles using a vehicle-track-coupled model, Transport problems = Problemy transportu. Gliwice: The Silesian University of Technology, 14, 27–39 (2019). https://doi.org/10.20858/tp.2019.14.3.3
17. Zhu, J.J., Ahmed, A.K.W., Rakheja, S.: An adaptive contact model for simulation of wheel-rail impact load due to a wheel flat, NaCoMM-20 (2007)
18. Andersson, C., Oscarsson, J.: Dynamic train/track interaction including state-dependent track properties and flexible vehicle components. Veh. Syst. Dyn. **33**, 47–58 (1999). https://doi.org/10.1080/00423114.1999.12063069
19. Johansson, A., Nielsen, J.C.O.: Out-of-round railway wheels - Wheel-rail contact forces and track response derived from field tests and numerical simulations. In: Proc. of the Institution of Mechanical Engineers, Part F: Journal of Rail and Rapid Transit, vol. 217, pp. 135–146 (2003). https://doi.org/10.1243/095440903765762878
20. Nielsen, J.C.O., Oscarsson, J.: Simulation of dynamic train–track interaction with state-dependent track properties. J. Sound Vib. **275**, 515–532 (2004). https://doi.org/10.1016/j.jsv.2003.06.033
21. Maskeliūnaitė, L., Sivilevičius, H.: The model for evaluating the criteria describing the quality of the trip by international train. Technol. Econ. Dev. Econ. **20**, 484–506 (2014). https://doi.org/10.3846/20294913.2014.949333
22. Sivilevičius, H., Maskeliūnaitė, L.: The numerical example for evaluating the criteria describing the quality of the trip by international. E&M Econ. Manag. **2**, 73–86 (2014). https://doi.org/10.15240/tul/001/2014-2-006
23. Technical description of ATLAS-LG system, RAKP project, ATLAS-LG Requirements for Installation, VAE Eisenbahnsysteme Gmbh. Document Number 012 44 90769, 8 p.
24. Kukėnas, V., Kharitonov, B., Levinzon, M., Jasevičius, R.: Improvement of diagnostic parameters of a rolling wheel with flat spot and experimental test on lithuanian railways. Appl. Sci. **10**, 7148 (2020). https://doi.org/10.3390/app10207148
25. Danilov, V.N.: Calculation of railway path in the zone of the joints, Works of the JSC VNIIZHT, 70, 116 p. (1952)
26. Andreev, G.E.: The work of the superstructure in straight sections during high-speed train traffic. The dissertation for the degree of Doctor of Technical Sciences. St. Petersburg State Transport University (1974)
27. Guidelines for the calculation of the vertical unsteady stress-strain state of the track on wooden and reinforced concrete sleepers in the area of the rail junction, Russian Federation Ministry of Railroads, Department of Railways, 43 p. (2005)

Stone Paved Road Digital Reproduction: A Workflow

Mattia Intignano[1], Salvatore Antonio Biancardo[1], Cristina Oreto[1], Nunzio Viscione[1], Rosa Veropalumbo[1], Francesca Russo[1], Francesco Abbondati[2], and Gianluca Dell'Acqua[1](✉)

[1] Department of Civil, Construction and Environmental Engineering, Università di Napoli Federico II, 80125 Napoli, Italy
{salvatoreantonio.biancardo,gianluca.dellacqua}@unina.it

[2] Department of Engineering, Università di Napoli Parthenope, 80133 Napoli, Italy

Abstract. The growing necessity to design and digitally represent historical pavements has led specialists to use different Building Information Modelling (BIM) tools to control the road design and construction phases. In this research paper, a Heritage BIM (H-BIM) approach was developed to recreate an archaeological road to accomplish the structural analysis of stone pavements. In detail, starting from the Digital Elevation Model (DEM) generated with Autodesk Infraworks, the road design process was performed. The corridor was modelled with Autodesk Civil 3D using a parametric road section created with the Subassembly composer tool. Subsequently, a visual programming application (Dynamo) based on Python language was adopted to extract corridor information (such as areas, volumes, subassembly parameters) and apply changes based on conditions established through code scripts. In detail, a workflow was developed to implements a disruption analysis of road stone pavements. As preliminary results, a tool is proposed to support the authorities and experts during the managing process.

Keywords: Heritage-BIM · Stone paved road · Procedural Modelling · Point Cloud

1 Introduction

The continuous search for new methods and approaches for increasing process efficiency in the AEC industry (Architecture, Engineering and Construction) has found in digitisation and computerisation an effective answer for the reduction of costs and production times issues.

BIM is defined as a process involving the generation and management of digital representation of a facility's physical and functional characteristics [1]. It is now recognised as the most appropriate methodology to undertake building and infrastructure projects [2, 3], having revolutionised the field of architecture and construction

O. Prentkovskis et al. (Eds.): TRANSBALTICA 2021, LNITI, pp. 131–139, 2022.
https://doi.org/10.1007/978-3-030-94774-3_13

for some years now [4], being both a technology and a methodology [5]. Although it remains a challenge for the AEC industry, since it requires a shift to a new way of working, there is a strong legislative endeavour to flat any differences in its adoption level among the various states of the EU [6].

The immense road heritage of European stone road pavements, starting from the Italian capital city, has never been the subject of wide-ranging policies aimed at enhancing recent times. The reconstruction of models managed in BIM tools can help the conservation of cultural heritage. Therefore, BIM has been declined in the so-called Heritage-BIM (H-BIM). It is defined as "a new way of modelling the existing structures, generating intelligent models that can contain and manage information, and concern all components of the project, including their geometric and identification information described in detail" [7].

Therefore, along with the consolidated experience in the architectural field [8–10], it has been proven that H-BIM is suitable also for infrastructures. Most of the historical urban areas in western countries are paved with stone elements [11], and Italy has one of the highest numbers of UNESCO Heritage sites, as the case of the Via Duomo in Naples [12].

H-BIM approach needs cutting-edge technologies, such as VR (Virtual Reality) [13] and UAV (Unmanned Aerial Vehicle) photogrammetry. The latter is an intelligent survey technology that can be used to automatically detect stone pavement's patterns [14, 15]. It was used effectively to the production of the 3D H-BIM for archaeological excavations in Hierapolis in Phrygia, Turkey [16], and Via del Vesuvio in Pompeii, Italy [17].

This research study aims to define a new approach to use the H-BIM methodology for the analysis and the management processes of archaeological sites focusing on the study of a stone pavement road located in Pompeii. The case study involves the terminal part of Via Stabiana on Porta di Stabia in the Archaeological Site of Pompeii. The paved area is 12 m long with a 5.20% average slope, that guarantees the outflow of the rainwater into an underground canal on the right side (North-South direction) and then outside the walls thanks to a change in the road's cross inclination.

The main idea of this work is to integrate archaeology and BIM, offering archaeological heritage protection agencies an innovative management and archiving tool. The proposed methodology, which will be discussed below, represents an element of novelty in the current panorama. It is the integration of engineering, digitisation and archaeology and poses a starting point for important questions about the possibility of studying the deep layers of the archaeological road infrastructure. This would help the investigation of the past construction techniques allowing for analysing their differences and similarities with present ones.

Since any H-BIM application should include an informed preliminary analysis of construction criteria, previous research works have been investigated to obtain an in-depth picture of knowledge on the case study, such as for Roman road building technologies and practices [18] and their design criteria [19]. Particularly significant is the study conducted in the archaeological site of Pompeii by Poehler and Crowther [20] on the paving techniques hypothesised from the surveys carried out.

2 Methodology

The pivotal methodology shown in Fig. 1 is the so-called Scan-to-BIM, which allows for the generation of a 3D BIM model from a point cloud [21, 22], involving two main work phases: 3D surveying through Laser Scanner processing the information surveyed with several BIM authoring software.

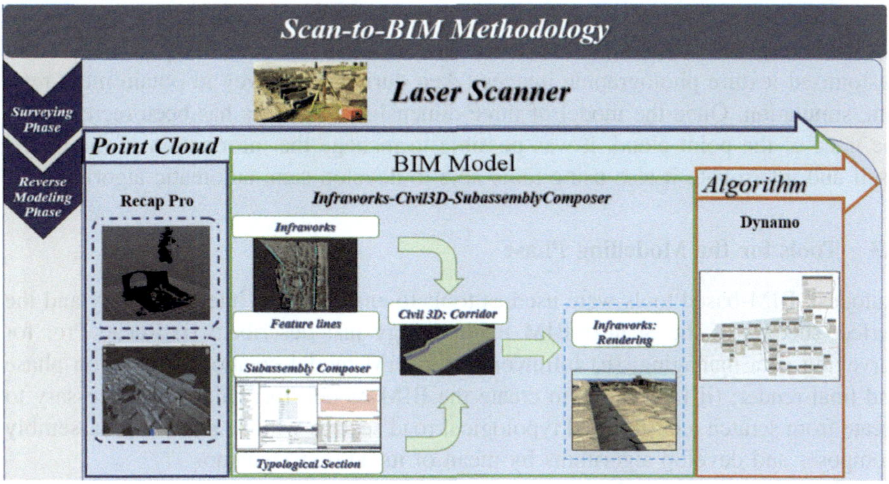

Fig. 1. Methodology workflow.

2.1 Scan-To-BIM Surveying Phase

The surveying phase of the Scan-to-BIM process relies on Laser Scanners technology. Electro-optical instruments are capable of surveying the surface of any object by mapping points in 3D. A Leica RTC360 model Laser Scanner has been used for the present work: it operates by emitting a laser pulse and calculating the distance from the object detected by measuring the round-trip time of the laser pulse itself. After setting the resolution parameter, connected to precision and scanning speed, a single 360° scan was performed in one minute and 50 s, obtaining a point cloud composed of 2 million points whose position is identified by coordinates [X, Y, Z] defined by the distance of the point from the instrument, the inclination angle formed by the conjunction of the point and the instrument concerning the vertical axis of the instrument itself and the azimuth angle formed by the conjunction of the point and the instrument with respect to a horizontal axis taken as reference. These coordinates were georeferenced through the parallel use of a GPS detection system to record the scanner's position at the time of scanning in a Gauss-Boaga system.

2.2 Scan-To-BIM Reverse Modelling Phase

The information thus obtained represents raw data, requiring a subsequent "reverse engineering" process that consisted of creating a 3D model shaping its surfaces on the points of the point cloud. 3D surfaces are generally divided into two main categories: (i) NURBS (Non-Uniform Rational Basis-Splines); (ii) Polygonal Meshes. In the first case, surfaces are defined by mathematical equations, very precise and stable. The latter are easier to model but they are highly dependent on the degree of simplification of the mesh itself usually providing rougher images compared to Nurbs ones. The final step of 3D modelling was the assignment of textures to Mesh surfaces using as a source of customised texture photographic images taken during the survey to obtain more realistic simulation. Once the model of three-dimensional surfaces has been recreated on the trace of the point cloud, it was possible to manage the information of the model itself and interrogate it also using tools able to develop semi-automatic algorithms.

2.3 Tools for the Modelling Phase

Autodesk BIM-based tools were used as tools to ensure 100% interoperability and the perfect success of the Scan-to-BIM methodology just described: (i) Recap Pro, for surveying data managing; (ii) Infraworks, for BIM model creation preparation phase and final render; (iii) Civil 3D, to create the BIM model, for which was necessary to create from scratch a customised typological road section with its add-on Subassembly Composer and develop algorithms by mean of its add-on Dynamo.

After the laser scanner-based survey, the starting point of the subsequent modelling phase was a point cloud in PTX format. This is an ASCII based format for saving point cloud data, that associates each point with 4 (X, Y, Z, intensity) to 7 values depending on whether there is colour information (RGB - Red Green Blue). Colourization and illumination tools were used to identify the parts of the cloud considered of interest, i.e., the road pavement. Normal and Elevation colourations were used to identify all those undesirable vertical elements, and Selection Tools were used to delete them, resulting in a lighter point cloud.

Recap Pro was also fundamental to ensure interoperability with other software used. Indeed, it was used to convert the original file format from PTX to RCP in order to import it in Autodesk Infraworks. Infraworks is a civil infrastructure design software that enables AEC professionals to model, analyse and visualise their projects in a real-world context, improving decision-making and project outcomes. In the first phase of the work, Infraworks was used for the DEM creation, that is, the model of the Elevation of a surface compared to a plan of referment. In this case, DEM represents a surface adherent to the archaeological pavement under study. Characteristic lines have been drawn on this surface at the road axis, road margins, and sidewalk edges to coordinate horizontal and vertical information. Then, they were exported in an SHP file format, extracting the spatial coordinates of the vertexes by which they are defined. In this way, 3D and georeferenced feature lines were obtained, ready to be imported into a BIM modelling tool, such as Autodesk Civil 3D. These feature lines are the base to develop the BIM model itself.

The 3D model of the road was obtained using Civil 3D. Usually, the corridor representing the infrastructure is obtained by extruding an assembly along an alignment, drawn on a topographic surface, made three-dimensional by the coordination with a design profile. In this application, the three-dimensional axis was obtained simply by importing into Civil 3D what has been produced with the work done previously with Recap Pro and Infraworks, that is, the characteristic lines consistent with the information reported by the point cloud.

Civil 3D has a well-stocked default library of structural and functional elements, but the assembly of the ancient Roman road does not correspond to any of those available since these ones follow principles and materials of modern conception, so it was necessary to build it from scratch using Subassembly Composer that is automatically bundled with the Civil 3D license and is an extremely useful tool for customising typological road sections. To create the geometry, one must proceed by selecting from the Toolbox the necessary elements, i.e., nodes, connections and surfaces enclosed by them, and add them to the working scheme on the Flowchart screen through a drag&drop operation. The position of the points in the section plane can be set from an initial point, by default the origin, by means of Delta-X and Delta-Y, or by imposing a slope combined to a Delta-X or a Delta-Y. Links are defined by indicating the starting and ending nodes. Shapes are defined by indicating the links that constitute their edges. For each element, point, link or shape, it is possible to associate a code, which is of vital importance for subsequent operations in Civil 3D, that happens to be a strongly hard-coded software. As the geometry is formed, the preview of what has been coded is obtained. By defining parameters in the Input/Output Parameters section, one can then use them to characterise the geometry and thus make the section parametric. Due to the conformation of the infrastructure and the variable horizontal-vertical trend of the three-dimensional polygonal lines used as reference, it is essential to set targets among the other parameters of the typological section itself.

Two other steps are needed to complete the corridor: build the extrusion surfaces and finally assign materials to characterise the render. It is possible to create the corridor surfaces through the Corridor Properties panel, click on Surfaces, and then specify the referment code; in this case, links defined in the assembly were used. Surfaces are created by triangulation – Mesh – using the vertexes of the links used as referment codes. Some errors and inaccuracies occurred but then these were fixed manually. Regarding the texture, it must reflect the component material of the 3D object to which it refers. Meta-data about materials are already contained in the parameters assigned to the section during the creation process with Subassembly Composer, but to give a visually appreciable result, it was chosen to customise a material-texture in Infraworks through photographic evidence taken during the survey.

Once the H-BIM model was obtained, Dynamo, an add-on of Civil 3D useful for creating an algorithm to automatise and speed up any process, was used to query the BIM model itself and develop side operations. The programming environment is a VPL - Visual Programming Language - that is based on the construction of a graph, in which the nodes are the elements of the logical content of the algorithm and the arrows connect them in input-output relationships. Moreover, the Dynamo feature for the creation of custom nodes by writing scripts in Python language was exploited to write an algorithm able to read the elevations of the profile of the characteristic baselines of the corridor. Further operations were developed in the script for the calculation of the local and total deviations between the linear interpolation of the elevations and the elevations themselves, in order to obtain an index designed to highlight the state of instability and irregularity of the pavement, useful information in order to plan the accessibility of weak users, with walking difficulties or other special needs, to the archaeological site.

3 Results and Discussion

Starting from the georeferenced point cloud obtained by Laser Scanner, the use of Autodesk Recap Pro allowed resizing the cloud itself by selecting the specific area of interest (Fig. 2a), resulting in an effective procedure for data manipulation (Fig. 2b). Brought from PTX format to RCP format, the point cloud file has been imported into Autodesk Infraworks, where DEM was obtained, i.e., a mesh surface consisting of the sum of the many triangles' surfaces vertices correspond to the points on the cloud.

Usually, the concept of DEM can be overlapped to DTM, especially when using Infraworks which helps the engineer to contextualise the work by dropping it in its territorial context. This time, the DEM faithfully depicted an already built surface, but a very irregular one, given the archaeological nature of the stone road under study.

Based on the paved road DEM, three feature lines (Fig. 2c, cyan colour) have been drawn. Afterwards, the spatial information has been extracted from the feature lines and imported in Civil 3D. Here the actual 3D model has been created: first the typological road section was created using Subassembly Composer Civil 3D add-on and then it was extruded along the feature lines previously imported from Infraworks (Fig. 2d). In this step, there is another element of innovation: where usually for the creation of the three-dimensional model of the road infrastructure Civil 3D requires a topographical surface on which to draw the alignment, having made the operations previously described on Infraworks, it was possible to realise the corridor using the spatial information obtained directly from the point cloud, i.e., from the survey of the existing artefact. The final render of the 3D model was then created by means of Infraworks using a customised texture (Fig. 2e). The H-BIM model obtained was then used to operate side automatised operations through Civil 3D add-on Dynamo, which allows for coding in VPL based on Python Language (Fig. 2f). The algorithm reads the elevation information of the H-BIM model, and after calculating a coefficient of disruption and irregularity of the pavement, it associates different levels of risk and practicability of the road stretch itself.

The uniqueness of the archaeological context is a major issue of the whole procedure since the software used have been conceived for the realisation of new structures whose typical feature is seriality and repetition. Therefore, it was necessary to adapt the available tools to the need to create unique three-dimensional elements to be inserted into a complex BIM paradigm composed of a multiplicity of these elements. This case study was an attempt to take another step forward in this direction, integrating established approaches with innovative modelling and scripting methods.

To conclude, the integration of advanced technological instrumentation, BIM software and innovative digitisation methodologies constitue a technical tool to produce the H-BIM of an archaeological reality. The model's core is the fidelity to the reality detected in the field and its purpose is flexible at user needs.

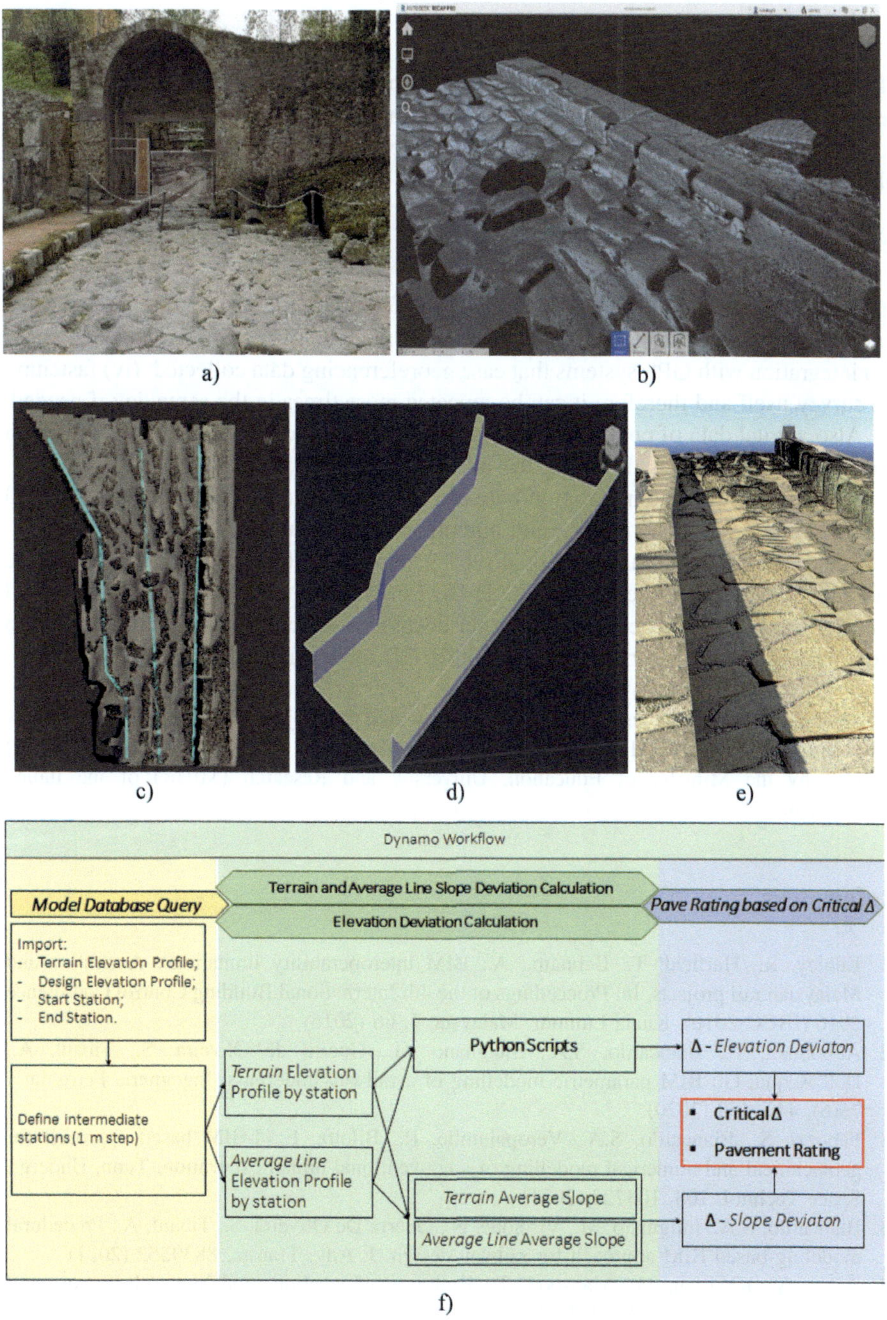

Fig. 2. Results: a) case study location; b) RCP output c) Infraworks EDM output; d) Civil 3D corridor output; e) Infraworks render; e) Dynamo algorithm in VPL.

4 Conclusion

The research paper has shown how the use of specific methodologies allows accurate detection of the geometric elements and their state of preservation for further evaluations for planning and managing maintenance treatments in the field of archaeological sites.

To date, the Scan-to-BIM methodology can be recommended for H-BIM digital reproduction, proving important advantages.

In fact, the use of laser scanner allows for (i) more precise surveys of the artefact than traditional techniques ones; (ii) collection of data in the form of a point cloud in a single file or in multiple files that can be merged in one, easing data managing; (iii) integration with GPS systems that ease georeferencing data collected; (iv) fastening the survey itself and therefore it can be repeated more times in the same day if needed.

Also, digital data of point cloud allows for BIM approach implementation to design and managing processes, lowering the overall costs.

Finally, it allows the integration between the use of industry-leading software and coding skills to automate queries and computational operations.

Some critical issues also emerged during this research work, represented by the difficulty of perfectly matching the 3D H-BIM model and the surveyed archaeological artefact. Therefore, it is desirable to further develop research in this direction, trying to use a more versatile and perhaps open-source 3D modelling BIM-based tool.

Acknowledgements. The research was developed within the Project of National Interest - PRIN 2017 "Stone Pavements. History, Conservation, Valorization and Design" (20174JW7ZL) financed by the Ministry of Education, University and Research (MIUR) of the Italian Government.

References

1. Kenley, R., Harfield, T., Behnam., A.: BIM interoperability limitations: Australian and Malaysian rail projects. In: Proceedings of the 4th International Building Control Conference 2016 (IBCC 2016), Kuala Lumpur, Malaysia, 7, 66 (2016)
2. Abbondati, F., Biancardo, S.A., Sicignano, G., Guerra de Oliveira, S., Tibaut, A., Dell'Acqua, G.: BIM parametric modelling of a railway underpass. Ingegneria Ferroviaria **75**(6), 443–459 (2020)
3. Fabozzi, S., Biancardo, S.A., Veropalumbo, R., Bilotta, E.: I-BIM based approach for geotechnical and numerical modelling of a conventional tunnel excavation. Tunn. Undergr. Space Technol. **108**, 103723 (2021)
4. Biancardo, S.A., Intignano, M., Viscione, N., Guerra De Oliveira, S., Tibaut. A.: Procedural modeling-based BIM approach for railway design. J. Adv. Transp., 8839362 (2021)
5. Azhar, S., Khalifan, M., Maqsood, T.: Building information modeling (BIM): now and beyond. Constr. Econ. Build. **12**(4), 15–28 (2012)
6. Charef, R., Emmit, S., Alaka, H., Fouchal, F.: Building Information Modelling adoption in the European Union: an overview. J. Build. Eng. **25**, 100777 (2019)
7. Murphy, M., McGovern, E., Pavia, S.: Historic building information modelling (HBIM). Struct. Surv. **27**(4), 311–327 (2009)

8. Martinez, S., Ortiz, J., Gil, M.: Geometric documentation of historical pavements using automated digital photogrammetry and high-density reconstruction algorithms. J. Archaeol. Sci. **53**, 1–11 (2015)
9. Conti, A., Fiorini, L., Massaro, R., Santorini, C., Tucci, G.: HBIM for the preservation of a historic infrastructure: the Carlo III bridge of the Carolino Aqueduct. Applied Geomatics (2020)
10. Donato, V., Biagini, C., Bertini, G., Marsugli, F.: Challenges and opportunities for the implementation of H-BIM with regards to historical infrastructures: a case study of the Ponte Giorgini in Castiglione della Pescaia (Grosseto – Italy)". In: ISPRS International Archives of the Photogrammetry, Remote Sensing and Spatial Information Sciences, XLII-5/W1, pp. 253–260 (2017)
11. Galli, E., Giuliani, F.: Stone pavement materials and construction methods in Europe and North America between the 19th and 20th century. Int. J. Architectural Heritage **13**(5), 742–768 (2018)
12. Biancardo, S.A., Oreto, C., Viscione, N., Russo, F., Ausiello, G., Dell'Acqua, G.: Heritage-BIM approach for Historical Urban Pavements: the case study of Via Duomo in Naples. In: Proceedings of the Transportation Research Board 100th Annual Meeting, Virtual Event, TRBAM-21-00817 (2021)
13. Banfi, F., Brumana, R., Stanga, C.: Extended reality and informative models for the architectural heritage: from Scan-to-BIM process to virtual and augmented reality. Virtual Archaeol. Rev. **10**(21), 14–30 (2019)
14. Grilli, E., Bruno, N., Autelitano, F., Roncella, R., Giuliani, F.: Automatic detection of stone pavement's pattern based on UAV photogrammetry. Autom. Constr. **122**, 103477 (2020)
15. Autelitano, F., Garilli, E., Giuliani, F.: Criteria for the selection and design of joints for street pavements in natural stone. Constr. Build. Mater. **259**(30), 119722 (2020)
16. Chiarabrando, F., D'Andria, F., Sammartano, G., Spanò, A.: UAV Photogrammetry for Archeological Site Survey. 3D Models at the Hierapolis in Phrygia (Turkey). Virtual Archaeol. Rev. 9(18) (2017)
17. Biancardo, S.A., Russo, F., Veropalumbo, R., Vorobjovas, V., Dell'Acqua, G.: Modeling roman pavements using heritage-BIM: a case study in Pompeii. Baltic J. Road Bridge Eng. **15**(3), 47–46 (2020)
18. Knapton, J.: The Romans and their roads – the original small element pavement technologists. In: Proceedings of 5th International Concrete Block Paving Conference, Tel-Aviv, Israel, pp. 17–52 (1996)
19. Garilli, E., Autelitano, F., Giuliani, F.: A study for the understanding of the Roman pavement design criteria. J. Cult. Heritage **25**, 87–93 (2017)
20. Poehler, E.E., Crowther, B.M.: Paving pompeii: the archaeology of stone-paved streets. Am. J. Archaeol. **122**(4), 579–609 (2019)
21. Capone, M., Lanzare, E.: SCAN-TO-BIM vs 3D ideal model. In: The International Archives of the Photogrammetry, Remote Sensing and Spatial Information Sciences, XLII-2/W9, pp. 219–226 (2019)
22. De Luca, M., Mauro, M., Russo, F., Dell'Acqua, G.: Before-after freeway accident analysis using cluster algorithms. Procedia Soc. Behav. Sci. **20**, 723–731 (2011)

Issue on Movement Stability of Three Sections Trailer Bus Train

Volodymyr Sakhno[1], Igor Murovanyi[2], Viktor Poliakov[1], and Valerii Dembitskyi[2(✉)]

[1] National Transport University, Kyiv 01001, Ukraine
[2] Lutsk National Technical University, Lutsk 43018, Ukraine
dvm2@meta.ua

Abstract. The article considers the stability of three sections passenger bus train that consists of three buses of the same type, during its operation in BRT system, which is usually used by the buses of particularly large capacity with the total length of up to 25 m. The efficiency in operation of such buses closely deals with the passenger traffic, which may be changed many times during the day. Therefore, a bus train that consists of two (or three) buses or trolleybuses, which operate in a clutch, the passenger capacity of which is similar to the paired buses and trolleybuses, may be promising. A three sections bus train operates during rush hours, and during the off-peak period - each bus separately (it is possible to park one or two buses on a dedicated site). The indices of cross-section stability and the factors that influence on their numerical values are defined in order to generalize the scheme.

Keywords: Bus · Metro bus · Stability · Angle of rotation · Angle of assembly · Layout · Speed · Acceleration

1 Relevance of Research

Within the International Summit of Mayors (IMC), which took place on October 9, 2018 in Kyiv, Daimler proposed that Ukrainian cities implement BRT (Bus Rapid Transit) transport system, which already operates in 200 cities around the world. On BRT line, as a rule, there are the buses of especially large capacity (18 or 22, 24, 25 m) that run on a separate (dedicated) lane with small intervals [1]. In addition, the efficiency of such buses is closely related to passenger traffic, which can change many times during the day. Therefore, a bus train that consists of two (or three) buses or trolleybuses, which operate in a clutch, the passenger capacity of which is similar to the paired buses and trolleybuses, may be promising. A three sections bus train operates during rush hours, and during the off-peak period - each bus separately (it is possible to park one or two buses on a dedicated site). The feasibility in establishment of trailer bus train is also confirmed by German company Göppel Bus. So, now, in the XXIst century, they have taken and designed the most common trailer. The trailer has the length of 8,350 mm or 11,090 mm, and together with the bus it formed a bus train with the length from 17.7 to 22.65 m, which, depending on the passenger flow, may be operated in different modes, as a bus train and as a regular bus [2]. Along with the

O. Prentkovskis et al. (Eds.): TRANSBALTICA 2021, LNITI, pp. 140–150, 2022.
https://doi.org/10.1007/978-3-030-94774-3_14

undeniable advantages of three-section articulated buses and trolleybuses, they also have some disadvantages - worse maneuverability and stability, compared to two-section ones. The peculiarity in design of functional systems that ensure the stability and maneuverability of any vehicle, including passenger trains, hereinafter is referred to as MVs, are the parallel processes of design, optimization and modeling of their movement dynamics, taking into account the multicriteria of sometimes contradictory tasks [3]. The characteristics of stability and maneuverability of MVs, as it is known, are determined by combination of operational, mass-geometric and design parameters of its modules (for a passenger bus train, these are three buses that operate in the clutch). In common case, the desired ratios of these parameters in terms of stability and maneuverability, even for the same vehicle in the range of operating loads and speeds are different. As a result, there is a difficulty at the early stages of MTs to obtain the accurate design parameters and quantitative indicators on the criteria of stability and maneuverability [3]. Since the curvilinear motion of MVs determines the design of trailer link control systems, at the first stage it is necessary to determine the parameters of the vehicle at circular motion [3, 4] and to perform various maneuvers [5]. At the second stage, the verification of parameters, being received before is held as to satisfaction of stability indices both in rectilinear movement, and when it makes various maneuvers. The selection and optimization of MVs' parameters for the whole range of operating speeds and loads requires the presence of differential equations of motion. Such equations with various degrees of complexity are given in [6, 7] etc. For two-section articulated bus trains, the equations to determine the maneuverability and stability indices are given in [8, 9], and for three sections ones - in [10–13]. In the cited works, the influence of design and operational factors on the value for critical speed of rectilinear motion is analyzed, in particular, it is established that its increase can be achieved by increasing the base of the bus, the first and second trailer, as well as the bus mass and drag coefficients, axles of buses and axles of trailers. At the same time, increasing the distance from the center of mass to the rear axle of the bus, increasing the distance from the center of mass to the coupling point of the bus with the first trailer, increasing the weight of trailers and drag coefficient of the steered bus axle reduces the critical speed. At the same time, these factors can affect differently at creation of three-sections trailer bus trains. The purpose of this paper is to determine the stability of the bus train that consists of three standard middle-class buses (for example, MAZ-206 buses).

2 Results of Research

[5] provides with the system of differential equations of motion for a three-section bus train that consists of a bus and two trailers both with spaced and converging axles. This system can also be used for a bus train that consist of three buses with the same type, [5].

The stability of the road train in research is considered, as a rule, as the plane-parallel movement of its links. It is believed that the normal reactions of the support surface on the wheels of the starboard and port sides are the same. In this case, the stability of the movement is considered for a flat model train. However, such approach to the extreme modes of movement can retraction to significant change in reactions of support surface to the wheels of the bus train. Therefore, it is necessary to consider the movement of the bus train in the longitudinal vertical and transverse planes. The connection between the sprung and unsprung masses of the actual bus construction is carried out by elastic and damping devices, and between unsprung masses and the road - through the tires, which are characterized both by elastic and damping properties. In the case under consideration, the forces of interaction in the traction couplings do not affect the redistribution of loads on the sides of the links in bus train. Therefore, a rather complex system - a three-link trailer can be considered as three systems - the first bus, the second and third buses, which move on their own. It is also assumed that the roll axis of each link is parallel to the support surface, and the movement of the links in bus train in the vertical plane at the corners of the roll affect the lateral movement, primarily by changing the vertical loads on the wheels, thereby changing the vertical reactions.

Let's make the equation of movement for buses in the cross plane, Fig. 1.

– for roll of body in the first bus

$$\begin{aligned} I_{X0}\ddot{\gamma} = & -Y_1^{(1)}h - 2Hq_{11}[H + \frac{\varepsilon}{2}(\cos\theta_0 + \cos\theta_0^{/})tg\gamma] - 2q_{21}(H+\varepsilon)^2 tg\gamma \\ & + z_0[(Y_{11}\cos\theta + Y_{11}^{/}\cos\theta^{/} - X_{11}\sin\theta_0 - X_{11}^{/}\sin\theta_0^{/}) + Y_{21} + Y_{21}^{/})] + \varepsilon\{F_a^o(\cos\theta_0 \\ & - \cos\theta_0^{/}) + [q_{11}z_a^o(\cos\theta_0 - \cos\theta_0^{/}) + Z_{a0}^{/}\cos\theta_0^{/} - Z_{a0}\cos\theta_0]\}; \end{aligned}$$

– for roll of body in the second bus

$$\begin{aligned} I_{X1}\ddot{\gamma}_1 = & -Y_1^{(1)}h - 2H_1q_{12}[H_1 + \frac{\varepsilon_1}{2}(\cos\theta_1 + \cos\theta_1^{/})tg\gamma_1] - 2q_{22}(H_1+\varepsilon_1)^2 tg\gamma_1 \\ & + z_2[(Y_{12}\cos\theta_1 + Y_{12}^{/}\cos\theta_1^{/} - X_{12}\sin\theta_1 - X_{12}^{/}\sin\theta_1^{/}) + Y_{22} + Y_{22}^{/})] + Y_2^{(1)}h_2 + \varepsilon_1\{F_{a2}^o(\cos\theta_1 - \cos\theta_1^{/}) \\ & + q_{12}[z_{a2}^o(\cos\theta_1 - \cos\theta_1^{/}) + Z_{a2}^{/}\cos\theta_1^{/} - Z_a\cos\theta_1]\}; \end{aligned} \tag{1}$$

– for roll of body in the first bus

$$\begin{aligned} I_{X2}\ddot{\gamma}_2 = & -Y_2^{(1)}h_2 - 2H_2q_{13}[H_2 + \frac{\varepsilon_2}{2}(\cos\theta_2 + \cos\theta_2^{/})tg\gamma_2] \\ & - 2q_{23}(H_2+\varepsilon_2)^2 tg\gamma_2 + z_3[(Y_{13}\cos\theta_2 + Y_{13}^{/}\cos\theta_2^{/} - X_{13}\sin\theta_2 \\ & - X_{13}^{/}\sin\theta_2^{/}) + Y_{23} + Y_{23}^{/})] + \varepsilon_1\{F_{a3}^o(\cos\theta_2 - \cos\theta_2^{/}) \\ & + q_{13}[z_{a3}^o(\cos\theta_2 - \cos\theta_2^{/}) + Z_{a3}^{/}\cos\theta_2^{/} - Z_{a3}\cos\theta_2]\}; \end{aligned}$$

Equation system (1) uses the following symbols:

H_i – the distance between the elastic elements of suspension in bus axle; i = 1…3;
I_{x0}, I_{x1}, I_{x2} – moments of buses inertia, relative to the longitudinal axis;

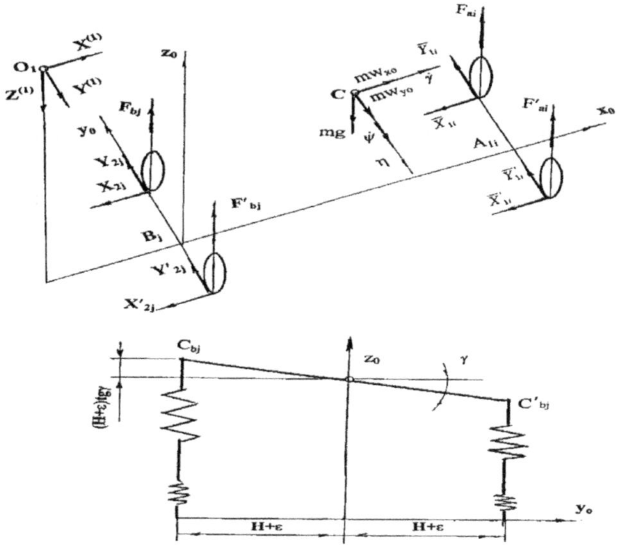

Fig. 1. Diagram of forces that act on the bus in the transverse plane.

ε_i – displacement of the center of bus mass in the transverse plane under action of lateral forces;
$\theta_0, \theta_0', \theta_1, \theta_1', \theta_2, \theta_2'$ – angles for rotation of external and internal wheels of buses;
$Y_1^{(1)}, Y_2^{(1)}$ – side reactions in traction couplings of buses;
h, h_2 – the distance from the support surface to the axis of the coupling device;
q_{11}, q_{12}, q_{13} – reduced rigidity of bus suspensions;
$F_{ai}^o\ F_{bj}^o$ – vertical reactions of the i-th front and j-th rear supports on the bus frame;
$Y_{11}, Y_{11}', X_{11}, X_{11}', Y_{21}, Y_{21}', X_{21}, X_{21}', Y_{21}, Y_{21}', X_{21}, X_{21}'$ – lateral and longitudinal reactions of the bearing surface on the axles of buses;
$Z_{a0}, Z_{a2}, Z_{a3}, Z'_{a0}, Z'_{a2}, Z'_{a3}$ – vertical reactions of the bearing surface on the left and right wheels of buses;
$z_{a0}, z_{a2}, z_{a3}, z'_{a0}, z'_{a2}, z'_{a3}$ – deformation in suspension of the right and left wheels in buses;
z_0, z_2, z_3 – height of the center of buses mass

The solution of equation system (2) allowed determining the roll angles of buses:

$$\gamma = \frac{\gamma^{(0)} 0,5H^2(q_{A1}+q_{B1}) + F_0(Z_o+h)}{-F_0\varepsilon + 0,5H^2(q_{A1}+q_{B1})};$$

$$\gamma_2 = \frac{\gamma_o^{(2)} 0,5H_2^2(q_{A2}+q_{B2}) + F_2(Z_o^{(2)}+h_2)}{-F_2\varepsilon_2 + 0,5H_2^2(q_{A2}+q_{B2})};$$

$$\gamma_3 = \frac{\gamma_o^{(3)} 0,5H_3^2(q_{A3} + q_{B3}) + F_3(z_3 + h_3)}{-F_3\varepsilon_3 + 0,5H_3^2(q_{A3} + q_{B3})}. \quad (2)$$

where F_0, F_2, F_3 – lateral force that acts in the center of mass of the buses; q_{Ai}, q_{Bi} – reduced rigidity in the front and rear suspensions of buses.

Dynamic components of vertical reactions in supports due to roll angles γ and $\gamma^{(0)}$, γ_2 and $\gamma^{(2)}$, γ_3 and $\gamma^{(3)}$ (loading and unloading), are defined as it follows:

– for the left side

$$\Delta G_1 = q_1(\gamma - \gamma^{(o)})H/2;\ \Delta G_2 = q_2(\gamma_2 - \gamma_o^{(2)})H_1/2;\ \Delta G_3 = q_3(\gamma_3 - \gamma_o^{(3)})H_3/2;$$

– for the right side

$$\Delta G_1^1 = q_{B1}(\gamma - \gamma^o)H/2;\ \Delta G_2 = q_{B2}(\gamma_2 - \gamma_o^{(2)})H_1/2;\ \Delta G_3 = q_{B3}(\gamma_3 - \gamma_o^{(3)})H_3/2;$$

Therefore, the dynamic loads, taking into account the onboard redistribution, will take the form:

$$\begin{aligned} G_1 &= G_1^o - \Delta G_1;\ G_2 = G_2^o - \Delta G_2;\ G_3 = G_3^o - \Delta G_3; \\ G_1^1 &= G_1^o - \Delta G_1^1;\ G_2^1 = G_2^1 - \Delta G_2^1;\ G_3^1 = G_3^1 - \Delta G_3^1. \end{aligned} \quad (3)$$

Equations (2, 3) are the basis to calculate the values of roll angles, as well as wheel loads on one side of the bus. As an example,

Figure 2a shows the graphs of roll change, and .

Figure 2b – graphs of changes in the loads on the wheels of one side of the bus in circular motion with radius of R = 50 m at speed of 15 m/s.

The obtained values of roll angles and loadings (unloading) are the basis to calculate the correction for resistance coefficients of bus wheels and to determine the lateral forces that act on the bus wheels, further to calculate the stationary motion parameters, critical rectilinear velocity, radius.

The lateral force on the wheels of the axles in bus train can be represented as [14]:

$$Y_i = k_i \delta_i. \quad (4)$$

The studies of dependence (5) in [12] found that the most acceptable for calculations can be considered as D.A. Antonov's dependence.

The benefits of D.A. Antonov's dependence is able to take into account the changes in the amount of diversion when oscillating and tilting the wheels to the bearing surface, when the body rolls, etc., despite a number of inconveniences due to the presence of a large number of adjustment factors. In addition, with increasing diversion angles, the error of the calculated data in comparison with the experimental ones is increased significantly, which also requires the adjustment.

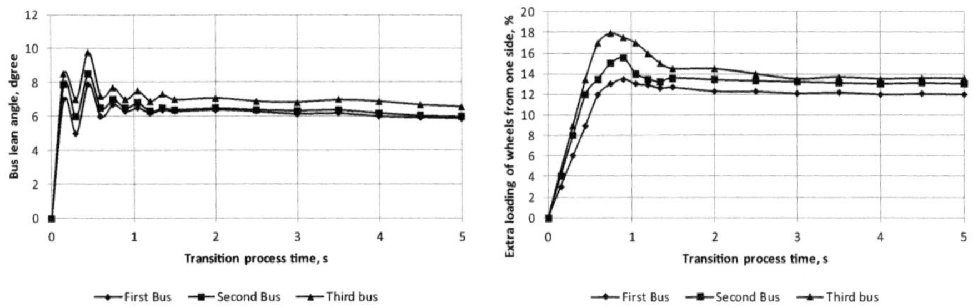

a) b)

Fig. 2. Change in roll angles and loading of the wheels from one side during the circular motion of buses.

D.A. Antonov's dependence is recorded as:

$$Y = qk_{yo}\delta, \tag{5}$$

where k_{yo3} – extreme value of coefficient k_{yo} depending on the normal reaction in contact of the wheel with the support surface; δ – the angle of the side of the wheel; q – coefficient that takes into account the change in the extreme value of coefficient $k_{YO_Э}$ depending on the modes of wheel movement, its installation and the properties of support surface.

The coefficient q for operation conditions of bus train, can be represented as:

$$q = q_N q_T q_\phi, \tag{6}$$

where q_N – the coefficient of influence on diversion of normal support reactions redistribution; q_T – coefficient of influence of traction and braking forces; q_φ – coefficient for quality of road surface.

The coefficients, included in expression (6), are determined by the following dependences [12], for vertical loading

$$q_N = 1 - 0{,}6\left(\frac{\Delta R_{ZЭ}}{R_{ZЭ}}\right)^2 + 0{,}4\left(\frac{\Delta R_{ZЭ}}{R_{ZЭ}}\right)^3 - 0{,}1\left(\frac{\Delta R_{ZЭ}}{R_{ZЭ}}\right) - 0{,}1\left(\frac{\Delta R_{ZЭ}}{R_{ZЭ}}\right)^2 \operatorname{sgn}\frac{\Delta R_{ZЭ}}{R_{ZЭ}}, \tag{7}$$

where $\Delta R_{ZЭ} = R_Z - R_{ZЭ}$, R_Z– normal reaction of the bearing surface on the wheels of the road train; $R_{ZЭ}$ – normal reaction of the support surface, which corresponds to the extreme value for coefficient of resistance to the side retraction.

The normal response to the wheels of bus train, taking into account the roll of the body can vary from zero to twice the nominal value.

The normal reaction of the support surface corresponding to the extreme value of the resistance to the lateral retraction is defined as [12]:

$$R_{Z э} = R_{Z ном} - 0{,}0726\left(\frac{HD}{B}\right)^2\left(\frac{d}{B}\right)^2\left(\frac{n_{сл}+8}{n_{сл}}\right)H(D^2-d^2)^2 \times \left[3{,}75R_{Z ном} + 15 - 1000\left(\frac{n_{сл}}{n_{сл}+10}\right)DB^2\sqrt{\frac{B}{H}}\right], \quad (8)$$

where $R_{Z\text{НОМ}}$ – normal reaction of the bearing surface that corresponds to the nominal air pressure in the tire according to the manufacturer's catalog, H; H – tire profile height, m; D, d – outer and inner diameters of the tire, m; B – tire profile width, m; $n_{сл}$ – the number of layers in tire cord.

Correction coefficient qф is defined by dependence [12]:

$$q_\varphi = \frac{arctg\left[\frac{q_N k_{YOэ}\pi}{2\varphi R_Z}(\delta - 0{,}026)\right]}{\frac{q_N k_{YOэ}\pi}{2\varphi R_Z}(\delta - 0{,}026)}, \quad (9)$$

where φ – the adhesion coefficient of the wheel to the support surface;

$$k_{уоэ} = 1000\left(\frac{n_{сл}}{n_{сл}+10}\right)DB^2\sqrt{\frac{B}{H}} + \frac{6{,}2}{d^3\left(n_{сл}+8\right)}\left[\frac{DHB}{\left(D^2-d^2\right)d}\right]p_{ш}, \quad (10)$$

where $p_{ш}$ – tire pressure.

MAZ-206 buses are equipped with 245/70R19.5 tires. For this type of tires,

Figure 3a shows the dependence of the influence coefficient on removal of the wheel of the bus and redistribution of normal support reactions qN and the coefficient for quality of the road surface $q_{ф}$, and in .

Figure 3b – the influence coefficient of traction and braking forces q_T.

As it follows from.

Figure 3, the correction coefficients of influence on resistance of redistribution of normal support reactions q_N, the quality of road surface $q_{ф}$ and the influence of traction and braking forces q_T from the normal reaction of the support surface on the bus wheel varies in a very narrow range – from 1.0 to 0.935, but this change is necessary to be taken into account in calculations of bus train stability when integrating the movement equations of bus train.

The stability of bus train in different modes can be judged by magnitude of the lateral speeds and accelerations that act in the center of mass at individual links. These velocities and accelerations were determined by solving the original nonlinear system of differential equations [5]. The magnitude of accelerations is determined by mode of movement. The mode of movement was considered stable if the acceleration of any bus did not exceed 4.5 m/s^2.

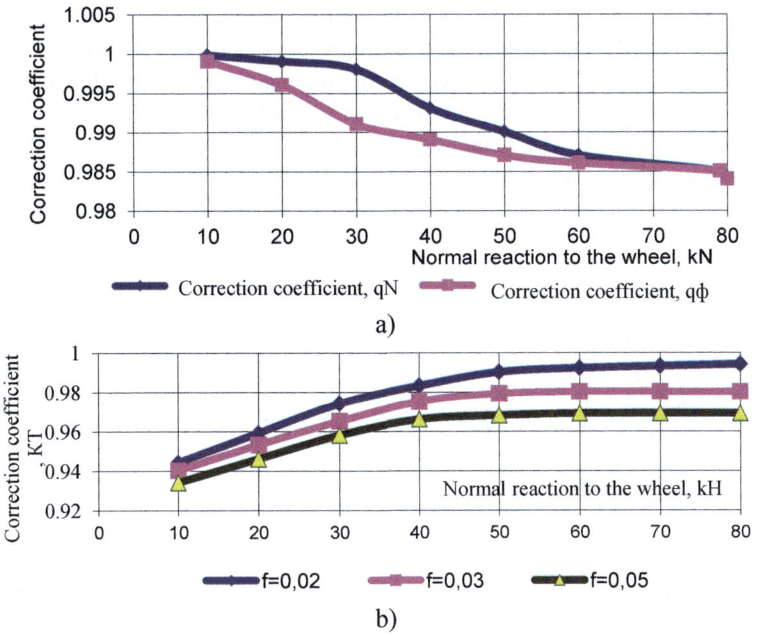

Fig. 3. Dependence of correction factors on resistance of retraction q_N, q_ϕ and q_T on the normal reaction of the support surface on the wheel.

Figure 4 as an example shows the results from calculation of magnitude of the lateral acceleration that act in the center of mass of all buses when performing their maneuvers “jerk of the steering wheel” and “shift” at speed of 15 m/s^2.

As it follows from the figures, the stability of the bus train when performing the maneuver “jerk of the steering wheel” is on the verge of permissible value. The lateral acceleration of the third bus is 4.3 m/s^2 with a permissible value of 4.5 m/s^2.

The solution of the linearized equations system [5] allowed to determine the critical speed of rectilinear movement at three-link trailer bus train, i.e. 29.78 m/s^2, which is more than allowed for buses in BRT system (25 m/s^2), i.e. in this case the stability of the bus train is provided. Thus, it can be considered expedient to design a three-link trailer.

3 Conclusions

1. The prospects to use a three-link bus train of three buses during its operation in BRT system are shown.
2. The spatial model of the three-link trailer bus train is designed. The roll angles and loading (unloading) of wheels in one board of buses at movement by circular trajectory are defined. The obtained dependences are the basis for calculation of correction factors to determine the lateral forces on the wheels of bus trains.

3. By integrating the initial system of equations, the lateral velocities and lateral accelerations in the center of mass at each bus are determined. It is established that when performing the maneuver "jerk of the steering wheel" at speed of 15 m/s^2, the stability of the bus is on the limit. The maximum lateral accelerations of the third bus are on the limit. The lateral acceleration of the third bus is 4.3 m/s^2 with a permissible value of 4.5 m/s^2.

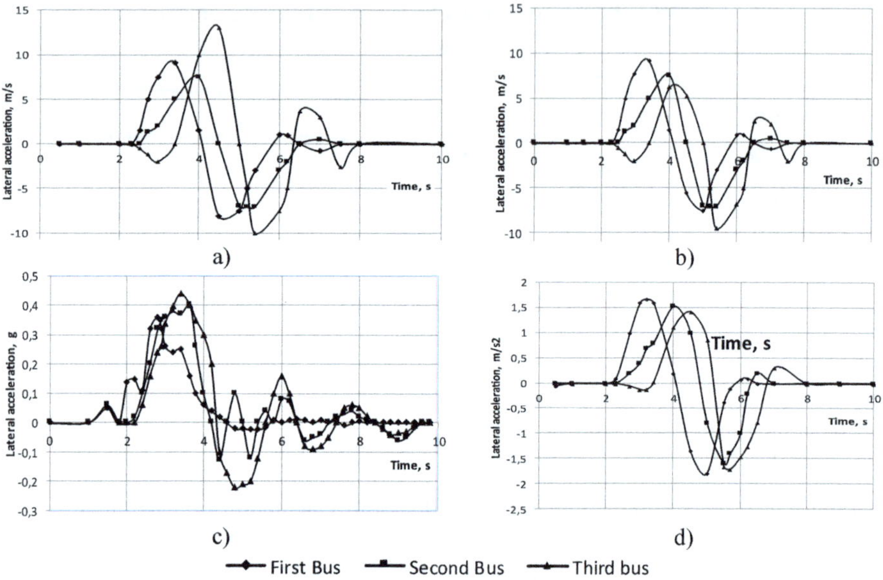

Fig. 4. Lateral speed and lateral acceleration that act in the center of buses mass, when performing the maneuver "shift" (a, c) and "jerk of the steering wheel" (b, d).

4. The critical speed of rectilinear movement of a three-link trailer bus train was determined, which amounted to 29.78 m/s, which is more than allowed for buses in BRT system (25 m/s), i.e. the stability of the bus train is ensured.

4 Discussion of Research Results

Today, BRT (Bus Rapid Transit) transport system operates in 200 cities around the world. On BRT line, as a rule, the buses of especially big capacity (18 or 22, 24, 25 m) run on a separate (allocated) lane with small intervals run. The efficiency in operation of such buses is closely related to passenger traffic, which can change many times during the day. Therefore, a bus train that consists of two (or three) buses or trolleybuses operate in a clutch, the passenger capacity of which is similar to articulated buses and trolleybuses, may be promising. During rush hours, a three-link bus train operates, and during the off-peak period - each bus separately (it is possible to park one or two buses on a dedicated site). Another indisputable advantage of such bus train is that the engine power of two buses is enough for its movement with the allowed maximum speed of 25 m/s, i.e. the third bus can be a simple two-axle trailer. This will significantly improve the economic and environmental performance of such bus train but this issue still needs to be addressed.

References

1. LiveJournal Home Page. https://griphon.livejournal.com/222403
2. Verkehrsgemeinschaft Osnabrück VOS Home Page. https://www.vos.info/ueber-uns/geschichte.html
3. Sakhno, V., Murovany, I., Sharai, S., Seleznev, V.: Comparative evaluation of maneuverability of large and extra large class buses Mobile machines. International scientific conference, Kaunas, Lithuania. Aleksandras Stulginskis University, pp. 38–47. (2017)
4. Sakhno, V.P., Marchuk, M.M., Marchuk, R.M.: Study of long haul truck movement along the curvilinear trajectory while steering a carryall semi-trailer - container by braking the wheels of one axle. INMATEH –. Agric. Eng. **52**(2), 107–112 (2017)
5. Sakhno, V.P., Bilichenko, V.V., Polyakov, V.M., Omelnitsky, O.E.: Maneuverability of metro buses. In: Bulletin of Mechanical Engineering and Transport: scientific journal/ Ministry of Education and Science of Ukraine, Vinnytsia National Technical University - Vinnytsia: VNTU , No. 2(6), pp. 131–140 (2018)
6. Sakhno, V.P., Verbytskyi, V.G., Bondarenko, A.Ye., Englezi, A.O.: The equation for motion of model of a four-link trailer road train. Motorway of Ukraine. Special issue. Bulletin of the Northern Scientific Center of TAU. Issue 10, pp. 117–120 (2007)
7. Sakhno, V.P., Bondarenko, A.Ye., Glinchuk, V.M., Englezi, O.A.: Equation of motion of a three-link road train of the "B-Double" type Motorway of Ukraine. Special issue. Bulletin of the Northern Scientific Center of TAU. Issue 12, pp. 164–169 (2009)
8. Sakhno, V.P., Poljakov, V., Myrovanyi, I., Seleznov, V.: Determination of movement stability of specialty large class hybrid bus with active trailer. INMATEN - Agricultural engineering, vol. 49/No2, pp. 107–118 (2016)
9. Sakhno, V.P., Polyakov, V.M., Murovany, S., Seleznyov, V.E.: Prior to the preliminary assessment of the stability of the hybrid bus of a particularly large class with an active trailer. Bulletin of the Kharkiv National Automobile and Road University. Vip.75, pp. 75–82 (2016)
10. Sakhno, V., Gerlici, J., Poliakov, V., Kravchenko, A., Omelnitsky, O., Lask. T.: Bus train motion stability in BRT system. XXIII Polish-Slovak Scientific Conference Machine Modeling and Simulation. MMS 2018.- Book of abstracts, September 4–7, Rydzyna Poland, pp. 49–57 (2018)
11. Sakhno, V., Omelnytskyi, O., Poliakov, V.: To issue about maneuverability of the bus. Systemy and means of self-transport. Badania, construction and technology. Selected items. - Monograph No. 12, Series: Transport. - Rzeszow, pp. 101–108 (2018)
12. Sakhno, V.P., Polyakov, V.M., Omelnytsky, O.Ye.: Maneuverability of a three-link metro bus. SCIENTIFIC PAPERS of the International scientific-practical conference "New technologies of motor transport development". - Kharkiv National Automobile and Road University. Research forensic center of the Ministry of Internal Affairs of Ukraine, pp. 305–308 (2018)

13. Sakhno, V., Gerlici, J., Poliakov, V., Kravchenko, A., Omelnitsky, O., Lask, T.: Bus train motion stability in BRT system. XXIII Polish-Slovak Scientific Conference Machine Modeling and Simulation. MMS 2018. Book of abstracts, September 4–7, Rydzyna Poland, p. 49 (2018)
14. Verbitsky, V.G., Sakhno, V.P., Kravchenko, A.P., Kostenko, A.V., Danilenko. A.Ye.: Cars. Stability: Monograph. Lugansk: Knowledge Publishing House, 176 p. (2013)

Research of Vehicle Airbag Non-deployment Cases

Robertas Pečeliūnas[1(✉)] and Tomas Pasaulis[2]

[1] Vilnius Gediminas Technical University, J. Basanavičiaus 28, LT 03224 Vilnius, Lithuania
robertas.peceliunas@vilniustech.lt
[2] Forensic Science Centre of Lithuania, Lvovo 19a, LT 09313 Vilnius, Lithuania

Abstract. In this paper, the conditions, criteria and the algorithms controlling the deployment of vehicle airbags are analysed. The sequence of the algorithm and specific thresholds used in airbags control are analysed as well. The means are provided for airbags testing to ensure their reliability and proper functioning. The types of road accidents including specific cases are described in terms of possible airbag non-deployment. The computer simulations of critical impact scenarios with a finite element model of light vehicle are done in LS DYNA software. The analysis of achieved impact force results show that even relatively small speed (30 and 50 km/h) causes different conditions for airbag deployment.

Keywords: Airbag deployment · Frontal impact · Vehicle collision · Passive safety

1 Introduction

The World Health Organization announces about 1.35 mln. road deaths and from 20 to 30 mln. injuries every year [1]. The 21% reduction of fatal road accidents was estimated in Europe Union (EU) during last eight-year period [2]; however, road accidents is still a relevant problem worldwide.

Airbags are one of the most important and complicated system of vehicle passive safety and the engineers are responsible for its relevant design and safety performance during road accident. According to U.S. National Highway Traffic Safety Administration (NHTSA) 2,790 lives were saved just because of front airbags in 2017 and more that fifty thousand lives since 1978 when the front airbags became widely used in road vehicles [3]. The front airbags reduce the risk of death for drivers by 29%, and by 32% for front passengers over 13 years of age [4]. According to NHTSA, the combination of front airbag and seatbelt reduces the severity of death during frontal impact by 61% comparing with 50% using only a seatbelt. Side airbags protects only mainly a head, but still reduce the severity of by 37% during side impact, and the side airbags that protect only the torso area reduce the severity of by 26% [5]. However, the cases of non-deployed airbags during road accident are also possible and sometimes unfortunately happens causing deaths or injuries.

O. Prentkovskis et al. (Eds.): TRANSBALTICA 2021, LNITI, pp. 151–160, 2022.
https://doi.org/10.1007/978-3-030-94774-3_15

The objective of this research is to analyse the causes and algorithms of airbag deployment, and to do a computer simulation of vehicle-obstacle impact in different moving cases with the purpose of estimating the deployment capabilities of car airbags.

2 Conditions, Criteria and Algorithms for Airbag Deployment

Vehicle airbag systems must function properly and be reliable; therefore, various analyses including different impact cases of computer simulations using Madymo, LS DYNA and other software are performed [6, 7]. A real driving tests of different vehicle driving modes also tests in test benches are organized for development of airbag systems.

The deployment of vehicle airbags during the impact depends on few thresholds. For some vehicles, the airbags are designed to deploy during a frontal collision with an undeformed wall when the speed of the vehicle is at least 23 km/h. Airbag deployment limits are generally governed by various vehicle design regulations, for example, U.S. law requires airbags to deploy at a speed of at least 22.5 km/h in the event of a frontal collision [8]. According to the NHTSA, the front airbags must deploy during a frontal or near-frontal impact with an undeformed wall at a speed of 12.8 to 22.5 km/h what is equivalent to a collision with a stationary vehicle at a speed of 26–45 km/h [3]. Seatbelts are usually sufficient to protect the driver and passengers at low collision speeds and the deployment of the airbags is unnecessary; however, at higher collision speeds, only the combination of seat belts and airbags can effectively protect [9].

The initial technologies of the airbags have been designed to deploy in all cases where certain thresholds were exceeded during a collision, which is not applied for now. It's been a while as the deployment of the airbags is adapted to the nature of the collision, its severity, the size and weight of the occupant, as well as the use of the belts and the position of the occupant in relation to the airbag. Vehicle body sensors transmit the signals to the airbag control unit, which determines in which mode the airbags must be deployed and whether they need to be deployed at all [8].

Advanced airbags have a dual-level inflation system which regulates the force of the deployment depending on the severity of the collision. This is necessary because passengers move forward at different speeds and for different periods of time; therefore, it is important that the airbag inflates at the right time and with the right force. All airbag systems are designed so that they do not deploy during sharp braking or when driving on uneven and potholed roads. The accelerations during these driving cases do not reach the thresholds of airbags deployment [10]. Also the deactivation of front airbag in cases of children is possible [8].

The airbag control unit receives a continuous signal from each sensor or MEMS (micro-electro-mechanical systems) and records data for a period of time after a specific event fixed [11]. The function of the vehicle's airbag system is to record the data about an event close to the deployment of the airbags which is also defined as Event Data Recorder (EDR). In the case of an airbag deployment, the collision data is permanently stored in the control unit. The data can be erased after 250 vehicle start cycles or when another event occurs close to the deployment of the airbags. The airbag control unit records data about 5 s before the collision and shortly after the collision [12].

The collision algorithmic process is performed in sequence shown in Fig. 1. Initially, the acceleration signals received from the sensors are processed by a signal filter and if the initial condition is met, the algorithm is excited. Normally, the algorithm does not work continuously, when the signal is received, the algorithm is excited within 4–5 ms. Then, if the results of the algorithmic solutions meet the various set thresholds, the airbags are deployed [13]. The algorithm usually has a very short time to process the received signal and make airbag deployment decisions; usually around 10–50 ms depending on collision type. Algorithms have to perform two main tasks: i) evaluate the received information and perform calculations based on airbag deployment, ii) determine the appropriate time for the airbags to deploy [9].

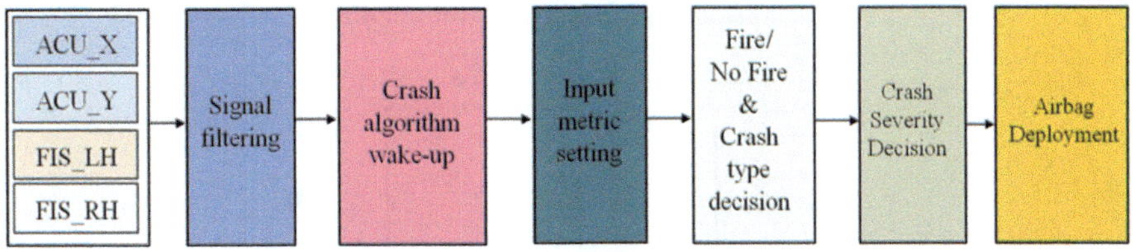

Fig. 1. Crash algorithm block diagram [13].

The control unit uses the appropriate algorithm to determine the severity of the collision by estimating one or more of the variables listed in Table 1 [11].

Table 1. Variables used in algorithms of airbag system [11].

Variable	Expression
Acceleration	$a = \frac{dv}{dt}$
Velocity	$v = \int adt = \frac{dx}{dt}$
Displacement	$x = \int vdt = \iint adtdt$
Energy density	$e = \int_{x_0}^{x} adx = \int_{v_0}^{v} vdv = \frac{1}{2} \cdot (v^2 - v_0^2)$
Energy	$E = \frac{1}{2} \cdot m(v^2 - v_0^2)$
Power	$\overline{p} = \frac{dE}{dt} = mva$
Power density	$p = \frac{\overline{p}}{m} = va$
Power rate density	$p' = \frac{dp}{dt} = vj + a^2$
Jerk	$j = \frac{da}{dt}$

Almost all algorithms used today are based on certain set thresholds where the acceleration is used mostly and the other variables are calculated from it [9]. Each airbag control unit has variable value limits against which algorithms estimate the need for airbag deployment. For a frontal collision, the longitudinal acceleration limits required to activate the airbags may be different: the lower acceleration limit varies between 4.7–7.5 g, and the upper acceleration limit value varies between 8.2–13.2 g [11]. An example of an algorithm scheme is shown in Fig. 2.

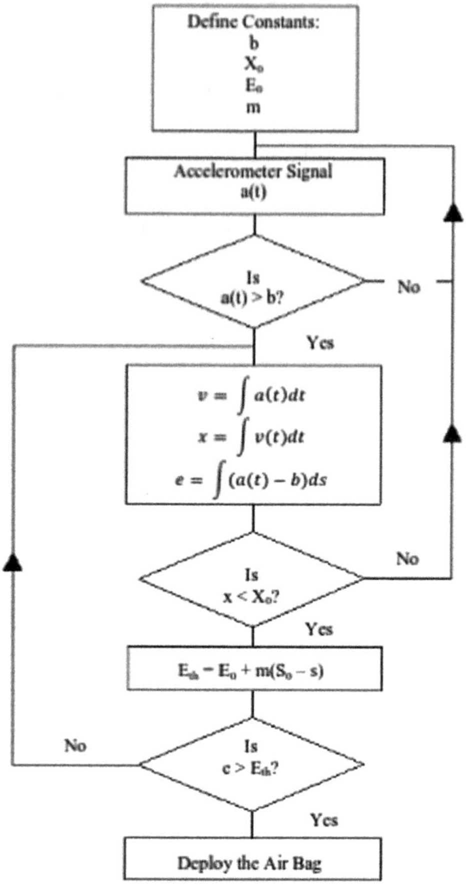

Fig. 2. An algorithm scheme for airbag deployment [11].

The presented algorithm estimates the acceleration (a), velocity (v), displacement (x) and energy density (e). If the acceleration (a) does not exceed a certain acceleration limit (b) during the collision, the algorithm is returned to its initial state. If the limit is exceeded, further calculations are performed (v, x, e). If the displacement (x) is less than the set limit, the limit energy density (E_{th}) is calculated. If the calculated energy density (e) is higher than the calculated limit value (E_{th}), the airbags are deployed, otherwise not deployed.

Airbag deployment schemes vary widely in different patents where the variables of acceleration and its jerk are used while new systems also perform an estimation of the presence of passengers and the position of the passenger in relation to the air bag.

New airbag systems are currently being widely developed to make the decision about the type of the vehicle collision before it happens. The system evaluates information before and during the collision, utilizing vehicle surrounding and impact sensing correspondingly. In this type of algorithm, the control unit independently uses the interface of the two main algorithms between pre-collision and collision cases (Fig. 3) [13].

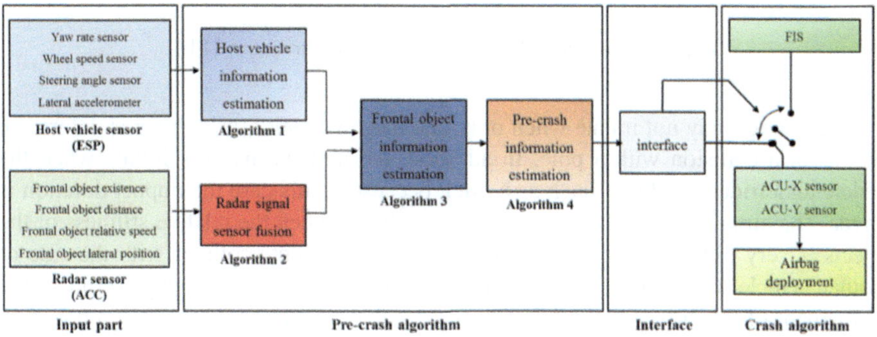

Fig. 3. Block scheme of an algorithm based on pre-collision information [13].

The input data used in the algorithms is derived from vehicle stability control sensors, radar and acceleration sensors. This algorithm is based on obtaining information from a pre-collision algorithm, the purpose of which is to predict the probability of a collision, the time before the collision, information about the type of collision and possible collision scenarios. Based on the information obtained, the airbags can be deployed so that their deployment corresponds to a particular defined collision configuration; however, the interface between the collision algorithm and the pre-collision algorithm is very important in this case. If the performance of the pre-collision algorithm fails, the overall performance of both algorithms deteriorates, unlike using only one, simple collision algorithm [13].

3 Critical Road Accidents Conditions for Airbag Deployment

Real traffic accidents are not regular collisions to steep solid obstacles at a certain set angle as it is evaluated during vehicle safety assessment tests; therefore, it is necessary to analyze cases where the deployment of the airbag is not fully ensured [11]. The deployment of vehicle airbags depends on the type of the collision. Statistically, the most common collisions occur in the front of the vehicle or in the front corner while the impact with the rear is least common (Fig. 4).

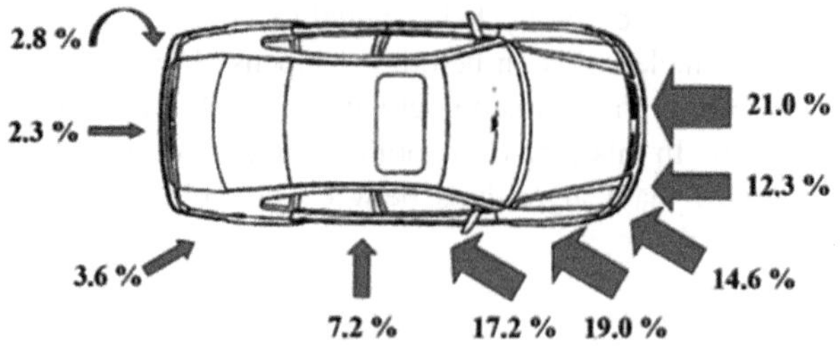

Fig. 4. Statistically based vehicle impact directions [14].

The airbags may not inflate when only one narrow area of the vehicle is deformed, such as in a collision with a pole. In addition, the airbags may not inflate when the vehicle underrides an object, such as a truck because at this case the impact duration is longer. The airbags may not deploy during a collision when the relative stiffness of the objects is very different, such as when the front of a vehicle collides with the side of another vehicle. Collisions that occur at an inclined angle do not always result in the deployment of the airbag, as in this case the deceleration does not occur in a direction parallel to the sensor [11]. These and similar cases are also described by manufacturers. As an example, Fig. 5 shows the cases in the Mazda CX-3 manual that do not allow the airbags to deploy.

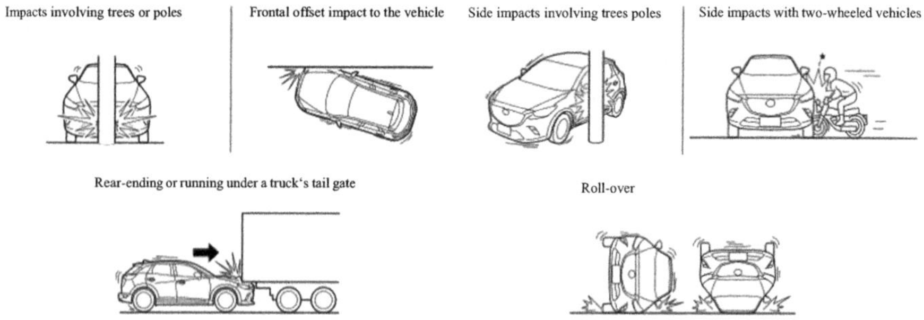

Fig. 5. Cases when the deployment of the airbags is not ensured [15].

Thus, the nature of the collision during an accident is an important factor on which the deployment of the airbags depends. As the main purpose of the airbags is to protect the occupants and the driver, a great attention must be paid to the design of the airbag systems and the arrangement of the system elements avoiding the cases of non-deployed airbags when it can mitigate the consequences of road accident.

4 Computer Simulation of Airbag Deployment

The finite element model of light vehicle Toyota Yaris is used for computer simulations. This model is created in Center for Collision Safety and Analysis of George Mason University (U.S.) with the purpose of various vehicle impact analysis into LS DYNA software (Fig. 6). The model was validated according to frontal impact performed by NHTSA [16]. The model is free available (https://www.ccsa.gmu.edu/models/2010-toyota-yaris/).

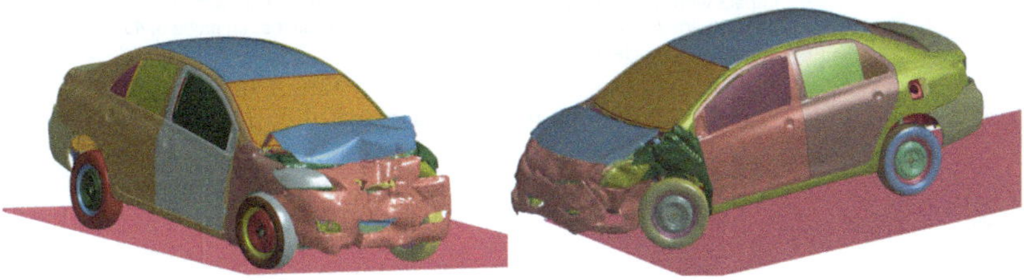

a) b)

Fig. 6. The view of Toyota Yaris vehicle after the impact with a rigid wall at 50 km/h: a) with 50% overlap; b) at an angle of 25°.

The finite element model of Toyota Yaris consists of 378,376 elements and 393,165 nodes. The construction of vehicle body, chassis and interior elements with their physical and mechanical properties are detail designed in this model.

Determining whether the airbags will deploy in a given collision requires having set of previously described acceleration limits. According [11], Toyota uses 5.8–10.2 g acceleration limits for airbag deployment during frontal impact. If the recorded maximum acceleration is over 10.2 g the airbag must deploy; however, the deployment of the airbags is likely if the acceleration is within the specified values.

At this research vehicle speed of 30 km/h and 50 km/h are used for collisions with rigid wall. The duration of computer simulation in LS DYNA was 150 ms. The cases of these impacts is analysed further: i) the impact with full overlap, ii) the impact with 50% overlap, iii) the impact with 25% overlap, iv) the impact at an angle of 45°, v) the impact at an angle of 25°.

5 Results and Discussion

The Toyota Yaris crash simulations provided the forces achieved during the impact by an obstacle collision (Fig. 7 and Fig. 8). These forces causes vehicle body deformations and previously described accelerations.

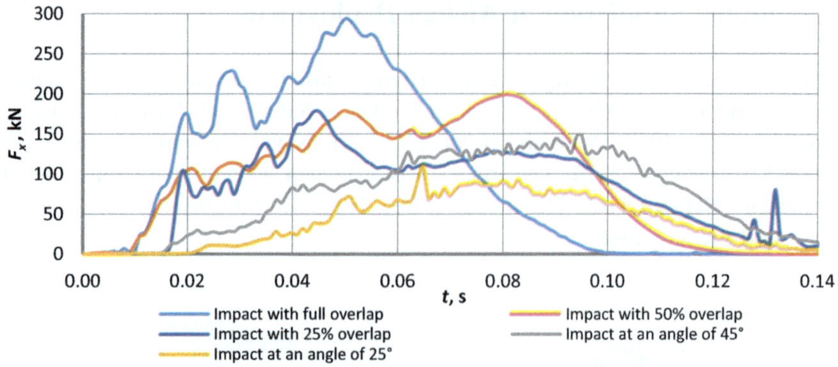

Fig. 7. Forces during the impact at a speed of 30 km/h.

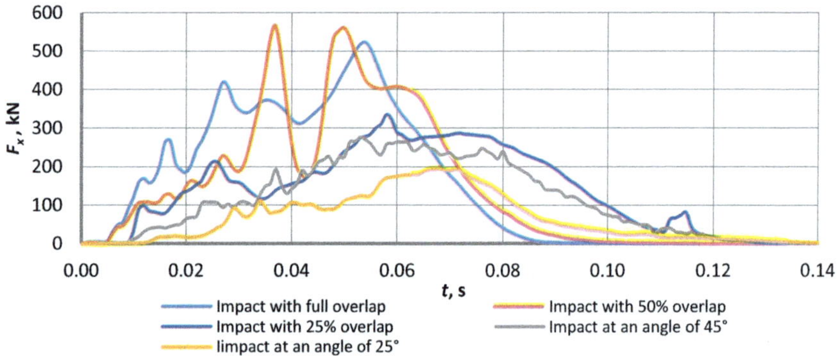

Fig. 8. Forces acting during the impact at a speed of 50 km/h.

The graphs show that the forces generated during a collision depend on the speed of the vehicle and the configuration of the collision. Table 2 presents the maximum values of the forces for simulation of each impact.

Table 2. The maximum values of the forces for simulation of each impact.

The case of impact	Speed before the impact, km/h	Maximum force, kN
The impact with full overlap	30	293.4
	50	522.5
The impact with 50% overlap	30	199.8
	50	561.3
The impact with 25% overlap	30	178.6
	50	334.5
The impact at an angle of 45°	30	147.6
	50	276.6
The impact at an angle of 25°	30	108.5
	50	193.5

It is recognised from the data table, that the maximum force is reached for vehicle impact with 50% overlap at 50 km/h speed, while the lowest force occurs for an impact of 25° angle at 30 km/h speed.

During the simulation of frontal impact with a rigid wall with 50% overlap at 30 km/h and 50 km/h speeds the longitudinal acceleration of 18.8 g was found for 30 km/h speed and it was reached after 0.083 s from the beginning of the impact. For simulation of 50 km/h the maximum acceleration was 58.5 g and it was reached after 0.041 s. At these cases the airbag deployment cause was exceeded (acceleration window 5.8–10.2 g), therefore airbags should be deployed for both analysed speeds. The upper acceleration limit for airbag deployment was reached after 0.0140 s at 30 km/h speed and after 0.0085 s at 50 km/h. It was also recognised a negative vehicle speed values just after the impact. This is because of the vehicle was blown back.

During the simulation of frontal impact with of a rigid wall at an angle of 25° at 30 km/h, the maximum longitudinal acceleration of 10.0 g was reached after 0.084 s from the beginning of the impact. For 50 km/h the maximum acceleration is 19.1 g and it was reached after 0.066 s. The limits of longitudinal acceleration for airbag deployment (5.8–10.2 g) is reached for 50 km/h case. For 30 km/h only the lower limit for airbag deployment was reaches; however, the upper limit was not reached. So the deployment of vehicle front airbag is guaranteed only for 50 km/h, and for 30 km/h the deployment is possible, but not guaranteed. The lower acceleration limit for airbag deployment was reached after 0.0565 s at 30 km/h speed and after 0.0439 s at 50 km/h. At this case the vehicle interacts with the obstacle (barrier) by sliding contact therefore vehicle's movement duration is longer.

The graphs also show that the force changes significantly during the impact duration. This is because of different shapes and stiffness of vehicle body. The highest forces was estimated when the impact reached front bumper beam, engine, front chassis legs.

6 Conclusions

After the research of vehicle airbag deployment, its related issues and performed computer simulations, the following conclusions are made:

1. The main variable for control of vehicle airbag is acceleration which activation thresholds varies between 4.7–13.2 g. The values of longitudinal acceleration used for airbags deployment analysis in computer simulation with light vehicle is between 5.8–10.2 g.
2. After computer simulation with software LS DYNA using Toyota Yaris finite element model, the largest longitudinal acceleration (58.5 g) was found during the impact with rigid wall with 50% overlap at 50 km/h speed. Changing the overlap of vehicle's collision with an obstacle from 50% to 100%, the longitudinal acceleration at a speed of 30 km/h increased 1.3 times. However, the largest increase (1.68 times) was found changing the overlap of vehicle's collision with an obstacle from 25% to 50% at 50 km/h speed.

3. Changing the angle of vehicle's impact with the obstacle from 25° to 45°, the longitudinal acceleration increased 1.4 times at both 30 km/h and 50 km/h speeds.
4. During the impact with rigid wall at an angle of 25° at a speed of 30 km/h, the resulting longitudinal acceleration (10.0 g) is within the limits of the airbag activation, in which case the deployment of the airbags is likely, but not mandatory. In all other simulated cases, the deployment of the airbags is mandatory because the longitudinal acceleration values obtained exceed the specified longitudinal acceleration limits.

References

1. Road traffic injuries. https://www.who.int/news-room/fact-sheets/detail/road-traffic-injuries. Accessed 17 Apr 2020
2. European Commission. Road safety. https://ec.europa.eu/commission/news/road-safety-2019-apr-04_en. Accessed 15 May 2020
3. Air Bags. https://www.nhtsa.gov/equipment/air-bags. Accessed 17 Oct 2020
4. Kahane, C.J.: Lives saved by vehicle safety technologies and associated Federal Motor Vehicle Safety Standards, 1960 to 2012 – Passenger cars and LTVs – With reviews of 26 FMVSS and the effectiveness of their associated safety technologies in reducing fatalities, injuries, and crashes. (Report No. DOT HS 812 069), Washington, DC: National Highway Traffic Safety Administration (2015)
5. McCartt, A., Kyrychenko, S.: Efficacy of side airbags in reducing driver deaths in driver-side car and SUV collisions. Traffic Injury Prevention (2007)
6. Hesseling, R.J., Steinbuch, M., Veldpaus, F.E., Klisch, T.: Identification and control of a vehicle restraint system. Proc. Inst. Mech. Eng. Part D: J. Automobile Eng. **220**(4), 401–413 (2006)
7. Lu, Y., Shen, J., Wang, C., Lu, H., Xin, J.: Studying on the design and simulation of collision protection system between vehicle and pedestrian. Int. J. Distrib. Sens. Netw. **16**(1), 1–9 (2020)
8. Nayak, R., Padhye, R., Kanesalingam, S., Arnold, L., Bahera, B.K.: Airbags, Textile Progress. Taylor & Francis, pp. 209–301 (2013)
9. Leschke, A.: Automobile Algorithm Concept for Crash Detection in Passenger Cars. Springer Fachmedien Wiesbaden GmbH (2020)
10. Ashley, B.: Defective Designs in Airbag Technology, The Forum: A Tennessee Student Legal J. ournal **2**(1), 49–80 (2015). Article 4
11. Kendall, J., Solomon, K.A.: Air Bag deployment criteria. The Forensic Examiner (2014)
12. Haladuick, T., Dyck, T., Maclnnis, S., Foster, C., Chapman, D.: Event data recorder use in Canadian criminal & civil collision reconstruction, European Association for Accident Research and Analysis (EVU) (2015)
13. Cho, K., Choi, S.B., Lee, H.: Design of an airbag deployment algorithm based on precrash information. IEEE Trans. Veh. Technol. **60**(4), 1438–1452 (2011)
14. Amin, M.S., Nasir, S.S., Reaz, M.B.I., Ali, M.A.M., Chang, T.: Preference and placement of vehicle crash sensors. Technical Gazette **21**(4), 889–896 (2014)
15. Mazda Cx–3 Owners Manual. Mazda Motor Corporation (2015)
16. Toyota Yaris Finite Element Model Validation Coarse Mesh. https://www.ccsa.gmu.edu/wp-content/uploads/2016/11/2010-toyota-yaris-coarse-validation-v1.pdf. Accessed 22 Apr 2021

Side Impact Analysis of the Bus Frame Structure

Tautvydas Pravilonis(✉) and Edgar Sokolovskij

Vilnius Gediminas Technical University, Vilnius, Lithuania
{tautvydas.pravilonis, edgar.sokolovskij}@vilniustech.lt

Abstract. Vehicle bodies and structures should be designed to absorb and redistribute as much impact energy as possible, thus protecting the driver and passengers from injury. The deformation zone of the vehicle in the case of a side impact is smaller than in the case of a front impact. A side impact collision directly affects the structure of the vehicle. Composite materials may be used in the structure to reduce vehicle weight and improve vehicle deformation. Bodies and structures where composite materials are used have a higher SEA value. Using Ansys software package according to UN / ECE Regulation No. 95, a side impact simulation, when a deformable barrier moving at a speed of 50 km/h impact to a structure was performed. The simulation was performed with a steel and hybrid bus frame structure. The simulation results show that the deformation of the hybrid frame structure is close to the deformation of the steel frame structure and the SEA value is 12% higher.

Keywords: Bus frame · Side impact · Kinetic energy · SEA

1 Introduction

Side impact accidents are one of the most dangerous accidents that happen every day. Although according to Liang and Le [1] the most dangerous accident for buses is a rollover. According to the sources that provide accident statistics by type of accident side impact accidents account for about 20% of all accidents. The incidents of corner, front and rear collisions respectively are as follows: 30%; 4%; 46% [2]. In 2018 9,670,000 traffic accidents related to vehicle collisions were recorded in the United States. The number of side-on-road accidents recorded in Europe is around 35–40% of all vehicle-related accidents [3].

The consequences of side impact accidents are painful not infrequently resulting in passenger deaths. The deformation zone of the vehicle in the case of a side impact is smaller than in the case of a front impact. The impact of a side impact is directly affected by the structure of the vehicle, which affects the safety of the occupants. Unlike in the case of a frontal or rear impact in the case of a side impact the impact energy is transmitted directly to the vehicle structure which directly affects the occupants. The structure of the vehicle should be constructed that in the event of a side impact accident, the structure absorbs the energy of the side impact and protects the occupants from injury

O. Prentkovskis et al. (Eds.): TRANSBALTICA 2021, LNITI, pp. 161–170, 2022.
https://doi.org/10.1007/978-3-030-94774-3_16

in the event of an accident. Side beams and reinforcement elements should withstand side impact force, absorb energy and protect occupants [4, 5].

The risk of side impact is distinguished by the fact that it is not possible to fit sufficient deformation zones in the lateral parts of vehicle structures as in the front and rear parts of vehicles. This limitation is even more pronounced in frame structures where square or rectangular profiles are used instead of shaped steel sheets. In order to minimize the consequences of a side impact vehicles shall be designed in a variety of ways to efficiently absorb the energy generated by the side impact so that it passes as little as possible to the occupants inside the vehicle.

Composite materials can be used in structures to increase the amount of impact energy absorbed. However, replacing steel with a fiberglass composite can lead to unacceptable deformations of the bus frame structure that may not ensure the safety of the driver or passengers in the event of a traffic accident leading to safety tests modeling various impact scenarios [6].

2 Methodology

Safety tests shall be performed to assess the strength and safety of the vehicle and the load on the driver and passengers. Side impact tests may be carried out in accordance with various requirements, directives, rules or standards:

- Federal Motor Vehicle Safety Standard (FMVSS 214). The barrier strikes a stationary vehicle at a speed of 54 km/h. In this test, the barrier is inclined by 27 degrees, so that the speed of the barrier splits into two components: a speed of 48 km/h in the direction of the Y axis and a speed of 25 km/h in the direction of the X axis. The width of the barrier is 1,676 mm and the ground clearance is 279 mm. The ground clearance of the deformable element is 330 mm. The weight of the barrier is 1,368 kg. The dummies are used in the front and rear seats;
- side impact test according to (UN / ECE) Regulation No. 95. A movable deformable barrier (MDB) strikes the test vehicle at a speed of 50 km/h. The barrier and the vehicle are arranged so that an angle of 90 degrees is formed between them. The barrier mass is 950 kg and a width of 1,500 mm. The ground clearance of the deformable element is 300 mm. The dummies are used only in the driver's seat [7].
- Euro-NCAP side impact test. This side impact test is very similar to (UN/ECE) Regulation No. 95 test. The difference is that according to the Euro-NCAP side impact test dummies are used in both the front and rear seats [8].

Such safety tests are quite expensive, so many manufacturers use numerical simulations based on the finite element method (FEM), which evaluated the strength and safety of the structure at the design stage to reduce these costs [9–11]. The results of the dynamic analysis make it possible to compare different structures and identify their weaknesses, which can deform under critical loads beyond the limit values. When a model is a tubular spatial system consisting of individual elements connected into one common node it is practically impossible to solve it by conventional methods, so in this case FEM are used [12].

In vehicle collision modeling using FEM the vehicle is divided into FE the contacts of the individual elements are described and the initial data (weight, energy absorption coefficient, body deformation depth, etc.) are entered [13].

The amount of energy absorbed by vehicles depends directly on the materials used in the structure of the vehicle, their density, yield strength, the wall thickness of the components, the geometry and the deformation zones.

When comparing the deformations of composite materials, they differ from metallic materials.

In metallic materials energy is dissipated through plastic folding, material hardening, and through heat loss. Composite materials in contrast dissipate energy due to external and internal friction and fiber bending and kinematic dispersion as the material decomposes. These properties give composite materials advantages in improving impact resistance. When assessing Specific Energy Absorption (SEA) it may be higher. Using composite materials, the same impact resistance of a structure can be achieved as with metallic materials but using composite materials the structure is lighter [14].

SEA also called absorbed energy per unit weight is a very important measure to reduce the weight of a structure and is used to assess the energy absorption capacity of structures [15, 16]. The absorption of specific energy can be calculated according to the equation [14]:

$$SEA = \frac{W}{\rho \cdot A \cdot \delta} = \frac{\int_0^{\delta} F \cdot d\delta}{\rho \cdot A \cdot \delta}. \quad (1)$$

where W – the total energy absorbed J; ρ – density, kg/m^3; A – cross-sectional area, m^2; δ – total compressive deflection mm; F – the compressive force, N.

According to the definition of impact suitability it is understood that the ideal, impact-friendly material used in vehicle structures must perform two main actions in the event of an accident: first, to absorb the kinetic energy of the impact; second, to dissipate this kinetic impact energy over time to ensure that the deceleration of the vehicle is less than the critical value above which passengers would suffer irreversible injury. According to Jacob [17] comparing two different materials with similar energy absorption properties a material with a higher specific energy absorption value and which has dissipated this energy over a longer period of time would be considered more impact resistant.

3 Side Impact Simulation and Results

The modeling of the side impact on the bus frame structure shall be performed in accordance with Regulation of the United Nations Economic Commission for Europe (UN/ECE) No 95. According to this Regulation a 950 kg barrier moving at a speed of 50 km/h strikes the structure of a bus frame [7]. The modeling process and accepted boundary conditions are presented below.

The geometric model of the bus frame structure and barrier was prepared by the Rhinoceros (Rhino version 6) program. In the next step, the model was uploaded to the

Ansys Workbench program. In side impact modeling were use mixed shell and solid FE. In the steel frame structure, all elements are connected using the bonded type and, in the hybrid, the bonded type and the cohesive zone model (CZM). Frictional was used to describe the contact of the frame structure with the base. The values of the coefficient of friction of a car tire with a dry road surface range from 0.8 to 0.9 [18]. A coefficient of friction of 0.8 was chosen for the model.

The properties of the materials used in the structure of the bus frame structure are given in Table 1.

After preparing the model it was broken down into FE. The FE frame structure of the bus frame structure consists of 425,495 elements and 430,682 nodes. The barrier in the model is accepted as a deformable body.

Table 1. Mechanical properties of pultruded glass fiber (GFRP) composite material and steel 1.4003 (Stala 400F).

Property	GFRP		Steel 1.4003 (Stala 400F)
	Directionality		
	longitudinal	crosswise	–
Density (kg/m^3)	2,000		7,700
Tensile ultimate strength (MPa)	400		450
Compressive ultimate strength (MPa)	400		–
Young's modulus (MPa)	39,000	4,875	220 · 103
Poisson's ratio	0.035	0.335	0.3
Shear modulus (MPa)	3,358	3,342	–
Yield strength (MPa)	–	–	400

To solve the problem of modeling the side impact to the frame structure the Ansys Workbench program supplement "Explicit dynamic" is used. Using the Velocity command, a speed of 50 km/h is entered into the program. It is at this speed that the deformable barrier (in the direction of the Y axis) collides with the frame structure. Barrier displacement was also limited in the direction of the Z axis. For all bodies the acceleration of free fall in the direction of the Z axis was added.

In the Ansys Workbench program selected for modeling, calculations are performed using the Newmark method. The equilibrium equation for this method is written as follows:

$$[M]\ddot{U} + [C]\dot{U} + [K]U = N, \tag{2}$$

$$\dot{U}_{i+1} = \dot{U}_i + \left[(1-\delta)\dot{U}_i\right] + \left[(1-\delta)\dot{U}_i + \delta\ddot{U}_{i+1}\right]\Delta t, \tag{3}$$

$$U_{i+1} = U_i + \dot{U}_i\Delta t + \left[\left(\frac{1}{2} - \alpha\right)\ddot{U}_{i+\alpha} \cdot \ddot{U}_{i+1}\right]\Delta t^2, \tag{4}$$

$$M\ddot{U}_{i+1} + C\dot{U}_{i+1} + KU_{i+1} = N_{i+1}. \tag{5}$$

where $\ddot{U}$ – acceleration; $\dot{U}$ – speed; U – displacement; $[M]$ – mass matrix; $[C]$ – damping matrix; $[K]$ – stiffness matrix; N – force; t – time; δ – change; α – number of individual elements.

Two bus frame structures were evaluated in the side impact modeling. Because the impact on the bus frame structure, which does not include exterior and interior elements, is modeled, only the mass of the bus frame structure is used, not the total mass of the vehicle. The first structure was made of steel and weighed 966 kg. The second bus frame structure was hybrid and weighed 814 kg. And it is a 15% lighter structure compared to a structure that is made only of steel.

In the developed model the deformable barrier moving at a speed of 50 km/h collides with the bus frame structure which deforms from the impact energy. The frame structure is built on an absolutely rigid surface and as written above contact with a coefficient of friction of 0.8 is described. Contacts are made where the front and rear axles of the bus are located. In solving the problem, it is assumed that the shock-absorbing properties of the surface on which the frame structure moves are not evaluated. The program also sets the numerical simulation of the collision to last 0.75 s. This impact time is sufficient to observe the deformations of the frame structure and to understand the impact resistance properties, as after 0.75 s there is no contact between the barrier and the structure.

The initial position of the structure and barrier is shown in Fig. 1. The following are the results of the numerical modeling of the side impact of the two bus frame structure.

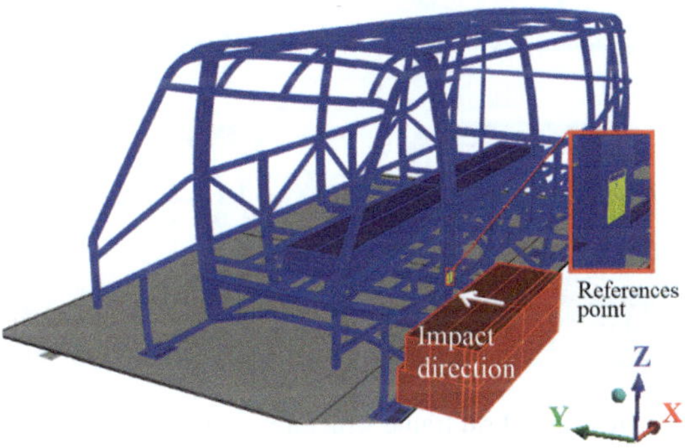

Fig. 1. Initial position of the side impact simulation.

The side impact modeling assumes that the steel frame structure meets safety requirements. It is treated as a reference and the differences in the obtained values are evaluated, which are recorded by side impact modeling with a hybrid frame structure.

Figure 2 and Fig. 3 show how steel and hybrid frame structures deform at different time points (four stages are distinguished).

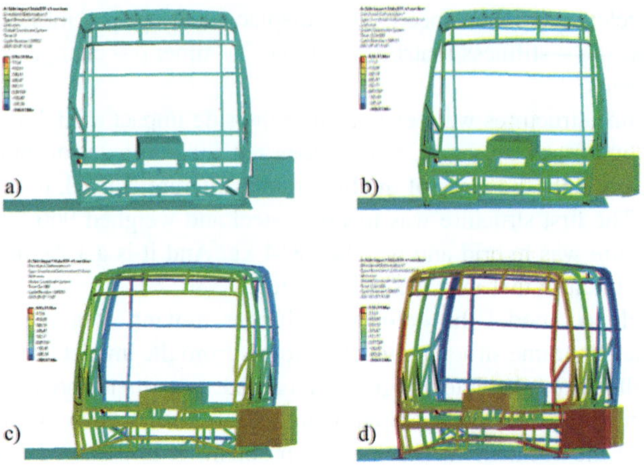

Fig. 2. Deformation progress in steel frame structure: a) start of contact; b) 25 ms from start of contact; c) 50 ms from start of contact; d) 75 ms from start of contact.

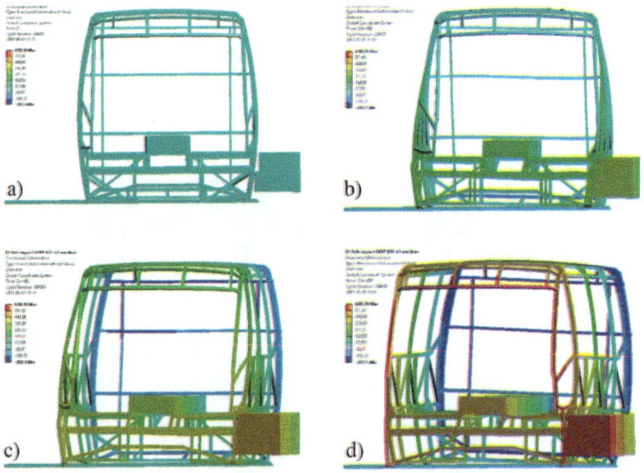

Fig. 3. Deformation progress in hybrid frame structure: a) start of contact; b) 25 ms from start of contact; c) 50 ms from start of contact; d) 75 ms from start of contact.

According to the presented figures of side impact simulations it is noticeable that both the steel and hybrid bus frame structure deforms in a very similar way. At the same moments of time the values of deformations are very close, the deformation zones are identical. The good strength of the hybrid frame structure is also confirmed by the fact that the maximum difference in displacements of the structures reaches 1% (in the direction of the Y axis) and the maximum concentrations of stress are formed in the same places. Also, these two different structures absorb the kinetic energy of the barrier in a very similar way. Figure 4 shows the change in the barrier's total kinetic energy over time. The initial total kinetic energy of the barrier is 91.64 kJ. This value corresponds to an initial speed of 50 km/h (13.89 m/s) for the moving barrier calculated as follows:

$$E_k = \frac{1}{2} \cdot m \cdot v^2 \tag{6}$$

where E_k – kinetic energy, J; m – mass of the structure, kg; v – initial barrier movement speed, m/s.

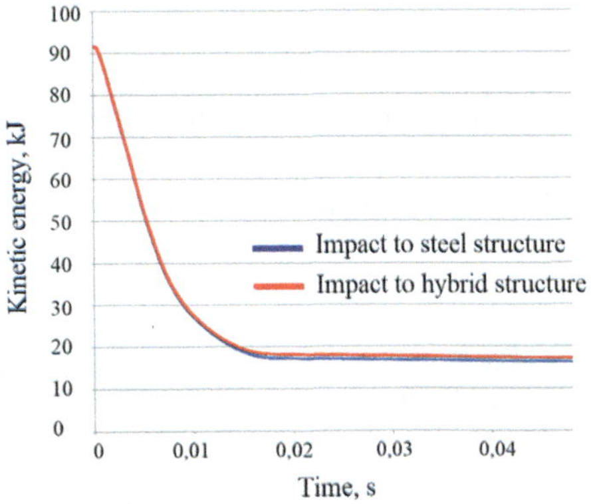

Fig. 4. Alteration of kinetic energy of barrier over time, J.

As shown in Fig. 4 the contact between the moving barrier and the bus frame structure occurs 0.015 s after the start of the test. Upon contact the kinetic energy of the barrier decreases to 18.8 kJ in the case of a steel frame structure and to 19.6 kJ in the case of a hybrid frame structure. As the test continues, the kinetic energy of the barrier remains the same until the end of the test, so the graph provides information for up to 0.04 s. The difference between the kinetic energy of the barrier after contact with the steel and hybrid frame structure is 4%. This difference in kinetic energies is due to a decrease in the mass of the hybrid frame structure. Due to the reduced mass the hybrid frame structure is less inert and therefore less dampens the speed of movement of the barrier and thus the kinetic energy.

Figure 5 shows a comparison of the total absorbed kinetic energy of a steel and hybrid frame structure when a deformable barrier with a speed of 50 km/h impact the structures. According to the presented graph, the value of absorbed energy at all time points in both structures is very close. The graph also shows that energy absorption does not follow a linear relationship, energy absorption is unstable.

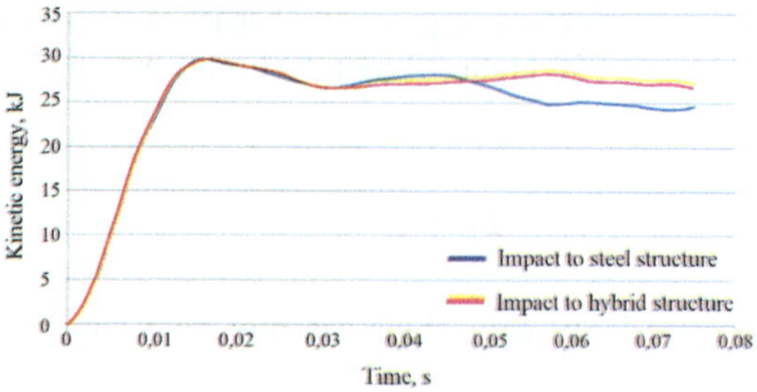

Fig. 5. Alteration of kinetic energy of frame over time, J.

The ratio of absorbed kinetic energy is similar in both steel and hybrid frame structure models. Therefore, it can be said that the hybrid frame structure absorbs impact energy almost in the same way as steel.

Figure 6 shows the change in geometry of different structures after side impact modeling.

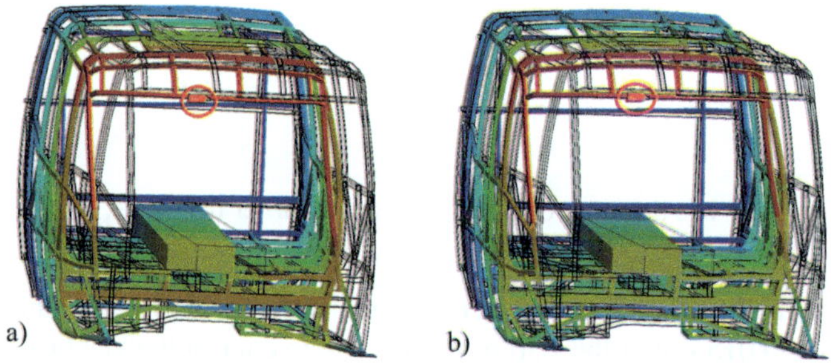

Fig. 6. The contour of the frame prior and after the side impact (black colour - frame contour before side impact): a) steel frame structure (front view); b) hybrid frame structure (front view).

In the steel frame structure, the maximum deformation is received at the top and amounts to 360 mm. Comparing the deformed steel frame structure with the undeformed (Fig. 6a), where the contour of the structure before deformation (black solid line) is shown after a side impact that was in a relatively low position, the maximum deformation is fixed in the upper part of the structure. This result was to be expected, as there are additional structural elements in the lower part of the structure over the entire floor area, which reinforce this part of the structure. The upper part of the structure shall have no additional reinforcements and the structural arches shall ensure sufficient strength of the structure. When evaluating a hybrid frame structure (Fig. 6b), the maximum deformation is also fixed in the upper part of the structure and reaches 387 mm. The difference between these deformations is about 7%. The greater deformation is reached in the hybrid frame structure due to the fact that it was in the upper part that the steel was replaced by a fiberglass composite, the mechanical properties of which are inferior to those of steel.

After performing a numerical simulation of the collision and determining the energy and deformation absorbed by the structures, the SEA can also be assessed. The SEA will be assessed according to Eq. (1). All the quantities required to calculate the specific energy absorption are known. The value of the deformation of the bus frame structure is used to calculate the SEA value, which is obtained by comparing the undeformed and deformed frame structure. In the case of a steel frame structure, they are: $W = 29.85 \cdot 10^3$ J; $\rho = 7,700\text{kg/m}^3$; $A = 0.125\text{m}^2$; $\delta = 0.360\text{m}$. In the case of a hybrid frame structure, they are: $W = 29.89 \cdot 10^3$ J; $\rho = 6,560\text{kg/m}^3$; $A = 0.125\text{m}^2$; $\delta = 0.387\text{m}$.

Calculating the SEA from the values given gives the following values: SEA_Steel = 86.15 J/kg; SEA_hybrid = 97.19 J/kg. The higher value of specific energy absorption is reached in the hybrid frame structure. The difference between the specific energy absorption values in different structures is about 12%. This difference is obtained because the amount of energy absorbed by the hybrid frame structure and the amount of deformation are close to the values of the steel frame structure but in the case of a hybrid frame structure the total structure density decreases significantly (about 15%). The higher the value of SEA, the better the structure is able to absorb energy. The resulting higher SEA value indicates that the use of composite materials significantly reduces the mass of the structure and at the same time maintains the impact resistance of the structure.

4 Conclusion

1. In the structure of a bus frame replacing steel with a fiberglass composite reduces the weight of the frame structure by about 15%. The reduction in mass would have a positive effect on the reduction of exhaust emissions.
2. Modeling of the side impact showed that the deformation of the hybrid frame structure is 7% higher compared to the steel frame structure.
3. Assessing the change of the kinetic energy of the barrier over time and its general change, it can be seen that the kinetic energy of the barrier decreases by about 80% after contact in 0.015 s. The change in the kinetic energy of the barrier in the different structures is very similar and the difference between them is 1.2 kJ. This shows that both structures absorb the kinetic energy of the barrier relatively equally.
4. Assessing how the kinetic energies of frame structures change, it can be seen that they also change uniformly. The difference in kinetic energy between different structures is about 5%.
5. After calculating and comparing the SEA values of different structures, it was found that a higher SEA value is obtained in the hybrid frame structure. This difference is 12%. The obtained result shows that the use of composite materials can ensure high impact resistance of the structure at a lower mass.

References

1. Liang, C.C., Le, G.N.: Analysis of bus rollover protection under legislated standards using LS-DYNA software simulation techniques. Int. J. Automotive Technol. **11**(4), 495–506 (2010)
2. Injury Facts, Type of Crash (2018). https://injuryfacts.nsc.org/motor-vehicle/overview/type-of-crash/
3. Breen, B.: Road safety study for the interim evaluation of Policy Orientations on Road Safety 2011–2020. (2015). https://ec.europa.eu/transport/road_safety/sites/roadsafety/files/pdf/study_final_report_february_2015_final.pdf
4. Pawar, A. P., Mankar, S. H.: Crashworthiness evaluation of low weight recyclable intrusion beam for side impact. Int. Eng. Res. J., 1894–1898 (2008)
5. More, K.C., Patil, G.M., Belkhede, A.: Design and analysis of side door intrusion beam for automotive safety. Thin-Walled Structures **153**(1–3), 106788 (2020)
6. Jeyakumar, P.D., Devaradjane, G.: Improvement of the frontal structure of a bus for crash accidents. Transp. Syst. **11**, 183–187 (2012)
7. Regulation No 95 of the Economic Commission for Europe of the United Nations (UNECE) — Uniform provisions concerning the approval of vehicles with regard to the protection of the occupants in the event of a lateral collision [2015/1093]., https://op.europa.eu/en/publication-detail/-/publication/a223bf6e-26c6-11e5-a342-01aa75ed71a1
8. Heiko, J., et al.: Review of the development of the iso side impact test procedure for child restraint systems. In: 20th International Technical Conference on the Enhanced Safety of Vehicles (ESV), Lyon, France, June 18–21 (2007)
9. Karlinski, J., Ptak, M., Dzialak, P., Rusinski, E.: Strength analysis of bus superstructure according to Regulation No. 66 of UN_ECE. Arch. Civil Mech. Eng. **14**(3), 342–353 (2014)
10. Li, Y., Lan, F., Chen, J.: Numerical study on the influence of superstructure configuration on coach rollover resistance performance. Int. J. Crashworthiness **17**(6), 637–654 (2012)
11. Na, J., Wang, T., Wu, C., Yan, Y.: A four-node membrane element model with bending modification for one-step algorithm for bus rollover impact. Eng. Comput. **32**(3), 607–620 (2015)
12. Marianski, M., Szosland, A.: Research on the strength of Standard bus bodies at rollover on the side. Technikal University of Lodz, pp. 29–41 (2012)
13. Noaimadudin, W.M.K., Nor, H.S., Amir, R.A.G., Nik, R.A., Khairul, I.A.J.: Modelling and simulation of a single deck bus subjected to rollover crash loading. Appl. Mech. Mater. **393**, 453–459 (2013)
14. Lukaszewicz, D.H.-J.A.: Automotive Composite Structures for Crashworthiness, BMW AG, Research and Innovation Centre, Knorrstrasse 147, 80788. Munich, Germany (2013)
15. Rafukka, I.A.: Numerical investigations in to crashworthiness characteristics of steel conical tubes. Nigerian J. Technol. **37**(2), 663–667 (2018)
16. Xu, T., Liu, N., Yu, Z., Xu, T., Zou, M.: Crashworthiness Design for Bionic Bumper Structures Inspired by Cattail and Bamboo. Applied Bionics and Biomechanics (2017)
17. Jacob, G. Ch.: Automotive Crashworthiness of Adhesively Bonded Carbon Fiber Polymer. Composite Structures, 245 p. (2006)
18. Cebon, D.: Handbook of Vehicle-Road Interaction. CRC Press. 616 p. (1999). ISBN 9789026515545

Assessment and Predictive Modelling of Transport and Operating Condition of Aerodrome Pavement: A Case Study of Zaporizhzhia International Airport Runway

Oleksandr Dubyk[1], Svitlana Timkina[1], Oleksandr Stepanchuk[1(✉)], and Olegas Prentkovskis[2]

[1] National Aviation University, Kyiv 03058, Ukraine
[2] Vilnius Gediminas Technical University, Vilnius, Lithuania

Abstract. The article presents the results of assessment and predictive modelling of the transport and operating condition of aerodrome pavement surface on the example of Zaporizhzhia International Airport runway. Domestic and foreign methods for assessing the transport and operating condition of aerodrome pavement have been analyzed. When assessing the transport and operating condition of aerodrome pavement, the method of signal evaluation is most widely used in Ukraine, while standard methods for determining the condition index of aerodrome pavement, *ACN – PCN* methods, and determining the complex *K* indicator are used abroad. The authors analyze Ukrainian and foreign literature in the field of assessment of transport and operating condition of aerodrome pavement. In the result of the analysis of literature sources, it was established that defects and damages of aerodrome pavement can be of three types: infringement of construction works technology; infringement of operating requirements; miscalculations and errors in construction made at the phase of preliminary research and designing. The reliability and suitability for further operation of the rigid aerodrome pavement of the runway with artificial turf of Zaporizhzhia International Airport have been assessed. A defective plan of the runway with artificial turf of Zaporizhzhia International Airport has been developed and the types of defects of paving slabs have been established. After calculating the value of the signal evaluation indicator, it was found that the pavement of the artificial runway of Zaporizhzhia International Airport is in the condition of unacceptable damage. The reliability of the aerodrome pavement is less than the allowable value of reliability. Measures have been developed to improve the transport and operating condition of the aerodrome pavement surface of the artificial runway of Zaporizhzhia International Airport.

Keywords: Runway · Transport and operating condition · Aerodrome pavement · Signal evaluation method · Reliability · Paving slabs

O. Prentkovskis et al. (Eds.): TRANSBALTICA 2021, LNITI, pp. 171–183, 2022.
https://doi.org/10.1007/978-3-030-94774-3_17

1 Introduction

Nowadays, both in Ukraine and abroad, the volume of air traffic has increased, which leads to an increase in the intensity of aerodromes operation. Of course, this adversely affects their transport and operating condition. Therefore, the development of aviation technology makes very high demands on both the geometric parameters and the quality and durability of aerodrome pavements, which necessitates the improvement of their design, construction, repair, and maintenance methods [1–5]. If we take domestic and foreign experience in the construction of aerodromes, it can be noted that a significant part of artificial aerodrome pavement has a significant level of wear. In other words, not all aerodromes can accommodate modern aircraft, as the pavement of aprons, taxiways, runways, and stands for long-term aircraft parking depleted their resource over time. Given the above, many aerodromes in Ukraine and around the world need a major overhaul. The runway with artificial turf is the main component of the aerodrome. The technical condition of aerodrome pavement does not meet flight safety requirements. The object of this article is rigid aerodrome pavement. The purpose of the study is to assess the transport and operating condition of aerodrome pavement (a case study of artificial turf of Zaporizhzhia International Airport runway) and provide an opinion on the possibility of their further operation. The scientific and practical value of the work lies in the proposed method of assessing the transport and operating condition of rigid aerodrome turf, where "reliability" is the main indicator. Also, the results of the research can be used for planing repair works during the overhaul of aerodromes with rigid reinforced concrete pavement. The article uses methods of signal evaluation and visual inspection of aerodrome pavement.

2 Analysis of Methods for Assessing the Transport and Operating Condition of Aerodrome Pavement

There is a qualitative and quantitative assessment of the transport and operating condition of aerodrome pavement. Qualitative assessment is performed to determine the suitability of the pavement for the bearing capacity of a given type of aircraft by comparing the aircraft classification numbers ACN and the bearing capacity of pavement PCN at the same strength of the soil base [4–6]. The classification number of pavement must not be lower than the classification number of the aircraft operated:

$$K \cdot ACN \leq PCN \tag{1}$$

where K is a coefficient that takes into account the intensity of aircraft movement.

Quantitative assessment of the transport and operating condition determine the suitability of pavement for operation based on the analysis of the nature and amount of damage. Criteria that determines the condition of the pavement surface are parameters that characterize the detected damage; quantitative indicators of transport and operating condition of the pavement surface; the category of the destruction of the aerodrome pavement.

When assessing the transport and operating condition, various methods are used:

- signal evaluation method S_k [7];
- standard method for determining the condition of aerodrome pavement *PCI*;
- determination of the complex *K* indicator.

Standard methods for determining the condition of the aerodrome pavement surface and the complex *K* indicator are foreign methods.

In Ukraine, the serviceability of rigid aerodrome pavement is determined by the indicator of the signal evaluation of the condition of the pavement S_k, which is determined by the formula [7]:

$$S_k = 5 - D, \tag{2}$$

where *D* is a composite indicator of damage rate calculated by the formula [7]:

$$D = (D_{cr} \cdot Q_{cr} + D_{j.s.} \cdot Q_{j.s.} + D_{ex} \cdot Q_{ex}) \cdot 100, \tag{3}$$

where D_{cr} is a composite indicator of through-cracks; $D_{j.s.}$ is an indicator of joint spalling; D_{ex} is a exfoliation index; Q_{cr} is a coefficient of through-cracks magnitude; $Q_{j.s.}$ is a coefficient of joint spalling magnitude; Q_{ex} is a magnitude factor of exfoliation.

The indicator D_{cr} is calculated by the following formula:

$$D_{cr} = \frac{n_{cr}}{n_{tot}} \cdot 100, \tag{4}$$

where n_{cr} is the number of slabs with through-cracks; $n_{заг}$ is the total number of slabs in the survey area.

Indicator $D_{j.s.}$ is calculated by the formula [7]:

$$D_{j.s.} = \frac{n_{j.s.}}{n_{tot}} \cdot 100, \tag{5}$$

where $n_{j.s.}$ is the number of slabs with joint spalling.

Indicator D_{ex} is calculated by the formula [7]:

$$D_{ex} = \frac{n_{ex}}{n_{tot}} \cdot 100, \tag{6}$$

where $n_{ш}$ is the number of slabs with exfoliation.

The aerodrome pavement index is calculated by the formula [8]:

$$PCI = 100 - MaxCDV, \tag{7}$$

where *Max CDV* is the largest value of the corrected reduced magnitude.

The complex *K* indicator is calculated by the formula [9]:

$$K = 100 - (K_A \cdot P_A + K_B \cdot_B + K_C \cdot P_C + K_D \cdot P_D), \tag{8}$$

where P_A, P_B, P_C, P_D are the magnitude coefficients for pavement areas; K_A, K_B, K_C, K_D are the aerodrome pavement quality value.

Table 1 shows the categories of transport and operating condition of the aerodrome pavement surface depending on the method of their evaluation.

Table 1. Categories of transport and operating condition of aerodrome pavement.

Evaluation method	Indicator values	Pavement condition
Signal evaluation S_k	3.5–5.0	Normal operation stage
	2.5–3.5	Critical stage
	Less than 2.5	Stage of unacceptable damage
Composite k indicator and index of pavement condition *PCI*	ss	Excellent
	71–85	Very good
	56–70	Good
	41–55	Satisfactory
	26–40	Bad
	11–25	Very bad
	0–10	Unsatisfactory

2.1 Theoretical Research Base

The works of many Ukrainian and foreign scientists [2, 3, 10–15] are devoted to the issue of classification of damages and defects of aerodrome pavement and the reasons of their formation. Analysis of Ukrainian and foreign literature allows us to conclude that the reasons of defects and damage of aerodrome pavement are [2, 3, 10–15, 17, 18]:

- infringement of construction works technology;
- infringement of operating requirements;
- miscalculations and errors in construction made at the phase of preliminary research and designing.

According to [2, 3, 10–15, 17, 18], processes that lead to the destruction of pavement are characterized as:

- mechanical (those processes that are caused by the application of force);
- complex (caused by exposure to chemical reagents, groundwater, freeze-thawing).

As mentioned by scientists [2, 3, 10–15, 17, 18], all destructions and damages to aerodrome pavement are the results of not one but a number of factors. According to the degree of destruction, there are three categories of damage:

Table 2. Allowable and critical reliability values.

Group of pavement areas	Allowable reliability value, $P_e^{\min}$	Critical reliability value, P_k
A	0.84	0.71
B	0.81	0.66
C	0.77	0.59
D	0.74	0.52

- emergency damage;
- damage of load-carrying structure;
- minor local damage, which is removed during operation maintenance.

Typical damage of rigid aerodrome pavement includes exfoliation of the uppermost layer of concrete, the formation of road bumps, and cracks, vertical displacement of slabs, spalling of corners and edges of slabs, joint spalling.

Monitoring of aerodrome pavement includes 2 stages [2, 3, 10–15, 17, 18]:

- preliminary examination carried out by the aerodrome service;
- detailed examination conducted by an expert commission.

The preliminary examination includes visual inspection; drawing up a defective plan; operational evaluation of technical and operation condition; conclusions on the operability of aerodrome parts; a decision to conduct a detailed examination. The detailed examination includes detection of hidden defects and damage; determination of material characteristics; calculation models of the "cover-base" system; estimation of loading on the pavement; assessment of slab flatness; degree of roughness; results of calculations and tests; conclusions on the aerodrome pavement condition. The recommendations for the restoration and strengthening of aerodrome pavement are being carried out basing on the conclusions on the aerodrome pavement condition. The risk of destruction of the aerodrome pavement overtime is calculated by the formula:

$$r(t) = \frac{n_{dam}}{n_{tot}}, \tag{9}$$

where $r(t)$ is the risk of the aerodrome pavement surface destruction; n_{dam} is the number of damaged slabs at the time t; n_{tot} is the total number of slabs of the aerodrome pavement [16–18].

Then, the probability of trouble-free operation (reliability) of the runway will be [16–18]:

$$P(t) = 1 - r(t). \tag{10}$$

The table shows the allowable and critical values of reliability depending on the group of aerodrome pavement areas:

During the operation of the aerodrome pavement, there comes a moment when the reliability of the pavement becomes less than $P_e^{\min}$. That is, there are defective slabs that have damage with values that are unacceptable for ensuring the safe performance of flights.

3 Assessment of Transport and Operating Condition of Aerodrome Pavement (A Case Study of Zaporizhzhia International Airport Runway)

The reliability and suitability for further operation of the rigid aerodrome pavement have been assessed on the example of Zaporizhzhia International Airport runway with artificial turf.

The existing artificial runway of Zaporizhzhia International Airport was built in 1955 and reconstructed in 1981. Zaporizhzhia International Airport accommodates:

- rectangular runway strip with dimensions of 3,100 × 270 m;
- ground runway measuring 2,100 × 85 m;
- artificial runway with dimensions of 2,500 × 42 m with fortified roadsides 9 m wide on both sides of the runway;
- taxiway T-1 of 1,031 m long, 21 m wide with fortified roadsides 2.5 m each;
- passenger apron measuring 253 × 210 m with parking stands for aircraft Yak-40, Tu-134, Tu-154, An-12, An-24, An-26, An-32, IL-76, B 737, DC 9, A 310, A 320, MD 82. The situational plan of Zaporizhzhia International Airport is shown in Fig. 1, and the general plan is presented in Fig. 2.

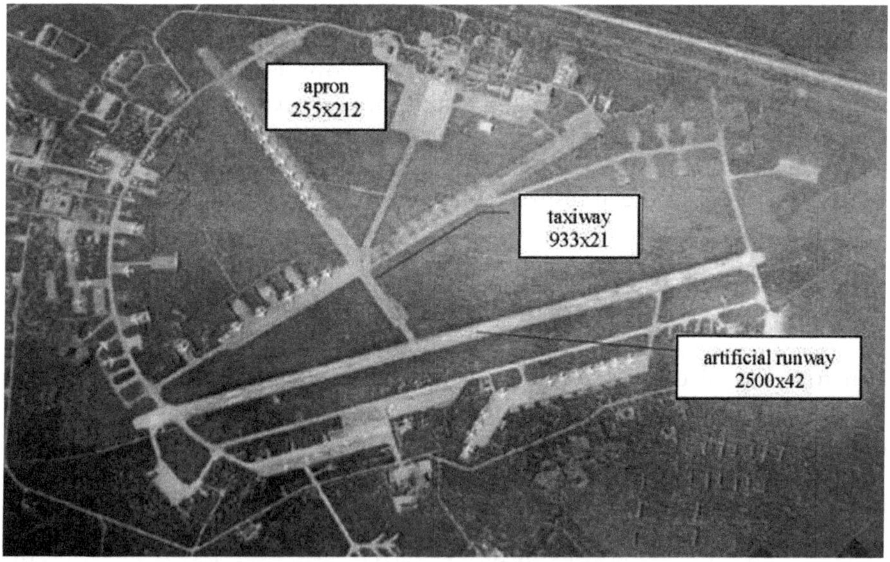

Fig. 1. Situational plan of Zaporizhzhia International Airport.

The artificial runway pavement has a two-layer structure: the upper layer is made of reinforced concrete 25–29 cm thick, the lower layer is made of cement concrete 20 cm thick, the base layer is blast-furnace slag 20 cm thick (Fig. 3).

Aerodrome pavements of the artificial runway, taxiway T-1, and the apron are in operation for a period that significantly exceeds the standard service life of rigid aerodrome pavement. The operation of aerodrome pavement for a long time and the aging of materials have led to their wear, the appearance of various deformations and fractures on the surface in the form of cracks, exfoliation and road bumps. Defects of Zaporizhzhia International Airport artificial runway are shown in Fig. 4.

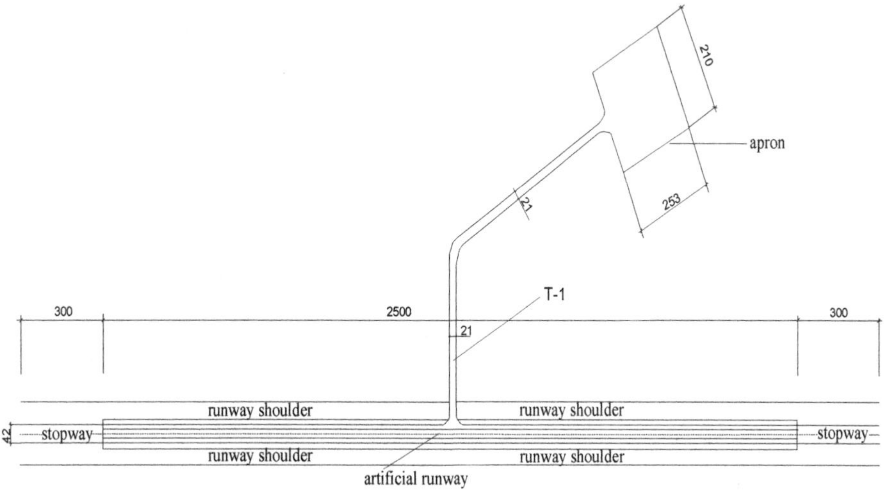

Fig. 2. General plan of Zaporizhzhia International Airport.

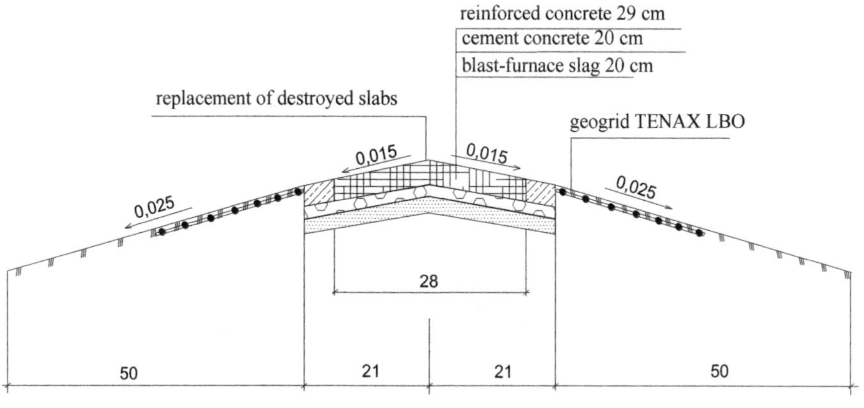

Fig. 3. Cross profile and construction of Zaporizhzhia International Airport pavement surface.

Among the main defects of the aerodrome pavement of the artificial runway there are exfoliation of slabs, longitudinal, lateral and diagonal through cracks, joint spalling, a grid of through cracks, and road bumps. As a result of inspection, it is given a conclusion on suitability of this aerodrome construction for the further operation and necessity of carrying out repair. Measures for strengthening the artificial runway are also developed. Aerodrome pavement continues to operate after a sufficient number of slabs have been damaged. Zaporizhzhia International Airport artificial runway is no exception.

a)

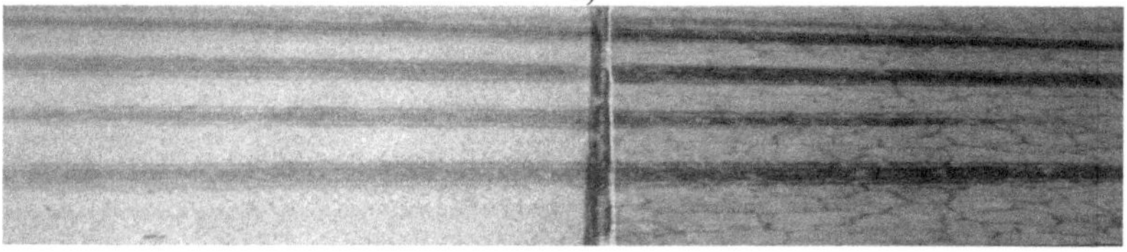

b)

c)

d)

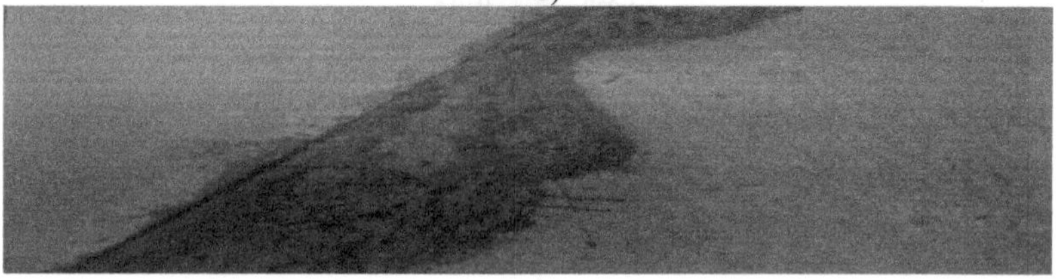

e)

Fig. 4. Defects and damages of Zaporizhzhia International Airport artificial runway: (a) is exfoliation of paving slabs; (b) is a grid of cracks; (c) is through longitudinal and lateral cracks; (d) is a damage to expansion joints; (e) is exfoliation of the upper layer of concrete.

Transport and operating condition of Zaporizhzhia International Airport artificial runway was assessed on the criterion using a reliability indicator, which includes trouble-free operation, durability, and maintainability. At the result of visual inspection of Zaporizhzhia International Airport artificial runway pavement, a defective plan for each slab has been developed, which is a plan of artificial pavement with numbering of longitudinal and lateral rows of slabs. A fragment of the defective plan of Zaporizhzhia International Airport artificial runway pavement is shown in Fig. 5. Each slab of the aerodrome pavement is 7 × 20 m in size, 42 m wide and divided into 6 rows of 7 m wide slabs.

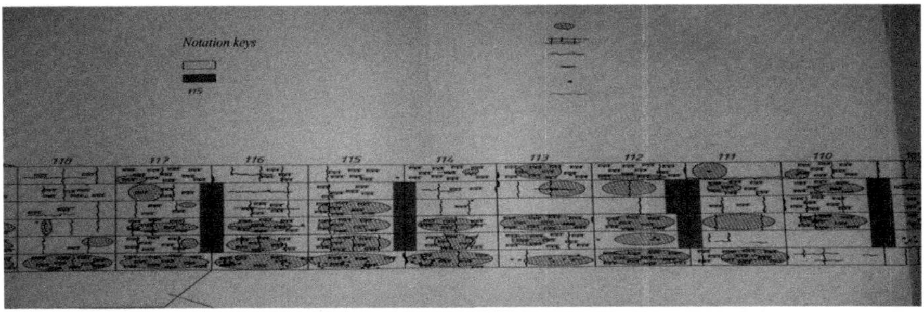

Fig. 5. A fragment of a defective plan of Zaporizhzhia International Airport artificial runway.

The total number of paving slabs at Zaporizhzhia International Airport is:

$$n_{tot} = \frac{l_{artil.runway}}{l_s} \cdot n_r = \frac{2500}{20} \cdot 6 = 750. \quad (11)$$

The result of visual inspection revealed:

- 300 slabs with through-cracks ($n_{cr} = 300$);
- 65 slabs with joint spalling ($n_{j.s.} = 65$);
- 269 slabs with exfoliation ($n_{ex} = 269$).

The total number of damaged slabs in a time of the aerodrome operation is:

$$n_{mad} = n_{cr} + n_{j.s.} + n_{ex} = 300 + 65 + 269 = 634. \quad (12)$$

The risk of destruction of the aerodrome pavement over time is calculated by the formula:

$$r(t) = \frac{634}{750} = 0.845, \quad (13)$$

Reliability of artificial runway operation at Zaporizhzhia International Airport is:

$$P(t) = 1 - 0.845 = 0.155.$$

That is, during the operation of the aerodrome pavement, its reliability is less than the allowable value of reliability for the aerodrome of 4C class:

$$P(t) = 0,155 \prec^{\min} = 0.77. \tag{14}$$

The generalized indicator of through-cracks is:

$$D_{cr} = \frac{300}{750} = 0.4. \tag{15}$$

The generalized indicator of slabs with joint spalling is:

$$D_{j.s.} = \frac{65}{750} = 0.087. \tag{16}$$

The generalized indicator of slabs with exfoliation is:

$$D_{ex} = \frac{269}{750} = 0.359. \tag{17}$$

We determine the generalized damage rate by the formula (3):

$$D_k = (0.4 \cdot 0.05 + 0.087 \cdot 0.1 + 0.359 \cdot 0.03) \cdot 100 = 3.95. \tag{18}$$

The suitability of the aerodrome pavement is assessed by the signal evaluation indicator:

$$S_k = 5 - 3.95 = 1.05. \tag{19}$$

Since the value of the signal evaluation is $S_k = 1.05$, according to Table 1, the pavement is in the stage of unacceptable damage.

4 Discussion and Interpretation of the Obtained Results

The analysis of the transport and operating condition of the artificial runway at Zaporizhzhia International Airport showed that the main defects of the aerodrome pavement are exfoliation of slabs, a grid of through-cracks, joint spalling, lateral and diagonal through-cracks, and road bumps. Among the causes of such defects there are temperature deformations in the pavement material during operational maintenance; loads from aircraft exceeding the calculated values. All defects of the aerodrome pavement are shown on the defective plan of the artificial runway. As the aerodrome pavement of Zaporizhzhia International Airport artificial runway is in the stage of unacceptable damage and has exhausted its calendar service life during a long period of operation, it is necessary to develop measures aimed at restoring its transport and operating condition. Therefore, the replacement of destroyed reinforced concrete slabs for the entire thickness of 29 cm must be performed on the artificial runway. The bottom layer and the base must remain unchanged. Almost 85% of the aerodrome pavement slabs of Zaporizhzhia International Airport need to be replaced. Overhaul of aerodrome pavement with replacement of slabs allows reducing costs by 20%. Having determined the indicator of the signal evaluation of aerodrome pavement, we found that the pavement is in the stage of unacceptable damage. To achieve this goal, we made theoretical research and visual inspections of the transport and operating condition of the artificial runway at Zaporizhzhia International Airport. Almost 9% of the artificial runway slabs at Zaporizhzhia International Airport have joint spalling, 40% have through-cracks, and 36% have exfoliation. The reasons for joint spalling of the slabs are that the expansion joints between the slabs are not completely filled with mastic and the top layer of concrete has insufficient strength. Slab edges may spall at the thermal expansion of concrete. When the edges of the aerodrome slabs spall, then the fittings expose. This can damage the tires of aircraft. Cracks develop most intensively in areas of pavement, that are repeatedly subjected to loads. The results of the assessment of transport and operating condition of Zaporizhzhia International Airport show us that cracks are the most common and at the same time dangerous type of destruction. Exfoliation of aerodrome pavement leads to loss of strength, reduction of thickness, formation of road bumps, and, consequently, reduction of bearing capacity of the pavement.

5 Conclusions

The article presents the results of the assessment and predictive modelling of the transport and operating condition of aerodrome pavement surface on the example of Zaporizhzhia International Airport runway. When assessing the transport and operating condition of aerodrome pavement in Ukraine and abroad, such methods are used: signal evaluation method, standard *PCI*methods for determining the condition index of aerodrome pavement, methods and determining the complex K indicator, $ACN - PCN$ method, etc. In Ukraine, the method of signal evaluation is very widely used, which includes the following types of defects and destruction of rigid aerodrome pavement as cracks, joint spalling, exfoliation. Both Ukrainian and foreign literature sources in the field of assessment of transport and operating condition of rigid airdrome pavement are analyzed. In the result of the analysis of literature sources, it was established that defects and damages of aerodrome pavement can be infringement of construction works technology; infringement of operating requirements; miscalculations and errors in construction made at the phase of preliminary research and designing. Typical damage of rigid aerodrome pavement includes exfoliation of the uppermost layer of concrete, the formation of road bumps, and cracks, vertical displacement of slabs, spalling of corners and edges of slabs, joint spalling. We made the assessment of transport and operating condition, reliability, and suitability for further operation of the hard aerodrome surface on the example of Zaporizhzhia International Airport, built in

1955 and reconstructed in 1981. A defective plan has been drawn up, where the artificial runway is divided into 750 slabs, each of which has dimensions of 7 × 20 m. The defective plan shows all the damage that each aerodrome pavement slab has. The prolonged operation of aerodrome pavements has led to their wear and tear, as well as the appearance of various deformations and destruction. More than half (almost 85%) of the slabs of Zaporizhzhia International Airport artificial runway are damaged. Therefore, the reliability of Zaporizhzhia International Airport operation is much less than its allowable value.

$(P(t) = 0,155 \prec^{\min} = 0.77)$. Its signal evaluation value is 1.05. This allows us to conclude that the artificial runway is in the stage of unacceptable damage. It necessitates taking urgent measures to improve the transport and operating condition of Zaporizhzhia International Airport artificial runway. In other words, all damaged slabs must be replaced with new ones. Unfortunately, this problem is not only at Zaporizhzhia International Airport, but mostly at a major part of airports in Ukraine, which were built in the middle of the XX century, where one or another element of the aerodrome has significant defects and damage.

References

1. Mezhdunarordnyye standarty i rekomenduyemaya praktika. Aerodromy. Prilozheniye 14 k konventsii o mezhdunarodnoy grazhdanskoy aviatsii (2009). http://www.gostrf.com/normadata/1/4293835/4293835980.pdf
2. Aerodromnyye pokrytiya. Sovremennyy vzglyad (2002). http://books.totalarch.com/aerodrome_cover_modern_look
3. Kulchitskiy, V., Makagonov, V., Romankov, N.: Otsenka tekhnicheskogo sostoyaniya iskusstvennykh aerodromnykh pokritiy. Aeroporty. Progressivnyye tekhnologii **4**, 12–14 (1999)
4. U.S. Department of Transportation Federal Aviation AC 150/5320-6F, Airport Pavement Design and Evaluations (2016). https://www.faa.gov/documentLibrary/media/Advisory_Circular/150-5320-6F.pdf
5. Tsykhanovskiy, V.: Raschet tonkikh plit na uprugom osnovanii metodom konechnykh elementov (2008)
6. Ushakov, V., Ageyev, B.: Naznacheniye remonta tsementobetonnykh pokrytiy na osnove vizual'noy otsenki ikh sostoyaniya. Nauka i tekhnika v dorozhnoy otrasli **2**, 41–43 (2006)
7. Nakaz Ministerstva transportu Ukrainy vid 01.07.2013 № 441 «Pro zatverdzhennia Instruktsiyi z ekspluatatsii aerodromiv derzhavnoyi aviatsiyi Ukrainy» (2013), zareyestrovanyi v Ministerstvi yustytsiyi Ukrainy 22.07.2013 za № 1229/23761 (2013). https://zakon.rada.gov.ua/laws/show/z1229-13#Text
8. Standard Test Method for Airport Pavement Condition Index Surveys (2018). https://www.techstreet.com/standards/astm-d5340-12-2018?product_id=2027838
9. Prusov, D.: Osoblyvosti doslidzhennia vzayemodii aerodromnyh pokryttiv z gruntovymy osnovamy zi slabkymy prosharkami. Visnyk natsionalnoho aviatsiinoho universytetu **2**, 129–133 (2009)
10. Tsykhanovskyi, V., Prusov, D.: Utochnennyi chyslennyi rozrakhunok zhorstkykh aerodromnykh pokryttiv na slabkykh gruntovykh osnovakh z urakhuvanniam neodnoridnosti materialu i osoblyvostei stykovikh elementiv plyt. Opir materialiv ta teoria sporud: Naukovo-tekhnichnyy zbirnyk **78**, 92–100 (2007)

11. Prusov, D., Beljatynskij, A.: The depending of embedded structures state of airports objects in terms of the boundary equilibrium state of soil semispace. Sci. Future Lithuania **3**(2),118–125 (2011)
12. Dubyk, A., Selenkov V., Talakh S.: Prochnostnoy raschet aerodromnykh pokrytiy pri slabykh gruntovykh osnovaniyakh. In: Proceeding of the 16th Conference for Junior Researchers Science – Future of Lithuania. Transport Engineering and Management, pp. 55–59 (2013)
13. Nerubenko, S.: Metodika prognozirovaniya shelusheniya betona pokrytiy. Materialy i tekhnologiya stroitelstva i remonta aerodromnykh pokrytiy **26**, 141–148 (2001)
14. Bileush, A., Berezivskyi, M., Serd, Y.: Patent of Ukraine 52548A. Method for determining bed coefficient of soil base. Kyiv: SE «Ukrainian Intellectual Property Institute» (2002)
15. Berezivskyi, M.: Osoblyvosti vplivu neodnoridnikh hruntovykh osnov na pruzhno-deformovanyi stan aerodromnoho pokryttiv (dys kand. Tekhn. Nauk: 05.22.11). Natsionalnyy aviatsiynyy universytet, Kyiv, Ukraina (2003)
16. Hameliak, I.: Osnovy zabezpechennia nadiinosti konstruktsiy dorozhnoho odyahu (dys ... dokt. Tekhn. Nauk: 05.22.11). Natsionalnyi transportnyi universytet, Kyiv, Ukraina (2005)
17. Shashkov, I.: Monitoring i prognozirovaniye rabotosposobnosti zhestkikh aerodromnykh pokrytiy (diss ... kand .. tekhn .. nauk). Voyennyy uchebno-nauchnyy tsentr voyenno-vozdushnykh sil «Voyenno-vozdushnaya akademiya imeni professora N.Ye. Zhukovskogo i YU.A. Gagarina», Voronezh, Rossiya (2012)
18. Fomin, V.: Otsenka i prognozirovaniye ekspluatatsionno-tekhnicheskogo sostoyaniya zhestkikh aerodromnykh pokrytiy (diss ... kand. Tekhn ... nauk). Moskovskiy avtomobilno-dorozhnyy institut, Moskva, Rossiya (2007)

Antiwear Coatings for Multi-material Composite Hydraulic Cylinder. A Tribological Study

Marek Lubecki(✉), Michał Stosiak, and Tadeusz Leśniewski

Faculty of Mechanical Engineering, Wrocław University of Science and Technology, Łukasiewicza 5, 50-371 Wrocław, Poland
{marek.lubecki,michal.stosiak,
tadeusz.lesniewski}@pwr.edu.pl

Abstract. Due to many advantages, composite materials are more and more widely used in the production of high-strength machine parts. An example of such an element is a hydraulic cylinder. Conventional hydraulic cylinders are made of steel, which, after appropriate treatment, ensures satisfactory conditions of cooperation between the inner surface of the cylinder and the piston seal. In the case of a cylinder made of composite material, it is necessary to check the tribological properties of the material surface and the potential selection of a material to act as an internal liner. Four materials were selected in the study - epoxy resin, polyurethane resin, CFRP and a commercial material used for sliding elements. Ball-on-disk tests were carried out and wear traces were observed using a non-contact profilometer.

Keywords: Composite materials · CFRP · Epoxy · Polyurethane

1 Introduction

Composite materials are gaining wide interest among mechanical designers due to their desirable properties, such as an excellent strength-to-weight ratio, the ability to manufacture complex shapes, corrosion resistance and resistance to magnetic fields [1]. The main industry branch driving progress in this area is the transport industry, especially the aviation and automotive industries. The use of composite materials allows to reduce the weight of vehicles, which directly translates into their efficiency, increased load capacity and, consequently, higher profitability [2, 3]. Other branches of the machine industry also see an opportunity for development in composite materials. One example is a hydraulic drive, where composite materials can be used to reduce the weight of accumulators and actuators.

One of the main advantages of the hydraulic drive is the high energy-to-weight ratio of the system. Conventional cylinders, however, are massive components made of steel. This is due to safety reasons and the necessity to withstand multiple, cyclic internal pressure loads [4, 5]. Making such an actuator from composite materials would allow for a radical reduction of its mass (even by 80%) while maintaining its operating parameters [6, 7].

O. Prentkovskis et al. (Eds.): TRANSBALTICA 2021, LNITI, pp. 184–193, 2022.
https://doi.org/10.1007/978-3-030-94774-3_18

A very important issue in the operation of a composite cylinder is to ensure appropriate tribological cooperation between the inner surface of the cylinder and the piston seal. In the case of composite cylinders, this is done by using an inner layer called a liner made of a material with good tribological properties. Another advantage of using a liner is the tightness of the cylinder, which cannot always be achieved with the composite only (weepage effect).

Plastics have long been used in the construction of gears and other elements, including gerotor pumps and the processes of their wear, are documented [8]. The usefulness of polymeric materials, and composite materials, in particular, has been proven many times not only in terms of design but also in terms of tribology. Combined with good mechanical properties, their high usefulness has been proven due to good wear resistance and self-lubricating properties. Thermoplastic materials, thermosets, and composites, [9, 10] are tested and used in various ways and their structure, composition, additives [11, 12] or the method of producing elements made of them [13] are modified. The promising results of these experimental studies are not clearly reflected in real objects, i.e. actuators, due to the focus on the material and not the real object, which is a much more difficult and time-consuming task. To transfer the results of laboratory research to real objects, it is necessary to choose the best material for this application, in the light of the adopted criteria, and then test the object made of the best material - the purpose of the research described in this paper is to choose the best material for further work, i.e. execution of the actual actuator and verification of assumptions. In the past, studies of materials for tribological coatings in composite actuators were carried out, but they did not take into account the influence of the composite substrate on the behaviour of materials during the test [14].

Four materials were selected in the study - epoxy resin, polyurethane resin, CFRP and a commercial material used for sliding elements (Belzona 1111 Super Metal). Ball-on-disk tests were carried out and wear traces were observed using a non-contact profilometer.

2 Materials and Methods

2.1 Materials

The materials were selected due to their availability, ease of processing and promising mechanical and tribological properties supported by a literature review. The first material was Huntsman Araldite LY1564 epoxy resin with Aradur 3487 amine hardener, the second material was Sika F180 two-component polyurethane, and the third was the commercially available Bezlona 1111 Super Metal. The basic mechanical and thermal properties of the materials are presented in Table 1. The substrate for the tested materials was an epoxy carbon laminate made by the vacuum bag method of 8 layers of carbon fabric with a twill weave and Araldite LY1564 epoxy resin with Aradur 3487 hardener. After the laminate was made, the surface was covered with coatings of the tested materials. Uncoated carbon epoxy laminate (CFRP) samples were also tested for comparison.

Table 1. Basic mechanical and thermal properties of tested materials.

Name	F180	Araldite 1564/3487	Super metal
Density [g*cm^{-2}]	1.06	1.11	2.5
Shore D hardness	74	–	84
Flexural modulus [MPa]	1,130	3,200	7,200
Flexural strength [MPa]	34	130	63
Pot life [min]	3.5	130–170	15
T_g [°C]	98	83	–

2.2 Methods

Tribological tests were carried out using the ball-on-disk method on the T-11 tribotester produced by the Łukasiewicz Research Network – The Institute for Sustainable Technologies in Radom. The samples were disks 1" in diameter and 6 mm in height. Bearing balls made of 100Cr6 steel with a diameter of 1/4" were used as counter-samples. The balls were made in accuracy class 10, and their hardness was 60–62 HRC. During the tests, the friction force was recorded. The temperature during the test was 21 °C. The test parameters were selected in such a way as to reflect the actual operating conditions of the piston-cylinder pair:

- Load - 9.81 N,
- Linear speed - 0.3 m/s (which corresponded to 573 rpm),
- Friction path - 2,160 m (corresponding to a test time of 2 h).
- Ambient conditions were controlled and unchanged during all tests.

After testing, all samples were subjected to friction trace observations using the Leica DCM8 contactless profilometer. Confocal measurements were made with the Leica Scan 6.5 software, at a magnification of 20x (Leica EPI 20x lens). For each sample, 4 measurements were made every 90 degrees. The post-processing of the results was performed with the Leica Map 7.4 software.

3 Results

Six samples of each type of material were tested. Figure 1 shows exemplary (representative) plots of the friction force during the test. For epoxy resin, a clear increase in friction force is visible after about 10 min, followed by a period of significant peaks reaching almost 6 N. This is due to the initial lapping of the surface necessary to stabilize the wear process. The friction force was the most stable for the F180 material. No sudden peaks were recorded for it, and with time the value of the friction force decreased slightly.

Figure 2 shows the maximum friction force recorded for individual materials during the test, while Fig. 3 shows the average friction force during the test. Comparing the two graphs, it can be noticed that the difference between the maximum and average friction force was: for Epoxy 63%, for F180 22%, for Super Metal 38% and for CFRP

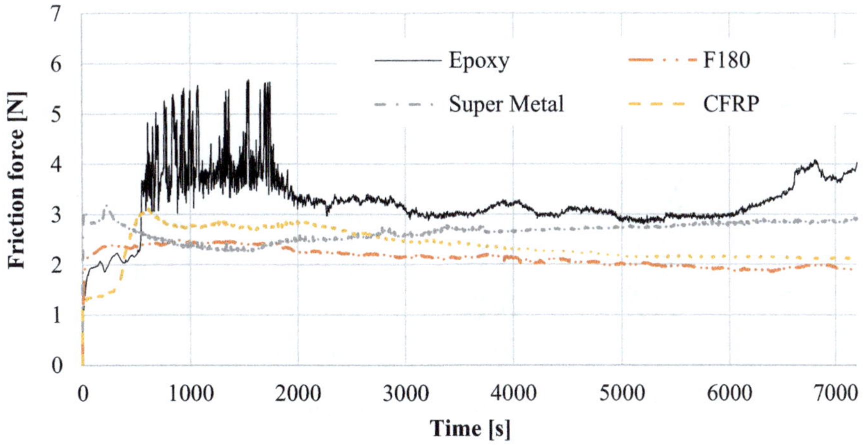

Fig. 1. Friction force during the test for samples from the tested materials.

32%. Both values were the smallest for F180 and the largest for Epoxy. After performing an analysis of variance with Tukey's post hoc test, differences were found between Epoxy and the rest, and between F180 and Super Metal. However, it cannot be said that at the 5% significance level there is a difference between F180 and CFRP.

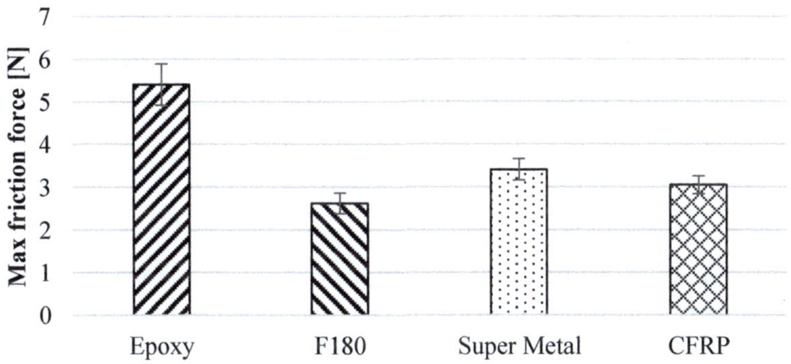

Fig. 2. Maximum friction force during the test.

Figure 4 shows the friction energy necessary to perform the test with the given parameters. This parameter was determined as the area under the diagram of friction force as a function of displacement. Numerical integration was performed with the Wolfram Mathematica program. After performing an analysis of variance with Tukey's post hoc test, differences were found between Epoxy and the rest, and between F180 and Super Metal. However, it cannot be said that at the 5% significance level there is a difference between F180 and CFRP.

Figure 5 shows sample cross-sections through the friction path, while Figs. 6, 7, 8 and 9 show 3D scans of the friction path surface.

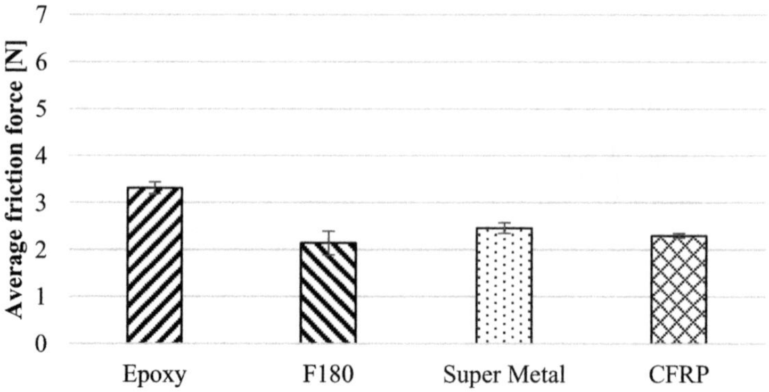

Fig. 3. Average friction force during the test.

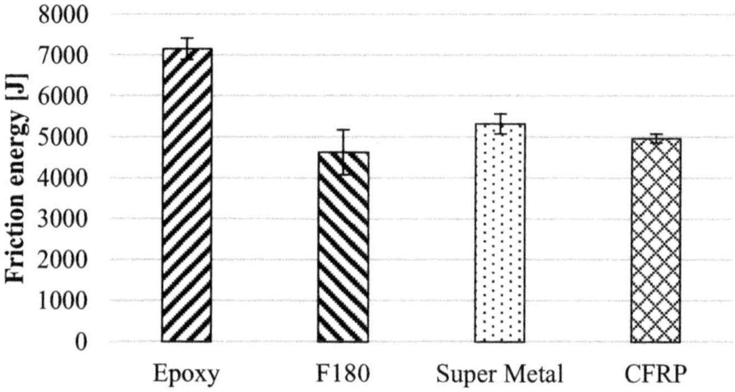

Fig. 4. Friction energy during ball-on-disk test.

Figure 10 shows the width of the friction trace for the tested materials, while. Figure 11 shows the depth of wear.

Comparing 3D scans, cross-section profiles and quantitative data on the depth and width of the abrasion marks, it can be seen that the Epoxy material is characterized by a very smooth surface of the friction mark without surface defects and micro craters. Despite this, it has the highest friction force and the greatest track depth and width. It is likely that the high wear in the first test phase resulted in a large contact area between the test material and the ball, which led to further surface smoothing in the second test phase. This hypothesis is supported by the results of the friction force over time.

The lowest wear was recorded for the F180 material, which was also characterized by the lowest values of friction force and friction energy. The 3D scan shows traces in the direction of the ball's movement over the sample, but the trace depth value itself is minimal. In addition to typical abrasive wear, one can also notice small plastic deformations manifested, among others, by elevations at the edge of the friction trace.

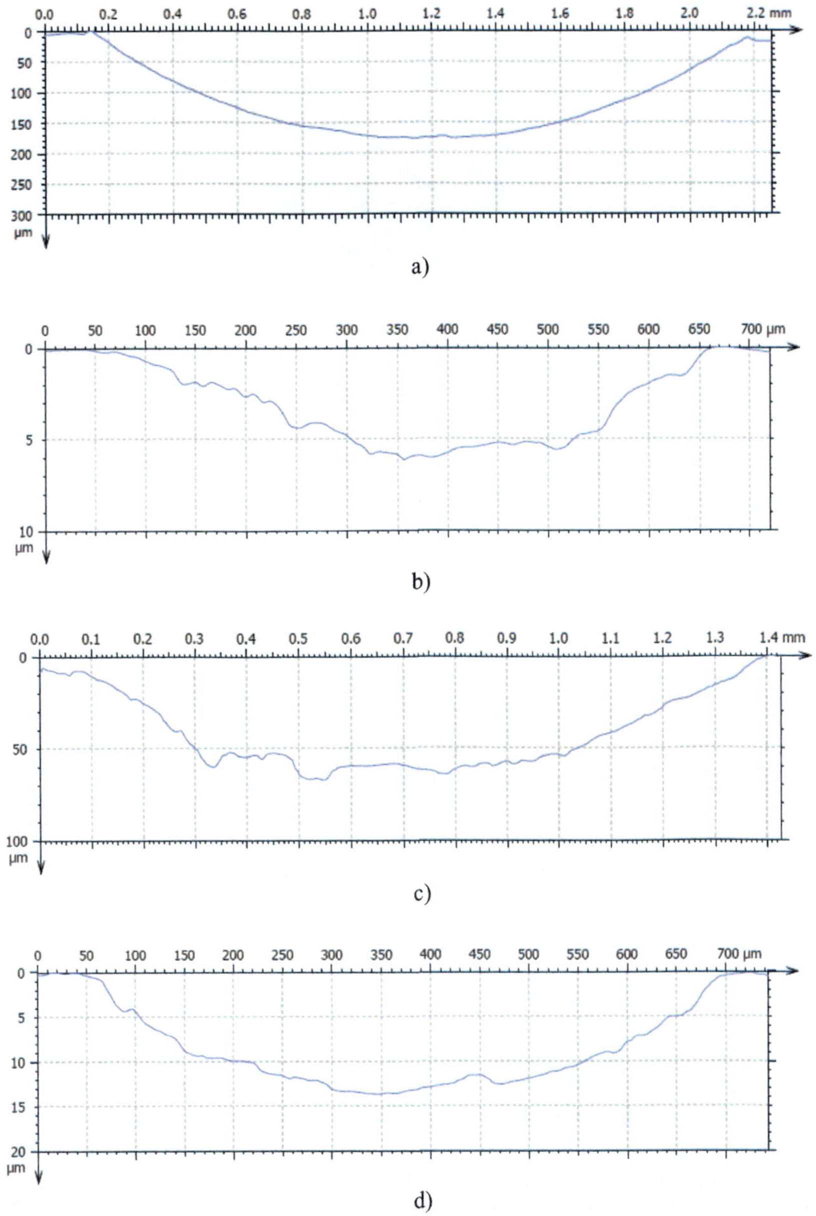

Fig. 5. Cross-sections through the friction path for materials: a) Epoxy, b) F180, c) Super Metal, d) CFRP.

In the case of the Super Metal surface, cracks and chipping can be noticed, leading to the deterioration of tribological properties. Both the friction force and the wear of this material were between Epoxy and F180.

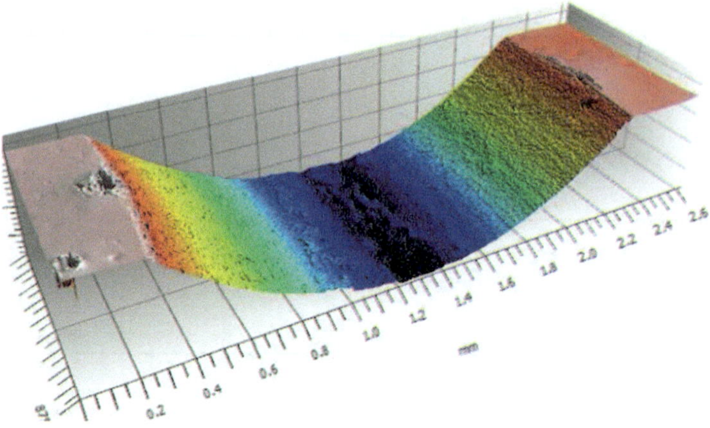

Fig. 6. 3D scan of the friction path surface for Epoxy material.

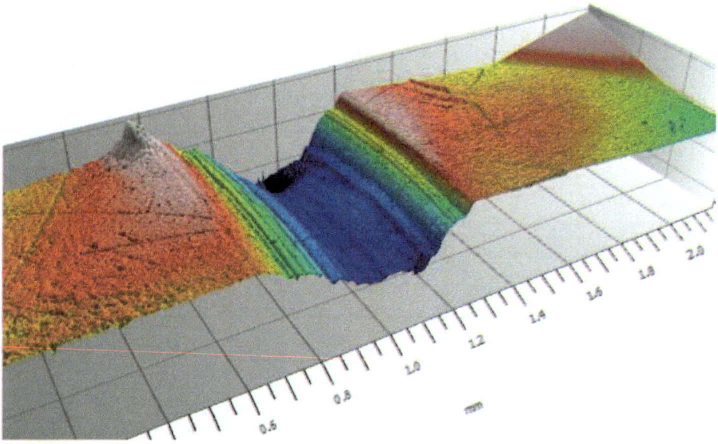

Fig. 7. 3D scan of the friction path surface for F180 material.

Similar, although smaller, chipping can be observed on the surface of the CFRP material. In this case, however, the friction force and wear were only slightly greater than for the F180.

In all cases, at the sliding contact, the tribological wear is additionally exacerbated by fatigue interactions. The ball sliding on the surface at any given moment is in contact with the coating at that point, similar to rolling a ball or a roller. Compressive stresses occur in front of the ball, and tensile stresses occur after the contact zone, and their size reaches very high values due to the stress concentration. This condition causes a continuous deformation of the coating-substrate system with a frequency related to the rotational speed.

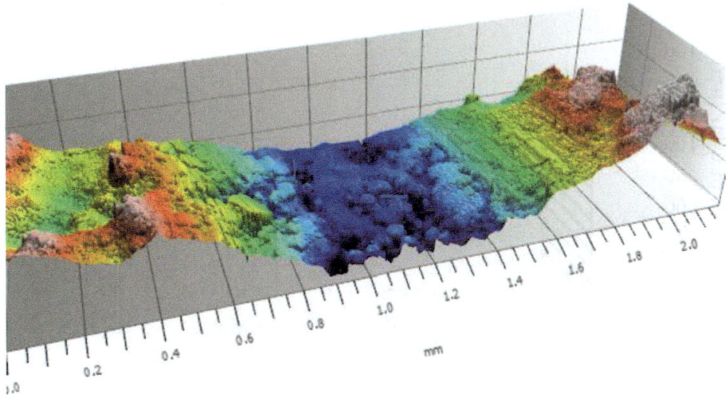

Fig. 8. 3D scan of the friction path surface for Super Metal material.

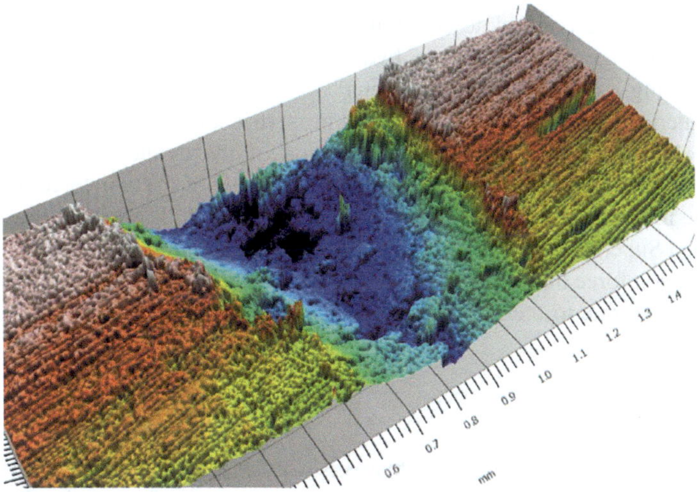

Fig. 9. 3D scan of the friction path surface for CFRP material.

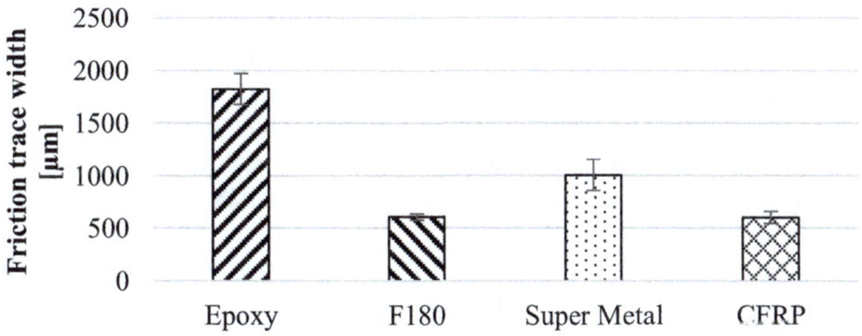

Fig. 10. Friction trace width.

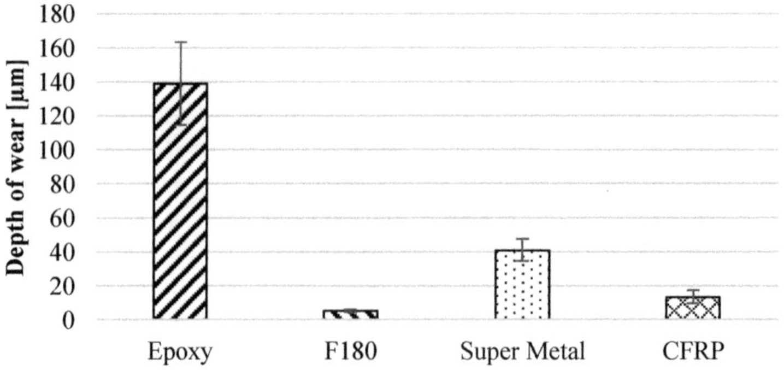

Fig. 11. Depth of wear.

4 Conclusion

The paper presents ball on disk tests of four materials potentially qualifying as materials for internal coatings in a composite hydraulic cylinder. A wear analysis was also carried out using a non-contact profilometer.

The material with the best tribological properties turned out to be F180 polyurethane, which will be used in the production of the cylinder prototype. It was characterized by the lowest wear and the lowest values of friction force and friction energy. The highest values were achieved by Epoxy, which disqualifies it as a sliding coating. Also Super Metal material was characterized by significantly higher values of wear and friction force than F180. Despite the good results obtained for CFRP material, it cannot be used due to the weepage effect in pressure loaded composite parts.

References

1. Kaw, A.K.: Mechanics of Composite Materials. CRC Press, Boca Raton (2005)
2. Kim, B., Kim, B., Kim, J., Joe, C.: Study on the development of composite CNG pressure vessels. Cryogenics (Guildf). **38**, 131–134 (1998). https://doi.org/10.1016/S0011-2275(97)00123-9
3. Banerjee, S., Messidoro, P., Ferrarese, A., et al.: Structural health monitoring technology for aerospace composite propellant tank. In: ASME 2011 Conference on Smart Materials, Adaptive Structures and Intelligent Systems, ASMEDC, vol. 2, pp. 563–570 (2011)
4. Baragetti, S., Terranova, A.: Limit load evaluation of hydraulic actuators. Int. J. Mater. Prod. Technol. **14**, 50 (1999). https://doi.org/10.1504/IJMPT.1999.036260
5. Marczewska, I., Bednarek, T., Marczewski, A., et al.: Practical fatigue analysis of hydraulic cylinders and some design recommendations. Int. J. Fatigue. **28**, 1739–1751 (2006). https://doi.org/10.1016/j.ijfatigue.2006.01.003
6. Mantovani, S.: Feasibility analysis of a double-acting composite cylinder in high-pressure loading conditions for fluid power applications. Appl. Sci. **10**, 826 (2020). https://doi.org/10.3390/app10030826
7. Solazzi, L.: Design and experimental tests on hydraulic actuator made of composite material. Compos. Struct. **232**,111544 (2020). https://doi.org/10.1016/j.compstruct.2019.111544

8. Krawczyk, J.: Functional verification and performance studies of the gerotor pump made of plastics. In: Stryczek, J., Warzyńska, U. (eds.) NSHP 2020. LNME, pp. 386–395. Springer, Cham (2021). https://doi.org/10.1007/978-3-030-59509-8_35
9. Chihai, R., Ungureanu, C., Cojan, A., et al.: Organic modified epoxy resin. tribologic aspects. In: Padgurskas, J. (ed.) The 10th International Conference BALTTRIB'2019 14–16 November 2019. Kaunas (2019)
10. Majdič, F., Strmčnik, E.: Tribological aspects of water hydraulics. In: SERBIATRIB '19 16 th International Conference on Tribology . Kragujevac (2019)
11. Li, G., Xu, H., Wan, H., et al.: Tribological properties and corrosion resistance of epoxy resin-polytetrafluoroethylene bonded solid lubricating coating filled with flake aluminum. J. Appl. Polym. Sci. **138**, 51003 (2021). https://doi.org/10.1002/app.51003
12. Zhang, E., Gao, F., Fu, R., et al.: Tribological Behavior of Phenolic Resin-Based Friction Composites Filled with Graphite. Materials (Basel). 14, 742 (2021). https://doi.org/10.3390/ma14040742
13. Kujawa, M., Kowalewski, P., Wieleba, W.: The Influence of Deformation under Tension on Some Mechanical and Tribological Properties of High-Density Polyethylene. Polymers (Basel) **11,** 1429 (2019). https://doi.org/10.3390/polym11091429
14. Lubecki, M., Stosiak, M., Leśniewski, T.: Comparative studies of tribological properties of selected polymer resins for use in hydraulic systems. Tribologia. **288**, 31–37 (2019). https://doi.org/10.5604/01.3001.0013.7764

Crash Test for the Evaluation of Vehicle Safety

Michal Ballay[1(✉)], Gustáv Kasanický[2], Pavol Kohút[2], and Ľudmila Macurová[2]

[1] Faculty of Security Engineering, University of Zilina, 1. mája 32, 010 26 Zilina, Slovakia
michal.ballay@fbi.uniza.sk

[2] Institute of Forensic Research and Education, University of Zilina, 1. mája 32, 010 26 Zilina, Slovakia
{gustav.kasanicky,pavol.kohut, ludmila.macurova}@uzvv.uniza.sk

Abstract. The article deals with the issue of vehicle safety. Impact tests are performed to evaluate the safety assessment of vehicles. Impact tests are also performed for the purpose of analysis and study of the mechanism of impact and movement of the vehicle after impact for the purpose of traffic accident analysis. The article contains an analysis of the dependence of the duration of the impact on the impact velocity, based on the results of impact tests. The aim was to find an approximation of the duration of the impact depending on the impact velocity and the approximation of the duration of the impact and in dependence on the overlap of the vehicle with a fixed obstacle.

Keywords: Crash test · Vehicle · Safety

1 Introduction

The theory of movement and impact of two vehicles for the needs of reconstruction and technical analysis of traffic accidents is described in many publications. However, only the current stage of development of computer technology has enabled its practical application in the analysis of traffic accidents in common expert practice.

Several programs are currently available for the reconstruction of vehicle accidents, which use a dynamic model of discrete forward simulation together with various shock models to simulate the movement of vehicles. Usually the simulation starts at the moment of impact, in which the vehicles are situated in their position just before the impact. Impact output parameters (vehicle speed, direction of movement vehicles, angular velocities of vehicles, etc.) are calculated using the impact model and then the required movement of the vehicles after the impact to their final positions is simulated.

O. Prentkovskis et al. (Eds.): TRANSBALTICA 2021, LNITI, pp. 194–201, 2022.
https://doi.org/10.1007/978-3-030-94774-3_19

2 Vehicle Safety Assessment in the Event of an Impact

Various impact tests are used to assess the safety of vehicles in an impact. In most cases, these tests are intended to simulate the collision of vehicles, such as frontal collision, side collision and rear collision. The nomenclature for vehicle collisions is given in the ISO standard. The EHK regulation – R 95 applies to frontal impact tests in Europe [1, 2].

The barrier according to Regulation R 33 is rigid (reinforced concrete) with a front width of at least 3 m and a height of 1.5 m. In the frontal test according to Regulation R 94, a barrier is used whose face is perpendicular to the direction of movement of the impacting vehicle, but the vehicle covers the face of the barrier by 40% of its width (Fig. 1). The barrier face is not rigid, but has a deformation structure with defined deformation characteristics [1, 3, 7].

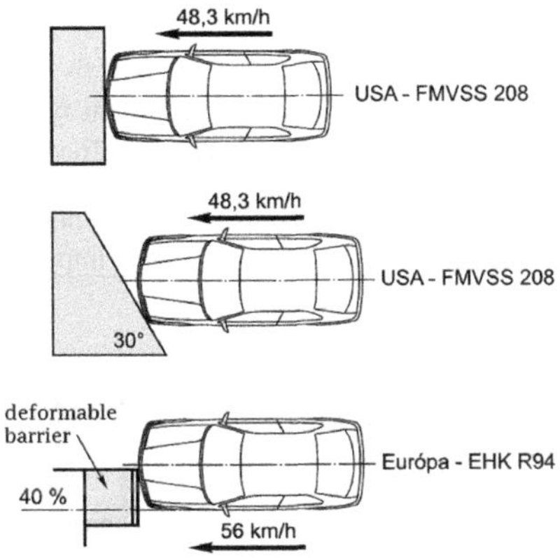

Fig. 1. Frontal crash test in the US and in Europe.

A movable barrier is used for the side impact according to Regulation R 95 (Fig. 2). The speed of the barrier at the moment of impact is 50 km/h.

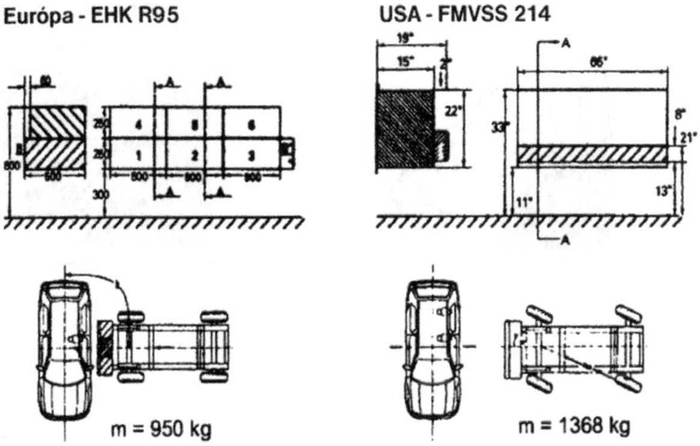

Fig. 2. Test equipment for a side collision between Europe and the US.

To test vehicles are also used in impact tests - is offset barrier (offset - crash, Fig. 3). The course of the collision and the appearance of the deformation for different types of impacts are shown in Table 1.

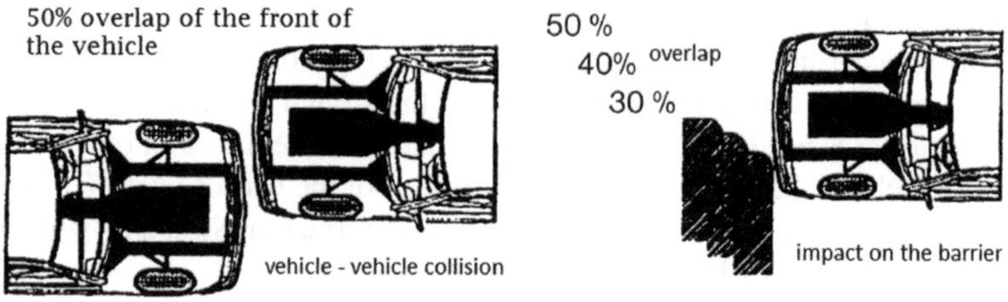

Fig. 3. Overlay vehicles [1].

Table 1. Collision process and deformation scale for different types of impact: A – vehicle – vehicle, B – vehicle – barrier (40% offset), C – vehicle – inclined barrier (30%) [1].

	Vehicles before and after impact	Orientation and forces on engine in impact
A – Collision of two vehicles	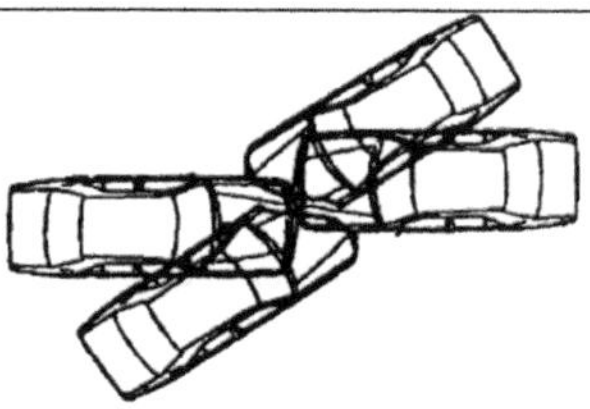	
B – Offset impact on a barrier	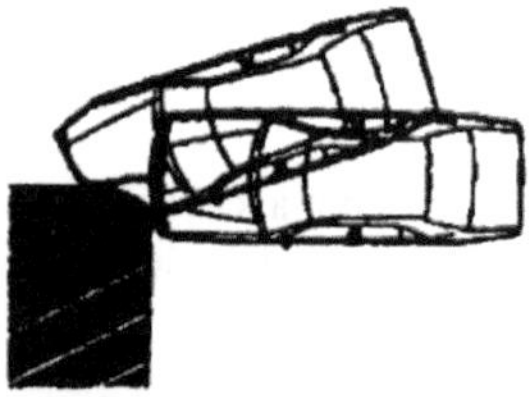	
C – Impact on an inclined barrier	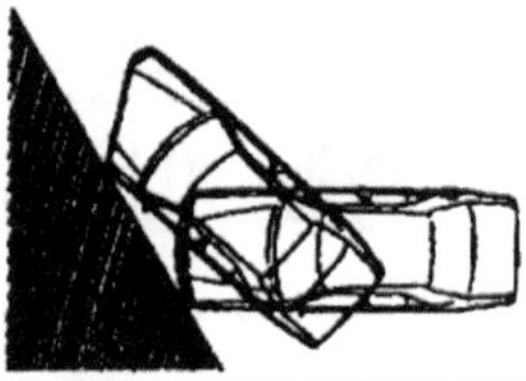	

3 Crash – Tests Simulating Traffic Accidents

We can derive the approximate time duration of impact according to impact speed and offset in impact on a non – deformable barrier from the analysis of results of crash – tests carried out by DSD Linz on November 2–3, 1996, in Graz, Austria [1, 2, 4, 8].

There were seventeen crash – tests in total. Table 2 shows the parameters used in finding the approximation relationship between time duration of impact and impact speed and in finding the approximation relationship between time duration of impact and impact offset. For a 100% offset of a Ford Escort we can state the following approximation relationship between time duration of impact and impact speed:

$$t = 0.609v + 51.932, \quad (1)$$

where t – time duration of impact (ms), v – impact speed (km/h).

Table 2. Crash – test parameters measured for a Ford Escort and Ford Sierra (impact on a non-deformable barrier) [1, 5, 6].

	Impact speed (km/h)	Impact offset on a non-deformable barrier (%)	Time duration of impact (ms)
Test no. 1	13	100	60
Test no. 3	37.5	100	75
Test no. 5	52	100	100
Test no. 8	95	100	110
Test no. 9	29.5	50	140
Test no. 10	40.5	30	170
Test no. 11	27.5	70	130

The measured values as well as the approximated graph are shown in Fig. 4.

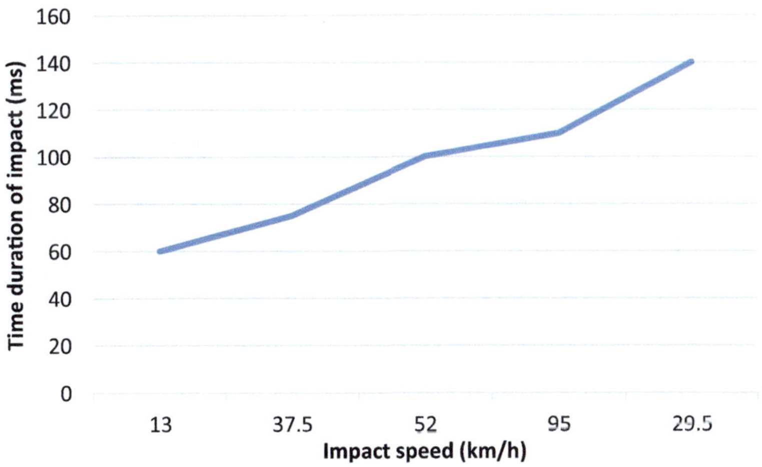

Fig. 4. Measured values and approximated graph of time duration of impact depending on impact speed for a Ford Escort (impact on a non-deformable barrier at full offset).

By analogy we can derive an approximation relationship between the time duration of impact and impact offset for a Ford Escort and Ford Sierra (in impact on a non-deformable barrier from speeds of 30–40 km/h):

$$t = -1.3617p + 211.36, \tag{2}$$

where t – time duration of impact (ms); p – impact offset (%).

The measured values as well as the approximated graph are shown in Fig. 5.

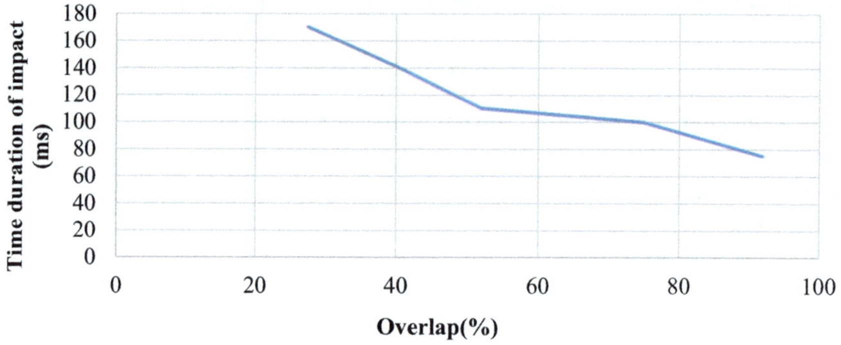

Fig. 5. Measured values and approximated graph of time duration of impact depending an impact offset for a Ford Escort and Ford Sierra (impact on a non – deformable barrier from speeds of 30–40 km/h).

Figures 6, 7, 8, 9, 10, 11 and 12 illustrate the graphs of speed and acceleration as recorded by the UDS as well as vehicle deformations.

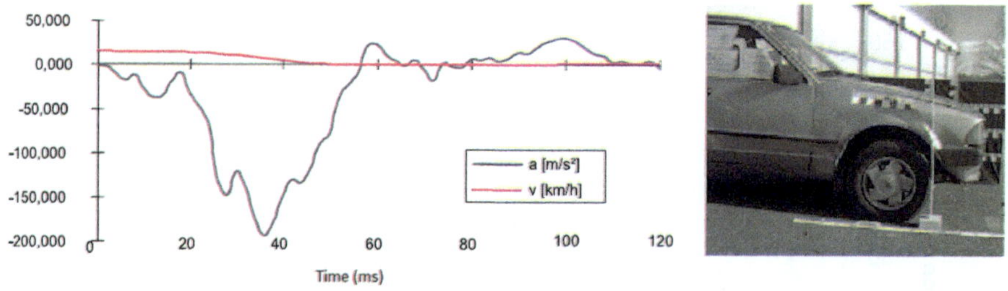

Fig. 6. Graphs of speed and acceleration in test no. 1 and vehicle deformations (on the right).

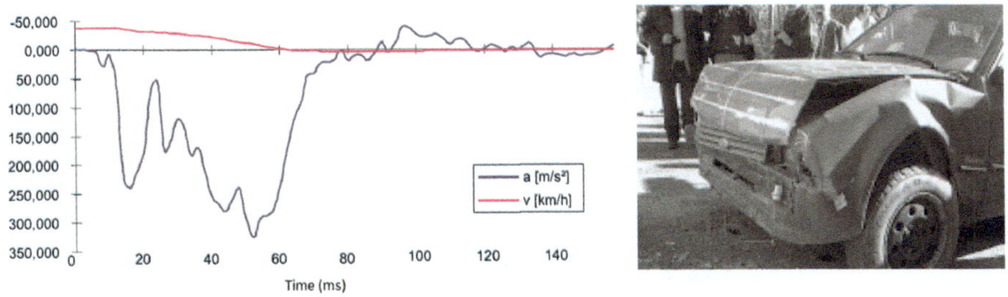

Fig. 7. Graphs of speed and acceleration in test no. 3 and vehicle deformations (on the right).

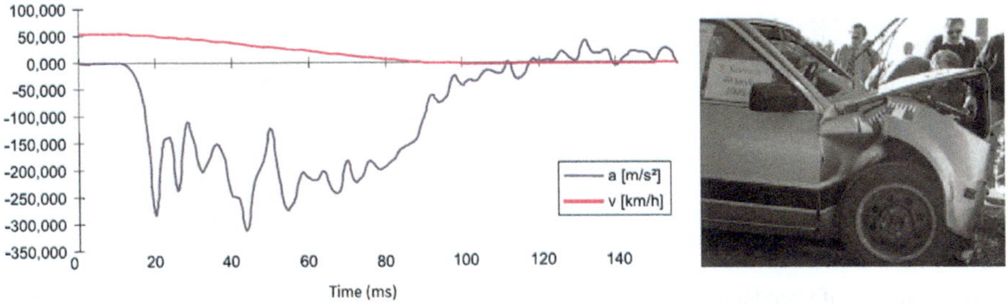

Fig. 8. Graphs of speed and acceleration in test no. 5 and vehicle deformations (on the right).

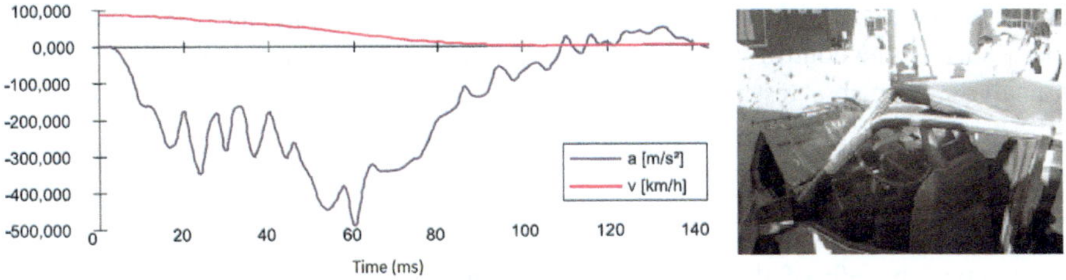

Fig. 9. Graphs of speed and acceleration in test no. 8 and vehicle deformations (on the right).

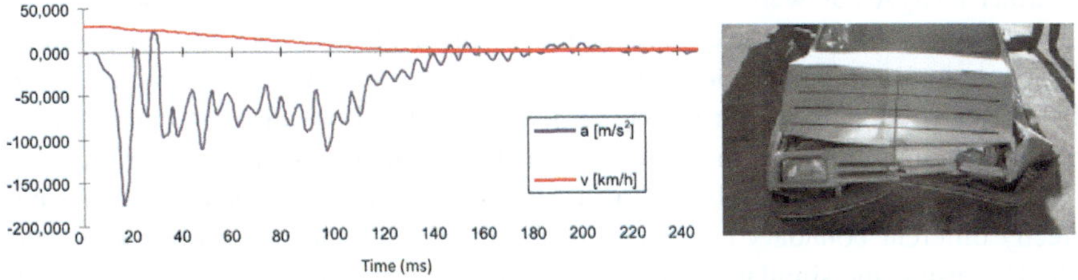

Fig. 10. Graphs of speed and acceleration in test no. 9 and vehicle deformations (on the right).

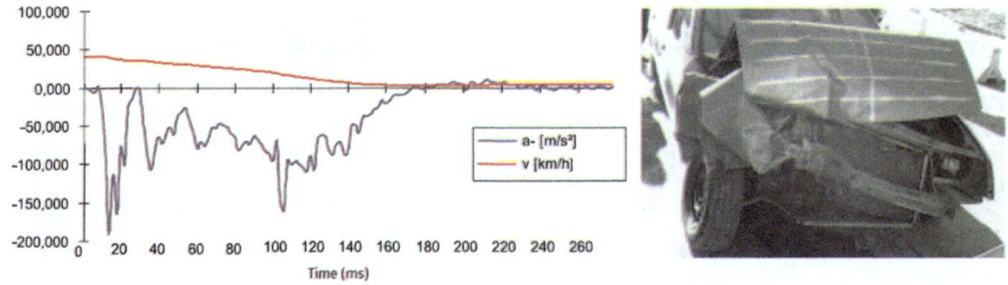

Fig. 11. Graphs of speed and acceleration in test no. 10 and vehicle deformations (on the right).

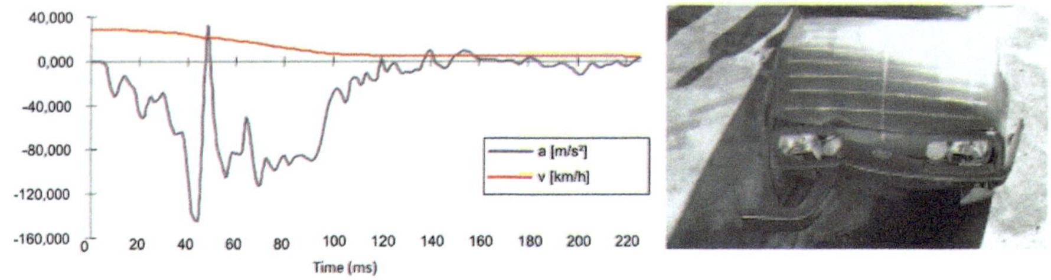

Fig. 12. Graphs of speed and acceleration in test no. 11 and vehicle deformations (on the right).

These crash – tests are carried out primarily to study the impact mechanism, its processes, vehicle deformations and vehicle motion depending on different impact parameters (velocity, impact offset, etc.). Some of the phenomena presented in this article, such as the control value of speed difference in the point of impact, were based on analysis of crash – test results.

4 Conclusions

The practical benefit of this article is mainly in the description of the dependences between the impact speed of the vehicle and the duration of the collision and also the dependence between the overlap of the vehicles and the duration of the collision. Knowledge of the duration of the collision depending on various parameters is essential for further analyzes as well as partial tasks in the process of technical analysis of the accident. This is, for example, solving the problem of passenger movement in the vehicle interior during a vehicle collision, where it is necessary to know the time of the collision, or solving the issue of the vehicle's distance from the first contact to the formation of digging tracks in the road - which is needed to analyze the crash.

The role of the expert in the technical analysis of a traffic accident is to determine correctly different boundary conditions of the simulation and perform the simulation itself. The results the simulations are then compared with reality. Experience helps the expert to find a solution to the real case by being able to determine correctly various technical parameters of the accident (road friction coefficient, range vehicle damage, the nature of skid tracks, the nature of vehicle movement after a collision, etc.). Nevertheless, several simulations have to be performed, that the values determined by the experts are confirmed and that the input and control parameters in each case they had the correct values. Some of these parameters they are sometimes entered only on the basis of their effect on the resulting error, which is possible performed only using computer algorithms.

Acknowledgement. The article was created with the support UNIZA grant system number 8010/2020.

References

1. Kasanický, G., Kohút, P., Lukášik, M.: Impact dynamics theory for the analysis and simulation of collisions. Pubslished by University of Zilina (2004). ISBN 80-8070-312-4
2. Moser, A., Steffan, H., Kasanicky, G.: Pedestrian model in PC-Crash –The Introduction of multi body system and its validation. In: Accident Reconstruction Technology and Animation IX (1999). SAE SP-1407
3. Kasanický, G., Kohút, P.: Impact dynamics theory for the analysis and simulation of collisions, Žilina: University of Žilina, EDIS, 350 pp. (2004). ISBN 80-8070-312-4
4. Vlk, F.: Motor vehicle bodies, Brno, 243 s (2000). ISBN 80-238-5277–9
5. Ballay, M., Macurová, L., Kohút, P., Čopiak, M.: Development of road safety status and the evaluation criterion causes of specific traffic accidents. In: Transport Means - Proceedings of the International Conference, 2018-October, pp. 765–770 (2018)
6. Steffan, H., Moser, A.: Unfallversuche -die Dynamik der Kollision, 6. Jahrestagung Europäischen Vereins fur Unfallforsuchung und Unfallanalysee.V., Bratislava (1997)
7. Leitner, B., Figuli, L.: Fatigue life prediction of mechanical structures under stochastic loading In: Machine modelling and simulations - ISSN 2261-236X (online). - Londýn: Édition Diffusion Presse Sciences. - s. [1–11] [online]. ISSN 2261-236X, MATEC Web of Conferences. Spôsob prístupu: (2017). https://www.matec-conferences.org/articles/matecconf/pdf/2018/16/matecconf_mms2018_02024.pdf
8. Leitner, B.: Autoregressive models in modelling and simulation of transport means In: Transport Means 2010: Proceedings of the 14th International Conference, 21–22 October, 2010, Kaunas University of Technology, Lithuania. Kaunas University of Technology, Kaunas, pp. 21–24 (2010). ISSN 1822-296X

Concept Idea of Engineer's Decision Support System

Andrius Macutkevičius and Raimundas Junevičius(✉)

Vilnius Gediminas Technical University,
Saulėtekio av. 11, 10223 Vilnius, Lithuania
{andrius.macutkevicius,
raimundas.junevicius}@vilniustech.lt

Abstract. Every vehicle has different driveline structure and different components; therefore, we must be mindful choosing one. In the early designing stages, deciding which structure and which component with specific requirements fits is not always that obvious as each engineer has its own experience and understanding about product it working with. Therefore, it is worthy to have an instrument, which can evaluate as many perspectives as possible of each drive and offer a solution to the decision-making task. The aim is to review possibilities using artificial intelligence in component selection, based on existing examples of the artificial intelligence uses. All processes in power drives are sufficiently, precisely and widely described, for this reason mathematical model creation is not the biggest challenge. Using design of experiment methods, mathematical model can be used as a data source for artificial intelligence (AI) training. Various structures of neuron network (NN) are used to reach higher accuracy of AI data processing. Many researches concentrate on particular component characteristics choice, but in a case of drive selection it's a complex combination of all drive components, so the possibilities of using AI (artificial intelligence) in solving interdisciplinary tasks are noted.

Keywords: Decision support system · Artificial intelligence · Engineering design

1 Introduction

Adopting a modern data architecture is vital for a digital transformation. A fundamental aspect with a significant impact on such transformation is to have a well-defined goal for modelling and handling the data in a system. The number of concepts in a system defines complexity. If this stays unmanaged, it can be seen as a major roadblock on the path towards an AI factory [2]. In this paper we are presenting methods and general structure of the tool which could help the engineer to make a specific decision in design process. The idea of the method is to develop an AI algorithm, which could make comparison of some specific design scenarios and propose set of parameters with some specific interval of values. This information will help the design team to develop product faster and to include human expectation to the design task not on assumptions but through the values of design parameters.

O. Prentkovskis et al. (Eds.): TRANSBALTICA 2021, LNITI, pp. 202–211, 2022.
https://doi.org/10.1007/978-3-030-94774-3_20

2 General Idea of Engineer's Decision Support System

In our method, we propose a tool, which will analyze possible new designing product capabilities by paying attention to some specific characteristics. These characteristics could be some physical parameters, or determined limitations based on human sensitivity.

Three type sets of data will be presented: 1. Data that will come from numerical simulation; 2. Data that will come from physical measurements; 3. Data that will come from generalized human expectation. The last data type could be created taking questionnaires or summarizing human responses to some specific vibration, acceleration or other irritant.

To solve the task, it's appropriate to acknowledge these few steps; one, Data preparation, two, AI algorithm development, three, Result verification. For the first generation model, only synthetic data will be used. To proceed, numerical simulations will be done and the set of significant parameters will be selected.

3 Data Preparation

Decisions and choices using known and calculated information usually leads to improvements. In this case it's AI information with data sets of known object parameters with important output of values. Nowadays emerging data sources include all information that can be detected and stored, this make more accurate prediction, forecasting and decisions makings through AI techniques [3]. But in other hand, the more data sets parameter will be used in model, the more time it takes to learn it and the more elaborate AI is needed. So parameters and data preparation is very important stage to create AI solution.

Data mining object can become mathematical or physical models with changeable parameters. The corresponding component characteristics are obtained by using different combinations of parameters and tests of various conditions. Design of experiment methods can be used in this step.

Factorial analysis is one of DOE (Design of Experiments) methods, which allow parameter analyses. Suppose, that we have parameters sets:

$$X = \begin{pmatrix} x_{01} & x_{11} & x_{21} & \cdots & x_{k1} \\ x_{02} & x_{12} & x_{22} & \cdots & x_{k2} \\ \vdots & \vdots & \vdots & \vdots & \vdots \\ x_{0N} & x_{0N} & x_{0N} & \cdots & x_{kN} \end{pmatrix}. \quad (1)$$

And the output parameters are:

$$Y = (y_1 y_2 \cdots y_N). \quad (2)$$

The following equation of parameters dependencies can be formed:

$$y = b_0 + b_1x_1 + b_2x_2 + \cdots + b_kx_k, \tag{3}$$

where b – weight coefficients.

Weight coefficients can be written as follows:

$$B = \left(X^TX\right)^{-1}X^TY,$$

$$b = \left(b_0b_1 \cdots b_k\right)^T. \tag{4}$$

Once the coefficients are set, it is possible to determine which parameter has a stronger effect on a particular output parameter. This will eliminate low significance parameters.

It is important to select the data input boundaries correctly because the method uses bordering values of parameters generating these sets of data. And the number of analysed simulations depends on the number of evaluated parameters:

$$N = 2^k.$$

Pareto graph is formed by analysing input parameters influence on output parameters. When results of the factorial analysis are calculated, it is possible to determine which parameters mostly affect the desired output parameters. As a result, in specific power drive design Pareto graph shows which parameters are more important than the others. This method is useful to determine model and experiment parameters and its ranges. Also it helps to simplify the complexity of the solution.

4 Artificial Intelligence

In actual world artificial intelligence (AI) is widely studied and developed. AI in mainly associated with the imitation of human thinking. The greatest achievement of AI development is digital computer program which could mimic and transcend a person's psychological capabilities, including reasoning, perception, creativity and emotions. Despite many philosophical and ethical issues, the application of knowledge in this area is expanding rapidly [1].

Advances in computing equipment during the last 21st century decades, expands the possibilities of artificial intelligence. In the second half of the 20th century began the development of advancing AI capabilities to replicate human reasoning. The following systems have been applied in different engineering processes. Since that, the classification of AI begun [6]:

1. Knowledge based systems.
2. Fuzzy logic.
3. Inductive learning.
4. Neural network.
5. Genetic algorithm.

Analysing artificial intelligence abilities, new trends, discoveries and the concept of AI changes with time. Initially, AI was perceived as any computer algorithm which can simulate human behaviour. Currently this view has expanded. Now mostly AI can use knowledge from learning. So AI creators doesn't need to code every AI action [8]. The term machine learning (ML), commonly used when talking about AI, refers to the ability of machine to "learn". In recent years, the term deep learning (DL) has been introduced in the field of AI. DL itself can be considered as subset of ML. DL is the next step in evolution of AI, developed to accomplish ML [11].

ML and DL learning used neural networks and deep neural networks algorithms. Neural networks (NN) based on natural neurons connections principle. Neuron can have any number of inputs x. Each input has weight coefficient w, which describe input and neuron interaction. So neuron state:

$$S = \sum_{i=1}^{n} x_i w_i. \tag{5}$$

Neurons state inertia is required, so output of neuron described using activation function:

$$Y = f(S). \tag{6}$$

Most popular activation function is sigmoid:

$$f(x) = \frac{1}{(1 + e^{-ax})}, \tag{7}$$

Neural network consists of layers with parallel neurons. Inputs and outputs connected with every layer of neurons. For this structure of NN it usually uses a back propagation method. Then data sets back propagating through NN a relation between inputs and output is formed by reducing the error value:

$$E(w) = \frac{1}{2}\sum_{j=1}^{p} (y_i - d_i)^2, \tag{8}$$

where y_i – current neuron output; d_i – current neuron output target; p – neuron number in output layer.

Gradient descent method is used for NN learning. In each iteration neuron weight changes by formula:

$$\Delta w_{ij} = -\eta \cdot \frac{\partial E}{\partial w_{ij}}, \tag{9}$$

where η – learning speed coefficient.

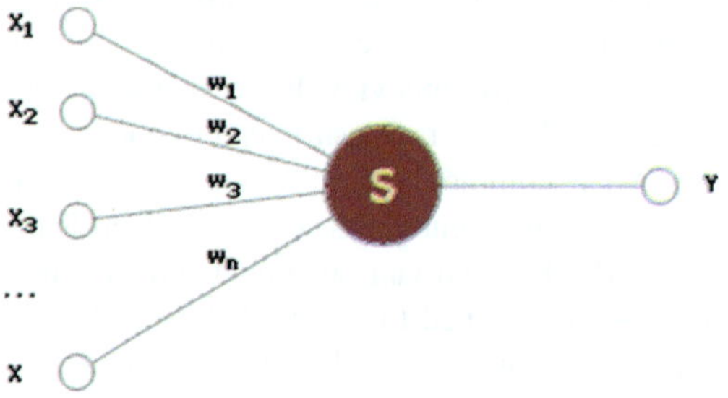

Fig. 1. Neuron mathematical model scheme.

Multi-stage trained groups of neural networks, that finally merge into one is offered in a features engineering methodology by M. Malin. Neural networks use function input parameter values to create interdependencies, which predict the best value of the output parameter. The essence of this method is to finalize the neural network model using accumulated data that neural networks have learned from interactions of input parameters. It is important to mention, that it does not matter what the output values of these neural networks are, because only the values of the last hidden layer is useful. These values with the input function parameters became the input parameters of final neural network model which shown in Fig. 2. Final model output parameters become result of the function [4].

Most interestingly is that such structure is able to ensure that the level of accuracy is equal to other popular methods like XGBoost and MLP with embedding.

The use of artificial neural networks (NN) is useful when there is a need to comprehend complex and hard to define dependencies. NN is used for the classification and processing of large data sets, where samples of classified data are already known [7]. Such systems have proven their worth in the processing of visual material. Also, it was proposed to train the neural network by generating different data sets of transmission types. But the latter system, presented in Fig. 3, repeat already designed sets [5].

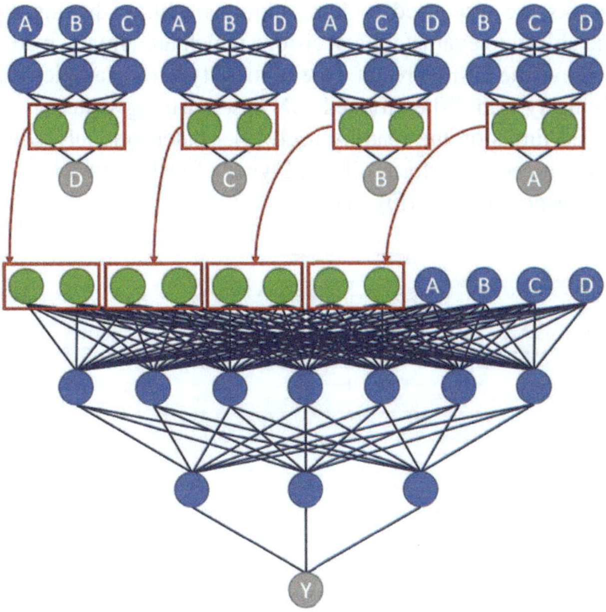

Fig. 2. Final neural network model [4].

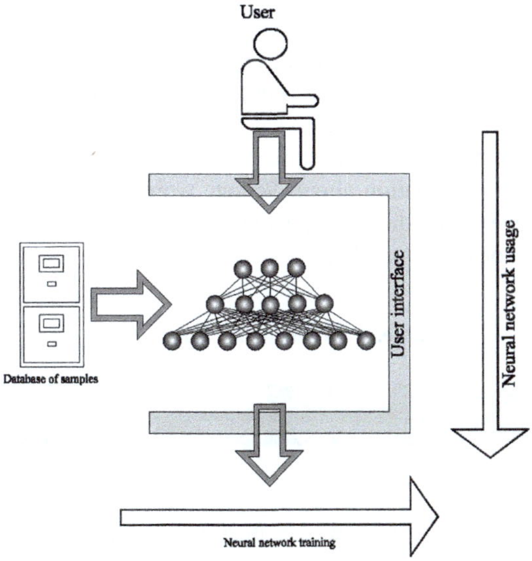

Fig. 3. Neural network learning and usage principal scheme [5].

Neural networks are also used to find solutions without having sternly defined methods with the ability to evaluate the decisions which are already made or are still in the making to achieve the desired result. One of these type of examples is offered by the autonomous driving system. When trained, NN based driving system is able to comply the lanes and speed limits on the road by itself [9]. Understandably, different types of NN have been developed to perform different types of tasks, but this is an example of AI decision-making possibilities (Fig. 4).

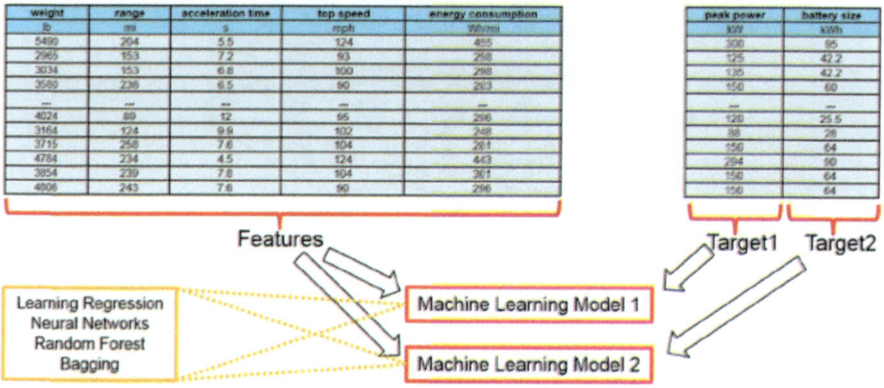

Fig. 4. Learning scheme [12].

Monitoring rising fuel prices lead to development of electric military vehicles. At the beginning of the design, there was a problem in selecting the optimal components initial parameters. Therefore, it is suggested to draw conclusions from the available information. This information from electric cars on the market have formed data set. The data set collected information about: vehicle weight, acceleration characteristics, maximum speed developed, distance traveled, maximum speed developed, power, battery size, energy consumption [12].

Table. 1. Methods accuracy [12].

	Linear Regression	Random Forest	Bagging	Neural Networks
Battery size mean error percentage	3.93%	9.34%	5.58%	5.50%
Battery peak power mean error percentage	20.72%	16.28%	16.96%	24.10%

Information processing and it's use is done with machine training technique. 80% of all available information is used to algorithm training and remaining 20% to test it.

In the initial stage, one must perform the selection of the most important parameters. Several methods are used to indicate the most important parameters. Their effectiveness is evaluated during testing. In this case, the advantage of machine training, is having a possibility to draw conclusions about vehicles being developed at the moment from already developed electric cars data. As example, the production and extraneous problems on the battery parameters is assessed using the solutions of the vehicles that are present on the market.

There are four methods used: linear regression, random forest, bagging and neural networks. The results are shown in the Table 1.

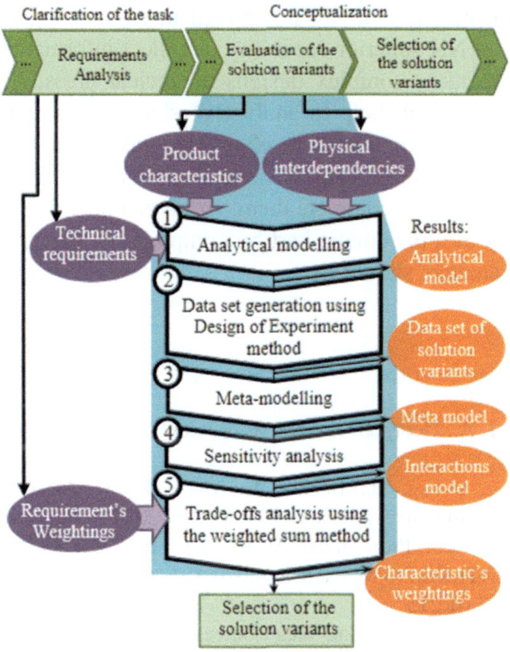

Fig. 5. Scheme of parameter analysis and modelling methods [10].

Results show that the size of the batteries is determined quite accurately according to desired parameters, however the maximum capacity of the battery can vary about 25%. Inaccuracies can appear due to insufficient data and different technology uses. Therefore, artificial intelligence methods are able to generate similar to data accuracy solutions.

Methodology which can solve optimization task, when optimization object is complexly dependent on different nodes and dimensions, but on purpose to optimize its parameters, it becomes difficult to evaluate impact of each parameter [10].

Therefore, it is proposed to create an analytical model, based on physical interdependence by testing a product using simulations, and adjusting the parameters to observe how optimized parameters change. A selection of successful combinations along with simulation output parameters, subsequently initiates a save for each of them as a data set. This data set can then be processed using neural networks or regression equations. As shown in Fig. 5 next stage is to assess the interrelationships between elements, so sensitivity analysis is performed. Thus, the most important data is extracted and used to evaluate the impact of each variable parameter on the output parameter by weight summation method. Then there will be no need to reuse other methods, only to make decisions based on the needs of weight.

5 Conclusion

AI is very popular technique and can be applied in engineering as decision support system. AI is based on NN superior capability to collect information from data which gives ability to create more complex designs in unclear areas. It will not solve all engineering problems but will help to analyze product design at its initial stage and compare different designs among themselves also taking in to account end user expectations. Considering these solutions are not finalized, for that reason future developments are necessary to complete them. As an example, it is planned to propose the design parameters of a pneumatic kart by comparing what type of components to use and what its parameters should be.

References

1. Alkrimi, J., Ahmad, A., George, L., Aziz, S.: Review of Artificial Intelligence **2**, 487–505 (2013)
2. Alvarez-Coello, D., Wilms, D., Bekan, A., Marx Gómez, J.: Towards a data-centric architecture in the automotive industry. CENTERIS 2020 - Int. Conf. Enterp. Inf. Syst. ProjMAN 2020 - Int. Conf. Proj. Manag. HCist 2020 - Int. Conf. Health Soc. Care Inf. Syst. Technol. 2020 CENTERISProjMANHCist 2020 181, pp. 658–663 (2021). https://doi.org/10.1016/j.procs.2021.01.215
3. Khan, A.I., Al-Badi, A.: Emerging data sources in decision making and AI. Procedia Comput. Sci., The 11th International Conference on Emerging Ubiquitous Systems and Pervasive Networks (EUSPN 2020)/The 10th International Conference on Current and Future Trends of Information and Communication Technologies in Healthcare (ICTH 2020)/ Affiliated Workshops 177, pp. 318–323 (2020). https://doi.org/10.1016/j.procs.2020.10.042
4. Malin, M.,: Automated Feature Engineering Using Neural Networks (2021)
5. Miltenovi, V.: Application of artificial intelligence methods in gear transmitters conceptual design UDC 007. 52, 62–233 (2001). https://www.semanticscholar.org/paper/APPLICATION-OF-ARTIFICIAL-INTELLIGENCE-METHODS-IN-.-Miltenovi/eb3940995e6461f0c8f0834716d45b83cc752ab7. Accessed 21 June 2021
6. Pham, D.T., Pham, P.T.N.: Artificial intelligence in engineering. Int. J. Mach. Tools Manuf. **39**, 937–949 (1999). https://doi.org/10.1016/S0890-6955(98)00076-5
7. Philip Chen, C.L., Zhang, C.-Y.: Data-intensive applications, challenges, techniques and technologies: a survey on Big Data. Inf. Sci. **275**, 314–347 (2014). https://doi.org/10.1016/j.ins.2014.01.015
8. Pohlink, C.: What is Artificial Intelligence without Machine Learning? (2019)

9. Sallab, A.E., Abdou, M., Perot, E., Yogamani, S.: Deep reinforcement learning framework for autonomous driving. Electron. Imaging **2017**, 70–76 (2017). https://doi.org/10.2352/ISSN.2470-1173.2017.19.AVM-023
10. Shabestari, S.S., Bender, B., Neumann, M., Song, Y.-W.: Decision support for design conflicts: a model-based method to analyze the interactions between technical requirements and product characteristics. Procedia Manuf. **52**, 203–208 (2020). https://doi.org/10.1016/j.promfg.2020.11.035
11. van Assen, M., Lee, S.J., De Cecco, C.N.: Artificial intelligence from A to Z: from neural network to legal framework. Eur. J. Radiol. **129**, 109083 (2020). Doi: https://doi.org/10.1016/j.ejrad.2020.109083
12. Xu, B., Rizzo, D., Onori, S.: Machine Learning Based Optimal Energy Storage Devices Selection Assistance for Vehicle Propulsion Systems (2020). https://doi.org/10.4271/2020-01-0748

Resistance to Impact of Loads of Shock Nature of Protective Coatings Intended for Protection of Boat Devices of Water Transport

Andriy Buketov[1], Oleksandr Sapronov[1(✉)], Abdellah Menou[2], Olha Syzonenko[3], Anna Sapronova[1], and Sergey Panin[4]

[1] Kherson State Maritime Academy, Kherson, Ukraine
[2] Moroccan Airports Authority, Mohammed V International Airport, Casablanca, Nouaceur, Morocco
a.menou@onda.ma
[3] The Institute of Pulse Processes and Technologies of NAS of Ukraine, Mykolaiv, Ukraine
sizonenko43@rambler.ru
[4] National Research Tomsk Polytechnic University, Tomsk, Russia
svp@ispms.ru

Abstract. The dynamics of fracture epoxy composites with different content of synthesized iron-carbidotitanic powder mixture containing the following phase composition of high-modulus compounds after treatment with high-voltage electric discharged synthesis is studied: Fe (70%) + Ti (5%) + TiC (20%) + Fe3C (5%).

It is proved that the introduction of the active additive into the epoxy binder at the optimal content ($q = 0.025$ pts.wt.) can increase the toughness of polymer composites by 1.7 times compared to the epoxy matrix. It is shown that the increase in impact resilience is associated with the orderliness of the structure of composites filled with high-modulus compound at crosslinking polymer composites. To determine the resistance of the developed composite materials to shock loads, changes in the propagation spectra of cracks under shock load over time are studied, which allows to control relaxation processes by changing the structures and, consequently, the properties of polymeric materials. The fracture surface of composite materials is investigated by optical microscopy. Based on complex researches of structure and mechanical characteristics, regularities of process of structure formation of composite materials are established. It is shown that to increase the mechanical characteristics it is necessary to increase the α-relaxation time, which in turn leads to a change in the nature of the destruction of polymeric materials, i.e. their structure.

Keywords: Synthesized iron-carbidotitanic powder mixture · Epoxy composite · Fracture surface · Crack propagation · Relaxation · Fracture structure

O. Prentkovskis et al. (Eds.): TRANSBALTICA 2021, LNITI, pp. 212–220, 2022.
https://doi.org/10.1007/978-3-030-94774-3_21

1 Introduction

During the operation of water transport, considerable attention is paid to the reliability of deck mechanisms, especially ship's boat devices, which are life-saving equipment. The operational term of life-saving equipment is affected not only by external factors (air with a temperature of 233 K to 323 K (±20 K), relative humidity up to 95% at a temperature of 293 K (±20 K), ultraviolet radiation, low atmospheric pressure, sea water, dew, fog), but also by mechanical factors (abrasive wear, impact of loads, friction, vibration) [1–4]. These and other factors lead to damage to the deck mechanisms, in particular the elements of the boat devices (cylinder, pulleys) and as a consequence to the failure of the mechanism or unit in general. For example, under the influence of moist air, the chemical composition of metal structures changes, corrosion occurs, which reduces the mechanical strength of the parts of the boat devices. Periodic change of the cold and warm seasons has a particularly detrimental effect on metal structures and parts of life-saving equipment, as a result of which not only their characteristics deteriorate, but the dimensions of working elements change, joints are destroyed, and deformations occur. With increasing temperature, the yield strength and modulus of elasticity of the metal decreases, aging accelerates, there is a risk of cracking, both in protective coatings and directly in metal structures and parts of life-saving equipment. For metals and alloys in the range of high loads and relatively low temperatures there is a mechanism of destruction due to the process of successive fluctuation of their bonds in the crystal lattice. Vibration leads to an increase in deformations and cracks, i.e. contributes to the accumulation of local defects, the emergence of fatigue [5–8]. Therefore, it can be stated that the influence of the above factors in the operation of life-saving equipment directly affects their speed in emergencies. One of the ways to ensure the reliability of metal structures and parts of boat devices is the development of new protective coatings with predictable properties, which will prevent the formation of deformations, cracks, rupture of strings in the crystal lattice [9, 10].

The purpose of the work is to develop a composite material resistant to impact loads.

2 Materials and Methods

The epoxy diane oligomer (ED-20 grade, State standard «GOST 10587-84») was chosen as the basic binder component under the Polymer Composite Material (PCM) formation. It is characterized with a set of improved properties in comparison with other known thermoset materials. Among them are: high adhesive bonding to metal substrates; ability of running curing process at low temperatures; lack of volatile organic emissions at product formation, easy processibility as well as low coating shrinkage at deposition onto long length surfaces of complex profile parts. The molecules of epoxy oligomers contain glycidyl and epoxy groups that can interact with curing agents to form cross-linked structure in the form of a net. All these result in substantially improving of the PCM characteristics [1, 3, 5].

The curing agent polyethylene polyamine PEPA (Industrial Standard «TU 6-05-241-202-78») was used for cross-linking of the epoxy composites. It allows running curing process at room-temperatures. As is known from the literature the PEPA is a low-molecular substance that consists of such structural monomeric units: $[–CH_2–CH_2–NH–]_n$. Different stages of cross-linking were varied and investigated under loading of the curing agent into composite at the following mixing rate of the components: 10 pts. wt. on 100 pts. wt. of epoxy oligomer ED-20 [1, 3, 5].

Synthesized iron-carbidotitanic powder mixture (SICPM) at a content of $q = 0.025\ldots5.000$ pts.wt. is used as an active additive for experimental studies. The filler is obtained by high-voltage electric discharge (HVED) synthesis. A mixture of powders of the following starting composition are used as the starting material: Fe (75%) + Ti(25%). During the synthesis, the accumulated energy of a single discharge (W_1) is 1 kJ, and the integrated specific energy of the processing (W_{spes}) is 25 MJ/kg. Due to HVED-processing, grinding of all treated particles is achieved and their phase composition changes in the direction of creating high-modulus compounds Fe (70%) + Ti(5%) + TiC(20%) + Fe_3C (5%) [11–13], which is established by quantitative X-ray phase analysis using a DRON-4-07 diffractometer under CuKα radiation.

Filled with SICPM epoxy composite, formed by the technology described in the works [14–17]. The topology of fracture surface and impact strength of CM were investigated by optical microscope MBS-10.

The impact resilience was measured with the use of notch less specimens with the sizes of 10 × 15 × 75 mm (ASTM D6110). The tests were carried out with the use of impact pendulum machine RKP-300 under high-rate loading (5.2 m/s). In doing so, the loading diagrams "loading-time" and "loading-bending reflection" were registered. A total work of impact fracture A is composed of two components: i) crack initiation A_i and ii) crack propagation A_p. The impact resilience was determined as [11–13]:

$$W = \frac{A}{bs}, \tag{1}$$

where A is the impact energy, consumed for the specimen fracture; b is a width of a specimen; s is a thickness of a specimen.

The use of the impact loading diagram processing software «VUHI-CHARPY» made it possible to determine fracture energy components by transforming «force – time» (P-t) dependence into «force – displacement» (P-s) one.

3 Experimental Results

It is found that the impact strength of the epoxy matrix is $W = 0.7$ J/cm^2 (Fig. 1, curve 1), which is consistent with the results of studies [14–16]. The introduction of SICPM at a content of $q = 0.025$ pts.wt. leads to a maximum increase in performance of 1.7 times ($W = 1.2$ J/cm^2).

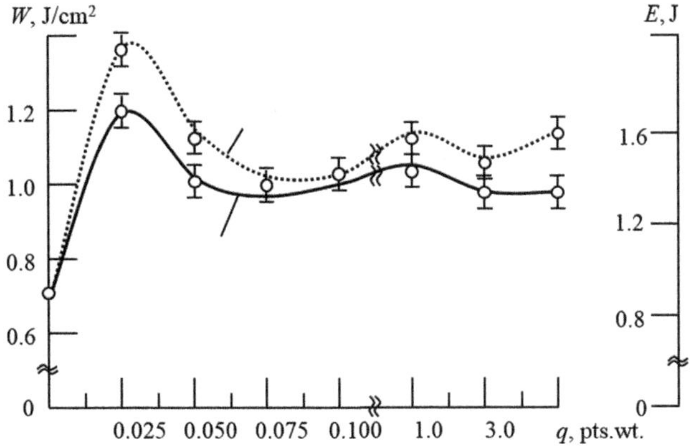

Fig. 1. Dependence of impact resilience (*W*) and energy (*E*) spent on the destruction of composite materials on the content of the synthesized iron-carbidotitanic powder mixture: 1 – impact resilience; 2 – energy expended on the materials destruction.

It is believed that such a content is the interaction of active hydroxyl OH and carbon C-C groups on the surface of the filler particles with macromolecules and segments of the epoxy oligomer, while providing uniform distances between the nodes of the structuring. The orderliness of the additive in the volume of the polymer contributes to the intensification of the crosslinking process and provides an increase in the mechanical properties [17] of the composite. Accordingly, the energy expended on the destruction of the composite on impact increases from $E = 0.90$ J to $E = 1.93$ J. Subsequent introduction of the synthesized iron-carbidotitanic powder mixture ($q = 0.050\ldots5{,}000$ pts.wt.) provides a slight reduction in impact resilience, which is $W = 0.9\ldots\ 1.1$ J/cm^2. The obtained values of impact resilience are 1.2…1.5 times higher than the epoxy matrix, which indicates the interfacial interaction of the components during polymerization. In this case, the decrease in impact resilience with increasing content of active additive is associated with the uneven distribution of particles by volume due to the supersaturation of the compositions with filler particles and a decrease in the sedimentation stability.

Taking into account that the mechanism of fracture of epoxy composites under impact loads is associated with the process of nucleation and propagation of microcracks [14] it is important to study these characteristics when the load (*P*) changes over time and analysis of the deformation of composite material from the load (*P*) (Table 1, Fig. 2, Fig. 3). The authors [15] showed that the propagation of cracks in polymeric material is divided into three discrete stages: crack growth, stable crack propagation, rapid and unstable crack growth, where the filler affects the trajectory of cracks in the polymer volume.

At the same time, the previously highlighted provisions are confined in the work and the mechanism of destruction of epoxy composite materials are considered [14–17], namely: the origin of the crack, its spread and direct fracture of materials. It should be noted that the stage of crack propagation (i.e. the time of crack propagation) can be compared with the relaxation processes of crosslinked polymers, which in turn will allow to control the processes of structure formation of filled polymer systems.

Table 1. Characteristic of the parameters P, τ, and l on destruction of epoxy composite materials filled with a synthesized iron-carbidotitanic powder mixture.

Content of a synthesized iron-carbidotitanic powder mixture, *q*, pts.wt	Maximum shock of destruction, *P*, kN	Time of crack propagation, *τ*, ms	Value of critical deformation *l*, mm
0.025	3.2	0.25	1.15
0.050	0.7	0.12	0.75
0.075	2.2	0.16	0.90
0.100	1.8	0.15	0.80
1.000	2.2	0.17	0.95
3.000	–	–	–
5.000	1.7	0.12	0.75

Therefore, it is found that the maximum load that leads to the destruction of the composite material is $P = 3.2$ kN (for CM with the content of SICPM of $q = 0.025$ pts. wt.). Such materials are characterized by branching in the polymer volume of micro-cracks with clearly defined straight chipping lines propagating in different directions (Fig. 4, a). This indicates the creation of barriers for the propagation of cracks by SICPM particles, which provides partial bending of the crack propagation trajectory and, as a consequence, an increase in crack propagation time.

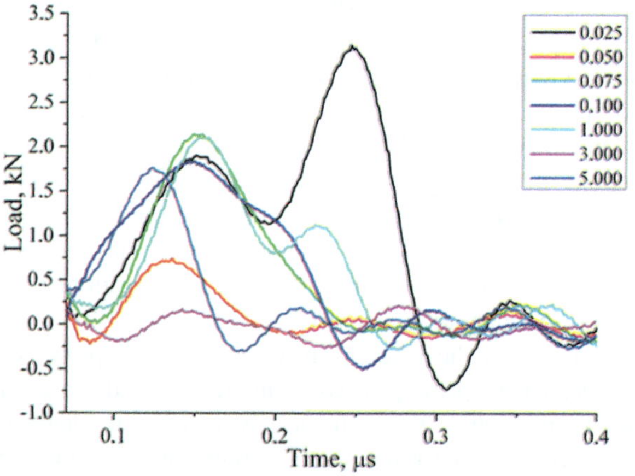

Fig. 2. Dependence of the crack propagation time of composite materials with different content synthesized iron-carbidotitanic powder mixture, *q*, мас.ч.: 1 – 0.025; 2 – 0.050; 3 – 0.075; 4 – 0.100; 5 – 1.000; 6 – 3.000, 7 – 5.000.

Thus, considerable attention in quantitative analysis must be paid to the time of propagation of the crack until the moment of destruction. For composites filled with synthesized iron-carbidotitanic powder mixture with a content of $q = 0.025$ pts.wt. the the maximum value of crack propagation time is $\tau = 0.25$ ms (among the studied materials).

It should be noted that the stage of crack initiation (Fig. 2) in terms of time parameters (areas where there is a noticeable pulsation) coincides with the so-called β-relaxation in polymers, i.e. it is the movement of individual lateral functional groups. And the characteristic time before the beginning of fracture (time of crack propagation) coincides in order of magnitude with α-relaxation. That is, it is the main relaxation process, the excitation of molecular motions of individual segments of the polymer network and the main chain. Increasing the α-relaxation time from τ = 0.10 ms [14] to τ = 0.25 ms (Fig. 2, Table 1) increases the mobility and flexibility of epoxy macromolecules, which in turn leads to increased mobility groups and segments of the polymer grid, and provides an increase in impact resilience from W = 0.70 J/cm^2 to W = 1.2 J/cm^2 (which is 2.0 times higher than the matrix value).

At the same time, the value of critical deformation increases to the maximum l = 1.15 mm (with the content of SICPM q = 0.25 pts.wt.), which indicates a fairly elastic properties of epoxy composites due to the ability of macromolecules to deform as a result of cyclic loads.

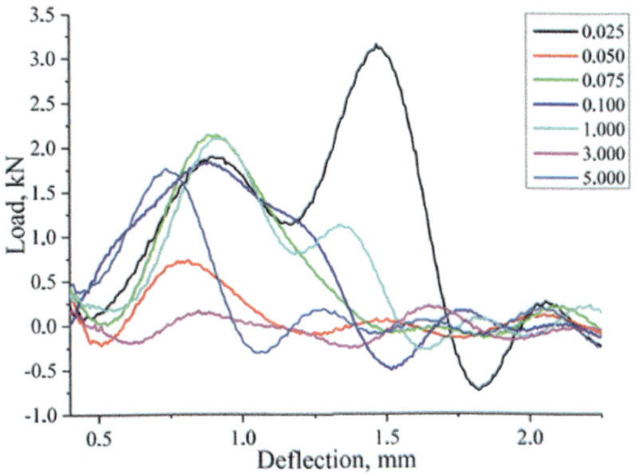

Fig. 3. Dependence of the deflection on the shock in composite materials with different content of particles of the synthesized iron-carbidotitanic powder mixture, q, мас.ч.: 1 – 0.025; 2 – 0.050; 3 – 0.075; 4 – 0.100; 5 – 1.000; 6 – 3.000, 7 – 5.000.

Increasing the content of SICPM to q = 0.050…5.000 pts.wt. leads to a nonlinear decrease in the maximum value of the load at the time of failure of the CM to P = 2.2… 0.7 kN and, accordingly, the crack propagation time decreases to τ = 0.17…0.12 ms (Table 1), which is associated with the formation of a significant number of microcracks that pass into the main cracks (Fig. 4, b–e) in the volume of the polymer. Thus, the presented analysis of macrodegradation of epoxy composites (Fig. 4) filled with particles of synthesized iron-carbidotitanic powder mixture allows stating that for the developed materials it is necessary to consider the theory of brittle fracture which has a transient character. The brittle fracture at the top of the crack is thought to break the chemical bonds, which are accompanied by the consumption of energy to overcome potential barriers. In this case, the relaxation time (Fig. 2, curve 2, 6, 7) with brittle fracture is minimal, which is consistent with the results of experimental studies. Resilient fracture is characterized by an increase in relaxation time (Fig. 2, curve 1, 3, 4, 5), which creates conditions for increasing the flexibility of polymer macromolecules and leads to a decrease in brittleness and, consequently, an increase in impact resilient.

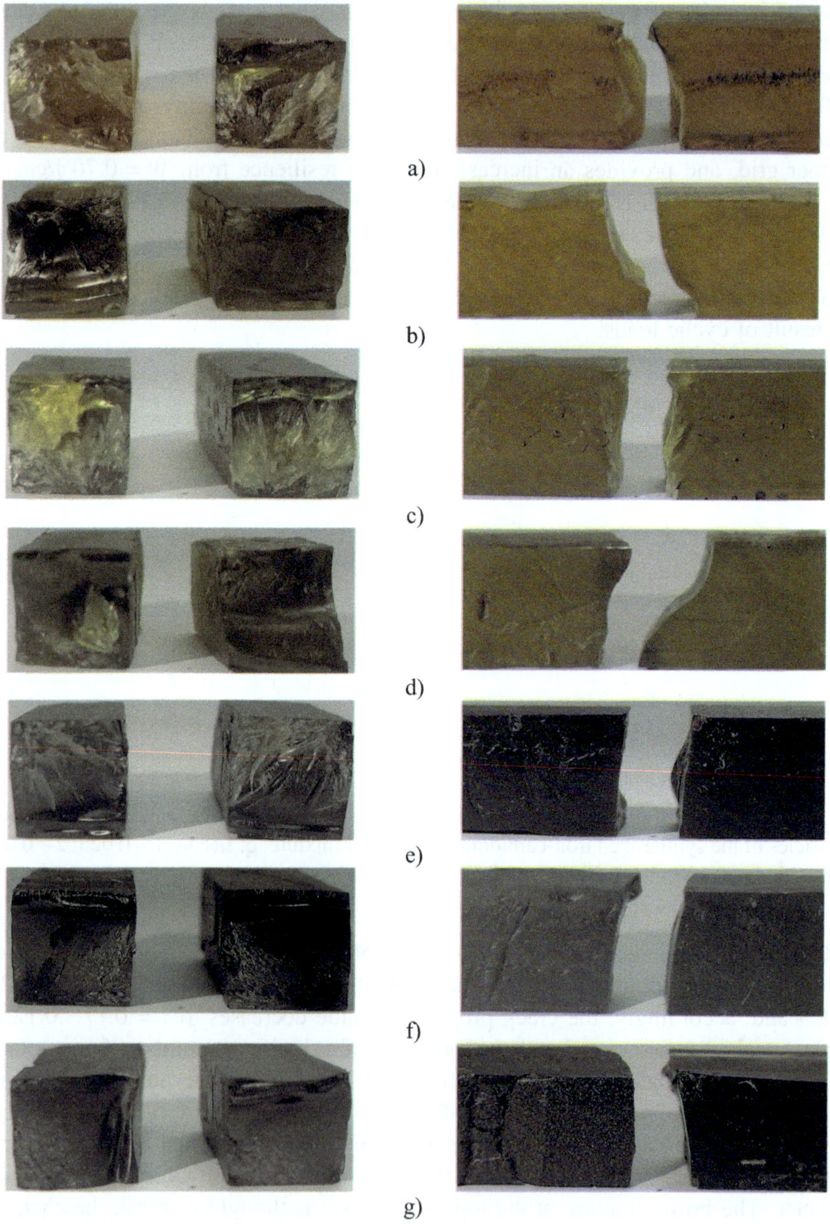

a)

b)

c)

d)

e)

f)

g)

Fig. 4. Type of macrodegradation of epoxy composites filled with synthesized iron-carbidotitanic powder mixture, q, pts.wt: a) 0.025; b) 0.050; c) 0.075; d) 0.100; e) 1.000; f) 3.000; g) 5.000.

4 Conclusion

The regularities of failure of epoxy composites filled with particles of synthesized iron-carbidotitanic powder mixture under impact loading are obtained in the investigations, which allows determining the composition of the protective coating intended for protection of boat water transport devices. Also:

It is found that the fractions of the synthesized powder iron-carbidotitanium charg synthesized iron-carbidotitanic powder mixture at the optimal content $q = 0.025$ pts. wt. contribute to the formation of the material with the maximum degree of crosslinking. This provides a maximum increase in the impact resilience of composites compared to the matrix from $W = 0.7$ J/cm^2 to $W = 1.20$ J/cm^2, and the energy expended on the destruction of chemical bonds increases from $E = 0.9$ J to $E = 1.93$ J.

It is proved that to increase the impact strength of epoxy composites it is necessary to increase the α-relaxation time by increasing the mobility of the segments of the polymer network and the main chain of the epoxy polymer, it is shown that the use of synthesized powder iron-carbidotitanic powder mixture in the amount of 0.025 pts.wt. allows to increase the time of α-relaxation by 2.5 times, and accordingly the value of impact strength by 2.0 times.

The micromechanisms of failure of epoxy composites with different content of the synthesized powder iron-carbidotitanic powder mixture is investigated by the method of optical microscopy. Based on the complex researches the theory of fracture of the developed composite materials is developed. It is proved that resilient fracture is characterized by an increase in α-relaxation time and, accordingly, an increase in cohesive strength, which ensures the stability of the developed materials during operation. Whereas brittle fracture is characterized by a decrease in the α-relaxation time and, accordingly, mechanical characteristics.

References

1. Zayed, A., Garbatov, Y., Guedes Soares, C.: Reliability of ship hulls subjected to corrosion and maintenance. Struct. Saf. **43**, 1–11 (2013)
2. Buketov, A., et al.: New black-filled epoxy coatings for repairing surface of equipment of marine ships. Transport **35**(6), 679–690 (2020)
3. Buketov, A., Maruschak, P., Sapronov, O., Zinchenko, D., Yatsyuk, V., Panin, S.: Enhancing performance characteristics of equipment of sea and river transport by using epoxy composites. Transport **31**(3), 333–342 (2016)
4. Soares, C.G., Garbatov, Y., Zayed, A., Wang, G.: Influence of environmental factors on corrosion of ship structures in marine atmosphere. Corros. Sci. **51**(9), 2014–2026 (2009)
5. Buketov, A., et al.: The use of complex additives for the formation of corrosion- and wear-resistant epoxy composites. Adv. Mater. Sci. Eng. **2019**, 1–5 (2019). https://doi.org/10.1155/2019/8183761
6. Lysenkov, E.A., Stryutskiy, O.V., Gomza, Y.P., Klepko, V.V.: The influence of carbon nanotubes on the sensitivity of humidity sensors based on organic-inorganic polymer materials. Funct. Mater. **22**(1), 40–46 (2015)

7. Sapronov, O., et al.: Development and use of new polymer adhesives for the restoration of marine equipment units. J. Mar. Sci. Eng. **8**(7), 527 (2020). https://doi.org/10.3390/jmse8070527
8. Sapronov, O., Maruschak, P., Buketova, N., Leschenko, O., Panin, S.: Investigation of Pm-75 carbon black addition on the properties of protective polymer coatings. In: 2016 AIP Conference Proceedings. Advanced Materials with Hierarchical Structure for New Technologies and Reliable Structures, Tomsk, Russia, vol. 1783, pp. 020194-1–020194-4 (2016). https://doi.org/10.1063/1.4966488
9. Grytsenko, O., Spišák, E., Dulebová, Ľ, Moravskii, V., Suberlyak, O.: Sorption capable film coatings with variable conductivity. Mater. Sci. Forum **818**, 97–100 (2015)
10. Matkovska, L., et al.: Structural peculiarities of ion-conductive organic-inorganic polymer composites based on aliphatic epoxy resin and salt of lithium perchlorate. Nanoscale Res. Lett. **12**(1), 423 (2017)
11. Sapronov, O.O., et al.: Research of crack initiation and propagation under loading for providing impact resilience of protective coating. Funct. Mater. **26**(1), 114–120 (2019)
12. Buketov, A.V., et al.: Dynamics of destruction of epoxy composites filled with ultra-dispersed diamond under impact conditions. Mech. Adv. Mater. Struct. **27**(9), 725–733 (2020)
13. Buketov, A.V., et al.: Influence of the structure of epoxy composite filled with discrete fibers on impact fracture of vehicle parts. Compos. Mech. Comput. Appl. Int. J. **11**(2), 113–127 (2020)
14. Dittanet, P., Pearson, R.A.: Effect of silica nanoparticle size on toughening mechanisms of filled epoxy. Polymer **53**(9), 1890–1905 (2012)
15. Brodnik, N.R., Hsueh, C.-J., Faber, K.T., Bourdin, B., Ravichandran, G., Bhattacharya, K.: Guiding and trapping cracks with compliant inclusions for enhancing toughness of brittle composite materials. J. Appl. Mech. **87**(3), 031018 (2020)
16. Lajtai, E.Z.: A theoretical and experimental evaluation of the Griffith theory of brittle fracture. Tectonophysics **11**(2), 129–156 (1971)
17. Visser, H.A., Caimmi, F., Pavan, A.: Characterising the fracture toughness of polymers at moderately high rates of loading with the use of instrumented tensile impact testing. Eng. Fract. Mech. **101**, 67–79 (2013)

Vehicle ABS Braking Performance on Road with Pavement Obstacles

Vidas Žuraulis(✉) and Aayush Chugh

Vilnius Gediminas Technical University,
J. Basanavičiaus street 28, 03224 Vilnius, Lithuania
vidas.zuraulis@vilniustech.lt,
aayush.chugh@stud.vgtu.lt

Abstract. Efficient and reliable vehicle braking is one of the most important part of vehicle constructional safety. The implementation of active safety, such as Anti-Lock Braking System (ABS), should improve the performance of vehicle braking; however, there are some cases when unfavourable driving conditions complicate the provided operation. The ABS braking performance is investigated in this paper when the fluctuations of wheel angular and peripheral movement are caused by road with pavement obstacles. The issue of ABS is investigated organising hardware-in-the loop (HIL) system with eight-phase algorithm for control of wheel braking pressure. The opportunity to restrict a complete control cycle after detection of rough pavement braking avoiding protracted wheel locking and a dynamic correction of the thresholds is estimated.

Keywords: Anti-lock braking system · ABS phase · ABS threshold

1 Introduction

The safety of passengers travelling in vehicles is in relatively high risk as an estimated 22,800 road traffic fatalities were recorded in the 27 European Union (EU) member states in 2019 [1]. However, the effectiveness of braking system of modern vehicles has always been one of the primary factors that can save the passenger's life or reduce the consequences. The EU has set a goal – no deaths and serious injuries on European roads by 2050 [2]. However, to reach "Vision Zero" the safer vehicles, safer infrastructure, better use of protective equipment, lower speeds, and better post-crash care is necessary.

Various Study reveal that vehicle Anti-lock Braking System (ABS) works effectively on even roads but on rough terrains the performance of ABS starts to deteriorate due to a number of factors such as axle and tire oscillations, wheel speed fluctuations and deficiencies in the ABS algorithms [3]. This causes the decrease of vertical load between tyre road pavement that results in lower contact force and undesirable vibrations for wheel - road interaction.

This paper aims to investigate the performance of ABS on rough terrains by developing eight-stage ABS algorithm in MATLAB/Simulink. An experimental

O. Prentkovskis et al. (Eds.): TRANSBALTICA 2021, LNITI, pp. 221–229, 2022.
https://doi.org/10.1007/978-3-030-94774-3_22

research in laboratory conditions using vehicle quarter car test rig is done taking into account the fluctuations in wheel vertical and angular dynamics at different road excitations.

2 Literature Review

With the increase in Sport Utility Vehicles (SUV), which conceptually can be used in unpaved road conditions, a new challenge appears to the braking efficiency. The specific issues also occurs for braking of heavy commercial vehicles [4]. ABS performance deteriorates on rough road conditions which further leads to increase in stopping distance and reduced lateral control of the vehicle [5]. Vibrations in wheel angular speed, as for example the wheel enters an obstacle, will also result in fluctuations in contact, since the generated tyre friction directly dependends on wheel slip ratio, normal load and speed [6, 7]. Controller malfunction occurs because of non-accurate signals and faults in the algorithm that leads to poor control decisions [8].

Road unevenness cause variation of friction between the tyre and the rough surface, which directly affects ABS braking and the design trends of a new control algorithms [9, 10]. A scientific reason of larger braking distance on rough terrains is well defined by Hamersma and Els as fluctuations in wheel loads because of rough surface and vehicle body motions [11]. Vertical force variation negatively influences longitudinal and lateral force thus increasing the braking distance. Wheel rotation also varies due to terrain excitation from pavement unevenness and this causes why ABS braking deteriorates on rough terrains [3]. These factors mainly causes a problems of incorrect inputs to the ABS control logic [12, 13]. The deterioration of ABS braking on rough terrain was analysed by both on-road testing and computer simulation [3]. General issues causing the deterioration of ABS at this case was the speed measurement during on-road tests and the existing response delays during modelling. Flat road pavement is usually normally the optimal choice for input of ABS logic at the design stage, because of simplifies simulation in various operating conditions [14]. However, it was also found a shorter braking distance using ABS when the tyre operates closer to its peak force on a hard terrain [15].

3 ABS Control Implementation

For this research, an ABS controller based on wheel rotation kinematics (mainly acceleration) has been modelled utilising ABS controller by the automotive manufacturer Bosch [16]. The ABS algorithm operates through several control phases and is designed around a set of pre-determined thresholds. When vehicle brakes are activated, the controlled parameters enters the first phase of the algorithm, where the brake pressure increases until the peripheral wheel acceleration crosses the threshold ($-a$). Then a holding state is (Phase 2) resulting that brake pressure would not increase more. Once the wheel slip threshold (λ_1) has been reached, the brake pressure is reduced (Phase 3) until the wheel peripheral acceleration exceeds the threshold ($-a$). At Phase 4 a pressure is held resulting wheel's acceleration. If the peripheral wheel acceleration again reaches the threshold ($-a$) then the tire slip is assumed to be in the unstable region and the controller returns to Phase 3, where the pressure is reduced, and the cycle is repeated. For driving velocity lower minimal (usually 5–10 km/h), the ABS logic is disengaged and the brake pressure is allowed to increase until the vehicle reaches a complete stop. One of the issue with this algorithm is that it only works at specific values of acceleration at any given velocity. Therefore, if the velocity of vehicle is changed, the values of acceleration thresholds also needs to change accordingly.

The main task of ABS is to prevent wheel locking up with the control of longitudinal wheel slip during emergency braking manoeuvres. To achieve this function, wheel sensors monitors its' peripheral velocity, and the logic calculates its' deceleration/acceleration. The operational modes (pressure hold, decrease or increase) depend on these values as well as the threshold levels (exceeded or not). The logic exploits the previously described Bosch algorithm and for further analysis is incorporated in MATLAB/Simulink (Fig. 1).

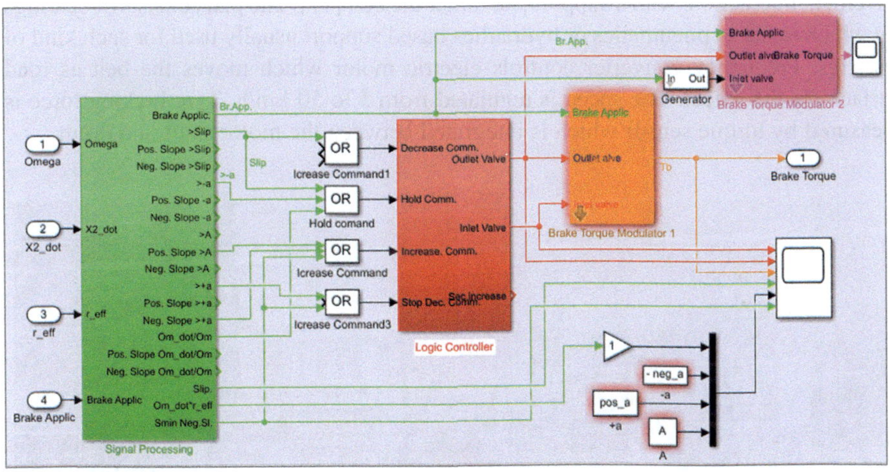

Fig. 1. MATLAB/Simulink model of ABS controller.

Model of the controller in Simulink contains three blocks:

- Signal Processing (SP) block which estimates the wheel deceleration ($\dot{\omega}$), wheel slip ratio (λ) and the ratio between wheel deceleration and angular velocity ($\dot{\omega}/\omega$). Calculated values are compared the respective thresholds and output signals for control are generated. This block incorporates signal processing including wheel peripheral acceleration and wheel slip, the threshold levels of the monitored parameters, and the outputs of the threshold-comparators.
- Logic Control (LC) block which uses a predefined sequence of actions to modulate the brake torque via the next block (Brake Torque Modulator) and to keep the wheel slip ratio below a present level (prevent wheel lock).

– Brake Torque Modulator (BTM) block uses brake torque values from LC. The maximal deliverable brake torque is assumed proportional to the brake pressure, therefore; modulating the brake pressure is equivalent to modulating the brake torque. A model of a two valve BTM is developed which has one inlet and one outlet valve. If there is no tendency of wheel locking, the BTM is in its initial state – not actuated valve (build-up cycle). For actuated inlet valve the wheel brake cylinder is closed from the master cylinder (hold cycle). When both valves are actuated, the wheel-brake cylinder is connected to an accumulator (reduce cycle).

4 Experimental Setup

Quarter car test rigs are useful measure for detail analysis of specific parts or systems in vehicle research ensuring the same operating conditions. The hardware-in-the loop (HIL) system was designed for analysis of ABS performance in case of wheel-obstacle interaction (Fig. 2). The test rig has three degrees of freedom with a moving belt for tyre-pavement interaction, with a support pad under the belt [17]. The pad consists of bearings set which replaces pneumatics or hydraulics based support usually used for such kind of test rigs. Frequency converter controls electric motor which moves the belt as road surface for the wheel. The speed is regulated from 5 to 50 km/h. Tyre braking force is measured by torque sensor which is integrated between the motor shaft and drum.

Fig. 2. Quarter car test rig: a) main components: 1 - sprung mass, 2 - unsprung mass, 3 - hydro.-pneumatic suspension, 4 - rolling surface, 5 - frame of the test rig, 6 - ABS hydraulic modulator with electronics, 7 - brake master cylinder, 8 - Arduino microprocessor, 9 - data acquisition system, 10–12 V battery, 11 - test rig frequency (velocity) controller, 12 - suspension pressure gauges, 13 - unsprung mass displacement sensor, 14 - PC with MATLAB/Simulink controller; b) sensors attached: 1 - wheel radius sensor, 2 – wheel angular speed sensor, 3 - unsprung mass acceleration sensor, 4 - velocity sensor, 5 - laser sensor for profile of rubber belt, 6 - brake hose pressure sensor (on the right), 7 - torque sensor (behind the drum), 8 - suspension deflection sensor (behind the wheel), 9 - cleats as road obstacles.

Three cleats height were used as obstacles per whole rubber belt to make the surface rough. The laser sensor was used for checking the surface displacement including cleats. The variation of 4–6 mm for input of wheel vertical displacement was used during tests. All the data from these sensors is logged in the data acquisition system (Fig. 2a, position 9) by remote control device.

The peripheral wheel acceleration is an input variable for ABS algorithm. The obtained values are shown in Fig. 3.

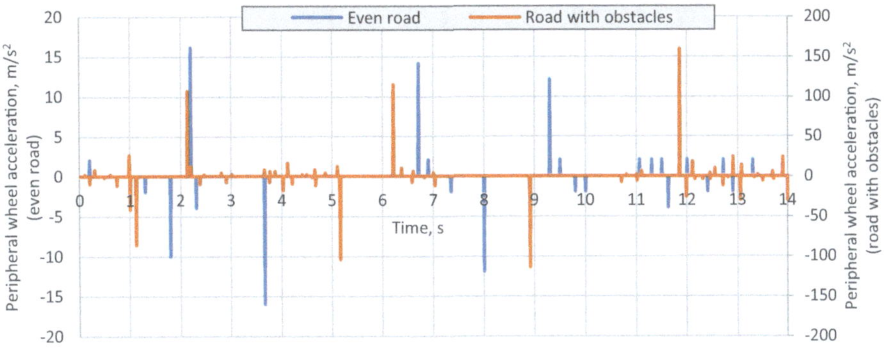

Fig. 3. Peripheral wheel acceleration during ABS braking on even road (left side scale) and on surface with obstacles (right side scale).

Different scale for vertical axis of wheel acceleration is used to represent the influence of even road and pavement with obstacle; however, the significant different in magnitudes between these two case is recognised – an acceleration on pavement with obstacles is up to ten times higher. The peaks corresponds to change in brake pressure and wheel rotational fluctuations. The influence of wheel-road interaction on ABS performance is analysed in next chapter.

5 Results and Discussion

After initial preparation, the performance of ABS with road obstacles was evaluated considering the following control thresholds: $A = 5$ m/s^2, $+a = 50$ m/s^2, $-a = -50$ m/s^2 and slip value was kept constant $\lambda = 0.2$ for all other testing parameters.

Main indicators of ABS braking performance comparing wheel interaction with even surface and interaction with selected obstacles are shown in Fig. 4. Left side graphs show ABS control indicators: ABS phases (according to used eight-phase algorithm) and brake states for pressure control (1 – pump, 2 – dump, 3 – hold). Right side graphs show wheel slip ratio and pressure into brake lines.

The braking on even surface was firstly done which show that ABS phases go through complete control cycle (all eight phases). This corresponds to efficient braking when wheel slip is limited up to 0.1 during high enough brake pressure (over 80 bars) (Fig. 4a). All three brake states activates corresponding to ABS phase.

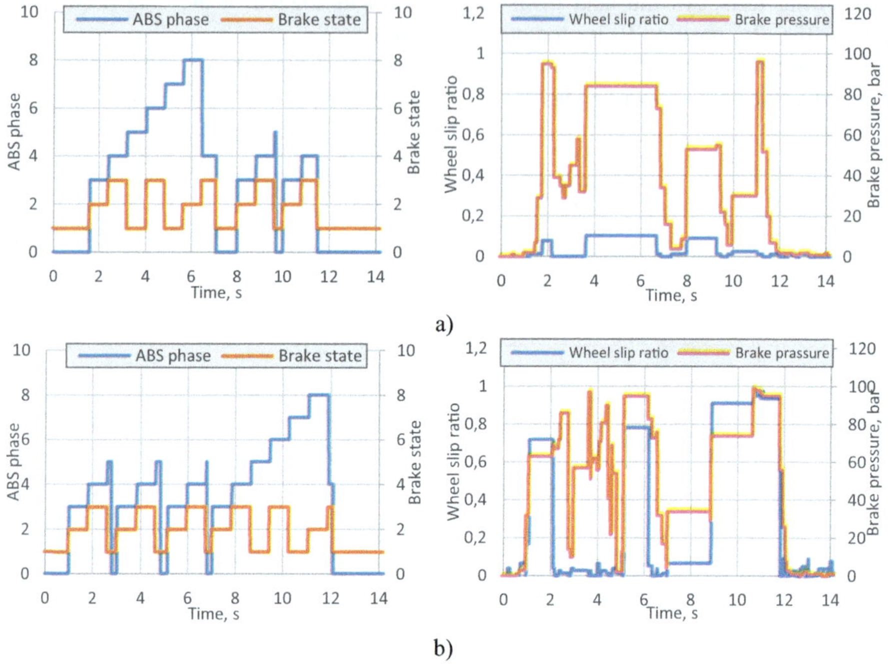

Fig. 4. ABS braking performance indicators: a) on even road; b) on surface with obstacles.

The same ABS performance by activating all tree brake states was estimated on surface with obstacles (Fig. 4b); however, wheel slip when ABS cycle goes through all eight phases is too high (around 0.9). This case does not ensures efficient vehicle braking or steering during braking as the tire is almost locked.

Wheel tend to lock when ABS goes to the 8th phase but instead of returning for certain time period to the 4th phase, it goes directly to the initial 0 (Fig. 4b). This movement of ABS algorithm from 8th to 0 phase contradicts the presented eight-phase algorithm as the phase should go from 8th to 4th phase, hold it there and then move to next phase or 0 stage where it could again start to build up the pressure for the next ABS activation. This ABS algorithm complies all the states of braking: pump, dump and hold but it still does not work efficiently enough in case of wheel interaction with road obstacle.

Wheel-obstacle interaction is a specific case for ABS braking therefore some corrections in control are necessary. Vehicle braking on rough road (including obstacles or high unevenness) can be identified from high wheel angular speed fluctuations during initial stage of braking (Fig. 3). In such a case an adaptive threshold levels instead of using fixed values is suggested. One of the purpose is to avoid higher that 5th ABS phase then pressure increase is maintained. This is not allowed on a pavement with high unevenness because of too high wheel slip at 6–8th phases (Fig. 4b). The switching between 3–4th phases will prevent from long wheel slip duration. The case of such conception by using Look-up table in Simulink is shown in Fig. 5. Here 3th phase (dump) and 4th phase (hold) just after, stops high wheel slip and keeps tight vehicle braking and possible steering. The fluctuations of wheel angular speed are obtained automatically.

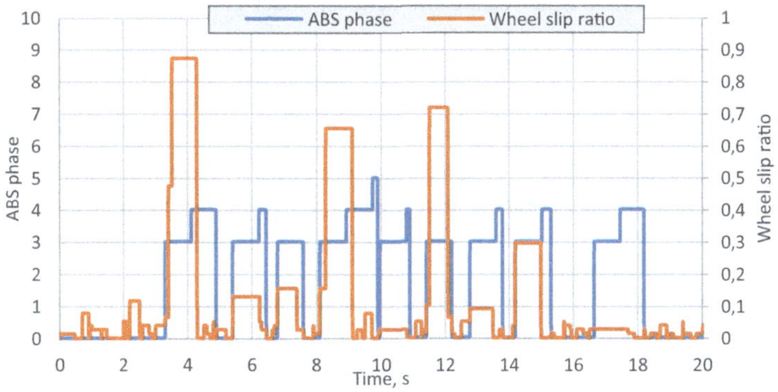

Fig. 5. ABS braking performance after implementing limited wheel slip duration.

The fluctuations of wheel angular speed are obtained automatically. The ABS does not lock the wheel when it reaches the 4th phase and then returns to zero phase, what is a contradiction to initial ABS algorithm.

The prevention of wheel lock during braking on road with obstacles also keeps the tire and suspension from unexpected damage. Locked wheel is more exposed by hard impact to pavement obstacle as border, transverse crack of the pavement or a haphazard stone.

6 Conclusion

Vehicle anti-lock braking system (ABS) performs its function good enough on the even road surfaces but the performance of ABS starts to deteriorate considerably when there are road obstacles or larger pavement unevenness. ABS performance on uneven surface is mainly evaluated by fluctuations of wheel rotation and the ability to complete all eight phases from the control algorithm. ABS is able to complete all eight phases when tested on even surface but as soon as the obstacles are put in it's path or the velocity is increased, it tends to keep wheel locked for too long. This is because the constant values of acceleration thresholds A, $+a$, $-a$ are used. One way of reducing the stopping distance and locking the wheel instantly when ABS is enforced is by introducing dynamic thresholds instead of using the fixed values. This is valid for all acceleration thresholds and also for slip conditions. All the analysed variables depends on velocity so dynamic its adaptation will led to the optimal choice of the variable and better ABS performance. The limited sequence of full cycle of presented ABS algorithm enable to stop the undesirable pump and the excess of braking pressure. This prevent from long wheel slip duration and keeps effective vehicle braking and possible steering in complicated driving conditions. The prevention of wheel lock during braking on road with obstacles (as border, transverse crack of the pavement) also keeps the tire and suspension from unexpected damage.

Acknowledgements. This research was funded from the European Union Horizon 2020 Framework Program, Marie Skłodowska-Curie actions, under grant agreement No. 872907.

References

1. European Commission. 2019 road safety statistics: what is behind the figures? https://ec.europa.eu/commission/presscorner/detail/en/qanda_20_1004. Accessed 06 Jun 2021
2. European Commission: Road safety: Europe's roads are getting safer but progress remains too slow. Accessed 20 May 2021
3. Penny, W.C.W., Els, P.S.: The test and simulation of ABS on rough, non-deformable terrains. J. Terrramech. **67**, 1–10 (2016)
4. Challa, A., Ramakrushnan, K., Gaurkar, P.V., Subramanian, S.C., Vivekanandan, G., Sivaram, S.: A 3-phase combined wheel slip and acceleration threshold algorithm for anti-lock braking in heavy commercial road vehicles. Veh. Syst. Dyn., 1–22 (2021)
5. Song, J., Che, W.S.: Comparison between braking and steering yaw moment controllers considering ABS control aspects. Mechatronics **19**, 1126–1133 (2009)
6. Schmeitz, A.J.C., Alirezaei, M.: Analysis of wheel speed vibrations for road friction classification. Veh. Syst. Dyn. **54**(4), 492–509 (2016)
7. Salehi, M., Noordermeer, J.W.M., Reuvekamp, L.A.E.M., Blume, A.: Parameter optimization for a laboratory friction tester to predict tire ABS braking distance using design of experiments. Mater. Des. **194**, 108879 (2020)
8. Jaiswal, M., Mavros, G., Rahnejat, H., King, P.D.: Influence of tyre transience on anti-lock braking. Proc. Inst. Mech. Eng Part K J. Multi-body Dyn. **224**(1), 1–17 (2009)
9. van der Merwe, N.A., Els, S., Žuraulis, V.: ABS braking on rough terrain. J. Terramech. **80**, 49–57 (2018)
10. Arrigoni, S., Cheli, F., Gavardi, P., Sabbioni, E.: Influence of tire parameters on ABS performance. Tire Sci. Technol. **45**(2), 121–143 (2017)
11. Hamersma, H.A., Els, P.S.: Improving the braking performance of a vehicle with ABS and a semi-active suspension system on a rough road. J. Terrramech. **56**, 91–101 (2014)
12. Marian, V., Luminita, S.M., Sorin, V.: The influence of forced steering vibrations on a wheel and dynamic effect of a wheel with ABS braking on undulated road II. In: Proceedings of the 3rd International Conference on Computational Mechanics and Virtual Engineering, Brasov, Romania (2009)
13. Kang, J.: Robust estimation method of tire torsional resonance frequency to detect decrease in tire inflation pressure. Veh. Syst. Dyn., 1–17 (2021)
14. Sivaramakrishnan, S., Singh, K.B., Lee, P.: Influence of tire operating conditions on ABS performance. Tire Sci. Technol. **43**(3), 216–241 (2015)
15. Eriksson, T.: Co-simulation of full vehicle model in Adams and anti-lock brake system model in Simulink. Master's thesis. Chalmers University of Technology (2014)

16. Day, T.D., Roberts, S.G.: A simulation model for vehicle braking systems fitted with ABS. SAE Technical Paper 2002-01-0559, pp. 11–25 (2002)
17. Žuraulis, V., Kilikevičius, A.: Quarter car test rig for extended dynamics research in laboratory conditions. In: Klomp, M., Bruzelius, F., Nielsen, J., Hillemyr, A. (eds.) IAVSD 2019. LNME, pp. 1425–1434. Springer, Cham (2020). https://doi.org/10.1007/978-3-030-38077-9_164

Comprehensive Analysis of Multilayer Coating Obtained by Ion-Plasma Spraying on Aircraft Structural Elements Made of Composite Materials in Order to Improve Their Physical and Technical Properties

Jevgēņijs Garbuzs(✉), Konstantins Sovkovs, Ruslans Garbuzs, and Jurijs Feščuks

Institute of Aeronautics,
Riga Technical University, Ķīpsalas iela 6B, Rīga 1048, Latvia
{jevgenijs.garbuzs,ruslans.garbuzs}@rtu.lv

Abstract. Method vacuum ion-plasma sputtering has been widely used in aviation for the protection of gas turbine blades, combustor components, flaps variable nozzle turbine engines of aircraft.

Use of ion-plasma sputtering, together with application of materials of chrome, nickel, yttrium, aluminum and other has made it possible to increase the service life and reliability of gas turbine engines.

All of the above indicators are relevant for aircraft parts and assemblies operating at high temperature conditions (about 1,700 °C).

The goal of the Multifunctional Nanostructured Coatings for Aircraft Structures (NANOCOAIRS) project is to investigate the use of nanostructured multilayer coatings for use in an aircraft anti icing system.

Keywords: Vacuum ion-plasma sputtering · Composite materials · Aircraft

1 Introduction

The main problems in creating perspective aviation technology are the development and implementation of new materials and technologies for the manufacture and application of protective coatings.

For the manufacture of structural elements of aircraft and similar machines, non-metallic composite materials based on various resins are widely used, using glass or carbon fibers as well as artificial polymer high-strength fibers as reinforcing elements. The resin that serves as the basis for such composites is characterized by low electrical conductivity, high plasticity, low wear resistance, as well as increased sensitivity to environmental influences (high humidity, ultraviolet radiation, high temperature). Therefore, in the manufacture of structural elements of aviation technology in order to protect against adverse environmental factors, its surface is covered with a protective layer. The protective layer is a variety of paint and varnish and metal coatings.

O. Prentkovskis et al. (Eds.): TRANSBALTICA 2021, LNITI, pp. 230–239, 2022.
https://doi.org/10.1007/978-3-030-94774-3_23

Metallic coatings are electroplated and obtained by sputtering. Coatings obtained by the method of vacuum ion-plasma sputtering have comparatively better properties.

The development and creation of thin, high-strength and wear-resistant metal coatings will help solve many of the problems connected with the negative impact of environmental factors on composite materials.

One of the negative factors for aircraft structures is icing, which could lead to an accident or disaster of the aircraft.

In order to prevent icing, various systems are used: chemical, mechanical and thermal.

In this study, the mechanical properties of nanostructured coatings created by the method of vacuum ion-plasma spraying were researched.

At the final stage of manufacturing parts, the method of vacuum ion-plasma sputtering is used. The coating created by this method helps to protect various parts and their surfaces from adverse factors [1, 2].

Most often, this type of coating is used to protect parts of gas turbine engines of aviation equipment that operate at high temperatures and under conditions of abrasive wear.

The operational reliability and economy of the engine depend on the normal operation of the combustion chamber and the gas temperature in front of the turbine. [3–6].

To increase the service life of aircraft, the metals of yttrium, titanium, zirconium, chromium, aluminum, nickel, etc. are used in ion-plasma spraying.

The goal of the Multifunctional Nanostructured Coatings for Aircraft Structures (NANOCOAIRS) project is:

1. development of technology for obtaining nanostructured multicomponent anti-icing coatings;
2. managing an experimental research of the characteristics of the created coatings;
3. research of anti-icing properties of the coating on samples of aircraft structures.

The coatings created by ion-plasma sputtering are often very thin (from 1 to 2 μ in this project), so they cannot be researched by themselves separately. One example is erosion, which actively destroys a given coating by abrasive wear. Therefore, the created coating - bases system must be researched together. Together with the hardness and flexibility, it is also necessary to take in to account properties such as plasticity and impact resistance, electrical conductivity [7].

This article examines the research stage, where the main goal is:

1 development of nano coatings, with the use of Al - Ti metals;
2 to investigate the obtained properties of the created coatings.

The created coating must have maximum wear resistance, low wettability, be resistant to corrosion, must have sufficient electrical conductivity to be used as a heat-generating element and have good adhesion to the base material.

2 Methods and Materials

The basis for sputtering the coating was taken a composites (carbon fibers based on epoxy resin).

Coating thickness 1–2 μm.

Before applying the coating, the surface of the samples was prepared: the surface of one sample was not processed, the surface of the second sample was processed with P1200 GRIT sandpaper, the surface of the third sample was processed with P2000 GRIT sandpaper, and the surface roughness of the prepared samples was measured, after spraying, the surface roughness of the created coating was checked. Roughness was measured with a TR100 Surface Roughness tester (see Fig. 1).

Using the modernized vacuum ion-plasma installation NNV6.6-I (see Fig. 2) [8–10], the necessary coatings were created. During the experiment, we used two independent plasma sources - a magnetron (Al) and an arc evaporator (Ti), as well as the supply of a reaction gas (N). This made it possible to create a coating with the required parameters.

Fig. 1. Measurement of the surface roughness of the sample with the TR100 Surface Roughness tester.

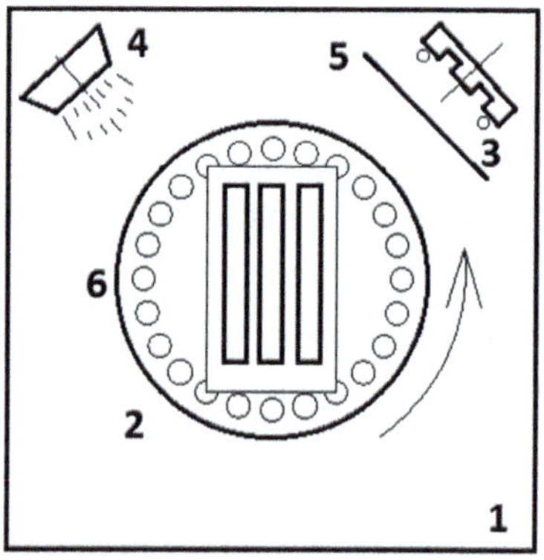

Fig. 2. Sputtering chamber diagram: 1 – vacuum chamber, 2 – turntable, 3 – arc evaporator (Ti), 4 – magnetron (Al), 5 – protective screen, 6 – samples.

Three coating programs were carried out for 3 prepared samples. The spraying program included the coating of three different layers of multicomponent coatings based on the Al - Ti - N system. One layer was an intermetalide, one layer was a conglomerate, and one layer was a nitride. During the spraying, the time of coating of the layers was changed. The ratio of Al to Ti by weight was chosen to be 32% Al to 68% Ti in the analysis of previous experiments [7].

3 Analysis of the Properties of the Created Coatings

When analyzing the obtained coatings, the following parameters were investigated:

1. Al and Ti content in the composition of the new coating,
2. mechanical properties of the coating,
3. electrophysical properties of the coating.

The dimensions of the samples were different for different experiments. The initial dimensions of the samples were: 250 × 30 × 2 mm (see Fig. 3a) Electrical conductivity was tested on them. Then fragments of 30 × 30 × 2 mm were cut from the samples (see Fig. 3b).

a) b)

Fig. 3. Simples with created coating: a) samples before cutting, b) fragments of samples after cutting.

HITACHI S-3000N Scanning Electron Microscope and BRUKER Quantax microanalyzer at 9,6 kV was used to conduct a chemical analysis of the created coating (see Fig. 4).

To research the mechanical properties of the created coating, was used:

1. Device PMT-3M1 was used to study the microhardness of the created coating.
2. Device Anton Paar Revetest Xpress RSX device was used to study the adhesion of the created coating.

Measurement of microhardness and adhesion is necessary to determine the hardness and strength of adhesion with material of the created coating.

An example of a graphical display of adhesion measurements is shown in Fig. 5.

Taking measurements with the Anton Paar Revetest Xpress RSX device, it is possible to take photographs of the degree of destruction of the created layer.

The process of destruction of the coating during the measurement of adhesion and its stages are described in the article [7].

Analyzing the graph of acoustic emission, it is possible to determine the initial moment of the next stage of destruction, which is characterized by a sharp increase in the graph line.

Compared to early stage samples [7]:

1. a visible scratch mark appeared later,
2. complete detachment of the created layer did not occur, despite the degree of destruction of the base itself. And in some examples, the layer remained until the final phase of the test (see Fig. 6).

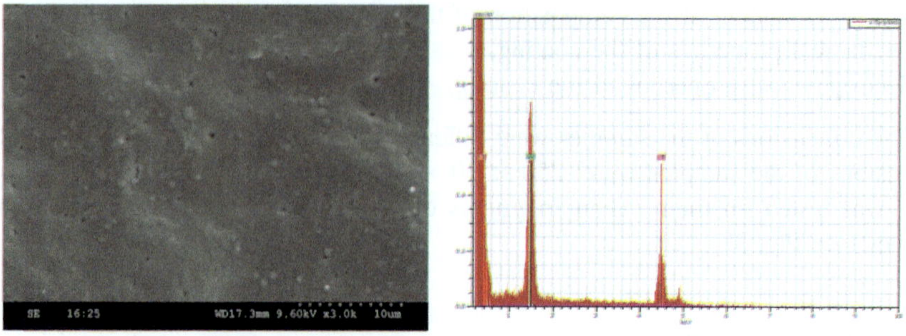

a) b)

Spectrum: 175/0/045

El	AN	Series	Net	unn. C [wt.-%]	norm. C [wt.-%]	Atom. C [at.-%]	Error [%]
Al	13	K-series	17520	28.94	33.51	47.21	1.4
Ti	22	K-series	6554	57.42	66.49	52.79	2.3
			Total:	86.36	100.00	100.00	

c)

Fig. 4. Display of data of electronic microscope HITACHI S - 3000N: a) created coating view with 3,000 magnification; b) graphical representation of the chemical composition of the surface of the created metal layer; c) table with the relative value of the required metals by mass and atoms.

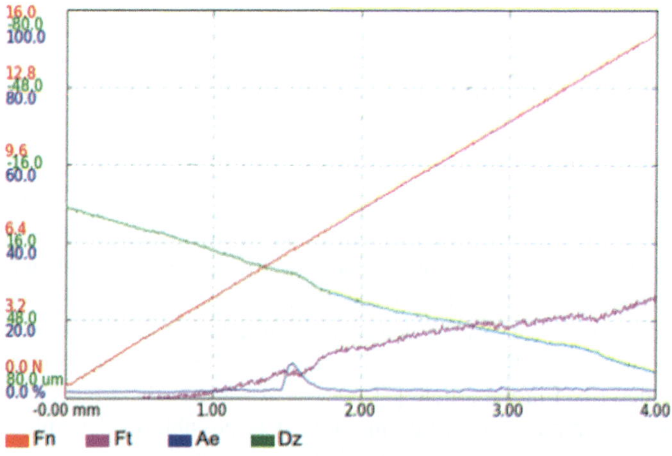

Fig. 5. An example of a graphical display of adhesion parameters: Fn – vertical load; Ft – tangential load-friction force; Ae – acoustic emission; Dz – indentation, depth.

	LC1	LC2	LC3	LC4	LC5
Pictures					
Values	1.64 N (0.32 mm)	2.73 N (0.62 mm)	7.07 N (1.80 mm)	8.21 N (2.14 mm)	14.55 N (3.88 mm)

Fig. 6. An example of saving the created layer until the final phase of testing.

Resistance was measured using a PeakTech 3320 multimeter. This value was necessary to determine the possibility of using the created layer in the anti-icing system of the aircraft.

4 Results

Chemical Composition

The created coatings were divided into 3 main groups, depending on the spraying program (G1, G2, G3) and 3 subgroups according to the degree of surface preparation of the samples before spraying (N1 – the original surface, N2 – the surface prepared with P1200 GRIT sandpaper, N3 – surface prepared with P2000 GRIT abrasive paper).

Figure 7 shows graphs with the obtained results of the chemical composition of the created coating.

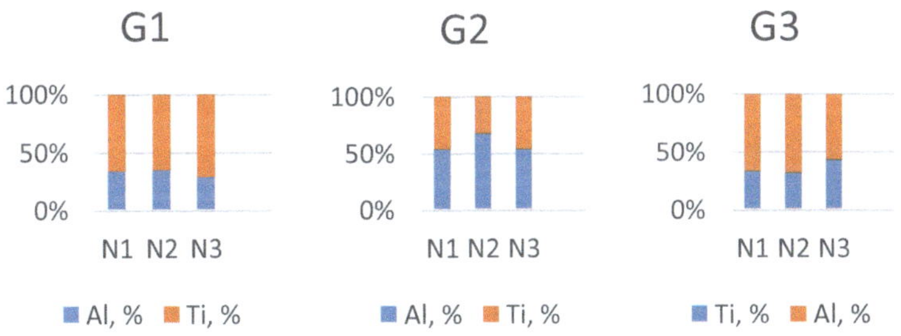

1) 2) 3)

Fig. 7. Results of analysis of Al and Ti content in the created coating: 1) 1st sputtering program; 2) 2nd sputtering program; 3) 3rd sputtering program; G1, G2, G3 - sample groups; N1, N2, N3 - sample subgroups.

Adhesion

Figure 8 shows graphs with the obtained results of measuring the adhesion of the created coating.

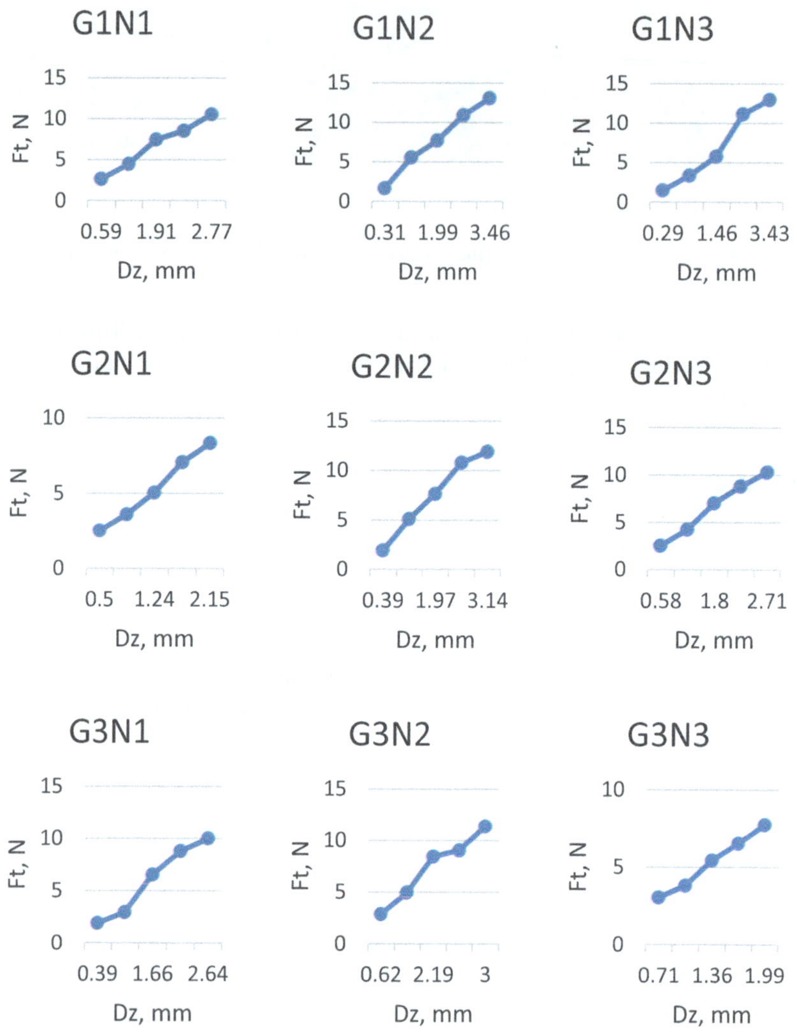

Fig. 8. Graphical results of measurements of the adhesion of the created coating: G1, G2, G3 – sample groups; N1, N2, N3 – sample subgroups.

Microhardness

Figure 9 shows graphs with the obtained results of measuring the microhardness of the created coating.

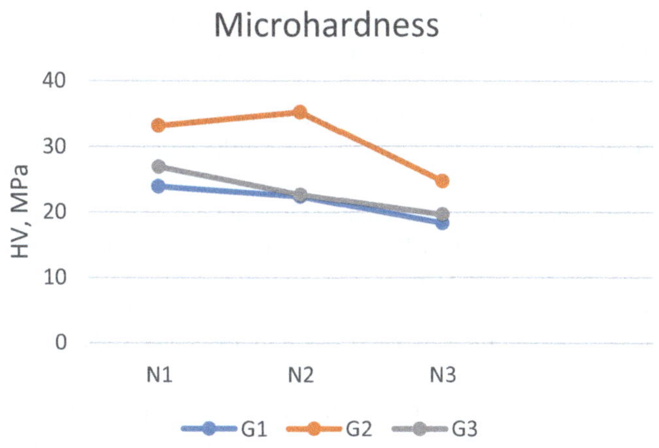

Fig. 9. Graphical results of measurements of the microhardness of the created coating: G1, G2, G3 – sample groups; N1, N2, N3 – sample subgroups.

Electrical Conductivity

Figure 10 shows graphs with the obtained results of measuring the electrical conductivity of the created coating.

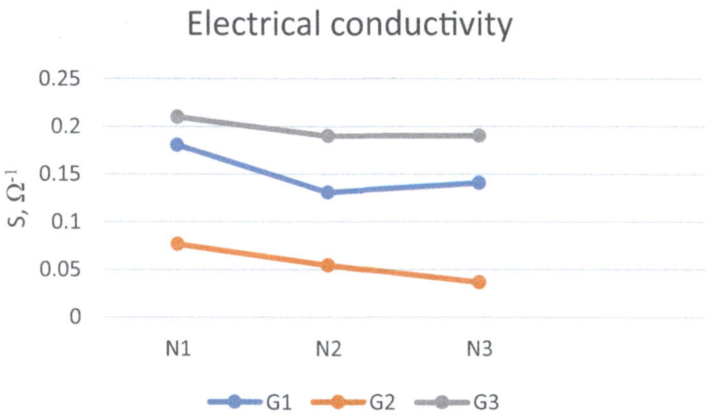

Fig. 10. Graphical results of measurements of the electrical conductivity of the created coating: G1, G2, G3 – sample groups; N1, N2, N3 – sample subgroups.

5 Conclusions

1. The necessary parameters of the settings of the vacuum ion - plasma installation were developed for applying the specified parameters of the ratio by mass Al - Ti for the creation of new coatings.
2. Investigating the properties of the created coatings, it was determined:

a. according to the results of measurements of hardest, the best results were obtained for samples G2N2 with the value of the ratio by mass Al - Ti 67.02%/ 32.98;
b. according to the results of measurements of adhesion, the best results were obtained for samples G1N2 with the value of the ratio by mass Al - Ti 34.60%/ 65.40%, while:
    i. complete detachment of the created layer did not occur, despite the degree of destruction of the base itself. And in some examples, the layer remained until the final phase of the test;
c. according to the results of measurements of electrical conductivity, the best results were obtained for samples G3N1 with the value of the ratio by mass Al - Ti 32.92%/67.08.

3. New coatings have a different range of mechanical properties and can be used in an anti-icing system; they can be used as a protective layer for the surface of structural elements and other various parts of aircraft.

Acknowledgements. This work has been supported by the European Regional Development Fund within the Activity 1.1.1.2 "Post-doctoral Research Aid" of the Specific Aid Objective 1.1.1 "To increase the research and innovative capacity of scientific institutions of Latvia and the ability to attract external financing, investing in human resources and infrastructure" of the Operational Programme "Growth and Employment" (No. 1.1.1.2/VIAA/1/16/176 "Multifunctional Nanostructured Coatings for Aircraft Structures (NANOCOAIRS)".

PostDoc Latvia NATIONAL DEVELOPMENT PLAN 2020 EUROPEAN UNION European Regional Development Fund

INVESTING IN YOUR FUTURE

References

1. Банных, О., Будберг, П., Алисова, С.: Диаграммы состояния двойных и многокомпонентных систем на основе железа. Металлургия, Москва (1986)
2. Лякишева, Н.: Диаграммы состояния двойных металлических систем. Машиностроение, Москва (2000)
3. Базылева, О., Аргинбаева Э., Туренко Е.: Высокотемпературные интерметаллидные сплавы для деталей ГТД. Авиационные материалы и технологии (3), с. 26–31 (2013)
4. Каблов, Е.: Стратегические направления развития материалов и технологий их переработки на период до 2023г. Авиационные материалы и технологии (5), с. 7–17 (2012)
5. Оспенникова, О.: Стратегия развития жаропрочных и теплозащитных покрытий. Авиационные материалы и технологии (5), с. 19–36 (2012)
6. Скибин, В.: Научный вклад в создание авиационных двигателей. Машиностроение, Москва, Т.2, (2000)

7. Garbuzs, J., Blumbergs, I., Garbuzs, R., Savkovs, K., Feščuks, J.: Applying of ion-plasma spraying on aircraft structures in order to improve the physical and technical properties of the glider surfaces. In: 2021 Proceedings of the 6th International Scientific and Practical Conference Transport, Education, Logistics and Engineering, pp. 5–12. Riga Aeronautical Institute, Latvia (2021)
8. Urbahs, A., Savkovs, K., Urbaha, M., Kurjanovich, I.: Nanostructured intermetal-ceramic coatings for blades of gas turbine engines. J. NATO Sci. Peace Secur. Ser. Ser. B Phys. Biophys. Nanodevices Nanomater. Ecol. Secur. Part 2 **28**, 307–314 (2012)
9. Urbahs, A., Urbaha, M., Savkovs, K., Bogdanova, S.: Wear resistant nanostructured multi-component coatings. J. NATO Sci. Peace Security Ser. Ser. B Phys. Biophys. Nanodevices Nanomater. Ecol. Secur. Part 1, 161–170 (2012)
10. Urbahs, A., Urbaha, M., Savkovs, K., Andrejeva, D.: Evaluation of the Physical and Mechanical Characteristics of Ion-Plasma Antifriction Coatings Based on Ti-Cu. Key Eng. Mater. **788**, 59–67 (2018)

Engine Combustion, Energy Management and Emissions

Determination of the Power Factor of Electric Rolling Stock of Alternating Current Consumption

Sergey Goolak[1(✉)], Borys Liubarskyi[2], Svitlana Sapronova[1], Viktor Tkachenko[1], and Ievgen Riabov[2]

[1] The State University of Infrastructure and Technologies, Kyiv 04071, Ukraine
[2] Kharkiv Polytechnic Institute, National Technical University, Kharkiv 61002, Ukraine
lborisl911@ukr.ne

Abstract. The analysis of the influence of operating modes of electric rolling stock of alternating current consumption and non-deterministic non-stationary process of changing the voltage of the contact network on the energy indicators of the traction drive of the electric rolling stock was carried out. As a result of the analysis, it was found that to improve the energy efficiency of the traction drive, to optimize the energy performance of the traction drive. It is proposed to introduce an optimal control system for the sampling frequency of the control signal when implementing pulse-width modulation, on the basis of which the algorithm of operation of the power transistors of the input converter of the traction drive is organized. As an objective function in the implementation of the optimal control algorithm, it is proposed to use the power factor of the drive as a function of time. It is mathematically proven that the power factor as a function of time is a mathematical convolution of the time functions of the efficiency and the use of the active power of the traction drive. Algorithms for determining the time functions of the power factor are proposed for the case of deterministic discrete, deterministic continuous, stochastic continuous and stochastic discrete signals of the efficiency and the use of the active power of the traction drive.

Keywords: Power factor · Objective function · Traction drive · Convolution of functions · Optimal control

1 Introduction

The operation of electric rolling stock of railways is associated with high power consumption. For the operation of electric rolling stock, the issue of reducing losses in traction drive systems is relevant [1]. On electric rolling stock of alternating current consumption, the efficiency and power factor are the energy indicators characterizing losses [2]. The efficiency characterizes the loss of active power in the traction drive system, the power factor – reactive. Thus, the task of optimal control of the traction drive should be associated with an increase in both the efficiency and the power factor, that is, a decrease in losses in the quality of active and reactive power [3].

O. Prentkovskis et al. (Eds.): TRANSBALTICA 2021, LNITI, pp. 243–252, 2022.
https://doi.org/10.1007/978-3-030-94774-3_24

But with such a formulation of the problem, certain difficulties arise associated with the fact that with an increase in the power factor, the efficiency decreases, and vice versa [4]. In addition, during the operation of the electric rolling stock, its operating modes are constantly changing – starting off, accelerating, braking, and the like. In the study [5], it is shown that the magnetic losses in the steel of the traction motor is a function of the frequency of magnetization reversal, which, in turn, depends on the rotational speed of the traction motor shaft. That is, the value of both the efficiency and the power factor changes all the time.

Another fact that affects the value of both coefficients is the non-stationary nature of the voltage change in the overhead contact network [6]. The random nature of the process of changing the voltage in the contact network is associated with the operation of the traction power supply system, namely the presence of one or several units of electric rolling stock in one feeder zone, the passage of the electric rolling stock of the feeder zone, and the like. The random nature of the voltage change in the contact network when analyzing the processes in the traction and auxiliary systems of the electric rolling stock requires their consideration of the traction drive as stochastic systems [7]. This means that when analyzing processes in traction and auxiliary systems, the methods of correlation analysis should be used.

To build an optimal control system for the traction drive and auxiliary systems in order to obtain optimal values of power factors and efficiency, optimal control algorithms should be applied. With a large number of existing algorithms for optimal control of both linear and stochastic systems, the common objective function in these algorithms is the objective function, on the basis of which one or another parameter is optimized.

Thus, the determination of the objective function for optimizing the energy performance of alternating current electric locomotives is an urgent task.

2 Problem Statement

The analysis of the energy efficiency of the traction power supply system requires the study of electrodynamic processes in all its components – traction substation, contact network and electric rolling stock [8]. An increase in the energy efficiency of an alternating current electric traction system as a whole means a decrease in total power losses in it, that is, a decrease in total power losses in each element of the system.

Electric rolling stock, as a consumer of electricity, introduces into the traction power supply system such distortions as dips and surges in the voltage of the contact network and the higher harmonic components of the voltage of the contact network [9]. These distortions lead to an increase in reactive power losses.

Voltage dips and surges of the contact network caused by the conditions of the passage of the electric rolling stock of the feeder zone [7], and the higher harmonic components – nonlinear volt-ampere and weber-ampere characteristics of the traction drive elements and auxiliary systems of the electric rolling stock [10].

On electric rolling stock of alternating current consumption with an asynchronous traction electric drive with both vector control [11] and direct torque control [12], the power section of the traction drive consists of an input converter – a controlled rectifier, a DC link – a smoothing reactor and an output converter – an autonomous inverter voltage. The algorithm for opening the power transistors in these converters is based on pulse width modulation (PWM). Since the input and output converters are separated by a DC link, only the operation of the input converter affects the active power utilization of an electric locomotive. Moreover, the higher the sampling frequency of the control signal when implementing PWM, the less high-frequency distortions the electric locomotive introduces into the network, that is, there is a high active power utilization factor [13]. With a sufficiently correct formulation of the problem, the study [13] did not take into account the fact that with an increase in the sampling frequency of the control signal during the PWM implementation, the heating of the power transistors of the converter increases. This leads to an increase in losses in semiconductor switches and, as a consequence, to a decrease in its efficiency [14]. That is, in order to increase the energy efficiency of the traction drive of the AC electric rolling stock, it becomes necessary to build a system for optimal control of the sampling frequency when implementing PWM.

To construct an optimal control system, an objective function is required, finding the extreme value of which will allow the optimal control algorithm to be implemented. Since it is necessary to obtain the maximum value of the power factor during the operation of the traction drive, it is obvious that such a function can be the power factor. Optimization of the power factor of the traction drive of electric rolling stock can be found in the study [15]. But the study [15] does not consider the issue of optimizing active power consumption. In addition, this study does not take into account the influence of the nature of the operation of the electric locomotive on the controlled energy indicators. This factor imposes certain restrictions on the use of the proposed solutions. The question of the influence of the operation nature of the electric rolling stock on the power indicator of the electric locomotive is considered in [16]. It shows that in order to optimize the operation of the traction drive system, the power factor should be considered as a temporary function, depending on the nature of the operation of the electric rolling stock. But it does not consider the effect of the operation of an electric locomotive on the efficiency.

The issue of full power consumption is considered in the study [17], where the use of energy efficiency indicators is proposed to take into account the efficiency of full power consumption. The proposed indicators of the energy efficiency of traction drives of AC electric locomotives take into account the total power consumption in the drive during the operation of the electric locomotive. But, the proposed indicators only record consumption. The study does not indicate how they can be used to build an optimal traction drive control system in order to reduce the total power consumption. Another important factor not considered in [15–17] is the random nature of voltage changes in the overhead contact network [7]. The random nature of the voltage change in the overhead contact network requires a fundamentally different approach to the construction of an optimal traction drive control system [18]. In such systems, the traction drive is considered as a low-pass filter and, in addition to optimizing the efficiency and power factor, the filtering error should be minimized. In other words, the problem of optimization of power losses in a traction drive with a single criterion becomes multi-criteria [19]. Another important factor not considered in [15–17] is the random nature of voltage changes in the overhead contact network [7]. The random nature of the voltage change in the overhead contact network requires a fundamentally different approach to the construction of an optimal traction drive control system [18]. In such systems, the traction drive is considered as a low-pass filter and, in addition to optimizing the efficiency and power factor, the filtering error should be minimized. In other words, the problem of optimization of power losses in a traction drive with a single criterion becomes multi-criteria [19].

Determination of the energy efficiency of a traction drive of an electric rolling stock is possible based on the results of numerical modeling based on simulation models created using the techniques used in [20], for a brushless inductor motor and a model of a motor with a transverse magnetic field given in [21].

The systematization of the results of the above studies allows to assume that the existing approaches to solving the problem of optimization of the consumed power in the traction and auxiliary systems of the electric rolling stock of alternating current consumption is to control the maximum possible value of the efficiency, or - the power factor. Obviously, such approaches make it possible to find acceptable solutions only if it is necessary to optimize the consumption of either active or reactive power. It follows from this that such approaches do not make it possible to obtain optimal solutions for traction drive control systems in order to reduce the total power consumption. This part of the problem can be solved by synthesizing the objective function of optimal control in order to reduce the consumption of full power, based on obtaining the optimal values of both energy indicators of the traction electric drive of AC electric locomotives.

3 Determination of Power Factor on the Basis of Transfer Functions of Power Factor and Efficiency of Consumption of Electric Rolling Stock of Alternating Current

3.1 Determination of the Power Factor for Deterministic Discrete Signals

The power factor as a function of time is given by:

$$\gamma(t) = \frac{P_1(t)}{S(t)}, \tag{1}$$

where $S(t)$ – total power consumption of electric rolling stock from the network; $P_1(t)$ – active power consumed by electric rolling stock from the network.

Efficiency as a function of time, action is determined by the expression:

$$\eta(t) = \frac{P_1(t)}{P_2(t)}, \tag{2}$$

where $P_2(t)$ – net power on the shafts of traction motors.

Since when implementing the PWM algorithm, discrete signals are used, Eqs. (1) and (2) are written in the transformation. The processing of the signals depicted in the transformation corresponds to the processing of signals in the frequency domain. For this case, the power factor:

$$\gamma(z) = \frac{P_1(z)}{S(z)}. \tag{3}$$

Efficiency:

$$\eta(z) = \frac{P_1(z)}{P_2(z)}. \tag{4}$$

In accordance with Eqs. (3) and (4), the structural diagram of the traction drive and auxiliary systems has the form (Fig. 1).

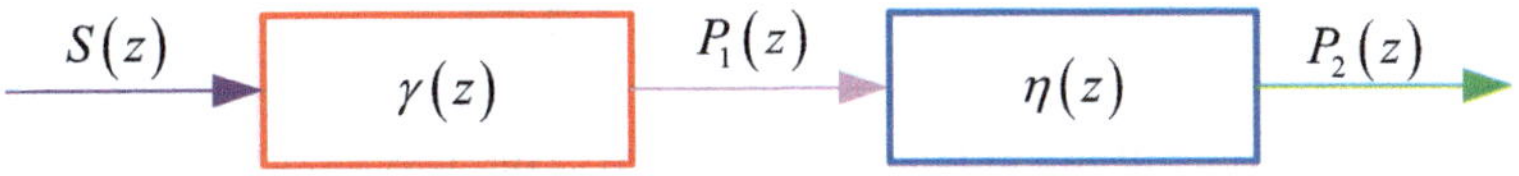

Fig. 1. Block diagram of the traction drive.

Transfer functions $\gamma(z)$ and $\eta(z)$ in Fig. 1 are pulse functions. For the convenience of further calculations, the variable z for the impulse function of the power factor is written as z_1 for the efficiency – as z_2. I.e.

$$z_1 = z_2 = z. \tag{5}$$

Then the impulse function of the power factor has the form:

$$\gamma(z_1) = \sum_{i=0}^{N_1} a_i \cdot z_1^{-i}, \tag{6}$$

where a_i – value of the power factor taken with a time delay $i \cdot \Delta T$; N_1 – the number of power factor readings.

Then the impulse function of the efficiency has the form:

$$\eta(z_2) = \sum_{j=0}^{N_2} b_i \cdot z_2^{-j}, \tag{7}$$

where b_j – value of the efficiency taken with a time delay $j \cdot \Delta T$; N_2 – the number of readings of the efficiency.

So, the objective function is defined as the transfer function of the electric rolling stock, where the input value is the total power consumption of the electric rolling stock from the network, and the output is the useful power (tangential power and payload of auxiliary systems):

$$k_p(z_1, z_2) = \gamma(z_1) \cdot \eta(z_2) = \sum_{i=0}^{N_1} a_i \cdot z_1^{-i} \cdot \sum_{j=0}^{N_2} b_i \cdot z_2^{-j} = \sum_{i=0}^{N_1} \sum_{j=0}^{N_2} a_i \cdot b_{N_2-j} \cdot z_1^{-i} \cdot z_2^{-(N_2-j)}. \tag{8}$$

Expression (8) is an operation of convolution of two functions γ and η, written in the form of Z – transformations [22].

After substitution of expression (5) into expression (8), the expression of the objective function takes the form:

$$k_p(z) = \gamma(z) \cdot \eta(z) = \sum_{i=0}^{N_1} a_i \cdot z^{-i} \cdot \sum_{j=0}^{N_2} b_i \cdot z^{-j} = \sum_{i=0}^{N_1} \sum_{j=0}^{N_2} a_i \cdot b_{N_2-j} \cdot z^{-(i+N_2-j)}. \tag{9}$$

The higher the value of the efficiency of the drive, the less the loss of active power in it. Accordingly, the higher the power factor value, the lower the total power loss. I.e.

$$k_p(z) = \gamma(z) \cdot \eta(z) = \sum_{i=0}^{N_1} \sum_{j=0}^{N_2} a_i \cdot b_{N_2-j} \cdot z^{-(i+N_2-j)} \to max. \tag{10}$$

Thus, when using any optimality criterion to determine the optimal values of the power factor and the efficiency of the rolling stock, it is necessary to control the maximum objective function.

When implementing the optimality criteria, operations such as integration and differentiation should be performed. These operations are more convenient to use in the spatial domain, where they can be implemented in the form of difference equations. Paserval's theorem [23] establishes a connection between the energies of discrete signals depicted in the frequency and spatial domains. In accordance with it, expression (8) for signals depicted in the spatial domain has the form:

$$k_p(t - NT) = \sum_{i=0}^{N_1} \sum_{j=0}^{N_2} \gamma(t - iT) \cdot \eta(t - (N_2 - jT)) \to max. \tag{11}$$

where $N = N_1 + N_2$; T – sampling period.

Based on similar considerations, as in the case of deterministic discrete signals displayed in the frequency domain, the maximum value of the objective function should be monitored.

3.2 Determination of the Objective Function for Deterministic Continuous Signals

As noted above, the objective function is the convolution of the power factor and the efficiency. For continuous signals, the convolution function is determined by the Duhamel's integral [24]:

$$k_p(t) = \gamma(0) \cdot \eta(t) + \int_0^t \gamma'(t) \cdot \eta(t - \tau)d\tau \rightarrow max, \tag{12}$$

where $\gamma(0)$ - power factor value at time zero; $\gamma'(t)$ - derivative of the power factor over time.

As in the case for deterministic discrete signals and, guided by similar considerations, for the case of deterministic continuous signals, the maximum value of the objective function should be monitored.

4 Determination of the Objective Function for Stochastic Signals

4.1 Determination of the Objective Function for Stochastic Continuous Signals

When processing stochastic signals, it is necessary to observe the minimum of the root-mean-square error between the actual value of the signal at the output of the system and its estimate. This problem can be solved using the Wiener-Hopf theorem [25]:

$$\int_0^\infty K_x(\tau - \lambda) \cdot k_p(\lambda)d\lambda = K_{xs}(\tau), \tag{13}$$

where $K_x(\tau)$ – autocorrelation function of the full power signal; $K_{xs}(\tau)$ - function of cross-correlation of signals of total power and useful power; λ - parameter characterizing the time interval for which the optimal value of the objective function is determined; $k_p(\lambda)$ - value of the objective function, determined using expression (12).

The solution of Eq. (13) made it possible to obtain a linear system that gives an estimate of the useful signal $\widehat{S}(t)$ at the output that is optimal in the sense of the minimum mean square error (variance):

$$J = M\left[\tilde{E}^2\right] \rightarrow min, \tag{14}$$

where

$$\tilde{E} = S(t) - \widehat{S}(t), \tag{15}$$

where $S(t)$ – stationary random process.

In other words. The task of determining the objective function is a two-criteria task. Using the first criterion, the maximum value of the objective function (12) is determined, with the help of the second (14), the value of the root-mean-square error between the actual signal at the output of the system and the estimation of this signal is minimized (13).

4.2 Determination of the Objective Function for Stochastic Discrete Signals

The Wiener-Hopf theorem for deterministic signals has the form [26]

$$\begin{cases} K_{xs}(t-kT,t-nT)=\sum_{i=0}^{N_1}\sum_{j=0}^{N_2} c_{i,j}\cdot K_x(t-(k-i)\cdot T,t-(n-j)\cdot T), \\ k\in N_1, \quad n\in N_2, \end{cases} \tag{16}$$

where $K_{xs}(t-kT,t-nT)$ - function of cross-correlation of signals of total power and useful power; $K_x(t-(k-i)\cdot T,t-(n-j)\cdot T)$ - autocorrelation function of the total power signal.

The solution of Eq. (16) with respect to all unknowns $c_{i,j}$, the total number of which will be N_1+N_2+2, makes it possible to find the impulse function of the traction drive system, minimizing the root-mean-square errors of filtering a random value of the total power.

5 Discussion of Results

As a result of the analysis, it is found that to improve the energy efficiency of electric rolling stock, to optimize the energy performance of its traction drive and auxiliary systems. It is proposed to introduce an optimal control system for the sampling frequency of the control signal when implementing pulse-width modulation, on the basis of which the algorithm of operation of the power transistors of the input converter of the traction drive is organized.

As an objective function in the implementation of the optimal control algorithm, it is proposed to use the power factor of the drive as a function of time.

It is mathematically proven that the power factor as a function of time is a mathematical convolution of the time functions of the efficiency and the use of the active power of the traction drive.

Algorithms are proposed for determining the time functions of the power factor for the case of deterministic discrete (11), deterministic continuous (12), stochastic continuous (13) and stochastic discrete signals of the efficiency and the use of active power of electric rolling stock (16).

6 Conclusions

As a research result, it is established:

1. The objective function for determining the optimal values of the energy indicators of the traction drive and auxiliary systems of the electric rolling stock of alternating current consumption is the convolution of the time functions of the power factor and the efficiency.

2. The maximum value of the objective function should be determined for deterministic processes, both discrete and continuous.

For stochastic processes, both discrete and continuous, the determination of the objective function is a two-criterion problem. The first criterion allows to find the maximum value of the objective function, the second – to minimize the root-mean-square error between the actual value of the signal at the output of the system and its estimate. The corresponding convolution functions should be used as the first criterion, and the corresponding forms of the Wiener-Hopf equation as the second.

References

1. Klimenko, I., Kalivoda, J., Neduzha, L.: Influence of parameters of electric locomotive on its critical speed. In: Gopalakrishnan, K., Prentkovskis, O., Jackiva, I., Junevičius, R. (eds.) TRANSBALTICA 2021. LNITI, pp. 531–540. Springer, Cham (2020). https://doi.org/10.1007/978-3-030-38666-5_56
2. Plakhtii, O., Nerubatskyi, V., Sushko, D., Ryshchenko, I., Tsybulnyk, V., Hordiienko, D.: Improving energy characteristics of AC electric rolling stock by using the three-level active four-quadrant rectifiers. Восточно-Европейский журнал передовых технологий **4**(8), 6–14 (2019)
3. Petrenko, O., Liubarskiy, B., Pliugin, V.: Determination of railway rolling stock optimal movement modes. Электротехника и электромеханика **6**, 27–31 (2017)
4. Chen, F., Feng, W., Ding, H., Lee, S., Jahns, T.M., Sarlioglu, B.: Comprehensive efficiency analysis of current source inverter based SPM machine drive system for traction applications. In: Chen, F., Feng, W., Ding, H., Lee, S., Jahns, T.M., Sarlioglu, B. (eds.) 2020 IEEE Energy Conversion Congress and Exposition (ECCE), pp. 3002–3009 (October 2020). IEEE (2020). https://doi.org/10.1109/ECCE44975.2020.9236270
5. Goolak, S., Sapronova, S., Tkachenko, V., Riabov, I., Batrak, Y.: Improvement of the model of power losses in the pulsed current traction motor in an electric locomotive. East. Eur. J. Enterp. Technol. **6**(5(108)), 39–46 (2020)
6. Feng, D., Zhu, H., Sun, X., Lin, S.: Evaluation of power supply capability and quality for traction power supply system considering the access of distributed generations. IET Renew. Power Gener. **14**(18), 3644–3652 (2020)
7. Goolak, S., Tkachenko, V., Bureika, G., Vaičiūnas, G.: Method of spectral analysis of traction current of AC electric locomotives. Transport **35**(6), 658–668 (2020)
8. Shen, X., Wei, H., Wei, L.: Study of trackside photovoltaic power integration into the traction power system of suburban elevated urban rail transit line. Appl. Energy **260**, 114177 (2020). https://doi.org/10.1016/j.apenergy.2019.114177
9. Wu, S., Wu, M., Wang, Y.: A novel co-phase power-supply system based on modular multilevel converter for high-speed railway AT traction power-supply system. Energies **14** (1), 253 (2021)
10. Xiao, F., Mai, Z., Liu, J., Lian, C.: Rotation-frequency oscillations suppression strategy for AC drive system with large inertia rotating load. IET Electr. Power Appl. **14**(12), 2412–2421 (2020)
11. Deryabin, E.I., Zhuravleva, L.A.: Electric traction drive of an agricultural tractor. In: IOP Conference Series: Earth and Environmental Science, vol. 548, no. 3, p. 032037. IOP Publishing (August 2020). https://doi.org/10.1088/1755-1315/548/3/032037

12. Ferestade, I., Ahmadian, M., Molatefi, H., Moaveni, B., Bokaeian, V.: Integrated sliding mode and direct torque controls for improving transient traction in high-speed trains. J. Vib. Control **27**(5–6), 629–650 (2020)
13. Tao, L., Guo, W.: Simulation study on influence of AC-DC locomotive mixed operation on grid-side. In: 2020 IEEE 2nd International Conference on Civil Aviation Safety and Information Technology (ICCASIT), pp. 336–340. IEEE (2020). https://doi.org/10.1109/ICCASIT50869.2020.9368674
14. Wang, W., Song, Q., Zhang, S., Li, Y., Ahmad, M., Gong, Y.: The loss analysis and efficiency optimization of power inverter based on SiC MOSFETs under the high-switching frequency. IEEE Trans. Ind. Appl. **57**(2), 1521–1534 (2020)
15. Martins, A., Morais, V., Ramos, C., Carvalho, A., Afonso, J.L.: Optimizing the train-catenary electrical interface in AC railways through dynamic control reconfiguration. EAI Endorsed Trans. Energy Web **7**(25), 1–15 (2019)
16. Wang, K., Hu, H., Zheng, Z., He, Z., Chen, L.: Study on power factor behavior in high-speed railways considering train timetable. IEEE Trans. Transp. Electr. **4**(1), 220–231 (2017)
17. Никифоров, М.М., Вильгельм, А.С., Сергеев, Р.В.: Применение индикаторов энергетической эффективности электровозов для оптимизации использования тяговых ресурсов. Известия Транссиба **1**(33), 88–98 (2018)
18. Okamoto, K., Goldshtein, M., Tsiotras, P.: Optimal covariance control for stochastic systems under chance constraints. IEEE Control Syst. Lett. **2**(2), 266–271 (2018)
19. Filyushov, Y.P., Simakov, G.M., Palagushkin, B.V.: The rule of electric drive multi criteria optimization solution choice. In: 2018 XIV International Scientific-Technical Conference on Actual Problems of Electronics Instrument Engineering (APEIE), pp. 71–75. IEEE (2018). https://doi.org/10.1109/APEIE.2018.8545926
20. Buriakovskyi, S., Liubarskyi, B., Maslii, A., Pomazan, D., Panchenko, V., Maslii, A.: Mathematical modelling of prospective transport systems electromechanical energy transducers on basis of the generalized model. In: 2019 9th International Conference on Advanced Computer Information Technologies (ACIT), pp. 76–79. IEEE (2019). https://doi.org/10.1109/ACITT.2019.8779998
21. Riabov, I., Liubarskyi, B.: Determination of phase flux-linkage of flux switching motor with spatial magnetic system. In: International Conference on Industrial Engineering, Applications and Manufacturing (ICIEAM), pp. 1–7. IEEE (2018). https://doi.org/10.1109/ICIEAM.2018.8728773
22. Gama, F., Marques, A.G., Leus, G., Ribeiro, A.: Convolutional neural network architectures for signals supported on graphs. IEEE Trans. Sig. Process. **67**(4), 1034–1049 (2018)
23. Lu, S.D., Sian, H.W., Wang, M.H., Liao, R.M.: Application of extension neural network with discrete wavelet transform and Parseval's theorem for power quality analysis. Appl. Sci. **9**(11), 2228 (2019)
24. Boev, V.M.: Calculation of transients in electrical circuits with «Incorrect» initial conditions with the help of the Duhamel integral and discontinuous functions. Электротехника и электромеханика **4**, 40–44 (2018)
25. Сгибнев, М.С.: Уравнение Винера-Хопфа в мерах с вероятностным ядром. Сибирские электронные математические известия **16**, 609–617 (2019)
26. Nakamori, S.: Robust RLS Wiener signal estimators for discrete-time stochastic systems with uncertain parameters. Front. Sig. Process. **3**(1), 1–18 (2019)

Comparing Analysis of Standard and Compact Versions of Two Metal Braid High-Pressure Hoses Basing on Experimental Research

Mykola Karpenko(✉), Olegas Prentkovskis, and Šarūnas Šukevičius

Department of Mobile Machinery and Railway Transport, Vilnius Gediminas Technical University, 10223 Vilnius, Lithuania
{mykola.karpenko, olegas.prentkovskis, sarunas.sukevicius}@vilniustech.lt

Abstract. The results of comparing analysis of behaviour and energy efficiency between standard (regular) and compact versions of high-pressure hoses used in the machinery hydraulic drives are presented in this article. The approach in this research is based on the experimental measurement of high-pressure hose vibration, inside fluid flow pressure measurement and frequency analysis. By experimental measuring's, compared a standard (regular) and compact versions of two metal braid composite high-pressure hoses. The current research was performed to describe a different between standard (regular) and compact versions of high-pressure hose behaviour from depend on fluid flow inside. In current research disclosed that by using different type of the two metal braid high-pressure hoses can be achieved a reduce of power losses, damped fluid pulsation and increasing hydraulic drive efficiency by replacing a high-pressure hoses on another type of high-pressure hose, during machinery hydraulic drive maintenance or on design stage.

Keywords: Energy-saving · High-pressure hose · Frequency · Composite · Damping

1 Introduction

Improving the stability of mobile machinery hydraulic drives is required a researching of the hydrodynamic processes, where an important role played transient phenomena associated with the development of pulsation and vortex structures of fluid flow, inside lines of the hydraulic drive and influence of that processes on hydraulic drive elements according to [1]. According to [2], the hydraulic drive of any modern mobile machinery, shown on Fig. 1a, is a complex and multi-criteria model, which includes a variety of different hydraulic components. For hydraulic components connecting into a single system used special elements what named – high-pressure hoses. According to [3] usually all hydraulic high-pressure hoses made from a composite materials based on the rubber, and it's an important structural parts of any hydraulic drive. Hydraulic high-pressure hose – high strength, lightweight, corrosion and impact resistance what make

O. Prentkovskis et al. (Eds.): TRANSBALTICA 2021, LNITI, pp. 253–262, 2022.
https://doi.org/10.1007/978-3-030-94774-3_25

them widely utilized in aviation industry [4] (Fig. 1b), transport engineering [5], and even in medical and sports fields [6].

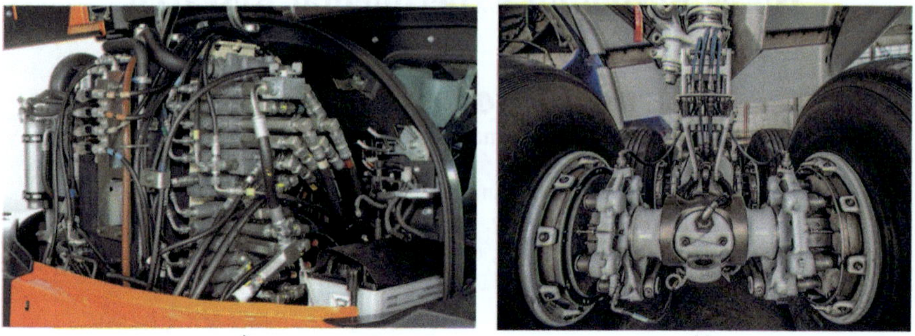

a) b)

Fig. 1. The typical components of hydraulic drive: a – the typical composition of mobile machinery hydraulic system; b – the typical composition of airplane chassis hydraulic system.

During machinery hydraulic drive operation a fluid with a high pulsation flowing from pump to working equipment's (hydraulic cylinders, motors, etc.) inside a high-pressure hoses. That fluid flow is associated with a variety of transient phenomena, according to [7, 8], which can be divided into two groups: high-frequency pulsations – resulted from pumps and low-frequency pulsations – excited in the high-pressure hoses. However, an accurate dynamic analysis is often difficult to achieve for the following reason that hydraulic driving process is strongly nonlinear due to both the compressibility of oil and the nonlinear characteristics of the pump and the high-pressure hoses. The task of researching a influences of fluid flow pulsation on high-pressure hose behaviour include a solutions on improving the stability of hydraulic drive, reliability, energy-saving and hydrodynamic processes inside all hydraulic drive.

The review of fluid flow pulsation influence inside high-pressure hoses by [9, 10] was considered more than a hundred articles on various aspects of the problematic. At the same time, for correcting investigation of composite materials behaviours should be taking in account the loads under what object belong during operation by [11]. That why experimental researches with combination of computational as well as advanced constitutive modeling a wide range used for investigation and improving using composite high-pressure hoses during machines design. The current research directed on investigation and comparing standard and compact versions of two metal braid high-pressure hoses, their possibility of fluid damping and energy parameters with taking in account frequency responses.

2 Research Objects

Hydraulic high-pressure hose (Fig. 2a) – high strength, corrosion and impact resistance with superior performance composite pipeline based on the rubber. The high-pressure hoses consists of three main elements: the inner rubber layer, reinforcing layers and the outer rubber layer (protective cover). 3D models of high-pressure hose with two braid layers (2SN/2SC) are presented in Fig. 2b. The two layers braid high-pressure hose is one of the most frequently used type of high-pressure hose in the hydraulic drives, according to [12].

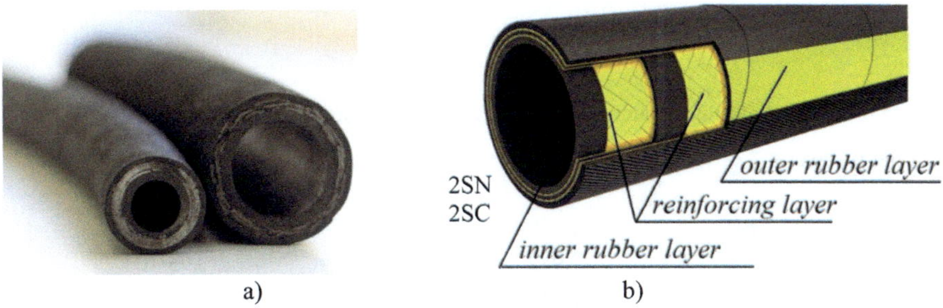

Fig. 2. Research Objects: a) – example of composite metal braid pipeline; b) – 3D view model of two metal braid high-pressure hoses.

In current research, for comparing analysis between using two metal braid high-pressure hoses types, deformation and frequency response of high-pressure hoses, depend from the fluid flow inside, was used two types of high-pressure hose:

1. Two metal braid reinforced hydraulic hose (2SN) according to European Standard [13].
2. Two metal braid reinforced hydraulic hose (compact version) (2SC) according to European Standard [14].

The high-pressure hose with conditional passage 08 DASH (which is equal to the diameters 1/2″ or 12.7 mm) is one of the most frequently used diameters in hydraulic drives [12] accepted for research. Parameters of used high-pressure hoses in the research is presented in Table 1. The material of high-pressure hoses layers presented in Table 2 and the material property of layers presented in Table 3.

Table 1. Physical and geometrical parameters of the high-pressure hoses.

Type	Inner diameter, mm	Outer diameter, mm	Max working pressure, bar	Min brake pressure, bar	Weight, kg/m
2SN	1/2″ or 12.7	23.2	275	1,100	0.63
2SC	1/2″ or 12.7	20.4	275	1,100	0.54

Table 2. High-pressure hoses layers material.

Type	Internal layer	Reinforcement	Layer between reinforcement	External covering
2SN	Synthetic rubber (NBR), extruded whole without joints	Steel wire (braid)	Synthetic rubber, uniform thickness	Anti-abrasive synthetic rubber NBR/PVC
2SC	Synthetic rubber (NBR), extruded whole without joints	Steel wire (braid)	Synthetic rubber, uniform thickness	Anti-abrasive synthetic rubber NBR/PVC

Table 3. Material property of high-pressure hoses layers.

Material of layer	Density, kg/m^3	Tensile Strength, MPa	Young`s modulus, MPa
Synthetic rubber, NBR	1,350	24.1	4
Steel wire braid	7,982	215	193•103
Anti-abrasive synthetic rubber NBR/PVC	1,235	22	35

Presented high-pressure hoses standards are International Certificate and are used in the most mobile machinery hydraulic drives. The length of high-pressure hoses, used in the current research is 1 m.

3 Experimental Research

Experimental research was conducted as a first step towards establishing the behaviour between standard and compact versions of two metal braid high-pressure hoses. The experimental part of the conducted research included the measurement and analysis of fluid pressure inside a high-pressure hoses and vibration analysis of high-pressure hoses behaviour depends from the fluid flow inside. The experimental investigation performed via Two Sample Measurement Design and based on One-Sample Statistical Method with Estimating Uncertainty in Repeated Measurements of data processing [15]. Test bench for the experimental research shown in Fig. 3.

The main parameters of operated hydraulic pump: Fluid pressure $\sim 2.48 \cdot 10^6$ Pa; Fluid flow rate ~ 35 l/min; Rotation $\sim 1{,}500$ rev/min.

The experiment measurement include a 3^d measuring's of fluid pressure inside high-pressure hose on inlet and outlet and 3^d measuring's of outer surface velocity deformation changing during inside fluid pulsation and frequency response spectrum.

For decrease during measurements was accepted and presented averages results of several measuring's. The percentage error value between non-contact laser scanning measurements is presented in Table 4 and the percentage error value between fluid pressure measurements is presented in Table 5.

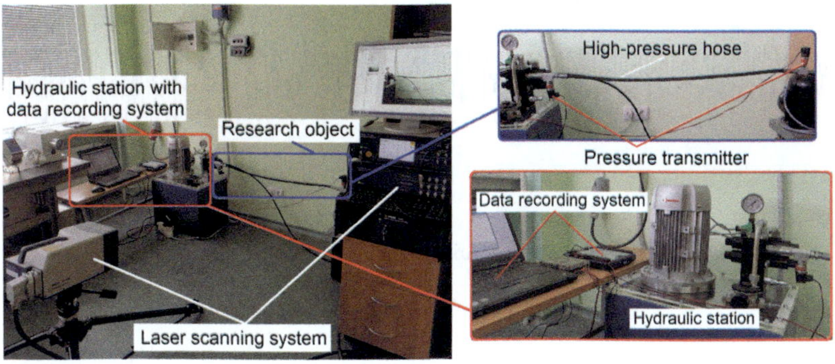

Fig. 3. The test bench for experimental research of high-pressure hoses.

Table 4. Percentage error of non-contact laser scanning measurements.

№ of non-contact measurement	Error at 2SN measurement, %	Error at 2SC measurement, %
1st measurement	1.36	1.54
2nd measurement	1.14	1.28
3rd measurement	1.29	1.34

Table 5. Percentage error of fluid pressure measurements inside high-pressure hoses.

№ of fluid pressure measurement	Error at 2SN measurement, %	Error at 2SC measurement, %
1st measurement	0.93	1.54
2nd measurement	1.09	1.28
3rd measurement	1.12	1.34

The measurement of the fluid pressure on inlet and outlet of high-pressure hoses shown in Fig. 4. Velocity measuring by sensor head of laser scanning system on the outer surface of the high-pressure hoses, during test, shown on Fig. 5.

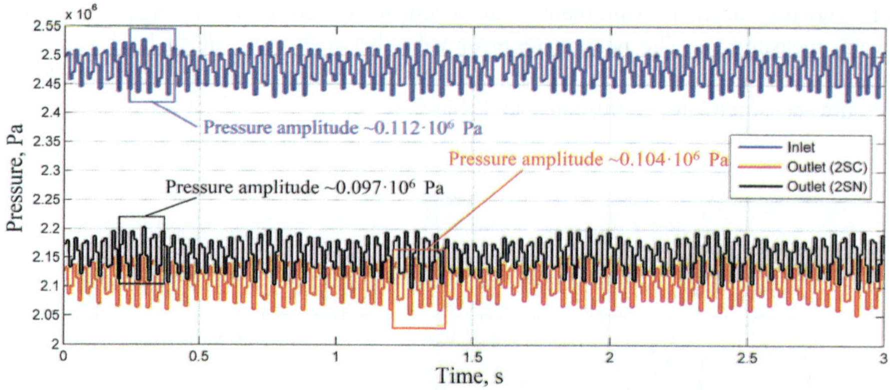

Fig. 4. The measurement of fluid pressure inside high-pressure hoses.

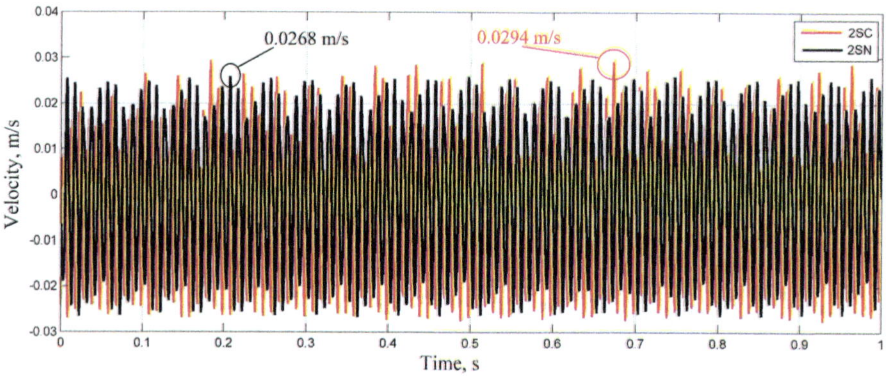

Fig. 5. The measurement of high-pressure hoses velocity deformation.

The nominal fluid pressure drop between inlet and outlet of high-pressure hoses is $\sim 0.332 \cdot 10^6$ Pa for 2SN and $\sim 0.381 \cdot 10^6$ Pa for 2SC was obtained by measurements. The fluid pressure amplitude reduce from inlet ($0.112 \cdot 10^6$ Pa) to $0.097 \cdot 10^6$ Pa on 2SN; $0.104 \cdot 10^6$ Pa on 2SC outlets of the high-pressure hoses. That confirm that different types of high-pressure hoses can be used for reducing dynamic loads inside hydraulic lines. From the comparing of measuring by PSV sensor head on surface of high-pressure hoses, an amplitude of displacement velocity by measuring was 0.0268 m/s on 2SN and 0.0294 m/s on 2SC, what confirm a high-pressure hoses deformation during a fluid pulsation inside. By graph can be point the less displacement velocity on standard version (2SN) of high-pressure hose and more displacement velocity on the compact version (2SC) of high-pressure hose, that is due to the fact that in the standard version of high-pressure hose, in radial direction the rubber layers is thicker than in compact version of the same two metal braided high-pressure hose.

The additional result obtained from experimental research comparing – fluid frequency response inside high-pressure hoses, based on spectrum analyses (Fig. 6).

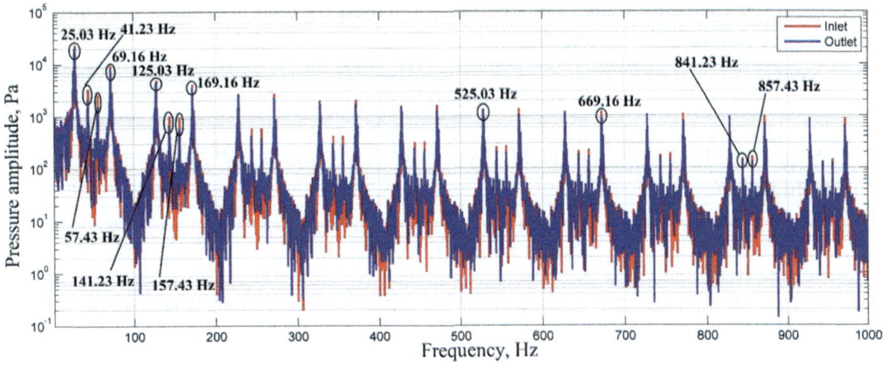

a)

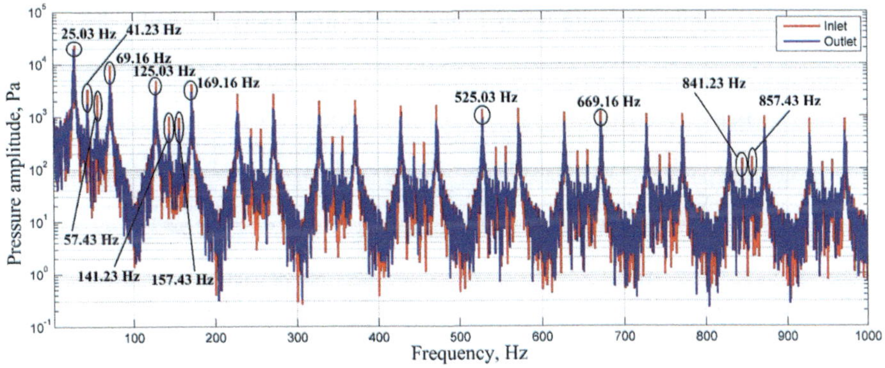

b)

Fig. 6. Frequency response from fluid flow inside high-pressure hoses: a) – 2SC; b) – 2SN.

By frequency's response of fluid flow inside the high-pressure hoses it is seen that the main fluid pressure amplitudes on frequency 25.03 Hz, 41.23 Hz, 57.43 Hz and 69.16 Hz with harmonic steps (for each harmonic step – 100 Hz) – 125.03 Hz, 141.23 Hz, 157.43 Hz, 169.16 Hz etc. The pressure amplitude on pointed frequencies and on they harmonic steps by standard version of high-pressure hose (2SN) were more damped, comparing to compact version (2SC). Since, the thickness and deformation of investigated high-pressure hoses are different, the damping of fluid pulsation between inlet and outlet amplitude constitute ~13.4% for 2SN and ~7.2% for 2SC of high-pressure hose. The results showed that by using standard version of two metal braided high-pressure hose more damping of fluid pulsation can be reached, compared to the compact version of the same high-pressure hose.

4 Analysis of the Energy Efficiency

The comparing analysis of the energy efficiency between standard and compact versions of two metal braided high-pressure hoses is presented on example of archived results from the measurements. The energy consumed by using different high-pressure hose can be calculated by equation:

$$E_{con.} = \Delta P_i(t) \cdot Q_i(t), \tag{1}$$

where $\Delta Q_i(t)$ – a fluid flow rate inside hydraulic drive, l/min; $\Delta P_i(t)$ – pressure losses by using high-pressure hoses, W:

$$\Delta P_i(t) = P_{inlet}(t) - P_{outlet}(t), \tag{2}$$

where $P_{inlet}(t)$, $P_{outlet}(t)$ – fluid pressure at the inlet and outlet of HPHs, respectively.

The energy flow charts is presented according to the archived results and showed in Fig. 7 (results performed for high-pressure hose length – 1 m). The energy flow charts is applicable for illustration energy transformation visually and quantitatively during using or replacing high-pressure hoses in the mobile machinery hydraulic drive.

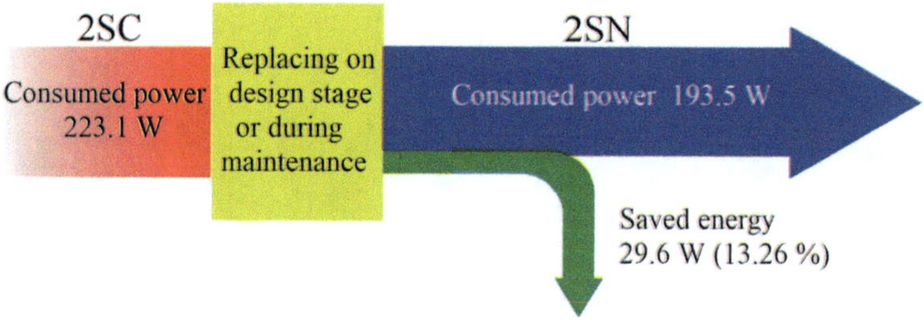

Fig. 7. The energy flow chart from comparing analysis.

The obtained results showed that the better options for energy saving in the hydraulic drive could be reached using standard version of two metal braid high-pressure hose. Compared to 2SC type, using a 2SN high-pressure hose an energy around 29.6 W, on length of hose one meter, can be saved. Alternatively, in percentage form 13.26% by each meter of high-pressure hose. The carried out research demonstrated that use 2SN high-pressure hose performed in the most efficient way than compact version.

5 Conclusions

In the present research, by experimental measuring's, compared a standard (2SN) and compact (2SC) versions of two metal braid high-pressure hoses. The approach presented in this paper of experimental measurement, based on fluid pressure measuring, high-pressure hose vibration measuring and frequency's analysis.

By experimental investigation, founded, that the nominal fluid pressure different between inlet and outlet of high-pressure hoses is $\sim 0.332 \cdot 10^6$ Pa for 2SN and $\sim 0.381 \cdot 10^6$ Pa for 2SC. By measurements, the fluid pressure amplitude reduce from inlet ($0.112 \cdot 10^6$ Pa) to $0.097 \cdot 10^6$ Pa on 2SN; $0.104 \cdot 10^6$ Pa on 2SC outlets of the high-pressure hoses. That confirm that different types of high-pressure hoses can be used for reducing dynamic loads inside hydraulic lines. From the comparing of measuring by PSV sensor head on surface of high-pressure hoses, an amplitude of displacement velocity by measuring was 0.0268 m/s on 2SN and 0.0294 m/s on 2SC, what confirm a high-pressure hoses deformation during a fluid pulsation inside. By obtained results can be point the less displacement velocity on standard version (2SN) of high-pressure hose and more displacement velocity on the compact version (2SC) of high-pressure hose, that is due to the fact that in the standard version of high-pressure hose, in radial direction the rubber layers is thicker than in compact version of the same two metal braided high-pressure hose.

The difference in the amplitude of fluid pulsation on the inlet and outlet of high-pressure hoses confirmed that they can be used to damp fluid pulsation and reduce fluid pressure losses inside hydraulic lines. Since, the thickness and deformation of investigated high-pressure hoses are different, the damping of fluid pulsation between inlet and outlet amplitude constitute $\sim 13.4\%$ for 2SN and $\sim 7.2\%$ for 2SC of high-pressure hose. The results showed that by using standard version of two metal braided high-pressure hose more damping of fluid pulsation can be reached, compared to the compact version of the same high-pressure hose. By frequency's response of fluid flow inside the high-pressure hoses it is seen that the main fluid pressure amplitudes on frequency 25.03 Hz, 41.23 Hz, 57.43 Hz and 69.16 Hz with harmonic steps (for each harmonic step – 100 Hz) – 125.03 Hz, 141.23 Hz, 157.43 Hz, 169.16 Hz etc. The pressure amplitude on pointed frequencies and on they harmonic steps by standard version of high-pressure hose (2SN) were more damped, comparing to compact version (2SC).

The obtained results showed that the better options for energy saving in the hydraulic drive could be reached using standard version of two metal braid high-pressure hose. Compared to 2SC type, using a 2SN high-pressure hose an energy around 29.6 W, on length of hose one meter, can be saved. Alternatively, in percentage form 13.26% by each meter of high-pressure hose.

In final step, research disclosed that the better options for energy saving in the hydraulic drive could be reached using standard version of two metal braid high-pressure hose by replacing compact version of high-pressure hose on its standard version high-pressure hose, during machinery hydraulic drive maintenance or on design stage.

References

1. Shen, W., Jiang, J., Su, X., Karimi, H.: Control strategy analysis of the hydraulic hybrid excavator. J. Franklin Inst. **352**(2), 541–561 (2015)
2. Karpenko, M., Bogdevičius, M.: Investigation of hydrodynamic processes in the system – "pipeline-fittings." In: Gopalakrishnan, K., Prentkovskis, O., Jackiva, I., Junevičius, R. (eds.) TRANSBALTICA 2021. LNITI, pp. 331–340. Springer, Cham (2020). https://doi.org/10.1007/978-3-030-38666-5_35
3. Cho, J.: Anisotropic large deformation and fatigue damage of rubber-fabric braid layered composite hose. Procedia Eng. **173**, 1169–1176 (2017)
4. Parvaresh, A., Mardani, M.: Model predictive control of a hydraulic actuator in torque applying system of a mechanically closed-loop test rig for the helicopter gearbox. Aviation **23**(4), 143–151 (2019)
5. Bogdevičius, P., Prentkovskis, O., Bogdevičius, M.: Transmission with cardan joint parameter influence to centrifugal pump characteristics. Mokslas – Lietuvos Ateitis/Sci. Fut. Lith. **9**(5), 559–564 (2017). (in Lithuanian)
6. Peters, S.T. (ed.): Handbook of Composites. Springer, Boston (1998). https://doi.org/10.1007/978-1-4615-6389-1
7. Dörfler, P., Sick, M., Coutu, A.: Flow-Induced Pulsation and Vibration in Hydroelectric Machinery: Engineer's Guidebook for Planning, Design and Troubleshooting. Springer, London (2013). https://doi.org/10.1007/978-1-4471-4252-2
8. Minakov, A., Platonov, D., Dekterev, A., Sentyabov, A., Zakharov, A.: The analysis of unsteady flow structure and low frequency pressure pulsations in the high-head Francis turbines. Int. J. Heat Fluid Flow **53**, 183–194 (2015)
9. Paidoussis, M., Li, G.: Pipes conveying fluid: a model dynamical problem. J. Fluids Struct. **7**(2), 137–204 (1993)
10. Paidoussis, M.: Fluid–Structure Interactions: Slender Structures and Axial Flow, vol. 2, 192 p. Elsevier Academic Press, London (2003)
11. Bogdevičius, M., Karpenko, M., Bogdevičius, P.: Determination of rheological model coefficients of pipeline composite material layers based on spectrum analysis and optimization. J. Theor. Appl. Mech. **59**(2), 265–278 (2021)
12. Karpenko, M., Bogdevičius, M.: Investigation into the hydrodynamic processes of fitting connections for determining pressure losses of transport hydraulic drive. Transport **35**(1), 108–120 (2020)
13. European standard. EN 853 2SN:2015. Rubber hoses and hose assemblies. Wire braid reinforced hydraulic type. Specification (2015)
14. European standard. EN 857 2SC:2015. Rubber hoses and hose assemblies. Wire braid reinforced hydraulic type. Specification (2015)
15. Karpenko, M.: Investigation of energy efficiency of mobile machinery hydraulic drives. Ph. D. dissertation, Vilniaus Gedimino technikos universitetas, 164 p. (2021)

Emissions of Polycyclic Aromatic Hydrocarbons by Road Transport into Roadside Areas

Valentina Iurchenko[1], Ievgeniia Ugnenko[2(✉)], Oksana Melnikova[1], Larysa Mykhailova[3], Elena Lebedeva[1], and Gintas Viselga[4]

[1] Kharkiv National University of Civil Engineering and Architecture, 40, Sumskaya Street, Kharkiv, Ukraine

[2] Ukrainian State University of Railway Transport, Feierbakh Square 7, Kharkiv 61050, Ukraine
ugnenko@kart.edu.ua

[3] Brandenburg University of Technology at Cottbus, Konrad-Wachsmann-Allee, 6, Cottbus, Germany

[4] Department of Mechanics and Materials Engineering, Vilnius Gediminas Technical University, J. Basanavicius str. 28, Vilnius, Lithuania
gintas.viselga@vilniustech.lt

Abstract. In experimental studies, the composition of polycyclic aromatic hydrocarbons (PAHs) in motor vehicle emissions was studied based on the analysis of PAHs in roadside soils of urban motor roads and suburban motor road. The concentration of PAHs in soils (1 m from the road) exceeded this indicator in the control sample by 8.7–152 times. The concentration of petroleum products in the investigated soils exceeded the specified permissible level in Ukraine (200 mg/kg), especially in the soils of the roadside area of the suburban motor road. The roadside area of suburban motor roads was contaminated with PAHs almost an 10 times higher than the roadside area of urban motor roads. The concentration of aliphatic compounds in the soils of the roadside area of urban motor roads was approximately 8 times higher than the concentration of aromatic compounds. While in the soils of the roadside area of the suburban motor road, the concentrations of aliphatic and aromatic compounds were very close. Among the studied PAHs, fluoranthene predominated in the soils of a suburban road, and phenanthrene slightly prevailed in the soils of urban roads.

Keywords: Polycyclic aromatic hydrocarbons · Motor vehicle emissions · Roadside soils · Environmental problem

1 Introduction

Polycyclic aromatic hydrocarbons (PAHs) are classified as hazardous and persistent pollutants of the natural environment; many of them are mutagens and carcinogens that pose a threat to human health. The study of PAHs is a priority in environmental studies. Natural surfactants do not pose a real threat; the environmental problem is created by

O. Prentkovskis et al. (Eds.): TRANSBALTICA 2021, LNITI, pp. 263–269, 2022.
https://doi.org/10.1007/978-3-030-94774-3_26

man-made surfactants [1, 2]. Among a large number of surfactants, benzo[*a*]pyrene, along with benzo[*b*]fluoranthene, benzo[*j*]fluoranthene, benzo[*k*]fluoranthene, dibenzo [*a, h*]pyrene, holantrene, have the greatest danger on the environment in terms of the degree of impact [3–5].

PAHs (surfactants) are organic compounds characterized by the presence in the chemical structure of three or more condensed benzene rings. Carcinogenic surfactants can be small (less than three rings) or large (more than four rings). Almost all man-made sources of surfactants are based on thermal processes associated with the combustion and processing of organic raw materials at temperatures of 650 to 900 °C and lack of oxygen. Anthropogenic surfactants enter the atmosphere in the form of soot particles (the product of incomplete combustion of fuel) and in the adsorbed state on the surface of solid particles. To date, hundreds of surfactants and their analogues have been identified that have more or less carcinogenic and mutagenic activity and fall into the group of anthropogenic harmful substances [1, 3, 4, 6, 7].

Mobile anthropogenic sources of surfactants are mainly land, road transport, aviation, and water transport. Twenty types of surfactants were detected in the exhaust gases of automobiles. In addition to 3,4-benzopyrene (3,4-BP), they contain other, no less dangerous surfactants that exhibit high mutagenic and carcinogenic activity such as benz[a]anthracene, benzfluoranthene, dibenzanthracene, and others [8–12].

Soils play the role of a kind of "depot", where petroleum products (PPs) including PAH enter due to anthropogenic emissions and natural inflows [13–15]. The concentration of 3,4-BP is 7 to 22 times higher at a depth of 0 to 5 cm than at a depth of 5 to 20 cm. Surfactants can move from the soil to plants, water and air. Many soil microorganisms have been found to be highly sensitive to the action of surfactants, which alters the formed microbiocenoses and affects the biological productivity of the soil [15–17].

The content of surfactants in the upper layers of soil increases in direct proportion to the rate of continuous growth of the number of transport units. The content of 3,4-BP is on average 90 to 1,300 ng/h in the US, 8 to 82 ng/h in Germany, 2 to 170 ng/h in France, 0.1 to 350 ng/h in Russia, 0 to 785 ng/h in Iceland. The highest concentrations of surfactants were found in areas adjacent to cities; the lowest in remote forest areas.

Many years of toxicological research have made it possible to identify surfactant compounds that need to be constantly monitored in the environment, both in terms of their own toxicity and in terms of the most likely release into the environment. The US Environmental Protection Agency (US EPA) monitors 16 compounds from the group of surfactants in environmental samples; the EU monitors 6.

The purpose of the work is to determine the impact of exhaust gases of road transport on soil contamination with surfactants of the roadside area of urban and suburban roads.

Since the determination of the concentration of surfactants in car emissions is a certain technical complexity, soils of the roadside area, in which the products of road transport emissions are condensed, were used as the object of study of the composition of surfactants.

2 Objects and Research Methods

Soils in the territories, boarded on automobile road: urban motor roads (Pushkinskaya St., Akademika Pavlova St., Traktorobudivnykiv St.), suburban motor road (M03 Kiev-Dovzhanskaya), were explored in the work. Samples of soils were collected by the modified method "envelope" at a depth of 0.5 to 1.0 cm. In soil samples PP concentration was estimated gravimetrically after hexane extraction according to the methods, which are recommended by the regulatory documents [18].

Identification and quantification of particular hydrocarbons, polluting of scanned soils, were performed with Gas Chromatography (GC, Carlo Erba GC8000) with a flame ionization detector (Fisons EL980 FID) and a mass spectrometer (Fisons MD800 MS). PP was extracted by means of triple chloroform extraction (5 ml) from 5 g of soil with the foolowing solving of the rest with hexane and cleaning of the extract through the column with florisil. Solvent evaporation was performed by applying of hydrocarbon free nitrogen at 20 °C to the sample volume 10–100 mcl, depending on the results of gravimetrical determination of PP according to EN ISO 9377-2, 2001. Traded diesel fuel and alkane standard Hewlett Packard (part. No. 18710-60170) were applied for quantification of the total amount of hydrocarbons and N-alkanes, respectively. Electron impact ionization (EI+) at 70 eV and full mode of scanning with the range of masses (m/z) from 50 to 250 were applied for detection. Quantitative estimation of the average (C15–C27) and heavy fraction of PP (>C27) in chromatograms was performed by applying of total ion current (TIC).

3 Results and Discussion

The characteristics of the composition of Germany standard sample of gasoline fuel used for quantification of volatile organic compounds (VOCs) are given in Table 1. As can be seen, a standard sample of gasoline fuel has a higher relative content of aromatic compounds (up to 67%) compared to aliphatic hydrocarbons (about 33%). In the emissions of diesel-fueled ICE (internal combustion engines), the major part of VOCs (about 57%) are aliphatic compounds, and the minor part of VOCs are aromatic hydrocarbons (about 44%). In the emissions of gasoline-fueled ICE aromatic compounds prevail [12].

Table 1. The qualitative and quantitative composition of VOCs in the emissions of the exhaust gases of internal combustion engines running on various types of fuel.

Hydrocarbons	The content of hydrocarbons in the total mass of VOCs, %		
	Exhaust gases of the internal combustion engine running on [12]:		Gasoline fuel standard (Germany)
	Gasoline fuel	Diesel fuel	
Aliphatic:	1.48–52.94	57.64–75.45	33.33
Aromatic:	47.06–101.24	24.55–42.36	66.67

The results of determining the concentration of oil products, aliphatic and aromatic hydrocarbons, PAHs in the soils of the roadside area of urban and suburban motor roads are given in Table 2. As can be seen, the concentration of oil products in the investigated soils exceeded the specified permissible level in Ukraine (200 mg/kg), especially in the soils of the roadside area of the suburban motor road.

Table 2. The composition of PPs that contaminate soils of the roadside area (1 m from the road) of the investigated motor roads.

Road	Side of the road	PP content, mg/kg	Aliphatic hydrocarbon content		Aromatic hydrocarbon content		PAHs content	
			mg/kg	in PPs, %	mg/kg	PPs, %	mg/kg	PPs, %
Prospekt Traktorobudivny-kiv	Leeward	268.0	238.0	88.8	30.0	11.2	1.7	0.6
Pushkinskaya Street	Windward	1185.0	1057.0	89.2	128.0	10.8	7.8	0.7
M03 Kharkiv-Dovzhansky	Windward	3798.0	2144.0	56.5	1654.0	43.5	137.3	3.6
Reference value		197.0	192.0	97.5	5.0	2.5	0.9	0.5

Aliphatic hydrocarbons prevailed among the hydrocarbons found in soil (as with in the VOC emissions generated by diesel-fueled engines), but their quantitative ratio was significantly different from the ratio of these classes of compounds in the exhaust gases of cars and in fuel (Table 1). The concentration of aliphatic compounds in the soils of the roadside area of urban motor roads was approximately 8 times higher than the concentration of aromatic compounds. While in the soils of the roadside area of the suburban motor road, the concentrations of aliphatic and aromatic compounds were very close. In addition, in the investigated soils of the roadside area of urban motor roads, the concentrations of oil products, aliphatic and aromatic hydrocarbons, PAHs exceeded these values in the reference sample by (PP content from 1.4 to 6.1), (aliphatic hydro-carbon content from 1.3 to 5.5), (aromatic hydro-carbon content from 6 to 25.7), (PAHs content from 1.9 to 8.7) times. In the soils of the roadside area of a suburban motor road, this excess was multiple (by 19.3 times for oil products, by 11.2 times for aliphatic hydrocarbons, by 330 times for aromatic hydrocarbons, by 152 times for PAHs), which is probably due to the absence of curbs on these motor roads, the intensity and composition of the traffic flow, the type of fuels used, etc. The determined level of soil contamination of the roadside area with PAHs makes it possible to assess the role of emissions from road transport as an intense source of man-made pollution of natural environment objects with environmentally highly hazardous compounds.

Table 3 shows the identification data for PAHs in the investigated soils. As can be seen, PAHs with the number of carbon atoms in the chain making C_{12}–C_{18} were identified in the soils of the roadside area of the investigated highways. It is possible that lighter and longer-chain aromatic compounds (including benzopyrene) were not detected due to the peculiarities of the chromatography mode (the range of detected masses is from 50 to 250). Among the identified PAHs phenanthrene ($C_{14}H_{10}$) (hazard class 1) and fluoranthene ($C_{16}H_{10}$) (hazard class 2 for soil) [19] have the highest concentrations that are on average 57% and 39%, respectively, of the total content of aromatic compounds in the investigated roadside soils. In the meantime, fluoranthene significantly prevails in the soils of a suburban road, and phenanthrene slightly prevails in the soils of urban roads. The share of PAHs in the PPs of the nearest roadside area of a suburban road is 3 to10 times higher than the value for urban roads.

Table 3. The concentration of PAHs that contaminate soils of the roadside area (1 m from the road).

Road	Side of the road	Concentration of PAHs, mg/kg					
		Acenaphthy-lene	Phenan-threne	Pyrene	Fluoran-thene	Benzo(a) anthracene	Chrysene
Prospekt Traktorobudiv-nykiv	Leeward	–	0.786	0.256	0.503	0.058	0.08
Pushkinskaya Street	Windward	–	2.493	1.427	2.436	0.659	0.757
M03 Kharkiv-Dovzhansky	Windward	10.547	13.699	25.865	43.372	19.841	23.96
	Leeward	–	4.904	4.072	7.375	2.096	1.994
Reference value		–	0.393	0.140	0.287	–	0.078

The results of determining the concentration of surfactants in roadside soils at different distances from Pushkinskaya Street are shown in Fig. 1. As can be seen, with an increase in the distance from the road, the concentration of PAHs in soils sharply decreases (similar to the dynamics of a decrease in the total concentration [20]).

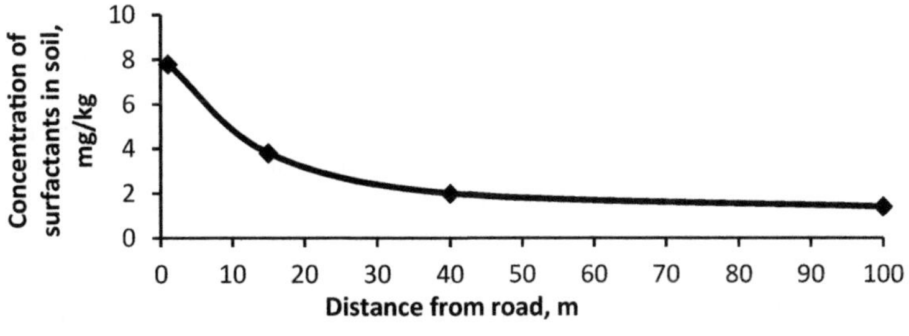

Fig. 1. Distribution of PAHs in the soils of the roadside area on Pushkinskaya Street.

The amount of PAHs in the soils of the roadside area of the investigated motor roads ranges within 0.11 to 3.61% of the total concentration of oil products in the soils (Table 2). However, they are extremely environmentally hazardous compounds - hazard class 1 [19]. In Ukraine, the concentration of PAHs in the soil is not standardized, but if we focus on the maximum permissible concentration (MPC) of the total PAHs (1 mg/kg) established for soils in Belarus [19], then their content in almost all samples of the investigated soils of the roadside area is approximately 2 to 137 of the MPC. In addition, in Belarus, approximate permissible concentrations (APC) have been established for certain PAHs of hazard classes 1 and 2 [19] (naphthalene, phenanthrene, fluoranthene, benzo(*a*)anthracene and chrysene), based on which, it can be concluded that in all investigated soil samples, an excess was observed: 3 to 249 of the APC in the roadside area of urban motor roads, and approximately 100 to 2,892 of the APC for a suburban motor road. The greatest excess of PAHs that is typical for the nearest roadside area of the M18 suburban road (without a curb) is most likely associated with the introduction of a significant part of PAHs with drop entrainment from the road bed and road washouts.

4 Conclusions

Road transport is an intensive source of soil contamination in the roadside area with environmentally hazardous compounds such as PAHs (hazard class 1), the concentration of which exceeded this value in the reference soil sample by 8.7 to 152 times, and exceeded the known MPC for soils (Belarus) by 2 to 137 times.

It has been found that the concentration of PAHs in the soil sharply decreases with an increase in the distance (approximately 15 m and more) from the road, reaching, in some cases, an environmentally safe level.

The roadside area of suburban motor roads was contaminated with PAHs almost an order higher than the roadside area of urban motor roads, which was caused by the absence of a curb (which leads to contamination by drop entrainment and road washouts), differences in the composition of the traffic flow, types of automobile engines and motor fuel.

References

1. Polycyclic aromatic hydrocarbons (PAHs): sources, pathways and environmental data. Environment Agency Horizon House, Deanery Road, Bristol BS1 5AH October, 44 p. (2019)
2. Rengarajana, T., Rajendranb, P., Nandakumarc, N., Lokeshkumard, B., Rajendrane, P., Nishigakib, I.: Exposure to polycyclic aromatic hydrocarbons with special focus on cancer. Asian Pac. J. Trop. Biomed. **5**(3), 182–189. http://www.sciencedirect.com/science/article/pii/S2221169115300034
3. Pies, C., Hoffmann, B., Petrowsky, J., Yang, Y., Ternes, T., Hofmann, T.: Characterization and source identification of polycyclic aromatic hydrocarbons (PAHs) in river bank soils. Chemosphere **72**(10), 1594–1601 (2008)

4. Rumney, H.S., Bolam, S.G., Law, R.J.: Polycyclic aromatic hydrocarbons in sediments at dredged material disposal sites around England: concentrations in 2013 and time trend information at selected sites 2008–2013. Mar. Pollut. Bull. **92**(1–2), 180–185 (2015)
5. Breedveld, G.D., Sparrevik, M.: Nutrient-limited biodegradation of PAH in various soil strata at a creosote contaminated site. Biodegradation **11**, 391–399 (2000). https://doi.org/10.1023/A:1011695023196
6. Semenov, M., Marynaite, Y.Y., Holobokova, L.P., Khuryhanova, O.Y.: Sources of polycyclic aromatic hydrocarbons in the surface air layer and surface water layer of Lake Baikal. Geoecology **3**, 56–64 (2018)
7. Kemelov, K., Maymekov, U., Sambaeva, D., Maymekov, Z.: Reducing concentrations of Benzo(a)pyrene in gas phase soot particles by using and burning water fuel emulsions. Pol. J. Environ. Stud. **29**(4), 2669–2677 (2020). https://doi.org/10.15244/pjoes/112367
8. Pshenyn, V.N.: Transport as a source of polycyclic aromatic hydrocarbons in the environment. Transport: Science, technology, management. All-Union Institute of Scientific and Technical Information, vol. 8, pp. 2–9 (1995)
9. Markov, V.A., Devianyn, S.N., Markova, V.V.: Environmental safety assessment of power plants with diesel engines. Saf. Technosphere **2**, 23–32 (2014). https://doi.org/10.12737/3668
10. Keyte, I.J., Albinet, A., Harrison, R.M.: On-road traffic emissions of polycyclic aromatic hydrocarbons and their oxy- and nitro- derivative compounds measured in road tunnel environments. Sci. Total Environ. **566–567**, 1131–1142 (2016)
11. Bryselbout, C., Henner, P., Carsignol, J., Lichtfouse, E.: Polycyclic aromatic hydrocarbons in highway plants and soils. Evidence for a local distillation effect. Analusis **4**, 290–293 (2000)
12. Gloria, J., Jacob, M., Reiko, H., Jennifer, C.: GS-MS determination of volatile organic compounds in gasoline and diesel emissions. Dartmouth Undergraduate J. Sci., 47–53 (2006)
13. Mykhailova, L., Fischer, T., Iurchenko, V.: Microbial activity and decomposition of soil organic matter in roadside soils contaminated with petroleum hydrocarbons. CLEAN - Soil Air Water **46**(6), 1800132 (2018). https://doi.org/10.1002/clen.201800132
14. Iurchenko, V., Melnikova, O., Mykhailova, L., Lebedeva, E., Mikhalevich, N.: Supporting of ecological safety of run-off from the territory of objects of road infrastructure, contaminated by petroleum products. In: Gopalakrishnan, K., Prentkovskis, O., Jackiva, I., Junevičius, R. (eds.) TRANSBALTICA 2021. LNITI, pp. 10–17. Springer, Cham (2020). https://doi.org/10.1007/978-3-030-38666-5_2
15. Hennadyev, A.N., Kozyn, Y.S., Shurubor, E.Y.: Dynamics of soil pollution by polycyclic aromatic hydrocarbons and indication of the state of soil ecosystems. Soil Sci. **10**, 75–85 (1990)
16. Vane, C.H., et al.: Polycyclic aromatic hydrocarbons (PAH) and polychlorinated biphenyls (PCB) in urban soils of Greater London, UK. Appl. Geochem. **51**, 303–314 (2014)
17. Horobtsova, O.N., Nazarenko, O.H., Mynkyna, T.M., Borysenko, N.Y.: The role of soil cover in the accumulation and migration of polycyclic aromatic hydrocarbons during technogenic pollution. Nat. Sci. **1**, 73–79 (2005)
18. Lurye, Yu.: Analytical Chemistry of Industrial Effluents, 448 p. Chemistry (1984)
19. The list of maximum permissible concentrations (MPC) and roughly permissible concentrations (APC) of chemicals in the soil: hygienic standards 2.1.7.12-1-2004. Resolution of the Chief State Sanitary Doctor of the Republic of Belarus dated February 25, vol. 28, p. 15 (2004)
20. Iurchenko, V., Melnikova, O., Mikhaylova, L., Lebedeva, E.: Contamination and "self-cleaning" of soils, boarded on the objects of automobile and road complex, from petroleum products. Procedia Eng. **187**, 783–789 (2017). https://doi.org/10.1016/j.proeng.2017.04.438

Liquefied Natural Gas Regasification Technologies

Vigailė Semaškaitė(✉) and Marijonas Bogdevičius

Department of Mobile Machinery and Rail Transport, Vilnius Gediminas Technical University, Plytinės St. 27, 10105 Vilnius, Lithuania
{vigaile.semaskaite, marijonas.bogdevicius}@vilniustech.lt

Abstract. Environment restrictions of global energy market lead to consider new energy resources. The important aspect is oil, coal transformation to others sources which could be changed by growth of natural gas market. Natural gas is getting as an alternative for changing oil and coal. Looking for better economical perspective, natural gas could be used as LNG, which contains a huge amount of cold energy. Conventionally, this energy is generated during regasification process of LNG and released as wasted energy into atmosphere or dump into seawater. In this paper, reviewed the comparative analysis of LNG regasification technology indicated, that possibility to recover LNG cold energy is the better solution with intermediate fluid vaporizer, which uses special cryogenic fluid. The second part of this paper, the 3-D numerical model for counter-flow of a single channel in the intermediate fluid vaporizer (Printed Circuit Heat Exchanger) is introduced with governing equations for mass, momentum and energy conservation.

Keywords: Liquefied natural gas vaporizers · Liquefied natural gas cold energy · Printed circuit heat exchanger · Governing equations · Fluid flow in tube

1 Introduction

Natural gas is getting as an alternative for changing oil and coal market until 2040 year. The growth of natural gas market influences extraction of natural gas from conventional and from unconventional supplies which accounts approximately (NG) supply form conventional reserves about 185.7 trillion m^3 and NG supply from non-conventional source (shale gas, tight gas, coal bed methane, aquifer gas, methane hydrates) about 845 trillion m^3 [1]. Commonly, natural gas is transferring by pipelines systems, which causes many restrictions especially in those places where are not convenient for construction. As better solution realization of NG is liquefied natural gas, which is 600-times less volume comparing with natural gas and could be stored underground, on the ground or by storage facility in the sea or open area [2]. Furthermore, LNG is cleaner product than natural gas doe to its extra purification that is performed during liquefaction. The liquefaction is thermodynamic process when natural gas is liquefied under boiling point (−162 °C) at atmospheric pressure. During this process, the energy is used to liquefy the natural gas and it is estimated more than

O. Prentkovskis et al. (Eds.): TRANSBALTICA 2021, LNITI, pp. 270–280, 2022.
https://doi.org/10.1007/978-3-030-94774-3_27

830 kJ/kg [3]. Looking for better economical perspective, this energy could be reused during regasification process and reduce the production of LNG. Conventional LNG regasification systems release wasted energy into atmosphere or dump into seawater, which not only consumes power to drive pumps or blowers but also exerts influence on the environment and ecosystem nearby the LNG receiving terminal [4].

In this research work analyses LNG regasification technologies, which have main impact on LNG cold energy generation process. The object was selected LNG Floating Storage Regasification Unit (FSRU). The comparative analysis of LNG vaporizers was done to compare possibility to change regasification technology in LNG FSRU. The second part of research concerns on LNG heat exchanger analysis.

2 Review of LNG Regasification Technologies

2.1 Literature Review

The regasification technology consists of special heat exchangers, that the main work is circulation of special fluid, water or air in tubes, pipes, etc. and heating of LNG under gaseous state. The typical heat exchanger is characterized as vaporizer in LNG regasification process. Construction of vaporizer could be several types: ORV (Open Rack vaporizer), AAV (Ambient Air vaporizer), IFV (Intermediate Fluid vaporizer), STV (Shell Tube vaporizer), SCV (Submerged Combustion vaporizer) [5–7] (see Fig. 1). According to statistics of LNG receiving terminals, 70% use ORV&Super-ORV, 2% use SCV, 5% use IFV [8].

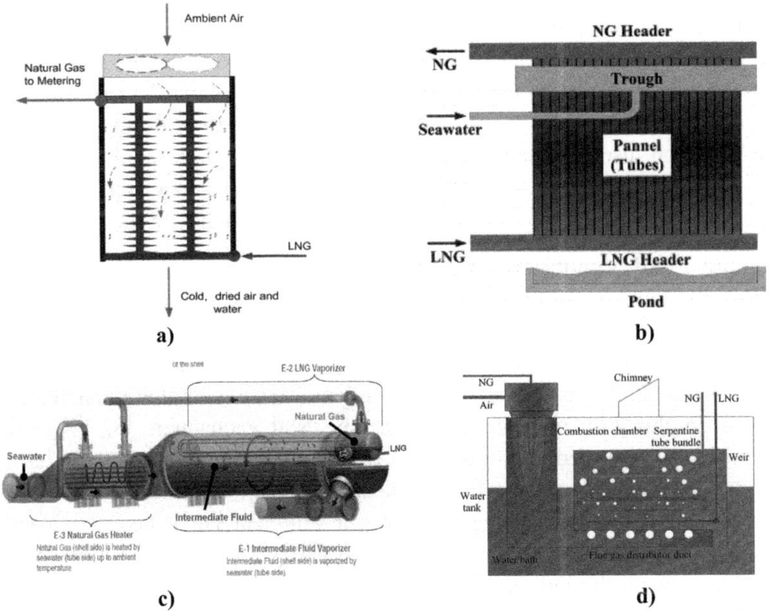

Fig. 1. LNG vaporizers: a – Ambient Air Vaporizer [6]; b – Open Rack Vaporizer [15]; c – Intermediate Fluid Vaporizer [17]; d – Submerged Gas Vaporizer [18].

Ambient Air Vaporizer. (AAV) is suitable in those places with warmer ambient temperatures and it is mainly used in peak shaving plants [8, 9]. The main heat source of AAV is the energy, which is extracted from the ambient air. The heat is absorbed directly from the surrounding air to heat the LNG by natural convection [9]. A typical AAV design configuration consists of long parallel or in serial fin tubes, that could allow air to exchange heat over a large area [6, 10]. Principle of vaporization process starts when LNG is vaporized directly with air passing through a number of interconnected tubes and then the air condenses and freezes forming frost. Looking for technical aspect, the frost is poor conductor and its generation reduces heat transfer coefficient which indicates effectiveness of vaporizer. Also, the AVV is commonly installed sections of vaporizers to prevent ambient air recirculation and to increase vaporization capacity [6, 9]. The vaporization efficiency of AAV depends on the heat transfer performance of AAV finned tube bundle, which greatly affects the stable and effective operation of LNG gas station [11]. The vaporization of AVV depends on frost growth and the deposition on the vaporizer wall because it causes limitation of working conditions [12]. So, some researchers are intensively working on frost formation solving this technical problem.

Open Rack Vaporizer. (ORV) is a type of commercial heat exchanger widely used in large regasification plants for base load Liquefied Natural Gas (LNG) receiving terminals. The mechanism of vaporizer is heat transfer tubes in which LNG flows from the bottom to the top inside the tube and other direction is seawater flow, which starts from the top to the bottom outside the tube [13]. The water spray equipment is installed on the top of vaporizer, which facilitates forming a uniform liquid falling film along the tube outside the heat transfer tubes [14]. The LNG is circulated in tubes and extracted heat transferred from seawater. The type of tubes is selected in this vaporizer with ribs, which is important for heat transfer area. The main challenge of this vaporizer is the heat transfer characteristics of the supercritical fluid flow inside the ribs tubes, which are different from that of a smooth tube [13]. The improved type of vaporizer is SuperORV, which also uses the sensible heat of seawater, but has different configuration of the heat transfer tube comparing with ORV. This new type SuperORV is characterized by its heat transfer tubes with double tube structure at the lower part [15].

Intermediate Fluid Vaporizer. (IFV) is a shell-and-tube vaporizer, which uses an intermediate fluid and it circulates by the gravitational force in the system. Before start work, intermediate fluid is evaporated by a heating source in the evaporator, then it is sent to transfer heat to LNG. After LNG heating, the intermediate fluid is cooled and condensed. A typical IFV is composed of a condenser, an evaporator and a thermolator, similar to a combination of three shell-and-tube heat exchangers [8, 16]. The heat transfer process of intermediate fluid occurs on the shell-side in which transfer heat to LNG inside the tube [8]. Looking for technical aspect, the intermediate fluid selection has to be considered according these criteria: a sufficient latent heat, environmental regulations such as ozone depletion potential (ODP) and global warming potentials (GWP) [17]. As well, the indirect heat transfers between the seawater and LNG allow to avoid the seawater freezing and improves heat transfer coefficient and operation reliability.

Submerged Combustion Vaporizer. (SCV) This type of vaporizer is used in peak shave terminals. The SCV structure is a water tank, a submerged burner, a flue gas distributor, a chimney, a serpentine tube bundle, and a cofferdam. The mechanism of this vaporizer mixture of natural gas and air is burned by the gas burner and after combustion through the sparge distributor at the bottom to generate an immersed jet to heat the water bath directly [18]. Then, LNG regasification process starts in the serpentine tube bundle, which function is to absorb heat from the turbulent water bath After LNG vaporization, steam from the combustion products condensates into liquid water and releases a lot of latent heat [8]. Even the sensible and the latent heat contents of the flue gas are recovered in the water, the overall thermal efficiency is enough high [19]. Based on many studies, the main challenge of this vaporizer remains huge thermal resistance of the ice layer, which causes the thermal behavior and has effect of regasification efficiency [18].

After LNG vaporization, steam from the combustion products condensates into liquid water and releases a lot of latent heat [8]. Even the sensible and the latent heat contents of the flue gas are recovered in the water, the overall thermal efficiency is enough high [19]. Based on many studies, the main challenge of this vaporizer remains huge thermal resistance of the ice layer, which causes the thermal behavior and has effect of regasification efficiency [18].

2.2 Comparative Analysis of LNG Vaporizers

The production of natural gas in regasification terminals depends on technology for regasification equipment. The following Table 1 shows the detailed analysis of LNG vaporizers, which was performed to verify 6 different criteria.

Table 1. Comparative analysis of LNG vaporizers.

Type	Environmental friendly technology	Construction cost	Technological simplicity	Maintenance	Operation cost	Capture of LNG cold energy
AVV	Low	Low	Simple operation	Low	Low	No
ORV	Medium	High	Simple operation	High	High	Possible
IFV	Medium	High	Complex technology	High	Low	Yes
SCV	High	Low	Need to observe	High	High	Possible

The Table 1 indicated, that the most suitable for small scale LNG plants is AAV with compact structure and low operating cost. Though, it is limited for large scale LNG terminals doe to the low efficiency. In those places, where are inland areas and no area for seawater or river water enough, the option is SCV vaporizer with large evaporation capacity. Also, the advantage of SCV that, unlike with ORV, AVV, IFV, it has fast start-up with high heat transfer efficiency. One of the key issues for SCV is high operating costs, which leads to consider about selection of ORV and IFV. In addition, ORV is mainly used in LNG terminals doe to low running cost, convenient maintenance and operation. However, the maintenance and operation cost of ORV is high and specific requirements for seawater. The main advantage of IFV, is possibility to utilize LNG cold energy for many cryogenic processes such as power generation, warehouse cooling, air separation, water desalination, carbon dioxide capturing and etc. For this reason, the many investigations were done, to adopt intermediated fluid between LNG and heat source to transfer heat and utilize LNG cold energy as a new product.

After introducing the LNG vaporizers, the theoretical research of LNG regasification process using IFV type is presented in the following sections.

3 Research Object

The main objective of this research work was concerned on Floating Storage and Regasification Unit. At present, FSRU is commercial type facility, which could regasify LNG to NG, storage LNG and reload LNG to another carrier. However, it is complicated system with variety of heat transfer mechanism, and the actual demand of improving efficiency and compactness [20]. The main heat exchange process starts in PCHE (Printed Circuit Heat Exchanger), where LNG extracts heat from transferred propane. Also, PCHE is characterized as intermediate fluid vaporizer type. The greatest advantage of this a new type of micro-channel heat exchanger is the compactness as main cryogenic heat exchanger on LNG carriers and LNG floating terminals for high efficiency [18, 20].

In the PCHE, LNG is heated from −155/−135 °C to −10 °C. Passages are between 1–2 mm. The PCHE is manufactured by SLS technology using SS316 stainless steel and constructed of stack of diffusion-bounded flat plates with flow channels. According to technical sheets, the core dimension of test PCHE is 646 * 552 * 1520 (L * H * W). At the start of a process propane enters as a saturated vapor to PCHE and after regasification it condenses as a liquid −20 °C. Meanwhile, the cross-flow heat exchange with the propane gas makes propane condense on the propane side as LNG is evaporated on the LNG side.

To conclude the previous sections above, the main issue of regasification process is described LNG cold energy waste. As solution could be adopted PCHE technology with LNG cold energy capture integration. To understand better cryogenics heat transfer mechanism and evaporation process in PCHE, the detailed theoretical approach is performed in following section.

4 Theoretical Approach of LNG Printed Circuit Heat Exchanger

Many studies have been done to analyze the evaporation and condensation processes for volatile liquids. Though, there many challenges for LNG regasification process as cryogenic conditions, determination of liquid and gaseous state which need to investigate.

Increasing regasification efficiency, it is important to observe heat transfer process, which depends on many variables such as pressure (p), temperature (T), mass flow (Q), volume fraction changes (α). In addition, the most equations are based on heat and mass transfer process. As solution, in this research, 3-D numerical model for counter-flow of a single channel in the PCHE is introduced with governing equations for mass, momentum and energy conservation for the entire computational domain. The governing equations indicate the alteration of surface area to volume ratio, that significantly impacts the heat transfer characteristics and internal flow dynamics of the system [21].

Momentum Equation. The flow pattern is described by momentum equation, which indicates how different forces influence fluid flow. The momentum equation consists of momentum source term (ρg) presenting gravity and ($F\sigma$) presenting surface tension [18, 22–24].

$$\frac{\partial(\rho\vec{v})}{\partial t}+\nabla(\rho\vec{v}\vec{v})=-\nabla p+\nabla\left[\mu\left(\nabla\vec{v}+\nabla\vec{v}^{T}\right)-\frac{2}{3}\mu\nabla v\right]+\rho\vec{g}+F_{\sigma}; \quad (1)$$

where F_σ – surface tension force ($N \cdot m^{-3}$); $\vec{v}$ – velocity vector shared by two phases ($m \cdot s^{-1}$); ρ – density ($kg \cdot m^{-3}$) (Tube density); μ – dynamic viscosity ($Pa \cdot s$); $\vec{g}$ – gravitational vector ($m \cdot s^{-2}$); p – pressure (Pa); ∇ – del operator.

Calculation of Density and Dynamic Viscosity. Mass and volume-averaged cell properties (ρ, μ) are used in a control volume containing more than one phases, are estimated [21]:

$$\rho=\alpha_q\rho_q+(1-\alpha_q)\rho_p, \quad (2)$$

where ρ – density ($kg \cdot m^{-3}$); α_q – volume fraction of qth phase (–); ρ_q – density of qth phase ($kg \cdot m^{-3}$); ρ_p – density of pth phase ($kg \cdot m^{-3}$).

$$\mu=\alpha_q\mu_q+(1-\alpha_q)\mu_p, \quad (3)$$

where μ – dynamic viscosity ($Pa \cdot s$); α_q – volume fraction of qth phase (–); μ_q – kinematic viscosity of qth phase ($Pa \cdot s$); μ_p – dynamic viscosity of pth phase ($Pa \cdot s$).

According to [18, 24] research, the volume of fluid model is the most suitable model for regasification process. Firstly, this model is used to separate the liquid and gas phases. The development of model was done by creating the propane condensation process in program Fluent [18, 21]. For this aspect, this model could be provided in

PCHE to verify liquid and gas volume at the same time. Using continuity equation, the LNG mass transfer rate (S_m) could be predicted as shown Eq. (4), (5), (6).

Continuity Equation

$$\frac{\partial}{\partial t}(\alpha_v \rho_v) + \nabla(\alpha_v \rho_v \vec{v_v}) = S_m, \tag{4}$$

$$\frac{\partial}{\partial t}(\alpha_l \rho_l) + \nabla(\alpha_l \rho_l \vec{v_l}) = -S_m, \tag{5}$$

$$\alpha_v + \alpha_l = 1, \tag{6}$$

where α – volume fraction (–); ρ – density $(kg \cdot m^{-3})$; t – time (s); S_m – the mass term source $(kg \cdot m^{-3} \cdot s^{-1})$; ∇ – del operator (–).

The mass source term is verified the rate of mass transfer due to evaporation and condensation, as follows [18]:

$$S_m = \dot{m}_{lv} - \dot{m}_{vl}, \tag{7}$$

where $\dot{m}_{vl}$ – mass transfer doe to condensation $(kg \cdot m^{-3} \cdot s^{-1})$; $\dot{m}_{lv}$ – mass transfer doe to vaporization $(kg \cdot m^{-3} \cdot s^{-1})$.

The evaporation (vaporization) starts, when liquid temperature is higher than the saturation temperature $T_l > T_{sat.}$. The rate of evaporation mass transfer is calculated by formula [18, 22]:

$$\dot{m}_{lv} = coeff * \alpha_l \rho_l \frac{(T_l - T_{sat.})}{T_{sat.}}, \tag{8}$$

where $\dot{m}_{lv}$ – mass transfer doe to vaporization $(kg \cdot (m^3 \cdot s)^{-1})$; coeff – configurable mass transfer parameter (s^{-1}); α_l – liquid volume fraction (–); T_l – liquid temperature (K); $T_{sat\cdot}$ – saturation temperature (K); ρ_l – liquid density $(kg \cdot m^{-3})$.

Condensation process starts, when vapor temperature is lower than saturation temperature $T_v < T_{sat.}$. The rate of condensation mass transfer is calculated by:

$$\dot{m}_{vl} = coeff * \alpha_v \rho_v \frac{(T_{sat.} - T_v)}{T_{sat.}}, \tag{9}$$

where $\dot{m}_{vl}$ – mass transfer doe to condensation $(kg/(m^3/s)^{-1})$; coeff – configurable mass transfer parameter (s^{-1}); α_v – vapor volume fraction (–); T_v – vapor temperature (K); $T_{sat\cdot}$ – saturation temperature (K); ρ_v – vapor density $(kg \cdot m^{-3})$.

Energy Equation. The energy source term (Q) presenting the latent heat transfer is added to reflect the heat transfer of phase transition) [22–24]:

$$\frac{\partial(\rho E)}{\partial t} + \nabla[\vec{v}(\rho E + p)] = \nabla(k\nabla \mathrm{T}) + \mathrm{Q}, \tag{10}$$

where Q – latent heat source term; k – heat conductivity coefficient (W/(m · K)); T – temperature (K); ρ – density ($\mathrm{kg \cdot m^{-3}}$); $\vec{v}$ – velocity vector shared by two phases ($\mathrm{m \cdot s^{-1}}$); E – Energy (J).

Above introduced Eqs. (1), (4–5), (10) present the Navier–Stokes equations, which are a set of partial differential equations and describe the motion of viscous fluid substances. The disadvantage of this method is the large number of equations, long solution time. Meanwhile, the equations could be solved using the method of the concentrated parameters to obtain a system of simple nonlinear differential equations. Medium (LNG flow in channel) is divided into the separated control volumes and the variable such as temperature could be determined in every moment in different control volume using Runge-Kutt method.

Heat Balance of Control Volume for Two Fluids and One Solid Plate. The main temperatures changes are described in below Eqs. (11–13), which indicate LNG flow during regasification process in channel, when the propane transfers the heat on PCHE plate.

For the hot-side fluid (propane):

$$\mathrm{m_h C_{p,h}} \frac{\mathrm{dT_h}}{\mathrm{dt}} + \dot{\mathrm{m}}_\mathrm{h} \mathrm{C_{p,h}} \left(\mathrm{T_h^{out}} - \mathrm{T_h^{in}}\right) = (\mathrm{hA})_\mathrm{h}(\mathrm{T_P} - \mathrm{T_h}), \tag{11}$$

where m_h – mass of propane (kg); $C_{p,h}$ – specific heat of propane ($\mathrm{J \cdot (kg \cdot K)^{-1}}$); $\mathrm{T_h}$ – temperature of propane (K); $\dot{m}_h$ – mass flowrate of propane kg $\cdot (\mathrm{s})^{-1}$; T_h^{out} – temperature of propane outlet (K); T_h^{in} – temperature of propane inlet (K); h_h – heat transfer coefficient of propane ($\mathrm{W \cdot (m^2 \cdot K)^{-1}}$); $\mathrm{T_P}$ – temperature of plate (K); A_h – heat transfer area ($\mathrm{m^2}$).

For the cold-side fluid (LNG):

$$\mathrm{m_c C_{p,c}} \frac{\mathrm{dT_c}}{\mathrm{dt}} + \dot{\mathrm{m}}_\mathrm{c} \mathrm{C_{p,c}} \left(\mathrm{T_c^{out}} - \mathrm{T_c^{in}}\right) = (\mathrm{hA})_\mathrm{c}(\mathrm{T_P} - \mathrm{T_c}), \tag{12}$$

where $\mathrm{m_c}$ – mass of LNG (kg); $\mathrm{C_{p,c}}$ – specific heat of LNG ($\mathrm{J \cdot (kg \cdot K)^{-1}}$); $\mathrm{T_c}$ – temperature of LNG (K); $\dot{\mathrm{m}}_\mathrm{c}$ – mass flow rate of LNG kg $\cdot (\mathrm{s})^{-1}$; $\mathrm{T_c^{out}}$ – temperature of LNG outlet (K); $\mathrm{T_c^{in}}$ – temperature of LNG inlet (K); $\mathrm{h_c}$ – heat transfer coefficient of LNG ($\mathrm{W \cdot (m^2 \cdot K)^{-1}}$); $\mathrm{T_P}$ – temperature of plate (K); $\mathrm{A_c}$ – heat transfer area ($\mathrm{m^2}$).

For the solid plate:

$$\mathrm{m_P C_{p,P}} \frac{\mathrm{dT_P}}{\mathrm{dt}} = (\mathrm{hA})_\mathrm{h}(\mathrm{T_h} - \mathrm{T_P}) - (\mathrm{hA})_\mathrm{c}(\mathrm{T_P} - \mathrm{T_c}), \tag{13}$$

where m_p – mass of plate (kg); $\mathrm{c_{p,P}}$ – specific heat of plate ($\mathrm{J \cdot (kg \cdot K)^{-1}}$); $\mathrm{T_P}$ – temperature of plate (K); T_P – temperature of plate (K); $\mathrm{T_h}$ – temperature of propane (K); T_c – temperature of LNG (K); A_c – heat transfer area of LNG ($\mathrm{m^2}$); $\mathrm{A_h}$ – heat transfer area of propane ($\mathrm{m^2}$).

The most difficult part, is to determine LNG temperature, density, pressure in different time during regasification process doe to the cryogenic conditions. From the literature review, pressure, density and temperature for vapor phase could be determined according Peng-Robinson equation state [2, 25]:

$$p = \frac{RT}{(v - b)} - \frac{a \cdot a_{ci}}{v(v + b) + b(v - b)}, \tag{14}$$

where p – pressure (Pa); v – molar volume ($m^3 \cdot mol^{-1}$); R – gas constant (8,314 J • (mol • K)$^{-1}$); T – absolute temperature (K); a, b, a_{ci} – effective parameters (–).

For cryogenic liquid phase, the specific molar calculation is made according to standard ISO 6976:1995 and Klosek-McKinley method [25].

LNG Density calculation is expressed by following equitation:

$$\rho_{LNG} = \frac{M_{mix}}{V_{mix}}, \tag{15}$$

where ρ_{LNG} – density of LNG ($kg \cdot m^{-3}$) by reference temperature, M_{mix} – molecular mass of mixture ($kg \cdot kmol^{-1}$), V_{mix} – molar volume of mixture ($m^3 \cdot kmol^{-1}$).

Molecular mass of mixture is expressed by following equitation:

$$M_{mix} = \sum M_i \cdot x_i, \tag{16}$$

where M_{mix} – molecular mass of mixture ($kg \cdot kmol^{-1}$), x_i – molar fraction of constituent i.

$$V_{mix} = \sum M_i \cdot V_i - \left[K_1 + (K_2 - K_1) \cdot \left(\frac{x_{N_2}}{0,0425}\right)\right] \cdot x_{CH_4}, \tag{17}$$

where V_{mix} – molar volume of constituent i at the temperature of LNG (m^3), V_i – molar volume of constituent i at the temperature of LNG (m^3), x_i – molar fraction of constituent i (mol), K_1, K_2 – volume correction factors (–); 0.0425 – this number is accepted for industry, that the nitrogen or butane content may not exceed 4% (–).

5 Conclusions

The comparative analysis was performed to verify possibility to change regasification technology in LNG FSRU. The results show, that the intermediate fluid vaporizer is the better option comparing with other types vaporizers. The main advantage of IFV, is possibility to utilize LNG cold energy for many cryogenic processes such as power generation, warehouse cooling, air separation, water desalination, carbon dioxide capturing and etc.

The main issue of regasification process is described LNG cold energy waste. As solution is investigated an adopted PCHE technology with LNG cold energy capture integration.

The evaporation and condensation process is analyzed in PCHE using the volume of fluid model. The model is focused on governing equations for mass, momentum and energy conservation. These equations are differential equation with partial derivatives. Conventionally, the equations could be solved using the method of the concentrated parameters to obtain a system of simple nonlinear differential equations (Runge-Kutta method).

References

1. Pfoser, S., Aschauer, G., Simmer, L., Schauer, O.: Facilitating the implementation of LNG as an alternative fuel technology in landlocked Europe: a study from Austria. Res. Transp. Bus. Manag. **18**, 77–84 (2016)
2. Ferrin, J.L., Perez, L.J.: Numerical simulation of natural convection and boil-off in a small size pressurized LNG storage tank. Comput. Chem. Eng. **138**, 106840 (2020)
3. Pospíšil, J., Charvat, P., Arsenyeva, O., Klimes, L., Spilacek, M., Klemes, J.J.: Energy demand of liquefaction and regasification of natural gas and the potential of LNG for operative thermal energy storage. Renew. Sustain. Energy Rev. **99**, 1–15 (2019)
4. Noh, Y., Kim, J., Kim, J., Chang, D.: Economic evaluation of BOG management systems with LNG cold energy recovery in LNG import terminals considering quantitative assessment of equipment failures. Appl. Therm. Eng. **143**, 1034–1045 (2018)
5. Egashira, S.: LNG vaporizer for LNG re-gasification terminal. KOBELCO Technol. Rev. **32**, 64–69 (2013)
6. Mokhatab, S., Poe, W., Mak, J.: Handbook of Natural Gas Transmission and Processing, 3rd edn., pp. 10–70. Elsevier, Walthan (2015)
7. Kanbur, B.B., Xiang, L., Dubey, S., Choo, F.H., Duan, F.: Cold utilization systems of LNG: a review. Renew. Sustain. Energy Rev. **79**, 1171–1188 (2017)
8. Han, C.L., Sun, H.Q., Li, Z.Y.: Thermal performance analysis and optimization design for LNG submerged combustion vaporizer. Cryogenics **95**, 47–56 (2018)
9. Agarwal, R., Thomas Rainey, S., Rahman, T.S., Perrons, R., Brown, R.: LNG regasification terminals: the role of geography and meteorology on technology choices. Energies **10**(12), 2152 (2017)
10. Lee, Y., Na, J., Lee, W.B.: Robust design of ambient-air vaporizer based on time-series clustering. Comput. Chem. Eng. **1**, 1–12 (2018)
11. Liu, S., Jiao, W., Zhang, P.: Conjugate numerical analysis on buddle effects for integrated heat transfer of LNG ambient air vaporizer. Appl. Therm. Eng. **180**, 115735 (2020)
12. Kuang, Y.W., Yi, C.C., Wang, W.: Numerical simulation of frosting behavior and its effect on a direct contact ambient air vaporizer. J. Nat. Gas Sci. Eng. **27**(1), 55–63 (2015)
13. Cheng, H., Ju, Y., Fu, Y.: Thermal performance calculation with heat transfer correlations and numerical simulation analysis for typical LNG open rack vaporizer. Appl. Therm. Eng. **149**, 1069–1079 (2019)
14. Cheng, H., Ju, Y., Fu, Y.: Experimental study on heat transfer characteristics of cooling falling film outside a vertical tube in open rack vaporizer. Appl. Therm. Eng. **172**, 115187 (2020)

15. Qi, C., Yi, C., Wang, B., Wang, W., Xu, J.: Thermal performance analysis and operation method with low temperature seawater of super open Rack vaporizer for liquefied natural gas. Appl. Therm. Eng. **15**, 61–69 (2019)
16. Wang, Z., Cai, W., Hong, W., Shen, S., Yang, H., Han, F.: Multi-objective optimization design and performance evaluation of a novel multi-stream intermediate fluid vaporizer with cold energy recovery. Energy Convers. Manage. **195**, 32–42 (2019)
17. Higashi, K., Kondou, C., Koyama, S.: Feasibility analysis for intermediated fluid type LNG vaporizers using R32 and R410A considering fluid properties. Int. J. Refrig. **118**, 325–335 (2020)
18. Pan, J., et al.: Numerical investigation of thermal-hydraulic performance of a printed circuit LNG vaporizer. Appl. Therm. Eng. **165**, 114447 (2020)
19. Honzowa, T., et al.: Numerical and experimental investigations on turbulent combustion fields generated by large scale submerged combustion vaporizer burners with water spray equipment. J. Nat. Gas Sci. Eng. **76**, 103158 (2020)
20. Bai, J., Pan, J., He, X., Wang, K., Tang, L., Yang, R.: Numerical investigation on thermal hydraulic performance of supercritical LNG in sinusoidal wavy channel based printed circuit vaporizer. Appl. Therm. Eng. **175**, 115379 (2020)
21. Saleem, A., Farooq, S., Karimi, I.A., Banerjee, R.: A CFD simulation study of boiling mechanism and BOG generation in a full-scale LNG storage tank. Comput. Chem. Eng. **115**, 112–120 (2018)
22. Ren, J.J., Shi, J.Y., Liu, P., Bi, M.S., Kai, J.: Simulation of thermal stratification and destratification in liquefied gas tanks. Int. J. Hydrogen Energy **38**, 4017–4023 (2013)
23. Hubert, A., Dembele, S., Denissenko, P., Wen, J.: Prediction liquefied natural gas rollovers using computational fluid dynamics. J. Loss Prev. Ind. **62**, 103922 (2019)
24. Migliore, C., Tuvilleja, C., Vesovic, V.: Weathering prediction model for stored liquefied natural gas (LNG). J. Nat. Gas Sci. Eng. **26**, 570–580 (2015)

Investigation of Gas Exchange in an Engine by Modeling

Ramis Zaripov[1], Nurbolat Sembaev[1], and Pavels Gavrilovs[2(✉)]

[1] Faculty of Engineering, Torighyrov University, 140000 Pavlodar, Kazakhstan
[2] Institute of Transport, Riga Technical University, Paula Valdena street 1, Riga 1048, Latvia
pavels.gavrilovs@rtu.lv

Abstract. The article reflects the issue of mathematical modeling of gas exchange processes. The features of modeling the processes of gas exchange based on the equations of gas dynamics are shown. The theoretical significance of the work lies in the methodology for constructing a computational model and individual results on modeling the process of gas exchange of an internal combustion engine. The mathematical model of the working process in the engine is created by the necessary conditions for calculating the entire gas exchange. The use of a computer allows a more accurate and complete description of all the main features. The filling of the dynamic phenomena in the intake system has the main purpose of optimizing the parameters of the engine design. The efficiency of work in the engine is determined by the perfection of the processes occurring during intake, mixture formation and combustion of a fresh charge, as well as by the perfection of the organization of exhaust gases. A sufficiently accurate assessment of the qualitative and quantitative relationships of design parameters with thermogasdynamic phenomena in the gas-air duct and gas exchange rates make it possible to fundamentally solve the problem of calculating the design parameters close to the optimal design parameters of the intake system and gas distribution of automotive engines, engines with injection and, to a certain extent, carburetor engines.

Keywords: Gas exchange · Law of conservation of mass · Momentum and energy · Mathematical model · Ideal gas equation of state · Energy equation

1 Analysis of Actual Problems of Engine Building and the Software Necessary for Their Solution

At the moment, any study of the theory of engine workflows during their design and modernization is not carried out without modeling programs. Creating models allows you to fully study the working processes, life support systems of engines, environmental safety of fuels, increasing the efficiency of power plants. The purpose of the work is to study the scientific foundations of the calculation of piston engines, universal mathematical models and applied programs for the thermodynamic calculation of two-stroke and four-stroke internal combustion engines with a refined consideration of the processes of mixing, combustion and gas exchange.

O. Prentkovskis et al. (Eds.): TRANSBALTICA 2021, LNITI, pp. 281–291, 2022.
https://doi.org/10.1007/978-3-030-94774-3_28

The more fully the mathematical model covers the considered physical processes and describes them more correctly, the more accurate and reliable the result can be obtained. Another very important property should have a software-the ability to solve complex optimization problems. It is not enough to get a good match with the experiment in the calculation; to solve practical problems, it is necessary to find effective ways to improve the design, to find the optimal values of many design factors that affect the workflow in different ways, and, sometimes, leading to ready for conflict situations.

For example, it is necessary to find a combination of the shape of the combustion chamber, the design of the fuel equipment and the injection characteristics that will simultaneously reduce both fuel consumption and NOx emissions. Optimization of one or even two parameters by iteration often does not allow you to get a good result for a problem with a large number of influencing factors. To increase the efficiency of solving optimization problems, it is rational to use formal search procedures of non-linear programming, which allow conducting an optimal search in automatic mode. To implement such a possibility, the program core, which implements a mathematical model of the process under study, must have, in addition to the necessary accuracy, also high speed, since when searching for the optimum, it is necessary to carry out calculations of many hundreds of design options.

Mathematical modeling of internal combustion engine workflows can be applied to the following list of tasks:

1. Forecasting and improving engine performance: high-speed, load, screw, diesel, high-altitude, characteristics with a change in the depth of immersion, etc.
2. Improvement of the turbocharging system, including:
   a) Selection of compressors and turbines for the piston part of the internal combustion engine.
   b) Optimization of the parameters of the compressor drive and power take-off of power turbines, the parameters of the additional combustion chamber in front of the power turbine.
   c) Optimization of the distribution of compression work between stages with multi-stage supercharging.
   d) Optimization of algorithms for controlling bypass devices and devices of the electric drive of the turbine.
3. Optimization of mixture formation and combustion to ensure the specified toxicity standards and minimum fuel consumption: optimization of the compression ratio, optimization of the injection advance, optimization of the injection characteristics (including the strategy of reusable injection), optimization of the shape of the combustion chamber and the parameters of the fuel equipment.
4. Improvement of the gas exchange system. Optimization of gas distribution phases, optimal design of inlet and outlet channels and windows, optimal design of pipelines (collectors).
5. Improving the operation of the internal combustion engine in unsteady modes and optimizing the control system algorithms.

To solve such a wide range of problems using mathematical modeling methods, software is currently used that implements mathematical models of two main classes:

1. Thermodynamic models of internal combustion engines, considering the engine as a combination of thermodynamic systems. (These programs are the most widespread).
2. Models based on solving problems of spatial hydrodynamics (in the English-speaking environment, called Computational Fluid Dynamic or CFD). The engine elements are divided into a large number (hundreds of thousands) of three-dimensional cells, for each of which a system of equations of conservation of energy, mass, momentum and state is solved in a three-dimensional formulation, (an actively developing direction.)

Models and programs of each class are designed to solve their specific tasks.

2 Thermodynamic Models of Internal Combustion Engines

The calculation schemes of typical thermodynamic models of internal combustion engines are presented in Fig. 1 and Fig. 2.

In the cylinder, due to its compactness, the pressure difference in volume is neglected, i.e., a 0-dimensional formulation of the problem is used. This assumption significantly simplifies the calculation and does not introduce noticeable errors in the results. In the thermodynamic approach, the cylinder is considered as an open thermodynamic system, or as a combination of several thermodynamic systems. The velocity field is not considered in it, the pressure and temperature inside each thermodynamic system are considered independent of coordinates, but dependent only on time (the angle of rotation of the crankshaft). In terms of numerical methods of fluid mechanics: the entire cylinder is a single cell for which a system of mass and energy balance equations is solved together with the equation of state:

$$\left\{\begin{array}{l} dU = -PdV + dQ_x - dQw + \sum\limits_{j=1}^{n} dI_j^* \\ dG = \sum\limits_{j=1}^{n} G_j \\ pV = GRT \end{array}\right\}. \tag{1}$$

where U – the internal energy, p – is the pressure; T – temperature; dQ_w – heat transferred to the walls; dQx – the heat from combustion; I_j^* – full enthalpy supplied to the working fluid with a mass of dGj, j-the source mass (enthalpy calculated based on parameters of braking); G – the mass of the working fluid, R is the gas constant.

The calculation of heat exchange with walls, as a rule, does not require large computing power. To calculate the amount of heat transferred to the walls, the Newton-Richman equation is used:

$$dQ_w = a_w F_w (T - T_w)\Delta\tau, \tag{2}$$

where $aw = f(\varphi)$ – the coefficient of heat transfer from the gas to the walls (there are a large number of formulas for calculating aw depending on the pressure and temperature in the cylinder, as well as on the cylinder size and operating parameters, the most common formula is Voshni [1]); $F_w = f(\varphi)$ – the current heat exchange area of the gas with the walls (in many programs, heat exchange is calculated separately for different surfaces: for the piston, for the cylinder cover, for the cylinder sleeve and even for the valve plates); T_w – the temperature of the heat-receiving surface; $\Delta\tau = \Delta\varphi/(6n)$ – the calculation step in time; φ φ – the calculation step by the angle of rotation of the crankshaft (usually no more than 1 degree by the crank gate); n – the speed of rotation of the crankshaft [15–17].

The calculation of combustion is currently the most complex and at the same time, the most relevant fragment of the mathematical model of the internal combustion engine, especially in the light of solving environmental problems. Its purpose is to determine the amount of heat released at each calculation step:

$$dQ_x = g_c H_u \frac{dx}{d\varphi} \Delta\varphi, \tag{3}$$

where g_c is the cyclic fuel supply, H_u is the lowest heat of combustion, $dx/d\varphi = f(\varphi)$ the heat release rate. The calculation of the heat release rate is the main difficulty due to the need to take into account the entire variety of influencing factors, such as the shape of the combustion chamber, fuel supply parameters, etc. The relevance of the correct determination of the heat release rate also increases due to the need to calculate the emission of harmful substances with exhaust gases of the internal combustion engine, which strongly depends on the characteristics of the combustion process.

Perfect mathematical models for the correct calculation of combustion simply did not exist until recently. Simple combustion models developed by V. I. Grinevetsky [2], I. I. Vibe [3], Watson [7], and also other authors [8–13] do not require any tangible computing resources, but they do not allow taking into account the design features of the combustion chamber of the internal combustion engine and its fuel equipment, as well as the influence of operating parameters. But the study of the influence of these aspects of the organization of the working process of the internal combustion engine is becoming particularly relevant in the current. The current time in the light of solving environmental problems. To take these factors into account, since the late 70s, methods for calculating combustion in a diesel engine have been developed, based on a multi-zone jet model. The most famous of them, which has received the most widespread in engineering practice all over the world, was the phenomenological model of Hiroyasu [4]. In the Russion, the Kavtaradze R.Z. model was developed and introduced into engineering practice [5]. Later, this model was developed, aimed at a more detailed account of the interaction of the fuel jet with the walls of the combustion chamber [6].

Modern phenomenological models of diesel combustion are logically complex, but they do not require large computing resources. The calculation time of combustion in a diesel engine with a single injection using the Hiroyasu model is about 1 min, the Razleitsev model requires 1–2 s, provided that the computer code is successfully implemented.

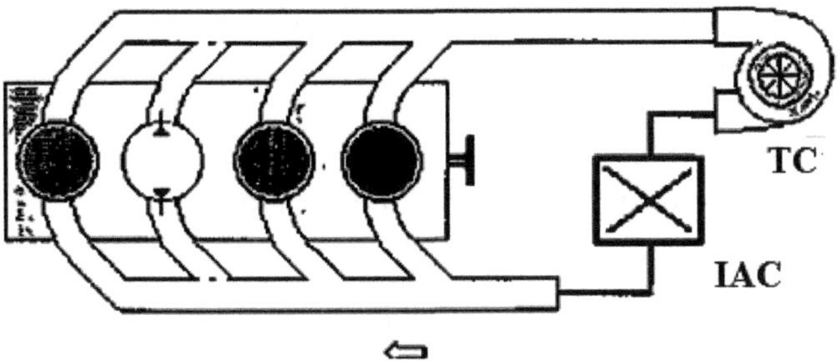

Fig. 1. Design diagram of an internal combustion engine with a 0-dimensional representation of the intake and exhaust pipelines.

3 Calculation of Gas Exchange

Calculation of gas exchange. Historically, the calculation of the gas exchange process caused the greatest difficulties in the past (now the calculation of the combustion process causes the greatest difficulties). The mathematical model of gas exchange has high requirements for accuracy and speed, since in thermodynamic models of the internal combustion engine, it accounts for most of the computer counting time. The calculation of the gas exchange process is extremely important for the correct calculation of the intra-cylinder parameters: it is necessary to correctly determine the mass of the working fluid, its pressure, composition and temperature at the time of closing the intake organs for the subsequent calculation of compression, combustion and expansion, as well as to evaluate the operation of the pump passages, the quality of cleaning and filling of the cylinder. The simplest method of calculating gas exchange is quasi-static [7–10], with the consideration of the inlet and outlet pipelines as open thermodynamic systems. The calculation scheme of this approach is shown in Fig. 1.

In this case, the same assumptions apply for both manifolds as for the cylinder:

- on the instantaneous propagation of disturbances;
- about instant mixing;
- about the absence of dependence of the gas parameters on the coordinates.

Collectors are considered as single volumes, the velocity field is not considered in them, pressure and temperature are considered independent of coordinates, but dependent only on time (the angle of rotation of the crankshaft). The gas parameters in the reservoirs are determined from a system of equations containing the equations of conservation of mass and energy, as well as the equation of state (1). The assumption that the gas parameters in the collector are independent of the coordinates makes it possible to calculate the working process in only one cylinder, assuming that the other cylinders work identically. The flows of mass dG_j and enthalpy dI^*_j entering the collectors from neighboring cylinders are taken along the averaged cylinder with a corresponding time shift. The selection of initial conditions is carried out by the establishment method: the calculation continues until the pressure and temperature in the collectors become periodic functions. In the simplest case, the mass dGj and the enthalpy dI^{*}_{j} of the gas flowing from the cylinder to the collector are determined by the equation of the stationary flow:

$$\begin{aligned} dJ_i &= W_0 \rho \mu f \Delta\tau; \\ dI_j^* &= dG_j C_p T_c - a_{wcr} F_{wcr} (T - T_{wcr}) \Delta\tau; \\ W_0 &= \left\{ \frac{\sqrt{\frac{2k}{k+1} RT_c} if \frac{p_c}{p_r} \geq \left(\frac{k+1}{2}\right)^{\frac{k}{k-1}}}{\sqrt{\frac{2k}{k-1} RT_c \left[1 - \left(\frac{p_c}{p_r}\right)^{\frac{k-1}{k}}\right]} if \frac{p_c}{p_r} \leq \left(\frac{k+1}{2}\right)^{\frac{k}{k-1}}} \right\} \end{aligned} \tag{4}$$

where p_c, Tc – gas pressure and temperature in the cylinder, p_r – gas pressure in the exhaust manifold, ρ – gas density in the cylinder, μf – area of the effective flow section at this calculation step; a_{wcr} – heat transfer coefficient from the gas to the channel walls, F_{wcr} – heat exchange area in the channel, T_{wcr} – the temperature of the channel wall, W_O is the velocity of the stationary flow.

However, the use of stationary dependencies (2) for calculating the flow velocity at each calculation step introduces significant errors in the results of calculating the free release period and causes instability of the calculation during purging, when the cylinder volume is small [11–14]. To increase the stability and accuracy of the calculation, the calculated step is reduced to 0.1–0.2 degrees of rotation of the engine crankshaft. However, this leads to a multiple increase in the counting time and only partially solves the accuracy problem. It is much more effective to use the dependencies obtained by A. S. Orlin for a non-stationary flow for calculating the flow rate [2]:

$$\begin{aligned} dC_{j,} &= 0,5(W_L + W_i)\rho\mu f \Delta\tau; \\ W_L &= \frac{(w_1 + w_0)\exp\left\{\frac{w_0 \Delta\tau}{l_{cr}}\right\} + (w_1 - w_0)}{(w_1 + w_0)\exp\left\{\frac{w_0 \Delta\tau}{l_{cr}}\right\} - (w_1 - w_0)}, \end{aligned} \tag{5}$$

where W_L is the speed at the end of the channel at the end of the calculated time interval, W_l is the speed at the end of the channel at the beginning of the calculated time interval, l_{cr} is the channel length. This approach practically does not affect the counting speed, but it makes the calculation stable and much more accurate, and also significantly expands the scope of the gas exchange model based on the 0-dimensional representation of collectors. The experience of calculating different engines using the described approach shows good accuracy in modeling gas exchange, both in high-speed engines and in two-stroke internal combustion engines [15–18].

However, there are problems that cannot be solved in principle within the framework of 0-dimensional representations. These are tasks where the assumption of the independence of the gas parameters from the coordinates is unacceptable:

- the study of the uneven filling along the cylinders,
- the design of pipelines for dynamic supercharging.

The elements of the gas-air path in the general case can have a large the length along the length, the sources and drains of the mass operating in the pulsating mode can be separated by a considerable distance, pipelines can have a complex configuration. All this leads to the fact that when modeling processes in pipelines, sometimes it is impossible to use the assumption that the parameters of the working fluid in them are independent of the coordinates. In this case, the collectors must be considered in a non-stationary, 1-dimensional representation, i.e. the equation of conservation of mass and energy is added to the equation due to the length of the pipelines, this is no longer a thermodynamic, but a gas-dynamic system of equations has to be solved in each cell, into which the collectors are divided (Fig. 2).

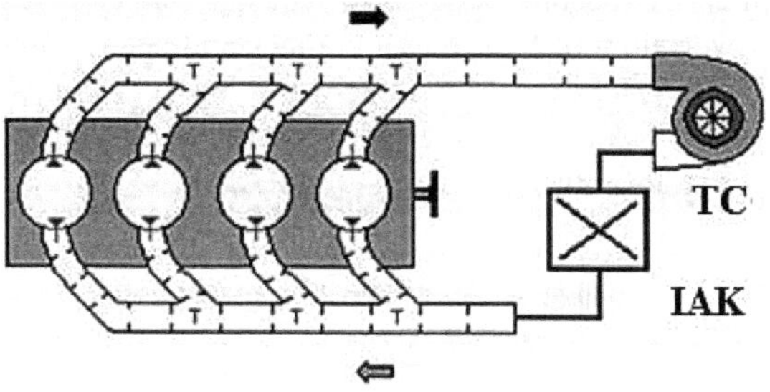

Fig. 2. The design scheme of the internal combustion engine with a 1-dimensional representation of the inlet and outlet pipelines.

Within the framework of this model, the gas parameters in the reservoirs are determined from a system of differential equations, including the equations of conservation of energy and momentum, as well as the equation of continuity and state, written for each of the cells into which the collector is conditionally divided:

$$\left\{\begin{array}{l} W = \dfrac{\partial W}{\partial x} + \dfrac{\partial W}{\partial \tau} = -\dfrac{1}{\rho}\dfrac{\partial p}{\partial x} \\ \dfrac{\partial \rho}{\partial \tau} = \rho\dfrac{\partial W}{\partial x} - W\dfrac{\partial \rho}{\partial x} \\ p = C\rho^{n} \end{array}\right\} \tag{6}$$

Special boundary conditions are recorded for the tees, the connection to the turbine and the inlet pipe from the inflatable air cooler (IAC). Most often, the system of equations is solved by:

- the method of characteristics (the most accurate method);
- by the Godunov method (the method of decay of an arbitrary gap);
- the Davydov-Belotserkovsky method of large particles (the Harlow method of particles in cells).

This approach is characterized by high accuracy. The parameters in each cylinder are calculated individually. The disadvantage of the non-stationary calculation method is a multiple increase in the volume of calculations, which makes it difficult to solve optimization problems. So the calculation of an 8-cylinder V-shaped diesel engine using a simple Vibe combustion model takes 20 min on a personal computer with a P4 processor.

For fruitful work, it is necessary to study a set of programs for modeling and optimizing work processes and determine a program for calculating internal combustion engines with a convenient user interface, a contextual help system, tools for automated data assignment that facilitate the identification of mathematical models [19–21]. The main parameters of the thermodynamic programs used for the study of internal combustion engines and which have become most widespread in Kazakhstan and abroad are presented in Table 1. The leader in popularity worldwide is the GT-Power program, which has become a de facto industry standard. All programs allow you to simulate the workflow of both diesel engines and spark ignition engines. Programs: BOOST, WAVE, GT-Power from three world leaders in software development: AVL, Ricardo Software, Gamma Technologies Inc. have almost equal opportunities. The cores of these programs are written in FORTRAN.

Table 1. An example of a table.

Program, developer	Gas exchange model	Combustion model	Note
IMPULSE	0-D	Vibe Razleitsev (1980)	Has not developed since 1985
DIESEL-PK [10] (Bauman Moscow State Technical University)	0D, ID	Vibe	Commercial model 2 and 4-stroke internal combustion engines
Lotus Engine Simulation [11] (Lotus Engineering)	1-D	Vibe	Commercial
VOLNA	1-D	Vibe Razleitsev (1980) Вибе	Has not developed since 1985
AMESIM [9] (LMS)	1-D	Model library: Vibe, CFD model, etc	Commercial
BOOST [12] (AVL)	1-D	Hiroyasu; table assignment dx/sf, custom model, CFD model with the code KIA (GT Power) or FIRE (BOOST). или FIRE (BOOST)	Commercial
WAVE [13] (Ricardo Software)			
GT-POWER [14] (Gamma Technologies Inc.)			

General basic properties of programs:

- WINDOWS applications;
- advanced user interface;
- a high degree of integration of the thermodynamic calculation of the internal combustion engine with other applications, for example, the calculation of bearings, cooling systems, dynamics of the valve mechanism, etc.;
- the ability to use user routines designed as DLLs for calculating various processes (including heat release rates);
- possibility of joint work with programs of three-dimensional modeling of gas flow;
- the ability to work together with SIMULINK packages, etc. for modeling (designing) control systems not only for the engine, but also for the entire vehicle as a whole;
- the possibility of matching the reciprocating internal combustion engine with the characteristics of turbines and compressors imported from text files in the SAE standard;
- the ability to calculate the unsteady modes of operation of the internal combustion engine;
- the possibility of calculating the unevenness of filling on the cylinders of the internal combustion engine.

4 Conclusion

Of the above, these programs allow us to most effectively solve the problems of optimal design of the gas exchange system, including:

- selection of valve timing phases individually for each cylinder;
- selection of the valve timing control law for transient engine operation modes (when installing an actuator to control the phases "on the move");
- design of pipelines, including for providing dynamic supercharging, register supercharging and exhaust gas recirculation systems;
- comparison of different engine concepts.

The latter is only partially true, due to the complexity of the task and the large amount of necessary source data. What may well be acceptable in the conditions of a corporation with experience, a staff of qualified specialists, detailed drawings and the results of experimental studies of the corresponding engines, may cause difficulties for researchers who do not have such resources.

With less reliability, these programs allow us to study intra-cylinder phenomena associated with the processes of mixing and combustion:

- study of the influence of the compression ratio,
- study of the effect of injection advance,
- study of the effect of the injection characteristics,
- study of the influence of the vortex intensity,
- study of the influence of the design of the sprayer (diameter, number and orientation of fuel jets).

References

1. Woschni, G.: Die Berechnung der Wandeverluste und der thermichen Belasttung der Bauteile von Dieselmotoren. MTZ **12**, 491–499 (1970)
2. Vyrubov, D.N., et al.: Dvigateli vnutrennego sgoraniya: Teoriya porshnevykh i kombinirovannykh dvigatelei [Internal Combustion Engines: The Theory of Piston and Combined Engines], 372 p. Mashinostroenie Publ., Moscow (1983). (in Russian)
3. Wiebe, I.I.: Novoe o rabochem cikle dvigatelej. [New on the Working Cycle of Engines], 272 p. Mashgiz Publ., Moscow (1962)
4. Hiroyasu, H., Kadota, T., Arai, M.: Development and use of a spray combustion modeling to predict diesel engine efficiency and pollutant emissions: part 1 combustion modeling. Bull. JSME **26**, 569–575 (1983)
5. Kavtaradze, R.Z., Onishchenko, D.O., Zelentsov, A.A., Kadyrov, S.M., Aripdzhanov, M.M.: Computational and experimental study of influence of piston's and sleeve's thermal insulation on formation of nitrogen oxides in combustion products of high-speed diesel. Vestnik MGTU im. N.E. Baumana. Ser. Mashinostroenie = Herald of the Bauman Moscow State Technical University. Ser. Mechanical Engineering, vol. 4, pp. 83–102 (2011). (in Russian)
6. Zakharov, L.A., Mironov, A.A., Malakhov, A.V., Kraynov, A.A., Zimina, T.N.: Modeling of gas exchange processes of a four-stroke piston engine with ignition from an external source. J. Phys. Conf. Ser. **1177**, 012017 (2019)
7. Grishin, Y.A., Bakulin, V.N.: New calculation schemes based on the large-particle method for modeling gas-dynamic problems. Dokl. Phys. **60**(12), 555–558 (2015)
8. Ambarev, K.: Simulation model of the operating cycle of internal combustion engine with variable compression ratio. J. Tech. Univ. Sofia Plovdiv Branch **16**, 147–152 (2011). ISSN 1310-8271. "Fundamental sciences and applications", Plovdiv, Bulgaria, book
9. Aghasafari, P., Israr, B.M.I., Pidaparti, R.M.: Investigation of the effects of emphysema and influenza on alveolar sacs closure through CFD simulation. J. Biomed. Sci. Eng. **9**, 287–297 (2016)
10. Arndt, D., Braack, M., Lube, G.: Finite elements for the Navier-Stokes problem with outflow condition. In: Karasözen, B., Manguoğlu, M., Tezer-Sezgin, M., Göktepe, S., Uğur, Ö. (eds.) Numerical Mathematics and Advanced Applications ENUMATH 2015. LNCSE, vol. 112, pp. 95–103. Springer, Cham (2016). https://doi.org/10.1007/978-3-319-39929-4_10
11. Kuleshov, A.S.: Multi-zone DI diesel spray combustion model and its application for matching the injector design with piston bowl shape. Proc. Inst. Mech. Eng. Part A J. Power Energy **222**, 309–321 (2008)
12. McCaw, D.L., Horrell, W.A.: On-Road Development of John Deere 6081 Natural Gas Engine: Final Technical Report, Deere & Company, 46 p. (2001)
13. Brandstein, A., Nakash, Y., Efrati, Y., Perel, D.: F100PW-229I thermodynamic model simulation with 'gasTurb 9'. Presented at the 45th Israel Annual Conference on Aerospace Sciences (2005)
14. AMEsim LMS Imagine. Lab Internal Combustion Engine. http://www.lmsintl.com
15. DIESEL-RK is an engine simulation tool. http://www.diesel-rk.bmstu.ru
16. Lotus Engineering Software. http://www.lesoft.co.uk
17. AVL BOOST: [Электронный ресурс]. http://www.avl.com. Accessed 09 Dec 2008
18. Ricardo Software: [Электронный ресурс]. http://www.software.ricardo.com. Accessed 09 Dec 2008
19. Gamma Technologies Inc.: [Электронный ресурс]. http://www.gtisoft.com. Accessed 09 Dec 2008

20. Kolade, B., Morel, T., Kong, S.-C.: Coupled 1-D/3-D analysis of fuel injection and diesel engine combustion. SAE Tech. Pap. Ser. –N 2004-01-0928, pp. 1–10 (2004)
21. Dudareva, N.Yu., Butusov, I.A., Kalschikov, R.V., Grin, R.R., Alexandrov, I.V.: The investigation of the effect of micro-arc oxidation modes on the adhesion strength of coatings (conference article). J. Eng. Sci. Technol. Rev. **7**(5), 5–8 (2014). Special Issue on Simulation of Manufacturing Technologies

Comprehensive Methodology for Comparative Environmental Assessment of Vehicles

Edvinas Valiulis and Saugirdas Pukalskas(✉)

Vilnius Gediminas Technical University, J. Basanavičiaus str. 28, 03224 Vilnius, LT, Lithuania
edvinas.valiulis@stud.vgtu.lt,
saugirdas.pukalskas@vilniustech.lt

Abstract. The negative impact of transport on the environment and the ways of ecological mobility are analysed in this paper. It is also highlighted that publicly recognised ecological vehicles – electric cars – are not absolutely "clean". Thus, based on research conducted by other scientists, recommendations of the EU institutions and other sources have developed a universal and comprehensive methodology for vehicle environmental assessment, which allows estimating the vehicle, energy consumption, CO_2 footprint during production and operation, and travel costs. The collected car data allowed us to identify the most environmentally friendly cars that have become Nissan Leaf, Renault Zoe and Hyundai Ioniq Plug-in, Honda E Advance, Toyota Yaris Hybrid.

Keywords: Environmental assessment · CO_2 emissions · Travel cost

1 Introduction

Operation of the vehicle features various harmful pollutants carbon monoxide (CO), nitrogen oxides (NO_x), hydrocarbons (HC), sulfur oxides (SO_2), particulate matter (PM), benzo(a)pyrene, etc. [1]. All these pollutants are harmful to the environment and people, therefore health care, industry, energy, fauna and flora suffer significant losses [2]. Particulate matter from diesel cars can cause smog. The formation of smog has a negative impact on human health – it causes respiratory and lung diseases [3].

The problem of climate change is directly linked to the 'greenhouse effect', where the atmosphere lets in visible sunlight but traps heat from the earth's surface [4]. Gases that contribute to the greenhouse effect, such as carbon dioxide, methane, and so on called greenhouse gases (GHGs). A certain level of GHG is necessary for life on Earth, but rapidly increasing GHGs concentrations in the atmosphere can have many negative consequences of climate change [5].

1.1 Environmental Impact of Transport

EU transport emits as much as 30% of total CO_2 emissions, of which 72% are land vehicles. CO_2 emissions are still a major problem for which alternatives are being sought. The EU has set a target for 2050 60% reduction in vehicle emissions compared to 1990 level. Other sectors since 1990 have significantly reduced emissions, but CO_2

O. Prentkovskis et al. (Eds.): TRANSBALTICA 2021, LNITI, pp. 292–302, 2022.
https://doi.org/10.1007/978-3-030-94774-3_29

emissions in the transport sector are due to increasing mobility of people and vehicles steadily rising. Passenger cars are a major polluter, accounting for as much as 60.7% of the total EU CO_2 road transport emissions from [6]. However, modern cars could be less polluting if they were shared more intensively.

The transport sector is the largest emitter of GHG, so reducing transport emissions is a key way to mitigate climate change. Increased use of electric vehicles could help achieve the EU's CO_2 reduction targets [7].

Nowadays, a growing emphasis on electric vehicles as an important part of achieving global climate change goals. These objectives are clearly reflected in measures to mitigate climate change [8]. Electric cars do not directly emit GHGs, but they are powered by electricity, which in many parts of the world is still produced from fossil fuels such as coal. A lot of energy is also used to make these "clean" vehicles, and especially their batteries.

The European Commission is proposing to modernize EU legislation on batteries. Batteries placed on the EU market should be in good working order and safe throughout their service life. This means that batteries must be manufactured with the least possible impact on the environment, using materials obtained without violating human rights and complying with all social and environmental standards. Batteries must be durable and should be used for other purposes, remanufactured or recycled at the end of their useful life. All valuable materials must be returned for further use [9].

Major carmakers such as Toyota or Volkswagen also contribute to less polluting battery production. These companies are investing in an alternative to the old type of lithium-ion battery, in the production of solid-state batteries, which should be safer and more durable. Due to their high efficiency at high temperatures and capacity, solid electrolytes are technology that can further increase energy density and battery safety [10].

1.2 Reducing GHG in the EU

The EU has three different legal mechanisms to manage air pollution: defining general air quality standards for ambient concentrations of air pollutants; setting national limits on total pollutant emissions; and designing source-specific legislation, e.g. to control industrial emissions or set standards for vehicle emissions, energy efficiency or fuel quality [11].

All EU Member States by 2050 must become climate neutral. The current EU emissions reduction targets for 2030 reduce emissions by 40% compared to the 1990 level. The Commission has recently proposed to increase this target to "at least 55%" in its amended proposal for EU climate law [12]. However, in the EP vote, Members agreed that by 2030 emissions need to be reduced to 60%. Finally, EU Member States have agreed to phase out all direct and indirect fossil fuel subsidies by 31 December 2025 [13].

2018 December 11 the Renewable Energy Directive was adopted in 2030 renewable energy, such as solar, wind and hydropower, will account for at least 32% of total EU energy consumption for electricity generation, transport, heating and cooling. Each Member State shall adopt and commit to the implementation of its national renewable energy action plan. To integrate the use of renewable energy in the transport sector,

Member States must oblige fuel suppliers to ensure that the share of renewable energy is reduced by 2030 account for at least 14% of total energy consumption [14].

Newly registered passenger cars in the EU must meet CO_2 emission standards. The car fleet must reduce CO_2 emissions from 130 g CO_2/km (2015 value) to 95 g CO_2/km (2021 value). CO_2 emissions from trucks have also been limited until 2030 to reduce it by 30%, with an intermediate target of 2025 is a 15% GHGs reduction [14].

2 Vehicle Life Cycle Assessment

The life cycle assessment (LCA) of vehicles is performed by assessing vehicle size, energy (or fuel) consumption, electricity or fuel production pollution, driving habits and even weather conditions [8].

LCA is a method used to optimise CO_2 emissions [15]. This method, based on ISO standards, makes it possible to assess the car's impact on the environment throughout its life cycle. This means that it is calculated from the extraction of raw materials to the disposal of the vehicle [14]. Life cycle analysis consists of five main areas: extraction, processing or purchase of raw materials; materials processing; vehicle assembly and production; operation; car recycling [16]. The first area of analysis is the extraction of raw materials from the subsoil or its surface. After the materials are extracted, they travel to their processing for further use, which occurs at the production sites. This stage of production also includes the production of engines, accumulators, fuels or electricity [17].

Another essential point in LCA is the production of the vehicle, which uses all the components already produced and the machined materials. This stage also includes tests on the strength of vehicle structures. Another area is car uptime: vehicle maintenance, repairs. The last stage is car recycling. Most LCA are completed without even starting to look at the car-recycling point, as the recyclable materials are different in each car, for example, batteries are not recycled or only a small part is recycled [18–20].

3 Environmental Assessment of Vehicles

3.1 Methodology of Environmental Assessment

The following criteria are assessed during LCA [21, 22]: vehicle size; emissions of pollutants; fuel type for vehicles with an internal combustion engine (ICE); electricity generation; battery capacity and its secondary use; car mileage.

After evaluating these criteria, a detailed LCA of cars was performed: the estimated total cost of ownership (TCO) for 1 km of road and CO_2 emissions during the production and operation of cars.

During the LCA, data and prices of cars sold in Lithuania were collected. Fuel and energy consumption, CO_2 emissions, distance per charge (for electric and plug-in hybrid cars), car sales price, car maintenance and inspection costs, pollution taxes and insurance cost were analysed and compared with electric, hybrid and conventional cars with ICE. According to these criteria, TCOs were calculated and their dependence on the mileage of the car was formed.

A separate calculation method was developed for each type of car. For a conventional car with only ICE, only one type of fuel is calculated. For a rechargeable hybrid car, both fuel and electricity costs and their prices need to be assessed. The prices of periodic technical inspections of different types of cars also differ, the most expensive service for CNG-powered cars (24.60 Eur). Each car is also subject to an individual pollution tax, which is calculated on the basis of the car's CO_2 emissions when it exceeds 130 g/km.

Conventional Cars. A formula has been developed to calculate the TCO per 1 km K_{km_con} of conventional (including hybrid) cars:

$$K_{km\_con} = \frac{\left(\frac{K_{fuel} \cdot B_{fuel} \cdot R_d}{100} \cdot D_m\right) + \left(\frac{K_{tech}}{t_{ap} \cdot Sk_m}\right) + \left(\frac{K_{ins}}{Sk_m}\right) + \left(\frac{K_t + K_r}{Sk_m \cdot t_{total}}\right) + K_{tp} + K_v}{R_m}, \quad (1)$$

where K_{fuel} – fuel price, Eur/l; B_{fuel} – fuel consumption, l/100 km; R_d – mileage per day, km; D_m – working days per month; K_{tech} – technical inspection price for car, Eur; t_{ap} – period valid mandatory technical inspection, year; Sk_m – number of months in the year; K_{ins} – price of compulsory civil insurance, Eur; K_t – pollution tax price, Eur; K_r – registration price, Eur; t_{total} – car life, year; K_v – car cleaning price per month, Eur; R_m – mileage per month, km; K_{tp} – maintenance price, Eur.

Electric Cars. A formula has been developed to calculate the TCO per 1 km K_{km_e} of electric cars:

$$K_{km\_e} = \frac{\left(\frac{K_e \cdot B_e \cdot R_d}{100} \cdot D_m\right) + \left(\frac{K_{tech}}{t_{ap} \cdot Sk_m}\right) + \left(\frac{K_{ins}}{Sk_m}\right) + \left(\frac{K_t + K_r}{Sk_m \cdot t_{total}}\right) + K_{tp} + K_v}{R_m}, \quad (2)$$

where K_e – price of electricity, Eur/kWh; B_e – electric car energy consumption, kWh/100 km.

Plug-In Hybrid Cars. The formulas used to calculate the TCO per 1 km of plug-in hybrid cars depending on the type of fuel (petrol or diesel):

$$K_{km\_hyb} = \frac{\left(\frac{K_{fuel} \cdot B_{fuel} \cdot R_d}{100} \cdot D_m\right) + \left(\frac{K_e B_e \cdot R_d}{100} \cdot D_m\right) + \left(\frac{K_{tech}}{t_{ap} \cdot Sk_m}\right) + \left(\frac{K_{ins}}{Sk_m}\right) + \left(\frac{K_t + K_r}{Sk_m \cdot t_{total}}\right) + K_{tp} + K_v}{R_m}. \quad (3)$$

CNG Cars. A formula has been developed to calculate the TCO per 1 km of CNG-powered cars K_{km_CNG}:

$$K_{km\_CNG} = \frac{\left(\frac{K_{CNG} \cdot B_{CNG} \cdot R_d}{100} \cdot D_m\right) + \left(\frac{K_{tech}}{t_{ap} \cdot Sk_m}\right) + \left(\frac{K_{ins}}{Sk_m}\right) + \left(\frac{K_t + K_r}{Sk_m \cdot t_{total}}\right) + K_{tp} + K_v}{R_m}, \quad (4)$$

where K_{CNG} – CNG price, Eur/m^3; B_{CNG} – CNG consumption of a car, m^3/100 km.

Assessment of CO_2 Emissions. A methodology for calculating CO_2 emissions from passenger cars was developed based on Hausfather [8] and WorldAutoSteel.org [23] sources.

According to the methodology [23], CO_2 emissions from cars are estimated as follows: 15% of total CO_2 emissions are emitted during the production of a conventional vehicle, and the remaining 85% are emitted during the operation of the car; the production of a plug-in hybrid emits the same amount of CO_2 as during the operation; meanwhile, as much as 85% of pollutants are emitted during the production of an electric car and the remaining 15% during its operation.

The formula used to calculate CO_2 emissions during the operation of conventional cars is:

$$CO_{2\,opr}^{conv} = \frac{M_{CO_2}.R_m.t_{total}}{1000}, \tag{5}$$

where M_{CO2} – CO_2 emissions of the vehicle, g/km; R_m – mileage per year, km.

Knowing that this represents 85% of the total CO_2 emissions, the CO_2 emissions from the production of the car are calculated:

$$CO_{2\,prod}^{conv} = \frac{CO_{2\,opr}.0,15}{0,85}. \tag{6}$$

The production of plug-in hybrids emits the same amount of CO_2 as during the operation of the vehicle [23]:

$$CO_{2\,prod}^{h} = CO_{2\,opr}^{h}. \tag{7}$$

The CO_2 emissions of electric cars during production can be calculated according to the recommendations of Hausfather, i.e. 175 kg CO_2/kWh multiplied by the battery capacity [8]:

$$CO_{2\,prod}^{e} = T_{bat}.175, \tag{8}$$

where T_{bat} – battery capacity, kWh.

The production of an electric car emits 85% of the total amount of CO_2, and during operation – 15%, therefore [23]:

$$CO_{2opr}^{e} = \frac{CO_{2\,prod}.0,85}{0,15}. \tag{9}$$

3.2 Results

According to the formulas, a price is calculated for each type of car per km (Fig. 1), which shows which type of car is the cheapest and which is the most expensive to operate. The graph shows that the maximum price per 1 km is for conventional and powerful petrol-powered cars. This is because this type of car not only has a relatively high fuel consumption, but also significantly higher maintenance and insurance costs.

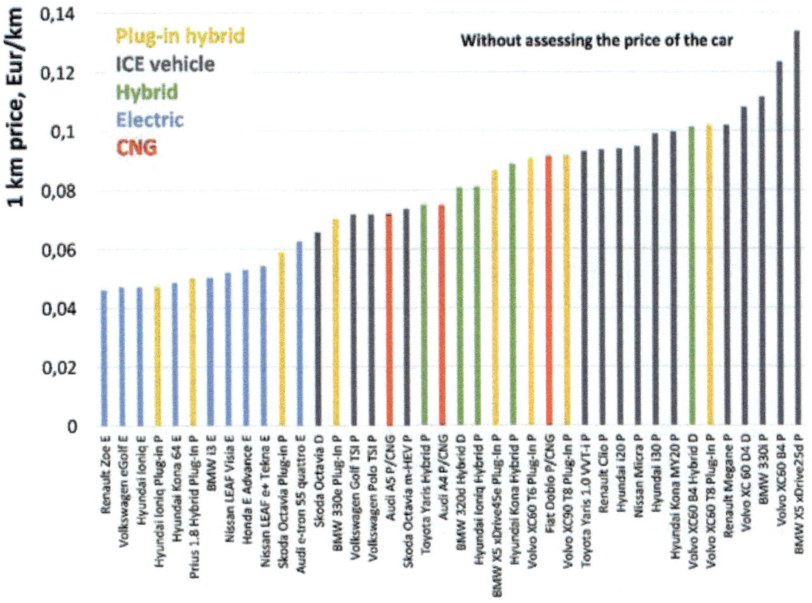

Fig. 1. The cost of various cars per 1 km of road, without estimating the initial price of the car.

Figure 1 shows that electric cars, and in particular small electric cars (blue columns), are ahead in terms of cost per 1 km, while conventional petrol and diesel cars have the highest costs (dark blue columns). Hybrid cars are concentrated in the middle of the evaluated cars (green columns), but Plug-in hybrids are scattered throughout the scale as their cost per 1 km is highly dependent on the size of the car: smaller cars like the Hyundai Ioniq Plug-in or Toyota Prius Plug-in have small enough costs ($\sim$5 ct./km) and is on par with the lowest cost electric cars, while large ones like the Volvo XC60 T8 Plug-in ($\sim$10 ct./km) appear at the bottom of the list, with the most costly petrol and diesel cars.

The graph in Fig. 2 shows the TCO in terms of operating costs and the initial cost of the car. For example, the graph shows that the Renault Clio petrol car costs the least (starting price € 11,822), but its long-term TCO are higher than the Volkswagen Polo TSI, which has a significantly higher starting price (€ 17,276) but lower fuel consumption and therefore the TCO of this car is lower than that of the Renault Clio eventually.

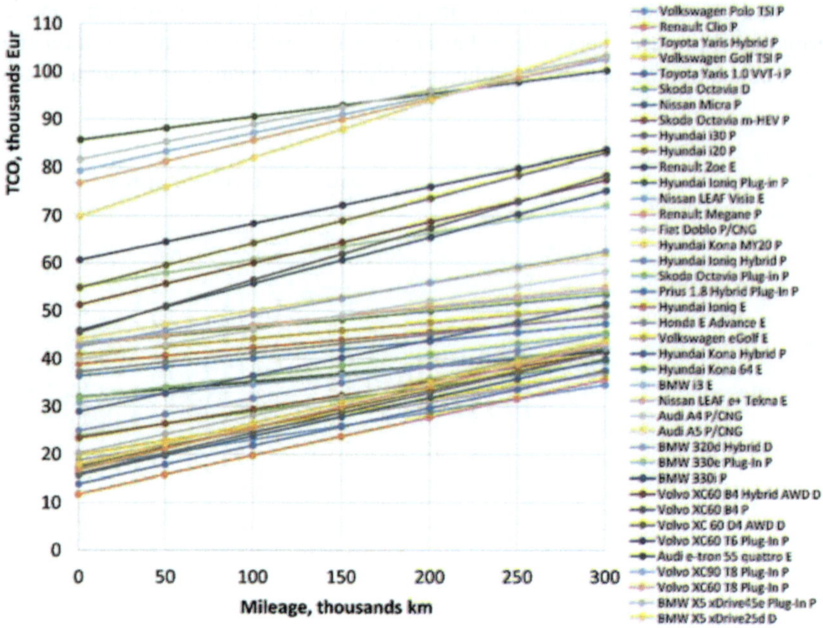

Fig. 2. Dependence of TCO of various cars on mileage, estimating the initial price car.

At the top of the chart are the big luxury cars BMW X5 xDrive25d, Volvo XC60 T8 Plug-in, BMW X5 xDrive45e Plug-in and Audi e-tron 55 quattro. The initial value of these cars is high, and the operating costs (due to high fuel consumption) are quite high, except for the electric Audi e-ton 55 quattro, which has similar operating costs for other smaller electric cars. Furthermore, although Audi e-ton 55 quattro is the most expensive of all cars valued, after 300 thousand km, its TCO will be lower than other luxury cars mentioned.

According to the formulas provided in the methodology, the CO_2 emissions of the assessed cars during their production and operation are analysed. The period of 8.3 years is analysed, assuming that each car travels 36,000 km (a total of 300,000 km) in one year (Fig. 3).

The graph shows (Fig. 3) that the least polluting cars are small compact electric cars like the Honda E Advance, Volkswagen eGolf, Nissan Leaf, BMW i3 and others. Their total CO_2 emissions during production and after 300 thousand km in 8.3 years reaches 6,000–8,000 kg. Meanwhile, looking at the CO_2 emissions of large luxury cars (BMW X5 xDrive 25d, Volvo XC60 B4, Volvo XC60 B4 Hybrid AWD or Volvo XC60 D4 AWD*)*, it can be seen that the values reach 60,000–70,000 kg, which is 10 times more.

The graph shows the three cars in the middle, whose columns rise above the others – *Volvo* plug-in cars. The CO_2 emissions of these cars during the operation are not high, but the emissions during the production of the cars are significantly higher than other cars because they have not only powerful ICE, but also a large capacity battery. This is in line with the results of other studies [24, 25].

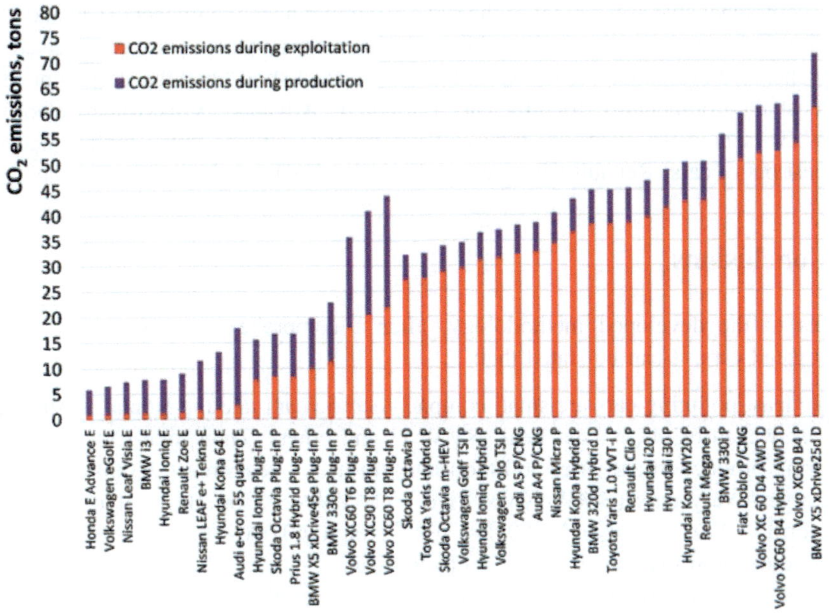

Fig. 3. CO_2 emissions of cars during their production and operation (the period 8.3 year).

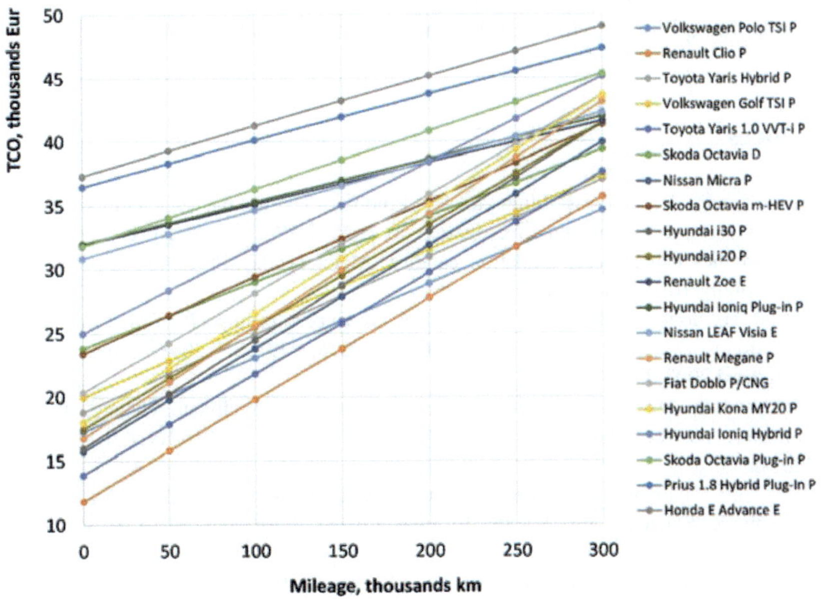

Fig. 4. Dependence of TCO of various cars on mileage, estimating the initial price car (small and compact cars).

A comprehensive assessment of the car's minimum negative impact (TCO and CO_2 emissions) shows that only small and compact cars are suitable. The cars selected for comprehensive evaluation are shown in Fig. 4. Figure 4 shows that the starting price of the most environmentally friendly cars (Honda E Advance, Volkswagen eGolf or BMW i3) is high, so even 300 thousand km mileage will be much higher than the cheapest petrol cars (Renault Clio, Toyota Yaris 1.0 VVT-i or Volkswagen Polo TSI).

4 Conclusions

Thanks to the developed methodology, after a comprehensive assessment of the investigated cars, it can be stated that:

1. The least polluting cars are small compact electric cars like the Honda E Advance, Volkswagen eGolf, Nissan Leaf, BMW i3 and others. Their total CO_2 emissions during production and after 300 thousand km in 8.3 years reach 6,000–8,000 kg;
2. Large luxury cars (BMW X5 xDrive 25d, Volvo XC60 B4, Volvo XC60 B4 Hybrid AWD or Volvo XC60 D4 AWD) emit 60,000–70,000 kg of CO_2, i.e. 10 times more than small compact electric cars;
3. The initial price of the most environmentally friendly cars (Honda E Advance, Volkswagen eGolf, Nissan Leaf, BMW i3), which emits the least amount of CO_2, is high, so after 300 thousand km mileage TCO will be higher than the cheapest small petrol cars;
4. The electric car Nissan Leaf Visia has the least negative impact. Renault Zoe and Hyundai Ioniq Plug-in, Honda E Advance, Toyota Yaris Hybrid continue to line up.

References

1. Nadanakumar, V., Jenoris Muthiya, S., Prudhvi, T., Induja, S., Sathyamurthy, R., Dharmaraj, V.: Experimental investigation to control HC, CO & NOx emissions from diesel engines using diesel oxidation catalyst. Mater. Today Proc. **43**, 434–440 (2021). https://doi.org/10.1016/j.matpr.2020.11.964
2. Demirel, H., Sertel, E., Kaya, S., Zafer Seker, D.: Exploring impacts of road transportation on environment: a spatial approach. Desalination **226**, 279–288 (2008). https://doi.org/10.1016/j.desal.2007.02.111
3. Kirk-Davidoff, D.: The greenhouse effect, aerosols, and climate change. In: Green Chemistry, pp. 211–234. Elsevier (2018). https://doi.org/10.1016/B978-0-12-809270-5.00009-1
4. North, G.R.: Climate and climate change | Greenhouse Effect. In: Encyclopedia of Atmospheric Sciences, pp. 80–86. Elsevier (2015). https://doi.org/10.1016/B978-0-12-382225-3.00470-9
5. Vanek, F.M., Albright, L.D., Angenent, L.T.: Energy Systems Engineering: Evaluation and Implementation. McGraw Hill Education, New York (2016)
6. European Parliament: CO_2 emissions from cars: facts and figures (infographics). https://www.europarl.europa.eu/news/en/headlines/society/20190313STO31218/co2-emissions-from-cars-facts-and-figures-infographics. Accessed 07 Oct 2021

7. European Environment Agency: New registrations of electric vehicles in Europe. https://www.eea.europa.eu/data-and-maps/indicators/proportion-of-vehicle-fleet-meeting-5/assessment. Accessed 07 Oct 2021
8. Hausfather, Z.: Factcheck: how electric vehicles help to tackle climate change. https://www.carbonbrief.org/factcheck-how-electric-vehicles-help-to-tackle-climate-change. Accessed 07 Nov 2021
9. European Commission: Green Deal: Sustainable batteries for a circular and climate neutral economy. https://ec.europa.eu/growth/content/green-deal-sustainable-batteries-circular-and-climate-neutral-economy_en. Accessed 07 Nov 2021
10. Lee, L.: How competition is driving innovation in the EV battery market. https://www.counterpointresearch.com/competition-driving-innovation-ev-battery-market/. Accessed 07 Nov 2021
11. Kurrer, C.: Air and noise pollution. https://www.europarl.europa.eu/factsheets/en/sheet/75/air-and-noise-pollution. Accessed 07 Aug 2021
12. European Commission: Regulation of the European Parliament and of the Council on establishing the framework for achieving climate neutrality and amending Regulation (EU) 2018/1999 (2020)
13. Haahr, T.: EU climate law: MEPs want to increase 2030 emissions reduction target to 60%. https://www.europarl.europa.eu/news/en/headlines/priorities/climate-change/20201002IPR88431/eu-climate-law-meps-want-to-increase-2030-emissions-reduction-target-to-60. Accessed 30 Jun 2021
14. Amanatidis, G.: Combating climate change. https://www.europarl.europa.eu/factsheets/en/sheet/72/combating-climate-change. Accessed 07 Nov 2021
15. Ternel, C., Bouter, A., Melgar, J.: Life cycle assessment of mid-range passenger cars powered by liquid and gaseous biofuels: Comparison with greenhouse gas emissions of electric vehicles and forecast to 2030. Transp. Res. Part D Transp. Environ. **97**, 102897 (2021). https://doi.org/10.1016/j.trd.2021.102897
16. Messagie, M.: Life cycle analysis of the climate impact of electric vehicles. Transport & Environment/Vrije Universiteit Brussel, Belgium (2018)
17. Verma, S., Dwivedi, G., Verma, P.: Life cycle assessment of electric vehicles in comparison to combustion engine vehicles: a review. Mater. Today Proc. (2021).https://doi.org/10.1016/j.matpr.2021.01.666
18. Walker, C.M., Tawfik, A.M.: Vehicle emissions and life cycle analysis models of gasoline and electric vehicles (2015). https://www.epa.gov/sites/production/files/2015-09/documents/awalker.pdf
19. Silvestri, L., Forcina, A., Arcese, G., Bella, G.: Recycling technologies of nickel–metal hydride batteries: an LCA based analysis. J. Clean. Prod. **273**, 123083 (2020). https://doi.org/10.1016/j.jclepro.2020.123083
20. Dong, Y., Zhao, Y., Hossain, M., He, Y., Liu, P.: Life cycle assessment of vehicle tires: a systematic review. Cleaner Environ. Syst. **2**, 100033 (2021). https://doi.org/10.1016/j.cesys.2021.100033
21. Zheng, G., Peng, Z.: Life Cycle Assessment (LCA) of BEV's environmental benefits for meeting the challenge of ICExit (Internal Combustion Engine Exit). Energy Rep. **7**, 1203–1216 (2021). https://doi.org/10.1016/j.egyr.2021.02.039
22. Helmers, E., Dietz, J., Weiss, M.: Sensitivity analysis in the life-cycle assessment of electric vs. combustion engine cars under approximate real-world conditions. Sustainability **12**, 1241 (2020). https://doi.org/10.3390/su12031241
23. World Steel Association AISBL: LCA – Life Cycle Assessment of Vehicle Emissions. https://www.worldautosteel.org/life-cycle-thinking/lca-videos/lca-life-cycle-assessment-vehicle-emissions/. Accessed 07 Nov 2021

24. Tiano, F.A., Rizzo, G., De Feo, G., Landolfi, S.: Converting a conventional car into a hybrid solar vehicle: a LCA approach. IFAC-PapersOnLine **51**, 188–194 (2018). https://doi.org/10.1016/j.ifacol.2018.10.035
25. Zhu, Y., Skerlos, S., Xu, M., Cooper, D.R.: System level impediments to achieving absolute sustainability using LCA. Procedia CIRP **90**, 399–404 (2020). https://doi.org/10.1016/j.procir.2020.02.137

Reviewing the Concept of Acoustic Agglomeration in Reducing the Particulate Matter Emissions

Sai Manoj Rayapureddy(✉) and Jonas Matijošius

Faculty of Transport Engineering, Department of Automobile Engineering, Vilnius Gediminas Technical University, J. Basanavičiaus Str. 28, 03224 Vilnius, Lithuania
{sai-manoj.rayapureddy, jonas.matijosius}@vilniustech.lt

Abstract. Air pollution is one of the world's major growing concern. Particulate Matter, which is a complex mixture of solid or liquid airborne particles, is one of the predominant emissions that shows a drastic impact on human health. While we observe a constant decline in these emissions from the transport sector over the years, the current number still exceeds the WHO & EU standards. The particulate matter of size less than 2.5 μm is found to be damaging human health by penetrating into the blood, heart and respiratory systems. In this article, we reviewed the technique Acoustic agglomeration and its application in reducing the Particulate Matter emissions. It involves a process of using sound waves to manipulate the motion of airborne particles to collide and agglomerate so that they can be easily filtered through the conventional filters. This process is found to be applied in various industries like chemical, air filtration in refrigerators and coal combustion plants. Several other experiments that involve acoustic agglomeration technique in reducing the particulate pollution coming from various sources including automobiles was reviewed. The success of these researches and the implementation of this technique in automobile might solve the problems associated with the particulate pollution.

Keywords: Particulate matter · Acoustic agglomeration · EURO 6 emission standards · Internal combustion engine

1 Introduction

"Air Pollution is Europe's biggest environmental health risk", says European Environment Agency. Although the quality of air is being improved over the years, experts say the pollutant concentration levels still exceed the WHO & EU standards making the air quality problem persist [1]. World Health Organization (WHO) estimates that nearly 7 million people all over the world die due to air pollution [2]. Nearly 90% of people who live in Europe are exposed to high concentrated pollutants that are considered harmful to human health [3]. Being exposed to this pollution for a longer period of time damages the respiratory system and might also lead to premature deaths [4]. Industries, Transport, agriculture, waste management and households are among the predominant

O. Prentkovskis et al. (Eds.): TRANSBALTICA 2021, LNITI, pp. 303–311, 2022.
https://doi.org/10.1007/978-3-030-94774-3_30

sources of pollution. Transport sector contributes for more than 10% of total emissions and accounts for nearly 30% of the total NOX emissions [2]. A detailed overall percentage of each pollutant released by the transport sector including non-exhaust is given in Table 1.

Table 1. Percentage of each pollutant released by the transport sector.

Pollutant	Road Transport exhaust, %	Non exhaust, %
CO	17.97	0
NO_X	28.12	0
PM_{10}	2.88	4.81
$PM_{2.5}$	5.39	4.5
SO_X	0.11	0

The European union aims to reduce the air pollution by joining hands with the industries responsible for its growth and by strengthening the regulations. In the case of transport sector, EU passes a set of regulations under the name Euro Emission Standard to control and contain the pollution that is being released from the automobiles. The latest revision which was done in 2014 under Euro 6 emission standards was presented in Table 2 [2, 5].

Table 2. Euro 6 vehicle emission standards.

Euro 6 emission limits (petrol), g/km	Euro 6 emission limits (diesel), g/km
CO – 1.0	CO – 0.50
HC – 0.10	HC + NO_x – 0.17
NO_x – 0.06	NO_x – 0.08
PM – 0.005 (direct injection only)	PM – 0.005
PM – 6.0×10^{11}/km (direct injection only)	PM – 6.0×10^{11}/km

Particulate Matter, nitrogen oxide and ozone are the top three pollutants that significantly affect the human health [6]. Of all the main components of Air Pollution, Particulate Matter (PM), due to their high toxicological effects, gained the attention of researchers [7–9]. Scientists are finding ways to reduce this emission which adversely affect human health ranging from minor coughing to severe heart and respiratory diseases [10, 11].

2 Particulate Matter

Particulate Matter is a complex mixture of solid particles and liquid droplets that are found in the air. The size of inhalable PM particles typically varies from 2.5 μm to 10 μm. When inhaled, these small sized particles enter into our body through our nostrils passing through the upper airway penetrating the vulnerable tissues and even damaging the lungs. The tiny particles also possess the threat of being transported to other organs by entering into the blood stream [12].

PM is emitted through a number of sources like wildfires, power plants, heating systems and automobiles in various sizes and shapes. Most part of this particle pollution is formed in the atmosphere as a result of complex chemical reactions between sulfur dioxide, nitrogen oxides and volatile organic compounds [13, 14]. Particles such as smoke, dirt, dust are large enough to be visible to the naked eye. These matter that contains microscopic solids and liquid droplets causes serious health problems when inhaled [15].

Based on the cause of emission, PM is categorized into exhaust and non-exhaust (primary PM) emissions. Primary PM accounts for the emissions that are released from tire and brake wear, road abrasions etc. In 2017, the transport sector accounted for more than 45% of non-exhaust $PM_{2.5}$ emissions, the value was 18% in 2000. Similarly, automobiles contributed for more than 60% of non-exhaust PM_{10} emissions, it was 32% in 2000 [16].

From the research data of National emissions reported to the Convention on Long-range Transboundary Air Pollution (LRTAP Convention) provided by European Environment Agency (EEA) the following graph has been plotted (see Fig. 1) [16].

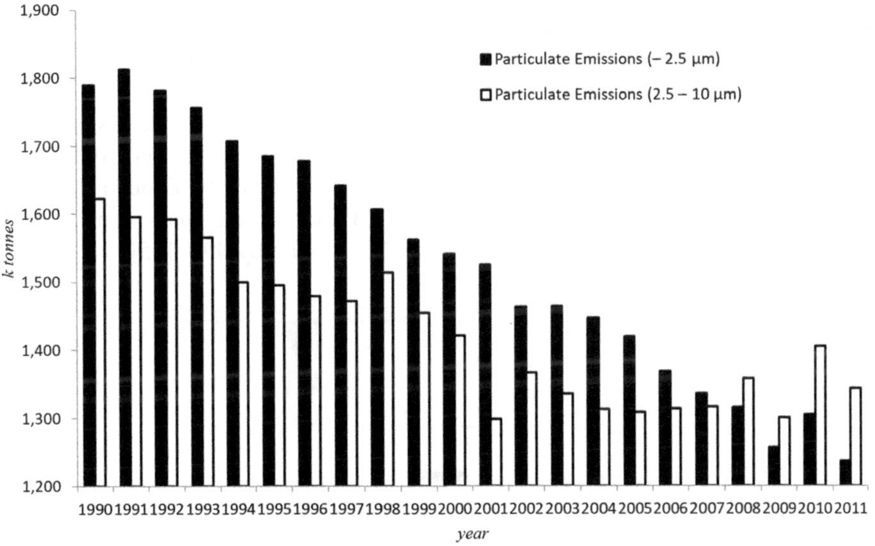

Fig. 1. Particulate Matter emissions (PM_{10} and $PM_{2.5}$).

Automobiles and industrial plants are equipped with PM filters to prevent exhaust particles from getting released into the atmosphere. The particles which are 2.5 µm in diameter are more dangerous as it is difficult to capture them through the conventional filters. In EU, $PM_{2.5}$ has been estimated to reduce life expectancy by more than 8 months [2, 14]. Particularly the small sized particles remain airborne for days after the release and can be carried to long distances from the source of origin [17]. Studies show these particles of particulate matter are responsible for a considerable amount of damage for human health [15]. Experiments and researches have been carried out to find new techniques such as Acoustic Agglomeration that help capture these tiny particles and reduce particulate pollution from effecting human health.

3 Acoustic Agglomeration

Acoustic agglomeration is the process of manipulating the motion of airborne particles with the help of high intensity, high frequency acoustic waves. These waves promote collision between the particles that leads to the formation of agglomerates. These clustered particles continue to agglomerate leading to enormous growth of size of the particles [18]. After collision, the particles are likely to adhere together due to meshing of irregular structures as well as inter-molecular attraction forces [19].

The concept of acoustic agglomeration was first observed by Patterson and Cawood by the name 'Phenomena in a Sounding Tube' in 1931 [20]. Several independent experiments were further developed leading to new theories and ways to estimate particle growth [18].

Unconditionally small size particles that are difficult to be captured using the conventional methods has been the primary focus behind the application of acoustic agglomeration technique [21, 22]. They contribute to reducing a significant amount of particle pollution largely from chemical industries, food processing industries, oil and gas production [23, 24].

Understanding the potential of the concept, Bing Feng Ng have successfully used acoustic agglomeration in Air-Conditioning and Mechanical Ventilation (ACMV) for filtering small size air particles (see Fig. 2) [18]. In another study by Zhou, the acoustic agglomeration operated at 1,400 Hz and 148 dB the particle mass concentration removal efficiency of in-house developed bag filter was up by 99% [25].

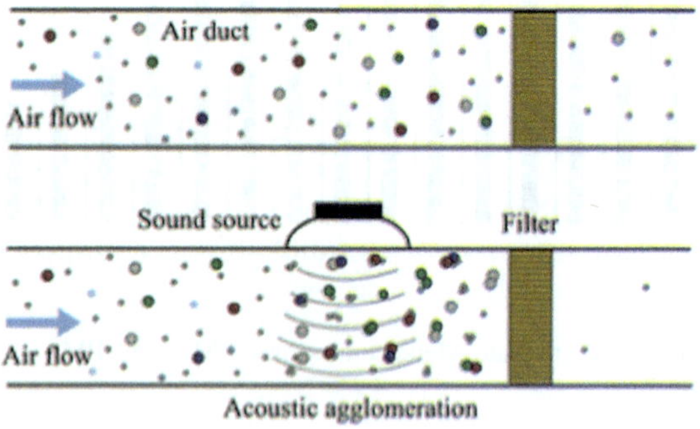

Fig. 2. Change in particle size while applying sound source.

3.1 Industrial Application of Acoustic Agglomeration in Reducing Particle Pollution

Gallego-Juárez conducted an experiment where they applied this phenomenon to reduce the particle emissions in coal combustion fumes from a semi-industrial pilot plant [26].

Their experimental setup basically consists of a circulating fluidized bed combustion plant of 0.5 MWt, a conventional electrostatic precipitator (ESP) and the acoustic agglomeration chamber driven by four high power and high directional acoustic transducers of 10 and/or 20 kHz (see Fig. 3).

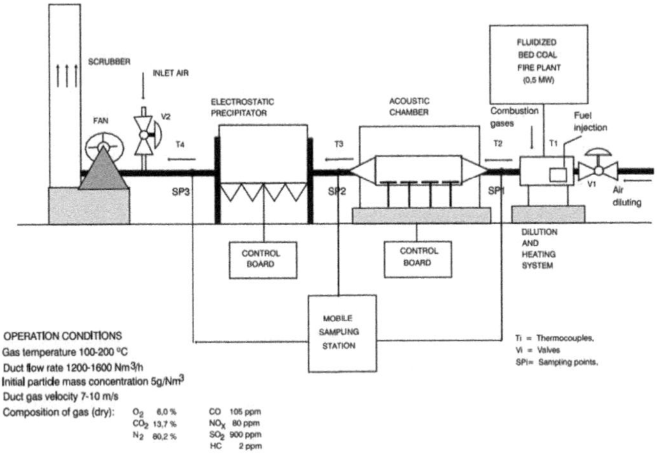

Fig. 3. Experimental setup of Gallego-Juárez.

Experiments were carried out with fume flow rates up to about 2,000 Nm^3/h, gas temperatures of about 150 °C and mass concentrations in the range 1–5 g/Nm^3. The fine particle reduction produced by the acoustic filter is of about 40% of the number concentration. The resultant mass reduction percentage (values when the acoustic chamber is on) given in the Table 3 is calculated with respect to the particle mass when the acoustic chamber is off. They found a significant reduction of total mass and number concentration of 37% and 40% respectively when the system worked at its peak power and frequency (4T $\times$ 400 W and 20 kHz).

Table 3. Summary of their findings.

Test	Mass reduction, %	Particle number reduction, %	
		Micron	Submicron
1	32	37	39
2	29	33	21
3	26	26	16
4	37	42	39
5	29	31	31
6	24	14	24

The principle of acoustic agglomeration has been broadly studied but its practical application has been limited without any real-time process. The reasons for its restricted application despite the tremendous successful experimentation might be due to the lack of suitable high intensity, powerful sound sources and development of its equipment on full scale.

4 Discussion

The primary objective of this review is to check the possibilities and limitations of using acoustic agglomeration technique in reducing particulate pollution from the exhaust of automobiles. Very limited experimentations have been performed implementing the agglomeration technique in transport sector.

4.1 Application of Acoustic Agglomeration in Reducing Particle Pollution from Diesel Engine

One such recent research has been carried out investigating the effect of acoustic agglomeration technique on the diesel engine exhaust particles by creating an acoustic chamber [27]. This experimental setup consists of diesel engine with load 45,000 Nm and speed 2,000 rpm, chamber where acoustic agglomeration takes place and measuring equipment. The sound pressure levels are designed to reach 140 dB at frequency 24 kHz. The reason behind the selection of this frequency range is to maximize the orthokinetic agglomeration and availability of acoustic generators with frequency 24 kHz. PM concentrations have been recorded with and without the acoustic waves. The sound levels have been continuously monitored for every 30 s.

Acoustic chamber which investigates the particle agglomeration was designed with a piezoelectric excitement source and an T – form pipe. Based on the design calculations, an acoustic camera was designed and manufactured. Particle Concentration Analyzer 4 APC ErgoTouch Pro was used along with Bruel & Kjær measuring software system Type 9727. During the experimentation, sound pressure levels are found to reach 136.6 dB at 24 kHz frequency [27].

As we can see from Fig. 4, there is considerable decrease in the total amount of particles when the acoustics is on. But when comparing the difference from Table 4, it can be observed that particles with diameter from 0.3–0.5 μm shows a decreased percentage of particles, while the area between 0.5–3 μm is found to be with no major change and the particles from 3–10 μm is shown to be increasing.

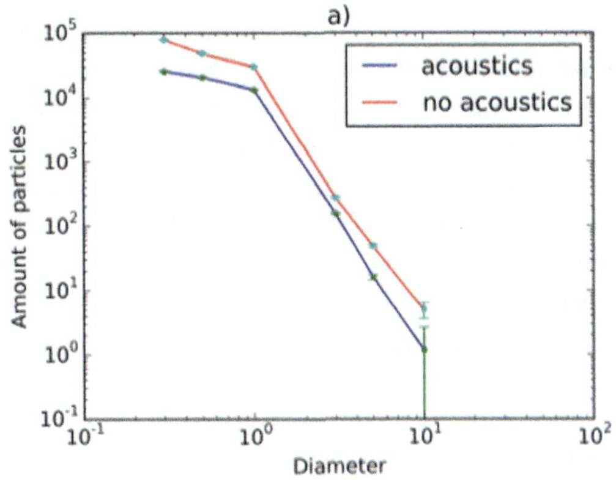

Fig. 4. Graph indicating the number of particles with and without acoustics along their diameter.

Table 4. Results of their findings.

Particle diameter, μm	Total amount of particles in 166.645 (cm^3) volume, number					
	0.3	0.5	1.0	3	5	10
With acoustic influence	25,391	20,381	13,269	152	16	1
Without acoustic influence	77,680	47,973	30,097	269	48	5
Difference, %	306	235	227	177	300	500

The increasing number of particles in the large size area can be explained by the increase in the size of particles after the agglomeration process. This experiment has been successful in showing that there is a scope of decreasing the total particulate matter (especially the small sized particles) emissions from the engine exhaust when the acoustic agglomeration technique is used.

5 Conclusion

1. Air Pollution is one of the primary concerns of all the European Countries and the world. Particulate Matter, NO_x and ozone are found to be the top three pollutants that significantly affect the human health.
2. Road transport accounts for more than 5% of total $PM_{2.5}$ emissions from exhausts and nearly 10% of the total $PM_{2.5}$ emissions combined.
3. It is almost impossible for a conventional filter to capture the particles with size less than 2.5 μm. These particles which are released into the environment are penetrated

directly into the lungs, blood and heart through the respiratory system causing serious health hazard.

4. Acoustic agglomeration is the process of manipulating the motion of airborne particles with the help of high frequency acoustic waves. This movement promotes collision between the particles that leads to the formation of agglomerates.
5. We reviewed the previous application of the principle of Acoustic Agglomeration in improving the air filtration efficiency in Air-Conditioning and Mechanical Ventilation (ACMV), particle mass concentration removal efficiency of in-house developed bag filter, reducing the particle emissions in coal combustion fumes from a semi-industrial pilot plant and found the results to be satisfying.
6. We achieved the primary objective of this review by understanding the possibilities of using the acoustic agglomeration technique in reducing particulate matter (with size less than 2.5 μm).
7. Reviewing the previous experimental conditions, it is revealed that sound waves with frequency of at least 140 db is required for the particles to agglomerate and sufficient time to find notable increase in their size.
8. By choosing acoustic generators with frequency 24 kHz we can maximize the orthokinetic agglomeration and they are also easily available in the market.

References

1. Oxley, T., Elshkaki, A., Kwiatkowski, L., Castillo, A., Scarbrough, T., ApSimon, H.: Pollution abatement from road transport: cross-sectoral implications, climate Co-benefits and behavioural change. Environ. Sci. Policy **19–20**, 16–32 (2012). https://doi.org/10.1016/j.envsci.2012.01.004
2. Emissions of air pollutants from transport—European Environment Agency. https://www.eea.europa.eu/data-and-maps/indicators/transport-emissions-of-air-pollutants-8/transport-emissions-of-air-pollutants-8. Accessed 21 Jun 2021
3. Lelieveld, J., Pozzer, A., Pöschl, U., Fnais, M., Haines, A., Münzel, T.: Loss of life expectancy from air pollution compared to other risk factors: a worldwide perspective. Cardiovasc. Res. **116**, 1910–1917 (2020). https://doi.org/10.1093/cvr/cvaa025
4. Brunekreef, B., Holgate, S.T.: Air pollution and health. Lancet **360**, 1233–1242 (2002). https://doi.org/10.1016/S0140-6736(02)11274-8
5. Williams, M.. Minjares, R.A.: Technical Summary of Euro 6/VI Vehicle Emission Standards 12
6. Rodriguez-Alvarez, A.: Air pollution and life expectancy in Europe: does investment in renewable energy matter? Sci. Total Environ. **792**, 148480 (2021). https://doi.org/10.1016/j.scitotenv.2021.148480
7. Baloch, R.M., et al.: Indoor air pollution, physical and comfort parameters related to school children's health: data from the European SINPHONIE study. Sci. Total Environ. **739**, 139870 (2020). https://doi.org/10.1016/j.scitotenv.2020.139870
8. Yi, B.-W., Zhang, S., Wang, Y.: Estimating air pollution and health loss embodied in electricity transfers: an inter-provincial analysis in China. Sci. Total Environ. **702**, 134705 (2020). https://doi.org/10.1016/j.scitotenv.2019.134705
9. Chen, Z., et al.: The association between high ambient air pollution exposure and respiratory health of young children: a cross sectional study in Jinan, China. Sci. Total Environ. **656**, 740–749 (2019). https://doi.org/10.1016/j.scitotenv.2018.11.368

10. Cao, W., et al.: NLRP3 inflammasome activation determines the fibrogenic potential of PM2.5 air pollution particles in the lung. J. Environ. Sci. **111**, 429–441 (2022). https://doi.org/10.1016/j.jes.2021.04.021
11. Zhang, X., Chen, X., Zhang, X.: The impact of exposure to air pollution on cognitive performance. PNAS **115**, 9193–9197 (2018). https://doi.org/10.1073/pnas.1809474115
12. Kinney, P.L.: Interactions of climate change, air pollution, and human health. Curr. Environ. Health Rep. **5**(1), 179–186 (2018). https://doi.org/10.1007/s40572-018-0188-x
13. Prospero, J.M., Lamb, P.J.: African droughts and dust transport to the Caribbean: climate change implications. Science **2003**(302), 1024–1027 (2003). https://doi.org/10.1126/science.1089915
14. US EPA, O. Particulate Matter (PM) Basics. https://www.epa.gov/pm-pollution/particulate-matter-pm-basics. Accessed 24 Jun 2021
15. Rajé, F., Tight, M., Pope, F.D.: Traffic pollution: a search for solutions for a city like Nairobi. Cities **82**, 100–107 (2018). https://doi.org/10.1016/j.cities.2018.05.008
16. Emissions of primary PM2.5 and PM10 particulate matter—European environment agency. https://www.eea.europa.eu/data-and-maps/indicators/emissions-of-primary-particles-and-5/assessment-3. Accessed 22 Jun 2021
17. Sapkota, A., et al.: Impact of the 2002 Canadian forest fires on particulate matter air quality in Baltimore City. Environ. Sci. Technol. **39**, 24–32 (2005). https://doi.org/10.1021/es035311z
18. Ng, B.F., Xiong, J.W., Wan, M.P.: Application of acoustic agglomeration to enhance air filtration efficiency in air-conditioning and mechanical ventilation (ACMV) systems. PLoS ONE **12**, e0178851 (2017). https://doi.org/10.1371/journal.pone.0178851
19. Sheng, C., Shen, X.: Simulation of acoustic agglomeration processes of poly-disperse solid particles. Aerosol Sci. Technol. **41**, 1–13 (2007). https://doi.org/10.1080/02786820601009704
20. Patterson, H.S., Cawood, W.: Phenomena in a sounding tube. Nature **127**, 667 (1931). https://doi.org/10.1038/127667a0
21. Song, C., et al.: Health burden attributable to ambient PM2.5 in China. Environ. Pollut. **223**, 575–586 (2017). https://doi.org/10.1016/j.envpol.2017.01.060
22. Reethof, G.: Acoustic agglomeration of power plant fly ash for environmental and hot gas clean-up. J. Vib. Acoust. Stress. Reliab. Des. **110**, 552–557 (1988). https://doi.org/10.1115/1.3269565
23. Riera, E., Gallego-Juárez, J.A., Mason, T.J.: Airborne ultrasound for the precipitation of smokes and powders and the destruction of foams. Ultrason. Sonochem. **13**, 107–116 (2006). https://doi.org/10.1016/j.ultsonch.2005.04.001
24. Riera-Franco de Sarabia, E., Gallego-Juárez, J.A., Rodríguez-Corral, G., Elvira-Segura, L., González-Gómez, I.: Application of high-power ultrasound to enhance fluid/solid particle separation processes. Ultrasonics **38**, 642–646 (2000). https://doi.org/10.1016/S0041-624X(99)00129-8
25. Zhou, D., Luo, Z., Jiang, J., Chen, H., Lu, M., Fang, M.: Experimental study on improving the efficiency of dust removers by using acoustic agglomeration as pretreatment. Powder Technol. **289**, 52–59 (2016). https://doi.org/10.1016/j.powtec.2015.11.009
26. Gallego-Juárez, J.A., Sarabia, E.R.-F.D., Hoffmann, T.L., Gálvez-Moraleda, J.C.: Application of the acoustic agglomeration to reduce fine particle emissions in coal combustion plants. Environ. Sci. Technol. **33**, 3843–3849 (1999)
27. Kilikevičienė, K., et al.: Experimental investigation of acoustic agglomeration of diesel engine exhaust particles using new created acoustic chamber. Powder Technol. **360**, 421–429 (2020). https://doi.org/10.1016/j.powtec.2019.09.057

Research of Automatic Rotational Frequency Control Systems of Automobile Diesel Engine

Anatolii Lisoval[1] and Alfredas Rimkus[2(✉)]

[1] Faculty of Automotive and Mechanical Engineering, Department of Engines and Thermal Engineering, National Transport University, Mykhaila Omelianovycha-Pavlenka Str. 1, Kyiv 01010, Ukraine
li-dvz@bigmir.net

[2] Vilnius Gediminas Technical University, J. Basanavičiaus str. 28, 03224 Vilnius, Lithuania
alfredas.rimkus@vilniustech.lt

Abstract. The purpose of the research is to compare the quality indicators of speed control of the automobile diesel engine with all-mode mechanical and microprocessor regulators in steady state conditions and in transient. Comparative motor research conducted on the same automobile diesel engine (125 kW, 4-cylinder, 120 mm cylinder diameter and 140 mm stroke). The object of experimental research is the automatic rotational frequency control systems of crankshaft diesel engine developed by the Authors. The results of the comparative analysis show that the parameters of PID controller (proportional-integral-differential) need to adjust according to the operating modes of the diesel engine. The stability of the rotational speed was achieved, developed by the microprocessor regulator at various steady speed modes, not exceed 5%. It was found that the values of the engine speed deviation and the duration of transients as good as using a serial mechanical regulator.

Keywords: Internal combustion engine · Diesel engine · Automatic rotational frequency control system · Microprocessor regulator

1 Introduction

Issues of development, implementation and operation of microprocessor control systems and diesel engine control remain relevant for the engine industry. Diesel engine reliability, fuel efficiency and emission concentrations depend on their proper tuning. Elemental base, digital information technology and, accordingly, software for such systems are constantly being improved [1]. The software for such control systems is the property of the company-developer. Electronic access keys of a certain level are transmitted to configure the operation of microprocessor control systems under operating conditions [2].

Automatic rotational frequency control systems (ARFCS) of the diesel engine are a mandatory part of the electronic control systems of the diesel vehicle. All-mode or dual-mode ARFCS are used on automotive diesel engines [3]. This depends on the conditions and operating modes of the diesel engine. Torque rise rapidly as

O. Prentkovskis et al. (Eds.): TRANSBALTICA 2021, LNITI, pp. 312–318, 2022.
https://doi.org/10.1007/978-3-030-94774-3_31

Turbocharged-diesel-engine engine speed increases above about 1,000 rpm, then remain relatively flat through the mid-speed range to close to the rated speed at which maximum torque is realized [4].

Many transport diesels engines are being converted to gas-diesels or pure gas engines in modern conditions [5]. Such measures to use more environmentally friendly gas fuel require a change in the setting of the microprocessor-based ARFCS accordingly [6].

For the creation and further research of new automatic systems of diesel engines and their gas modifications PID regulators are used because of their versatility [7, 8]. However, it is known that two-mode ARFCS in transport operation have better fuel efficiency indicators than all-mode regulators [3]. The PID controller can be used for coordinated control of the exhaust gas recirculation valve and intake valve phases in heavy duty diesel engines [9]. The PID controller has been installed on a standard electronic control unit of spark ignition engine, and experimental results on a test engine show that it meets the performance requirements in a wider range of operation [10].

New and improved ARFCS of diesel engines are investigated the theoretically and experimentally. The theoretical research includes mathematical simulation of ARFSS stability in steady-state and transient modes [11].

At the department of engines and thermal engineering of the National Transport University developed microprocessor dual-mode and all-mode diesel engines speed regulators. The dual-mode microprocessor regulator (MR) on microcontroller firm Microchip and an actuator of the proportional electromagnet was developed. Actuator operates on the rail of the fuel pump. This dual-mode speed MR was tested on a SMD-23.07 (125 kW, 4-cylinder, 120 mm cylinder diameter and 140 mm stroke) automobile diesel engine [12].

The all-mode MR designed by Heinzmann components – an electronic control unit and a rotary solenoid as an actuator. The electronic control unit and the actuator are shown in Fig. 1.

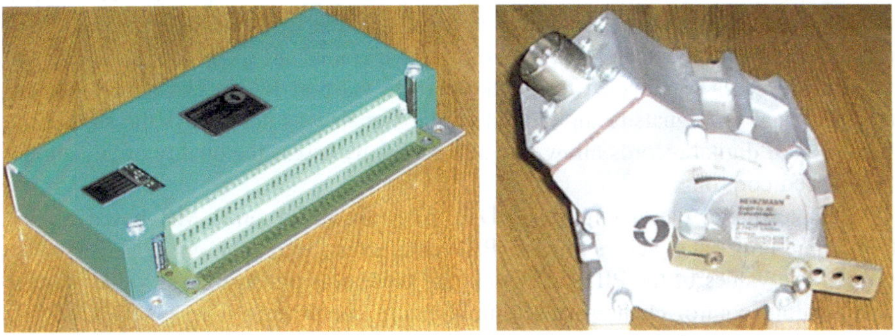

a) b)

Fig. 1. Components of the microprocessor regulator: a) electronic control unit; b) actuator.

In the electronic control unit, the PID speed control algorithm is implemented [13–15]. The all-mode MR on a diesel engine SMD-23.07 was tested, as well as on 6 and 8 – cylinders YMZ diesels engines (120 mm cylinder diameter, 130 mm stroke). The ultimate goal of researches of the all-mode MR was the development of universal recommendations for adjustment the PID parameters for an automobile diesel engines [14]. In order to ensure stable operation and high-quality transients of automobile diesel engine, an individual choice of the actuator and PID parameters is necessary taking into account their operating conditions. The effect of the PID parameters of MR of the diesel engine on fuel consumption is noticeable during transients. Fuel savings can be 2.25% when properly adjusted [12].

2 The Purpose and Methodology of Research

The results of the research presented in the article are the continuation of work on adjustment an all-mode MR for automobile diesel engine. The literature on the automatic speed regulation of a diesel engines basically describe the advantages and disadvantages of different type regulators [3].

The purpose of the research is to compare the quality indicators of the speed control of an automobile diesel engine with all-mode mechanical and microprocessor regulators at steady-state conditions and transients.

Motor comparatives researches performed on the same four-cylinder automobile diesel engine SMD-23.07 with turbocharger. The mechanical (serial) and the developed microprocessor control systems of a diesel engine with a rated power of 125 kW at 2,000 rpm are taken as a subject of experimental researches. Adjustments of all other diesel engine systems and mechanisms were the same.

3 The Results of Comparative Research

To compare the different types of speed controllers, only those engine test results that conducted under the same conditions and modes selected. For all types of diesel engines speed controllers, a digital recording of the crankshaft speed at steady state and transient processes was maded to a computer file via port R232. The degree of conversion of analogy signals to digital and digital signals was not always the same [12]. The results of digital records allow you allow visually assess the quality of adjustment of the speed control.

When testing a diesel engine SMD-23.07 with an all-mode MR based on the Heinzmann units, the values of the PID parameters were chosen: P = 5, I = 10, D = 12. The values of the PID parameters are present as dimensionless values for the corresponding control channels and amplification coefficients. These values of PID parameters are basic values.

The alignment controlling quality conducted at steady-state conditions of a diesel engine at 800 rpm (Fig. 2), 1,100 rpm (Fig. 3, a) and 1,800 rpm (Fig. 3, b) with regulators of various types. In the frequency range of 800…1,800 rpm, the all-mode

MR had the reference values of the PID parameters, and after 1,850 rpm P-component was increased 2.5 times.

With the stability of maintaining the rotational speed by the developed all-mode system MR does not exceed 5%. In steady-state modes, all-mode MR works no worse than the mechanical and dual-mode microprocessor regulators.

Transients were simulating on the motor-brake stand KS-56 by “momentary” movement of the fuel control lever from one extreme position to another. The rejection of the rotational speed during the transient process of the all-mode MR system was 1.3%, while the mechanical regulator was 2.3%. Transient time is 0.7 s in both cases.

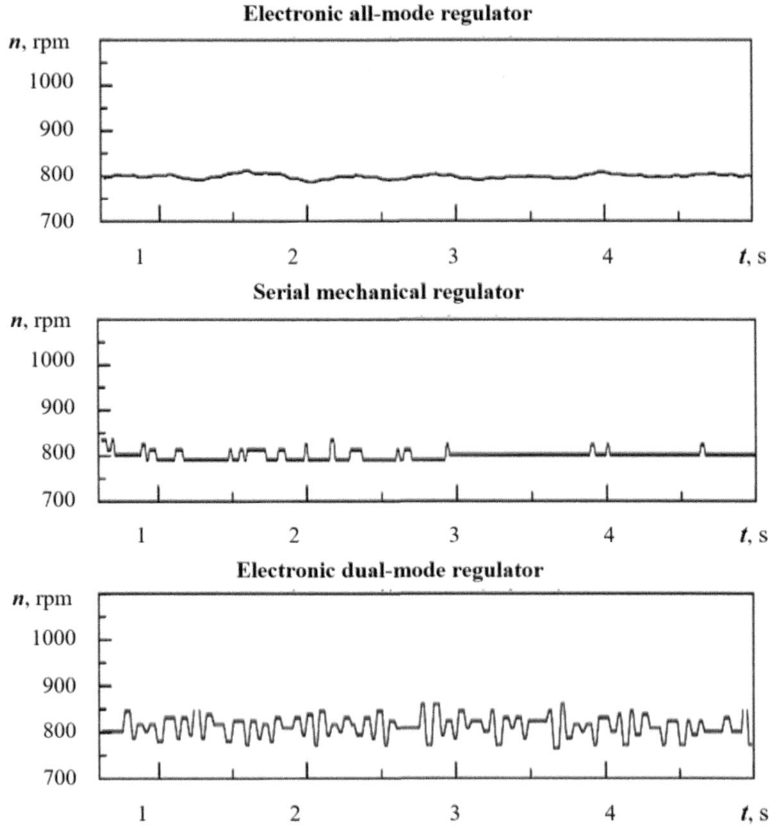

Fig. 2. Comparison of diesel engine operation SMD-23.07 with various regulators in idle mode 800 rpm: n – rotational speed of the engine crankshaft, t – time.

If the duration of the free acceleration mode on a SMD-23.07 diesel engine with a mechanical all-mode regulator does not exceed 5 s, then the rail of the fuel pump does not reabsorb the fuel supply. The same free acceleration mode was creating on the diesel engine with the all-mode system MR. During the free acceleration period, the MR kept the speed of fuel pump rail at 0.3 mm/s.

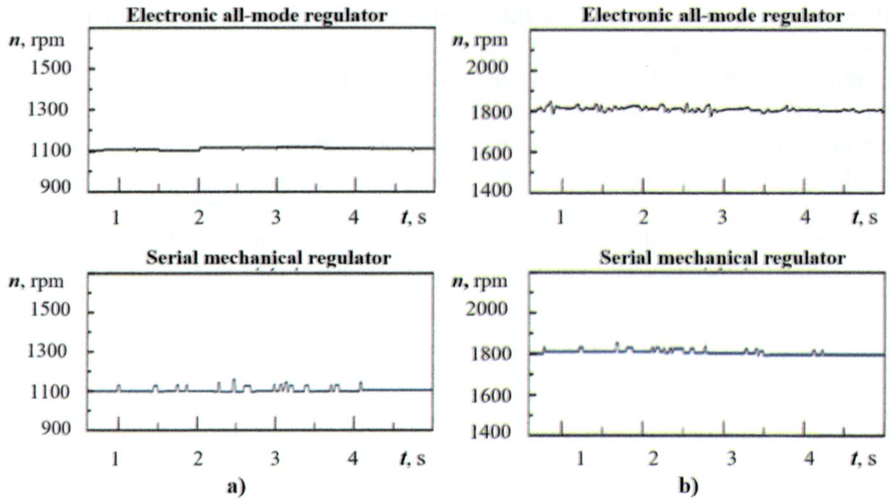

Fig. 3. Comparison of diesel engine operation with various regulators at: a) n = 1100 rpm, b) n = 1,800 rpm.

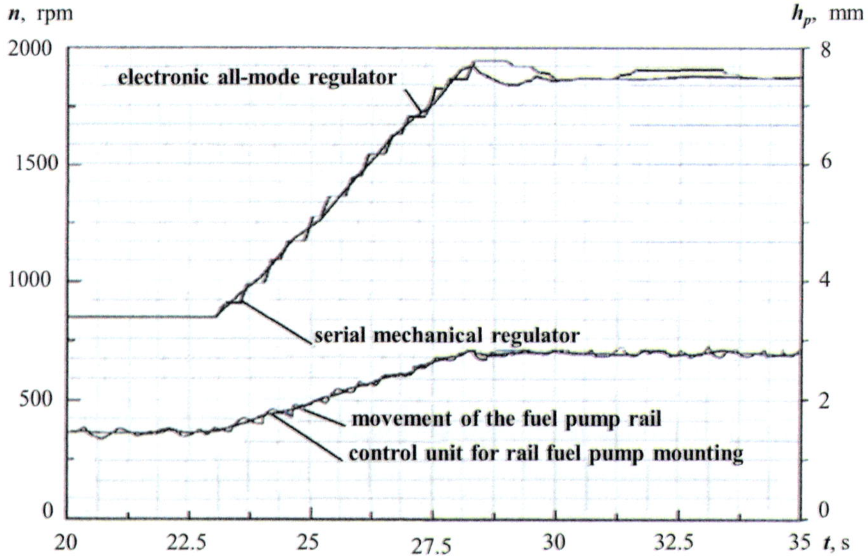

Fig. 4. Comparison of transients of a diesel engine with mechanical and electronic regulators by free acceleration mode: n – rotational speed of the engine crankshaft; h_p – movement of the fuel pump rail; t – time transition.

Figure 4 shows a comparison of transients of a diesel engine SMD-23.07 with a mechanical and MR (electronic) regulators by free acceleration mode. The results show satisfactory performance of the developed all-mode MR in dynamic modes.

4 Conclusions

An all-mode speed regulator for a diesel engine has been developed and tested, which based on Heinzmann components – a microprocessor control unit and a rotary solenoid as an actuator:

1. The test results showed that the parameters (proportional-integral-differential) and gain factors of the PID controller require individual adjustment in accordance with the operating modes of the diesel engine.
2. The results of comparative analysis show the satisfactory performance of the developed all-mode microprocessor controller in steady state and dynamic modes.
3. The developed all-mode microprocessor controller maintains the diesel engine crankshaft speed with an accuracy from 0% to 5%. Control quality indicators – values of speed overshoot and duration of transients, do not exceed similar values of the serial regulator.
4. The developed microprocessor speed controller can be using not only on diesel engines, but also on their gas modifications and gas-diesel engines.

References

1. Aleksandrov, A.A., Ivashchenko, N.A.: Encyclopedia, vol. 4. Internal combustion engines ["Entsiklopediya. Tom IV. Dvigateli vnutrennego sgoraniya"]. Machine Building, Moscow (2013)
2. Wargula, L., Krawiec, P.: The research on the characteristic of the cutting force while chipping of the Caucasian Fir (Abies Nordmanniana) with a single-shaft wood chipper. IOP Conf. Ser. Mater. Sci. Eng. **776** (2020). https://doi.org/10.1088/1757-899X/776/1/012012
3. Krutov, V.I.: Automatic control and management of internal combustion engines ["Avtomaticheskoe regulirovanie i upravlenie dvigateley vnutrennego sgoraniya"]. Machine Building, Moscow (1989)
4. Heywood, J.B.: Internal Combustion Engine Fundamentals, 2nd edn. McGraw-Hill Education, New York (2018)
5. Rimkus, A., Pukalskas, S., Matijošius, J., Sokolovskij, E.: Betterment of ecological parameters of a diesel engine using Brown's gas. J. Environ. Eng. Landsc. Manag. **21**(2), 133–140 (2012). https://doi.org/10.3846/16486897.2012.679661
6. Markov, V.A., Furman, V.V., Bebenin, E.V.: Improvement of diesel and gas-diesel engine speed control system ["Sovershenstvo-vaniye sistemy regulirovaniya chastoty vrashcheniya dizel'nogo i gazodizel'nogo dvigateley"]. Autogas Fuelling Complex + Altern. Fuel **4** (2016)
7. Denisenko, V.V.: PID-controllers: the issues of realization ["PID-regulyatory: voprosy realizatsii"]. Mod. Autom. Technol. **4**, 60–65 (2007)
8. Kutrubas, V.A., Sycheva, E.E.: Effective PID controller ["Effektivnyy PID- regulyator"]. Ind. ACS Controllers **5**, 60–65 (2013)
9. Wahlström, J., Eriksson, L., Nielsen, L., Pettersson, M.: PID controllers and their tuning for EGR and VGT control in diesel engines. IFAC Proc. Vol. **38**(1), 212–217 (2005). https://doi.org/10.3182/20050703-6-CZ-1902.01923

10. Kwiatkowski, A., Werner, H., Blath, J.P., Ali, A., Schultalbers, M.: Linear parameter varying PID controller design for charge control of a spark-ignited engine. Control Eng. Pract. **17**(11), 1307–1317 (2009). https://doi.org/10.1016/j.conengprac.2009.06.005
11. Markov, V.A., Pozdnyakov, E.F., Furman, V.V., Plakhov, S.V.: Simulation of the diesel engine rotational speed automatic control system. Proc. High. Educ. Inst. Mach. Build. **7** (712), 35–46 (2019). https://doi.org/10.18698/0536-1044-2019-7-35-46
12. Kostritsya, S.V.: The choice of rational parameters and the development of electronic speed regulator of the diesel engine ["Vibir ratsionalnih parametriv i rozrobka elektronnogo regulyatora chastoti obertannya dizelya"]. National Transport University, Kyiv (2014)
13. Heinzmann: Basic information on digital control ["Bazovaya informatciya po tcufrovomu upravleniu"]. DG 95–105, Kyiv (1997)
14. Heinzmann: Basic digital system Pandaros II ["Tcufrovue regulyatoru skorosty"]. DG 95-105, Kyiv (1997)
15. Heinzmann: Digital speed controllers ["Tcufrovue regulyatoru skorosty"]. DG 95-105, Kyiv (1997)

Influence of the Addition of Alcohol Compounds to Gasoline on the Performance of a Modern Spark Ignition Engine

Gintaras Valeika[1(✉)], Yuriy Gutarevych[2], Yevheniy Shuba[2], Jonas Matijošius[1], Oleksandr Dobrovolskyi[2], and Dmytro Ovchynnikov[2]

[1] Vilnius Gediminas Technical University, 10223 Vilnius, Lithuania
{gintaras.valeika, jonas.matijosius}@vilniustech.lt
[2] National Transport University, Kyiv 01010, Ukraine
{shuba90, dobrovolskyioleksandrntu, ovchynnikovdmytrontu}@i.ua

Abstract. The article presents the results of experimental studies of fuel efficiency and environmental indicators of an engine with a fuel injection system, feedback and a two-stage exhaust gas neutralization system running on gasoline with different content of alcohol compounds, in particular bioethanol. The study was carried out on an engine installed on a brake stand, equipped with measuring equipment, which made it possible to load the engine and determine the indicators in all possible operating modes. To assess the effect of the addition of alcohol compounds in a wide range of speed and load modes, the studies were carried out using three factorial experiments. The concentration of pollutants in the exhaust gases of the engine was determined after the neutralizer.

Keywords: Gasoline engine · Alcohol compounds · Bioethanol · Fuel efficiency · Environmental performance · Neutralization system · Feedback · Bench tests

1 Introduction

Automobile transport is the largest consumer of light oil products [1]. The main sources of energy for vehicles are internal combustion engines: spark ignition engines and diesel engines [2]. The constant increase in the number of cars requires an expansion of their fuel base [3]. Given the limited oil reserves and the environmental problem, one of the possible solutions to expand the fuel base is the use of fuels from renewable energy sources [4]. These fuels include biofuels. For spark ignition engines, the closest substitutes for regular gasoline are alcohol fuels, in particular bioethanol [5]. The use of alcohols as a motor fuel has been known for a long time, but only in recent years [6], due to the shortage of gasoline [7], bioethanol began to be widely used in spark-ignition automobile engines, mainly as an additive to gasoline fractions [8]. The performance of spark ignition engines using a bioethanol additive to gasoline has been investigated in many studies [9, 10]. But the amount of the additive was insignificant,

O. Prentkovskis et al. (Eds.): TRANSBALTICA 2021, LNITI, pp. 319–328, 2022.
https://doi.org/10.1007/978-3-030-94774-3_32

since during the operation of cars, gasolines with small amounts of the additive were widely used [10]. The amount of bioethanol added to gasoline is constantly increasing. Today, gasoline with the addition of alcohol compounds (up to 40%), based on bioethanol, is available at filling stations. This gasoline is actually a gasoline-alcohol mixture. One of the topical issues when using additives of alcohol compounds to gasoline is to determine the effect of their value on the fuel efficiency and environmental performance of a modern engine. The conducted experimental studies are aimed at solving this issue [11].

These studies [12–14] have shown that the content of NO_x in the engine exhaust gases is practically the same, the concentrations of CO and HC also differ insignificantly, although there is a tendency to their decrease when working on a gas-alcohol mixture. An increase in the concentration of acetaldehyde CH_3CHO is observed when operating on a mixture, the content of formaldehyde HCHO changes in different ways under different load conditions. As a result of the studies carried out, it was concluded that the use of ethanol compounds does not lead to a deterioration in the environmental performance of a spark ignition engine.

At the same time, some studies have established a significant effect of bioethanol additives in gasoline on the performance of a spark ignition engine [15, 16].

At partial load modes, where the internal combustion engine operates on lean mixtures, when switching to mixed gasolines, an increase in specific fuel consumption is also observed with practically unchanged NO_x concentrations and a decrease in HC concentrations in the exhaust gas [17].

Based on the results of experimental studies, it was established in the work that in the presence of vehicles with different adjustments in operation, it is advisable to limit the amount of bioethanol added to gasoline in an amount of 10% by volume [18].

As the analysis shows, studies to determine the performance indicators of gasoline engines with injection systems with feedback and a neutralizer in the exhaust system when operating on mixed gasolines with an alcohol content of more than 20% have been practically not carried out [19].

The main aim of this study is to determine how bioethanol alcohol affects the performance of a spark ignition engine by evaluating energy and environmental presentation.

Since the additive of bioethanol to gasoline is widely used, the value of this additive is constantly increasing. The main power systems of modern cars are injection systems, it is advisable to determine how the additive value affects the fuel efficiency and environmental performance of a modern engine. This determines the essence of the experimental research carried out.

2 The Methodology of the Experiment

The purpose of experimental studies is to determine the effect of the content of alcohol compounds in gasoline on energy, environmental and fuel efficiency indicators of the engine.

Experimental studies were carried out on an engine with an injection system and a feedback system on gasoline with an alcohol content of 0%, 9%, 18%, 27% and 36%.

The object of experimental research is a Volkswagen BBY spark ignition engine (Fig. 1), equipped with a fuel injection system and feedback and a two-stage exhaust gas after treatment system with accelerated warm-up and exhaust gas recirculation.

Fig. 1. Volkswagen BBY engine on the brake stand.

The technical characteristics of tested engine are shown in Table 1.

Table 1. Technical characteristics of the Volkswagen BBY engine.

Indicator name, units	Value
Engine volume, dm^3	1.39
Cylinder arrangement type	Inline
Number of cylinders	4
Piston stroke, mm	75.6
Cylinder diameter, mm	76.5
Compression ratio	10.5
Fuel injection system	Magneti Marelli 4MV
Power, kW	55 at 5000 rpm
Maximum torque, Nm	126 at 3800 rpm
Fuel	Gasoline 95 RON

Fuel for the experiment with different concentrations of alcohol compounds was obtained by mixing available gasolines from the network of 95 RON and 95 RON E40 filling stations. The physicochemical composition of 95 RON and 95 RON E40 gasolines was preliminarily determined (Table 2).

It is known that changes in engine operating parameters, such as torque or vacuum in the intake manifold, engine crankshaft speed and the amount of alcohol compounds added to gasoline, depend on the values of the main performance indicators of a gasoline engine. Namely - hourly consumption of fuel and air, concentrations of harmful substances in exhaust gases, excess air ratio, ignition timing. Therefore, in the future, for reasons of reducing the time and energy resources, it was decided to cover all the speed and load ranges of the engine by carrying out three factorial experiments.

Table 2. Physicochemical composition of gasolines 95 RON and 95 RON E40.

No.	The name of indicators	Units	Analysis results		Test method
			95 RON	95 RON E40	
1	Density at 20 °C	kg/m^3	737.3	749.2	GOST 3900–85
2	Mass fraction of aromatic hydrocarbons	%	35.0	15.8	ASTM D 6733–01
3	Mass fraction of benzene	%	0.72	0.75	ASTM D 6733–01
4	Volume fraction of oxygen-containing compounds - methanol - ethanol - isopropyl alcohol - isobutyl alcohol - tertiary butyl alcohol - ethers (C_5 i above) - other oxygen-containing compounds with a boiling end point not exceeding 210 °C	%	< 0,1 0 0 0 0 7.4 0	0 27.8 0 8.7 0 0.2 0	ASTM D 6733–01

During the experiment, sequentially, for each point, the value of the torque, engine crankshaft speed, and the percentage of alcohol compounds in gasoline were set. The following parameters were measured: time of consumption of fuel and air doses, vacuum in the intake manifold, temperature of oil and coolant, opening angle of the throttle valve and ignition timing, excess air ratio, concentration of CO, HC, CO_2, NOx, O2 in exhaust gases before and after the neutralizer.

The main indicators of the robots of the engine were measured as follows:

Fuel consumption was determined by the mass method (electronic scales MERA – VM-2/3, scale division – 0.5 g) by measuring the consumption time of the set fuel dose using an electronic stopwatch (TAKSUN-TS-613A, graduation – 0.01 s). The fuel dose was determined depending on the engine operating conditions, while the minimum measurement time was at least 30 s. The air flow rate was measured with rotary gas meters RG-100 (the scale division of the meter is 0.01 m^3). Depending on the operating mode of the engine, the air dose for measurement was set. Time was measured with an electronic stopwatch (TAKSUN-TS-613A, graduation 0.01 s) [22]. The metrological characteristics of the equipment are shown in Table 3.

Table 3. Metrological characteristics of instruments and equipment used in the experimental studies of the engine.

No.	The quantity that is measured	Dimension	Measuring	Measurement range	Measurement error
1	Crankshaft speed	rpm	Automated brake stand SAK-670	Up to 3,200	± 1 rpm
2	Torque	Nm	Dynamometric link with strain gauge	0…800	± 1 Nm
3	Fuel weight	kg	Electronic scales MERA VM - 2/3	0.0005…0.25 0.25…1.0 1.0…1.5 1.5…2.0 2.0…3.0	± 0.0005 kg ± 0.0010 kg ± 0.0015 kg ± 0.0020 kg ± 0.0030 kg
4	Air volume	Cubic meters	Rotary gas meter RG - 100	0…1	± 0.01 m^3
5	Time	Seconds	Stopwatch TAKSUN-TS-613A	0…180	± 0.01 s
6	Intake manifold vacuum	kPa	Exemplary vacuum gauge	−100…0	± 0.5 kPa
7	Ignition timing	Degree of crankshaft rotation	OBD II diagnostic interface with special adapter	−40…70	± 1°
8	Ignition timing	Degree of crankshaft rotation	Stroboscope SNAP – ON MT1241, scale	−40…70	± 1°
9	Coolant temperature	°C	Temperature sensor with thermistor	0…100	± 2 °C
10	Atmosphere pressure	kPa	Aneroid barometer	81…106	± 0.17 kPa
11	Air temperature	°C	Mercury thermometer		± 0.1 °C
12	Relative humidity	%	Hygrometer		± 2.5%
13	Excess air ratio	–	Gas analyzer Bosch BEA-060	0.5…9.999	–
14	Content of components in exhaust gas: CO CO2 O_2	%	Gas analyzer Bosch BEA-060	0…10% 0…18% 0…21%	± 0.03% ± 0.5% ± 0.1%
	HC calculated as hexane (C6H14) NOx	ppm	Gas analyzer Bosch BEA-060	0…9999 0…5000	± 10 ppm ± 25 ppm

The volumetric content of harmful substances in the exhaust gas and the excess air ratio were determined with a Bosch BEA 060 gas analyser.

3 Research Results and Their Analysis

Using the results of the experiment, mathematical models of the engine were obtained in the form of dependences of engine performance on engine speed and load mode (torque, power) for various contents of alcohol compounds in mixed gasolines. Figure 2a shows the engine load characteristics as the dependence of the engine performance on gasoline without the addition of alcohol compounds and the addition of 36% of the effective power for the engine speed of 2,300 rpm.

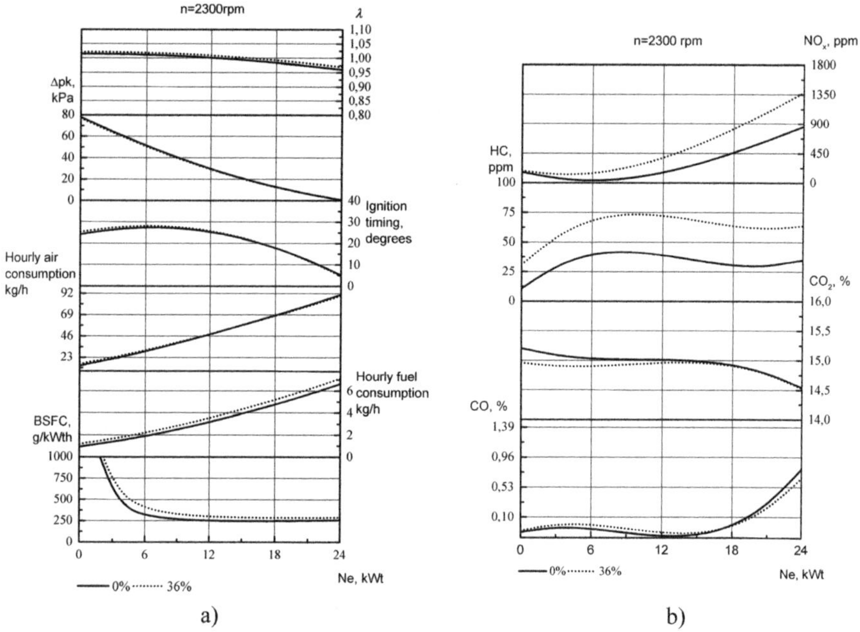

Fig. 2. Loading characteristics of the VW BBY engine.

The analysis of the characteristics shows that the feedback injection system ensures the operation of the engine when the fuel-air mixture is close to stoichiometric, although when operating on a gas-alcohol mixture, the excess air coefficient, in particular in the zone of high loads, is slightly higher. Figure 2b shows the dependences of the content of pollutants in the exhaust gases after the neutralizer for the same operating conditions of the engine.

This composition of the fuel-air mixture is maintained in all load and speed modes (Fig. 3). The same engine power in all load modes is obtained not by changing the position of the throttle valve (ΔPk is the same for both gasolines), but by the necessary change in the cyclic fuel supply, as evidenced by large values of the hourly fuel consumption at the same air flow rate (Fig. 2a, Fig. 4). The type of fuel has practically no effect on the value of the ignition timing (Fig. 5).

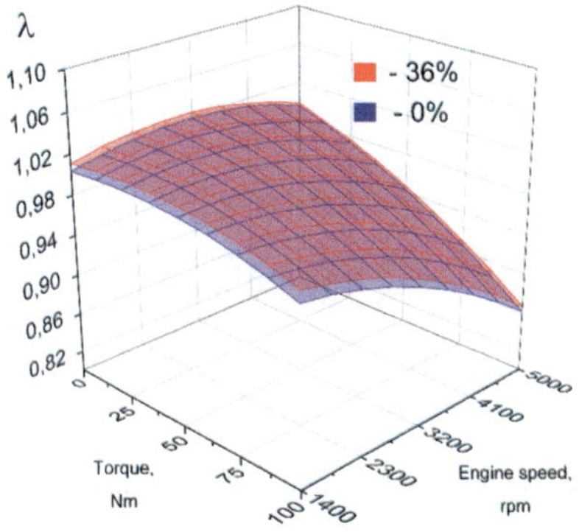

Fig. 3. Dependence of the excess air ratio of the VW BBY engine when running on A-95 gasoline and a gas-alcohol mixture.

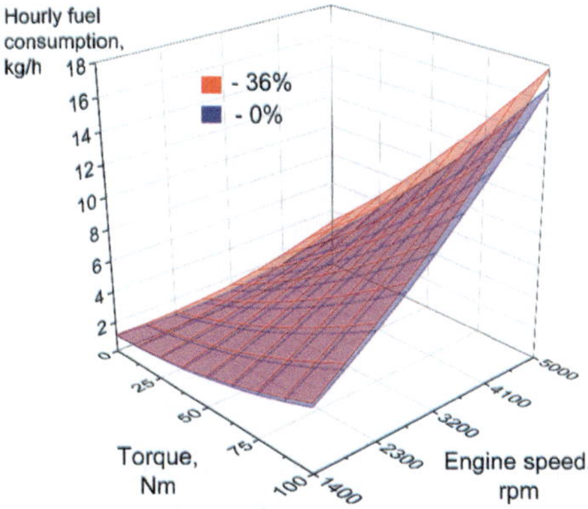

Fig. 4. Dependence of the hourly fuel consumption of the VW BBY engine when running on A-95 gasoline and a gas-alcohol mixture.

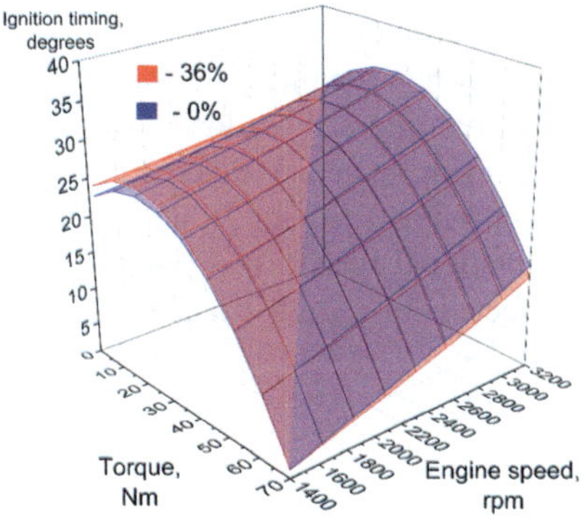

Fig. 5. Dependence of the ignition timing of the VW BBY engine when running on gasoline with different content of alcohol compounds.

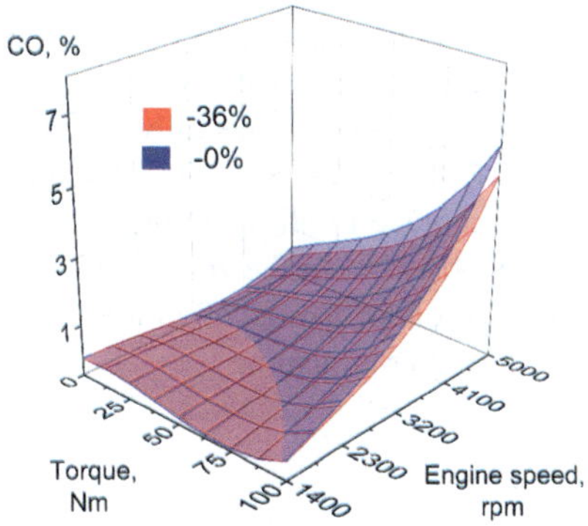

Fig. 6. Dependence of CO concentrations in the exhaust gases after the neutralizer the VW BBY engine when running on gasoline with different content of alcohol compounds.

Since the control system maintained the composition of the fuel-air mixture close to stoichiometric for all types of fuel, the patterns of changes in the concentrations of pollutants in the exhaust gases are the same. In this case, the concentrations of carbon monoxide are close in value (Fig. 2b, Fig. 5). At the same time, the concentration of hydrocarbons when working on a gas-alcohol mixture is much higher (Fig. 2b, Fig. 6), although their absolute value is insignificant (Fig. 8).

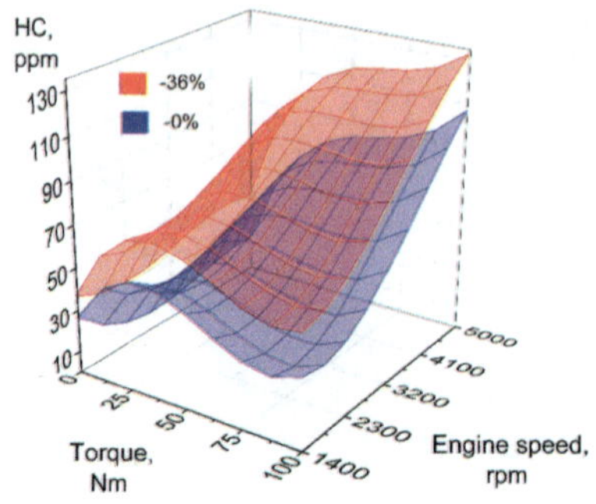

Fig. 7. Dependence of HC concentrations in the exhaust gases after the neutralizer the VW BBY engine when running on gasoline with different content of alcohol compounds.

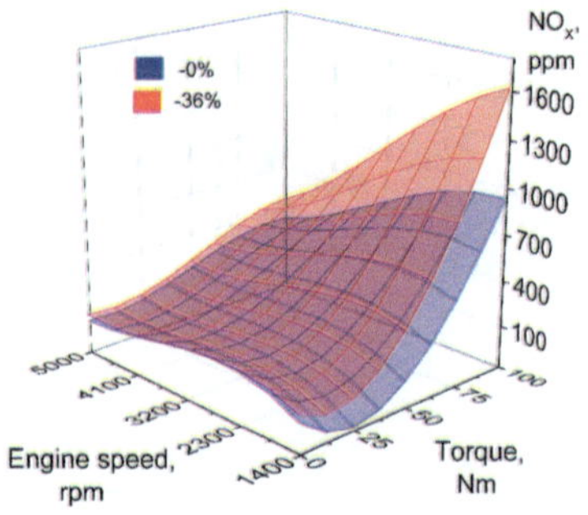

Fig. 8. Dependence of NO_x concentrations in the exhaust gases after the neutralizer the VW BBY engine when running on gasoline with different content of alcohol compounds.

This is due to the influence of various factors, in particular the chemical composition of the fuel. The concentration of nitrogen oxides when running on a gas-alcohol mixture is also higher, especially when the engine is running at high load, where, as noted above, the excess air ratio is slightly higher. An increase in emissions of nitrogen oxides takes place in all operating modes of the engine (Fig. 2b, Fig. 7), a significant increase is observed in the zone of high loads and rotational speeds, it is possible that the intensity of formation of nitrogen oxides is affected by a high concentration of oxygen during combustion.

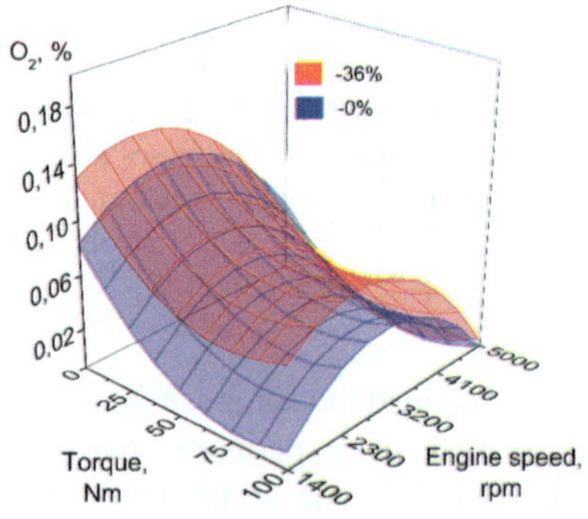

Fig. 9. Dependence of O_2 concentrations in the exhaust gases after the neutralizer the VW BBY engine when running on gasoline with different content of alcohol compounds.

Figure 9 shows the oxygen concentration in the exhaust gases after the catalytic converter in all possible high-speed and load modes of engine operation. It can be seen that the oxygen concentration is higher when working on gasoline-fuel mixtures.

4 Conclusions

1. From the data obtained on the operation of the engine with the injection system and feedback, it is seen that the use of gas-alcohol mixtures with a high content of alcohol compounds in such engines does not lead to significant changes in the operating processes of the engine.
2. The control system maintains the air-fuel mixture close to stoichiometric in all speed and load modes.
3. When replacing gasoline without alcohol compounds with a gas-alcohol mixture, an increase in fuel consumption is observed as a result of the low heat of combustion of the mixture.
4. Conversion of an engine with an injection system and a feedback system to a gas-alcohol mixture can lead to a slight depletion of the fuel-air mixture.
5. Improved fuel efficiency will lead to increased nitrogen oxide emissions. Adjusting the richer air/fuel mixture to reduce NOx emissions will result in lower fuel efficiency.

References

1. Alla, S.A., Bianco, V., Tagliafico, L.A., Scarpa, F.: Pathways to electric mobility integration in the Italian automotive sector. Energy **221**, 119882 (2021). https://doi.org/10.1016/j.energy.2021.119882
2. Bereczky, Á.: The past, present and future of the training of internal combustion engines at the department of energy engineering of BME. In: Jármai, K., Bolló, B. (eds.) Vehicle and Automotive Engineering. LNME, pp. 225–234. Springer, Cham (2017). https://doi.org/10.1007/978-3-319-51189-4_22
3. Zoldy, M., Hollo, A., Thernesz, A.: Butanol as a diesel extender option for internal combustion engines. In: Proceedings of SAE 2010 World Congress & Exhibition, April 12 (2010). https://doi.org/10.4271/2010-01-0481
4. Fuc, P., Lijewski, P., Sokolnicka, B., Siedlecki, M., Szymlet, N.: Analysis of exhaust gas flow through a particulate filter in the exhaust of the spark ignition direct injection engine. In Proceedings of the Computational Technologies in Engineering (TKI 2018); Military Univ Technol Warsaw, Fac Mech Engn, Dept Mech & Appl Comp Sci; EC Test Syst; SYMCOM, vol. 2078 (2019)
5. Alalwan, H.A., Alminshid, A.H., Aljaafari, H.A.S.: Promising evolution of biofuel generations. Subj. Rev. Renew. Energy Focus **28**, 127–139 (2019). https://doi.org/10.1016/j.ref.2018.12.006
6. Rimkus, A., Zaglinskis, J., Stravinskas, S., Rapalis, P., Matijosius, J., Bereczky, A.: Research on the combustion, energy and emission parameters of various concentration blends of hydrotreated vegetable oil biofuel and diesel fuel in a compression-ignition engine. Energies **12** (2019). https://doi.org/10.3390/en12152978
7. Çay, Y., Korkmaz, I., Çiçek, A., Kara, F.: Prediction of engine performance and exhaust emissions for gasoline and methanol using artificial neural network. Energy **50**, 177–186 (2013). https://doi.org/10.1016/j.energy.2012.10.052

8. Ershov, M.A., Grigoreva, E.V., Habibullin, I.F., Emelyanov, V.E., Strekalina, D.M.: Prospects of bioethanol fuels E30 and E85 application in russia and technical requirements for their quality. Renew. Sustain. Energy Rev. **66**, 228–232 (2016). https://doi.org/10.1016/j.rser.2016.07.054
9. Alptekin, E., Canakci, M., Ozsezen, A.N., Turkcan, A., Sanli, H.: Using waste animal fat based biodiesels–bioethanol–diesel fuel blends in a DI diesel engine. Fuel **157**, 245–254 (2015). https://doi.org/10.1016/j.fuel.2015.04.067
10. Thangavelu, S.K., Ahmed, A.S., Ani, F.N.: Review on bioethanol as alternative fuel for spark ignition engines. Renew. Sustain. Energy Rev. **56**, 820–835 (2016). https://doi.org/10.1016/j.rser.2015.11.089
11. Tutak, W.: Bioethanol E85 as a fuel for dual fuel diesel engine. Energy Convers. Manage. **86**, 39–48 (2014). https://doi.org/10.1016/j.enconman.2014.05.016
12. de Almeida, L.Q., Sales, L.C.M., Sodré, J.R.: Fuel consumption and emissions from a vehicle operating with ethanol, gasoline and hydrogen produced on-board. Int. J. Hydrogen Energy **40**, 6988–6994 (2015). https://doi.org/10.1016/j.ijhydene.2015.03.167
13. Alptekin, E.: Evaluation of ethanol and isopropanol as additives with diesel fuel in a CRDI diesel engine. Fuel **205**, 161–172 (2017). https://doi.org/10.1016/j.fuel.2017.05.076
14. da Costa, R.B.R., Valle, R.M., Hernández, J.J., Malaquias, A.C.T., Coronado, C.J.R., Pujatti, F.J.P.: Experimental investigation on the potential of biogas/ethanol dual-fuel spark-ignition engine for power generation: combustion, performance and pollutant emission analysis. Appl. Energy **261**, 114438 (2020). https://doi.org/10.1016/j.apenergy.2019.114438
15. Altın, İ., Bilgin, A., Sezer, İ.: Theoretical investigation on combustion characteristics of ethanol-fueled dual-plug SI engine. Fuel **257**, 116068 (2019) https://doi.org/10.1016/j.fuel.2019.116068
16. Li, Y., Chen, Y., Wu, G., Liu, J.: Experimental evaluation of water-containing isopropanol-n-butanol-ethanol and gasoline blend as a fuel candidate in spark-ignition engine. Appl. Energy **219**, 42–52 (2018). https://doi.org/10.1016/j.apenergy.2018.03.051
17. Li, Y., et al.: Combustion, performance and emissions characteristics of a spark-ignition engine fueled with isopropanol-n-butanol-ethanol and gasoline blends. Fuel **184**, 864–872 (2016). https://doi.org/10.1016/j.fuel.2016.07.063
18. Kiani, M.K.D., Rostami, S., Eslami, M., Yusaf, T., Sendilvelan, S.: The effect of inlet temperature and spark timing on thermo-mechanical, chemical and the total exergy of an SI engine using bioethanol-gasoline blends. Energy Convers. Manage. **165**, 344–353 (2018). https://doi.org/10.1016/j.enconman.2018.03.066
19. Gutarevych, Y., et al.: Improving fuel economy of spark ignition engines applying the combined method of power regulation. Energies **13**, 1076 (2020). https://doi.org/10.3390/en13051076
20. Unni, J.K., Govindappa, P., Das, L.M.: Development of hydrogen fuelled transport engine and field tests on vehicles. Int. J. Hydrogen Energy **42**, 643–651 (2017). https://doi.org/10.1016/j.ijhydene.2016.09.107
21. Heywood, J.B.: Internal Combustion Engine Fundamentals; Second edition.; McGraw-Hill Education, New York (2018). ISBN 978-1-260-11610-6.
22. Gritsuk, I., Volkov, V., Mateichyk, V., Gutarevych, Y., Tsiuman, M., Goridko, N.: The evaluation of vehicle fuel consumption and harmful emission using the heating system in a driving cycle. SAE Int. J. Fuels Lubr. **10**, 236–248 (2017). https://doi.org/10.4271/2017-26-0364
23. Fuc, P., Lijewski, P., Ziolkowski, A., Dobrzynski, M.: Development of a method of calculation of energy balance in exhaust systems in terms of energy recovery. In: Proceedings of the ASME International Mechanical Engineering Congress and Exposition, 2017 8. ASME (2018)

Evaluation of Fault Impact on Reliability Scenario of Ship Distribution Network by Statistical Analysis

Gintvilė Šimkonienė(✉) and Stasys Donėla

Lithuanian Maritime Academy, I. Kanto Street 7, 92123 Klaipėda, Lithuania
{g.maslauskaite, s.donela}@lajm.lt

Abstract. Nowadays, in accordance with all kinds of requirements, with the increasing dimensions of ships, electrical systems have become more complex. In the absence of proper maintenance more difficult to control, maintain and support the electric systems of vessels. There are a lot of models created to solve problems of electrical power systems to evaluate network reliability index. The article provides an overview of the concept of electrical reliability, analyses of reliability calculation models. Researchers present the weakness sides, the main problems of vessel electric system, which means it should receive more attention. A created structured model shows the main failures, evaluate the location, cause of failure, may affect the complete liquidation of the device, system breakdown its damage. Considering the operating time of device, its location, causes of its failure or not, installation, and deterioration lifetime, it is possible to improve reliable operation of the electrical system and network, in conjunction reducing the probability of failures and network chain failure reactions.

Keywords: Reliability · Reliability indices · Ship distribution network · Fault · Reliability calculation methods

1 Introduction

Nowadays, the need for electricity is very important. But even more important is the uninterrupted and reliable operation of the entire electrical system. Reliable supply of electricity is and will be an essential factor to produce electric energy without interruption, with an acceptable degree [5] of reliability. Reliability and security of electricity supply is the main factor to ensure that the ship's electrical equipment and electrical system components works adequately, and it perform their intended functions. Faults, theirs rate, electrical stress, mechanical stress, temperature, short circuits, overvoltage, disconnections, and force majeure has big influence for the network reliability and lifetime of network components and all of system. Electric energy reliability concern with appropriate equipment isolation, rapid response fault removal and reaching avoid all of this.

Reliability is currently the subject of numerous studies, which interested many scientists and researchers around the world [2, 18, 21]. Reliability plays important roles in the normal operation [8, 13] because, in the event of a loss (of electricity supply, the

O. Prentkovskis et al. (Eds.): TRANSBALTICA 2021, LNITI, pp. 329–342, 2022.
https://doi.org/10.1007/978-3-030-94774-3_33

ship's machinery is unable to perform its intended functions, even in the absence of full operation of the ship. Mathematically, reliability can be defined as probability that part of equipment or system will function in normal operation over time [14]. Reliability shows the overall relationship between power system quality and safety [19].

Electric network reliability also potentially significant adverse effects not only the equipment but also for people who is working with electrical equipment on board. The use of any electrical equipment carries a risk of potentially dangerous electric shock or even death. Anyone working with live parts can come into direct or indirect contact with electrically conductive materials and be injured. Electrical networks: More than 50 V of alternating current (AC) or 120 V of direct current (DC) are considered hazardous.

All kind of regulations and laws require that the supply of electricity should be ensuring the quality and continuity of service [11]. All Statutory regulations are based on the experience and the following basic precautions which can help to avoid people deaths, danger for equipment, injury and damages. Scientists creating methods how to improve network reliability, how to show the weakness nodes, parts of network, calculating and trying to be able to anticipate before a fault occurs, to predict it and to eliminate perishable parts of the network in a timely manner.

To minimize the safety risk of engineers and equipment, all electrical system should be created, designed, and installed to comply with all international standards and requirements. This is necessary to minimize the potential adverse effects of electricity, the risk to the engineer working, the equipment, and the electrical system.

2 Reliability Literature Review

Electric network reliability is defined as a device suitability, functionality, continuous working capacity, meeting all operating characteristics within the specified frame of time. Reliability is the probability of a component or system to operate as expected or not [7]. Also, this factor of the efficiency and operability of the electrical system associated with sustained and momentary supply interruptions [5]. The frequency and duration of the device failure [10] are self–contained indicators that measure the reliability of a device or system.

Known that all devices have reliability impact if it is working correct and it is good exploitation. The reliability of the device also depends on the operation of the whole system, operating modes, voltage and current level, compatibility of devices, principle of operation, device manufacturer, and so on. Equipment failure and its rate effect of service life not even just for damage device but could be for the entire system.

The reliability problem become so important when on board electricity become the main source to work for the main engines, cooling, and ventilation systems, to meet the needs of people, during technical works. The reliability of the electrical system is very important depending on the type of vessel. Reliability has a direct effect on the outage costs and even power outage is for several moments, may cause huge damage. Outage costs means to damage equipment or system.

Vessel electrical network consists of many individual elements which the capacity has influence of not interrupted power supply to electric network customers (pumps, electric machines etc.). However, when any one device is not working in normal conditions and its influence may affect other elements that are dependent on the failing unit not working.

There are a lot of methods, network configurations, involving automations devices that improves network reliability [3, 5]. To know which method is useful and would be the best for ship electrical system, it is necessary to find out what are the main methods are useful to assess the reliability of the electric network to prevent further damage progress. To know what types of failures that have the greatest impact on power outages and lack of power supply, complete damage to the power grid.

3 Factors Affecting Electric Network Reliability

Reliability of electric network is influenced by many external and internal factors [20]: overvoltage, overloads, damages, not correct projecting cable lines configurations, incorrectly distributed power, affection from nature, cable laying not adequately operated monitoring system are the main factors affecting the ship electric distribution network. The failure of conventional power equipment is one of the main causes of the power outage. A lot of research analyzing the major faults phenomena: lightning, short circuit, vibration, temperatures, human failures – such as the major cause of outages on power distribution lines. Most of the outages are unexpected disturbances or other interruptions which can lead to the blackout of the whole distribution network. It can damage line connected equipment such as instrument transformers, asymmetry surge, arresters, shunt reactors and circuit breakers, disconnect the main ship customers.

Outages in the network show weakness of the ship electric grid. Voltage dips, flickers and swells cause the fluctuations in electric low voltage network. Faults and its duration can show the weakness sides of ship electric network. Unexpected factors may occur, which may damage the electric network or parts of it. When a fault occurs in the electrical system it damages both thermally and mechanically. These factors can cause fires, explosions, and the details of the equipment melt. The main factors which affect the ship's electrical network are shown in Fig. 1. This model shows how the types of fault affect the reliable operation of the electrical system.

Ships electric network reliability and quality of service will be ensured only by investing in worn-out electrical equipment. Shipping companies can suffer huge losses due to a power outage. Each outage or disconnection event should be recording in the system in a separate record with the following primary data (planned and unplanned disconnection):

1. fault location;
2. date and time;
3. reason;
4. interruption duration;
5. number of users, affected due to fault;
6. number of disconnections between transmission and distribution systems affected;
7. environmental impact.

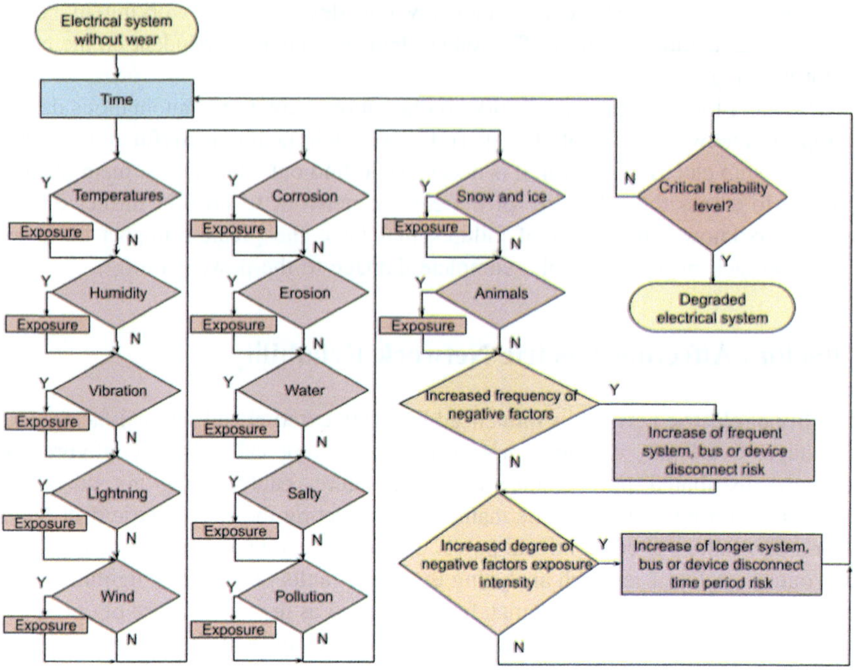

Fig. 1. Factors of ship's electricity distribution system degradation.

Reliability and quality of service will be guaranteed only by investing in old electrical equipment. Allow only those persons who have special education and permission, permits to work with electrical system. Even for the smallest fault without electricity for several or even several hours, the vessel may suffer very large losses of downtime.

4 Reliability Index Computation

4.1 Distribution Network Reliability Evaluation

Reliability is the probability that an electrical system or equipment will operate without failure for some time. The system reliability usually is evaluating of each year with annual indexes which will be taken up and the ships electrical side. By the time to failure of an item it is mean the time elapsing from when the item is put into operation until it fails for the first time. Assuming that the starting point is 0. The time to failure is a random variable T. Functions shown in Fig. 2 depict how failure rate change with the amount of time.

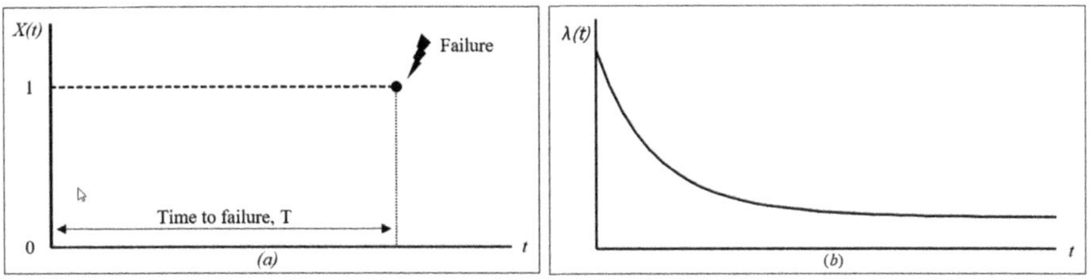

Fig. 2. Examples of hazard functions: (*a*) operation level of device before failure; (*b*) aging after failure (negative aging).

The time of operation of the device until its failure is random and depends on the random nature of the electrical resistance of the insulation, structural parts, material, the influence of environmental factors (ambient temperature, pressure, humidity, radiation, mechanical disturbances, etc.) and random surges. The operation time of device from installation or regular repair to failure, after which additional repair would be required, is divided by (gamma) exponential statistical law, the section function of which has the expression:

$$F_T(t) = 1 - \exp\left(-\frac{t}{\tau}\right), \quad (1)$$

where T and τ random and average operating time; t – independent function argument (e.g., one year). This method called proportional hazard method [16].

Reliability value is the probability of an event that the actual operating time will be higher than the set (agreed) value t:

$$p = 1 - q = P(T > t) = 1 - F_T(t), \quad (2)$$

where q – unreliability size.

Equivalent failure rate component can be determining:

$$\lambda_{eq} = \sum\nolimits_{i=1}^{n} N_i\left(\lambda_{flri}\pi_{Qi}\right), \quad (3)$$

where λ_{flri} – simple failure rate of the i component; π_{Qi} – quality coefficient of i component; N_i – of i component number; n – number of component types. There are three basic load-point reliability indices used to predict the reliability are the average load point failure rate (l_s), the average load point outage duration (r_s), the average annual load point unavailability (U_s).

$$\lambda_j = \lambda_{sj} + \sum\nolimits_{i=1}^{n} \lambda_{ij} + \sum\nolimits_{k=1}^{n} p_{kj}\lambda_{kj}, \quad (4)$$

$$U_j = \lambda_{sj}r_{sj} + \sum\nolimits_{i=1}^{n} \lambda_{ij}r_{ij} + \sum\nolimits_{k=1}^{n} p_{kj}\lambda_{kj}r_{kj}, \quad (5)$$

$$r_j = \frac{U_j}{\lambda_j}. \tag{6}$$

Failure rate of network components can be calculated as a weight using the following equation:

$$\lambda_a = W_a \sum_i W_{F,i} \cdot \lambda_{F,i}, \tag{7}$$

where p_{kj} – the control parameter of lateral section k which depends on the fuse-operating model; λ_{ij}, λ_{kj}, λ_{sj} – the failure rates of the main section i; r_{sj},r_{ij}, r_{kj}– the outage durations (switching time or repair time) for the free elements, respectively.

There are two dimensions which are used usually to say about reliability: frequency of interruption – how often or how many interruptions customers (or devices) experience (SAIFI, SAIDI or MAIFI).

System average interruption frequency index:

$$SAIFI = \frac{number\ of\ customer\ interruptions}{total\ number\ of\ customers\ in\ system}\ during\ the\ period \tag{8}$$

This is the number of customer interruptions per year, divided by the total customers on the system.

System average interruption duration index:

$$SAIDI = \frac{sum\ of\ duration\ of\ all\ customer\ interruption}{total\ customers\ in\ system}\ during\ the\ period \tag{9}$$

Momentary average interruption index (MAIFI):

$$MAIFI = \frac{number\ of\ customer\ momentary\ interruption}{total\ customers\ in\ system}\ during\ the\ period \tag{10}$$

In regulated environments, the duration threshold is set by the regulator and is a key measure for counting duration-based interruptions, resulting in the application of penalties. These indices indicate the annual average performance of the network in terms of interruption frequency and duration. They are weighted by the number of customers or energy supplied and is either presented on a system wide or customer basis [15].

The reliability of ship distribution network is assessed by the availability and quality of electrical network. To protect the power grid as effectively as possible, scientists offer many variations that could protect against frequent power outages.

4.2 Model Analyses of Ship Distribution Network

For comparative reliability state analysis of the situation is possible when interruption and reliability indicators should be collecting, sorting, systematizing in the same or similar principles. In these days there are no accumulate all historical data interruptions and interruptions are not reliable and representative in the data.

There are a lot of methods to simulate electric network reliability, estimate the probability of failure. Methods vary in complexity and the number of benchmarks. The influence to choose the right model has a setting which parts of the network estimated, evaluate. To choose the right method, it is important to understand the underlying reasons that affect the electric network reliability or unreliability of the electrical power supply. It is important to understand what the basic devices running one or another condition, which occur faults.

To simulate reliability usually applied analytical or strategic approach, Weibul distribution "Bath bulb" [15], Markov and Monte Carlo simulations. Also, reliability can be calculate using different hierarchical levels method [17].

One of the most popular is the topological matrix, which is based on the Monte Carlo method. This method evaluates the electrical component of the network's dependency on each other for work or damage. Monte Carlo simulation (stochastic modeling method) the most widely used approach for calculating failure probability and reliability assessment [6, 14]. This modeling method is simple compared to other methods [1, 11, 19]. One of the most disadvantages of this model is: require a long computation point while using analytical part requires considerably less computation.

So, what kind of method to choose – depends on network topology and depending on what is achieved.

5 Results of Experimental Research

5.1 Reliability Evaluation

Failure analysis and prognosis are very important tools for the network reliability. There are two kinds of interruptions: short interruptions and long interruptions. It depends on fault and difficulty of failure. Faulty network and its components can also act other components surroundings. Faults should be collecting at the same time to evaluate all networks and its components data: fault type; types of cable, transformer, generator, commutation apparatus; lifetime, exploitation age; how many times faulty; voltage; power; weather conditions; vibrations; temperatures. For the power quality of ships has influence bad quality of device, network construction, materials also human factors: how with electric network and its components working, how following the rules.

7 ships of one company were evaluated for the study of the article. A principal scheme (Fig. 3) was chosen for the study, which will be used to assess the reliability of the network.

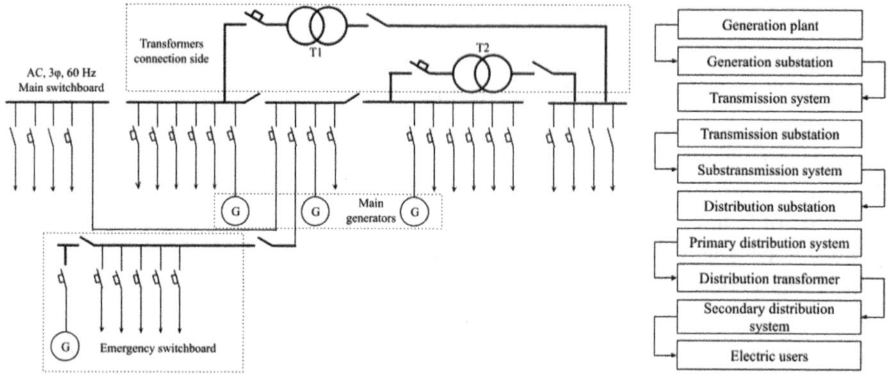

Fig. 3. Ship distribution network.

The statistics shows that ships MV (440 V–690 V) has the highest impact on reliability, because more than 90% dominated by the power supply disconnected failures.

Based on real data from ship company in 2018–2019, made table with the main factors which were recorded. Took one year, all faults are rated in 100%, the percentage size is distributed according to the event. The data of failures are converted to percentages and shown in Table 1. Known that 7% of all outages was in High Voltage side (above 1,000 V) and 93% of all outages was in lowest side (until 690 V).

Table 1. Faults distribution – based on ship company real data case.

Fault name	Percentage of all failures, %
Environmental factors (nature conditions etc.)	17
The same parameters having devices but created of different companies (internal characteristics are different)	21
The human factors	30
The period of installation and capacity of load, lifetime	5
Load	12
Force majeure	15
Total:	**100**

Each termination event is recorded in the system as a separate item, a reference to those primitives (planned and unplanned interruption): termination systems in place; date and time; termination of the cause and its group (only unscheduled interruptions); termination of the nature of the termination of duration; the number of users affected by the termination; termination of affected joints between the transmission and distribution systems number; the impact on the environment. It is also influenced by the accuracy of the indicators of statistical data points.

From table data we see that this company for ships using relatively few estimates. It is not enough to know about the basic problems of electrical distribution network. From the data know that the weakness ships side is Medium Voltage (until 690 V) (see Fig. 4).

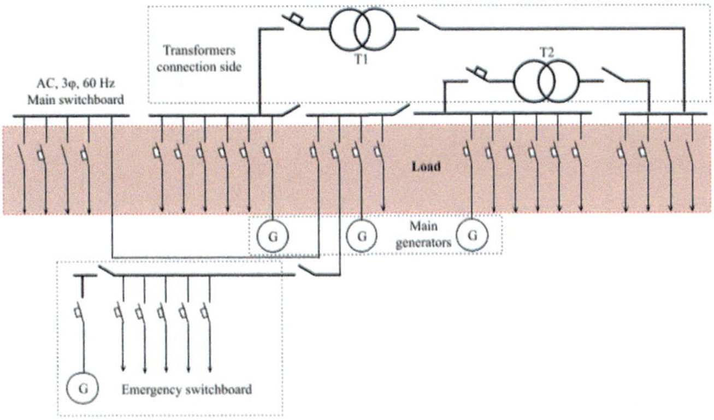

Fig. 4. Ship distribution network weakness sides.

Electric network's reliability also depends on network topology, component's information. All networks can be shown as nodes or divided into the block schemes (Fig. 6). Ship electrical network is meshed network and, switchboard mostly is divided more than into two sections (Fig. 3). Figure 3 show that all systems (generation, transmission, distribution) and its elements are very closed linked. If something happen to the main generators electric energy could be supplied from an emergency generator or battery secured side. Electric network configuration has big influence in electric supply reliability. Configuration and network type depends also on the ship type (tanker, passenger ship, ship with electric propulsion etc.).

5.2 Ships Electrical Network

To evaluate reliability of network can be estimated using created model (Fig. 5).

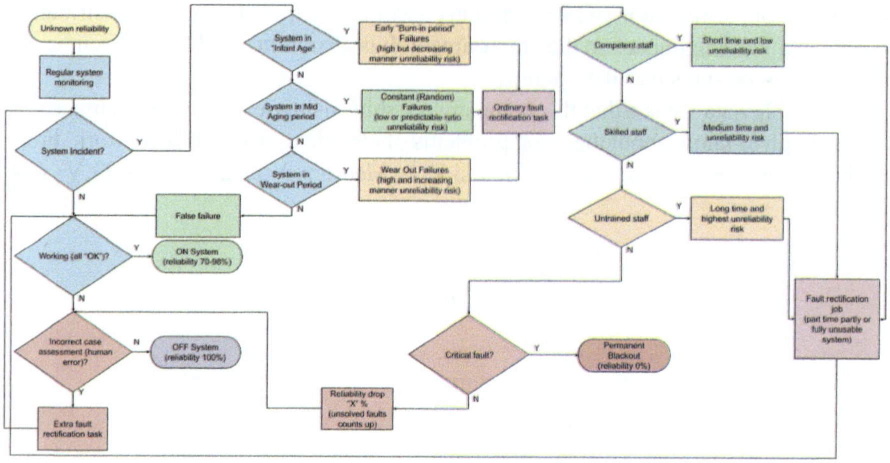

Fig. 5. Reliability model.

When selecting the model, it is acceptable that the distribution network operates normally, without any failures. The main purpose of network reliability evaluation models is to identify the main factors that have the most negative impact on the stability, for operation when the electrical network is not operating in the normal operating state, therefore, requires a quotative assessment of network reliability by assigning a user as an attribute value. Reliability assessment models are based on equipment failure rate (intensity), recovery time, scheduled and unscheduled repair times, average scheduled shutdown frequency, and so on.

The suitability of the evaluation technique and model, the abundance of the evaluated indicators, the evaluation of the respective system, the suitability and quality of their input are the main and most important aspects in choosing the most appropriate solutions and the number of models.

Quantifying the reliability of ships distribution networks (Fig. 6) has now evolved into a set of defined reliability indices that are recognized throughout the industry. Calculating the reliability performance of a distribution network is relatively simple but the quantity of data for any network other than extremely small ones means that a software solution is the only practical way. Nevertheless, it is vitally important that the principles of the calculation be understood by the utility because it is always wise to be able to check that the software calculations agree broadly with the experience of engineers and their quick calculations. The major advantage of using software for the calculation is that it is simple to study the effect of changes in the network.

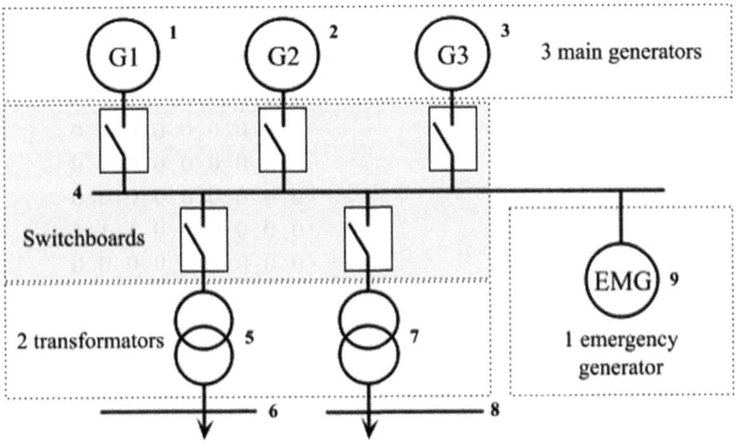

Fig. 6. Ship electric scheme with grouping elements.

Making calculation using node scheme (Fig. 6) showing in the Table 2 when electric energy working with normal condition. Making matrices with numbers: 2 – this element is working; 1 – element is working when 2 numbering element is working; 0 – element is not working. The matrix of elements (devices) states shown in table 2 whereby numbering by row corresponds of all scheme or part of it states, numbering by column – connection method. Topological matrix describes scheme in Fig. 6, which is divided into buses and numerical values. Buses present relationship of electrical system devices, connection, and continuous operation of the system.

Table 2. Electrical network working in normal condition, all generators working constantly operating conditions.

Links of scheme	Buses of the network	Connection method								
		1	2	3	4	5	6	7	8	9
Link 1	1	2	1	1	1	1	1	1	1	0
Link 2	2	1	2	1	1	1	1	1	1	0
Link 3	3	1	1	2	1	1	1	1	1	0
Link 4	4	1	1	1	2	1	1	1	1	0
Link 5	5	1	1	1	1	2	1	0	0	0
Link 6	6	1	1	1	1	1	2	0	0	0
Link 7	7	1	1	1	1	0	0	2	1	0
Link 8	8	1	1	1	1	0	0	1	2	0
Link 9	9	0	0	0	0	0	0	0	0	0

Table 3 represent state the scheme when is vessel in black out, whole system is not working, turning emergency generator and electricity supplies only to the main electricity customers. Calculation results presents in Fig. 7.

Table 3. Working just emergency generators.

Links of scheme	Buses of the network	Connection method								
		1	2	3	4	5	6	7	8	9
Link 1	1	0	0	0	0	0	0	0	0	0
Link 2	2	0	0	0	0	0	0	0	0	0
Link 3	3	0	0	0	0	0	0	0	0	0
Link 4	4	0	0	0	0	0	0	1	1	1
Link 5	5	0	0	0	0	0	0	0	0	0
Link 6	6	0	0	0	0	0	0	0	0	0
Link 7	7	1	1	1	1	0	0	2	1	1
Link 8	8	1	1	1	1	0	0	1	2	1
Link 9	9	1	1	1	1	0	0	1	1	2

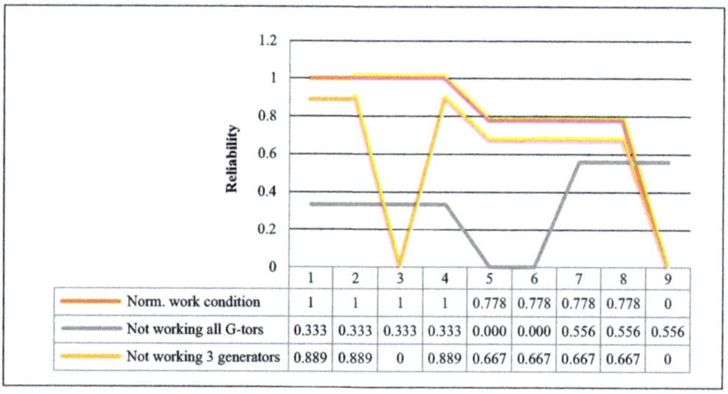

Fig. 7. Comparing of calculation results.

The results of calculation are shown in Fig. 7. Results show that reliability strongly depends how electric network is constructed. Thus, the design and reconstruction of the power grid, it is important to investigate and analyze reliability. A reliability analysis will make it possible to assess the extent to which the system meets the performance criteria, to quantify possible variations and to help make an economic decision. The obtained results of the reliability analysis will allow to determine the frequency of failures of a specific equipment or group of devices, the average duration of failures. The information obtained and analyzed when describing and analyzing the databases can be important for the distribution network operators: the most common causes of failures are given; areas where there is insufficient load distribution; the weakest areas of protective equipment; what contributes to power outages and failures.

6 Conclusions

To ensure more reliable and secure electric energy supply to the main devices of ships, it is necessary to choose program, method or algorithm to increase reliability, electricity efficiency of electricity distribution network. To ensure the safety of electricity facilities for the environment, environment requirements, at the same time considering the needs of consumers. Main failures in the weakness parts of the network can be avoided, increase reliability and quality of service will be ensured only by investing in worn-out electrical equipment. Device modeling is a very important factor that affects the reliability of the whole system. The modeling should be as simple as possible, but at the same time empowered and accurate in capturing and assessing all possible risks. Only by understanding the model and its level of limitations can the results be interpreted correctly without fetishizing them. Reliability indexes and it is using can show the weakness places of ships electric network.

References

1. Brown, R.E.: Electric Power Distribution Reliability, 1st edn. CRC Press, Boca Raton (2002)
2. Chernenko, I.V.: Effect of switchgear failures in calculations of structural reliability of power supply circuits at industrial facilities. In: 2018 International Conference on Industrial Engineering, Applications and Manufacturing (ICIEAM), pp. 1–4. IEEE, Moscow, Russia (2018)
3. Falaghi, H., Haghifam, M., Tabrizi, M.R.O.: Fault indicators effects on distribution reliability indices. In: CIRED 2005 - 18th International Conference and Exhibition on Electricity Distribution, pp. 1–4, IET, Turin, Italy (2005)
4. Ge, S., Xu, L., Liu, H., Zhao, M.: Reliability assessment of active distribution system using monte carlo simulation method. J. Appl. Math. **2014**, 1–10 (2014)
5. Giron, C., Rodriguez, F.J., Urtasum, L.G., Borroy, S.: Assessing the contribution of automation to the electric distribution network reliability. Int. J. Electr. Power Energy Syst. **97**, 120–126 (2018)
6. Haakana, J., Kaipia, T., Lassila, J., Partanen, J.: Simulation method for evaluation of the challenges in the reliability performance of medium-voltage networks. In: 17th Power Systems Computation Conference (PSCC), Stockholm, Sweden (2011)
7. Heydt, G.T.: Future renewable electrical energy delivery and management systems: energy reliability assessment of FREEDM systems. In: IEEE PES General Meeting, pp. 1–4. IEEE, Minneapolis, USA (2010)
8. Huang, H., Zhang, H.: Application and analysis of long-distance EPON in transmission lines monitoring system. Adv. Mater. Res. **317–319**, 1583–1589 (2011)
9. Jianping, G., Guopei, W., Zhenhai, L.: Power supply reliability analysis and improvement measures of regional power grid. In: 2008 China International Conference on Electricity Distribution, pp. 1–7. IEEE, Guangzhou, China (2008)
10. Lakehal, A., Chelli, Z., Djeghader, Y.: A hybrid Bayesian network based method to as-sess and predict electrical power network reliability. In: 2019 4th World Conference on Complex Systems (WCCS), pp. 1–5. IEEE, Ouarzazate, Morocco (2019)

11. Liu, H., Hu, H., Huang, J.: A method of reliability evaluation for complex medium-voltage distribution system based on simplified network model and network-equivalent. In: 2008 China International Conference on Electricity Distribution, pp. 1–6. IEEE, Guangzhou, China (2008)
12. Liu, H., Wang, Z., Han, M.: Reliability analysis of solenoid valve power supply based on time-varying fault rate. In: 2019 4th International Conference on Mechanical, Control and Computer Engineering (ICMCCE), pp. 154–1543. IEEE, Hohhot, China (2019)
13. Liu, Z., Wang, Y., Li, W., Xu, S., He, W., Sun, K.: A planning method of clustering ONUs based on reliability. In: 2017 16th International Conference on Optical Communications and Networks (ICOCN), pp. 1–3. IEEE, Wuzhen, China (2017)
14. Narimani, A., Nourbakhsh, G., Ledwich, G.F., Walker, G.R.: Impact of electric energy storage scheduling on reliability of distribution system. In: 2015 IEEE 11th International Conference on Power Electronics and Drive Systems, pp. 405–408, IEEE, Sydney (2015)
15. Pasha, G.R., Khan, M.S., Pasha, A.H.: Empirical analysis of the Weibull distribution of failure data. J. Stat. **13**(1), 33–45 (2006)
16. Pylvänäinen, J., Verho, P., Järvinen, J., Kunttu, S., Sarsama, J.: Advanced failure rate and distribution network reliability modelling as part of network planning software. In: CIRED 2005 - 18th International Conference and Exhibition on Electricity Distribution, pp. 1–5. IET, Turin, Italy (2005)
17. Tripathi, A., Sisodia, S.: Comparative study of reliability assessment techniques for composite power system planning & applications. In: International Conference on Emerging Trends in Mechanical and Electrical Engineering, pp. 8–13, Ghaziabad: International Organization of Scientific Research, Tekanpur, India (2014)
18. Weiqing, M., Huiming, Z., Chaoqun, K., Ye, Z.: Reliability analysis of power supply based on recursive algorithm. In: 2014 IEEE 5th International Conference on Software Engineering and Service Science, pp. 327–330. IEEE, Beijing, China (2014)
19. Yuan, Y., Jingyan, Y., Jianhua, Z., Xiangning, X.: Reliability analysis of power distribution systems based on hybrid method into account of voltage sags. In: 2005 International Power Engineering Conference, pp. 945–949. IEEE, Singapore (2005)
20. Zhang, X., Bie, Z., Li, G.: Reliability Assessment of Distribution Networks with Distributed Generations Using Monte Carlo Method **12**, 278–286 (2011)
21. Zhu, G., Qin, F., Zhou, T., Shang, X., Qian, Y.: Failure distribution of nuclear power station DC power supply modules and maintenance benefit analysis. In: 2019 International Conference on Quality, Reliability, Risk, Maintenance, and Safety Engineering (QR2MSE), pp. 489–494. IEEE, Zhangjiajie, China (2019)

Comparative Analysis of Energy and Ecological Indicators of a Spark Ignition Engine Running on Different Amount of Petrol and Bioethanol

Gabrielius Mejeras(✉) and Alfredas Rimkus

Vilnius Gediminas Technical University, J. Basanavičiaus Street 28, 03224 Vilnius, Lithuania
{gabrielius.mejeras,alfredas.rimkus}@vilniustech.lt

Abstract. Results presented in experimental study, which extends of numerical modeling studies of petrol mixed with 10% bioethanol (E10), bioethanol mixed with 15% petrol (E85). During the test AVL BOOST program with modeled HR16DE spark ignition engine parameters were used. Bench tests carried out using the stoichiometric mixture $\lambda = 1.0$ and the lean mixture $\lambda = 1.1$. Results for $\lambda = 1.2$ were obtained by numerical modeling. The addition of bioethanol to petrol increases the oxygen content of the fuel therefor stoichiometric mixture requires less air. This allows the engine control unit to increase the amount of fuel injected. Lower bioethanol calorific value limits engine torque. After determining brake torque, break specific fuel consumption and ecological data, results of various fuels were determined. It was found that the highest engine brake torque was developed using mixture E85. E85 gives higher brake thermal efficiency (BTE), energy efficient increases when the mixture is leaner. However, it is not advisable to use lean mixture ($\lambda > 1.1$), as engine torque and power is reduced.

Keywords: Spark ignition engine · Petrol · Bioethanol · Fuel mixtures · Combustion

1 Introduction

Environment and ecology are one of the most relevant topics today. First European (EU) standard for emissions was set in 1993, and in 1997 the Kyoto Protocol was signed. The second more important Paris agreement was signed in Paris at 2016 which aims to reduce pollution level all over the world [1]. The search for alternative fuels in EU and all over the world began that could replace convectional fuels. One of the ways to reduce greenhouse gas (GHG) impact from internal combustion engines and to reduce energy dependence on petroleum-derived fuels is to use bioethanol (E85) (Table 1). E85, which contains 85% ethanol and 15% petrol, has a higher density, higher octane number compared to pure petrol. E85 maximum combustion pressure in the cylinder is lower than that of petrol because it has a lower net calorific value. E85 reduces CO, CH, CO_2 emissions due to lower C/H ratio and has more oxygen in the

O. Prentkovskis et al. (Eds.): TRANSBALTICA 2021, LNITI, pp. 343–352, 2022.
https://doi.org/10.1007/978-3-030-94774-3_34

fuel [2]. Yusoff carried out tests on pure petrol by mixing certain parts of ethanol and butanol with pure petrol. The above alcohols were added at 10% and 20% by volume. The test was performed with a 4-cylinder engine. Ethanol-containing fuels showed higher break thermal efficiency (BTE) values than iso-butanol containing fuels. The increase in combustion efficiency is due to the presence of higher oxygen content in ethanol, which had a direct impact on the increase in BTE. The increase in the latent heat evaporation of ethanol is also higher than that of iso-butanol. Taking all load conditions into account, the increase in BTE with E10, E20 fuel mixtures increased by 8.0%, 13.6%, while with butanol B10 and B20 increased by 5.5% and 6.4% [3]. Gupta study of E5 fuel found that the cylinder pressure and temperature of the test engine increased as the engine load was increased with E5 fuel. The thermal efficiency value increased by an average of 0.14% compared to regular petrol [4].

Table 1. Fuel properties comparison.

Indicator	Fuel mixture	
	E10	E85
Chemical formula	C4…C12	C2H5OH
Molecular weight μ	100…105	46.07
Elemental composition %		
Carbon	85…88D	52.2
Hydrogen	15…12	13.1
Oxygen	~0.025	34.7
Density (20 °C) ρ, kg/m^3	738.2	782.9
Boiling point, t_s, °C	27…225	78
Freezing point, °C	−40	−114
Stoichiometric air to fuel ratio l_O, kg air/l kg fuel	14.7	10.5
Cetane number	8…14	8
Octane number	95	108
Lower heating value (*LHV*), MJ/kg	41.73	30.88

Meanwhile, Akansu also performed a study with 20% ethanol fuel (E20). The test was performed with a high displacement engine. The experiment was performed at constant engine speed – 1,500 rpm. The test showed that the BTE was reduced by 1.94% and 5.16% with E20 fuel when engine load was 50% and 100% respectively, compared to petrol. This decrease in energy efficiency is thought to be due to a decrease in energy density when petrol is blended with ethanol [4, 5]. Renzi conducted a study with a spark ignition engine where ethanol blended with petrol fuel and mixtures of E0, E50 and E80 are used. The study found that E50 fuels produced higher BTE values than other fuel blends. The author of the study indicates that the air to fuel ratio was close to stoichiometric when such results were obtained. Thermal braking efficiency values were also found to decrease with the use of E80 fuel. In other words, it had the opposite effect on power growth due to poorer fuel combustion [6].

Yusoff performed test using pure petrol, ethanol and iso-butanol fuel blends (E10, E20, iB10, iB20) in a four-cylinder spark ignition engine. The test results showed that due to the higher specific heat of alcohol evaporation, the coefficient of filling (volumetric efficiency) of the cylinder increased. Higher oxygen content in alcohol improved combustion quality. However, taking into account all load conditions during the test, the torque using E10, E20, iB10, iB20, E5iB5 and E10iB10 decreased by 2.33%, 3.88%, 2.56%, 3.06%, 1.06% and 3.77% compared to petrol [3]. Masumas test shows that using a mixture of petrol fuel with 15% ethanol (E15) increased the torque by 2.73% compared to petrol at 4,000 rpm. The increase in torque can be explained by the higher-octane number of ethanol, the latent heat of evaporation, and the oxygen concentration in the fuel [7]. The aim of the study is to investigate using AVL numerical analysis, the change in combustion, in a SI engine by changing the fuel mixture and air to fuel ratio.

2 Research Methodology

Influence of bioethanol on the performance of an internal combustion engine was evaluated based on bench tests and numerical modeling of the HR 16DE spark ignition engine from Nissan Qashqai (Table 2). The test was performed at a constant engine speed (n = 2,000 min^{-1}), throttle open position (15%) and ignition angle Θ = 22 Crank Angle Degree (CAD) before Top Dead Center (TDC). Bench tests were performed using E10 (petrol 90% / bioethanol 10%) and E85 (petrol 15%/bioethanol 85%) with an excess air ratio of λ = 1.0 and λ = 1.1. Based on the test results, analysis of combustion process was performed, and the trends of changes in the combustion (start of combustion, intensity of combustion, duration of combustion) were determined. Numerical modeling of the combustion process was performed using the analysis data, the results of which were validated with an experiment when λ = 1.0 and λ = 1.1. Based on the identified combustion process trends, modeling was performed for E10 and E85 fuels in the presence of a lean mixture (excess air ratio λ = 1.2), for which no experimental studies were performed.

Table 2. Engine Nissan HR 16 technical data.

Parameter	Value
Number of cylinders	4
Cylinder bore, mm	78
Piston stroke, mm	83.6
Number of valves per cylinder	4
Displacement, cm^3	1,598
Compression ratio, ε	10.7
Nominal power, kW (min^{-1})	84 (6,000)
Maximum engine torque, Nm (min^{-1})	156 (4,400)

The stand tests were performed in the initial stage of the research, using the test equipment of the laboratory of internal combustion engines of the Department of Automobile Transport at VILNIUS TECH. The tested engine is controlled by the programmable ECU MoTeC M800 and AMX 200/100 load stand. The pressure in the cylinder is determined using a pressure sensor AVL ZI31 with a built-in spark plug and recorded with the AVL DiTEST DPM 800. The hourly consumption of E85 is measured with an electronic fuel consumption meter AMX 212F. Emissions were tested using an AVL DiSmoke 4000 emission tester.

Numerical engine operation simulation was implemented using AVL Boost, creating an internal combustion engine model using visual icons. The following boundary conditions were selected for modeling: inlet air pressure The following boundary conditions were selected for modeling: inlet air pressure, temperature, fuel properties and amount of fuel injected, excess air ratio, indicators required to calculate heat release according to Vibe function (start of combustion, duration of combustion and intensity of combustion). It was chosen that E10 and E85 fuel mixtures are injected into the intake manifold with four injectors.

3 Results and Their Analysis

Experimental research and AVL BOOST modeling shows Fig. 1 and Fig. 2 that when the engine is running and the fuel mixture is attenuated with E10 and E85 fuel, the brake torque M_B and power P_B are reduced as the fuel supplied to the cylinders is reduced. E85 has 2…9% higher torque and power depending on the lambda setting. The leaner the mixture, the bigger the difference.

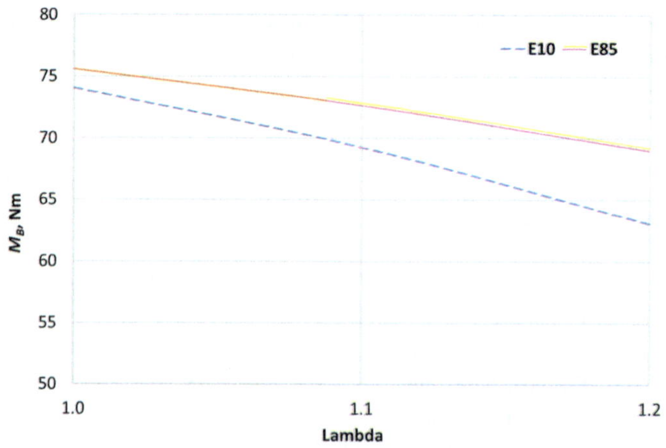

Fig. 1. Brake torque *MB*.

Higher brake torque and power can be explained because E85 stoichiometric mixture requires less air ~28% compared to E10 (Table 1). To the same air quantity in the engines cylinder Electronic Control Unit (ECU) sprays ~28% more fuel. Although E85 lower calorific value is ~26% smaller compered to E10. Therefore, much more torque and power is not obtained when using E85. The enthalpy of evaporation differs significantly for E85 and E10. E85 consumes more energy during evaporation, more cooling the intake air, this increases density and cylinders volumetric efficiency. The higher air mass entering the cylinders makes it possible to increase the amount of E85 fuel, this increases the torque and power.

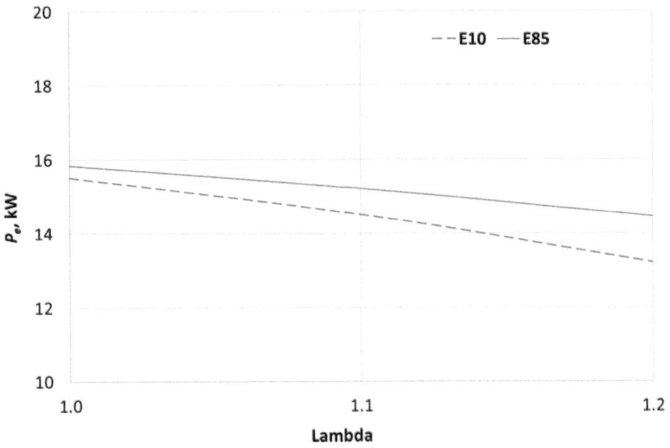

Fig. 2. Break power P_B.

Indicated Specific Fuel Consumption (ISFC) study graph (Fig. 3) shows, that by leaning the mixture to $\lambda = 1.2$, the comparative fuel consumption decreases running on both E10 and E85 fuels, more air increase combustion efficiency. However, the ISFC of E85 fuel is ~26% higher and this results in a lower calorific value of this fuel.

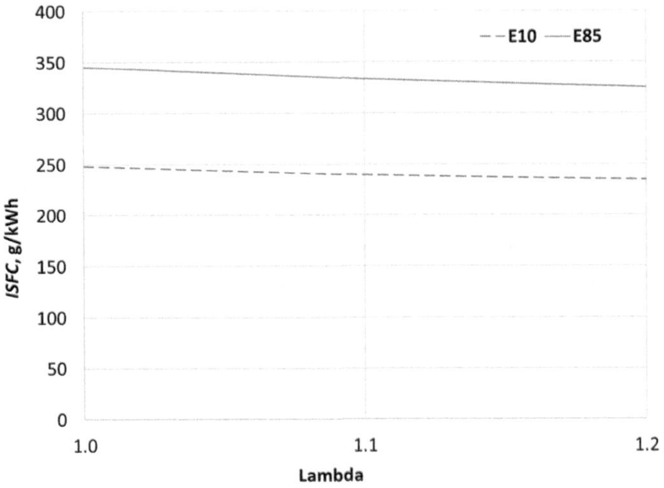

Fig. 3. Indicated specific fuel consumption.

The Break Thermal Efficiency (BTE) calculation graph (Fig. 4) shows that when the mixture is leaned to λ = 1.2, the BTE increases with the engine running on both E10 and E85 fuels, as higher air volumes increase combustion efficiency. However, thinning the mixture more then λ = 1.1 is not rational, as the engine torque and power decrease. The BTE of E85 fuel mixture is ~2% higher and this results in more efficient combustion of bioethanol fuels, which improves due to the additional oxygen in the fuel.

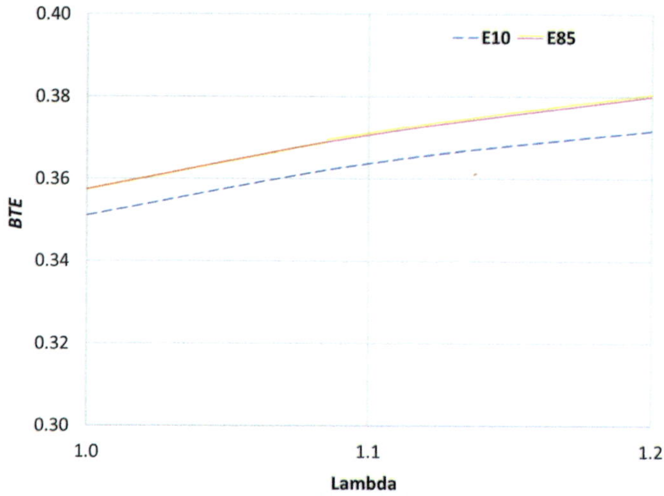

Fig. 4. Break thermal efficiency of engine.

Peak Fire Temperature results obtained by numerical simulation (Fig. 5) show that as the mixture is leaned to λ = 1.2, the combustion temperature decreases with the engine running on both E10 and E85 fuel mixtures due to the lower cyclic fuel injection into the cylinders. Peak Fire Temperature of fuel mixture E85 is ~5.5% lower and this is due to a larger temperature drop due to ethanol evaporation and a slightly slower combustion process. This is confirmed by the subsequent peak of temperature. When λ = 1.1, the Peak Fire Temperature of E10 fuel reaches 14.8 CAD ATDC fuel E85 – 15.5 CAD ATDC. This reduces emissions of nitrogen oxides (NO_x), which forms in high temperatures. Numerical modeling has shown that the use of E85 fuel reduces NO_x emissions by up to 1.5 times (Fig. 6).

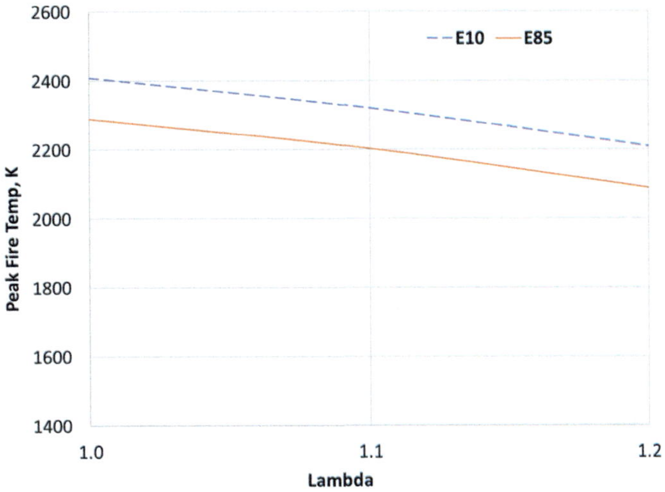

Fig. 5. Peak Fire Temperature of fuel mixtures E10 and E85.

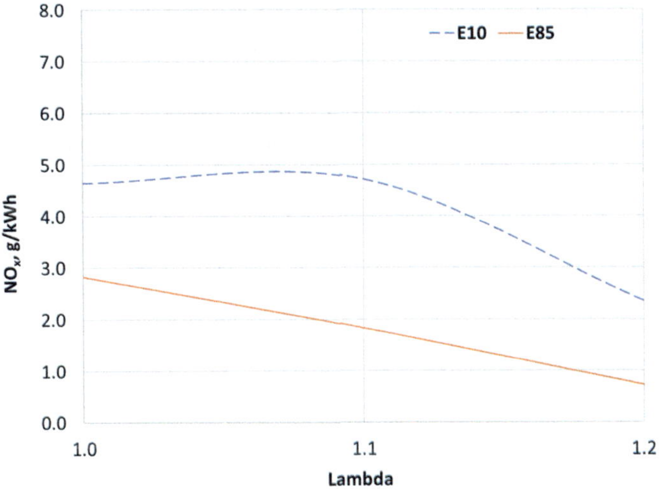

Fig. 6. Nitrogen oxide emissions of fuel mixtures E10 and E85.

The Indicated mean effective pressure (IMEP) graph (Fig. 7) shows that when the mixture is leaned to $\lambda = 1.2$, the IMEP decreases with the engine running on both E10 and E85 fuels due to the lower cyclic fuel injection into the cylinders. However, we see that thinning the mixture more $\lambda = 1.1$ IMEP decreases more intensively and therefore the engine torque and power decrease. By leaning E85 fuel mixture, the reduction in IMEP is smaller because the ethanol evaporates more of the air (the enthalpy of evaporation) and increases volumetric efficiency.

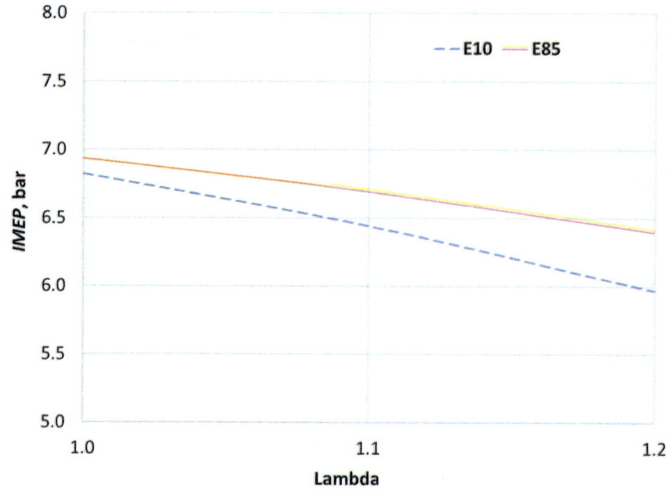

Fig. 7. Indicated mean effective pressure of fuel mixtures E10 and E85.

Carbon monoxide (CO) emissions (Fig. 8) are reduced by ~84% and ~78% by leaning the mixture from $\lambda = 1.1$ to $\lambda = 1.2$, by running the engine on E10 and E85 fuel mixtures, due to increase in oxygen involved in combustion. It has also been found that the CO emission of fuel mixture E85 is ~65% lower compared to E10, although combustion takes place at lower temperatures. The better combustion process is due to the higher oxygen and hydrogen content in bioethanol.

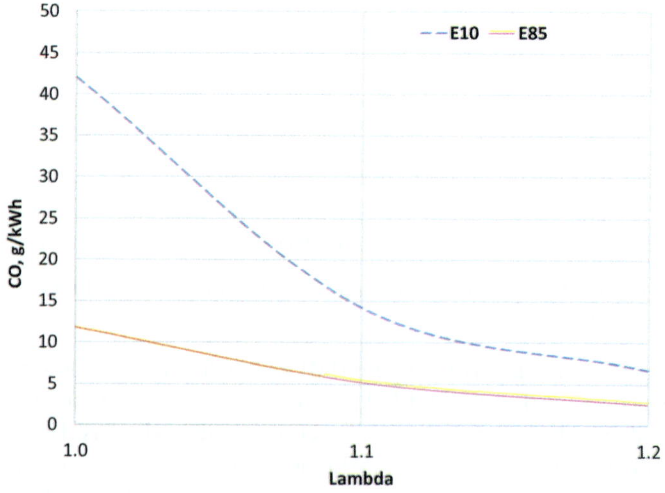

Fig. 8. Carbon monoxide emissions of fuel mixtures E10 and E85.

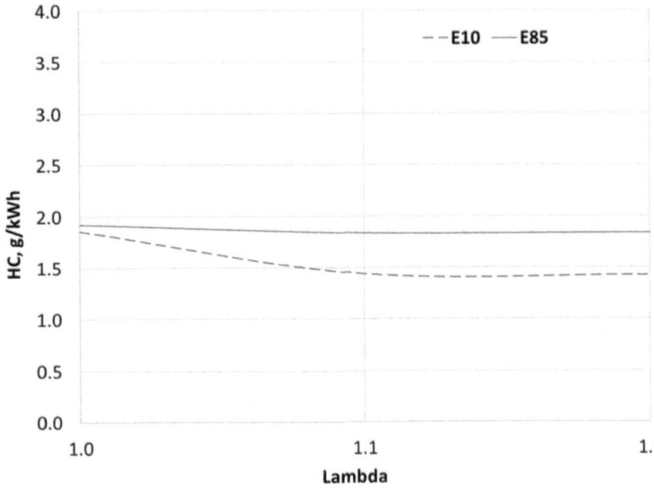

Fig. 9. Hydrocarbon emissions of fuel mixtures E10 and E85.

Bench tests (Fig. 9) and numerical simulations have shown that the emissions of unburnt hydrocarbons (HC) of E85 and E10 fuel mixtures decrease from $\lambda = 1.0$ to $\lambda = 1.1$ during leaning of the mixture, and changes little if leaned more. A small difference in HC emissions was found using E85 and E10 fuel mixtures. HC emissions during the burning of fuel mixture E85 increased because of lower temperature (Fig. 5).

4 Conclusion

Differences in energy and ecological performance were determined when the engine was running on E10 and E85 fuel with stoichiometric ($\lambda = 1.0$) and lean mixtures of ($\lambda = 1.1$, $\lambda = 1.2$):

1. When the engine is running on E85 fuel, the torque and power is 9% higher than on the E10. The increase in power is due to the mass of E85 injected, which is higher due to the lower air demand for the combustion of this fuel and better volumetric efficiency of the engine. However, when leaning the mixture, torque and power decrease and it is inappropriate to lean more then $\lambda = 1.1$ because then the power of E85 becomes less than E10 when $\lambda = 1.0$.
2. The Indicated Specific Fuel Consumption of E85 fuel is ~26% higher compared to E10 due to the lower calorific value of this fuel, but the Break Thermal Efficiency is ~2% higher because the oxygen in bioethanol improves the combustion process.
3. Peak Fire Temperature with E85 fuel is ~5.5% lower compared to E10 due to colder intake mixture and slower combustion. This reduces NO_x emission.
4. CO emissions from E85 were found to be ~65% lower compared to E10 due to the higher oxygen and hydrogen content of E85.

Acknowledgment. To obtain the results, a virtual internal combustion engine simulation tool AVL BOOST was used. It was acquired by signing an agreement between AVL Advanced Simulation Technologies and Faculty of Transport Engineering of VILNIUS TECH.

References

1. Campbell, P., Zhang, Y., Yan, F., Lu, Z., Streets, D.: Impacts of transportation sector emissions on future U.S. air quality in a changing climate. Part II: air quality projections and the interplay between emissions and climate change. Environ. Pollut. **238**, 918–930 (2018). https://doi.org/10.1016/j.envpol.2018.03.016
2. Davis, G., P.E, W.: Ethanol vehicle cold start improvement when using a hydrogen supplemented e85 fuel, Collect. Tech. Pap. 35th Intersoc. Energy Convers. Eng. Con-Ference Exhib. 32
3. Yusoff, M.N.A.M. et al.: Comparative assessment of ethanol and isobutanol addition in gasoline on engine performance and exhaust emissions. J. Clean. Prod. **190**, 483–495 (2018). https://doi.org/10.1016/j.jclepro.2018.04.183
4. Gupta, P., Das, P.P., Mubarak, M., Shaija, A.: Performance and emission analysis of single cylinder SI engine using bioethanol-gasoline blend produced from Salvinia Molesta. In: IOP Conference Series Materials Science Engineering, vol. 297, p. 012005 (2018). https://doi.org/10.1088/1757-899X/297/1/012005
5. Göktaş, M., Balki, M.K., Sayin, C., Canakci, M.: An evaluation of the use of alcohol fuels in SI engines in terms of performance, emission and combustion characteristics: a review. Fuel **286**, 119425 (2021). https://doi.org/10.1016/j.fuel.2020.119425
6. Renzi, M., Bietresato, M., Mazzetto, F.: An experimental evaluation of the performance of a SI internal combustion engine for agricultural purposes fuelled with different bioethanol blends. Energy **115**, 1069–1080 (2016). https://doi.org/10.1016/j.energy.2016.09.050
7. Masum, B.M., Masjuki, H.H., Kalam, M.A., Palash, S.M., Habibullah, M.: Effect of alcohol–gasoline blends optimization on fuel properties, performance and emissions of a SI engine. J. Clean. Prod. **86**, 230–237 (2015). https://doi.org/10.1016/j.jclepro.2014.08.032

Overview of Problematic Aspects of Passenger Car Hybrid Technologies

Tadas Vipartas[1,2](✉), Alfredas Rimkus[1,2], and Máté Zöldy[3]

[1] Vilnius Gediminas Technical University, J. Basanavičiaus Street 28, 03224 Vilnius, Lithuania
{tadas.vipartas,alfredas.rimkus}@vilniustech.lt

[2] Vilnius College of Technologies and Design, Antakalnio Street 54, 10303 Vilnius, Lithuania
{t.vipartas,a.rimkus}@vtdko.lt

[3] Budapest University of Technology and Economics, Műegyetem Street 3, Budapest 1111, Hungary
zoldy.mate@kjk.bme.hu

Abstract. Hybrid electric vehicles (HEV) with Atkinson cycle engines typically use variable valve timing (VVT) technology that reduces fuel consumption up to 30% due to the delay in closing intake valve. Engine thermal efficiency can be raised even up to 40%, however it is necessary to reduce heat loss through the cooling system and to avoid knocking combustion. Electric Vehicle (EV) mode operates at very low loads and the traction force is generated only by the electric motor, while internal combustion engine (ICE) is off to avoid an inefficient zone. Parallel Hybrid (PH) mode operates at higher loads, ICE is on and depending on accelerator pedal and on the State of Charge (SOC) of high voltage batteries, the powertrain can operate in Smart Charge (SC) and Electric Boost (E-Boost). HEVs emission profile is not always improved due to cold start events. NO_x, HC, CO and particle number (PN) increase significantly after cold start. Moreover, the amount of emissions depends on SOC of batteries. At high SOC fuel consumption, CO and NO_x emissions are reduced, while at low SOC emissions are increased significantly enough. Analysis show that cold start CO_2 emissions along the New European Driving Cycle (NEDC) and Worldwide Light duty vehicle Test Procedure (WLTP) procedures have differences even up to 30%.

Keywords: Hybrid vehicles · Emissions · CO_2 · NEDC · WLTP · Atkinson cycle

1 Introduction

It is now widely acknowledged that the increase in temperature in the Earth's atmosphere is due to greenhouse gas emissions from a variety of human activities. 2015 The United Nations Climate Change Conference agreed to keep temperature rises below 2 °C, which meant significant changes in the global energy supply system [1].

The sharp rise in global energy demand has led to the search for alternative energy sources, and the uncertain situation over hydrocarbon prices, international agreements

O. Prentkovskis et al. (Eds.): TRANSBALTICA 2021, LNITI, pp. 353–360, 2022.
https://doi.org/10.1007/978-3-030-94774-3_35

and economic turmoil, such as the Kyoto Protocol and Agenda 21, has forced countries to adopt unconventional and renewable energy strategies [2]. Thus, energy production from alternative sources is essential to reduce environmental problems phasing out fossil fuels [3].

The transport sector is identified as one of the largest polluter, accounting for 25% of total global greenhouse gas emissions [4]. The situation is very similar in European countries. 2015 the transport sector accounted for 23% of greenhouse gas (GHG) emissions [5], while road transport accounted for more than 70% of this amount [6].

The European Union is aiming for 2050 reduce GHG to zero. The evolution of society is a one of the main difficulties and an opportunity looking ahead. To meet this, EU countries must allocate more funding for environment-friendly processes [7].

To achieve ambitious goals, more attention is being paid to hybrid electric vehicles (HEVs). Their direct emissions are typically lower than conventional vehicles (CVs) due to reduced fuel consumption. The problems of limited distance and self-charging batteries are solved by using internal combustion engine (ICE) in the vehicle [8]. At low loads only an electric powertrain is used, while at high loads it supports ICE [9]. ICE can operate efficiently in all zones due to electric motor. Most HEVs with Atkinson cycle engines use VVT technology, which delays the closing of the intake valve. The 20–30% reduction in fuel consumption is due to Atkinson cycle engines in HEVs [10]. Takakashi et al. state that engine thermal efficiency can be raised even up to 40%. Application of Atkinson cycle, EGR with cooling, electrical water pump and reduction of friction loss would allow to reach that value. As well as CO_2 emissions would also be reduced. However, main task is to reduce heat loss through the cooling system and to avoid knocking combustion.

The aim of the article is to highlight the problems of automotive hybrid technologies based on the analysis of scientific articles.

2 Analysis of the Operating Modes and Architectural Structure of a Hybrid Electric Vehicle Powertrain

Different hybrid technologies may improve gases from tailpipe and reach fuel efficiency in comparison with petrol and diesel vehicles. However, in order for the ICE to work most effectively in all ranges, manufacturers have faced significant problems [11].

Cubito et al. notes that hybrid vehicle can operate in two modes. The main factors are vehicle speed, energy request and state of charge (SOC) of high voltage batteries.

1. Electric Vehicle (EV) mode: at very low loads, the ICE is off, avoiding an inefficient range, and the traction force is generated only by the electric motor MG2 (Fig. 1a).
2. Parallel Hybrid (PH) mode operates at high loads, ICE is on and depending on vehicle load and the battery SOC, the powertrain can operate in Smart Charge (SC) and Electric Boost (E-Boost)

   - During SC, ICE operating points are shifted at higher load levels, although the power requirement is less than need to drive the vehicle. ICE operates near the optimal efficiency zone; the battery is charged via the generator MG1 (Fig. 1b).

- During E-Boost mode additional power is supplied to the MG2 from high voltage batteries. ICE receives support only at very high loads (Fig. 1c).

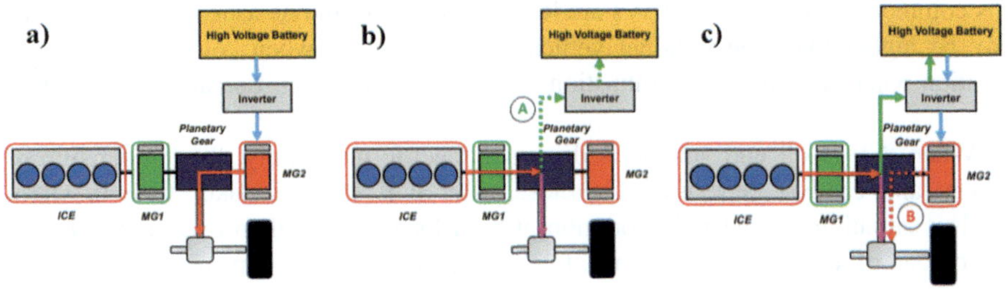

Fig. 1. Hybrid vehicle operating modes: a) EV mode; b) SC mode; c) E-Boost mode [12].

The powertrain of HEVs could be classified into three main groups: series hybrid electric vehicle, parallel hybrid electric vehicle and a series-parallel hybrid electric vehicle. Each of them has its advantages and disadvantages. The structure of series hybrid electric vehicle is the simplest and typically consists of an ICE, generator and electric motor. However, the efficiency of this vehicle is the lowest [11]. The series-parallel system requires more power composite devices, therefore considered more complicated [13].

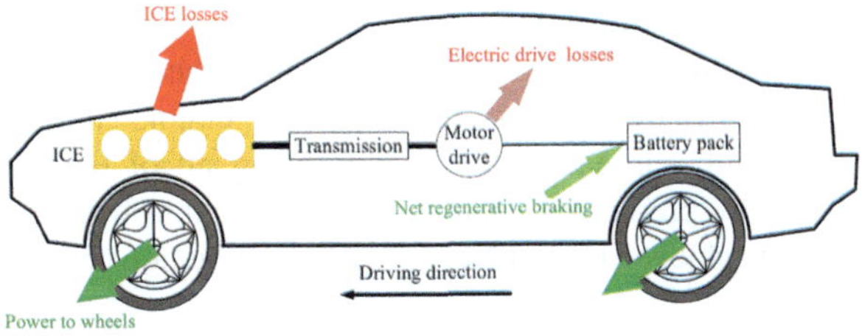

Fig. 2. Energy apportion in hybrid vehicle [14].

Typical energy apportion in a parallel hybrid electric vehicle is shown in Fig. 2. Most of the energy loss (from 63% to 72%) is in an internal combustion engine, while electric powertrain loss ranges between 5 and 20%. Power to wheels is approximately 27–38%, energy returned to the batteries and subsequently to the road 5–9% [14].

3 Pollutant Emission Studies of Hybrid Electric Vehicles

3.1 Change in Pollutant Emissions of Hybrid Vehicles During "Cold Start" Event

It is common to assume that HEVs are less polluting, but in one aspect their emissions may be higher than conventional vehicles. This happens when cold start occurs, a situation where the internal combustion engine does not reach operating temperature [15].

When ICE is off and vehicle is driven in electric mode only, the temperature of coolant and catalyst decreases. Therefore, after ICE starts, it takes time to warm up. This situation is especially common at high loads. HEV often encounters several ICE on and off events in a single trip, this is related to an energy management strategy that is responsible for reducing fuel consumption by using an electric motor and stored energy in the batteries [16].

Typically, two methods are used to evaluate the change in emissions: chassis dynamometer test and Real Driving Emissions (RDE). Bagheri et al. analysis of studies by other researchers show the direct cross comparison between HEVs (after cold start) and CVs emissions performance results. Percentage difference for pollutants are shown in Fig. 3. The positive values presents increasing emissions comparing HEVs to CVs, negative indicate reduced emissions compared HEVs to CVs.

There is sufficiently significant change in emissions. Based on chassis dynamometer studies the average nitrous oxides (NO_x), hydrocarbons (HC), carbon monoxide (CO) and particle number (PN) expanded to 104%, 6.4 times, 22% and 17.9 times. The average RDE results show NO_x and HC decreased 5% and 39%, while CO and PN increased 13% and 5 times [17].

However, to reduce of HEVs after cold start emissions there are few control options under three categories: after treatment system warm-up, engine calibration adjustment and filter installation with sorbent materials. Assessing all of them, after treatment warm-up with electrically pre-heated catalytic converter control strategy is one of the most effectives and can achieve HC and NO_x emissions reduction about 40%–50% [18].

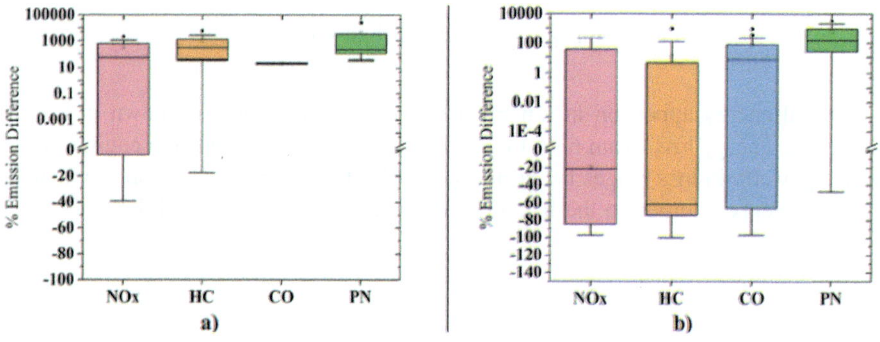

Fig. 3. Ecological parameters compared HEVs to CVs. Emissions NO_x, HC and CO (units mass per distance), PN – (particle number per distance): a) Chassis dyno testing results; b) RDE testing results [17].

Duarte et al. emphasizes that battery SOC is considered the main parameter influencing fuel consumption and exhaust emissions under low power driving modes. When the SOC of the battery is less than 50%, fuel economy reduced 57%, carbon monoxide increased 27%, nitrous oxides increased 56%. However, if the battery charged at least 70%, fuel economy improved to 40%, carbon oxides and nitrous oxides reduced to 34% and 61% [19].

At the moment, energy management system is responsible for ICE control (on or off). It depends on load and battery SOC without evaluation its effect on emissions. High voltage batteries usually maintained when SOC is ~45%–70% [20]. Similar trends are presented by Frey et al., emphasized that state of charge is important on nitrous oxides concentration in exhaust system. When SOC is high NO_x are reduced about 42% [21].

3.2 Differences in Emissions from Various Hybrid Vehicle Research Technologies

However, to estimate the exact CO_2 emissions of vehicles there are type approval (TA) procedures: Worldwide Light duty vehicle Test Procedure (WLTP) and New European Driving Cycle (NEDC). From 2017 September European Commission has decided to replace the previously applicable NEDC to WLTP attempting to close the gap between TA procedures [12] also reported emissions differences.

The European Commission's report presents that ICE passenger cars have an average 1.21 difference ratio of CO_2 between the procedures, considering the certification values for carbon dioxide emissions.

Evaluating the real amount of $CO_{2,}$ the correction coefficient applied and is named K-Factor. Tile pipe emissions are calculated according to the following equation:

$$M_{CO2\_cor} = M_{CO2} - K_{CO2} \cdot Q, \tag{1}$$

where $M_{CO2\ cor}$ – corrected concentration of carbon dioxide; M_{CO2} – corrected concentration measured in chassis dynamometer; K_{CO2} – K-Factor and Q is the battery integral due to M_{CO2} twice measurement [22].

Differences between NEDC and WLTP procedures also note Cubito et al. and present analyses on carbon dioxide of a Euro-6 hybrid electric vehicle (with NiMH batteries) from the chassis dynamometer tests. The carbon dioxide measured when a cold start occurs are shown in Fig. 4. Terms "High SOC" correspond to 70% and "Low SOC" to 30% of battery charge. CO_2 concentration increased ~30% in WLTP (Fig. 4a). Tests were also performed by changing the coolant temperature from 25 °C ("Cold") to 70 °C ("Hot"), the results are presented in Fig. 4b.

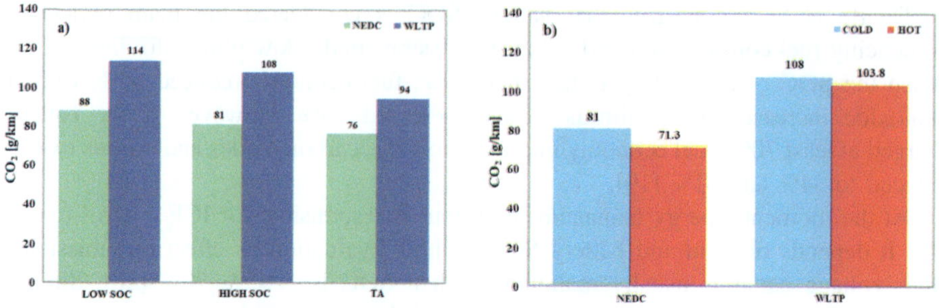

Fig. 4. Carbon dioxide concentration between procedures [12].

CO_2 emissions increased ~14% at coolant temperature 25 °C compared to 70 °C, while carbon dioxide increased ~4% according to the WLTP procedure. Comparing the differences between procedures there is a noticeable a significant variation. NEDC procedure indicates ~33% CO_2 reduction compared to WLTP at coolant temperature 25 °C. Increased temperature to 70 °C this change is even greater, respectively to ~46%.

Reducing coolant temperature from 25 °C to −7 °C causes ~16.5% increase of about 19 g/km of CO_2 emissions at low SOC and ~21% increase of CO_2 (~23 g/km) at high SOC [12].

4 Conclusions

HEVs with Atkinson cycle engines typically use VVT technology that reduces fuel consumption up to 30% due to the delay in closing intake valve. Air recirculation during compression results the reduction of the actual gas involved in compression. Adopting cooled EGR, electrical powertrain and reducing friction losses engine thermal efficiency can be raised up to 40%. This results CO_2 emissions proportionately. However, it is necessary to reduce heat loss through the cooling system and prevent detonating combustion.

Parallel Hybrid (PH) mode operates at higher loads, ICE is on and depending on vehicle load and the battery SOC, the powertrain can operate in Smart Charge (SC) and Electric Boost (E-Boost). During SC, ICE operates near the optimal efficiency zone because the engine operates medium and higher loads.

HEVs emission profile is not always improved due to cold start events compared to conventional vehicle. Results from chassis dynamometer tests show the average increase: NO_x – 104%, HC – 6.4 times, CO – 22% and PN – 17.9 times. RDE tests results show a smaller increase in pollutants: NO_x − 5%, HC − 39%, CO − 13% and PN – 5 time. Different results obtained using different test methods were due to different test conditions. Increased emissions could be explained due to cold start event, where the ICE and exhaust system catalysts do not reach normal temperature. The mechanical losses of the engine also increase; the thermal efficiency is reduced. However, to improve emissions, could be used electrically pre-heated catalytic converter where reduction of HC and NO_x up to 50%.

The amount of emissions also depends on the SOC of high voltage battery. When the SOC is less than 50% both fuel consumption ($\sim$57%) and pollutant emissions (CO $\sim$27%, NOx $\sim$56%) increase.

In order to reduce transport emissions more quickly, research on using of fossil fuels should be continued, as well as more research into renewable biofuels needed.

References

1. Ferella, F., Puca, A., Taglieri, G., Rossi, L., Gallucci, K.: Separation of carbon dioxide for biogas upgrading to biomethane. J. Clean. Prod. **164**, 1205–1218 (2017)
2. Nizami, A.S., et al.: Developing waste biorefinery in Makkah: a way forward to convert urban waste into renewable energy. Appl. Energy **186**, 189–196 (2017)
3. Quereshi, S., Pant, K.K., Dutta, S., Naiya, T.K.: Unfolding the role of molybdenum disulfide as a catalyst to produce platform chemicals from biorenewable resources. Biomass Convers. Biorefin., 1–14 (2020).https://doi.org/10.1007/s13399-020-00888-7
4. International Energy Agency. https://www.iea.org/reports/co2-emissions-from-fuel-combustion-overview. Accessed 05 Jul 2021
5. European Commission: Directorate General for Mobility and Transport. EU transport in figures: statistical pocketbook 2017 (Publications Office, 2017)
6. European Environment Agency. https://www.eea.europa.eu/themes/transport/intro. Accessed 04 Jul 2021
7. European Union. https://ec.europa.eu/clima/policies/strategies/2050_en. Accessed 07 Jul 2021
8. Rahman, I., Vasant, P.M., Singh, B.S.M., Abdullah-Al-Wadud, M., Adnan, N.: Review of recent trends in optimization techniques for plug-in hybrid, and electric vehicle charging infrastructures. Renew. Sustain. Energy Rev. **58**, 1039–1047 (2016)
9. Un-Noor, F., Padmanaban, S., Mihet-Popa, L., Mollah, M., Hossain, E.: A Comprehensive study of key electric vehicle (EV) components, technologies, challenges, impacts, and future direction of development. Energies **10**, 1217 (2017)
10. Takahashi, D., Nakata, K., Yoshihara, Y., Ohta, Y., Nishiura, H.: Combustion development to achieve engine thermal efficiency of 40% for hybrid vehicles. SAE Technical Paper 2015-01-1254 (2015). https://doi.org/10.4271/2015-01-1254
11. Das, H.S., Tan, C.W., Yatim, A.H.M.: Fuel cell hybrid electric vehicles: a review on power conditioning units and topologies. Renew. Sustain. Energy Rev. **76**, 268–291 (2017)
12. Cubito, C., et al.: Impact of different driving cycles and operating conditions on CO_2 emissions and energy management strategies of a euro-6 hybrid electric vehicle. Energies **10**, 1590 (2017)
13. Banshoya, H., et al.: Hybrid operable in series mode and in series - parallel mode. U.S. Patent Application 10/214,093 (2019)
14. SteelShield Homepage. http://www.steelshieldtech.com.hk/Applications/Automotive_Energy_Requirement.html. Accessed 12 Jul 2021
15. Huang, Y., et al.: Re-evaluating effectiveness of vehicle emission control programmes targeting high-emitters. Nat. Sustain. **3**, 904–907 (2020)
16. Varella, R.A., Gonçalves, G., Duarte, G., Farias, T.: Cold-running NO_x emissions comparison between conventional and hybrid powertrain configurations using real world driving data. SAE Technical Paper 2016-01-1010 (2016). https://doi.org/10.4271/2016-01-1010

17. Bagheri, S., Huang, Y., Walker, P.D., Zhou, J.L., Surawski, N.C.: Strategies for improving the emission performance of hybrid electric vehicles. Sci. Total Environ. **771**, 144901 (2021)
18. Laurell, M., Pace, L., Ekström, F., Konieczny, K.: Strive for zero emissions impact from hybrids. SAE Technical Paper 2019-24-0146 (2019). https://doi.org/10.4271/2019-24-0146
19. Duarte, G.O., Varella, R.A., Gonçalves, G.A., Farias, T.L.: Effect of battery state of charge on fuel use and pollutant emissions of a full hybrid electric light duty vehicle. J. Power Sources **246**, 377–386 (2014)
20. Yang, Z., et al.: Real driving particle number (PN) emissions from China-6 compliant PFI and GDI hybrid electrical vehicles. Atmos. Environ. **199**, 70–79 (2019)
21. Frey, H.C., Zheng, X., Hu, J.: Variability in measured real-world operational energy use and emission rates of a plug-in hybrid electric vehicle. Energies **13**, 1140 (2020)
22. European Commission. Joint Research Centre. From NEDC to WLTP: effect on the type approval CO2 emissions of light duty vehicles (Publications Office, 2017)

Logistics and Transportation

Algorithm for Determining the Optimal Length of the Rail Line by Current Automatic Locomotive Signaling

Ravshan Aliev(✉) and Marat Aliev

Department of Information Systems and Technologies,
Tashkent State Transport University, Tashkent, Uzbekistan

Abstract. The article proposes a mathematical model for determining the optimal parameters of coding sections for transmitting reliable information to a locomotive, taking into account the absence of rail circuits on lines with semi-automatic blocking. On such areas, the length of the rail line should be selected taking into account the transfer of reliable information to the locomotive, depending on many parameters that affect the information transfer path. To determine the optimal parameters of the information transmission line, the problems of deriving equations from a locomotive receiver were considered, taking into account the absence of boundaries, a mathematical model of a rail line was developed, and rail line equivalent schemes along a locomotive receiver of a coding section on a high-speed line were presented. As a result of the work, analytical expressions were determined to determine the current flowing under inductive coils, a simulation model was developed, and studies were carried out to select the optimal parameters. The proposed method of transmitting information to a locomotive, the condition of the track sections, clearly determines the length of the coding section and thereby allows to increase the safety of train traffic.

Keywords: Rail circuits · Rail line · Locomotive receiver · Automatic locomotive signaling system · Track receiver · Train shunt

1 Introduction

Every year, with the growth of train speeds in railway transport expanding the introduction of modern equipment automation and telemechanic. Auto block, automatic locomotive signaling, dispatching and electrical centralization, crossing signaling devices. Are under construction the most effective means of intensifying the work of railway transport, improving the use of its basic complex and expensive technical means is the automation of the transportation process control with help complex systems of interval regulation of train traffic and ensuring its safety [1–4]. From reliability the work of these systems, especially for high speed lines, depend a lot the rhythm of rail transportation and traffic a train safety.

JSC "Uzbekistan Temir Yollari" has taken a course towards modernizing its railways where the priority is to increase the speeds passenger train. In this direction, a high-speed line has already been put into operation on the Tashkent-Samarkand section

O. Prentkovskis et al. (Eds.): TRANSBALTICA 2021, LNITI, pp. 363–374, 2022.
https://doi.org/10.1007/978-3-030-94774-3_36

with a train speed of 160–260 km/h. A number of measures were taken to ensure the safety of high-speed trains, such as the strengthening of railways, updating the dimensions of buildings, changing the configuration of railroad switches, as well as train control systems, which should include sensors for monitoring the condition of sections of the pathway and the evaporation condition of the rail filaments [8, 11].

2 Mathematical Model of a Rail Line of a Coding Section for High-Speed Lines

For sections of the road with low train traffic, on the railway of Uzbekistan is used semi-automatic blocking, control of employment and free road segment between the stations which is carried out counters axes [13].

A promising plan for the introduction of innovative technologies in railway transport on one of the lines between the stations designed the coding section between the input traffic light H and the warning traffic light PN Fig. 1, having provided a transmitting point (indicated by the letter T) at a distance of 1,000 m from the input traffic light, at it the total length of the section of approaching was 3,125 m.

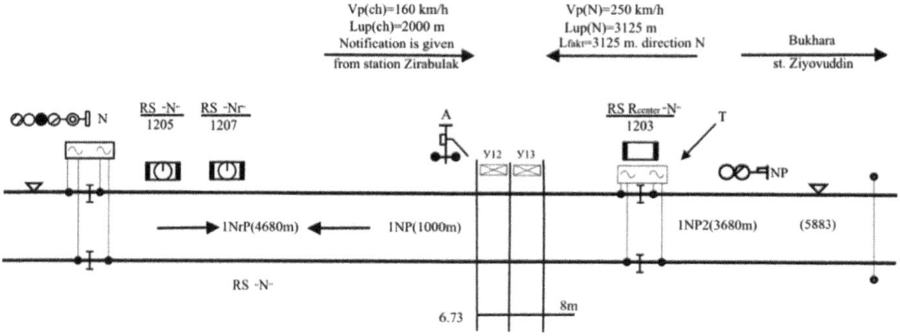

Fig. 1. Scheme area of the coding on the stage.

A significant drawback of such a code translation scheme is that broadcasting devices (cupboard RS RC «N») transmitting devices are installed, that distort the code signal, which can lead to an emergency.

The length of such sections should be selected taking into account the transfer of reliable information to the locomotive, depending on many parameters that affect the information transfer path.

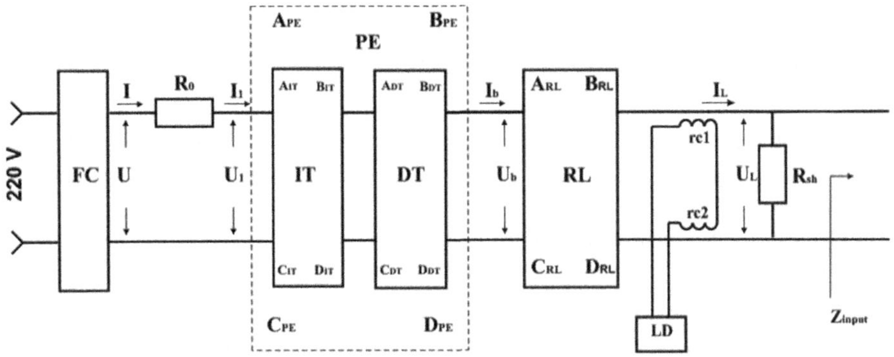

Fig. 2. Scheme of substitution of the encoding section on the high-speed line (current at the beginning of the rail line (I_b), voltage at the beginning of the rail line (U_b), rail line power equipment (PE), locomotive receiver coil (rc), frequency converter (FC)).

where Ait, Bit, Cit, Dit are the coefficients of the four poles of an isolation transformer. A_{dt}, B_{dt}, C_{dt}, D_{dt} are the coefficients of the four poles of the throttle transformer 2TT-1-150 with a transformation ratio n = 3.

Z_{input} is input resistance of the rail line behind the train. It can be taken equal to the wave impedance of the rail line $Z_{input} = Z_v$.

The coefficients of the four pole N are determined by the following expression:

$$\left\| \begin{matrix} A_{pe} & B_{pe} \\ C_{pe} & D_{pe} \end{matrix} \right\| = \left\| \begin{matrix} A_{it} & B_{it} \\ C_{it} & D_{it} \end{matrix} \right\| * \left\| \begin{matrix} A_{dt} & B_{dt} \\ C_{dt} & D_{dt} \end{matrix} \right\|. \tag{1}$$

To implement the task, the initial stage in creating the software was the development of a mathematical model of the rail line.

To develop methods for analysis, calculation and synthesis of the rail line, it is necessary to draw up a general equivalent circuit shown in Fig. 2.

The coefficients of the four poles of the rail line are determined by known formulas:

$$A_{rl} = ch\gamma l; B_{rl} = Z_v * sh\gamma; C_{rl} = \frac{1}{Z_v} * sh\gamma l; D_{rl} = ch\gamma l.$$

where

$$\gamma = \sqrt{Zg_{эmax}} \text{ is spread coefficient} \tag{2}$$

$$Z_V = \sqrt{\frac{Z}{g_{эmax}}} \text{ is wave impedance rail line} \tag{3}$$

$$g_{эmax} = \frac{(2g_{иmax}+g_o)2g_{иmax}}{4g_{иmax}+g_o}, \tag{4}$$

where z is the resistivity of the rail line; g_{max} is conductivity of insulation of the rail line 1 sim/km; g_o is maximum conductivity of grounded contacts support = 0.5 sim/km.

Voltage and current flowing under the receiving coils of a locomotive alarm: I_L is current under the receiving coils of the locomotive.

$U_L = I_L * \frac{R_{sh} * Z_{v2}}{R_{sh} + Z_{v2}}$ is voltage value at the receiving end.

Voltage and current at the beginning of the rail line:

$$U_n = A_{rl}U_L + B_{rl}I_L, \tag{5}$$

$$I_n = C_{rl}U_L + D_{rl}I_L, \tag{6}$$

$$U_n = A_{rl} * I_L * \frac{R_{sh} * Z_{v2}}{R_{sh} + Z_{v2}} + B_{rl} * I_L, \tag{7}$$

$$I_n = C_{rl}I_L * \frac{R_{sh} * Z_{v2}}{R_{sh} + Z_{v2}} + D_{rl}I_L. \tag{8}$$

The voltage on the primary winding of the isolation transformer and the current in it:

$$U_1 = A_N * \left(A_{rl} * I_L * \frac{R_{sh} * Z_{v2}}{R_{sh} + Z_{v2}} + B_{rl} * I_L\right) + B_N * \left(C_{rl}I_L * \frac{R_{sh} * Z_{v2}}{R_{sh} + Z_{v2}} + D_{rl}I_L\right), \tag{9}$$

$$I_1 = C_N * (A_{rl} * I_L * \frac{R_{sh} * Z_{v2}}{R_{sh} + Z_{v2}} + B_{rl} * I_L) + D_N * (C_{rl}I_L * \frac{R_{sh} * Z_{v2}}{R_{sh} + Z_{v2}} + D_{rl}I_L). \tag{10}$$

Voltage and current at the output of the frequency converter (FC):

$$I = I_1 = C_N * (A_{rl} * I_L * \frac{R_{sh} * Z_{v2}}{R_{sh} + Z_{v2}} + B_{rl} * I_L) + D_N * (C_{rl}I_L * \frac{R_{sh} * Z_{v2}}{R_{sh} + Z_{v2}} + D_{rl}I_L), \tag{11}$$

$$\begin{aligned} U = {} & A_N * \left(A_{rl} * I_L * \frac{R_{sh} * Z_{v2}}{R_{sh} + Z_{v2}} + B_{rl} * I_L\right) + B_N * \left(C_{rl}I_L * \frac{R_{sh} * Z_{v2}}{R_{sh} + Z_{v2}} + D_{rl}I_L\right) \\ & + R_0 * \left[C_N * \left(A_{rl} * I_L * \frac{R_{sh} * Z_{v2}}{R_{sh} + Z_{v2}} + B_{rl} * I_L\right) + D_N * \left(C_{rl}I_L * \frac{R_{sh} * Z_{v2}}{R_{sh} + Z_{v2}} + D_{rl}I_L\right)\right]. \end{aligned} \tag{12}$$

To determine the current flowing under the receiving coils of the locomotive, we represent Eq. (12) in the following form:

$$U = I_L * \left\{ A_N * \left(A_{rl} * \frac{R_{sh} * Z_{v2}}{R_{sh} + Z_{v2}} + B_{rl} \right) * B_N \left(C_{rl} * \frac{R_{sh} * Z_{v2}}{R_{sh} + Z_{v2}} + D_{rl} \right) + R_0 * \left[C_N * \left(A_{rl} * \frac{R_{sh} * Z_{v2}}{R_{sh} + Z_{v2}} + B_{rl} \right) + D_N * \left(C_{rl} * \frac{R_{sh} * Z_{v2}}{R_{sh} + Z_{v2}} + D_{rl} \right) \right] \right\}, \tag{13}$$

where from:

$$I_L = \frac{U}{A_N * \left(A_{rl} * \frac{R_{sh}*Z_{v2}}{R_{sh}+Z_{v2}} + B_{rl} \right) + B_N * \left(C_{rl} * \frac{R_{sh}*Z_{v2}}{R_{sh}+Z_{v2}} + D_{rl} \right)} \rightarrow \frac{U}{+ R_0 * \left[C_N * \left(A_{rl} * I_L * \frac{R_{sh}*Z_{v2}}{R_{sh}+Z_{v2}} + B_{rl} * I_L + D_N * \left(C_{rl} I_L * \frac{R_{sh}*Z_{v2}}{R_{sh}+Z_{v2}} + D_{rl} I_L \right) \right) \right]} \tag{14}$$

In addition, it becomes necessary to determine the locomotive signaling current when the train approaches the supply end.

As the train approaches the supply end, the current in the rails increases significantly, the nature of its change is determined by the dependence of the transmission resistance main schemes replacement Z_{rt} from the length of the rail line $Z_{rt} = f(l)$.

The change in the ALS current along the rail line can be determined by the formula:

$$\frac{I_{nl}}{I_{kl}} = \frac{Z_{rte}}{Z_{rtb}}, \tag{15}$$

where I_{nl} is the current under the coils of the locomotive at the beginning of the rail line, I_{kl} is current under the coils of the locomotive at the end of the rail; Z_{rte} is the resistance of the transmission of the rail line by the locomotive receiver when the train is on the supply end of the rail line. Z_{rtb} is the resistance of the transmission of the rail line along the locomotive receiver when the train is at the end of the rail line.

At finding the train in the end of the rail line, when $l_1 = l$, and $l_2 = 0$;

$$Z_{rte} = = Z_v \sinh\gamma l_1 + \frac{Z_{ch} * Z_{ire} * \cosh\gamma l}{Z_{sh} + Z_{ire}} + Z_{ib}^I * \left(\frac{1}{Z_v} \sinh\gamma l * \frac{Z_{sh} * Z_{ire}}{Z_{sh} + Z_{ire}} + \cosh\gamma l \right).$$

At finding the train in the beginning of the rail line, when $l_1 = 0$, and $l_2 = l$.

$$Z_{rtb} = \frac{Z_{ch} * (\cosh\gamma l_2 Z_{ire} + Z_v \sinh\gamma l_2)}{Z_{sh}\left(\frac{1}{Z_v} \sinh\gamma l_2 Z_{ire} + \cosh\gamma l_2 \right) + \cosh\gamma l_2 Z_{ire} + Z_v \sinh\gamma l_2} + Z_{ib}^I. \tag{16}$$

Using the above expressions, the conducted analysis current of the automatic locomotive signaling and determine out the required power supply capacity of the rail line.

3 Algorithm for Determining the Optimal Parameters of the Rail Line of the Coding Section

For research and determine the optimal length of the coding section on the base above expressions an algorithm was developed and a program was compiled, the results of calculations on a computer are shown in Table 1 and in the graphs of Fig. 4, 5 and 6.

The program for determining length of the additional shunting zone on approach of the train.

```
Programdopzona;
label 1,2,3,4,5,6,7;
var
z,fz:array[1..5] of real;
i,j:integer;
ri,zvm,azv,bzv,fzv,gm,ag,bg,fg,ach,bch,chm,
charg,ash,bsh,shm,sharg,are1,aim1,am,fa,are,
aim,bm,fb,bre,bim,cm,fc,cre,cim,dm,fd,dre,dim,
azn,bzn,k1,k1re,k1im,k1m,fk1,k2m,fk2,
k2re,k2im,k3m,fk3,k3re,k3im,k4m,fk4,k4re,k4im,
zp,zpm,fzp,zlm,fzl,zlre,zlim,zldshnm,ldshn,
zldshnre,zldshnim,fzldshn,k5re,k5im,k5m,fk5,
k6re,k6im,k6m,fk6,k7m,fk7,k7re,k7im,k8m,fk8,
k8re,k8im,k9m,fk9,k9re,k9im,k10m,fk10,k10re,
k10im,k11m,fk11,k11re,k11im,zpshre,zpshim,zpshm,
fzpsh,kshn,l,zvxnm,fzvxn,azvxn,bzvxn,zvxn,fn,
zpm1,fzpm1,achn,bchn,chnm,chargn,ashn,bshn,
shnm,shargn,k12re,k12im,k12m,fk12,
k13re,k13im,k13m,fk13,k14re,k14im,k14m,fk14,
k15re,k15im,k15m,fk15,k16re,k16im,k16m,fk16,
zpn1re,zpn1im,zpn1m,fzpn1:real;
begin
for i:=1 to 2 do
readln(z[i],fz[i]);
 i:=1; {vvod Chastoti Z i fz}
 {z[i]:=3.84;fz[i]:=78.45;}
1: ri:=1.3;
...
....
end.
```

Program for research of a jointless track circuit with a potential receiver on train departure.

```
program uxod;
{$APPTYPE CONSOLE}
uses
 SysUtils;
{begin
 { TODO -oUser -cConsole Main : Insert code here }
label 1,2,3,4,5,6,7,8;
var
z,fz,zvxlsh,zvxpsh,fzvxlsh,fzvxpsh:real;
i,j:integer;
ri,zvm,azv,bzv,fzv,gm,ag,bg,fg,ach,bch,chm,
charg,ash,bsh,shm,sharg,are1,aim1,am,fa,are,
aim,bm,fb,bre,bim,cm,fc,cre,cim,dm,fd,dre,dim,
azn,bzn,k1,k1re,k1im,k1m,fk1,k2m,fk2,
k2re,k2im,k3m,fk3,k3re,k3im,k4m,fk4,k4re,k4im,
zp,zpm,fzp,zlm,fzl,zlre,zlim,zldshnm,ldshn,
zldshnre,zldshnim,fzldshn,k5re,k5im,k5m,fk5,
k6re,k6im,k6m,fk6,k7m,fk7,k7re,k7im,k8m,fk8,
k8re,k8im,k9m,fk9,k9re,k9im,k10m,fk10,k10re,
k10im,k11m,fk11,k11re,k11im,zpshre,zpshim,zpshm,
fzpsh,kshn,l,zvxnm,zvx1m,fvx1,fzvxn,azvxn,bzvxn,zvxn,fn,
zpm1,fzpm1,achn,bchn,chnm,chargn,ashn,bshn,
shnm,shargn,k12re,k12im,k12m,fk12,zxm,fzx,zxre,zxim,
k13re,k13im,k13m,fk13,k14re,k14im,k14m,fk14,
k15re,k15im,k15m,fk15,k16re,k16im,k16m,fk16,
zpn1re,zpn1im,zpn1m,fzpn1,Rsh,x,k17re,k17im,k17m,fk17,
k18re,k18im,k18m,fk18,k19re,k19im,k19m,fk19,ksh,
k20re,k20im,k20m,fk20,k21re,k21im,k21m,fk21:real;
begin
{for i:=1 to 2 do
readln(z[i],fz[i]);}
i:=1; {vvodChastoti Z ifz}
z:=4.26;fz:=78.99;Rsh:=0.06;x:=0;
1: ri:=0.15;
2: l:=0.25;
3: fn:=80.0;
..
end;
end. (the program contains 540 sheets, therefore only part of the program is shown).
```

At startup the program, a window appears on the computer screen Fig. 3.

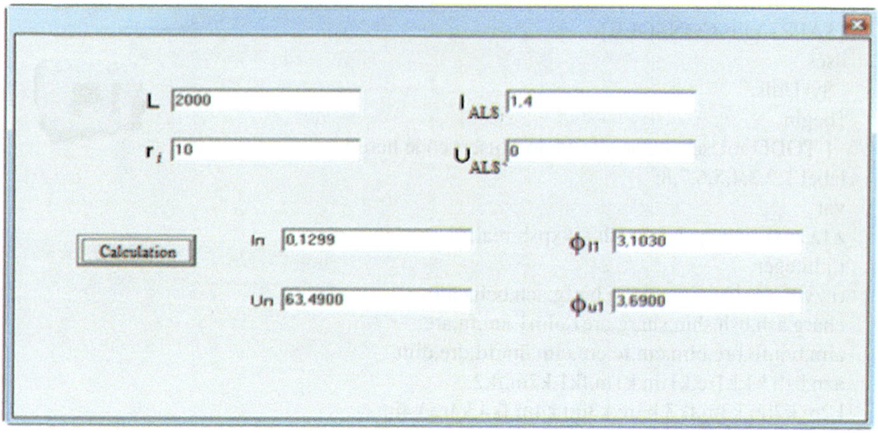

Fig. 3. Calculation of voltage and current depending on the resistance of the ballast and the length of the section.

Where the top part of the window shows the parameters, which necessary introduce it is the length of the rail line L, ballast resistance r, ALS current I_{ALS} and voltage under the receiving coils of the locomotive U_{ALS}, after entering this data, the "calculation" button is pressed and in the windows of In and Un, that is, we get the current and voltage at the output of the frequency converter, in this case, the current is In = 0.1299 A, and the voltage is Un = 63.49 V.

Table 1. The dependence of voltage and current at the output of the frequency converter from the resistance of the ballast and the length of the rail line.

n/p	*L*, м	R_i = 1 Om/км		R_i = 2 Om/км		R_i = 5 Om/км		R_i = 10 Om/км	
		U_b, volt	I_b, ampere	U_b, volt	I_b, ampere	U_b, volt	I_b, ampere	U_b, volt	I_b, ampere
1	1,000	40.54	0.0985	38.5	0.0931	37.25	0.09	36.79	0.0862
2	1,500	60.29	0.1348	54.46	0.12	51.02	0.1116	44.77	0.1086
3	**3,400**	**223.7**	**0.4649**	153.4	0.3	118.6	0.2231	107.3	0.1992
4	4,200			**228.2**	**0.4931**	158.5	0.2918	137.8	0.2471
5	5,200					**223.9**	**0.4074**	181.8	0.321
6	6,000							**224.9**	**0.3929**

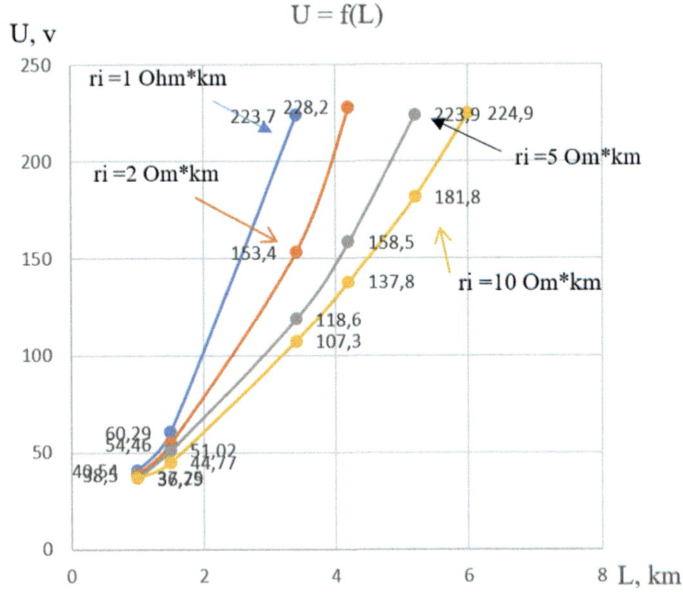

Fig. 4. Graphs of the dependence of the voltage of the power source on the length of the rail line at a current of locomotive signaling I = 1.4 A.

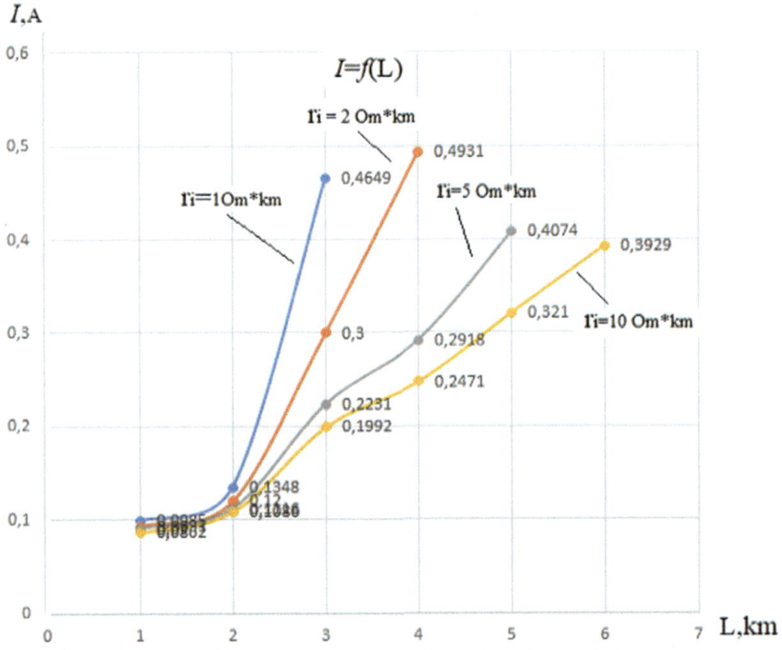

Fig. 5. Graphs of the dependence of the current of the power supply on the length of the rail line at a current of locomotive signaling I = 1.4 A.

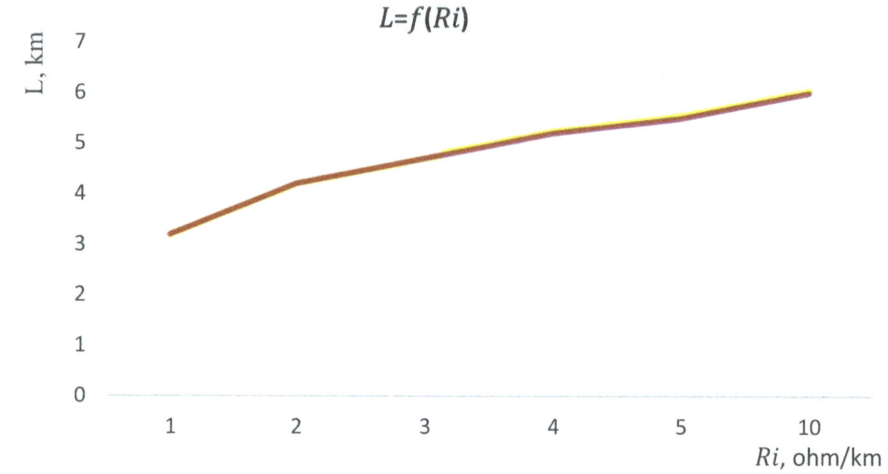

Fig. 6. The graph of the dependence of the length of the rail line on the insulation resistance at a current of locomotive signaling I = 1.4 A.

Can also get a graph of changes in the supply voltage from resistance in depending change resistance ballast Fig. 7.

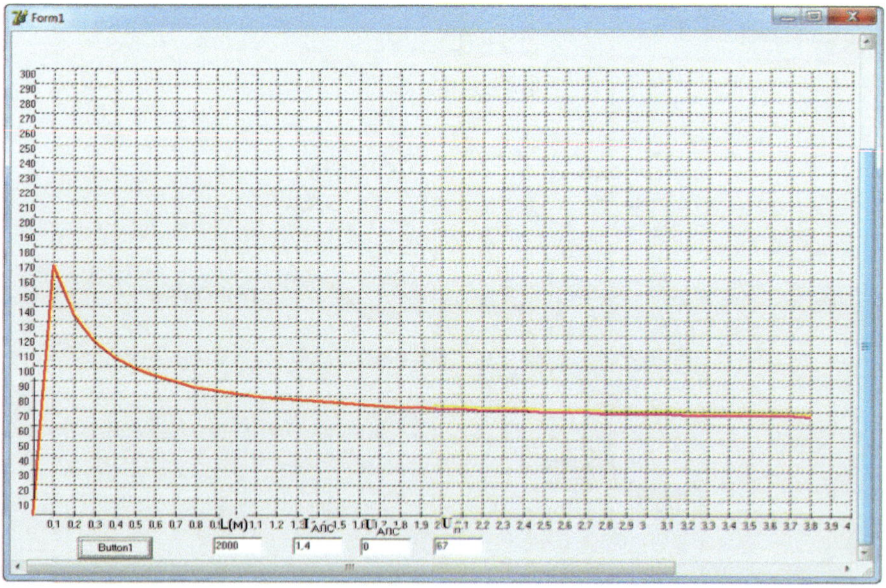

Fig. 7. Graph of the change in the supply voltage from the resistance depending on the change in the resistance of the ballast.

4 Conclusion

The proposed mathematical model for determining optimal parameters rail line of the coding section allows conduct research and calculations for a wide range of changes in the ballast resistance rail line.

The calculations performed to determine the length of the coding section and experimental studies allow us to assert that at condition the ballast resistance is equal to 10 Ω * km, the length of the coding section can reach up to 5,500 m and there is no need to set a translating point for broadcasting code signals.

References

1. Tilk, I.G.: New automatic and telemechanical devices for railway transport, 168 p. UrGUPS, Yekaterinburg (2010)
2. Современные системы интервального регулирования и обеспечения безопасности движения поездов / Е. Н. Розенберг [и др.] // Автоматика, связь, информатика: Научно-популярный производственно-технический журнал. - No 12. - С. 22 - 23. (2005) [In Russian: Modern systems of interval regulation and ensuring safety train traffic / E. N. Rosenberg [et al. // Automatics, communication, informatics: Popular industrial and technical journal]
3. Aliev, R.M., Aliev, M.M., Tokhirov, E.T.: Methodology for determining the optimal values of resistance at the ends of the jointless track circuit with considering twofold shunting. Int. J. Emerg. Trends Eng. Res., IJETER **8**(9), 5048–5052 (2020)
4. Aliev, R.M., Tokhirov, E.T., Aliev, M.M.: The mathematical model of the sensor for monitoring the state of the track section with current receivers. IJRTE **8**(5), 1–4 (2020). ISSN 2277-3878. https://doi.org/10.35940/ijrte.E5936.018520
5. Rosenberg, E.N., et al.: Exclusion of the passage of the forbidden traffic light signal: new technique and technology. Autom. Commun. Inf. Popular Ind. Tech. J. **2**, 10–11 (2008)
6. Tilk, I.G., Lyanoy, V.V.: System development perspectives interval regulation of train traffic. Autom. Commun. Inf. **8**, 7–9 (2007)
7. Zorin, V.I., Shuhina, E.E., Titov, P.V.: Microprocessor-based locomotive safety systems for new generation trains. World's Railroads **7**, 61–69 (2003)
8. Fedorov, N.E.: Modern autoblocking systems with tonal rail circuits, 130 p. Samara (2004)
9. Rosenberg, E.N., Voronin, V.A.: Intelligent systems interval regulation. Autom. Commun. Inf. **2**. 23–24 (2011). [In Russian: Розенберг, Е.Н. & Воронин, В.А. Интеллектуальные системы интервального регулирования // Журнал «Автоматика, связь и информатика»]
10. Intelligent Transport Systems (ITS) for sustainable mobility. UN, Economic Commission for Europe, UNECE. Geneva, 120 p., February 2012
11. Aliev, M., Aliev, R., Tokhirov, E., Nurmuhamedov, T.: Four-pole rail coefficients of the jointless track circuit in the presence of one of the ends track circuit insulating joints. Chem. Technol. Control Manage. **2019**(4) , 89–92 (2019). Article 6. https://uzjournals.edu.uz/ijctcm/vol2019/iss4/6
12. Aliev, R.M., Aripov, N.M.: Methods for calculating the coefficients of the four-pole rail track circuit without insulating joints for locomotive receiver at service able rail threads. European science review #7-8 Vienna 2015, pp.146–148 (2015)
13. Theeg, G., Vlasenko, S. (eds.): Railway Signalling & Interlocking. International Compendium, 448 p. A DVV Media Group Publication. Eurailpress (2009)

14. Нижниченко, Д. А.: Системы обеспечения безопасности движения высокоскоростных электропоездов. // Журнал «Автоматика, связь и информатика». No 10. - С. 18–21 (2009). [In Russian: Nizhnichenko, D.A. Systems ensuring safety traffic of high-speed elektrotrain // «Automation, communication and informatics»]
15. Алиев, Р. М., Алиев, М. М., Акбаров, У.: Устройство контроля состояния перегона. Патент на полезную модель. Агентство по интеллектуальной собственности Республики Узбекистан. Ташкент. № FAP 01155 от 21.07.2015 г. [In Uzbekistan: Aliev, R.M. & Aliev, M.M. & Akbarov, U. Utility Model Patent FAP 01155 Device control condition track section between stations]

Inclusion of Cyber Security Mechanisms in the Development of the Telematics System

Jana Handriková[1], Júlia Mihoková Jakubčeková[1], Darina Stachová[1(✉)], and Eleonóra Benčíková[2]

[1] Faculty of Security Engineering, University of Žilina, Univerzitná 8215/1, 010 26 Žilina, Slovak Republic
{jana.handrikova,julia.mihokova, darina.stachova}@fbi.uniza.sk

[2] Faculty of Logistics and Crisis Management, Tomas Bata University in Zlín, nám. T. G. Masaryka 5555, 760 01 Zlín, Czech Republic
bencikova@utb.cz

Abstract. The increase in demand for the transport of people and goods is leading to a rapid development of transport systems. Data protection must include the privacy and security of end users, service providers and application providers. In this document, we propose a new data protection framework based on privacy and security technologies. Privacy technology allows users and service providers to define a flexible data model. Security technology provides traditional features such as encryption, authentication, and denial. In addition, it provides a secure environment for secure execution, which is necessary to restrict access to data for specific purposes. The increase in the number of system components places high demands on the management of the entire system. Intelligent control systems used by computer architectures such as cloud computing, fog computing, SDN or NFV architecture. The safety of this system is essential for safe and efficient transport. They also use development systems from research projects that improve the individual functions of telematics systems. Necessary solution already in the development phase, where the requirements of cyber security must be implemented. This study examines the weaknesses of cyber security. Threats and attacks exploiting these vulnerabilities are identified and classified. The final part proposes and discusses the development guidelines and mitigation strategies to be used in the development of autonomous and unmanned systems.

Keywords: Cyber security · Cyber risk · Cyber threat · Cyber terrorism · Cyber-attack · Critical infrastructure · Cyber space

1 Introduction

Key sectors such as transport, energy, health and finance are increasingly dependent on digital technologies for the performance of their core functions. Although digitalization offers enormous opportunities and solutions to many of the challenges facing Europe, not least during the COVID-19 crisis, it also exposes the economy and society to cyber threats. Cyber-attacks and cybercrime are on the rise across Europe in terms of numbers

O. Prentkovskis et al. (Eds.): TRANSBALTICA 2021, LNITI, pp. 375–384, 2022.
https://doi.org/10.1007/978-3-030-94774-3_37

and sophistication. This trend is set to continue in the future, with 22.3 billion devices worldwide expected to be connected to the Internet of Things by 2024. In December 2020, the European Commission and the European External Action Service (EEAS) presented a new European Union (EU) cyber security strategy [12]. The aim of this strategy is to strengthen Europe's resilience to cyber threats and to ensure that all citizens and businesses can make full use of trusted and reliable services and digital tools. The new strategy contains concrete proposals for the introduction of regulatory, investment and policy instruments. Certification plays a key role in ensuring high standards of cyber security for ICT products, services and processes. The fact that different EU countries currently use different safety certification systems creates market fragmentation and regulatory barriers. The Network and Information Security (NIS) Directive was introduced in 2016 as the first ever EU-wide legislative measure to intensify cooperation between Member States in the crucial area of cyber security. It set out security obligations for basic service operators (in key sectors such as energy, transport, healthcare and finance) and digital service providers (online markets, internet search engines and cloud services). In December 2020, the European Commission proposed a revised Network and Information Security (NIS2) Directive to replace the 2016 Directive. The new proposal responds to the evolving threat environment and takes into account the digital transformation of EU society accelerated by the COVID-19 crisis [1] (Fig. 1).

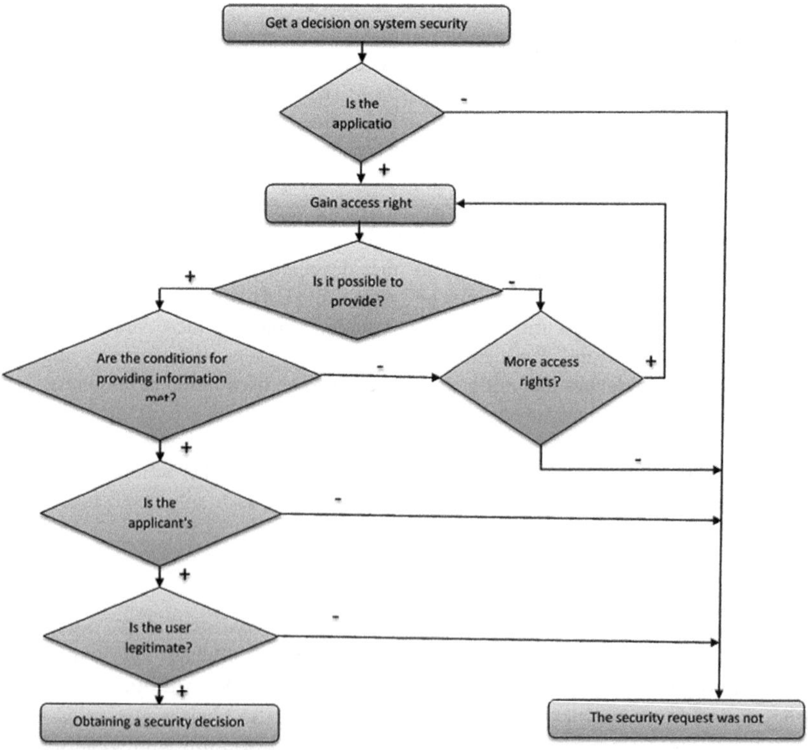

Fig. 1. Functional diagram for obtaining system security information.

Security solutions also include the transport sector. The elements that enter this system and affect each other are human, vehicle and infrastructure. The types of input data with which it is possible to address safety criteria and which affect the safety of the road transport system can be defined according to the authors [2] to:

- Type of data (Vehicle speed, Vehicle position, Communication state, SPaT information, RSU position, Distance to the object, Road/weather conditions),
- Source data (GPS, Temperature sensor, Laser scanner, Ultrasonic sensor, Wireless card, Camera, On-board unit (UBO), On-board diagnostics (OBD)).

Each of these components of the transport system interacts with each other. The interaction takes place through the exchange of information between man and vehicle, between vehicle and infrastructure (vehicle to infrastructure systems), between vehicles to each other (vehicle to vehicle infrastructure). Using any information system brings with it risks. In [3], the authors identified security challenges in information transport systems. In other articles, the authors focused on solving cyber safety issues in a specific type of telematics system, in [4] on Cyber Security and Risk Analysis on Connected Autonomous Vehicles, in [5] the authors present a classification of warning application for road safety. Other articles focus on a specific aspect of cyber security, for example in [6] the authors focused on privacy risk for vehicles during un-encrypted and periodic communication in Vehicular Edge Computing (VEC) systems. Information flows play an important role in road safety. The unavailability of information systems can have an impact on: Impact on people, Impact on the economy, Impacts on the environment, Political impacts, Synergic and cumulative effect of individual impacts. However, as well as ensuring the correct operation, it is important that the CIA triad is maintained - confidentiality, data integrity, availability in the whole process of information exchange - in their acquisition, transmission and processing. Authors in [7] presents an extensive survey of all types of IoT architectures prevalent in the industry. System security, including cyber security, must cover all layers of the system architecture Devices in the Perception layer, Protocols used to connect these devices to the middle-ware, Application interface used to manage these devices and the data generated out of these [7]. Many telematics systems use central cloud data processing. Terminal devices use cloud services over the network (Internet). The advantages of cloud data processing include Terminal devices are only used to record data, they do not transform it into information, Terminal devices are cheaper, less software requirements for terminal equipment, less demands on the management and maintenance of terminal equipment, Cloud allows you to dynamically change allocated resources, Ability to use foreign computing resources in the cloud (data storage, virtual servers, applications), Resilience of the application in terms of the availability of hardware and software resources, Flexibility in the type of device used as an access point (workstations, tablets, mobile phones, etc.), Ability to access this technology through various platforms, Ability to access data without geographical restrictions, Cloud management is provided by its operator. Disadvantages of cloud data processing:

- High demands on the capacity of the transmission path.
- High demands on transmission speed.
- High demands on the security of the transmission path.

The increasing computing power of end devices allows the transfer of part of the data processing to these devices. The authors at [8] focused on investigate challenges of an IoT data stream processing at the edge computing layer. The division of processing between end devices and the cloud brings with it a division of responsibilities for application and data security between the cloud operator and the owner/owners of the end devices. Division of responsibilities is described by the shared security model at Fig. 2.

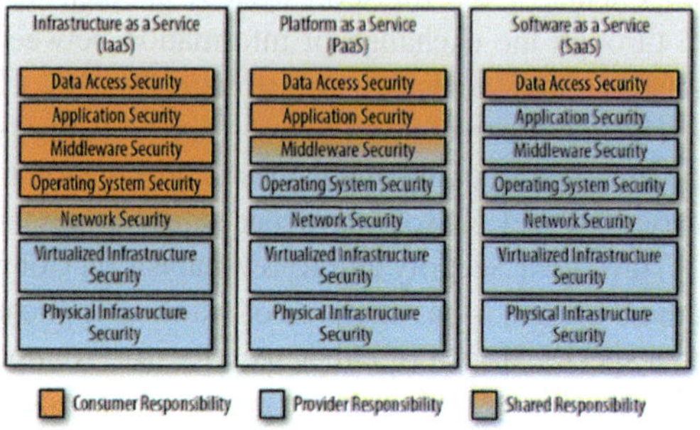

Fig. 2. Division of responsibilities for security in the cloud environment [9].

On the one hand, delegating part of the responsibility to the cloud operator has its undeniable advantages - the operator often invests far more resources in security technologies and security management than a customer using cloud services could afford. However, the data gets out of the owner's control. The terminal accesses the services provided by the cloud over the network, often over the Internet. Terminal equipment, hardware and software are managed by the organization, respectively. A private person, as the owner is therefore responsible for all components of the cyber security of the terminal equipment. Thus, the cyber security of applications using cloud services includes the cyber security of end devices, the data transmission path, including the security of data transmitted over it, and the security of the cloud. The cyber security of the terminal can be described by RBD diagrams as a sequence of subsystems in series, i. e. failure of any subsystem will cause failure of the whole system - terminal equipment. Due to the effort to minimize the price of the terminal equipment, the redundancy of hardware or software components is used to a negligible extent.

$$R_{SYS} = R_1(t) \times R_2(t) \ldots \times R_n(t), \tag{1}$$

where R_{SYS} – reliability of the system, $R_1(t)$ ... $R_n(t)$ – reliability of the subsystems.

The subsystem of connection of the terminal device to the global network can be divided into a subsystem dependent on the terminal device and a subsystem that ensures the transmission of data over the network, their routing, finding the optimal route of their delivery. The terminal may use one or more alternative connection technologies to connect to the global network - at least one connection method must be available for the system as a whole to function properly. Also, data routing over the global network (Internet) can take place in several possible routes depending on the active elements of the network, their routing protocols, network topology - the individual subsystems of possible transmission are in parallel in the RBD diagram. Parallel probabilities are calculated by multiplying the unreliability (Q) of the series components where Q = (1 R) if only one unit needs to function for system success:

$$Q_{SYS} = Q_1(t) \times Q_2(t) \ldots \times Q_n(t) \tag{2}$$

where Q_{SYS} – unreliability of the system ($1-R_{SYS}$), where R_{SYS} is reliability of the system), $Q_1(t) \ldots Q_n(t)$ - unreliability of the subsystems.

Due to the ever-increasing computing capacity of terminal equipment, it is possible to transfer part of the data processing to terminal equipment. This concept uses edge computing. The objectives of such data processing are Reduce the latency (delay) resulting from the need for communication between the cloud and end devices, Reduce the limitations of available network bandwidth, Increase network throughput.

2 Goals and Designed System

Prerequisite for ensuring the availability of the information system, i. e. of all tangible and intangible assets of the owner, which are processed in an automated system, is the identification of threats and vulnerabilities of the system. Security threats have their origin in the external and internal environment of the asset owner. The risk of a security incident can be reduced by identifying system vulnerabilities and taking appropriate security measures. System vulnerabilities can be detected based on previous experience - for example, by performing an attack tree analysis - or by detecting an abnormal state or behaviour of the system. Machine learning algorithms are playing an increasingly important role in recognizing atypical system security phenomena. The aim of the authors was to identify the vulnerabilities of the research project development and to implement security measures to increase cyber security into the development and deployment phase of the research project. 1. Design a system using edge computing features, i. e. a system in which the terminal, in addition to recording data from the environment, also partially processes it in order to limit the volume of data transmitted for further centralized processing. 2. Identify system development vulnerabilities. 3. Identify and implement those security mechanisms that reduce the risk that the research project will not be completed and operated. The aim is to reduce the risks identified as large and medium to acceptable. The created system uses a single-chip Raspberry Pi 2 computer as end devices, with the Debian (Raspbian Stretch with desktop) operating system installed. The license plate recognition application was chosen as a model application. The Open Source library OpenALPR, written in C++, was used. Library functions analyze individual frames of a video to determine the most likely textual representation of license plate characters. The software platform BOINC (Berkeley Open Infrastructure for Network Computing) was used for centralized processing, primarily intended for volunteer computing and computer grids. It is free software distributed under the Lesser General Public License (LGPL), version 3 and higher. Communication between these basic elements of the system took place over a fixed computer network, the communication was not encrypted.

3 Methodology

The taxonomy listed in [10] was used to detect the vulnerability of the system. Based on the fact that it was a research project, with a simple architecture and a small number of components, the risks arising from the identified vulnerability were assessed only qualitatively, i.e. verbal expression. Risk is a function of the probability of an event occurring and its effect.

1. Vulnerabilities resulting from external events: The system proposed by us is intended for research purposes, therefore the risk resulting from natural disasters was classified as small. It envisaged its development, testing and operation in a controlled environment. The Raspberry Pi terminal was chosen, which is suitable for outdoor use due to its characteristics. Due to current events, the COVID 19 crisis, the classification of this risk has been changed to high risk. The risk arising from the conflict with the legislative environment was classified as large. In order to reduce this risk to acceptable, several safety measures have been taken. Image recordings of the vehicles of the members of the research team were used to obtain test data. Due to the use of a private cloud, members of the research team were responsible for security in the cloud. The risk arising from insufficient financial security of the project was classified as a large risk. In order to minimize financial costs, a Raspberry Pi terminal was chosen, with a purchase price of approximately $50. The system designed for central processing of data obtained from the perception layer was installed on a regular PC, which was available to members of the research team. Open Source software was chosen. The risk arising from the dependence on the availability of services of other providers was assessed as small. If data transmission via the mobile network were used, this risk would increase. It would be assessed as an acceptable risk.

2. Failed Internal Processes: The number of members of the research team was limited. Therefore, only a very small number of processes related to the design or operation of the system have been identified. These processes mainly covered the area of communication between members of the research team, determining responsibility for the design and operation of each subsystem, creating documentation, description of detected system error states, possible causes, setting incident resolution procedures to put the system into operation. The members of the research team are also its operators, the processes related to control processes were mainly related to the stated research objectives - for example, the creation of metrics related to image processing speed in case of process parallelization. Due to the introduction of security measures, the risk assessment has been changed from large to acceptable.

3. System and Technology failures: Each automatic data processing system consists of hardware, software resources and the provision of communication between these individual components. Hardware vulnerabilities result from the development and operation of the system on components with insufficient performance, capacity, or high maintenance requirements. Due to the goals of the research project, we chose the

Raspberry PI as a terminal. Characteristics Raspberry Pi are Availability, The price, ARM processor, Battery powered option, Low power consumption and the resulting low need for cooling, MicroSD card slot, Possibility to connect a permanent storage medium via USB. The BOINC system, which provided centralized processing of the analysed camera recording, was installed on a standard PC. Neither in the case of the terminal equipment nor in the case of the centralized data processing equipment has cyber security been increased by the use of hardware component redundancy. If a cloud solution is chosen, the redundancy of hardware resources in the cloud would increase the level of cyber security, but would disproportionately increase the cost of the system and the requirements for its maintenance, including increased demands on human resources. The BOINC system was chosen for central processing, which confirmed its ability to process even large-scale tasks by processing many volunteer calculations. They are listed on the pages of this software, as shown in Table 1.

Table 1. The website [11] describes some types of security problems and built-in options for their elimination.

Vulnerability	Precautions
Falsifying results	Setting validator rules - for example, a result is considered correct if at least two out of three results are the same according to the set metric. Adaptive replication of tasks, fuzzy logic can be used to determine the limits of tolerance of the validity of the result
Credit forgery (ratings). Attackers return results and declare higher CPU usage time than they actually were	BOINC tries to avoid some problems: for example, it detects when applications are using too much disk space, memory, or CPU time and interrupts the execution of those applications
Distribution of malicious executable files. Attackers infiltrate the BOINC server and, by modifying the database and files, try to distribute their own executable files (such as virus programs) that are disguised as BOINC applications	BIONC uses a public signature procedure based on cryptography using a public key to authenticate executables and the BOINC project server
Data server capacity exceeded. Attackers repeatedly send large files to BOINC data servers, filling their disks and degrading them	BOINC provides an optional certificate mechanism to prevent attacks on data servers. Each output file is assigned a maximum size. Each project has a pair of authentication keys to upload
Theft of a participant's account information by a network attack. Attackers use BOINC protocols to steal account information	BOINC does not provide any mechanisms to reduce the risk of obtaining project participant account information by attacking a computer network. Security can be enhanced by using high-security communication protocols - such as https, not http

(*continued*)

Table 1. (*continued*)

Vulnerability	Precautions
Theft of project files. Attackers steal input and/or output files	The input and output files used by BOINC applications are not encrypted Use high-security communication protocols - such as https, not http
Intentional misuse of host computers by projects. The project intentionally distributes an application that abuses host participants, e.g. by stealing sensitive information stored in files	Projects can minimize the likelihood of causing problems by pre-testing applications. Projects should thoroughly test their applications on all platforms and with all input scenarios before upgrading them to production status A user who plans to participate in a project should always check the institution that created the project and request to participate in it
Unintentional misuse of host computers by projects. The project releases an Application that inadvertently abuses the host, e.g. deleting files or causing failures	A user who plans to participate in a project should always check the institution that created the project and request to participate in it. If such behavior is suspected, the application should terminate its participation in the project and inform the project creator

With regard to the security of the communication, it is necessary to assess in particular the availability of the transmission path, its capacity and the use of transmission protocols encrypting the communication and verifying the identity of the subscribers, for example by using public and private keys. One of the basic security measures is the control of physical and logical access to individual components of the system. Logical access control was performed only through one-step authentication, using a password. Failure of any subsystem will cause the whole system to fail, so the risk related to System and Technology failures was assessed as large. Following the introduction of security measures, it has been changed to acceptable.

4 Actions of People

The group of people who interact with the system is very small - it consists only of members of the research team. The risk of intentional damage due to a very small team can be considered small. We can consider the risk resulting from insufficient knowledge and experience of the persons involved as medium risk. The risk of errors and delays is about the same. In order to minimize the risk arising from these vulnerabilities, the processes of mutual regular inspection and partial testing of the system were set up. The vulnerabilities before and after the introduction of security measures are shown in Table 2.

Table 2. Vulnerability before and after the introduction of security measures.

Vulnerability	Before implementing security measures	After the introduction of security measures
External events		
Disaster (except pandemic)	Small	
Pandemic	Big	
Legal issues	Big	Acceptable
Business Issues	Big	Acceptable
Service dependency	Small, resp. acceptable	
Failed Internal Processes		
Process design or execution	Big	Acceptable
Process control	Big	Acceptable
Supporting Process	Big	Acceptable
System and Technology failures		
HW	Big	Acceptable
SW	Big	Acceptable
Systems	Big	Acceptable
Action of people		
Inadvertent	Medium	Acceptable
Deliberated	Small	
Inaction	Medium	Acceptable

5 Conclusion

Cyber security helps to identify, assess and address threats in cyberspace, reduce cyber risks and eliminate the impact of cyber-attacks through information crime, cyber terrorism and cyber espionage by strengthening the confidentiality, integrity and availability of data, systems and other elements of the information and communication infrastructure. Connected devices, including the machinery, sensors and networks that make up the Internet of Things, will be key to further shaping Europe's digital future, and their security will be just as important. 5G networks are key for critical industries such as energy, transport, banking and healthcare. Ensuring the resilience of 5G networks is therefore very important for our company. The technology is key to Europe's competitiveness in the global market, and its cyber security is crucial to achieving the Union's strategic autonomy. It is crucial to find innovative solutions that will protect us from the latest, most advanced cyber threats. For this reason, cyber security is an important part of the EU's framework programs for funding research and innovation. An interesting direction for further research is metrics to monitor the effectiveness of the proposed elimination tactics. Cyber security is becoming one of the most important challenges today and it is therefore essential that states at national level adopt binding legislation in the field of national cybersecurity protection, which would ensure an adequate level of protection of critical infrastructure and basic security areas of the state.

References

1. European Commission. Revised Directive on Security of Network and Information Systems (NIS2). https://ec.europa.eu/digital-single-market/en/news/revised-directive-security-network-and-information-systems-nis2. Accessed 16 Mar 2020
2. Malik, R.Q., et al.: Mapping and deep analysis of vehicle-to-infrastructure communication systems: coherent taxonomy, datasets, evaluation and performance measurements, motivations, open challenges, recommendations, and methodological aspects. IEEE Access **7**, 126753–126772 (2019). http://ieeexplore.ieee.org/stamp/stamp.jsp?tp=&arnumber=8758170&isnumber=8600701. Accessed 04 Mar 2021
3. Zhao, M., Walker, J., Wang, C.-C.: Security challenges for the intelligent transportation system. In: Proceedings of the 1st International Conference on Security of Internet of Things, SecurIT 2012, pp. 107–115. Association for Computing Machinery, New York (2012). https://www.researchgate.net/publication/262239533_Security_challenges_for_the_intelligent_transportation_system. Accessed 12 Jan 2021
4. Jonnalagadda, S.K., Jakkala, P.R.: Cyber security and risk analysis on connected autonomous vehicles. Solid State Technol. (2020). https://www.researchgate.net/profile/Surya-Kiran-Jonnalagadda/publication/347909272_Cyber_Security_and_Risk_Analysis_on_Connected_Autonomous_Vehicles/links/5fe6c635299bf14088440b1a/Cyber-Security-and-Risk-Analysis-on-Connected-Autonomous-Vehicles.pdf. Accessed 15 Feb 2021
5. Trager, J., Kalová, L., Pagany, R., Dorner, W.: Warning apps for road safety: a technological and economical perspective for autonomous driving – the warning task in the transition from human driver to automated driving. Int. J. Hum. Comput. Inter. **37**(4), 363–377 (2021). https://www.tandfonline.com/doi/full/10.1080/10447318.2020.1860545. Accessed 25 Feb 2021
6. Shit, R.C., et al.: Privacy-preserving cooperative localization in vehicular edge computing infrastructure. Concurr. Computat. Pract. Exp., e5827 (2020). Early View. https://onlinelibrary.wiley.com/doi/full/10.1002/cpe.5827. Accessed 18 Oct 2020
7. Acharya, V., Hegde, V.V.: Security frameworks for internet of things systems - a comprehensive survey. In: 2020 3rd International Conference on Smart Systems and Inventive Technology (ICSSIT), Tirunelveli, India, 2020, pp. 339–345. https://doi.org/10.1109/ICSSIT48917.2020.9214127. http://ieeexplore.ieee.org/stamp/stamp.jsp?tp=&arnumber=9214127&isnumber=9214075. Accessed 18 Mar 2021
8. Xhafa, F., Kilic, B., Krause, P.: Evaluation of IoT stream processing at edge computing layer for semantic data enrichment. Fut. Gener. Comput. Syst. **105**, 730–736 (2020). ISSN 0167-739X. https://www.sciencedirect.com/science/article/pii/S0167739X19321296. Accessed 22 Nov 2020
9. Dotson, C.: Practical Cloud Security. O´Reilly. https://www.oreilly.com/library/view/practical-cloud-security/9781492037507/ch01.html. Accessed 10 Nov 2020
10. Pardini, D.J., Heinisch, A.M.C., Parreiras, F.S.: Cyber security governance and management for smart grids in Brazilian energy utilities. JISTEM J. Inf. Syst. Technol. Manag. 14(3), 385–400 (2017). ISSN 1807-1775. http://www.scielo.br/scielo.php?script=sci_arttext&pid=S1807-17752017000300385&lng=en&nrm=iso. Accessed 31 Mar 2021
11. Boinc. Security issues in volunteer computing. https://boinc.berkeley.edu/trac/wiki/SecurityIssues. Accessed 31 Mar 2021
12. European Council: Council of the European Union: Cybersecurity: how the EU tackles cyber threats. https://www.consilium.europa.eu/sk/policies/cybersecurity/

Interaction Between Marketing and Logistics in Transport Customer Service

Kristina Čižiūnienė[1(✉)], Ieva Meidutė-Kavaliauskienė[2], and Diana Višneveckaja[1]

[1] Faculty of Transport Engineering, Department of Logistics and Transport Management, Vilnius Gediminas Technical University, Plytinės g. 27, 10105 Vilnius, Lithuania
kristina.ciziuniene@vilniustech.lt, diana.visneveckaja@stud.vgtu.lt

[2] Department of Business Technologies and Entrepreneurship, Vilnius Gediminas Technical University, Saulėtekio al. 11, 10223 Vilnius, Lithuania
ieva.meidute-kavaliauskiene@vilniustech.lt

Abstract. Increasing globalization of international trade and transnationalization of companies require free movement of resources and goods, which is not possible without transport. Here consistent coordination of the movement of materials and goods offering a solution to many problems becomes vital. In order to successfully develop their activities, businesses need to focus on customers, searching for new ways to create added value and to establish long-term relationships with them. The transport sector is no exception, as this is where both marketing and logistics interests intersect and must be coordinated successfully. The essential idea of marketing, which promotes the establishment of long-term relationships between a customer and an organization through the benefits of services provided to the consumer, goods sold and other marketing elements, has become increasingly popular among businesses. Competition is a distinctive feature of any activity where interests clash, but logistics and marketing have the greatest impact on competitiveness. However, in order to fully exploit efficiency of both logistics and marketing, analyzing these areas and their tools both separately and from the perspective of their points of interaction in transport companies is necessary. This article examines the opportunities offered by logistics and marketing integration, also presenting the results of a quantitative research (by questionnaire) in the transport sector.

Keywords: Transport sector · Logistics · Marketing

1 Introduction

Increasing globalization of international trade requires free movement of resources and goods, which is not possible without logistics and transport services. According to [1, 2], consistent coordination of movement of materials and goods, which offers a solution to many problems, becomes vital. Thus, in order to successfully develop their activities, businesses need to focus on customers, searching for new ways to create added value and establish long-term relationships with them [3].

O. Prentkovskis et al. (Eds.): TRANSBALTICA 2021, LNITI, pp. 385–393, 2022.
https://doi.org/10.1007/978-3-030-94774-3_38

Logistics services are exceptional, so the assessment of service quality is a difficult task, as services have absolutely different features compared to tangible goods (such as intangibility or inseparability from the consumer, which becomes a subjective factor in assessing quality, short-term nature, etc.) [4]. It should be noted that the provision of services can be complicated as it involves different processes and participants in different geographic locations and with different distribution [5–7]. Assessment of the quality of services provided by one or another logistics company requires choosing respective quality indicators, which are not always directly related to the level of customer satisfaction. Misuse of information, improperly selected and evaluated customer service model or random decisions can lead to significant financial losses, poor quality of services and a defeat in competition for a viable place in the market.

The aim of this article is to assess the interaction between marketing and logistics in customer service in the transport sector.

The article consists of 3 parts: introduction, Sect. 2 with three subsections, where the first one presents relevant research, the second offers a research methodology, and the third presents the results of the quantitative research, and the conclusions of the article. Quantitative research was used to determine the significance of logistics and marketing tools in transport companies.

Methods used in the article include analysis and empirical research (questionnaire).

2 Customer Service as a Point of Interaction Between Marketing and Logistics

This part of the article analyses two key aspects: integration of logistics and marketing from a theoretical point of view, and results of a quantitative research of the transport sector.

2.1 Logistics and Marketing Integration

Under current conditions, when the success of companies is measured based on their focus on customers, service process takes on a different meaning, where order processing of each individual customer rather than increasing the volume of services becomes the decisive factor. This promotes interaction between logistics and marketing. The main directions of logistics and marketing are the development of customer service policy, creation of logistics service standards and development of relationships between suppliers and customers [8, 9]. The author believes that the solution of the task of customer service policy development first requires conducting an analysis of the company's capabilities, segmenting the consumer market by consumer group, the level of service use and customer approach to the services provided.

Consumers set their parameters for the service and its characteristics. This is why identifying groups of potential consumers is very important. Marketing complements and develops logistics, integrates everything into a mobile system, therefore, in order to identify priority elements of logistics, consumer behavior must be analyzed [10]. Parameters, such as loyalty, image, returns, number of orders, comprehensiveness of purchased services, and costs of losing a customer are used in the analysis. Each group is subject to an individual incentive programme to increase customer loyalty. Consumer behavior and its forecasts have been analyzed using such programmes as discounts, credits, deferrals and bonuses for new purchases. The interaction between logistics and marketing was observed to allow ensuring a higher level of service and cutting costs in the supply chain. The common tasks of logistics and marketing include research of demand and supply, provision of information and economic analysis [11].

Marketing is perceived as a system of approaches and a possibility to create strategic tasks in marketing processes, while logistics is a marketing mechanism [12]. The scheme below (see Fig. 1) has been used in the analysis of interaction between logistics and marketing the ultimate goal of which is customer service [13].

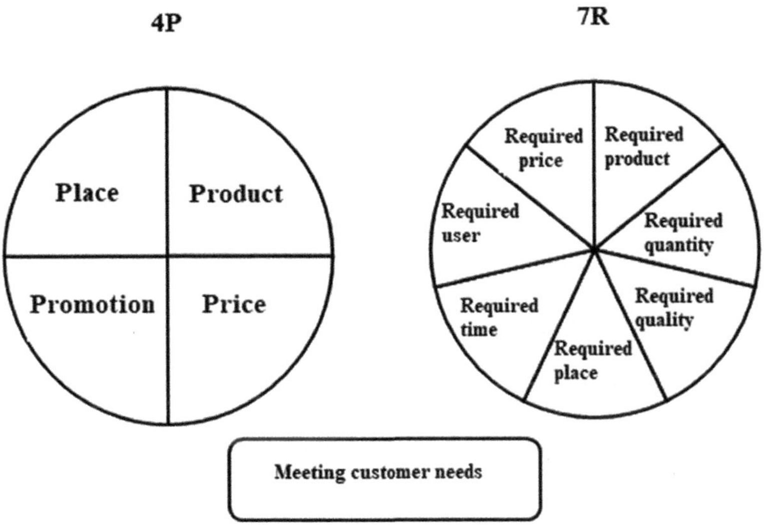

Fig. 1. Interaction between logistics and marketing [13].

The assessment of marketing and logistics often leads to a controversy over which one of them is more important. Marketing shapes and defines the demand, while logistics ensures traffic flow to the consumer. The two of them are inseparable. In order to be successful, integrating them is absolutely necessary, because marketing is a concept of management (planning, organization, control) and sales focused on demand, while logistics is focused on managing material and information flows and their effective use in solving complex tasks inside and outside the company and seeking to meet customer needs.

Marketing is focused on market research, advertising and psychological impact on the consumer, while logistics focuses on the development of technological systems in the supply chain [14]. One of the main ways to create value for customers through interaction between marketing and logistics is to deliver goods on time or to reduce the delivery time. Also, integration of marketing and logistics is essential when it comes to customer service. From the customer's point of view, quality, speed (time) and reliability are the requirements which customers raise for interaction between marketing and logistics [15].

Therefore, when planning their activities, organizations must include all elements of the marketing complex in them, but they can implement them using different strategies [16]. A combination of marketing mix elements must be determined taking into account both end users and distribution channels [17]. The marketing mix consisting of product, price, place and promotion (4Ps) must be analyzed as marketing tools that need to be optimally combined in order to achieve marketing goals [18], while pre-transaction, transaction and post-transaction elements of customer logistics must respond thereto. Increasing the demand for services is impossible without the marketing concepts [19]. The relationship between logistics and marketing can be seen in the assessment of types of services and goods, pricing, promotion, sales forecasting, analysis, and the development of customer service policies [20]. Marketing is one of the areas which companies should focus on the most.

2.2 Research Methodology

Research object was the application of logistics and marketing measures in the customer service of transport companies.

Aim of the research was to determine the possibility of the assessment of quality of customer service of transport companies.

Hypotheses raised:

H1 – companies operating in the transport sector devote less attention to marketing tools;

H2 – companies operating in the transport sector conduct too few customer service quality surveys;

H3 – focus of companies operating in the transport sector has been equally divided among quality assessment criteria.

Research method – a questionnaire survey, which was conducted in companies providing logistics and transport services. Representatives of logistics and transport companies were interviewed in the survey. In order to obtain the most accurate data possible and to ensure the reliability of the research, correct sample of the research object had to be chosen. It was calculated using the Paniott formula:

$$n = \frac{1}{\Delta^2 + \frac{1}{N}}, \tag{1}$$

where n – sample size, Δ – permitted error value, N – size of the entire population.

According to the data of the public register of companies *rekvizitai.lt*, having included in the search such areas of activity as transport services, logistics services and forwarding, 11,506 companies providing these services in Lithuania during the research period were found.

The sample size is the number of respondents to be interviewed during a research in order to have the results reflect the opinion of all respondents (population) with the chosen probability and error. The sample must be representative in order to be able to decide on the entire population based on the results. Having chosen a probability of 95% and an error of 7, the sample size of 193 was used.

2.3 Results of the Quantitative Research

78.5% of all responses came from small businesses, 16.9% - from medium-sized enterprises and 4.6% - from large enterprises. The average age of the respondents ranged from 26 to 35 years.

Representatives of companies were asked to answer how they rate the quality of their customer service from 1 to 10. The overall average of the answers received was 8.07 points.

Respondents had to choose which criterion determined competitive advantage of their company (see Fig. 2).

Fig. 2. Criteria reflecting why services of the company have been chosen.

Timely service, prompt response to inquiries, and awareness of the company's brand were among the most popular answers. The level of professionalism of employees was a less popular option, chosen by 12.6% of respondents only. Such criteria as a great variety and uniqueness of services were also included in the research.

The research aimed to identify the criteria that affect customer dissatisfaction (see Fig. 3).

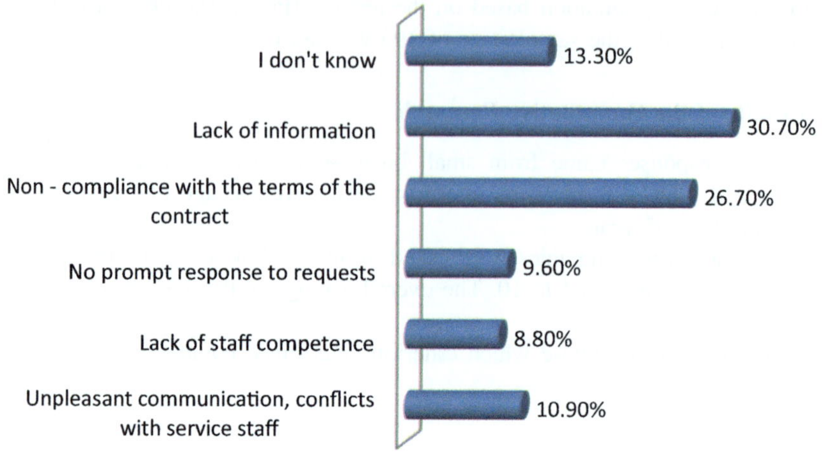

Fig. 3. Criteria that cause customer dissatisfaction.

30.7% of respondents mentioned lack of information as one of the main reasons why customers are dissatisfied. Non-compliance with contract conditions was named as having a significant impact, and was chosen by 26.7% of respondents. The third most popular answer was "I don't know" with 13.3% of respondents having chosen it. This may mean that these companies either do not conduct any customer service surveys or do not inform employees about processes ongoing in the company. The assessment of these answers was aimed at finding out whether companies should improve their customer service. 53% of respondents said improvements were still needed. This suggests that companies need a customer service assessment. Respondents' responses on the application of respective measures are presented in Table 1.

Table 1. Use of different measures.

Measure	Responses (%)	
	Yes	No
Quality measurement system	27	73
CRM	44	56
Customer service standards	41.7	58.3
Quality improvement research	22.5	77.5
Advertising	84.9	15.1

Advertising is one of the most popular measures among companies.

Customer service in the transport sector is of two types: logistics and marketing. The research revealed that logistics service was given a priority with 53.3% of respondents having chosen this answer (see Fig. 4).

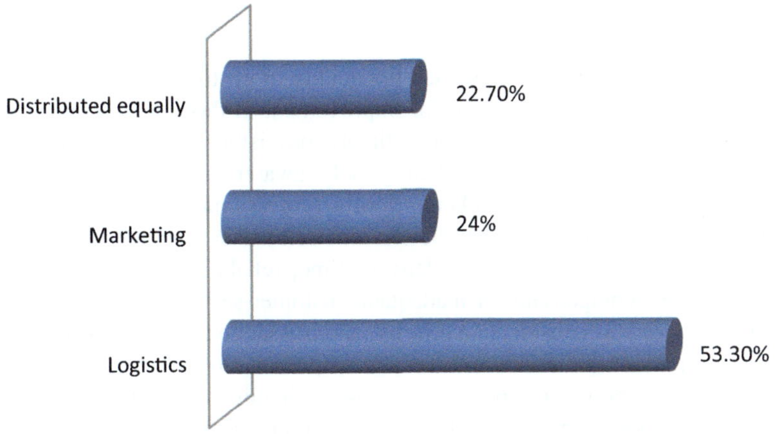

Fig. 4. Distribution of companies by priority type of customer service.

Respondents were also asked to assess which elements have been fostered more in their company (1 – most important; 3 – least important). Interestingly, the obtained data revealed that transaction elements came in first, followed by pre-transaction elements and post-transaction elements.

The primary focus in the processing of orders was on transaction elements, and, unlike in case of purchasing goods, customers in the customer service sector did not find post-transaction elements important.

During the survey, respondents also assessed the importance of quality assessment criteria, logistics and marketing tools:

- **Importance of quality assessment criteria in serving customers.** Respondents indicated that the criteria of materiality, responsiveness and reliability were most important to them. Criteria of comprehensiveness and persuasiveness were least popular.
- **Importance of logistics tools in customer service.** When assessing logistics tools, respondents chose planning and forecasting, analysis and cost optimization, and information technology as their top priority.
- **Importance of marketing tools in customer service.** When assessing marketing tools, respondents unanimously (96.9%) indicated the price as the most important marketing tool. Advertising was somewhat less popular referred to as important by 72% of respondents, and sales was indicated by 62.9% of respondents. A mere 24.9% of respondents referred to service standards as important.

Respondents also spoke on the impact of the marketing tool – staff and logistics tools – on customer service. Respondents were more likely to disagree with the fact that service staff was more important for customer service quality than the service itself. As many as 93.3% of respondents agreed that delivery time, reliability and flexibility compensate for poor customer service. However, 83.9% of respondents believed that frequent rotation of employees led to poorer customer service quality.

3 Conclusions

1. During the research, respondents rated the quality of customer service at 8.07, and they agreed that there still is room for improvement in this area in companies.
2. Customers found the logistics tool – timely provision of services and prompt response to inquiries –, and the marketing tool – awareness of the company's brand – most appealing, while the marketing tool – professional employees – was least important. Respondents think that the service itself rather than staff determines the quality of customer service, while delivery time, reliability and flexibility, rather than employees, compensate for inadequate customer service.
3. Logistics service was considered a priority, and transaction elements were named as the most important, which reconfirmed the importance of logistics tools – timely provision of the service and prompt response to inquiries. Reliability was considered to be the most important quality assessment criterion.
4. More than a half of the surveyed companies do not have a quality measurement system, CRM or customer service standards in place and do not conduct quality improvement surveys or customer surveys, which confirms that most companies do not know why their customers are dissatisfied. Areas for improvement, such as provision of information to customers, performance of contract terms and conditions and informing employees about emerging issues, were identified.
5. The hypotheses raised at the beginning of the research proved to be right: **H1** – companies operating in the transport sector devote less attention to marketing tools. **Substantiation:** 53.3% of companies which took part in the survey focus more on logistics service. **H2** – companies operating in the transport sector conduct few surveys of customer service quality. **Substantiation:** 77.5% of companies that took part in the research do not conduct any quality improvement surveys.
6. One hypothesis raised at the beginning of the research proved wrong: **H3** – companies operating in the transport sector divide their attention equally among different quality assessment criteria. **Denial:** 93.3% of the surveyed companies considered reliability (timely processing of orders and compliance with conditions) to be the most important, while a mere 13% of companies considered persuasiveness as relevant.

References

1. Meidute-Kavaliauskiene, I., Vasiliene-Vasiliauskiene, V., Vasilis Vasiliauskas, A.: Identification of sectorial logistics service quality gaps by applying SERVQUAL method. Transport **35**(4), 419–434 (2020)
2. Lambert, D.M., Cooper, M.C.: Issues in supply chain management. Ind. Mark. Manag. **29** (1), 65–83 (2000)
3. Fonseca, J.P.C., Ferreira, F.A.F., Pereira, L.F., Govindan, K., Meidute-Kavaliauskiene, I.: Analyzing determinants of environmental conduct in small and medium-sized enterprises: a sociotechnical approach. J. Cleaner Prod. **256**, 1–13 (2020)
4. Kilibarda, M., Andrejic, M., Popovic, V.: Research in logistics service quality: a systematic literature review. Transport **35**(2), 224–235 (2020)

5. Bazaras, D., Palšaitis, R., Čižiūnienė, K., Petraška, A., Kaminskas, K.: Assessment of the influence of social-cultural environment in the context of global logistics. In: Kabashkin, I., Yatskiv (Jackiva), I., Prentkovskis, O. (eds.) RelStat 2018. LNNS, vol. 68, pp. 647–657. Springer, Cham (2019). https://doi.org/10.1007/978-3-030-12450-2_62
6. Dahooie, J., Babgoharia, A.Z., Meidutė-Kavaliauskienė, I., Govindan, K.: Prioritising sustainable supply chain management practices by their impact on multiple interacting barriers. Int. J. Sust. Dev. World **28**(3), 267–290 (2021)
7. Restuputri, D.P., Masudin, I., Sari, C.P.: Customers perception on logistics service quality using Kansei engineering: empirical evidence from Indonesian logistics providers. Cogent Bus. Manag. **7**(1), 1751021 (2020)
8. Абрамова, Е.Р.: Взаимодействие маркетинга и логистики для формирования системы потребительского сервиса в цепях поставок. Вопросы структуризации экономики **3**, 112–115 (2010)
9. Şükrü Akdoğan, M., Durak, A.: Logistic and marketing performances of logistics companies: a comparison between Germany and Turkey. Procedia. Soc. Behav. Sci. **235**, 576–586 (2016)
10. Игрокова, К.А.: Сравнительный анализ деятельности служб маркетинга и логистики. *Вестник университета (государственный университет* управления), 197–200 (2016)
11. Шилько, ИС.: Совершенствование управления системой сбыта продукции предприятия на основе взаимоденйствия маркетинга и логи стики. Перспективы науки и образования **2**(30), 164–171 (2013)
12. Новгородова, Т.А.: Взаимодействие маркетинга и логистики. Актуальные проблемы авиации и космонавтики **2**(11), 320–322 (2015)
13. Макарова, ИВ., Семенова, Е.А.: Влияние логистики на конкурентоспособность предприятия. Молодий вчений **1**(28), 178–185 (2016)
14. Репич, Т.А.: Маркетинг и логистика – что важнее? Мясной бизнес, 86–89 (2008)
15. Beniušienė, I., Tijūnaitienė, R.: Marketingo logistika: laiko verte aptarnaujant klientus. **1**(5), 111–114 (2005)
16. Jarašūnienė, A., Batarlienė, N., Čižiūnienė, K.: Business risk management at transport companies: Lithuanian study case. In: Transport means 2018: Proceedings of the 22nd International Scientific Conference, Trakai, Lithuania. Part 1, 03–05 October 2018, pp. 297–340. Kaunas University of Technology, Kaunas (2018)
17. Rutkauskas, A.V., Ginevičius, A., Stasytytė, V.: Optimization of the structure of marketing complex costs as a means of business sustainable development. Ekonomika [Economics] **78**, 115–133 (2007)
18. Jakučionytė, K., Jakubavičius, A.: Tarptautinis marketingas plėtojant logistikos paslaugas: inovatyvūs sprendimai, iš Tarptautinės ekonomikos ir vadybos [International economics and management]: 17-osios Lietuvos jaunųjų mokslininkų konferencijos "Mokslas – Lietuvos ateitis", įvykusios Vilniuje 2014 m. Vasario 6 d., pranešimų medžiaga. Vilnius: Technika, pp. 145–148 (2014)
19. Хаирова, С.М.: Маркетинговое и логистическое обеспечение услуг транспортно-экспедиционных организаций региона. Вестник Сибирской государственной автомобильно-дорожной академии **2**(24), 136–140 (2012)
20. Kempa, E.: Logistics and marketing management as an element of creating value system. Adv. Logist. Syst. **4**(1), 109–113 (2010)

Sales Forecast and Layout Analysis of a Damper Manufacturing Industry for Autonomous Transport

Nalina Hamsaiyni Venkatesh(✉)

Kaunas University of Technology, Studentu g. 56, 51424 Kaunas, Lithuania
nalven@ktu.lt

Abstract. Technological advancements in the field of transport and mobility are numerous with advancements and improvements in every possible scope. The subsystems of vehicles are a constant area of research for improvement, by the fact a new development in the field of magnetorheological fluid dampers has been proposed in published previous research. The proposed design of the damper and its features can capture the market with proper background planning to establish a small-scale manufacturing industry. This paper deals with discussing and planning a layout to establish an industry and to analyze the break-even sales quantity based on the inquired costs. The market sales analysis is performed using a different parameter that affects sales figures. Employing the bass business model to the discussed conditions the different sales scenarios are predicted and analyzed to come up with strategic plans to alter the sales trends in the time of declension. The potential outcomes and suggestions are discussed to conclude a complete background study to establish a small-scale automotive industry.

Keywords: Magnetorheological damper · Industrial layout · Break-even analysis · Bass diffusion model · Sales forecast · Market scenario analysis

1 Introduction

The potential advancement in the suspension industry has been in research for a decade to develop various improvements. Several types of research are being carried to validate different alternatives. As suspensions play a vital role in vehicle stability and safety, the experimental testing and validation required to evaluate the performance are done carefully. The parameters exhibited during testing are analyzed to determine the performance. Certain designs proposed by researchers have been experimentally and numerically verified to exhibit expected outcome performance. A research model of a magnetorheological fluid damper was developed with a specifically formulated magnetorheological fluid to eliminate certain downsides commonly experienced with the conventional ones. The proposed design exhibited promising experimental results paving a way towards a future scope of converting the design into a product and some areas of research improvements [1]. This work concentrates on further discussion of the background work and business analytics of the proposed design as a small-scale original equipment manufacturing (OEM) industry.

O. Prentkovskis et al. (Eds.): TRANSBALTICA 2021, LNITI, pp. 394–406, 2022.
https://doi.org/10.1007/978-3-030-94774-3_39

There several practices on building an industrial layout based on different parameters and processes involved in the manufacturing of the products. The layout built using the scope of JAVA called the Computerized Relative Allocation of Facilities Techniques (CRAFT). The main aim of the implementation of these techniques is to obtain a facility layout having minimal material flow cost. The study was based on estimating the process step and the distance between the process station to determine a matrix to evaluate the best and most efficient layout to reduce the induced costs using mathematical equation [2]. The aisle structures in a facility layout design have a huge role and affect the cost involved in the movement, time, and efficiency associated with the movement of materials and goods. The aisle structures must carefully analyze and designed considering different parameters such as the usage of the entry and exits of a facility and its size, width, orientation, and so on. A mathematical model has been discussed and devised to determine the appropriate size, orientation, width, and several aisles in a particular facility to effectively use and cut down the wastes associated with it [3]. The facility layout design or industrial layout depends on numerous factors as indicated in different research. The parameters which decide on sustainable factors of the designed layout are termed as the sustainable facility layout which holds account of the social impacts such as the power consumed, pollution, and safety. These are discussed in detail for which a mathematical formula has been generated to determine the factor. This was built using numerous technologies such as big data analytics, machine learning, meta-heuristic which provide various data to analyze the sustainable compatibility of the layout [4]. A literature review of the parameters that drive the design of facility layout problems is studied and categorized based on the drivers. The study serves as a useful guideline tool for a designer as it guides and suggests some latest methodologies and their research to obtain data of process, method, and mathematical model [5]. The layout specifically customized to adopted for a foundry was proposed and studied to provide data on constraints and operation guidelines to cooperate with them [6]. There are several studies and predicting models to create a forecast of the adoption of new technologies in the market. In transportation and mobility sector are in research to develop CAM. Many researchers have contributed towards predicting the outcome of these various technologies under different scenarios using different prediction models. Adoption different latest technologies and their contribution towards enhancing the performance of autonomous vehicles are a constant area of research to scholars. A study was conducted using bass model to determine the adoption of autonomous truck for freight purpose [12]. The new car launch model and its sales prediction was carried out using previously launched similar model to evaluate the market statistics [13]. Different business models that are in use and their differences have been stated and discussed in this research to evaluate the electric vehicle adoption [16]. The research was dedicated to OEM manufacturing companies and the different business model and strategies to be adopted to capture market was discussed [17]. A review of the sustainable business models for automotive field was summarized [18]. A Delphi based scenario analysis to determine the most accurate business model for predict of automotive parts was covered in this research [19]. The development scope lying in the autonomous vehicle and its market has been analyzed and discussed by the author [21].

2 Main Objectives

The objective of this research is to create a simple facility layout design for the proposed OEM small-scale manufacturing industry to plan the functioning. With the data and parameters of planning from the facility layout, further approximation of the estimated costs can be built to analyze the business aspects of converting the new design into a potential commercial product. The estimated costs can be used to determine the break-even quantity of the damper. A bass business model can be created for the same to study the trend in sales with a predicted set of scenarios. As this prediction model can be employed to a product with no historic data, it can be used to estimate the trend. This model is based on two potential factors or categories of the market namely innovators and imitators. This will be a brief background analysis to estimate the outcomes of converting the invention into a commercial product.

3 Facility Layout Plan

The basic classification of the layouts in production based on the size of the industry and pattern of the production process carried out is discussed to categorize the large variety. The key focus of the research was to establish the easiest way to develop an effective and maintenance-free layout with minimal effort by classifying certain parameters [7]. With this research as a reference, a simple layout was built on a small-scale industry basis. The complexity of the designed layout model can be determined by using some simple denotation and parameter analysis of the layout. These are discussed and elaborated by researchers to understand the link between the factors or parameters that govern the complexity of the layout structure [8]. The potential geographical and market conditions were analyzed to determine the challenges in starting up a new manufacturing industry to manufacture the designed damper. The market analysis was done to determine the potential competitors, their features, and sales background to obtain data on their product and customer requirements. The product is altered and tailored to fit customers of different transport users. The layout of the manufacturing unit was designed in such a manner that the movement between departments is reduced as much as possible by keeping the process cycle a guideline to avoid rerouting. The manufacturing line is divided into three different departments with multiple stages of the process namely: fluid preparation department, manufacturing department, and assembly department respectively, and the process flow chart as shown in Fig. 1 and Fig. 2.

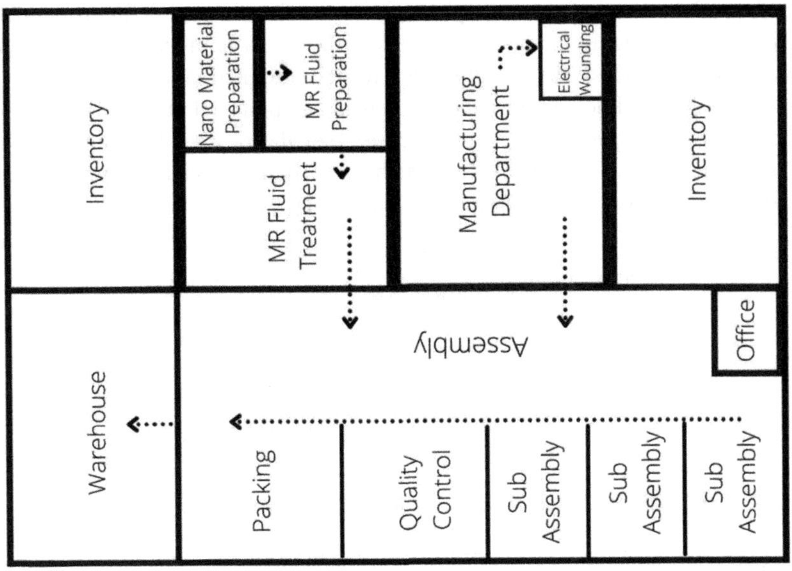

Fig. 1. Factory layout.

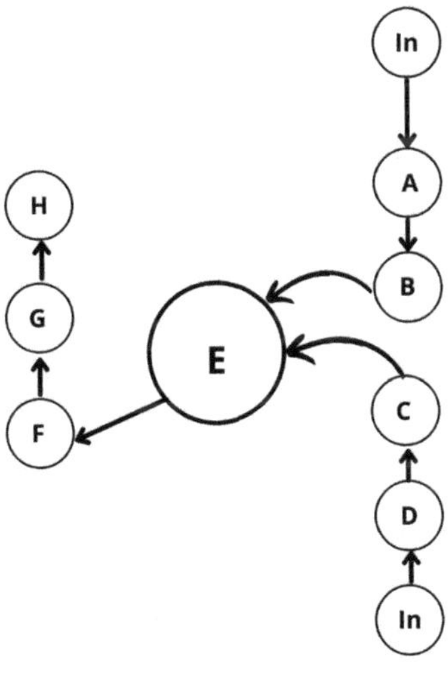

Fig. 2. Process flowchart.

The plant layout is converted into a process flow chart to ease the process of evaluating the complexity of the layout by certain parameters called density index and path index as explained and indicated by previous researches [8] The density index is the measure of the ratio of some graph nodes to the theoretical maximum number of those nodes connecting all the nodes, where k is the actual number of connections and n is the number of nodes. The density index for the designed layout is 0.16. The path index denotes the ideal deviation in the theoretical and practical number of paths in the layout. The obtained parametric values denote that the designed layout is a simple layout with the sequential process for which cycle index, decision point index, and redundancy distribution index are not necessary to be evaluated using Eqs. 1 and 2.

$$\textit{Density Index} = \frac{k}{n(n-1)} = \frac{9}{8(8-1)} = 0.160 \quad (1)$$

$$\textit{Path Index} = 1 - \frac{p}{N} = 1 - \frac{2}{2} = 0 \quad (2)$$

4 Cost Estimation and Break Even Analysis

There are several techniques based on which the manufacturing costs are estimated using artificial intelligence, enterprise resource planning, and computer-aided designed integrated cost estimation. Researchers have created a paper that deals or guides on how to select a cost estimation technique based on the innovation quotient factor and value of the manufacturing products using a decision tree. This has been formulated using several previous references and works of literature to support and validate different methodologies [9].

The break-even analysis is the method adopted to determine the point or quantity of sales wherein the establishment makes neither a profit nor a loss by breaking even with the costs inquired to develop and establish the industry or business. This is the most commonly used technique to determine the break-even point in a business. The parameters that govern this calculation are the fixed costs, variable costs, and selling price per unit. The fixed costs are otherwise known as the constant expenditure during business such as the rent, salary, equipment, and setup costs. The variable costs are the material costs, electricity, maintenance, and emergency expenditures. These are usually manually calculated to determine the approximate units to formulate business strategies to plan. Researchers have contributed to the digitalization of this calculation to ease the effort. The scope of artificial intelligence is being exploited to determine the break-even quantity and productivity of a textile industry that deals with more than 1,000 variants of fabrics and its products [10]. The break-even point for battery trucks in Latin American states is obtained using a model which involves varying factors that are geographical-based conditions. The obtained results have helped to conclude the trend in each state by considering the current status and predicted future scenarios along with exceptions and advertising [11].

Table 1. Estimated fixed costs (€).

Fixed cost (€)	
Rent	180,000
Salary	19,008
Equipment	115,464.6
Setup	1,760
Total	316,323.6

The firm is set to handle a global e-commerce market with a standard retail market price for single unit purchases and customize the product for bulk industrial orders or function as a supplier. The average costs of setting up the manufacturing industry, equipment, labor, material, electricity, advertising, and packaging charges were estimated. These estimated values are as tabulated in Table 1 and Table 2. These figures were used to calculate the break-even unit of the damper with a standard retail selling price of 600 €.

$$Break\ even\ time = \frac{FC}{SP - VC} \tag{3}$$

where FC – Fixed Costs; SP – Selling price per unit; VC – Variable price per unit.

The labour rates in Lithuania were defined as 8.8 € per hour in the Eurostat released on April 2019. This figure was used to determine the approximate labour charges enquired depending on the planned schedule of work. The average rent rate in Lithuania was found to be 3–5 € per m^2 [14, 15] The plant layout was estimated to be established for approximately 3,000 m^2. The approximate cost of the equipment and charges were estimated to contribute the equipment and setup charges. The variable cost of the damper is assumed to be 100 €. The break-even point of the sales as quantity of units were calculated using the Eq. 3. The estimated total fixed cost is approximately 317,000 €. These calculations were made with certain detailed planning for the functioning of the industry. The industrial layout was planned for 3,000 m^2 area of land with basic construction building and internal detailing. The break-even point was calculated to be 634 units using Eq. 3. The break-even graph is as shown in Fig. 3. The graph indicates the slow growth of revenue with an increase in sales to achieve a breaking point with the total expenses. Similar calculations were performed for different variable costs per unit of the damper to obtain a view of the units to be sold to break even. The obtained values are tabulated as shown in Table 2 and a graphical representation of the comparison of variable cost per unit that can affect the break-even point is shown in the graph in Fig. 4.

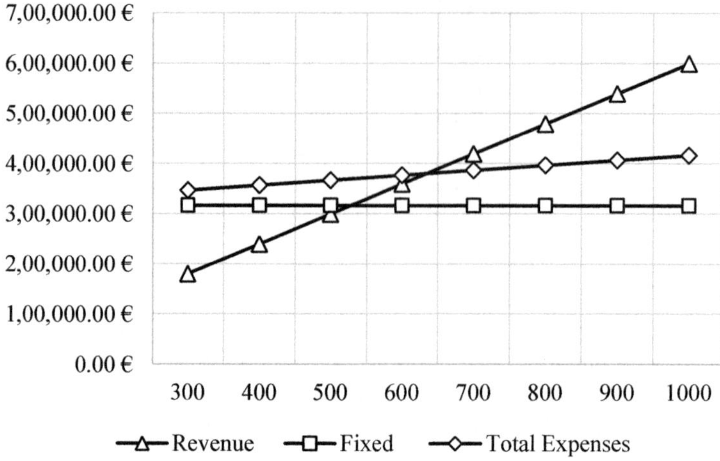

Fig. 3. Break-even analysis.

Table 2. Break-even units for different variable costs per unit (€).

Scenario	BE in units	Variable cost per unit (€)
Case 1	576.3	50
Case 2	634	100
Case 3	704.4	150
Case 4	792.5	200

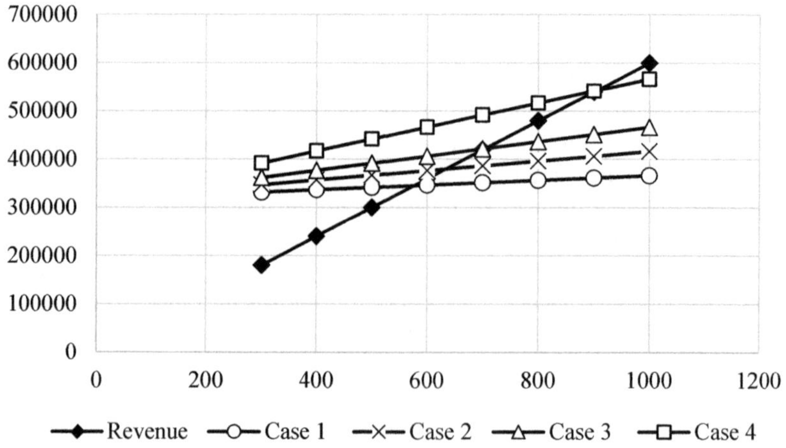

Fig. 4. BEP for different variable costs per unit (€).

5 Sales Prediction Using Bass Diffusion Model

The basic model is used to analyze considered the values of the coefficient representing the different sales scenarios. Several researchers have adopted this model to obtain forecasts of the outcome assuming different scenarios and how it could affect the market trend [12]. The method of analyzing the sales data history of similar technology in the market for its market trend has been adopted generally in several research papers with uncertainty factors. The Bass diffusion model is employed as a robust model with the available historic data to adapt to the new technology. In research, to launch a new call model which was similar to an existing model in the market was analyzed using the bass diffusion model to obtain the sales forecast outcomes [13]. Based on the diffusion of innovations theory, the Bass diffusion model predicts the number of adopters of an invention or device using different parameters such as relative benefit, observability, reinvent ability, trialability, compatibility, and difficulty. The innovators and imitators, which are the two consolidated parameters of the discussed variables, are used to decide the adopters. The innovators are a group of people who adapt to a new product without waiting for a review or input from existing customers, whereas the imitators are a group of people who are influenced by previous adopters. The formula below can be used to calculate the rate of acceptance and the number of adaptors in the nth year as in Eq. 4.

$$n(t) = p * [m - N(t)] + \left(\frac{q * N(t)}{m}\right) * [m - N(t)], \tag{4}$$

where n(t) is the number of adopters at time t with a market potential of m, N(t) cumulative adopters at time t, p is the coefficient of adoption of innovators and q is the coefficient of adoption of imitators. With this equation, the model prediction can be made using different values of p and q where it can be stated that $0 \geq p,q \geq 1$.

Table 3. Parameter description of futuristic scenarios.

Description	Assumed scenarios	p value	q value
Case 1	Steady advertising and marketing of the damper and it's features	0.01	0.34
Case 2		0.02	0.34
Case 3		0.03	0.34
Case 4	Steady advertising and marketing of the damper and it's features with additional discounts	0.01	0.36
Case 5		0.02	0.36
Case 6		0.03	0.36
Case 7	Accelerated advertising and marketing of the damper with added feature to enhance safety	0.01	0.38
Case 8		0.02	0.38
Case 9		0.03	0.38

For a business model analysis, p ranges from 0–0.03 and q ranges from 0.3–0.5 of which 0.38 is said to be an idealistic value. For this purpose of study three different values of each set is of p and q is considered as a case and the prediction for the sales outcomes are calculated to analyze the market trend as shown in Table 3.

For the above-mentioned set of parameters, the base model was used to calculate the damper's expected sales quantity in Lithuania. Despite external factors such as accelerating population growth and leading to improved vehicle demand for comfort and efficiency, the forecasted sales figures reflect the acceptance of the MR dampers by the consumer potential. The obtained forecasted revenue volumes were plotted on a table, along with the demand growth for each year. The bass model was applied to the MR damper for the nine cases in this analysis, as seen in Table 3. To assess the pattern of influence of the diffusion parameter values, the different forecast revenue figures were plotted against the year as a graph. From all the obtained graphs it is evident that the parameters greatly affect the capture of the market and the peak sales duration varies with the parameter and initial sales values. The forecasted sales quantity graph is as shown in Fig. 5.

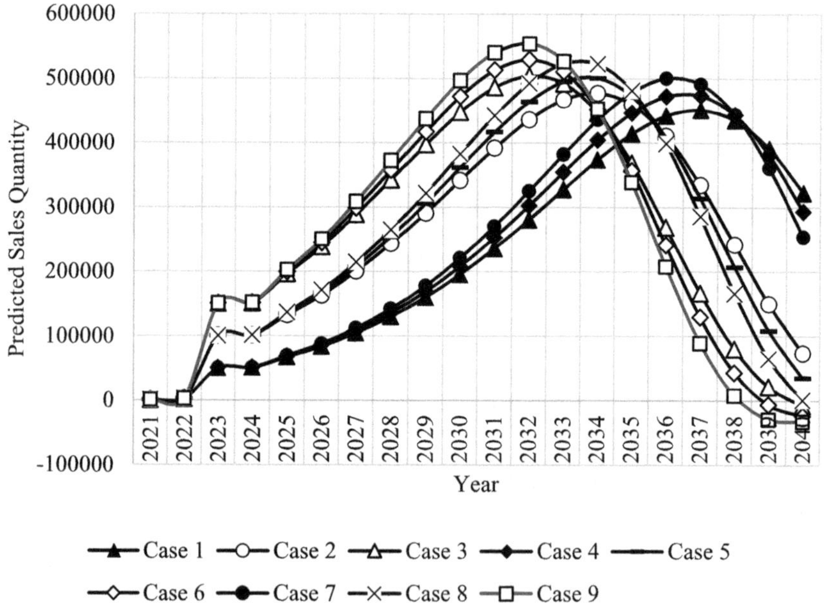

Fig. 5. Predicted sales.

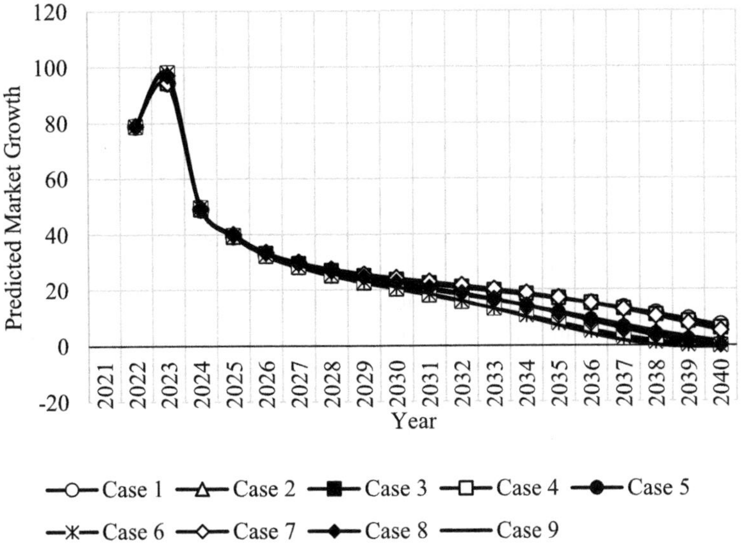

Fig. 6. Predicted market growth.

The predicted market growth is indicated in Fig. 6. The sales quantity curve depends on the p and q parameters and the effect of the parameters can be seen in the graph, the graph tends to have an early peak point when the q values are greater and tend to have a quicker sales fall. From the graph and trend in Lithuania concerning the sales observed, we can conclude that the most ideal values of the parameter are cases 1 and 2 for a steady business.

6 Discussion

The break-even analysis briefly indicates the year in which the break-even quantity will be achieved. The various cases and their outcomes are also indicated to avoid unexpected outcomes in the case of uncertainty. With these ahead calculations we can evaluate the sales and respective profits obtained. The change in selling price and contribution margin and greatly affect the duration at which the break-even point will be achieved. We can change the parameters dependently to alter the outcomes according to need. The ideal sales trend with an expected rate of 10% of the break-even quantity achieved is furnished from the study. The trend clearly explains how the bass model parameters affect the sales trend. The company can take necessary actions to accelerate the sales using different business strategies such as advertising, marketing through discounts and replacement offers. These show that there is an existing potential market to achieve and conquer. These discussed figures are an approximation that can be used as a methodology to estimate similar business establishments. This research focuses on how a new product can be potentially integrated into the new mobility launch in the country to impact the market to accelerate the market capture by eventually showing the difference in the performance of the product compared to the existing ones in the market. This research is to determine the break-even time and sales trend to obtain declension and peak periods to essentially plan business plans to alter the sales as required.

There definitely are certain threats that have to be given attention to before establishment. The existence of secondary automobile and parts market will definitely deviate the potential customers due to the lack of information or knowledge on the new product and its efficiency. The purchasing capacity of the population of the country will affect the expected sales outcomes [22]. The selling price of the new product may not be adhering to customers due to a variety of alternative options. Strategic business planning along with research and development of the product can alter potential threats. With these strategies combined along with the strategic planning of the launch and development of autonomous vehicles employing this damper as a ride comfort feature in Lithuania can greatly affect the efficiency and quality of the vehicles indeed increasing the factor of imitators in the Bass diffusion model contributing to improved market capture. The introduction of this MR damper into the market as a feature of the autonomous vehicles can prove to be a game-changing factor that elevates the establishment of the product and brand to obtain new retail and commercial customers which will be beneficial to build the business to slowly become a leading damper manufacturing industry globally. Further research can be carried out in the technical aspects of this damper and its integration into the autonomous vehicle. The data integration of the road conditions from various sensors such as the LIDAR and image sensing camera should be converted and given as data input to the variable damping altering unit in the damper system to continuously update the damping effects depending on the load conditions at that instant. This research scope can reform certain aspects in terms of the performance of vehicles considering the data transfer system on connected vehicle systems in the European Union (EU) [20]. The instant information exchange on the status of road conditions and vehicles in certain lanes can be utilized to instantaneously adapt vehicle systems or in this case, suspension systems to perform with the expected outcome considering the equipped scope of future vehicles and autonomous vehicles.

7 Conclusion

This research was focused to analyse the business aspect of the improved MR damper. This work is a continuation of the technical research to support background analytics to determine the outcomes of starting up a small OEM manufacturing industry. The industrial layout has been proposed by considering the governing parameters to reduce the complexity of the layout which in turn will save cost and time. The proof that the layout is simple is the value of the density index, for which the calculated value is nearly zero. The break-even analysis was carried out to determine the average quantity of sales to be made to break even and the time required to break even. The break-even quantity was obtained to be 634 units for variable cost per unit 100€ and selling price per unit 600€. The sales prediction was carried out to determine the peak sales and sales declension duration to formulate a strategic plan to implement to alter the sales. The inclining sales period for the assumed cases was from the year 2032–2037. Beyond which there occurs a definite declension which has to be addressed by strategic business planning to make the product sustainable in the market. Thus, the background business analytics for the proposed new product was carried out in this research.

References

1. Nalina, H.V., Andrius, D.: Research on semi active suspensions in passenger car. In: 24th International Scientific Conference. Transport Means, pp. 1010–1016 (2020)
2. Prasad, N.H., Rajyalakshmi, G., Reddy, A.S.: A typical manufacturing plant layout design using CRAFT algorithm. Procedia Eng. **97**, 1808–1814 (2014)
3. Pourvaziri, H., Pierreval, H., Marian, H.: Integrating facility layout design and aisle structure in manufacturing systems: formulation and exact solution. Eur. J. Oper. Res. **290**(2), 499–513 (2021)
4. Tayal, A., Solanki, A., Singh, S.P.: Integrated framework for identifying sustainable manufacturing layouts based on big data, machine learning, meta-heuristic and data envelopment analysis. Sustain. Cities Soc. **62**, 102383 (2020)
5. Al-Zubaidi, S.Q.D., Fantoni, G., Failli, F.: Analysis of drivers for solving facility layout problems: literature review. J. Ind. Inf. Integr. 100187 (2020)
6. Gamberini, R., Ruggerini, T., Lolli, F.: The plant layout of a foundry: constraints, operative guidelines and a case study. IFAC-PapersOnLine **51**(11), 1180–1185 (2018)
7. Gayam, N. R., Shanmuganandam, K., Vinodh, D.: Layouts in production industries: a review. In: Materials Today: Proceedings (2020)
8. ElMaraghy, H., AlGeddawy, T., Samy, S.N., Espinoza, V.: A model for assessing the layout structural complexity of manufacturing systems. J. Manuf. Syst. **33**(1), 51–64 (2014)
9. Schwabe, O., Shehab, E., Erkoyuncu, J.: An approach for selecting cost estimation techniques for innovative high value manufacturing products. Procedia CIRP **55**, 41–46 (2016)
10. Annamalai, S.: A multivariate analysis and investigation on break even and productivity measures using artificial intelligence: a post research work in textile process industries. In: Materials Today: Proceedings (2020)
11. Tanco, M., Cat, L., Garat, S.: A break-even analysis for battery electric trucks in Latin America. J. Cleaner Prod. **228**, 1354–1367 (2019)
12. Simpson, J.R., Mishra, S., Talebian, A., Golias, M .M.: An estimation of the future adoption rate of autonomous trucks by freight organizations. Res. Transp. Econom. **76**, 100737 (2019)
13. Ismail, Z., Abu, N.: New car demand modeling and forecasting using bass diffusion model. Am. J. Appl. Sci. **10**(6), 536 (2013)
14. Lithuania in Europe. Statistics Lithuania (2018)
15. Statistical Yearbook of Lithuania. Edition -73 (2018)
16. McManus, W.: Market models for predicting PHEV adoption and diffusion. University of Michigan, Ann Arbor, Transportation Research Institute (2009)
17. Kolar, E., Lindström, L.: Future Business Model for OEMs in the Automotive Industry Business Model Adaptation Based on the Role an OEM Takes in a Future Business Network (Master's thesis) (2018)
18. Wells, P.: Sustainable business models and the automotive industry: a commentary. IMB Manage. Rev. **25**(4), 228–239 (2013)

19. Fritschy, C., Spinler, S.: The impact of autonomous trucks on business models in the automotive and logistics industry–a Delphi-based scenario study. Technol. Forecasting Soc. Change **148**, 119736 (2019)
20. Monios, J., Bergqvist, R.: The transport geography of electric and autonomous vehicles in road freight networks. J. Transp. Geograph. **80**, 102500 (2019)
21. Kaltenhäuser, B., Werdich, K., Dandl, F., Bogenberger, K.: Market development of autonomous driving in Germany. Transp. Res. Part A: Policy Pract. **132**, 882–910 (2020)
22. Raslavičius, L., Azzopardi, B., Keršys, A., Starevičius, M., Bazaras, Ž, Makaras, R.: Electric vehicles challenges and opportunities: lithuanian review. Renew. Sustain. Energy Rev. **42**, 786–800 (2015)

An Assessment of Airports Logistics Capabilities in Africa: A Conceptual Framework

Aya Medany[1(✉)], Ilmars Blumbergs[1], Khaled Elsakty[2], and Raghda Bahaa[2]

[1] Transport Department, Riga Technical University/Aeronautics Institute, Riga, Latvia
aya.medany@olympian.org, ilmars.blumbergs@rtu.lv
[2] Transport Department, Arab Academy for Science, Technology and Maritime/Logistics and International Transportation, Giza, Egypt
{khaled.sakty, raghda.bahaa}@aast.edu

Abstract. Air cargo has become a key element of global supply chains, and it is expected to continue growing because of the increasing worldwide economic integration, shorter product life cycles, reductions in inventory stocks, and increasing competition between the airlines. The airlines competition has increased due to the demand for the speed and reliability benefits that air freight offers. The new industrial organizational concepts have made air cargo an important asset for world trade, supporting its dynamic including just-in-time and e-commerce, reduction of time to market, and enlargement of markets. The major challenge for future growth of air cargo transport is capacity limitations at major airports especially in Africa. The African countries are still politically unstable with regular revolutions, change of constitutions, and trucked elections to stay in power. Several factors in the next decades will give African countries the opportunity to turn the tide. This paper aims to assess the African Hub Airports (AHA) capabilities in terms of soft and hard infrastructures (SI & HI) and upgrade the most appropriate airport to be close to change requirements for an elite hub. The SI assessment includes transportation process, e.g., e-freight, airlines cooperation, electronic air waybill (e-AWB), and customs regulations. The HI assessment includes on-ground cargo handling, e.g., storage, warehousing facilities, handling equipment, specialized cargo aircrafts, and security measures.

Keywords: African airport · Air hub capabilities · Air transport logistics · Cargo handling · E-Freight · Hard infrastructure · Soft infrastructure

1 Introduction

Air cargo transportation has become a key element of the global logistics market, growing faster and wider every day, and it is expected to continue growing because of increasing worldwide economic integration, shorter product life cycles, reductions in

O. Prentkovskis et al. (Eds.): TRANSBALTICA 2021, LNITI, pp. 407–418, 2022.
https://doi.org/10.1007/978-3-030-94774-3_40

inventory stocks, and increasing competition among the airlines and airports. The global air freight market was valued at $270.2 billion in 2019 and is projected to reach $376.82 billion by 2027 registering a compound annual growth rate of 5.6% [8]. However, the world air cargo traffic growth is expected to be increased by 4.0% until 2039 (IATA, ICAO, Boeing 2020). There is a continued demand for the speed and reliability benefits that air freight offers. Industries that require transport of time-sensitive and high-value commodities such as perishables, consumer electronics, high-fashion apparel, pharmaceuticals, industrial machinery, and automobile components that recognize the value of air freight, and; consequently, this value will continue to play a significant role in their shipping decisions. Although the share of air cargo in the total freight transport is low, measured in tones lifted or ton kilometers carried, the share of values transported is high, which underlines its importance for the functioning of supply chains for high valued goods in the economy. Air cargo represents about 1% of world trade exchanged in volume of goods shipped, but represents 35% of world trade in value [10]. In addition, new industrial organizational concepts have made air cargo an important asset for world trade, supporting its dynamic:

1. Just in time and E-commerce.
2. Reduction of "time to market".
3. Enlargement of markets (firms serve larger markets from fewer points). Air cargo is also used for the most valuable and time-sensitive goods:
4. 30% of shipped goods are perishables (by nature: food, or by destination: press).
5. The other 70% is manufactured goods (spare parts, chemicals, pharmaceuticals, etc.).

It has a special place in modern supply chains, carrying the most valuable, most perishable, and most urgent shipments across the world. From necessities like pharmaceuticals to luxuries like exotic flowers or diamonds, air cargo services shrink-me and space to link customers to distant sources quickly. The development of air cargo networks was historically dominated by the air passenger hubs, because the major part of air cargo was transported in the bellies of passenger airplanes. With the extension of dedicated cargo airlines and the increasing environmental problems of big passenger hubs, the development of efficient future air cargo networks is an upcoming challenge. Basically, the network configurations of air cargo carriers can be developed towards pure cargo carriers, combined carriers, or integrated service providers. A major challenge for future growth scenarios is capacity limitations at major airports. Airport investment plans exist, but in terms of strained public budgets, airport investments directly compete with other public sectors. Therefore, profitable and necessary airport investment needs to be selected from an air transport perspective. Airlines which combine passenger and cargo services, such as Lufthansa, are concentrated around a small number of airports which are the airlines passenger hubs. Therefore, network configurations of pure cargo airlines, such as Cargolux, which focus their services on freight only, are much more diverse. There is a huge gap between the European and African air cargo hubs from the development side or the number of cargo fleet; however, the African countries have more land and labor to serve better, that's why structural changes and a new mind-set from African governments are desperately needed and more studies and researches should be done.

The African countries are still politically unstable with a lot of changes, and this is due to many factors. The first factor is demographics, where the rate of the average annual population growth of the African countries is 2.5% (www.worldometers.info, 2019) – more than two times faster than the world average. The world's population is expected to increase by 2 billion persons in the next 30 years, from 7.7 billion to 9.7 billion in 2050, according to the United Nations report. The new population projections indicate that nine countries will make up more than half this projected growth: India, Nigeria, Pakistan, the Democratic Republic of the Congo, Ethiopia, the United Republic of Tanzania, Indonesia, Egypt and the United States of America (in descending order of the expected increase) (un.org, 2019). Moreover, when most of the rest of the world will have to face an aging population and an increase of healthcare expense and pension expenditure, Africa, with its young and numerous inhabitants, will have a considerable advantage. In addition, the fast-growing population will represent an increasingly attractive economic market for businesses. Asia's need for natural resources will increase, and Europe's investment in fresh products from Africa will be necessary.

The Single African Air Transport Market (SAATM) is a flagship project of the African Union Agenda, where an initiative is introduced by the African Union to create a single unified air transport market in Africa to advance the liberalization of aviation in Africa and act as an impetus to the continent's economic integration of 2063 agenda. SAATM will ensure that aviation plays a major role in connecting Africa, promoting its social, economic and political integration and boosting intra-Africa trade and tourism as a result. According to IATA, Single African Air Transport Market (SAATM) aims to create a single unified air transport market in Africa to advance the liberalization of civil aviation in Africa, acting as an incentive to the continent's economic integration agenda [20].

The rest of the paper is organized as follows: The second section will discuss the literature review, the obstacles that Africa faces, and the air freight in Africa and the Middle East. The third section will review the statistics of Africa in airports. The fourth section will discuss the air freight in Ethiopia and the statistics of Ethiopia in airports. In the fifth section, air freight in Egypt will be tackled. The sixth section will be about the conceptual framework suggestion. The seventh section will discuss the research recommendations. And finally, the conclusion.

2 Literature Review

The impact of an airport on the global connectivity in response to network changes was evaluated at the route level and was limited to direct or one-stop connections. While these studies employed several different centrality indicators and network metrics as the proxies of different properties of air transport networks, they largely neglected the issue of whether those measures were able to (i) reflect the connection exemplified by the passenger flows, and (ii) explain how connectedness of an airport has increased or decreased over time [6]. Studies of Barros (2008), Gillen and Lall (1996), Sarkis (2000), Lin and Hong (2006), Fung et al. (2008), Malighetti et al. (2007), and Gitto and Mancusso (2012) support the efficiency advantages of the hub airports. Both studies of

Bel and Fageda (2010) and Bilotkach et al. (2012) claimed that hub airports set higher aeronautical charges probably due to either their increased market power or their effort to deal with congestion problems. Although in some studies (Lin and Hong, 2006) hub status is measured by the share of transfer/transit passengers, we align with Gillen and Lall (1996), Sarkis (2000), Malighetti et al. (2007), Bel and Fageda (2010), and Bilotkatch et al. (2012) who consider hubs to be the home-base for one or more airlines [16].

2.1 Research Methodology

The study depends on the qualitative-quantitative approach by applying the descriptive-analytical research methodology. It depends on previous researches, studies, periodicals, and reports related to this subject.

2.2 Obstacles Africa Faces

This section will discuss the studies of the problems that Africa faces. Local knowledge is apparently essential as the continent's largest problems are political instability, poor infrastructure and handling facilities, and excessive regulations and bureaucracy. Nieszner adds the need to improve overall customs regulations and cargo handling security, and Mebrate adds traffic rights to above mentioned problems. "One has to go through a long bureaucratic process before securing fifth and seventh traffic freedom rights," Merbrate says. "As we develop the air cargo industry on the continent; countries need to ease the stringent policies which are crucial to build a conducive operating environment for airlines" [12]. Air cargo has limited profit margin, and these obstacles can disturb smooth operations, which can easily lead an entrepreneur from profit to loss. There are more problems like the need to improve overall customs regulations and cargo handling security and traffic rights. This will be essential if Africa wants its success to be sustainable in the long-term. If the industry is based on non-African companies, it will collapse in case that they leave. Experts expect African carriers will try to take advantage of the "New Southern Silk Road" and capture a higher percentage of the passenger and cargo traffic by expanding domestically and internationally to China, India and Latin America through partnerships. Africa's airlines continue to struggle and collectively remain in the red while airlines in every other region in today's favorable environment are profitable,political interference and government meddling in airlines as well as protectionism & unnecessary high taxation are common problems.

According to 'Air Cargo World 2016', the African airports are absent among the top 30 airports in Cargo, and the African airlines are not among the top 25 airlines. Looking at the top 30 airports in terms of air freight, North American and Asian airports are dominating the rankings helped by the large domestic markets of individual countries in the regions. However, the major hubs of Europe and the Middle East, such as Frankfurt, Paris, Dubai, or Doha, do make it to the rankings as they provide access to a large market of end point customers for goods flown by air freight. Hermann (2013), the regional director of Lufthansa cargo for Africa, discussed that Lufthansa Cargo is experiencing strong growth, especially from Egypt, Ethiopia and Kenya. Herman added that Lufthansa is already working with the stakeholders in the supply chain to develop the air cargo business and improve quality and processes; in addition, his research shows that Africa's Air Cargo transportation and airports need to develop three main areas. First, faster development of e-freight, e-transactions and e-AWBs, i.e., most countries are still considerably lagging behind in their development and need to catch up in order to benefit from more transparent and cost-effective processes. Second, security and airport infrastructure need to improve, especially in West Africa. Third, there needs to be a wider spread of commodities being shipped from Africa. Knowing that the majority of exports from Africa are perishable goods with relatively low returns, the operation of shipping routes to Africa will remain a challenge (ATN, 2013).

2.3 Air Freight in Africa and the Middle East

The global cargo market is experiencing a revolution, and there are new trade corridors that are emerging linking the markets of Africa, Latin America, the Middle East and Asia, which are growing faster than western markets; while Africa is emerging middle class, infrastructure developments and increasing foreign direct investments, it will eventually drive cargo traffic growth. It is expected that in 2030 air freight in Africa and middle East will grow by 5.8 percent and the African fleet will double to 1,210 aircrafts, 60 percent being added to existing fleets. More specifically, just 0.7% of the total weight of African exports are transported by air, corresponding to 32.8% of the value. Africa is a region that is very rich in mineral wealth. This can be proved by some high-end products that source their materials in Africa (Air cargo world, 2013).

Saudi Arabia Vision 2030 program focuses on four key sectors: industry, mining, energy and logistics. The Kingdom hopes to tap on its strategic location and become a trading hub that connects Asia, Europe and Africa, Crown Prince Mohammed bin Salman said in his foreword for the kingdom's Vision 2030 blueprint, "Our geographic position between key global waterways makes the Kingdom of Saudi Arabia an epicenter of trade and the gateway to the world". It also sits on the Asia-to-Europe trade route, which is responsible for 12% of global container trade in 2019. Saudi Arabia will still need all hands-on deck to achieve Vision 2030. The Kingdom plans to develop itself into a powerhouse for global investment. Saudi Arabia is trying to effectively link with other countries in the region through enhanced logistic services and new cross-border infrastructure projects including land transport projects with Africa through Egypt. Logistical and trade exchanges will be streamlined for further cementing our pre-eminent position as a major trade hub [9].

Market-wise, the following graph shows a very strong air freight relationship between Africa and Europe, as more than half of the value and weight are heading there. Asia and North America follow in distance, with a combined value percentage of 32% and almost the same in terms of weight [19]. Asia is the only region where the air freight volumes with Africa show higher shares in value than in weight. The rest of the world regions show only fragments of the total Africa airborne trade, as it can be seen in Fig. 1.

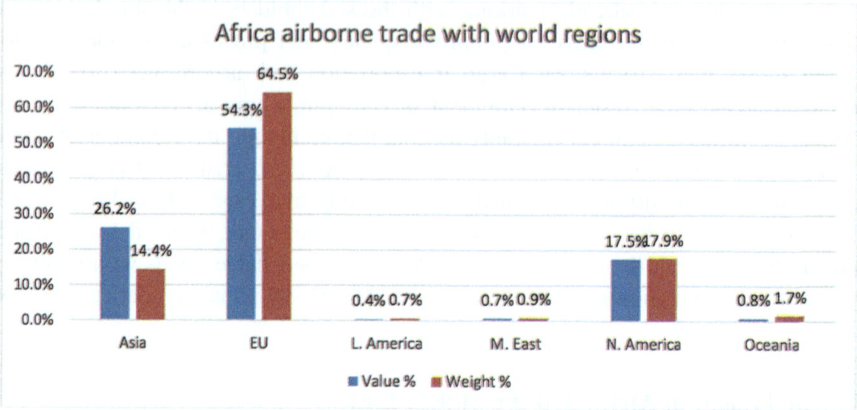

Fig. 1. Trade of Africa with world regions in weight and value IATA [10].

3 Statistics of Africa in Airports

The air transport industry in Africa directly generated an estimated 381,000 jobs in 2014. This number was mainly driven by on-site airport jobs (55%) and jobs with airlines or handling agents (35%). The air transport sector supported over 1 million jobs and contributed $26.5 billion to Africa's GDP. Africa was the region responsible for the lowest number of aircraft deliveries in 2015. In total, only 29 new airliners were delivered. This generated about $2.623 million for aircraft manufacturers. Ethiopian and Kenya airways were responsible for more than 50% of regions total acquisition. African airlines account for 5% of the total worldwide aircraft. However, most African carriers continued to struggle in 2015. Air traffic liberalization in the continent is needed for air traffic to develop to its full potential. Unfortunately, Angola and Nigeria, Africa's two largest oil exporters, have been hit particularly hard by the downturn [14]. Table 1 will show the global air cargo transportation in 2017.

Table 1. Annual world airport traffic report, 2018.

Region	Million Metric Tons
Asia-Pacific	47.1
North America	33.1
Europe	21.6
The Middle East	9.3
Latin America – Caribbean	5.4
Africa	2.2
Total	118.6

4 Air Freight in Ethiopia

Ethiopian airlines are the most successful in Sub-Saharan Africa with 19 million seats in 2019 surpassing the Emirates. In addition, it is the richest in Africa. The success of the Ethiopian Airlines can be attributed to two factors.

First, the airline is independent from the government and its vision is lucid. Nevertheless, the Ethiopian airlines is highly protected by the government. As a result, Ethiopian Airlines has already attained its goals of Vision 2025 in 2018 and is now pursuing its Vision 2035. Second, the geographical location of Ethiopia has enormously helped in preserving its position as an aviation hub connecting Africa with the Middle East, Asia, Europe, and the Americas. The Ethiopian Airlines has invested heavily in modernizing its fleet, which is now the youngest in the continent (International Finance, 2020).

Addis Ababa is the central shipping point of most/all perishable goods like live cattle, fish, oil, flowers, fruits, vegetables, coffee, and gas and mining equipment from Africa to Liege airport—Africa's gateway to Europe. Especially in the transport of perishables, professional storage, seamless documentation, and rapid transshipment play a key role in trading with industrial nations, and Addis Ababa provides all these criteria. In the Ethiopian airlines report, government intention to develop its capital to become a freight hub for Africa. In addition, they are planning to compete with hubs like Dubai to become an international cargo hub.

4.1 Statistics of Ethiopia in Airports

Ethiopian Cargo is one of the seven strategic business units of the Ethiopian Airlines Group and the largest cargo service provider in Africa. Its cargo network connects to more than 30 international destinations spread across four continents including 20 in Africa; 8 in Gulf, the Middle East and Asia; and 2 in Europe via its hub in Addis Ababa. The Ethiopian Cargo service has warehouse facilities with a capacity of 250,000 tons of cargo per annum with cold room temperature ranging from −21 to 10 °C (2,000 m^2). It has also additional cold room facility with a capacity of 65,000 tons of cargo with temperature ranging from 2 to 4 °C in 6,000 m^2 dedicated for export products. When the airline completes building a new cargo two phase terminal, it will make Ethiopian Cargo terminal one of the largest terminals in the world with a capacity of handling 1.2 mil. tons per annum. In table two, the major categories of owned and leased aircraft are stated (Table 2).

Table 2. Ethiopian Airlines fleet (www.cargo.ethiopianairlines), 2018.

Fleet	A/C Type (pass.)	Number	Fleet	A/C Type (cargo only)	Number
Q400 Bombardier	Q400 Bombardier	19	Boeing B777-200LRF	B777-200LRF	6
Boeing B737-700	B737-700	8	Boeing B757-260F	B757-260F	2
Boeing B737-800	B737-800	15	**Total Operating Fleet**		**8**
Boeing B767-300ER	B767-300ER	6			
Boeing B777-200LR	B777-200LR	6			
Boeing B777-300ER	B777-300ER	4			
Airbus A350-900XWB	A350-900XWB	2			
Boeing B787-8	B787-8	16			
Total Operating Fleet		**84**			

5 Air Freight in Egypt

Cairo, with its distinguished geographical location, is considered a favorable hub for transshipments via Egypt Air fleet to/from major cities for the Egyptian exports and imports. Egypt is working hard to overcome geopolitical demand softness. Accordingly, it has 11 destinations for air cargo and 73 for passengers with 9,000 employees working in Egypt Air cargo, 14 of them are representatives of all control authorities. Egypt Air has storage capacity of 120,000 tons and area of 55,000 m^2 and serves 37 foreign Airlines. It's the main gate for fruit and vegetables to Europe & the Middle East. The following table shows the major categories of owned and leased aircraft (Table 3).

Table 3. Egypt Air fleet (www.egyptair), 2018.

Pass. Aircraft	In service	Orders	Notes	Cargo aircraft	In service	Orders
Airbus A320-200	4	–	To be phased by 2019	Airbus A300-600RF	2	–
Airbus A320neo	–	15	Deliveries from 2020	Airbus A320-200/P2F	–	2
Airbus A330-200	4	–		Airbus A330-200/P2F	2	1
Airbus A330-300	4	1		**Total**	**4**	**3**
Boeing 737-800	29	–				
Boeing 777-200ER	2	–				
Boeing 777-300ER	6	–				
Boeing 787-9	2	4	Deliveries end 2019			
Total	**51**	**20**				

6 Conceptual Framework Suggestion

The framework (FW) of the upgraded African hub (UAH) is displayed in Fig. 2. It is created for facilitating the activities required to perform the assessment. From the bottom to the top, there are two pathways with arrows pointing upward and downward. This means that if the process is not stable on the top, more than one hub needs to be upgraded, or some more information is required, then the assessment is repeated. During monitoring, if any significant malfunction is found at any step, it is possible to restart the assessment, as an internal evaluation process. This FW makes all work visible, easy to track, and illustrates the logics of the assessment of selected UAH.

This framework is conceptualized in response to the required assessment of airport logistics capabilities in Africa. The overall target of the FW is to define and upgrade hubs. There are four parallel activities identified to reach the UAH. All activities start from the baseline of the assessment (BLA). The BLA of this FW mainly depends on all information available for the existing African sea and air ports, including the relative location, capacity, soft infrastructure, hard infrastructure, and the present processes of cargo handling. Once the baseline information is collected, the framework is divided to two main pillars: the first will be defining the needs for upgrading the selected hub/hubs, and the second pillar will be about selecting the best hub. Both pillars will be continuously exposed to monitoring and evaluation, and; thus, may lead to revising the assessment.

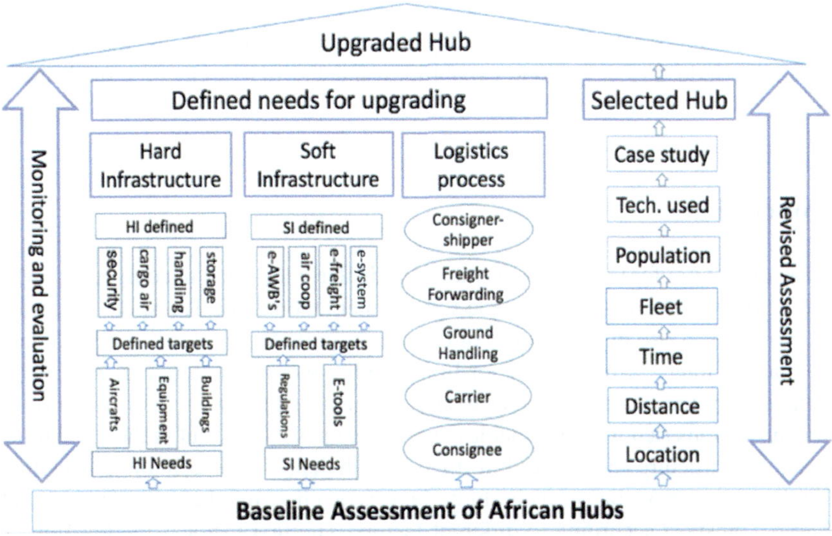

Fig. 2. The Proposed framework of the Air hub in Africa, source (developed by the researcher).

The first pillar points to the defined needs for UAH, and consists of: i) hard infrastructure (HI), ii) soft infrastructure (SI), and iii) logistics process (LP). The HI needs are defined for aircrafts, equipment and buildings. Consequently, HI is delineated through the information gathered regarding security, cargo air, handling and storage. From the defined HI, the requirements necessary for the UAH could be identified. On the other hand, SI needs are defined for the existing regulations and E-tools. This leads to the defined targets of e-AWB's, air coop, e-freight, and e-system. The logistics process for each hub under the assessment consists of five different operations: consignee, carrier, ground handling, freight forwarding, and consigner shipper.

The fourth pillar is used as a comparison between the selected UAH with a reference developed hub in the major global air-transport. This is done by taking into consideration the technology used in the process of handling and storage, the population in the country according to the demographics, the number of aircraft they have and if it includes any cargo aircraft, and finally the location of the hub compared to other cities for calculating the time and distance needed to decrease the cost and waiting time.

7 Research Recommendations

1. The Air-hub infrastructure is better to be owned and combine between the civil and military air traffic control especially in the region that the study focuses on.
2. One of the techniques that provides more evidence is "system wide information management" (SWIM), which switches from disparate concepts for handling data to a more harmonized and global infrastructure. This infrastructure will improve the efficiency of the Air-hub and analytics which are characteristic of 'big data' in other industries.
3. For maximizing the restructuring of logistic chains to serve the rapidly growing e-commerce industry, the unique capabilities and the new areas of growth that air cargo provides are required.

8 Conclusion

Air transport is vital for manufactures trade, particularly the trade in components which form a major part of cross border trade today. International Air Transport Association (IATA) forecasts a rise in the number of cargo transported around the world. Air cargo is essential for many facets of modern life. In other words, moving perishable goods from one side of the world to another would not be possible without air transport. One of the main regions rich with perishables and minerals is Africa, but in the same time, it has weak infrastructure and lack of information with delay in the technology techniques used in the transport operation. This study focuses on choosing the best African hub for Air Cargo Transport according to different factors. The first factor is the location, and choosing it depends on three main criteria that can be outlined at the local level: criteria related to airport environment, criteria related to airport infrastructure, and criteria related to airport operations. The second factor is the airlines strategies. A cargo alliance to/from Africa that unites airways from different contents needs to be established on the model of the existing cargo alliances which are developed mainly from existing passenger alliances such as Sky Team Cargo from Sky and WOW from Star Alliance members. A flagship project of the African Union Agenda 2063 an initiative of the African Union, signed by 28 countries. According to IATA, Single African Air Transport Market (SAATM) aims to create a single unified air transport market in Africa to advance the liberalization of civil aviation in Africa, acting as an impetus to the continent's economic integration agenda.

References

1. ATN2019, article ATN: Lufthansa Cargo continues to go from strength to strength. Accessed 18 Mar 2019 (2019)
2. Bardai, A.M., Er, A.Z., Johari, M.K., Noor, A.A.M.: A review of Kuala Lumpur International Airport (KLIA) as a competitive South-East Asia hub. IOP Conf. Ser. Mater. Sci. Eng. **270**(1), 012039 (2017). https://doi.org/10.1088/1757-899X/270/1/012039
3. Blachut, J.: The polish air HUB or the central airport in Poland, the solidarity port. IOP Conf. Ser. Mater. Sci. Eng. **471**, 112081 (2019). https://doi.org/10.1088/1757-899X/471/11/112081
4. Bofinger, H.C.: WIDER Working Paper 2017/36 Air transport in Africa A portrait of capacity and competition in various market segments (2017)
5. Cargonewswire2019, Liberalization & Modernization: the way to push frontiers of excellence in air freight industry in Africa. Accessed 20 Feb 2019 (2019)
6. Cheung, T., Wong, C., Zhang, A.: The evolution of aviation network: Global airport connectivity index 2006–2016. Transp. Res. Part E Logist. Transp. Rev. **133**, 101826 (2020). https://doi.org/10.1016/j.tre.2019.101826
7. Egyptian, T.H.E., Company, H., Navigation, A.I.R.: Arab Republic of Egypt the egyptian holding company for **31**(70) (2005)
8. Halmare & Mutreja: Air Freight Market by Servive: Global Opportunity Analysis and Industry Forecast, Mayank halmare, Sonia Mutreja (2020–2027) (2021) '
9. Han, E.S., Goleman, D., Boyatzis, R., Mckee, A.: 済無No Title No Title. J. Chem. Inf. Model. **53**(9), 1689–1699 (2019)
10. IATA (International Air Transport Association) 'IATA Cargo Strategy', IATA Cargo Strategy (2018). https://www.iata.org/whatwedo/cargo/Documents/cargo-strategy.pdf
11. ICAO: Second ICAO Meeting on the Sustainable Development of Air Transport in Africa (2017). https://www.icao.int/Meetings/SUSDEV-AT/CountryProfiles/Mauritius.pdf
12. In, L.O.G.: Africa's cargo boom: Is it sustainable? (2013)
13. Mo, Y., Li, S., Zhao, Y., Wang, L.: Analysis of China aviation network structure evolution based on ward system clustering. IOP Conf. Ser. Earth Environ. Sci. **189**(6), 062039 (2018). https://doi.org/10.1088/1755-1315/189/6/062039
14. Mott, M.: Annual Analyses of the EU Air Transport Market 2012, Transport, (December), p. 32 (2016)
15. Oxford Economics. Economic Benefits from Air Transport in Malaysia, p. 5 (2011)
16. Paraschi, E., Georgopoulos, A., Papatheodorou, A.: Abiotic determinants of airport performance: insights from a global survey. Transp. Policy **85**, 33–53 (2020). https://doi.org/10.1016/j.tranpol.2019.10.017
17. Redding, S.J., et al.: History and Industry Location: Evidence from German Airports (2010)
18. Report, T.: Air Freight Transport of August (2016). https://doi.org/10.13140/RG.2.2.32927.76965

19. Tzimourtos, G., et al.: Air freight Transport - A Strategic Modeling Approach on a Global Scale, Delft University of Technology (2014)
20. UNECA. The Single African Air Transport Market is launched, p. 2063 (2018). https://www.uneca.org/stories/single-african-air-transport-market-launched

Websites (Harvard system)

http://www.transport.gov.za
https://www.fraport.com
https://www.innofrator.com
www.pcf-frankfurt.de
www.liegeairport.com
https://cargo.ethiopianairlines.com/CargoNetwork
http://www.egyptair-cargo.com
https://aci.aero
https://www.iata.org
http://www.cargonewswire.com
https://www.fwd.news/new-asia-africa-el-dorado-myth/
https://www.icao.int
https://www.logisticsmiddleeast.com/transport/air-cargo/
http://www.cargonewswire.com/
https://aircargoworld.com/
https://www.internationalairportreview.com/
https://internationalfinance.com/how-did-ethiopian-airlines-make-it-happen/
https://www.worldometers.info/world-population/africa-population/
https://www.un.org/development/desa/en/news/population/world-population-prospects-2019.html
https://lot.dhl.com/vision-2030-the-future-of-logistics-in-saudi-arabia/
https://vision2030.gov.sa/en/programs/NIDLP

Concept of Resistance in the Railway Infrastructure Elements Protection

David Rehak, Lucie Flynnova(✉), and Simona Slivkova

Faculty of Safety Engineering, VSB – Technical University of Ostrava, Lumirova 630/13, 700 30 Ostrava, Czech Republic
lucie.flynnova@vsb.cz

Abstract. Railway transport is an important subsector of Europe's critical infrastructure. Railway infrastructure elements are under continuous risk from a wide range of disruptive events which might disrupt the functioning of railway infrastructure and threaten the safety of passengers. For these reasons, their protection must be guaranteed. A meaningful approach to the protection of critical infrastructure elements is the concept of resilience. In the current model, resilience consists of repressive and reinforcing factors but lacks preventive factors. Based on the above, the paper deals with the concept of resistance in the protection of railway infrastructure elements. In this context, resistance can be perceived as the ability of an element to prevent the occurrence of a disruptive event. These are preventive measures which determine structural resistance and resistance in the protection of railway infrastructure elements.

Keywords: Critical infrastructure · Railway infrastructure · Resilience · Resistance · Protection

1 Introduction

Railway transport is a key form of transport across the world. Its importance is indicated by the railway transport subsector's rank alongside the road, air and water transport subsectors as critical transport infrastructures of Europe [1]. According to statistics, railway transport is one of the safest modes of transport [2]. Despite this fact, railway infrastructure elements are constantly exposed to disruptive events of an anthropogenic or naturogenic character. These can have adverse effects on the functionality of infrastructure and the health and safety of transported persons. For these reasons, comprehensive studies of the protection of railway infrastructure and continual improvements are necessary. A possible approach to railway infrastructure protection is application of the concept of critical infrastructure resilience. Three factors currently determine resilience and are given the most consideration. These are robustness, recoverability and adaptability [3]. However, some current publications clearly state that resilience should include a fourth determining preventive factor, namely resistance [4, 5]. Based on this assertion, the paper addresses the concept of resistance in the context of resilience.

O. Prentkovskis et al. (Eds.): TRANSBALTICA 2021, LNITI, pp. 419–428, 2022.
https://doi.org/10.1007/978-3-030-94774-3_41

2 Description of Railway Infrastructure

The basis of secure and high-quality railway transport is reliable infrastructure. Transport infrastructure means all the roads and fixed transport equipment necessary for the operation of vehicles and the safety of such traffic. Railway infrastructure includes, namely, railway land, railway units, rails, switches, signalling equipment, level crossings (including road safety equipment), other civil engineering works (such as bridges, tunnels, overpasses, elevated track crossings) and associated station infrastructure (e.g. platforms, security equipment) [6].

These elements of railway infrastructure are classified into three groups according to their topological structure – line, point and areal elements [7]. Line elements facilitate the transmission, delivery or transport between two physically separate locations (i.e. the connections between individual elements/locations). In terms of importance, these constitute a fundamental group which relates to all point and area elements. In the context of railway transport, line elements are individual lines or line sections.

Point elements are closed units which fulfil their function for the needs of a specific line element. Mainly, these are locally defined points located in a small area. A point element can also work for more than one line element (e.g. a switch at a point where one line diverges or at the crossing of two lines). In the context of railway infrastructure, point elements are primarily railway equipment (e.g. communications equipment, signalling equipment, electrical equipment), railway structures, level crossings and stops.

Areal elements have the nature of a planar aggregate and include locations where several point and line elements can function simultaneously, where the very existence of so many elements in one place can have a critical character. Areal elements are the most complex group of elements. At least two point elements (possibly more) and at least one line element (can be more) operate within the areal element. In such an element, the effects of an outage can accumulate. These elements primarily include railway junctions, and level crossings and railway stations with signalling equipment.

3 Threats to Railway Infrastructure

Disruptive events of various type can have a significant adverse impact on the functionality of railway infrastructure, and the health and safety of transported persons.

A basic step in the protection of railway infrastructure is therefore always the consistent identification of threats that such disruptive events can raise. Threats to railway infrastructure can be divided into two basic groups according to their origin: (1) threats arising from railway transport and railway areas and (2) threats arising from beyond the railway area, i.e. from the surrounding environment.

3.1 Threats Which Arise from Railway Transport and Railway Areas

Under the railway infrastructure system, safety is based primarily on the Directive of the European Parliament and of the Council on Railway Safety [8], and its basic concepts in this area are accident, serious accident, and emergency.

The common safety indicators of the above-mentioned directive identify specific threats arising from railway transport and infrastructure (Table 1). The categorization of threats into threats related to railway transport or threats related to railway infrastructure is based on their primary cause.

Table 1. Threats arising from railway transport and infrastructure.

Threat category	Threat type
Related to railway transport	Collision of train with rail vehicle or obstacle within the clearance gauge; Derailment of a train; Level crossing accident, including accident involving pedestrians at a level crossing; Accident to persons involving rolling stock in motion; Fire in rolling stock; Dangerous chemical substance leakage during transport
Related to railway infrastructure	Broken rail; Track buckle and other track misalignment; Wrong-side signalling failure; Fire beyond the perimeter of the runway; Explosion beyond the perimeter of the runway

3.2 Threats Which Arise from the Surrounding Environment

The second area where threats to railway infrastructure may arise is beyond the perimeter of the track, i.e. the surrounding environment. For the purposes of the present paper, we deal primarily with the environment of the Czech Republic. These threats may vary from country to country, according to the nature of the territory.

The fundamental document used by the Czech Republic in this regard is the "Threat Analysis for the Czech Republic" [9]. Here, the authors identified a total of 72 types of threat for the entire territory of the country. These were categorized according to their nature into six groups: abiotic, biotic, cosmic, technogenic, sociogenic, and economic. In the context of railway infrastructure, abiotic, technogenic and sociogenic threats can be considered especially relevant (Table 2).

Table 2. Types of threat to railway infrastructure.

Threat category	Threat type
Abiotic	Flood; Flash flood; Extreme wind; Extreme high temperatures; Earthquake; Slope instability; Tornado; Extreme low temperatures; Atmospheric discharges; Black ice, glaze ice, and rime
Technogenic	Special flood; Fire beyond the perimeter of the runway; Explosion beyond the perimeter of the runway
Sociogenic	Huge legitimacy disturbance (terrorism); Other forms of legitimacy disturbance (vandalism)

4 Protection of Railway Infrastructure Elements

Due to their open nature and interconnection with other infrastructures, railway infrastructure elements are vulnerable to a number of threats and therefore require a high level of security. The protection of railway infrastructure elements is accomplished through measures which have both a preventive and repressive nature. The following section outlines possible methods for the protection of railway infrastructure elements.

4.1 Approaches to the Protection of Railway Infrastructure Elements

Physical protection in the critical infrastructure system is generally provided through individual measures which fall under the categories of physical security measures, regime and organizational measures, and technical security devices, especially mechanical barriers and alarm systems [10]. Railway infrastructure elements are thus protected through the integration of all three abovementioned categories of measures.

Many authors in their studies deal with the use of equipment for the detection and monitoring of threats to railway infrastructure. For example, Catalano et al. [11] devised an Intrusion Detection System which uses FBG sensors (Fiber Bragg Gratings Sensors). Cañete et al. [12] proposed the use of wireless sensor networks integrated into the construction of fixed lines for monitoring during installation and maintenance phases. Wang and Ni [13] proposed the use of wireless sensor networks for prompt detection of earthquakes and for use during safety inspections of railway networks. Scholten et al. [14] presented the use of wireless sensor networks for checking the integrity of freight trains, especially the detection of situations where the undesirable loss of a railway car may occur.

Hartmark [15] introduced a number of measures which can be used to protect train rails from frost. Richter [16] addressed protective measures against threats associated with high voltage in railway transportation. Capra [17] proposed five main measures, mainly from the category of physical security and systemic measures, to protect railway infrastructure from terrorist attack. El-Koursi and Bruyelle [18] described preventive technical measures to reduce the number of suicides, the death rate of people at level crossings and the crossing of railways outside designated areas. Several publications, most which deal with technical protection measures, have addressed the issue of inappropriate levelling [19, 20]. Starita and Scaparra [21] examined effective allocation of protective measures in railway infrastructure systems.

4.2 The Concept of Resilience in Critical Infrastructure Protection System

Another possible approach in protecting infrastructure elements is application of the concept of critical infrastructure resilience. The term resilience was first defined by Holling [22] in 1973 in the context of resistance and stabilization of ecosystems, as a measure of the persistence of a system and its ability to absorb disruptions without significantly altering the state of the system. Later, the term resilience began to be used in other scientific disciplines such as psychology, sociology and economics, later also in engineering. In the context of critical infrastructure, the term resilience was first used in 2009 in the document Critical Infrastructure Resilience Final Report and Recommendations [3], where it was defined as the ability to reduce the magnitude, impact, or duration of a disruption. Resilience is the ability to absorb, adapt to, and/or rapidly recover from a potentially disruptive event.

Based on the above definitions, in a critical infrastructure system, resilience must be viewed as a cyclical process of continuous improvement in prevention, absorption, recovery and adaptation in a system. Figure 1 presents the cycle of resilience, showing the strengthening of resilience from its initial level (represented by the black dashed line) to a new level (represented by the red dashed line). The difference between these levels (Δ) is the level of strengthening in resilience.

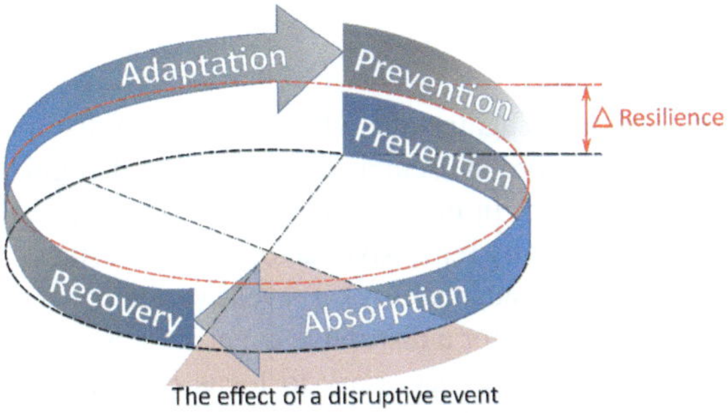

Fig. 1. Critical infrastructure resilience cycle [23].

The critical infrastructure resilience cycle begins with the prevention phase. In this phase, preventive measures are implemented to prepare the element for a potential disruptive event which may occur in the future. During the effects of a disruptive event, resilience passes into the absorption phase, which is determined by the level of robustness. Robustness is defined as the ability of an element to absorb the effects of a disruptive event, either through the structural properties of the buildings or the technologies applied (i.e. structural robustness) or through security measures (i.e. security robustness). When the disruptive event ends, the recovery phase begins. For the recovery phase, it is crucial that the element demonstrates the ability of recoverability, i.e. the ability to recover from a disruptive event and to restore its activity to the original or other required level of performance. The duration of the recovery phase then depends on the available resources and time pool of the individual recovery processes. This is followed by an adaptation phase, which ostensibly completes the critical infrastructure resilience cycle. Adaptation is characterized by adaptability, i.e. the ability of a critical infrastructure element to be prepared for the possibility that a past disruptive event repeats in the future. The adaptation phase is improved through the implementation of internal processes, such as risk management or innovation and educational processes [24].

Resilience is an important factor in the system of protection of critical infrastructure elements. Currently, the resilience of critical infrastructure is determined by three components: robustness, recoverability and adaptability [3]. These components are further determined by individual variables. An overview of these variables is given in Table 3. By strengthening the individual components of resilience and their variables, the level of resilience increases and thus reduces the vulnerability of the element.

Table 3. Components and variables which determine the resilience of critical infrastructure [24].

Components	Variables
Robustness	Redundancy; Detection ability; Responsiveness
Recoverability	Material resources; Financial resources; Human resources; Recovery processes
Adaptability	Risk management; Innovation processes; Educational and development processes

The protection of elements against the adverse effects of disruptive events must be continuously ensured, i.e. before the actual occurrence of a disruptive event, during the event, and after its termination. In the current model, however, resilience is perceived as a condition which is shaped by repressive factors (i.e. robustness and recoverability) and strengthening factors (i.e. adaptability). Preventive factors (i.e. resistance) which are useful in preparing elements for a disruptive event are absent in the current concept of resilience. For this reason, it is necessary to focus on this group of factors determining the resistance of critical infrastructure elements.

5 Resistance of Railway Infrastructure Elements

The term resistance in the context of resilience is defined by Sugden [25], who investigates the relationship between resistance and resilience in ecosystems. The author's work can be considered a key starting point in resistance research. The author defines resistance as a measure of how an ecosystem changes after a perturbation, such as the introduction of an alien species [26] or climate change [27]. Resilience then is the extent to which an ecosystem recovers after the removal of the source of change [25]. In addition to ecology, the term resilience is also used in a number of other scientific disciplines, such as sociology, where the term is used in connection with the resistance of society [28, 29], or in medicine in relation to the resistance of bacteria and viruses to antibiotics [30, 31].

5.1 View of Resistance in the Critical Infrastructure Resilience System

As mentioned above, in the current model, resilience is determined by three components: robustness, recoverability and adaptability. Resistance is not defined as a separate component in the concept of critical infrastructure resilience but is partly seen as a part of robustness [3]. The concept of resilience therefore currently lacks a preventive factor which would deal with measures that could prevent the actual occurrence of a disruptive event. For this reason, the present paper proposes viewing resistance as a separate component determining resilience in the prevention phase, building on current publications which explore resistance in the context of critical infrastructure resilience [4, 32].

In the context of critical infrastructure, resistance can be defined as the ability of an element to prevent the occurrence of a disruptive event. Resistance includes preventive measures determining structural resistance and security resistance. The first of these, structural resistance, is the ability of an element to withstand the effects of adverse factors based on its design, location in the system, and the technologies it applies.

In the case of security resistance, the element resists the effects of adverse factors using a number of security measures. The factors and variables determining the resistance of critical infrastructure elements are presented in Table 4.

Table 4. Factors and variables determining the resistance of critical infrastructure elements.

Component	Factors	Variables
Resistance	Crisis preparedness	Security planning, threat analysis
	Anticipation ability	Indication of resilience disruption, inspections, surveys, software applications
	Physical resistance	Structural properties of buildings and use of technologies
	Security measures	Organizational measures, regime measures, technical means

The essence of resistance is the prevention of disruptive events. Therefore, the variables determining resistance are also of a preventive nature, and individual measures which strengthen resistance are implemented in the prevention phase to prevent the occurrence of a potentially disruptive event.

Crisis preparedness is a set of measures to increase the preparedness of an element for disruptive events [24]. Crisis preparedness measures include, for example, security planning, which includes emergency plans, crisis preparedness plans, and threat analyses. The given measures serve primarily to raise awareness to the very existence of potential threats, to evaluate them, and to summarize the procedures, methods and measures which deal with disruptive events.

Anticipation ability means the ability of the system to predict the possible occurrence of a disruptive event, which creates a time interval that can be used to implement further preventive measures. Reliable prediction of growing threats (i.e. the effect of the threat on railway infrastructure) then enables a prompt response and prevention of the actual occurrence of a disruptive event. To achieve this, a resilience disruption indication system, which uses indicators to assess the level of resilience of the elements and the possibility of their disruption, can be applied [5]. Based on the assessment of a possible disruption of the element's resilience, preventive measures can be applied to prevent the occurrence of a disruptive event. Other measures which can be used to predict the occurrence of a disruptive event include, for example, regular inspections, surveys or the use of software applications which enable the prediction of disruptive events.

Physical resistance means the structural properties of buildings and applied technologies which resist the adverse effects of anthropogenic and naturogenic threats through their material and structural resistance, and thus prevent the occurrence of disruptive events.

Security measures are a set of organizational and regime measures and technical means applied to increase the security of the element against disruptive events [24]. These measures include the use of alarm systems, mechanical barriers, access controls, guarding of buildings, etc. [10].

Strengthening the individual factors determining resistance increases the level of resistance and thus the overall resilience. The allocation of an entity's resources in the application of individual measures should focus on the most critical areas of the system which are at risk and vulnerable.

5.2 Applying the Concept of Resistance in Railway Infrastructure

Resistance can be considered the most important component of resilience over time since it includes measures which can prevent the occurrence of disruptive events when the required level is reached. According to the European Railway Agency (ERA), the number of disruptive events in railway transport is declining [2], but a high level of protection of individual elements in railway infrastructure must continue. Because it is not possible to ensure the protection of all elements, selected elements are designated critical, and protecting these elements is prioritized. For such elements, the concept of resilience can be applied. Many authors have already addressed the issue of railway infrastructure resilience [33–35]. The term resistance is used in the professional literature dealing with railway transport, but in a context different from the context examined in the present paper. The concept of resistance in the system of railway infrastructure resilience therefore has great potential, and more research should be devoted to the issue.

6 Conclusion

The concept of resilience is an important factor in the system of protection of critical infrastructure elements. Resilience is seen as a cyclical process of continuous improvement of the four phases, which are prevention, absorption, recovery, and adaptation in a system. The prevention phase is determined by the level of resistance. This is determined by several factors whose strengthening increases the level of resistance and the actual resilience. The concept of resistance is an excellent topic for further research because a resistant system can prevent the occurrence of disruptive events. This would reduce not only the number of system failures and damage to property but also the number of deaths and injuries. Railroad infrastructure is one of the key subsectors of the national economy, and in the countries of the European Union, railroad transport is classified as a subsector of critical infrastructure. Because of its importance in the functioning of the state and the daily life of its inhabitants, it is important to continuously protect elements of railway infrastructure against the adverse impacts of disruptive events. To achieve this, the concept of resilience and resistance, which has great potential in this sector, can be applied.

Funding. This work was supported by the Technology Agency of the Czech Republic [grant number CK01000015] and by the VSB – Technical University of Ostrava [grant number SP2021/28].

References

1. Council Directive 2008/114/EC of 8 December 2008 on the Identification and Designation of European Critical Infrastructures and the Assessment of the Need to Improve their Protection (2008)
2. Report on Railway Safety and Interoperability in the EU. European Union Agency for Railways (2020)
3. National Infrastructure Advisory Council: Critical Infrastructure Resilience Final Report and Recommendations. U.S. Department of Homeland Security (2009)
4. Rehak, D., Hromada, M., Lovecek, T.: Personnel threats in the electric power critical infrastructure sector and their effect on dependent sectors: overview in the Czech Republic. Saf. Sci. **127**, 104698 (2020)
5. Splichalova, A., Patrman, D., Kotalova, N., Hromada, M.: Managerial decision making in indicating a disruption of critical infrastructure element resilience. Adm. Sci. **10**(3), 75 (2020)
6. Commission regulation (EC) No 851/2006, of 9 June 2006, specifying the items to be included under the various headings in the forms of accounts shown in Annex I to Council Regulation (EEC) No 1108/70 (Codified version) (2006)
7. Rehak, D., Slivkova, S., Pittner, R., Dvorak, Z.: Integral approach to assessing the criticality of railway infrastructure elements. Int. J. Crit. Infrastruct. **16**(2), 107–129 (2020)
8. Directive (EU) 2016/798 of the European Parliament and of the Council, of 11 May 2016, on railway safety (recast) (2016)
9. Ministry of the Interior of the Czech Republic: Threat Analysis for the Czech Republic: Final Report 2015. Approved by Resolution of the Government of the Czech Republic of 27 April 2016, no. 369, pp. 9 (2015)
10. Kampova, K., Lovecek, T., Rehak, D.: Quantitative approach to physical protection systems assessment of critical infrastructure elements: use case in the Slovak Republic. Int. J. Crit. Infrastruct. Prot. **30**, 100376 (2020)
11. Catalano, A., Bruno, F.A., Pisco, M., Cutolo, A., Cusano, A.: An intrusion detection system for the protection of railway assets using fiber Bragg grating sensors. Sensors **14**, 18268–18285 (2014)
12. Cañete, E., Chen, J., Díaz, M., Llopis, L., Rubio, B.: Sensor4PRI: a sensor platform for the protection of railway infrastructures. Sensors **15**, 4996–5019 (2015)
13. Wang, D., Ni, Y.: Wireless sensor networks for earthquake early warning systems of railway lines. In: Ni Y.Q., Ye X.W. (eds.) Proceedings of the 1st International Workshop on High-Speed and Intercity Railways. Lecture Notes in Electrical Engineering, vol. 148. Springer, Berlin (2012)

14. Scholten, H., Westenberg R., Schoemaker, M.: Sensing train integrity. In: Sensors, pp. 669–674. IEEE, Christchurch, New Zealand (2009)
15. Hartmark, H.: Frost protection of railway lines. Eng. Geol. **13**, 505–517 (1979)
16. Richter, B.: Overvoltage protection concept for DC railway systems. J. Electrostat. **65**, 356–362 (2007)
17. Capra, G.S.: Protecting critical rail infrastructure. Counterproliferation Paper No. 38 USAF Counterproliferation Center, Air University, Maxwell Air Force Base, Alabama (2006)
18. El-Koursi, E.M., Bruyelle, J.L.: Railway accident prevention and infrastructure protection. J. Civil Eng. Architect. **10**, 96–107 (2016)
19. Silla, A., Luoma, J.: Trespassing on Finnish railways: identification of problem sites and characteristics of trespassing behaviour. Eur. Transp. Res. Rev. **1**, 47–53 (2009)
20. Rådbo, H., Silla, A., Lukaschek, K., Burkhardt, J., Paran, F.: Current knowledge of railway suicides and trespassing accidents. Rrduction of Suicides and Trespasses on RAILway property Collaborative project (RESTRAIL) (SCP1-GA-2011–285153) (2012)
21. Starita, S., Scaparra, M.P.: Optimizing dynamic investment decisions for railway systems protection. Eur. J. Oper. Res. **248**(2), 543–557 (2016)
22. Holling, C.S.: Resilience and stability of ecological systems. Annu. Rev. Ecol. Syst. **4**, 1–23 (1973)
23. Rehak, D., Senovsky, P., Hromada, M., Lovecek, T.: Complex approach to assessing resilience of critical infrastructure elements. Int. J. Crit. Infrastruct. Prot. **25**, 125–138 (2019)
24. Rehak, D., Senovsky, P., Slivkova, S.: Resilience of critical infrastructure elements and its main factors. Systems **6**(2), 21 (2018)
25. Sugden, A.M.: Resistance and resilience. Science **293**(5536), 1731 (2001)
26. Knapp, R.A., Matthews, K.R., Sarnelle, O.: Resistance and resilience of alpine lake fauna tofish introductions. Ecol. Monogr. **71**(3), 401–421 (2001)
27. Mahardja, B., et al.: Resistance and resilience of pelagic and littoral fishes to drought in the SanFrancisco Estuary. Ecol. Appl. **31**(2), e02243 (2021)
28. Baaz, M., Lilja, M., Schulz, M., Vinthagen, S.: Defining and analyzing "resistance": possible entrances to the study of subversive practices. Altern. Glob. Local Polit. **41**(3), 137–153 (2017)
29. Hollander, J.A., Einwohner, R.L.: Conceptualizing resistance. Sociol. Forum **19**(4), 533–554 (2004)
30. European Centre for Disease Prevention and Control. https://antibiotic.ecdc.europa.eu/en/get-informedfactsheets/factsheet-experts. Accessed 02 Apr 2021
31. Ddzidic, S., Susković, J., Kos, B.: Antibiotic resistance mechanisms in bacteria: biochemical and genetic aspects. Food Technol. Biotechnol. **46**(1), 11–21 (2008)
32. Rogers, C.D.F., et al.: Resistance and resilience – paradigms for critical local infrastructure. Municipal Eng. **165**(2), 73–83 (2012)
33. Smoczyński, P., Gill, A., Kadziński, A.: Safety recommendations as a method of strengthening resilience of the railway system. In: Proceedings of 24th International Scientific Conference. Transport Means 2020, pp. 804–809 (2020)
34. Adjetey-Bahun, K., Birregah, B., Châtelet, E., Planchet, J.L.: A model to quantify the resilience of mass railwaytransportation systems. Reliab. Eng. Syst. Saf. **153**, 1–14 (2016)
35. Bababeik, M., Khademi, N., Chen, A.: Increasing the resilience level of a vulnerable rail network: the strategy of location and allocation of emergency relief trains. Transp. Res. Part E **119**, 110–128 (2018)

Threat Assessment of the Railway Infrastructure Soft Targets

Simona Slivkova[1(✉)], David Rehak[1], Lenka Michalcova[2], and Radim Pittner[3]

[1] Faculty of Safety Engineering, VSB – Technical University of Ostrava, Lumirova 630/13, 700 30 Ostrava, Czech Republic
{simona.slivkova, david.rehak}@vsb.cz

[2] Faculty of Transportation Sciences, Czech Technical University in Prague, Konviktska 20, 110 00 Prague 1, Czech Republic
brehovska@fd.cvut.cz

[3] Railway Administration, Department of Safety and Crisis Management, Dlazdena 1003/7, 110 00 Prague 1, Czech Republic
pittner@spravazeleznic.cz

Abstract. Railway transport is one of the significant areas for the functioning of human society. The task of transport is well-thought movement of goods and persons from one place to another. The essence of railway transport is the service of transporting people and goods by rail using railway infrastructure, railway vehicle and under the supervision of rail traffic management systems. However, elements of the railway infrastructure are constantly threatened by a wide range of different threats, which may cause a disruption of their functionality or a threat to the safety of transported persons. At present, attention is also significantly focused on the protection of so-called soft targets. This focus is also reflected in the field of railway transport and railway infrastructure. To ensure the protection of the soft targets of railway infrastructure, threats need to be assessed comprehensively so that an effective response to these threats can be ensured. The aim of this article is to present the procedure of assessing the threats of soft targets of railway infrastructure according to the level of their danger resulting from the level of expected impact. Various threats enter into this assessment in order to maintain the complexity of this assessment.

Keywords: Railway infrastructure · Soft targets · Railway stations · Threat assessment

1 Introduction

The essence of railway transport is the transport of people and goods using railway infrastructure, rail vehicles, energy, and labor on the railways. It is a complex system that must consider the means of transport, the transport medium, and other organizational requirements [1].

Railway transport is one of the most important ways of transporting people and cargo by land. Within the North Atlantic Treaty Organization (NATO), railway transport is one of the main modes of transport for the armed forces [2]. Since 2008,

O. Prentkovskis et al. (Eds.): TRANSBALTICA 2021, LNITI, pp. 429–438, 2022.
https://doi.org/10.1007/978-3-030-94774-3_42

railway transport has also been included in the European Critical Infrastructure Sector [3].

Transport as a sector itself is also an important condition for the development of the economy and the whole society of the state. Transport affects virtually all areas of public and private life and the business sector. It is a necessary condition for increasing the competitiveness of the state [4].

Due to their importance, transport systems are often ranked among the targets of terrorist attacks, mainly due to their high impact on the social, economic, psychological, and political levels [5, 6].

The security situation in the world is constantly deteriorating in terms of terrorism, and the incidence of violent attacks is increasing. Organizers of such attacks are increasingly motivated to target unprotected places with a high concentration of people, regardless of whether they are politically or religiously symbolic [7].

Protection of soft targets is becoming a priority for individuals, social groups, and the governments. Ensuring their protection minimizes property damage and reduces the number of injured or victims. The protection of soft targets is a long-standing problem, especially in developed countries, which are the primary target of terrorist attacks and similar threats. The key step for any protection is a comprehensive risk assessment of the element. In this assessment, it is necessary to use an adequate method of risk analysis and to correctly interpret the risks [8–10].

2 Description of Railway Infrastructure

Generally, any subsystem (including railway transport) can be understood as a group of interconnected and interactive parts that perform an important work or task and are part of a larger system. In principle, the railway transport system consists of the infrastructure, the services it provides, the service providers, the threats affecting the subsystem and the impacts caused by the disruption of the subsystem [11].

The function of railway infrastructure is primarily linked to the railway. It is intended for the movement of rail vehicles, including fixed equipment needed to ensure the safety and continuity of railway transport [12].

Railway infrastructure means all roads and fixed installations of railway transport which are necessary for the operation of railway vehicles and the safety of such operation. Specifically, these may include, for example, the following elements [13]:

- land and railway track,
- passenger platforms and loading ramps,
- construction of railway stations,
- equipment for heating railway switches,
- civil engineering buildings (bridges, overpasses, level crossings, tunnels),
- level crossings of the track (including equipment for ensuring road safety),
- railway superstructure (rails, switches, and crossings),
- access roads for passengers and goods, including roads,
- security, signaling and telecommunication equipment,
- infrastructure management service facilities.

Together, the railway infrastructure forms a comprehensive complex of elements that must be perceived as significantly functionally interconnected. All movement on the infrastructure is organized by a complex traffic management system, which must take into account several timetables on different transport routes [11, 14].

3 Determination of Soft Targets of Railway Infrastructure

The essence of determining the soft targets of railway infrastructure is to take into account three basic characteristics, the combination of which can identify those elements of railway infrastructure that are soft targets in the field of railway transport and are also assessed as systemically important. These characteristics are: (1) railway infrastructure, (2) soft targets, and (3) significance, criticality. Based on the integration of defined characteristics, it is possible to determine the elements of interest of the railway infrastructure (see Fig. 1).

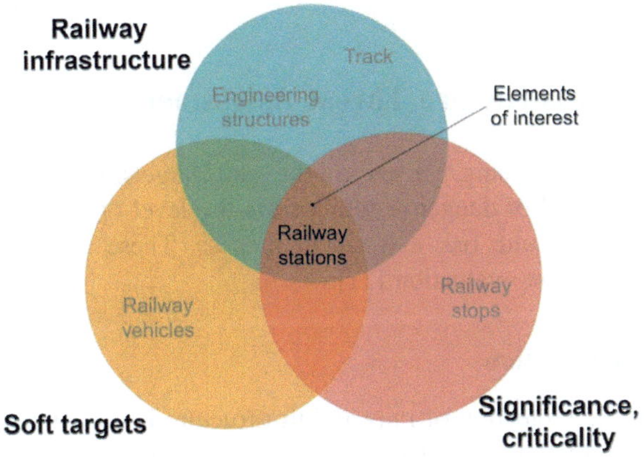

Fig. 1. Determination of interest elements of railway infrastructure.

The characteristics of the railway infrastructure have been outlined in the previous section. The term "soft targets" is usually used to refer to vulnerabilities that can be selected by terrorists in an effort to maximize losses and thereby cause population fear and significant media coverage [15].

These are usually publicly accessible places with a high concentration of people who have a low level of protection. These may be open or enclosed spaces or facilities to which the public has free access and which, due to the high concentration of persons, may be identified as a potentially suitable target for attackers or terrorists [16, 17].

In the field of railway transport, railway stations and railway vehicles can be considered as soft targets on the basis of the above definitions. However, with regard to the first characteristic of railway infrastructure, the list is selected only for railway stations.

The area of "significance, criticality" enters here as a criterion for the selection of those elements that are important for a given system (i.e. critical, or also significant).

The critical element can be the element of the system that is important for the operation of the system and whose malfunction would have serious effects on the monitored system, or such an element, the failure of which could cause large negative effects on the system's ability to provide the service. Significance often reflects the importance of a given element for a large part of society or for a given system, also with regard to the interdependencies in the given system [18–20].

Based on a more detailed specification of these characteristics, it is possible to identify important railway stations in a given railway transport system. These are primarily those stations that are important or critical elements of a given transport system (e.g. they form a significant railway junction in a given place). Smaller stations on side lines as well as railway stops will be excluded from the identification due to this aspect.

Railway stations determined in this way become elements of interest for the subsequent process of comprehensive threat assessment of soft targets of railway infrastructure.

4 Railway Infrastructure Threat Assessment

The procedure for assessing threats to the soft targets of railway infrastructure according to their level of danger resulting from the level of expected impact reflects the basic principles of the risk assessment process. These are threat identification, analysis, and subsequent evaluation [21].

4.1 Threat Identification

The key step in the protection of railway infrastructure is always the consistent identification of threats that may adversely affect the railway infrastructure. In the proposed procedure, threats are divided into two basic groups with regard to the nature of their origin: (1) threats from the area of railway transport and railways, and (2) threats arising outside the perimeter of the railway, i.e. arising from the environment.

Threats in the Field of Railway Transport and Railways. This group of threats is tied to the railway transport system and also to the elements of the railway infrastructure. It is thus possible to assume their occurrence within the entire railway system. This group of identified threats will be basically similar anywhere in the world, as the railway systems in different countries differ only with minimal deviations.

In an effort to assess the probability of the occurrence of these threats, some deviations will already arise, primarily with regard to the quality of the equipment, or their technical functionality or modernization. However, the present proposal does not work with probability.

There can be several sources for identifying such threats, they were primarily used for the proposed procedure [12, 22, 23]. The resulting list of identified threats contains:

- Collision of rolling stock (Label A1),
- Derailment of a railway vehicle (A2),
- Collision of a railway vehicle with an obstacle in the clear section (A3),
- Encounter of rail vehicles with road vehicles (A4),
- Rail break (A5),
- Track deflection (A6),
- Emergency at the pantograph-traction line interface (A7),
- Fire in the perimeter of the runway (A8),
- Explosion in the perimeter of the runway (A9),
- Failure of signaling (safety) systems (A10),
- Leakage of dangerous goods (substances) during its transport (A11).

Threats Arising Outside the Perimeter of the Runway, I.E. Arising from the Surrounding Environment. The group of threats arising from the environment is significantly affected by local conditions. Thus, the results may vary significantly across countries. The basis is the assessment of threats to the territory, i.e. the state.

When trying to assess the probability of the occurrence of these threats, it will be necessary to evaluate the local conditions, i.e. the immediate vicinity of the station. But as mentioned above, the present proposal is not working with the probability.

To identify threats to this group, it is possible to use existing threat analyzes of the country. For the territory of the Czech Republic, it is the Threat Analysis for the Czech Republic [24]. Here, the authors identified a total of 72 types of hazards for the territory of the Czech Republic, which they divided according to their character into abiotic, biotic, cosmic, technogenic, sociogenic and economic groups. For the needs of the proposed procedure, those threats were selected which were identified as risks unacceptable and risks conditionally acceptable in the authors' evaluation.

As the Threat Analysis for the Czech Republic [24] contains a complete list of threats for the entire state, it was necessary to select those threats that may have a negative effect on the railway infrastructure system. At the same time, the final list of threats in this group no longer includes those threats that have been identified within the group of threats from the area of railway transport and railways (e.g. leakage of a dangerous substance during transport, or a railway accident). The resulting list of identified threats to this group contains:

- Flood (Label B1),
- Flash flood (B2),
- Extreme wind (B3),
- Extremely high temperatures (B4),
- A special flood (B5),
- Violation of large-scale legality (terrorism) (B6),
- Other violations of the law (vandalism) (B7),
- Earthquake (B8),
- Slope instability (B9),
- Tornado (B10),
- Occurrence of extremely low temperature (B11),
- Atmospheric discharges (B12),

- Fire outside the perimeter of the runway (B13),
- Explosion outside the perimeter of the runway (B14),
- Black ice, glaze, and atmospheric icing (B15),
- Epidemics - mass infections of persons (B16),
- Leakage of hazardous chemical from stationary equipment (B17),
- Radiation accident (B18),
- Disruption of large-scale heat supply (B19),
- Disruption of large-scale electricity supply (B20),
- Disruption of large-scale drinking water supplies (B21),
- Snow calamity (B22).

4.2 Analysis of the Impacts of Identified Threats

The following text presents an analysis of the identified threats. It focuses on the possible impact of identified threats on railway stations. Both threats from the field of railway transport and threats arising outside the perimeter of the track are analyzed.

The analysis assesses the possible impact of the threat on certain elements of the railway station (see Table 1).

Table 1. Threats in the field of railway transport and railways.

Railway station element	Label
Station building including all necessary equipment and another station building	E1
Pre-station area	E2
Platforms, including underpasses, overpasses, and possible shelters	E3
Tracks for entry and departure of trains, including technical equipment	E4
Other tracks, e.g. for decommissioning or deposition, including technical equipment	E5
Other technological objects, e.g. transformer stations, security devices	E6

The assessment of all elements of the railway station is related to all identified threats. The "danger" of the threat is assessed. A hazard is a property that characterizes how dangerous a given threat is in general in relation to a given element of a railway station. This evaluation takes place only on a general scale, the real impact can be assessed only in the context of the specific location of the element, when it is possible to obtain a variable probability for the resulting degree of risk.

The hazard assessment reaches a total of four levels:

- 0 = negligible level of danger,
- 1 = low level of danger (disruption of the function of the element),
- 2 = high level of danger (failure of element function),
- * = only secondary impact (without physical disruption or damage to the element, it manifests itself only within the limitation of some function of the element, which causes a restriction of railway traffic in the given place).

The results of the evaluation are presented in the following tables (Table 2).

Table 2. Hazard assessment of railway transport threats.

Threat	E1	E2	E3	E4	E5	E6
A1	0	0	1	2	1	1
A2	0	0	1	2	1	1
A3	0	0	0	2	1	1
A4	0	0	0	2	0	1
A5	0	0	0	2	2	0
A6	0	0	0	2	2	0
A7	0	0	0	2	2	1
A8	2	1	1	2	2	2
A9	2	1	1	2	2	2
A10*	0	0	0	1	1	1
A11*	0	0	0	1	1	1

Table 3. Environmental hazard assessment.

Threat	E1	E2	E3	E4	E5	E6
B1	2	1	1	2	2	2
B2	2	1	1	2	2	2
B3	1	1	0	2	2	1
B4	0	0	0	2	2	0
B5	2	1	1	2	2	2
B6	2	2	1	2	2	2
B7	1	1	0	2	1	2
B8	2	1	1	2	2	2
B9	1	1	1	1	1	1
B10	1	1	0	2	2	1
B11	0	0	0	1	1	0
B12	1	1	0	1	1	1
B13	1	1	0	1	1	1
B14	1	1	0	1	1	1
B15	0	0	0	1	1	0
B16*	1	1	0	0	0	0
B17*	1	1	0	1	1	1
B18*	1	1	0	0	0	0
B19*	1	0	0	0	0	0
B20*	2	2	0	2	2	2
B21*	1	0	0	0	0	0
B22*	1	1	0	1	1	1

4.3 Threat Hazard Assessment

Three assessment categories were created for the final threat assessment (Table 3):

- I. category – threats with a high level of danger (threats that can significantly damage the element of the railway station, or completely destroy it and thus cause the failure of the element; the value "2" from the tables above),
- II. category – low-risk threats (threats that can cause minor impacts and partial damage, i.e. disruption of the function of the element; the value "1" from the tables above),
- III. category – threats that can cause only secondary impacts (impacts on the function of the element without physical damage, impacts in the form of traffic restrictions at the station).

The result of the implementation of the proposed procedure is a list of possible threats for a given element of a railway station with classification into individual categories according to the non-security of the threat. Within the presented example of evaluation in the Czech Republic, it is possible to present the final list of threats, e.g. for pre-station proctor stations (see Table 4).

Table 4. Threats in the field of railway transport and railways.

Threat	Category
Violation of large-scale legality (terrorism)	I
Fire in the perimeter of the runway	II
Explosion in the perimeter of the runway	II
Flood	II
Flash flood	II
Extreme wind	II
A special flood	II
Other violations of the law (vandalism)	II
Earthquake	II
Slope instability	II
Tornado	II
Atmospheric discharges	II
Fire outside the perimeter of the runway	II
Explosion outside the perimeter of the runway	II
Epidemics - mass infections of persons	III
Leakage of hazardous chemical from stationary equipment	III
Radiation accident	III
Disruption of large-scale electricity supply	III
Snow calamity	III

5 Conclusion

The aim of the paper was to present the procedure for assessing the threats of soft targets of railway infrastructure according to the level of their danger resulting from the level of expected impact.

Already in history, but also today, a number of experts have expressed the opinion that railway transport has an irreplaceable place in transport services and in the national economy. The area of railway transport is also significantly involved in the activities of protection of so-called soft targets. For effective protection of elements, a comprehensive threat assessment is always the first step. As part of the proposed procedure, an assessment of the danger of the given threat resulting from the level of the expected impact was used for this purpose.

The result of the process is a list of threats that can negatively affect an element of the railway station and which are classified according to their level of safety. These results can be a significant input for the process of protecting soft railway infrastructure targets. The proposed procedure was practically applied in the Czech Republic. Its adaptation for use in other countries consists only in supplementing the threat analysis of the country. The procedure is primarily prepared for application in the field of railway infrastructure, but its basic principles can be adapted to the needs of other sectors.

Funding. This work was supported by the Technology Agency of the Czech Republic [grant number CK01000015].

References

1. Zangani, D., Fuggini, C.: Towards a new perspective in railway vehicles and infrastructure. In: Procedia - Social and Behavioral Sciences, vol. 48, pp. 2351–2360. Genova, Italy (2012). https://doi.org/10.1016/j.sbspro.2012.06.1206
2. Vlkovsky, M., Ivanusa, T., Neumann, V., Foltin, P., Vlachova, H.: Optimizating cargo security during transport using dataloggers. J. Transp. Secur. **10**, 63–71 (2017). https://doi.org/10.1007/s12198-017-0179-4
3. Council Directive 2008/114/EC of 8 December 2008 on the identification and designation of European critical infrastructures and the assessment of the need to improve their protection. European Council, Brussels, Belgium (2008)
4. Transport policy of the Czech Republic for the period 2014−2020 with a view to 2050. Ministry of Transport of the Czech Republic, Prague (2013). Approved by Resolution of the Government of the Czech Republic No. 449 of 12 June 2013, p. 89 (2013)
5. Hedel, R., Boustras, G., Gkotsis, I., Vasiliadou, I., Rathke, P.: Assessment of the European programme for critical infrastructure protection in the surface transport sector. Int. J. Crit. Infrastruct. **14**(4), 311–335 (2018). https://doi.org/10.1504/IJCIS.2018.095616
6. Szatmári, M., Leitner, B.: Threat assessment of railway stations as a tool for increasing of soft targets security level. In: Proceedings of 24th International Scientific Conference. Transport Means 2020, pp. 119–125. Kaunas, Lithuania (2020)

7. Leitner, B., Luskova, M.: Assessing security of soft targets using complex systems analysis methods. In: Hofreiter, L., Berezutskyi, V., Figuli, L., Zvaková, Z. (eds.) Soft Target Protection. NSPSSCES, pp. 241–255. Springer, Dordrecht (2020). https://doi.org/10.1007/978-94-024-1755-5_19
8. Apeltauer, T., Apeltauer, J., Okrinova, P.: Soft target protection analysis using pedestrian simulation. In: Proceedings, Fire and Evacuation Modeling Technical Conference (FEMTC), p. 16 (2020)
9. Hofreiter, L., Halaj, M., Jankura, R.: Building a security culture as a tool for soft targets protection. In: Hofreiter, L., Berezutskyi, V., Figuli, L., Zvaková, Z. (eds.) Soft Target Protection. NSPSSCES, pp. 139–147. Springer, Dordrecht (2020). https://doi.org/10.1007/978-94-024-1755-5_11
10. Foltin, P.: Security of logistics chains against terrorist threats. In: 17th International Conference the Knowledge-Based Organization, Conference Proceedings 1: Management and Military Sciences. Sibiu, Romania (2011)
11. Rehak, D., Slivkova, S., Pittner, R., Dvorak, Z.: Integral approach to assessing the criticality of railway infrastructure elements. Int. J. Crit. Infrastruct. **16**(2), 107 (2020). ISSN 1475-3219
12. Act No. 266/1994 Coll., Railways Act, as Amended. https://www.zakonyprolidi.cz/cs/1994-266. Accessed 13 Apr 2021
13. Commission regulation (EC) No 851/2006, of 9 June 2006, specifying the items to be included under the various headings in the forms of accounts shown in Annex I to Council Regulation (EEC) No 1108/70 (Codified version)
14. Dvorak, Z., Luskova, M., Rehak, D., Slivkova, S.: Criticality assessment of railway bridges. In: Transbaltica XI: Transportation Science and Technology: Proceedings of the International Conference Transbaltica. Springer, Vilnius, Lithuania, Cham (2020). https://doi.org/10.1007/978-3-030-38666-5_50
15. Karlos, V., Larcher, M., Solomos, G.: Review on Soft target/Public Space Protection Guidance. EUR 29116 EN, Publications Office of the European Union, Luxembourg (2018). https://doi.org/10.2760/553545
16. Kalvach, Z.: Basics of Soft Targets Protection - Guidelines (2nd version), p. 46. Prague, Soft Targets Protection Institute (2016)
17. Soft targets protection concept for 2017–2020, p. 32. Ministry of the Interior of the Czech Republic, Prague (2017)
18. Fekete, A.: Common Criteria for the Assessment of Critical Infrastructures, p. 10. Bonn, German, Federal Office of Civil Protection and Disaster Assistance (2011)
19. Jönsson, H., Johansson, J., Johansson, H.: Identifying critical components in technical infrastructure networks. In: Proceedings of the Institution of Mechanical Engineers (2007), vol. 222, no. 2, pp. 235–243. https://doi.org/10.1243/1748006XJRR138
20. Leitner, B., Rehak, D., Kersys, R.: The New Procedure for Identification of Infrastructure Elements Significance in Sub-sector Railway Transport. Communications - Scientific Letters of the University of Žilina, vol. 20, no. 2, pp. 41–48. Žilina, Slovakia (2018)
21. International Organization for Standardization/Technical Committees Risk Management. ISO 31000 Risk Management-Guidelines. International Organization for Standardization/Technical Committees Risk Management, Geneva, Switzerland (2018)
22. Directive (EU) 2016/798 of the European Parliament and of the Council, of 11 May 2016, on railway safety (recast)
23. Decree No. 1 to the SZDC D17 regulation. Railway Administration, state organization, General Directorate, pp. 11. Czech Republic, Prague (2017)
24. Paulus, F., Kromer, A., Petr, J., Cerny, J.: Threat Analysis for the Czech Republic. Fire and Rescue Service of the Czech Republic, Czech Republic, Prague (2015)

Sensitivity Test of the STEM Method Modified to Prioritize the Allocation of Traffic Path Capacity

Pavel Purkart(✉), Jan Kruntorád, and David Vodák

CTU in Prague, Faculty of Transportation Sciences, Konviktska 20, 11000 Prague 1, Czech Republic
{purkapav, kruntjan, vodakdav}@fd.cvut.cz

Abstract. CTU in Prague Faculty of Transportation Sciences, Department of Transport Systems, deals in its research with the issue of allocating the capacity of a railway transport path. The capacity of any transport route is a limiting element in terms of its functionality. It is not always possible to meet all requirements and the allocator has to decide how to allocate capacity, so that it is as efficient as possible and achieves as many societal benefits as possible. The research team deals with this issue by using a modified STEM (Step Method) method for its research, which is a tool for linear optimization. The article presents the use of this method on the railway line Plzeň - Žatec in the Czech Republic. It deals with its sensitivity while changing the parameter of the number of passengers in the individual segments of the passenger rail transport. There are many requirements for the allocation of railway capacity, but the infrastructure is not able to satisfy all of them. The article presents a simulation of three variants of the number of passengers expected in regional expresses. This represents the possibility of allocating capacity to different train segments in situations when the infrastructure cannot satisfy all requirements for railway capacity allocation.

Keywords: Railway infrastructure capacity · Requirements of operators and regional public transport managers · STEM method · Plzeň – Žatec railway line · Optimization

1 Introduction

The capacity of the railway infrastructure is a parameter which influences its usability. Not only in the Czech Republic, but in all developed countries, where rail transport is used as the backbone of transport services in regions, its capacity is a major problem, often failing to meet all requirements. This brings the question of how to realize the operation of trains so that this is as effective as possible with regard to the infrastructure restrictions [1, 2]. There are no universal methods and practically every infrastructure manager solves this issue differently.

This issue is a part of the research, which is currently conducted by CTU in Prague, Faculty of Transportation Sciences, Department of Transportation Systems, which encounters this issue not only in scientific work, but also while solving practical studies

O. Prentkovskis et al. (Eds.): TRANSBALTICA 2021, LNITI, pp. 439–447, 2022.
https://doi.org/10.1007/978-3-030-94774-3_43

for various subjects. To solve the problem, the STEM (Step Method) method is currently being considered and tested.

2 Using the STEM Method

The STEM method can solve linear mathematical problems with more purpose functions. The aim of this method is to find compromise solutions, whose realization should bring the most benefits. The main principle of the method is the calculation of purpose function ideal values for individual cases. This calculation is followed by minimizing the compromise solution deviation from the ideal purpose function values. The basic of the method is an interactive procedure of searching the compromise solution.

Benefit of the STEM method is that there is only minimal need of communication between a submitter and a solver (compared to another methods). The scale method for individual criterion is set by calculation. The submitter must decide whether the result of the calculation is acceptable for him or not. Therefore, the method consists of a calculation and decision-making process. The calculation is stopped, if the submitter finds the result acceptable, otherwise the solver must be informed by the submitter in order to change the criterions or their numbers, the whole calculation is made again.

The STEM method consists of the following steps:

1. Solver calculates the optimal solution for individual criterions (purpose function) separately. The number of calculations fits the number of criterions.
2. Solver calculates the scales of individual criterions according to formula (1):

$$w_i = \frac{z_{ii} - \min\limits_{i=j...k} z_{ij}}{z_{ii}} \frac{\alpha}{\sqrt{\sum\limits_{i=1}^{n} c_{ij}^2}} \quad (1)$$

where z_{ij} – element of optimization criterion values matrix for optimization in the individual optimization criterion (z_{ij} is the value of optimization criterion $j = 1,...,k$ in the case of optimization according to the criterion $i = 1,...,k$), c_{ij} – element of the price matrix – element of individual optimization criterion coefficients matrix. Value α comes from Eq. (2):

$$\sum_{i=1}^{k} \frac{z_{ii} - \min\limits_{i=j...k} z_{ij}}{z_{ii}} \frac{\alpha}{\sqrt{\sum\limits_{i=1}^{n} c_{ij}^2}} = 1 \quad (2)$$

In reality, we have to calculate the coefficient alfa value first and then count the scales of individual criterions. If the scale fits the constraint $w_i \geq 0$ for more criterions, the solver adds a new variable $d \geq 0$ and solves the model with a new optimization criterion (3).

$$\min f(x,d) = d \tag{3}$$

There is a form (4) for variable *d*:

$$d = \max_{i=j...k} \left\{ w_i \left(z_{ii} - \sum_{j \in J} c_{ij} X_j \right) \right\} \tag{4}$$

We have to implement constraint (5) for the correct calculation:

$$w_{ii} \left(z_{ii} - \sum_{j \in J} c_{ij} X_j \right) \leq d \tag{5}$$

If the constraint $w_i > 0$ fits for only one value $i = 1, \ldots, k$, the solver can simplify the constraint (5) to (6):

$$\min f(x) = \sum_{i=1}^{k} w_{ii} \left(z_{ii} - \sum_{j \in J} c_{ij} X_j \right) \tag{6}$$

3. Solver presents the results to the submitter. The submitter must modify the criterions or add/remove some of them, if he does not find the results acceptable. Solver goes back to step 2. Solver has found a compromise solution if the submitter of is satisfied with the result. The solution is optimal if the value $d = 0$ is reached.

3 Model Modification and Determination of Stable Values

Established evaluation criterions for individual lines for the modified STEM method:

Daily Estimated Average Number of Passengers in the Limiting Railway Section in Thousands
This parameter presents the daily average number of passengers on the route in the limiting section - the section with the lowest capacity. The value presents the passenger numbers in the trains of the given line in this section.

Daily Estimated Average Number of Passengers on the Route in Thousands
This parameter presents the daily average number of passengers on the route or on the logically selected section of the route. This parameter provides an evaluation of the total route use. It is not sufficient to consider the potential only on the limiting section mentioned above, but it is also crucial to assess the potential of the whole route.

The Use of Maximal Line Speed in a Logically Selected Railway Section
Trains are often unable to reach the line's maximal speed so they cannot make the full use of railway line parameters – this is the reason for implementing this parameter. When the train is able to reach the railway line maximal speed in the chosen section, the ratio will be 1 (100%). If the speed is not reached, the ratio will decrease. If the

line's maximal speed is up to 100 km/h in the selected section and the train is able to reach a speed of only 80 km/h, the ratio will logically decrease to 0.8 (80%).

Evaluation of System Links on the Route in Selected Section
The parameter should evaluate the links to other lines, the aim is to identify the importance of route in the network. The overall value of the parameters is the sum of the following points for all transfer nodes/points in the selected section of the route. The transfer nodes/points are evaluated as follows:

- 2 points – a transfer node with system links to railway routes in at least three other directions (at least a crossroad station, but rather nodal station) with the possibility of system links to bus routes or city public transport,
- 1 point – a transfer point with system links to railway routes in at least one or two other directions with the possibility of system links also to bus routes/city public transport or a transfer point with frequent system links to bus routes or city public transport

If a route is routed through an important transfer node, it receives 2 points for that node. It receives 1 point for each transfer point (lower importance). The higher the sum of points, the more frequent and important the links are, and therefore the operation of the route is crucial for the public transport routes network.

Comparison of Travel Times Between Individual Car Transport and Railway Route in the Three Selected Sections with the Highest Passenger Numbers
This parameter is set in order to compare the ability of the train route to compete with individual car transport. In the selected section of the line, the three busiest connections will be selected and the ratio of the travel time of individual car transport in the given section to the travel time using the railway route will be determined. For the mentioned group of three connections, the value will be determined separately and then the average of the three values that will be included in the evaluation will be calculated. If the value exceeds number 1, public transport is on average in a selected connections faster than individual car transport.

The railway line Plzeň - Žatec was chosen for the model test. The line runs from the regional capital city of Plzeň (Pilsen) in the western part of the Czech Republic to the agglomeration in the Podkrušnohorská basin in the northern part of the country (cities of Žatec, Chomutov, Most and Jirkov). Especially in the Pilsen agglomeration, the capacity of the railway line is very restrictive, therefore it was chosen for model test. The STEM method has been modified. It was originally intended for project evaluation, providing evaluation and results for prioritization of projects in relation to a limited budget. It calculates the occupancy options of the selected section of the railway line in the selected time interval, so we can decide which trans should be allowed to pass the section in order to maximize benefits for society after the modification.

Following conflicts of regional public transport managers demands are expected:

- route R (fast train) Plzeň – Most in 120 min interval,
- route Sp (regional express) Plzeň – Žihle in 120 min interval,
- route Os no. 1 (regional train no. 1) Plzeň – Žihle in 60 min interval,
- route Os no. 2 (regional train no. 2) Nýřany – Plzeň – Plasy in 60 min interval.

This route schema ensures that the total interval of the fast segment of trains in the section Plzeň – Žihle will be 60 min and 30 min for regional trains in the peak period.

Given the fact that the basic interval of the most sparsely represented train segments is 120 min, this value was also chosen as the starting point for determining the length of the evaluation period. We consider even traffic in both directions, so for each direction in this period there are 60 min of the track capacity available, including all the operations (operation of railway signalling equipment, etc.), if expressed by the number of minutes, not the number of paths, as considered in the model. This value is reduced to *50* min in order not to reach the occupancy rate of 100%. We considered minimal operation of the freight trains on this line, so there are no requirements for their paths. This is based on the usual prioritization of passenger trains at the expense of freight trains during the peak hours in the railway sections with capacity restrictions.

According to the graphical timetable the most restrictive section is the *Horní Bříza – Kaznějov* section [5]. This section is considered for the calculation with the following occupancy time for individual routes:

- route R 8 min,
- route Sp 9 min,
- route Os 10 min.

The following parameters were stable for the STEM sensitivity test:

The Use of Maximal Line Speed in a Logically Selected Railway Section
For the use of maximal line speed, the determined parameters are set in the Table 1.

Table 1. The use of maximal line speed in a logically selected railway section.

Route	The use of maximal line speed in a logically selected railway section
R	1.00
Sp	1.00
Os no. 1	2 × 1.00 = 2.00*
Os no. 2	2 × 1.00 = 2.00*

* For regional trains, the value is multiplied by two, as two trains pass in each direction over a reference period of 120 min.

In the case of all connections, the full use of maximal line speed is planned.

Evaluation of System Connection Links on the Line in a Logically Defined Section
The below mentioned nodes are served by the individual routes. The evaluation of individual nodes according to their significance are summarized in the Table 2 and Table 3.

Table 2. Evaluation of system connection links of model routes in individual nodes. Part 1.

Route	Nodes – evaluation						
	Nýřany	Plzeň – Jižní P	Plzeň hl.n	Plzeň-Bolevec	Třemošná	Horní Bříza	Kaznějov
R	–	–	2	0	0	0	1
Sp	–	–	2	0	0	1	1
Os no. 1	–	–	2	1	1	1	1
Os no. 2	1	2	2	1	1	1	1

Table 3. Evaluation of system connection links of model routes in individual nodes. Part 2.

Route	Nodes – evaluation						
	Plasy	Mladotice	Žihle	Blatno	Žatec	Chomutov	Most
R	1	0	1	1	1	2	2
Sp	1	1	1	–	–	–	–
Os no. 1	1	1	1	–	–	–	–
Os no. 2	1	–	–	–	–	–	–

The values for the individual routes are summed into the model as one number, and the accumulation of these values is expressed in Table 4. For experimental reasons, the values of the routes that passes through the section twice during the evaluation interval are not multiplied by two.

Table 4. Cumulative evaluation of system connection links of model routes.

Route	Cumulative evaluation of system connection links of model routes
R	11
Sp	7
Os no. 1	9
Os no. 2	10

Comparison of Travel Times Between Individual Car Transport and the Railway Route in the Three Selected Sections with the Highest Passenger Numbers
The values of this parameter were determined for individual routes from the average values of the following important connections:

- R: Plzeň – Mostecko, Plzeň – Plasy, Plzeň – Žihle,
- Sp: Plzeň – Kaznějov, Plzeň – Plasy, Plzeň – Žihle,
- Os no. 1: Plzeň – Horní Bříza, Plzeň – Plasy, Plzeň – Žihle,
- Os no. 2: Plzeň – Plasy, Plzeň – Horní Bříza, Plzeň – Nýřany.

The results are summarized in Table 5.

Table 5. Comparison of travel times between individual car transport and railway route in the three selected sections with the highest passenger numbers.

Route	Comparison of travel times between individual car transport and railway route in the three selected sections with the highest passenger numbers
R	0.80
Sp	0.87
Os no. 1	0.80
Os no. 2	0.87

4 Sensitivity Test of the STEM Method Modified to the Transport Problem of Railway Capacity Allocation

The sensitivity of the STEM method was practically tested on changes in passenger numbers. The change was made in the regional express segment, which is not operated on the line today, but due to the fast and more frequent connection of the northern Pilsen region to the regional city of Pilsen, it is worth to consider operations of these trains. The aim of the test was to try to determine from roughly what border of passengers it is appropriate to run these trains at the expense of other segments of passenger transport.

The following daily values of the number of passengers were selected as input data (var. 0) summarized in Table 6.

Table 6. Passenger numbers of model lines – var. 0.

Line	Daily estimated average number of passengers in the limiting railway section in thousands [thousands of passengers per 24 h]	Daily estimated average number of passengers on the whole route of the line [thousands of passengers per 24 h]
R	0.9	1.4
Sp	0.8	0.9
Os no. 1	0.5	1.5
Os no. 2	0.3	2.5

Var. 0 model prefers *R, Sp and Os no. 2* lines, while $d = 0.167$. It is a compromise solution, which can be represented by the fact that the stated daily number of passengers for regional express is sufficient to conclude that it is purposeful to operate them at the expense of other routes.

For the next test, the daily numbers of passengers on regional express were reduced. In the variant A, the expected numbers of passengers in express trains were reduced according to Table 7.

Table 7. Passenger numbers of model lines – var. A.

Line	Daily estimated average number of passengers in the limiting railway section in thousands [thousands of passengers per 24 h]	Daily estimated average number of passengers on the whole route of the line [thousands of passengers per 24 h]
R	0.9	1.4
Sp	0.7	0.8
Os no. 1	0.5	1.5
Os no. 2	0.3	2.5

Var. A assumes $d = 0.1638$. It is also a compromise solution, not a global optimum.

According to the above stated, the switching value has still not been detected (a value that determines when another line is selected to be preferred). The third test of variant B was made, while the number of passengers in regional express was reduced by another 100 passengers per day. The inputs of variant B are summarized in Table 8.

Table 8. Passenger numbers of model lines – var. B.

Line	Daily estimated average number of passengers in the limiting railway section in thousands [thousands of passengers per 24 h]	Daily estimated average number of passengers on the whole route of the line [thousands of passengers per 24 h]
R	0.9	1.4
Sp	0.6	0.7
Os no. 1	0.5	1.5
Os no. 2	0.3	2.5

Variant B contains a final calculation and chooses lines *R, Os no. 1 and Os no. 2* for railway capacity assignment. *Value d reached 0.1695, so we have a compromise solution. The calculation results in the switching value for starting the operation of regional expresses being between 600–700 passengers per day in the restricted section Kaznějov - Horní Bříza and 700–800 passengers per day on the entire regional express line, assuming that the numbers of passengers on other lines are stable and do not change*. However, with such small changes, there were also minimal changes in the weights of individual evaluation criterions, so we can state that the modified method is also very well applicable in practice.

5 Conclusion

The above-mentioned form of the research shows that the STEM method is practically usable for solving problems in determining the preference for the allocation of capacity of the transport route, and that it works satisfactorily even with the change of selected parameters. The challenge is to test the parameters changes in a specific case.

It will probably be possible to modify the STEM method not only for the problem of railway capacity, but also for solving other tasks by computational path. In the above task, the method achieves relatively satisfactory results. Another task of the scientific team is to test other specific cases of railway lines as well as other methods and compare the results achieved by them.

It is obvious that the evaluation criterions must be chosen carefully and responsibly, otherwise the method will not give satisfactory results. If this is met, it can be a suitable tool for deciding or assessing situations whose optimal or suboptimal solution is not obvious.

Acknowledgment. All the facts presented in this article are based on the results of the research of the CTU Faculty of Transportation Sciences, Department of Transport Systems. This research was supported by the Grant Agency of the Czech Technical University in Praha, grant No. SGS18/150/OHK2/2T/16 Railway track parameters for transportation service optimization and No. SGS20/138/OHK2/2T/16 Design and optimal use of railway infrastructure parameters.

References

1. Hansen, I.A., Pachl, J.: Railway timetable and traffic. DVV Media Group GmbH – Eurailpress, Hamburg (2008). ISBN 978-3-7771-0371-6
2. Pachl, J.: Systemtechnik des Schienenverkehrs: Bahnbetrieb planen, steuern und sichern, mit Beispielen. 4., überarb. und erw. Aufl. B.G. Teubner, Stuttgart (2004). ISBN 3-519-36383-6
3. Purkart, P.: Optimal cooperation of public transport segments. Praha: Defense date 06 Dec 2019. Doctoral Minimum. CTU FTS. Department of Transportation Systems. Supervised by Týfa, L
4. Teichmann, D., Dorda, M.: Comparison of two selected methods in evaluating of investments in transport infrastructure. In: Finance and Performance of Firms in Science, Education and Practice: proceedings of the 7th International Scientific Conference: 23–24 April 2015, Zlín, Czech Republic, pp. 1524–1536. Tomas Bata University in Zlín, Zlín (2015). ISBN 978-80-7454-482-8
5. Public available operational documents (timetables, etc.) of Správa železnic, state organization (Czech state railway infrastructure manager)

"Green" Sector of the Air Transport of Ukraine Sustainable Development

Olena Sokolova, Mariya Grygorak, and Viktoriia Ivannikova(✉)

National Aviation University, Kyiv 03058, Ukraine

Abstract. Harmful impact of global air transport on the environment is evaluated in the article. Global environmental problems and directions for their reduction at the national level are identified. It is established that the most harmful influence for the ecosystem is done by the air transport during performance of logistics processes and operations at the airport. World experience of solving the environmental pollution problems in aviation sector on the basis of principles and methods of "green" logistics is studied. It has been found that Ukrainian airlines are not active enough to use environmental solutions in their activities; the main reason of this is the lack of effective mechanisms for implementation and state support of "green" initiatives. Main tools of sustainable environmental development, which should be implemented in the aviation industry of Ukraine, are proposed. It is proposed a mechanism for formation of a "green" airport in Ukraine. It represents a system of principles, methods and tools of logistics management of production and technological processes as well as airport infrastructure, to reduce their negative impact on the ecosystem and achieve a high level of resource efficiency.

Realization of the system of principles of "green" logistics allows to implement gradually technologies for de-carbonization of the airport depending on its spatial development.

Keywords: Environment · "Green" aviation · Sustainable development

1 Actually of the Subject Matter

Currently, at the global level a huge attention is paid to the searching of solution for energy conservation problem, fuel efficiency, as well as control over the level of harmful emissions in the atmosphere by various fields and sectors of the economy.

According to the conclusions, done by the UN experts, at the current temperatures rise, melting of permafrost will be a threat to the planet in 2050. Main reason of the rapid warming in the world is concentration of carbon dioxide CO2 in the atmosphere, which is increased due to the further expansion of industrial production [1]. 3% of the total CO2 emissions belong to the air transport, which remains the only means of quick transportation of passengers and cargo over long distances. Rapid development of air transport would inevitably cause enormous damage to the environment, but initiatives, realized by the industry, are designed to stabilize and then reduce greenhouse gas emissions. The COVID-2019 pandemic led to the reduction of business activity, which

O. Prentkovskis et al. (Eds.): TRANSBALTICA 2021, LNITI, pp. 448–455, 2022.
https://doi.org/10.1007/978-3-030-94774-3_44

allowed to obtain a short-term positive environmental effect. However, according to the ICAO forecast according to the various scenarios of aviation sector development up to 2050, CO_2 emissions are expected to be increase from 890 to 2800 Mt. [2]. Taking into account expected trends of climate changes as well as global requirements regarding environmental safety, the aviation sector is obliged to take active steps in this direction in order to save its role in the global environment and not to lose potential benefits for society, in general.

Ukraine, as a member of the world aviation community, must follow established international recommendations and requirements regarding achievement environmental effect. That's why, in order to achieve the set global aviation goals there is an urgent need to find and implement resource-efficient "clean" technologies for the Ukrainian aviation sector both at the level of individual aviation enterprises and the industry level, in general.

2 Problem Solution

According to the International Council On Clean Transportation report, it had been established that in 2019 aviation CO_2 emissions amounted to 920 million tons (Mt), and the largest number of emissions, almost 85%, was made by passenger aircraft [3].

In 2019, compared with 2013, the total amount of CO_2 emissions from this segment has been increased by 33% and equals to 785 million tons (Mt). CO_2 emissions from air passenger transportation by narrow-body aircraft comprise 43% of the total world's volume, by wide-body aircraft - 36%, and regional - 6%. The last 15% are CO_2 emissions from air cargo transportation, 8% of which was performed by passenger aircraft as additional loading, and 7% - by cargo aircraft. In 2020, due to the implementation of quarantine measures because of Covid-19 pandemic and blocking of air transport activity, global volumes of air passenger transportation have been decreased by almost 60%. This has led to a slight reduction of the energy intensity of commercial passenger aircraft and, in the result, increased environmental efficiency of the industry, in general.

Among the world's major pollutants are United States, China, Great Britain, Japan and Germany. Moreover, 55% of all global aviation CO_2 emissions belong to the markets of US, EU and China [4].

According to the annual global study Energodata, level of the Ukraine's GDP energy intensity is 2 times more than the average value in the other world countries. For example, level of Poland's GDP energy intensity is 2.5 times lower than in Ukraine, level of Germany's GDP energy intensity is 3.3 times lower than Ukrainian one [5].

In 2019 the amount of carbon dioxide emissions in Ukraine has been decreased by 4.03% compared with the previous 2018 and was equal to 121,300 thousand tons. At the same time, the total amount of pollutant emissions in 2019 has been equal to 4108.3 thousand tons (reduction rate - 0.31%). 1648.8 thousand tons of emissions have been generated by mobile devices, which is 2.3% higher than in 2018 [6].

In the transport sector, the largest amount of harmful emissions into the atmosphere belongs to the motor transport. It should be noted that information on CO_2 emissions by the Ukrainian aviation sector are not published in official statistical sources since

2016. According to the international experts, achieving of a short-term environmental effect in 2020 will not solve the problem of air transport negative impact on the environment and climate changing.

Therefore, implementation of the fiscal and regulatory measures are considered are to be the main priorities, which will ensure operational and technical efficiency, as well as risk management that may arise during creation of "clean" aviation infrastructure, new (energy efficient) aircraft engines, power plants, etc. (see Fig. 1).

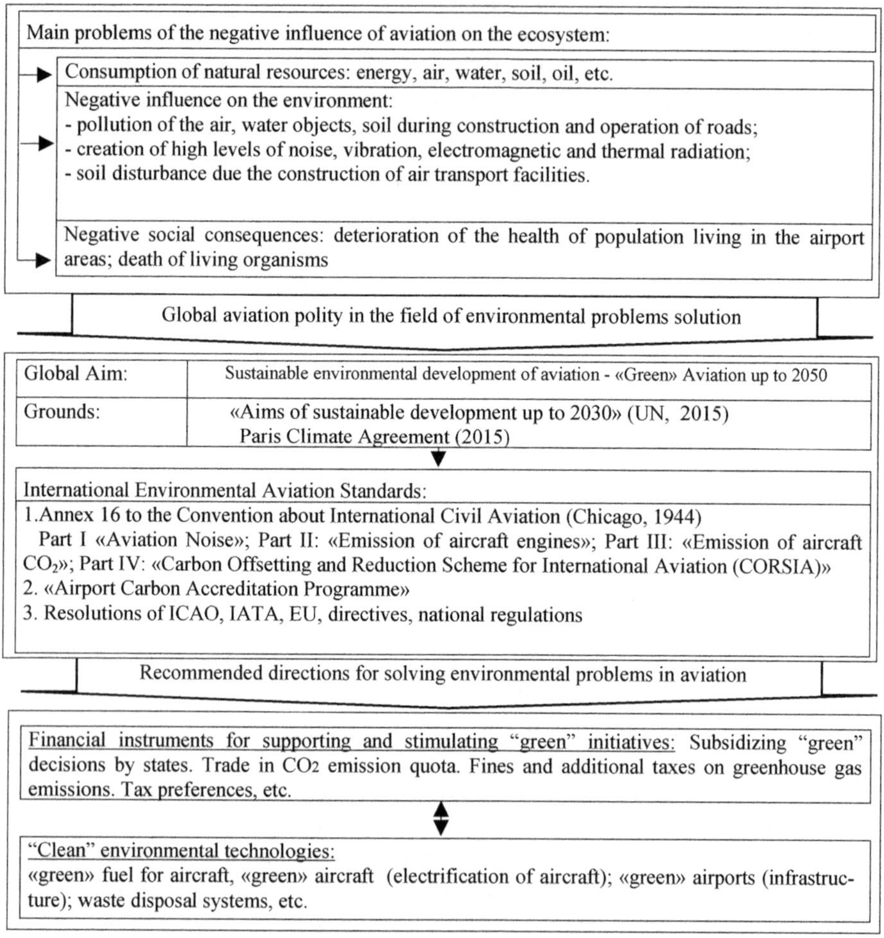

Fig. 1. Global policy for solving environmental problems in aviation.

In order to solve the global problem of reducing the negative impact of the aviation industry on the environment there is a necessity in transformation of its existing state, in accordance with the principles of sustainable development, i.e. its transformation into a sustainable air transport system [7–9].

Taking into account determined global trends in the development of "clean" aviation in the nearest future, Ukrainian aviation sector faces a number of problems and challenges, the solution of which requires appropriate changes both at the legislative level and foresee technological modernization of the air transport system, in general. Moreover, the key factor should be state support and stimulation of energy efficient solutions implementation, which will foresee organization of coordinated work of administrative, financial and legislative structures as well as research institutions.

Ukrainian aviation enterprises are not active enough to develop programs for "green" technological solutions implementation into their activity. First of all, it is observed because of uncertainty of financial mechanisms to stimulate "green" initiatives and lack of own resources for their implementation. In the most countries of the world, the incentives for reduction harmful emissions are an effective system of taxation and penalties for exceeding the established amount of emissions, which allows to invest received budget funds to the resource-efficient projects.

At this stage, Ukraine has introduced a carbon tax, the rate of which (10 UAH/t CO2 - 0.36 USD/t CO2) is the lowest one in comparison with the other countries. For example in Poland, having the economy, which is the most close to the Ukrainian one, the carbon tax rate is 1 US dollar/t CO2.

According to the international experience, a carbon tax can be a powerful tool to reduce harmful emissions if the rate is high. At the same time, in case of considerable increase of the carbon tax rate, there is a risk that, for the first, companies will provide inaccurate data to reduce tax costs, and for the second, this value will be included into the products prime cost and will cause increasing of its final cost. The low tax rate does not create incentives for any technological changes, but leads to the appearance of a fair concept, called "polluter pays".

It should be added that in the most countries of the world the carbon tax mechanism is narrower than in Ukraine. Also, the world practice provides exemption from carbon taxation if there are risks of CO2 leakage or the industry is vulnerable to high emission costs. As a rule, this category includes agriculture and forestry, as well as transport (by types).

That's why, in order to achieve the expected effect from the aviation sector, it is necessary to solve a wide range of existing problems, starting from the state regulation of the tasks and possible scenarios of their development. At the same time, the key functions here should be performed by the State Aviation Service of Ukraine. It should determine requirements, develop methods and implement necessary tools for gradual decarbonization of domestic aviation enterprises on the basis of regulatory and supervisory policy.

Main tools of sustainable environmental development to be implemented in the aviation industry of Ukraine should include: creation of an effective environmental management system; formation of a flexible "environmental" pricing policy; creation of an effective public procurement policy; reforming of the "environmental" taxation system; development and usage of effective mechanisms of state investment into the infrastructure, correspondent to the principles of sustainable development; implementation of state support mechanisms for the researches, connected with the development of "green" technologies; development and implementation of the strategies for "green" air transport development as a composite part of a strategy for the sustainable "green" society formation.

It is obviously, that the greatest damage to the ecosystem is caused by the air transport during implementation of logistics processes and operations at the airport. Therefore, reduction of harmful substances emissions during airport activities should be carried out in the technical, operational and logistics areas [10]. The environmental effect in technical area can be achieved due to the improvement of aircraft engine design, usage of alternative types of fuel (energy). Operational activities will include staff training and usage of information and communication technologies. Main activities directed to the reduction of the negative environmental impact in the logistics sector include: optimization of the distribution network, carrying out procurement and production activities taking into account CO2 emissions, usage of intermodal and multimodal technologies, "green" designing of warehouses, packaging management, waste management, etc. It means that environmental problems in aviation sector should be solved on the basis of principles and methods of "green" logistics. Implementation of the "green" logistics principles allows to balance economic, social and cultural as well as environmental efficiency of the logistics system (supply chain) [11].

As it was mentioned before, airport logistics processes have a high level of energy consumption. That's why, development and implementation of the measures for access to the "green" energy is an extremely important step in airports decarbonization, and it also gives possibility to replace fossil fuels in production by alternative energy sources. Airports can get "clean" energy by two ways: buy or generate it. In this case, the second option is more expensive one from a technical point of view, but it is the most attractive one from the side of business sustainability.

In order to achieve maximum socio-economic and environmental effects, Ukrainian airports should be developed not only from the standpoint of the aviation industry needs, but also should take into account the features and capabilities of the urban system to which they belong and have a significant impact. Thus, according to the world experience, if the airport is located within the city, then due to the territorial constraints, logistics and commercial infrastructure will be mainly concentrated on its territory. In case of airport location outside the city, business infrastructure will be developed around it, which allows you to create special economic zones and develop the airport on the bases of aeroscities or aerotropolises models.

This approach will provide an opportunity to resolve systematically environmental problems and challenges for ensuring sustainable functioning of both the air transport system and other economics sectors of individual regions and country in general.

Thus, taking into account the information, described above, the authors propose a mechanism for the formation of a "green" airport as one of the necessary step to achieve sustainable development of the aviation sector of Ukraine. It consists of the system of principles, methods and tools for management of production and technological processes and airport's infrastructure objects in order to reduce their negative impact on the ecosystem and achieve high level of resource efficiency (see Fig. 2).

According to the proposed mechanism, the airport's greenhouse gas emissions management subsystem should monitor and control the following processes: 1) direct emissions from the sources owned or controlled by the airport; 2) emissions from purchased electricity and 3) all other indirect emissions from other sources that are not controlled by the airport, but related to its activities.

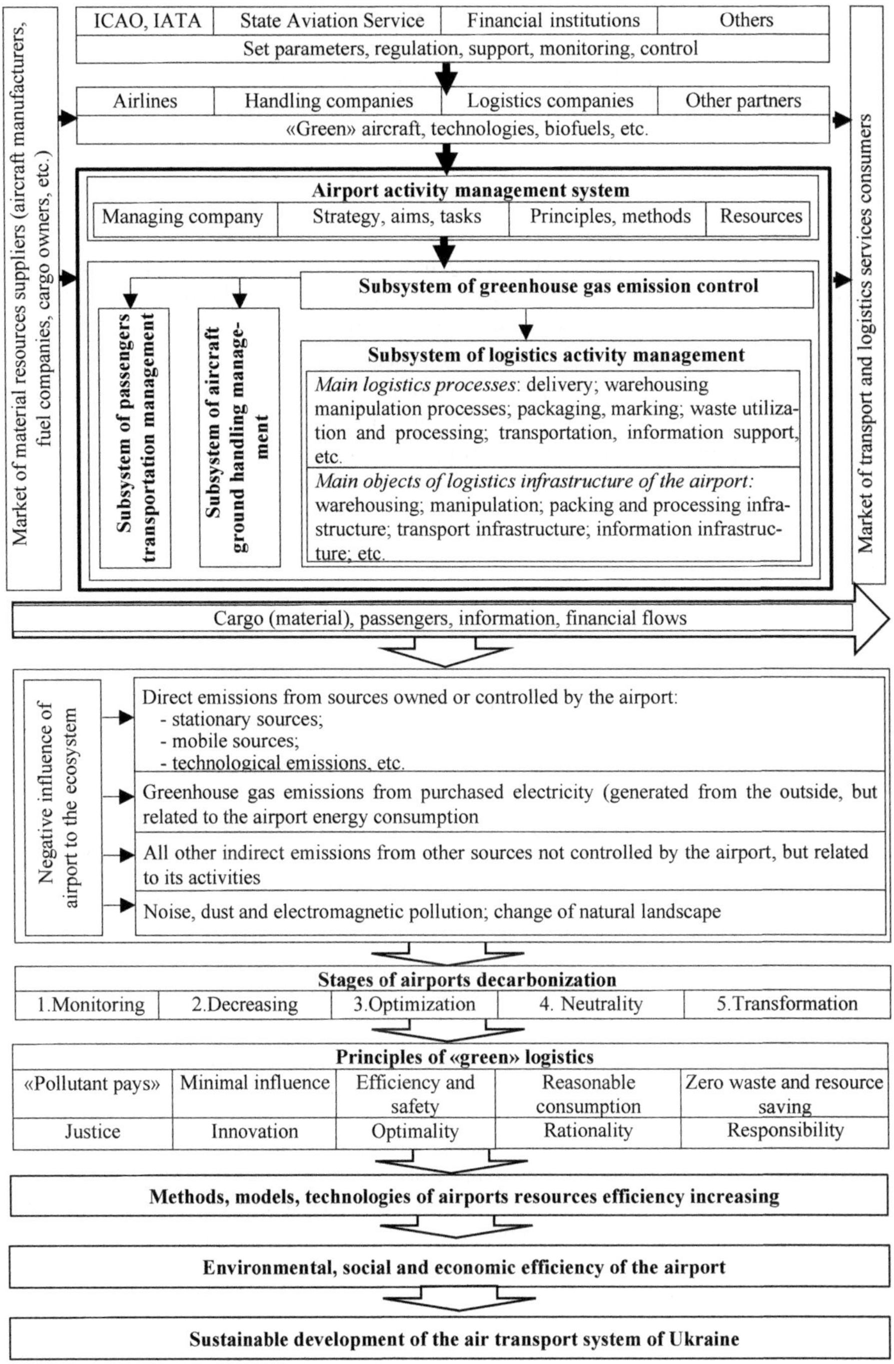

Fig. 2. Mechanism of a "green" airport formation in Ukraine.

Realization of the system of principles of "green" logistics allows to implement gradually technologies for airport de-carbonisation, depending on its spatial and territorial de-velopment as key objects of transport and logistics infra-structure at the national and regional levels [12].

3 Conclusions

In the context of global climate changing and the vector chosen by the international community to creation of a "clean planet", aviation sector of Ukraine must resolve the task, directed to the achieving of the set universal indexes of greenhouse gas emissions decreasing by 2050. Due to this fact, there is a necessity to improve legislation in or-der to provide state support, regulation and stimulation of "green" initiatives implementation, as well as realizing structural changes in the aviation industry, in general. Application of the "green" logistics principles allows to implement a systematic approach to solving problems of aviation emissions reduction and developing environmen-tally efficient solutions throughout the whole cycle of air transport products creation.

References

1. Climate Change. https://www.un.org/en/global-issues/climate-change. Accessed 12 Apr 2021
2. Environmental Trends in Aviation To 2050 – ICAO. https://www.icao.int. Accessed 18 Apr 2021
3. CO2 emissions from commercial aviation: 2013, 2018, and 2019. https://theicct.org/sites/default/files/publications/CO2-commercial-aviation-oct2020.pdf. Accessed 18 Apr 2021
4. Comi, A., Savchenko, L.: Last-mile delivering: analysis of environment-friendly transport. Sustain. Cities Soc. **74**, 103213 (2021)
5. Український зелений курс: декарбонізація або смерть. https://www.epravda.com.ua/projects/ekopromyslovist/5cde72e4e561c. Accessed 21 Apr 2021
6. The environment of Ukraine 2019. http://www.ukrstat.gov.ua/druk/publicat/kat_u/2020/zb/11/Dovk_19.pdf. Accessed 21 Apr 2021
7. Vashisth, A., Kumar, R., Sharma, S.: Major principles of sustainable transport system: a literature review. Int. J. Res. Appl. Sci. Eng. Technol. **6**(II), 1597–1605 (2018)
8. Ogryzek, M., Adamska-Kmiec, D., Klimach, A.: Sustainable transport: an efficient transportation network – case study, sustainability. MDPI Open Access J. **12**(4), 2–14 (2020)
9. Budd, T., Intini, M., Volta, N.: Environmentally sustainable air transport: a focus on airline productivity. In: Walker, T., Bergantino, A.S., Sprung-Much, N., Loiacono, L. (eds.) Sustainable Aviation, pp. 55–77. Springer, Cham (2020). https://doi.org/10.1007/978-3-030-28661-3_4

10. Smokers, R., Tavasszy, L., Chen, M., Guis, E.: Options for competitive and sustainable logistics. Sustain. Logistic Transp. Sustain. **6**, 1–30 (2014)
11. Sokolova, O., Soloviova, O., Borets, I., Vysotska, I.: Development of conceptual provisions to effectively manage the activities of a multimodal transport operator. Eastern-Eur. J. Enterp. Technol. **13**(109), 38–50 (2021)
12. Yudin, O., Ziubina, R., Buchyk, S., Matviichuk-Yudina, O., Suprun, O., Ivannikova, V.: Development of methods for identification of information-controlling signals of unmanned aircraft complex operator. Eastern-Eur. J. Enterp. Technol. **2/9**(104), 56–72 (2020)

Model of Transport System Optimization in Kyiv (Ukraine)

Viktoriia Ivannikova(✉) and Oleksii Nesterov

National Aviation University, Kyiv 03058, Ukraine

Abstract. Kyiv is a capital city of Ukraine having a developed transport infrastructure, which consists of the airports, railways, highways, waterways and bridges. In addition, Kyiv has a developed system of urban and suburban passenger transport. Subway, trams, buses, trolleybuses and other modes of transport carry thousands of passengers every day, mostly to the places of work or study from large sleeping areas on the outskirts of the city to the industrial downtown. However, unfortunately, the transport system of the city of Kyiv cannot be called ideal. Poor road conditions on most highways, underdeveloped subways and low capacity of many important roads lead to considerable congestion during rush hour. That's why, new project for the development of transport system of Kyiv, namely the subway, highway connection to other cities of Ukraine and helping to reduce traffic by constructing the detour road for transit transport, has been proposed in the article. The main aim of the project on transport system of Kyiv optimization is to overcome the problem of overcrowded vehicles during rush hour.

Keywords: Cargo transportation · Development project · Passenger transportation · Transport system · Urban transport

1 Introduction

Kyiv is the capital and the biggest city of Ukraine. The geographical location and political status of Kyiv led to the fact that for centuries the transport system of this city has been continuously improving to create the biggest and the most developed transport hub in modern Ukraine.

A city of Kyiv is located in a favorable position on the intersection of the main routes connecting western part of Europe with the eastern part and with Asia. Due to its convenient geographical location, from time immemorial Kyiv has been developing as an important transport hub.

The transport system of Kyiv consists of three airports, two of which are accommodated for handling passenger flights, railways with a hub at Kyiv-Pasazhyrskyi station [1], highways with bridges, among which seven bridges across the Dnipro River play a special role in transport communication, the subway in intra-city passenger traffic, and waterways, which now pay more historical than the practical role [2].

Urban transport system in Kyiv is represented by such transport modes, as subway, bus, trolleybus, tram, funicular, non-public minibuses and ring urban rail.

O. Prentkovskis et al. (Eds.): TRANSBALTICA 2021, LNITI, pp. 456–467, 2022.
https://doi.org/10.1007/978-3-030-94774-3_45

The Kyiv subway is not only one of the main transport arteries of the city. Currently, this network consists of three operating lines (see Fig. 1) with a total length of 70 km and 52 operating stations, the platforms of which are designed for five-car trains. The downtown lines have three transfer stations [3].

Fig. 1. Map of Kyiv subway with the urban rail and rapid trams lines [4].

The highest developed ground public transport in Kyiv is bus with 85 existing routes, one of which is working only at nights [5]. There are 22 routes of trams in Kyiv, 5 of which are called Kyiv Express Trams because they go by isolated rails and have specially accommodated stations that function like subway stations, having ticket desks and turnstiles. They also have more modern vehicles than the regular tram. Other tram routes function on free access rails and go slower than the rapid tram. Trolleybuses in Kyiv serve passengers on 51 routes – 47 of them function at days and other 4 routes are night ones [6]. The funicular in Kyiv is represented by only one route with two stations, which connects the Poshtova Square, located at the bottom near the Dnipro embankment, with Mykhailivska Square, which is located at the top. Minibus is a privately owned route taxi that uses short buses. Due to the special appearance and mostly bright yellow color, minibuses are always visible on the roads of Kyiv, which makes them the most popular transport in the city, even though the vehicles are often in poor condition and it is almost impossible to feel comfortable inside the cabin.

However, unfortunately, the transport system of the city of Kyiv cannot be called ideal. Poor road conditions on most highways, underdeveloped subways and low capacity of many important roads lead to considerable congestion during rush hour, which is complicated by the low driving culture of local drivers and the inability of law enforcement to monitor all traffic violations. There are many promising projects to improve this transport system, which due to lack of financial resources have not yet been implemented. The development of transport in the city of Kyiv has been standing still for several years now, which is why there are more and more problems with communication between different districts of the city.

This study is aimed to present the new ideas of the authors for the development of transport system of Kyiv, namely the subway, highway connection to other cities of Ukraine and helping to reduce traffic by constructing the detour road for transit transport. If at least part of the existing projects to improve the transport system of the city and its environs are implemented, there will be a significant change in the quality and efficiency of this network [7]. If all the ideas, including those proposed in this study, are implemented, the city of Kyiv will receive a modern and very convenient European-style transport system.

2 Background

The current transport system of the city of Kyiv faces many problems every day, one of which is traffic jams. This is due to the fact that the majority of the city's population lives in remote outskirts and has to travel to work located mostly in downtown by its own or public transport every morning (see Fig. 2).

One of the most populated areas of Kyiv is Troyeshchyna with a population of over 250,000 inhabitants, located in the northwestern part of the city. Every day at its exits, the biggest traffic jams occur because this district does not have the subway transport [8]. In addition, not every city resident has their own car, which leads to the fact that during rush hour public transport is also overcrowded and unable to carry all interested passengers.

Buses, trolleybuses and minibuses run quite often, but they cannot serve all the passengers who need them, and they often get stuck in traffic jams, often even becoming their cause. Trams do not get stuck in traffic jams [9], but they are often slower than road transport, and there is a considerable risk of a complete stop of the entire line, which makes it risky to use this transport during rush hour [10]. The subway in Kyiv is underdeveloped: there are only three lines that do not even connect the most populated areas with the downtown, so locals have to use land transport to get to the nearest station, and it takes money and time.

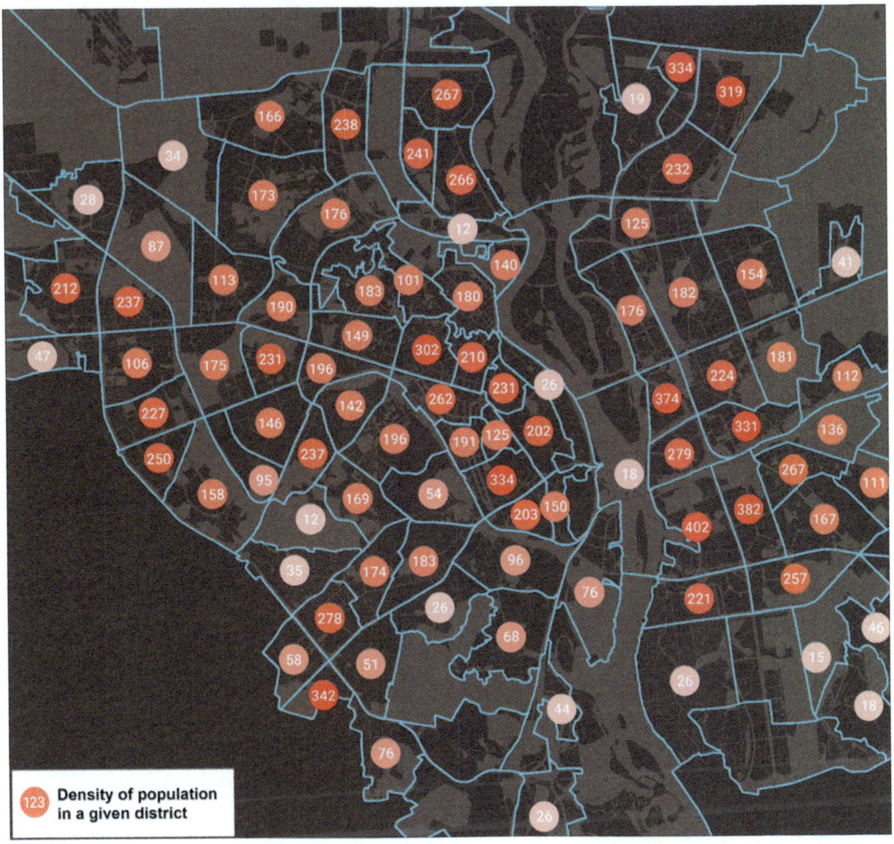

Fig. 2. Scheme of population of Kyiv by neighborhoods.

In addition, the road surface on many of Kyiv's main highways is of poor quality, forcing drivers to slow down and go around particularly dangerous potholes, leading to accidents. Sometimes even a newly repaired road due to the unprofessionalism of workers and drivers is in good condition for a short time.

Kyiv has two ring roads (see Fig. 3): the Small, which is the border for the downtown, and the Big, which currently runs only along the right bank of the Dnipro and is the immediate border for the cityline. Big Ring Road now exists only partly and runs along the western border of the city, mostly outside the city. It unites the Dnipro, Odessa, Zhytomyr and Hostomel directions. Small Ring Road is such only in fact, because it does not stand out as a separate street in Kyiv.

However, the map of Kyiv clearly shows the streets and bridges that merge in the downtown into a ring road with a total length of 38.9 km, which is on 4.5 km longer than the section of the Big Ring Road that exists today.

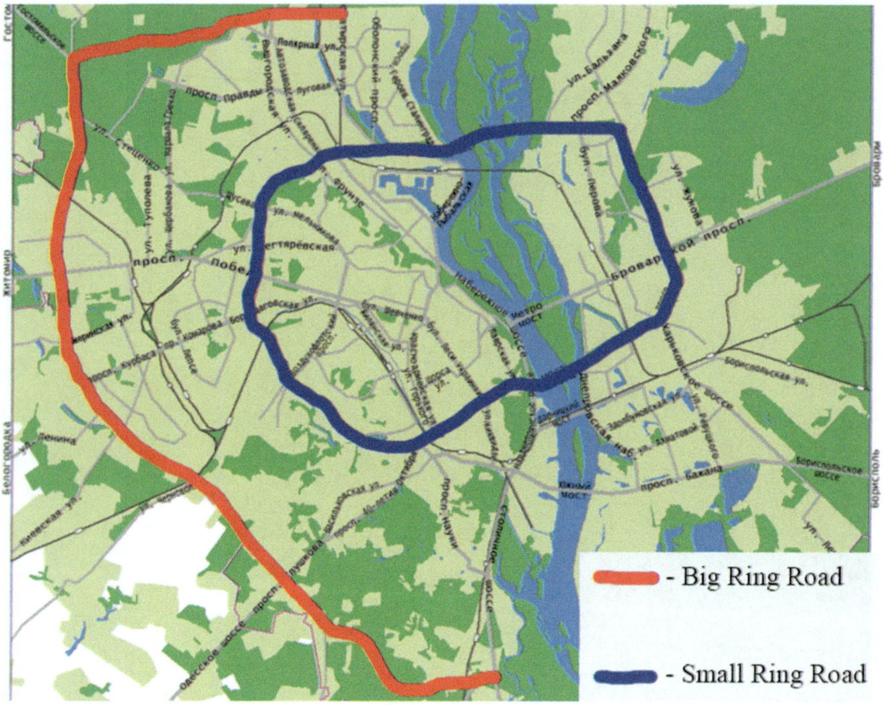

Fig. 3. Map of ring roads of Kyiv.

3 Project

Analyzing the identified opportunities of the transport system of Kyiv, the authors set the main directions of its development and create new projects for its optimization, which differ from the existing ones.

The project has been divided into two types. The first one is aimed to ensure the safe road passage of transit vehicles through the city without additional travel through Kyiv. The second project consists in improving Kyiv's urban and suburban transport system to make it primarily convenient for passengers: residents of the city's most populated areas and tourists.

3.1 Road Detour

Unfortunately, now it is impossible to avoid the arrival of transit transport, which can significantly exceed the permissible limits within the city limits, within the cityline of Kyiv. This is due to several important factors:

- The city stands on the banks of the Dnipro, which must be crossed by land transport; the bridges are built only within Kyiv, so to travel between the river banks it is needed to drive through the city.

- The Big Ring Road is not built on the left bank of Kyiv, so it is impossible for transit trucks to go around the residential areas located there.
- Most of the highways of interregional and international importance are a continuation of the large avenues of Kyiv, on which the vehicles of local residents move daily. For example, the Kyiv-Chop highway is a continuation of Peremoha Avenue, Brovarskyi Avenue serves as a continuation of the St. Petersburg-Sumy-Kyiv highway, and the highway to Boryspil Airport and Kharkiv within Kyiv is Mykola Bazhan Avenue.
- Unlike many other cities in Ukraine and Europe, Kyiv's ring roads serve to travel between neighboring areas of the city rather than for transit traffic through it.

The authors' projects to optimize the road network of the city of Kyiv include solving all the above problems. Continuation of the Big Ring Road on the left bank of the Dnipro is a very expensive idea, because it requires the construction of additional bridges in the northern and southern parts of Kyiv, as well as pave a wide highway where there are now villages or cottages, thermal power plant and natural obstacles such as forests and lakes.

In order to ensure the safe transit of large trucks through Kyiv, the only existing bridge across the Dnipro outside the city can be used. This bridge is a road built on top of the dam of the Kyiv Reservoir near the suburbs of Kyiv called Vyshgorod. The length of the entire dam is 40 km, and the bridge across the Dnipro – a little more than one kilometer. At the moment, the road surface of the dam consists of only two lanes, so the structure serves the value of part of the Kyiv hydroelectric power plant rather than the transport road. However, in the future, this bridge, like the dam itself, can be expanded to have more capacity for transit vehicles and better protect the city from heavy water flows from the Kyiv Reservoir.

At the same time, the problem due to which this bridge is still rarely used is that it is located on the opposite bank of the Desna River, which is a tributary of the Dnipro and flows into it in the northern part of Kyiv. The nearest existing bridge to Kyiv across the Desna is located near the town of Oster in the Chernihiv Region, namely 65.3 km from the taken dam bridge (see Fig. 4). As the distance from Kyiv borders to Oster is approximately the same, so the total length of performed detour is equal to more than 130 km.

That is why, given the above conditions, it is possible to build a new bridge across the Desna, which can become part of the future bypass road for transit freight or passenger transport. The northern border road of Kyiv on the left bank is Myloslavska Street. At its end, there is an abandoned building and a swamp that surrounds the place, where the Desna flows into the Dnipro. In the case of the final dismantling of this building and the construction of the overpass and the bridge over the Desna, a road can be built that can connect the dam bridge with Kyiv. Then there will be a single highway that will allow transit traffic to bypass Kyiv downtown, located on the right bank, from the north. In addition, this road will allow residents of the left bank of Kyiv to get faster to the northern outskirts of the city on the right bank without having to go through other districts where there is a risk of getting stuck in traffic jams. Thus, the highway, which will consist of an overpass over swampy soils and a bridge over the Desna River, will connect the existing Myloslavska Street and the dam of the Kyiv Hydroelectric Power Station, crossing it near the village of Oseshchyna.

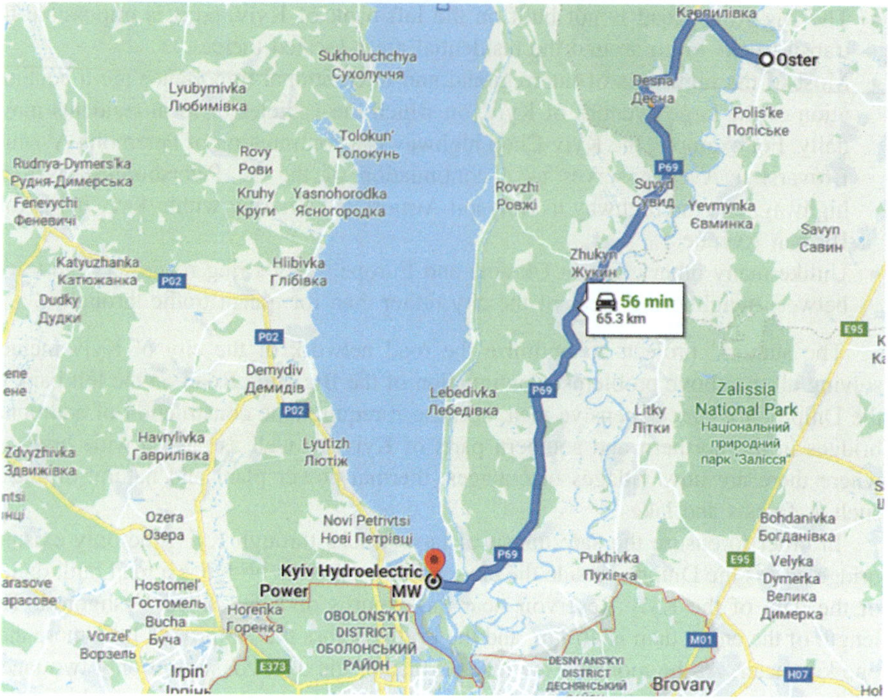

Fig. 4. Map of the existing road detour of Kyiv.

As for the continuation of this highway after the dam on the Right Bank, it is worth mentioning the existence of such an important transport hub in that part of the Kyiv agglomeration as Hostomel Antonov Airport. The option of building a highway from Vyshgorod to Hostomel is possible, and it could become a road directly to Antonov Airport. Route P69, which is a continuation of the dam on the right bank of the Dnipro north of Vyshgorod, rests on a dead end at the intersection with highway P02, while being approximately at the same latitude as Hostomel Airport. The space between these objects consists of forests and swamps without settlements, so it was proposed to build a highway there (see Fig. 5). It will be an access road to the airport from the east, where the railway station is located.

Thus, together with the existing P69 route and the previously proposed route from the Kyiv district of Troyeshchyna to the dam bridge across the Dnipro, there will be a highway that will connect the left bank of Kyiv with Antonov Airport. Its total length from intersection with Myloslavska Street to the border of Hostomel village will be equal to 30 km, 7.5 km of which exist now. The highway will solve the following problems:

- The highway can become a bypass road for Kyiv from the north for transit transport, so the need to build the Big Ring Road on the left bank with the construction of two additional bridges will be partially eliminated.

- This route will allow residents of the left-bank part of Kyiv to get to the north-western outskirts of the city faster without passing through the center.
- As part of the highway will be a new bridge across the Desna, it will allow Kyiv residents to reach the northern part of Ukraine faster.
- The dam bridge will be used more and will unload those bridges across the Dnipro, which are located further south within Kyiv, first of all the Pivnichnyi Bridge, which faces the problem of congestion on a daily basis.
- In case of opening passenger flights through Antonov Airport, it will be possible to get to it quickly and conveniently by road.

Fig. 5. Complete project of the highway from northern Kyiv to Hostomel.

3.2 Urban Transport

The main aim of the project on optimization of urban passenger transport is to overcome the problem of overcrowded vehicles during rush hour. However, land transport should not be called poorly developed, given the variety of vehicles in Kyiv and the sufficient number of their routes. That's why, the project is aimed at improving the subway itself, which is lacking in several prominent residential areas of Kyiv. There have been projects for decades to build a fourth subway line to Troyeshchyna district, but so far they have not been implemented due to the unfinished construction of the Podilsko-Voskresenskyi Bridge, which is to become a connecting subway bridge for this line.

But even the existing subway projects do not solve another issue for Kyiv's infrastructure: the connection with Kyiv's largest airport, Boryspil. That is why, it was decided to propose new subway development project in Kyiv, taking into account not only existing ideas, but also other factors, such as the population of different parts of the city and their importance for the city's economy. It includes the extension of two of the three existing lines and the construction of four new ones. Thus, Kyiv will have a developed subway system of seven lines, the scheme of which is shown in Fig. 6.

The priority projects that are being implemented now are the extension of the green subway line to the Vynohradar district and the construction of a yellow line from the Troyeshchyna district to Zhuliany airport. After modification of these projects to get a more convenient and clear subway scheme in the end, it is propose to extend the green line not to the northwest, but to the north of Kyiv with the final station on Taras Shevchenko Square. The yellow line after the Igor Sikorsky Airport can be extended to the Big Ring Road to implement the future project. In addition, the route of this line on the left bank has been changed: instead of heading east to Bratyslavska Street, it was proposed to turn the line directly to Troyeshchyna district and lay the subway on or under Vladimir Mayakovski Avenue, which is the most important transport artery in this area.

The next step is to extend the green line east to Boryspil airport from Chervonyi Khutir station with the construction of extra stations in the villages of Shchaslyve and Chubynske along this route. Thus, it will be convenient and cheap for passengers to get from the airport to the downtown, and residents of those villages that mainly work and study in Kyiv will be able to use this fast and relatively inexpensive transport.

The cyan left-bank line will allow direct connections from Troyeshchyna to the southern residential areas of Darnytskyi district. However, the construction of this line has not been considered as a priority due to the development of a developed tram system in this part of Kyiv, but this line has been kept in the scheme, because of its existence in the official projects of Kyiv subway development.

Fig. 6. Prospective project of subway development in Kyiv.

The two new, not represented in existing subway development projects lines, which have been added to the proposed scheme will run under two ring roads in Kyiv: the purple one will follow the Small Ring Road, and the lime line will follow the Big Ring Road. Since the Big Ring Road is built only on the right bank of the city, it was decided to connect its north and south ends in a lime line along the Dnipro embankment and Stalingrad Heroes Avenue, so that this line came out completely on the right bank. In order to connect all the terminals of the subway lines on the right bank with the lime line, it is proposed to extend the blue line one station north to the Pivnichna station, which is the same name as the street under which it will be located.

When creating this subway project, it was taken into account, that all proposed routes passed through existing roads and bridges: thus, the construction of a purple line involves laying rails for subway trains on the Pivnichnyi and Paton bridges. If the project is implemented, it will solve the following problems [11, 12]:

- Providing convenient, fast and cheap travel to two airports in Kyiv – Zhuliany and Boryspil. Given the current price of the Kyiv subway and ground transport to Boryspil airport, a trip there by subway will cost 10 times cheaper than by bus or train.
- Providing communication with the most populated areas of Kyiv: Troyeshchyna, Vynohradar and Borshchahivka.
- Creation of two ring lines that will provide fast communication between other lines and will run on two ring roads in Kyiv.

In total, the project consists of 4 new subway lines and 86 new stations, as well as 28 new transfer points between lines, which looks colossal compared to the current three. The detailed description of the proposed Kyiv subway development project is as follows:

1. The red line (Lisova-Akademmistechko) remains unchanged, except for the completion of new transfer points to new stations.
2. The blue line (now Teremky-Heroiv Dnipra) should be extended to the end of Obolonskyi Avenue to connect it with the Big Ring Road and the future lime line.
3. The green line (now Syrets-Chervonyi Khutir) should be extended to Taras Shevchenko Square after Syrets and by Boryspil Highway to Boryspil Airport after Chervonyi Khutir, as well as to receive the completed stations Lvivska Brama and Telychka, which will serve as a transfer station in the future.
4. The yellow line (Zakrevskyi Street-Vyshneve) will run from Myloslavska Street along Vladimir Mayakovski Avenue, the new Podilskyi Bridge, through the Podil district and the Central Railway Station to Povitroflotskyi Avenue to Sevastopol Square, Zhuliany Airport and Big Ring Road near the route to Vyshneve village. Compared to the existing project [11], the unnecessary Zatoka Desenka station has been removed, which would be located in a non-residential and informal area, so it would not be popular and could become unprofitable. In addition, the route on the left bank has been changed so that it turned into the Troyeshchyna district.
5. The cyan line (Myloslavska-Pivdenna) will run along Honore de Balzac Street, residential areas Rusanivka, Berezniaky and Osokorky to Kolektorna Street.

6. The purple line (small ring) will run along the Small Ring Road, namely Roman Shukhevych Avenue, Bratyslavska Street, Yuri Gagarin Avenue, Sobornist Avenue, Paton Bridge, Druzhby Narodiv Boulevard, Valeriy Lobanovskyi Avenue, Chokolivskyi Boulevard, Vadym Hetman Street, Oleksandr Dovzhenko Street, Olena Teliga Street, Stepan Bandera Avenue and Pivnichnyi Bridge.
7. The lime line (big ring) will run along the Big Ring Road, Poliarna Street, Pivnichna Street, Stalingrad Heroes Avenue and the embankment along the Dnipro.

In addition to the above advantages of the proposed subway project, a few other important locations in Kyiv that will be served by this mode of transport in case of implementation of all these ideas are described below:

- Both passenger airports in Kyiv will be connected to the subway network.
- The Electrotekhnichna and Sholem Aleichem stations of the purple line will be located near two important markets on the left bank of Kyiv – Troyeshchyna and Lisovyi, respectively.
- Yordanska Station will serve as a transfer station for the Pochaina subway station, which will help to unload the poorly organized stop of many land transport routes located there.
- Stations Babi Yar, Naberezhna, Spivoche Pole and Pyrohiv will be located near the famous sights of Kyiv and will help tourists to reach them easier.
- Stations Karavayevi Dachi, Vyshneve and Volynska will be located near important for suburban transport railway stations.
- Vadym Hetman and Borshchahivska stations will be located near major universities in Kyiv, so students and teachers will get there much faster.
- Rusanivka, Sadova, Dachna and Pivdenna stations will be located in cottage villages within Kyiv, where urban transport is currently poorly developed.
- Subway lines will run under all major highways in Kyiv, so this will help solve the problem of congestion.

4 Conclusions

The novelty of the presented research is the introduction of a new bypass for transit transport passing through Kyiv, as well as additional metro lines in the city of Kyiv, which were not implemented in previous projects for the development of urban transport.

In case of implementation of the two proposed projects for the development of the transport system, Kyiv will receive a modern and convenient system for all transport users. The semi-ring road on the Left Bank will allow vehicles traveling from Eastern Ukraine and the Middle East to the west to bypass Kyiv from the north. Thus, transit transport will not get stuck in traffic jams or create them on the most important highways of the city, and it will be able to go through the capital of Ukraine faster without passing through its downtown. It should be noted that most of the future highway has already been built: only one Vyshgorod-Hostomel highway and an overpass from Troyeshchyna district to the Kyiv hydroelectric dam with a bridge across the Desna near the village of Oseshchyna need to be completed to the existing roads on the left bank of Kyiv and on the outskirts of Vyshgorod.

Implementation of the project of this highway will open additional prospects for Hostomel Antonov Airport. In this way, it will be possible to finally open passenger terminals near this airfield, because then it will be much easier for people to get to it. Moreover, the existence of a freight railway station on the territory of the airfield together with the future highway will make Antonov Airport an important multimodal hub for air, rail and road transport.

As for the second project of subway development in Kyiv, it is primarily aimed at resolving an important issue that has been raised by Kyiv authorities for years: providing residential areas in Troyeshchyna, Vynohradar and Borshchahivka with this type of transport, which, importantly, does not get stuck in traffic jams. In addition, if this idea is implemented, both passenger airports in Kyiv will also have a connection with this relatively inexpensive mode of transport. After the final implementation of the project of the Kyiv subway development, which will consist of seven lines, the land transport network in the city will be significantly unloaded. This will be an improvement for the roads of Kyiv, because then the number of accidents and traffic violations will be reduced.

References

1. Kyiv-Pasazhyrskyi Railway Station. https://en.wikipedia.org/wiki/Kyiv-Pasazhyrskyi_Railway_Station. Accessed 17 Apr 2021
2. Transport in Kyiv. https://en.wikipedia.org/wiki/Transport_in_Kyiv. Accessed 8 Apr 2021
3. About subway. http://www.metro.Kyiv.ua/node/90. Accessed 8 Apr 2021
4. Scheme of subway lines. http://www.metro.kiev.ua/node/101. Accessed 8 Apr 2021
5. Transport modes and advantages of bus. http://svit-express.com.ua/perevozki-po-ukraine/. Accessed 17 Apr 2021
6. Trolleybus as a mode of public transport. https://ria.ru/20101122/299491309.html. Accessed 12 Apr 2021
7. Savchenko, V., Laptiev, O., Kolos, O., Lisnevskyi, R., Ivannikova, V., Ablazov, I.: Hidden transmitter localization accuracy model based on multi-position range measurement. In: 2020 IEEE 2nd International Conference on Advanced Trends in Information Theory, IEEE ATIT 2020, Proceedings, pp. 246–249 (2020)
8. 11 Infrastructural Projects that can make Kyiv better. https://www.epravda.com.ua/rus/publications/2019/12/11/654728/. Accessed 8 Apr 2021
9. How the tram is better than the bus. https://gre4ark.livejournal.com/678424.html. Accessed 19 Apr 2021
10. Disadvantages of tram. https://zen.yandex.ru/media/id/5d9e3e07bc251400b1190385/nedostatki-tramvaia-5dc6166172b73d07e87e14c0. Accessed 11 Apr 2021
11. Sokolova, O., Soloviova, O., Borets, I., Vysotska, I.: Development of conceptual provisions to effectively manage the activities of a multimodal transport operator. Eastern-Eur. J. Enterp. Technol. **13**(109), 38–50 (2021)
12. Yudin, O., Ziubina, R., Buchyk, S., Matviichuk-Yudina, O., Suprun, O., Ivannikova, V.: Development of methods for identification of information-controlling signals of unmanned aircraft complex operator. Eastern-Eur. J. Enterp. Technol. **2/9**(104), 56–72 (2020)

Opportunities for the Development of Digital Transport and Cargo Platforms in Asia

Kristina Čižiūnienė[1(✉)], Marija Kolosova[1], and Jūratė Liebuvienė[2]

[1] Faculty of Transport Engineering, Department of Logistics and Transport Management, Vilnius Gediminas Technical University, Plytinės g. 27, 10105 Vilnius, Lithuania
kristina.ciziuniene@vilniustech.lt

[2] Department of Transport Engineering, Klaipėda State University of Applied Sciences, Jaunystės g.1, 91274 Klaipėda, Lithuania
j.liebuviene@kvk.lt

Abstract. In the European Union, the transport sector generates over 20% of the total GDP, so this steadily growing indicator shows the importance of effectiveness of transport in operation. As the influence of globalization expands the boundaries of economic and transport markets, the need to optimize and streamline activities of the transport sector and procedures for logistic actions has become inevitable. Digital transport management platforms, which not only facilitate and speed up the work of logistics managers, but also significantly increase the security of the transport market, can help to tackle the challenges in this area. Digital freight platforms promote transformation and optimization of business models. In Europe, where road transport is highly developed, there are ample transport exchange platforms, and customer assistance, a transport and freight exchange, an interactive map and effective feedback are the key products which they offer. There are far more of them in Europe and they are significantly more developed than in other regions (e. g. Asia or Africa), but a narrow focus on clearly defined market segments currently limits and weakens the potential of the platforms. Researchers say that in the long-term perspective the impact on the market of such platforms can be significant, which the experience of other markets, such as the growth of e-commerce or the transition to the food industry market, has already shown. Therefore, this article will analyse most popular transport and freight platforms in Europe and assess possibilities of their integration into Asian markets.

Keywords: Transport and cargo platforms · Transport exchanges · Digital freight platforms · New markets

1 Introduction

Intelligent Transport Systems (hereinafter - ITS) is a name to describe transport systems where vehicles interact with the environment and with each other to provide enhanced driving experience, and where intelligent infrastructure improves safety and capacity of road systems [1].

O. Prentkovskis et al. (Eds.): TRANSBALTICA 2021, LNITI, pp. 468–476, 2022.
https://doi.org/10.1007/978-3-030-94774-3_46

Intelligent technology is a general term for the application of integrated communication, control and information processing technologies in the transport system [2]. ITS improves efficiency and productivity of the transport system, quality of services, increases mobility, reduces energy consumption and environmental impact [3]. Online electronic transport exchanges are very important for transport companies. Such exchanges can be found in both Eastern and Western Europe [2]. Thus, transport platforms are widespread not only in Europe, but also in other continents, such as North America or Asia [4].

An electronic transport exchange is an online information platform tailored to the needs of freight and transport owners and freight forwarders [2]. Electronic freight and transport exchanges allow to offer customers additional logistics services that increase the competitiveness of companies: help track cargo, collect data on the logistics process, compare alternative logistics options, serve customers individually, and enable managers to communicate with each other [5].

A dozen years ago, intelligent transport platforms also known as exchanges were designed to search for and offer available cargo and empty vehicles. With rapid development of computer technology and the Internet, the needs of users have changed, also changing software functions. As a result, today, freight and transport exchanges are multifunctional platforms that increase the efficiency of the entire transport sector.

Transport exchanges enable their users to receive orders more easily and quickly, reduce costs and find new business partners [2]. Exchanges significantly facilitate communication between transport professionals, help to establish new contacts, and make it easy to find customers and/or carriers [6]. There is a considerable amount of information on transport platforms used in Europe, but not enough on transport platforms used or operating in Asia or North America.

Changes in the transport and logistics sector have been driven by a number of external and internal factors, starting from a shortage of drivers to investment in logistics infrastructure, changing consumer behavior and changing international trade relations, which have opened new transport routes between Europe and Asia.

Exchanges are time and money savers, and an opportunity for a rapid expansion of companies driven by newly established connections [6]. Given the consequences of development of transport exchanges, it is safe to say that they have had a positive impact on the development of transport, forwarding and logistics sectors. Firstly, they made collaboration more secure, allowing to optimize costs, and making medium and small businesses more efficient. Secondly, they made establishing new international connections much easier.

The aim of the article is to investigate the need of Lithuanian logistics companies for the development of transport platforms in the Asian region.

The following tasks were set to achieve this goal:

1. To conduct a survey of Lithuanian logistics companies.
2. To evaluate the development of the transport platform in the Asian market in the context of material and human resources.

Methods used in the article:

- Analysis, generalization and synthesis of scientific literature.
- Empirical research (survey), analysis of results.

2 Development of Digital Transport and Cargo Platforms in Asian Market

2.1 Materials and Methods

The research process is complex and heterogeneous [7]. Such stages as research planning, empirical research stage, statistical and theoretical processing of the obtained material, and practical application of the results can be distinguished in the process [8]. In the process of cognition, empirical and theoretical cognition always occur together. Empirical data collection is a painstaking process, which requires knowing exactly what you want [9].

Grūnovienė [10] distinguishes two main types of research: quantitative and qualitative research. In quantitative research, the researcher discusses the statement under consideration in advance, drawing a conclusion on whether or not the research proved the statement right [11]. Quantitative research can be statistical or experimental. Quantitatively, the aim is to search for, measure and calculate external features for a single explanation, laws, rules, versality and universality [8].

In order to obtain unbiased and accurate results reflecting the current situation in the transport, logistics and forwarding sector, a written questionnaire was chosen as the method of research. In order to maximize questionnaire response rate, attention was paid not only to the content of the questionnaire, but also to its presentation to respondents, as interest in the questionnaire process is one of the most important factors in data collection.

The sample size was calculated using the formula [8].

$$n = \frac{z^2 \cdot \upsilon(1-\upsilon)}{\Delta^2}, \tag{1}$$

where n – sample; z – the quantile of the normal difference; υ – event frequency; Δ – accuracy of probability estimation.

The research focuses on companies providing transport services. According to the data of the Department of Statistics of Lithuania, there are 8,501 enterprises registered and engaged in transport and storage operations. To determine the frequency of an event, it is necessary to find out the number of companies that potentially use transport platforms. An assumption was made that 45% of transport companies use transport platforms. S Sample calculations showed that at least 99 respondents had to be interviewed in order to obtain satisfactory results for the survey. However, knowing the research methodology and the main problem, which is unreturned questionnaires, questionnaires were sent to 200 respondents, increasing the likelihood of reaching the needed number of returned questionnaires for the survey to reflect the most realistic situation possible. 150 completed questionnaires were returned.

2.2 Results and Discussion

In order to find out whether Lithuanian logistics companies would benefit from the establishment of a transport platform in the Asian market, a survey was conducted in participation of 150 respondents. The aim of this survey was to examine whether there is a need for the emergence of various logistics companies from the Asian region on transport exchange platforms.

The majority of respondents (75%) were medium-sized enterprises with up to 60 employees, while small businesses with up to 15 employees comprised the smallest share. This may be due to the fact that many small businesses do not use transport platforms and may be reluctant to participate in such a survey, while representatives of large companies often simply do not have time to answer various questionnaires, and this survey may be no exception, as only 17% of respondents represented corporations in the survey.

It is important to determine types of companies that were represented in the survey (Fig. 1).

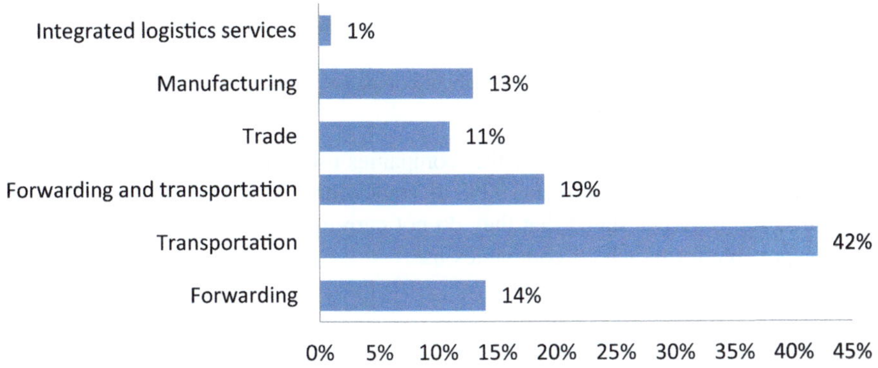

Fig. 1. Company types.

Figure 1 shows that companies whose principal activity is transportation (42%) and activities related to both forwarding and transportation (19%) accounted for the largest share of respondents, while 14% of respondents said that their companies were engaged in forwarding activities only. 1% of respondents (one company) represented companies engaged in the provision of integrated logistics services.

Having assessed the above aspects, analysing countries which the respondents usually cooperate with was expedient. The questionnaire contained open-ended questions, which respondents could answer the way they wanted (sometimes indicating more than one country). Figure 2 shows most popular answers.

As per Fig. 2, Lithuanian companies mostly work with Germany, the Netherlands and Denmark, and the least - with Georgia. In their response to this question, respondents also mentioned such countries as Poland, Ukraine, Lithuania, but all of them were mentioned up to 10 times only. Interestingly, the Scandinavian countries Norway, Sweden and Finland were also mentioned very rarely - up to 15 times each.

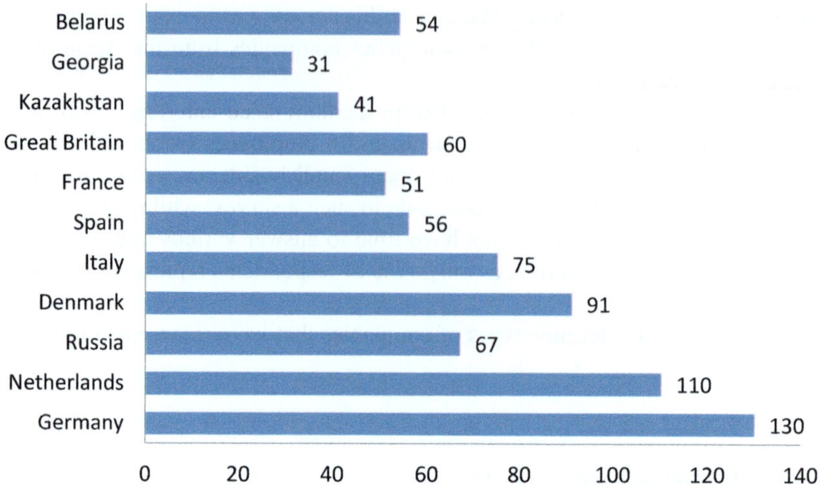

Fig. 2. Countries that Lithuanian companies work with the most.

The study aimed to find out whether companies use transport and freight exchange platforms to search for customers. 92% of respondents answered that they use such platforms, and only 8% stated that they do not currently use them. Figure 3 illustrates the transport platforms which companies use.

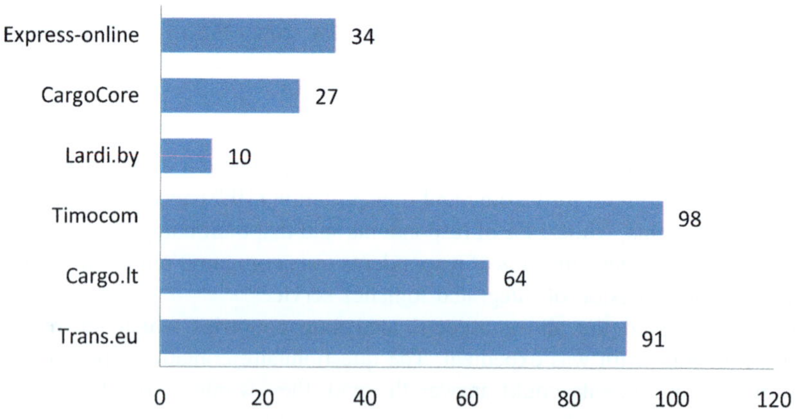

Fig. 3. Platforms that respondents use.

Figure 3 shows that transport platforms TimoCom and Trans.eu were mentioned most often, while the little-known free transport exchange platform Lardi.by established in Belarus received the fewest votes.

Since respondents were asked to name the platforms they use, the survey sought to find out why companies chose those particular platforms. The question was open-ended, and the most popular answers included: convenient, accurate data; many offers, many users; recommended by partners; reasonable price; the platform best meets the needs of the company.

These answers reveal that respondents choose a particular freight and transport exchange platform for a variety of personal reasons, and meeting customers' needs was one of the most important of them. Respondents, who answered that they use freight and transport exchange platforms, were asked to name why they do that, and those, who answered that they do not use such platforms, were asked why they do not use them.

The most common answers of **respondents who use platforms** were: faster and more reliable search; such a solution is necessary to fill empty shipments; platforms create a sense of security; reliable partners with whom long-term relationships have been established.

The most common answers of **respondents who do not use platforms** were: too expensive; no offers corresponding to the company's labor market can be found; the company tried using transport platforms, but it didn't work out; the company has enough business partners and is not looking for more.

The answers reveal that companies find the price of the transport exchange and the regions where it works important. Reliability and security which the platform guarantees to its customers was mentioned as the main advantage thereof.

The survey also sought to find out whether respondents think that there are enough customers from the Asian region on transport platforms. 90% of respondents said that the platforms they use do not have enough customers from the eastern region, and 10% said that this region was irrelevant for them, so they could not answer the question. None of the respondents answered that there are no freight or transport offers on the transport platforms which their companies use or that the number of offers is insufficient. In this context, the aim was to find out whether respondents' companies would benefit from more freight and transport offers from companies in the Asian region on transport platforms. All the respondents unanimously answered "yes" to this question, which shows great interest of companies in such an opportunity.

Respondents, who answered "no" to the question "Do you use transport and freight exchange platforms to search for customers/orders", were asked whether their companies would start or at least consider using services of transport platforms if they had more offers from logistics companies in the Asian region. As many as 96% of respondents answered "yes" to this question and a mere 4% said "no". This means that with the emergence of freight or transport offers from the Asian region on a transport platform, interest in such a platform would increase in Lithuania as well. It is therefore important to assess the aspects required for the establishment of a transport platform in the Asian market.

Given the findings of the study that it is important for transport/logistics companies to have freight from the Asian market on transport and freight platforms, it is important to assess what measures should be taken to implement this. Therefore, in order to attract customers from the Asian market to the transport platform, a new division needs to be established in this region.

Each project requires a variety of resources, including material, human, financial, time, and other resources.

Every new office established as a company expands needs a headquarters. A new office should have at least 5 separate air-conditioned rooms for the following teams:

- a separate sales department with 4 employees. This is the department which would have the most employees, as it is the main source of income;
- customer service and marketing department together. Since customer service department would be responsible for answering customer questions, it would initially have 1 employee only, as there would not be many customer questions when the office is just starting to run. Marketing department would be responsible for attracting customers and would have 2 employees;
- management office with 1 employee;
- human resource department specialist, accountant and IT specialist together, 3 persons in total;
- business security and debt collection departments together. Business security department would have 2 employees and would be responsible for customer due diligence, while debt collection department would have 1 employee dealing with customer issues.

Thus, each department is responsible for certain different tasks, so they need employees with different experiences. A new office must have representatives of all these departments. Sales department must have the most employees, as it generates direct revenue for the company and serves potential customers. Business security is another important department, which screens new customers who want to collaborate with the platform. This department must have 2 employees, with one employee screening up to 10 companies per day. Customer service department is also important. It needs one employee at the start, possibly growing to two employees after half a year. IT department needs one employee to maintain work computers, install the necessary programs and arrange equipment needed for other employees. Marketing department needs two employees, who would spread the word about the platform in the new market as widely as possible. There should be one person in human resource department to recruit employees, supervise their work and assist the company manager in organizational matters. At the beginning, one person would be enough at debt collection department, and after 6 months, another employee could be hired, if necessary. A company manager is needed to manage the newly opened office and an accountant – to oversee the company's finances. In total, the new office should have 14 employees, including the head of the company, and after 6 months, 2 more people could be hired, leading to a total of 16 employees. This number may increase in the course of business, depending on the workload and the available resources. EUR 29 106.12 is required to set up and maintain an office in Almaty in the first month, and EUR 210 794.34 would be needed to carry out its activities for at least one year.

3 Conclusions

1. The results of the survey revealed that many companies operating in the logistics sector in Lithuania use services of transport platforms, but they are not satisfied with the number of available transport and cargo offers from the Asian region. Having included such offers, 96% of respondents would start using transport platforms, although they have not been using them so far. Therefore, establishing a transport platform in the Asian market is necessary to remain the leader in the industry and to attract new customers.
2. The assessment of transport development in the Asian market revealed that the operating budget for the first year would be EUR 210,794.34 EUR, while a realistic assessment of return on investment showed that this project would pay off in 1.82 years and the company would generate a profit of EUR 15,000.3 at the end of the second year.
3. Transport platforms therefore need to pay more attention to activities of their competitors and to carry out a thorough analysis of them in order to remain the leader in the future and not to lose customers; they also need to use the opportunity to expand into the global market beyond Europe; not to be afraid of a flexible pricing policy, as if the starting price is too high, smaller companies will not even be able to try to apply for the services provided by the platform, immediately turning to competitors offering cheaper services.

References

1. Williams, B.: Intelligent Transport Systems Standards. Artech House, Bosto London (2008)
2. Batarlienė, N., Jarašūnienė, A.: Intelektinės Technologijos Transporte: Vadovėlis. Vilniaus Gedimino technikos universitetas, Vilniaus (2020)
3. Keršys, A.: Darni Transporto Plėtra. Lietuvos edukologijos universiteto leidykla, Kaunas (2013)
4. Čižiūnienė, K., Kolosova, M., Gomienė, Ž., Burinskienė, L.: Azijos ir Šiaurės Amerikos transporto platformų analizė. Sustainable Environ. Dev. Sci. Articles **1**(17), 32–41 psl. Klaipėda (2020)
5. Liebuvienė, J., Šileikienė, A., Čižiūnienė, K.: Elektroninių krovinių ir transporto biržų Lietuvos logistikos sektoriuje panaudojimo galimybių tyrimas. Investigation of electronic freight exchange using in Lithuanian logistics sector//Inžinerinės ir edukacinės technologijos: mokslinių straipsnių žurnalas = Engineering and educational technologies: scientific journal. Kaunas: Kauno technikos kolegija (2017)
6. Batarlienė, N.: Informacinės transporto sistemos. Technika, Vilnius (2011)
7. Kardelis, K.: Mokslinių tyrimų metodologija ir metodai. Mokslo ir Enciklopedijų leidybos centras, Vilnius (2017)
8. Kardelis, K.: Mokslinių tyrimų metodologija ir metodai. Vadovėlis (ketvirtas leidimas), Leidykla Liucijus, Šiauliai (2007)
9. Čėsna, B., Bagdžiūnaitė-Litvinaitienė, L., Jakubavičius, A.: Moksliniai tyrimai ir inovacijos inžinerijoje: Technologija, Vilnius (2012)

10. Grūnovienė, D., Macas, A., Pauliukėnas, L.: Mokslinio tyrimo procesas ir jo organizavimas. Kauno kolegijos leidybos centras, Kaunas (2005)
11. Bitinas, B., Rupšienė, L., Žydžiunaitė, V.: Kokybinių tyrimų metodologija. S. Jokužio leidykla-spaustuvė, Klaipėda (2008)

Transportation Problems and Their Application in Planning Transport Provision of Area Evacuation

Zuzana Gašparíková(✉) and Bohuš Leitner

Žilinská Univerzita v Žiline, Univerzitná, 8215/1, Žilina 010 26, Slovakia
{zuzana.gasparikova,bohus.leitner}@uniza.sk

Abstract. Transportation problems are one of the specific problems of linear programming. These are optimization tasks and their purpose is to minimize the total cost of transporting commodity or people in complex logistics processes. The solution is to optimize the delivery system of the certain entity (e.g. commodity, persons or materials) from sources (suppliers) to destinations (customers) in order to minimize the total transport costs associated with distribution. In the case of area evacuation, transportation problems are used in planning the optimal transport of evacuees from boarding points to evacuation centers, while the calculations take into account in particular the transport distance and the total cost of transporting one person. The aim of the article is to demonstrate the possibilities and advantages of optimization procedures with a focus on transportation problems in planning and managing transport provision in the logistics process – evacuation of the area. The article contains a draft transport plan of inhabitants of the micro-region endangered by floods during their evacuation to designated evacuation centers. In determining the initial solution were applied three methods: The North West Corner Method, The Index Method and The Vogel's Approximation Method. The best of them were described in the optimality test. Based on quantitative analysis of the problem, it was established of the optimal plan for the transport evacuees from the designated area.

Keywords: Area evacuation · Transport provision · Transport planning · Transportation problems · Distribution method · Software support · Optimal distribution plan

1 Introduction

Evacuation is one of the basic measures of collective protection and she is carried out to the necessary time of the stay the persons in the endangered area. We can say that evacuation is population protection measure, which reduced number of the people in the territory or in the area threatened by an extraordinary event.

Area evacuation mean the evacuation of the people from larger urban areas. Evacuated residents are all persons who do not perform rescue work. Evacuation management is a complex and extensive process, because the beginning and extent of

O. Prentkovskis et al. (Eds.): TRANSBALTICA 2021, LNITI, pp. 477–489, 2022.
https://doi.org/10.1007/978-3-030-94774-3_47

the evacuation is difficult to predict. Evacuation management solve a large number of tasks before, during and after the evacuation.

The activities that belongs to the evacuation management are divided into several phases (Fig. 1).

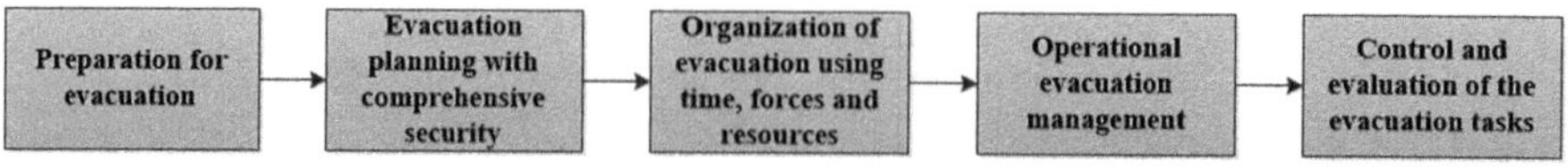

Fig. 1. Phases of the evacuation [1].

The basis of evacuation planning is analysis of situation, which analyzes the extent of the crisis phenomena, time constraints, evaluation of self-evacuation, evaluation of usable forces and resources for evacuation, forecast for emergency development, setting aims, proposing procedures to achieve aims and selecting the most appropriate option. Evacuation planning is an important process that identifies certain specific tasks to individual branches of public administration, aligns the evacuation of people, animals and property, controls the fulfillment of these tasks, processes documentation for evacuation management and lists them population before emergency situations occurs. Each evacuation management should also be an operational management in order to control and guide further planned activities during the performance of individual tasks.

2 Mathematical Support in Planning of Area Evacuation

Protection the population from emergencies is a system of activities, which are carried out based on analyzes of the endangered area and the probability of occurrence (frequency of occurrence) of emergencies.

Evacuation is a complex logistical process that requires thorough preparedness to minimize the damage caused by an emergency situation. Since saving the life and health of the population is mainly about optimizing time, it makes sense to use applicable mathematical disciplines and software support based on mathematical methods when planning and managing area evacuation.

Transport process management strategies and the optimal allocation of resources (entry points) and targets (evacuation centers) play a key role in evacuation planning. Several foreign authors are dealing with this domain. In *Emergency Planning Evacuation as a Network Design Problem* the authors [2] discuss the NDP (The network design problem), which includes the analysis and performance of the transport network, its optimality and transmission. The aim of the NDP is to optimize the measurement of system performance in order to minimize the total travel costs and at the same time to perform the behavior of road users when choosing a route [3]. João Coutinho-Rodrigues from the Portuguese University of Coimbra in his work *Design of evacuation plans for the densely urbanized city centers* [4] dealt with transport provision and creation of primary and secondary evacuation routes. Primary route takes into account the length of the route and its risk. The location of the threat and the evacuation center are known. Searching for the secondary route, only the location of the threat is known and the shortest route determines the place where evacuees arrive. In the work *Integration of spatial analysis and an agent-based model into evacuation management for shelter assignment and routing* [5] was created The Adaptive Evacuation Route Algorithm (AERA) where the authors consider the total evacuation time and safe evacuation route criteria to be the main optimization. The authors Janáček and Sibila deal with computer-assisted approaches to the design of the evacuation plan in the work *Optimal evacuation plan design with IP-Solver* [6]. They assign available means of transport to boarding points in endangered areas so that total time needed to evacuation was as short as possible.

2.1 Types and Application of Transportation Problems

Except for basic formulations of transportation problem are exist the other types of models for logistics processes. *General distribution problem* (GDP) differs from the classic transportation problem mainly in that the capacities of resources and customer requirements are not listed in the same measures. For their comparability, it is necessary to supplement certain “conversion factors”. For example, production line capacity (hours), customer request (pcs) – conversion factor (pcs/hour). *Container transportation problem* (CTP) changes the basic formulation of the transportation problem. In the case of CTP, transport between suppliers and customers is carried out means of containers, which have a certain defined capacity C. Transport costs are not related to one unit of transport goods, but they are considered as one container. There are costs associated with transporting one container, but they are the same regardless of the container is full or half empty. The optimal CTP solution should lead to the individual transported containers being used as efficiently as possible. Other types of transportation problems can arise from various combinations of solved problems, such as *allocation problems* – they solve the optimal grouping of resources into districts served from individual sources. The most famous type of transportation problem is the so-called *coverage task*. It is an optimization of the decision where to build service stations (rapid assistance stations, fire stations, consignment sorting centers, etc.) At the same time, it is necessary to assign them their territorial scope – identify the circuits to be served by these centers [7].

3 The Classic Transportation Problem in the Transport Provision of Area Evacuation

Transportation problems belong to a special category of linear programming problems - distribution problems. Transportation problems are historically one of the oldest tasks of linear programming and in addition to them, the distribution problems also include the so-called allocation problems, allocation problems for optimal distribution of resources and their various combinations [8].

The aim of transportation problems is to schedule the distribution of goods or materials from suppliers (sources) to customers (destinations) so that the total costs are minimized. In transportation problems, it take into account m suppliers $D_1, D_2, \ldots, D_m$

with quantify of goods produced $a_1, a_2, \ldots, a_m$ and n customers $O_1, O_2, \ldots, O_n$ with requests $b_1, b_2, \ldots, b_n$. The relationship between supplier and customer is evaluated by the rate $c_{ij} (i = 1, 2, \ldots, m; j = 1, 2, \ldots, n)$ usually determined on the basis of quantified total transport costs or only according to the kilometers of distance between sources and customers. The aim is to determine volume of transport x_{ij} between i-th source and j- th customer so that the capacities of the sources are not exceeded and customer requirements have been met.

Based on the number of customer requests and supplier capacities, transportation problems can be divided into balanced and unbalanced problems (Table 1).

Table 1. Basic division of transportation problems according to their balance

Balanced transportation problem	Unbalanced transportation problem	
$\sum_{i=1}^{m} a_i = \sum_{j=1}^{n} b_j$	$\sum_{i=1}^{m} a_i \neq \sum_{j=1}^{n} b_j$	
	Customer deficit	Resource deficit
	$\sum_{i=1}^{m} a_i > \sum_{j=1}^{n} b_j$	$\sum_{i=1}^{m} a_i < \sum_{j=1}^{n} b_j$

A *balanced transportation problem* satisfies the condition that the number of requests from customers and the capacity of resources is the same. In this case, all supplier's capacities are exhausted and customer demand is satisfied.

An *unbalanced transportation problem* is inconsistency between requirements and capacities. The solution of such a problem can take place only after the transformation of this problem into a balanced transportation problem, usually by adding the so-called fictitious supplier or customer. Transport costs in such a case are zero $(c_{ij} = 0)$, because neither a fictitious supplier or a customer not exist.

The resulting mathematical model of the transportation problem can be expressed in summative form (Fig. 2)

minimization

$$z = \sum_{i=1}^{m} \sum_{j=1}^{n} c_{ij} . x_{ij} \quad \rightarrow \quad MIN$$

subject to

$$\sum_{j=1}^{n} x_{ij} = a_i \qquad \text{for } i = 1 \ldots m$$

$$\sum_{i=1}^{m} x_{ij} = b_j \qquad \text{for } j = 1 \ldots n$$

$$x_{ij} \geq 0, \qquad \text{for } i = 1, 2, \ldots, m, \ j = 1, 2, \ldots, n.$$

Fig. 2. Mathematical model of the transportation problem.

4 Optimization of Transportation Problem in the Transport Provision of Area Evacuation

The North West Corner Method, the Index Method and the Vogel's Approximation Method are most often applied to obtain initial solutions to the transportation problem [8]. A "matrix of differences" is used to verify the optimality of the solutions, which we obtain by the methods for the initial solutions of transportation problems. Determining the elements of the "matrix of differences" D is part solution of the optimality of the transportation problems using the *potential method.* If all elements of the "matrix of differences" are larger or equal to zero $\left(d_{ij} \geq 0\right)$, the solution is optimal. If any element of the "matrix of differences" is less than zero, the solution is not optimal and it is necessary to change the element in which the optimality is disturbed by the so-called "*movement in a cycle*", but it is important that the solution is not disturbed (one variable is created, the other disappears). The result obtained by the "movement in the cycle" must be optimized again. In order to find the optimal solution, we can also apply the MODI method (modified distribution method), which is based on the relationship (1), where c_{ij} represents rate value, r_i and s_j represents row and column numbers:

$$H_{ij} = r_i + s_j - c_{ij}. \tag{1}$$

If all the value of H_{ij} is less than zero, the solution obtained is optimal.

In our case, we will be solved unbalanced transportation problem (the number of requests for evacuation of inhabitants is not equal to the number of capacities of the selected evacuation centers). The capacity of the evacuation centers exceeds the number of evacuees, so it was necessary to introduce a fictitious boarding point. The number of inhabitants needed for evacuation from individual places is real, obtained from a flood risk maps (Fig. 3), the number of capacities of the intended evacuation centers is based on the capacities of selected objects. The individual rates determine the average costs incurred for the transport of one person, where the cost including transport distance, the price of fuel, costs associated with activating and ensuring of evacuation centers and their services and transport back after the evacuation.

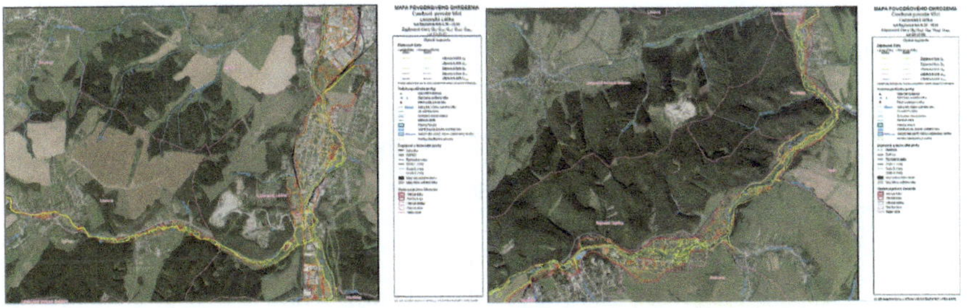

Fig. 3. Flood risk maps [9].

Transportation problem - In an extraordinary event caused by 100-years of water on the water flow Rajčanka, the number of potentially endangered is 606 inhabitants in Rajecké Teplice, 1,227 inhabitants from Lietavská Lúčka and 127 inhabitants of the village Porúbka. In case of evacuation, there are 4 evacuation centers Lietavská Lúčka - Primary school (LL PS), Lietavská Lúčka - House of culture (LL HC), Porúbka -House of culture (P HC) and Rajecké Teplice - Primary school (RT PS). There are 5 boarding places - 2 in Lietavská Lúčka (LL1, LL2), 1 in Porúbka (P) and 2 in Rajecké Teplice (RT1, RT2) in the endangered municipalities (Fig. 4). In the chosen model scenario, it is assumed that all residents from the endangered area will be transported to the evacuation centers only by designated means of transport and self-evacuation will not be considered.

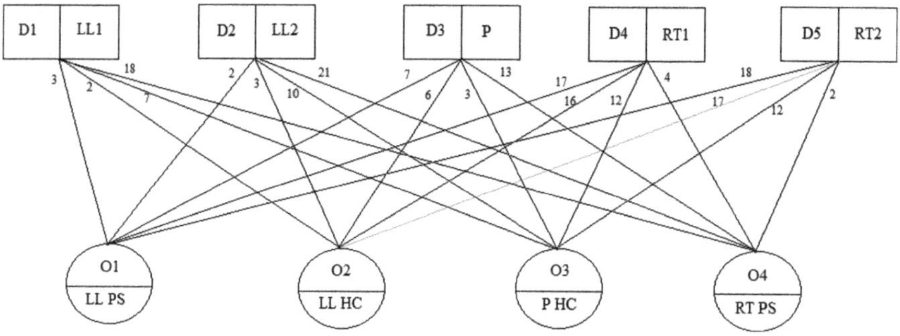

Fig. 4. Scheme of transportation problem.

In this case, it represents the boarding point of the supplier and the evacuation center of the customer. The individual rates represent the unit costs (Eur/person) for the transport of one person.

4.1 Obtaining an Initial Solution to the Transportation Problem

The North West Corner Method allows to obtain relatively quickly an initial solution (Table 2), which is acceptable, but not usually optimal. The method does not take into account the rates c_{ij}.

Table 2. Initial solution of The North West corner method.

	LL PS	LL HC	P HC	RT PS	Requirements of boarding places
LL 1	3 (670)	2 (38)	7	18	708
LL 2	2	3 (412)	10 (107)	21	519
P	7	6	3 (113)	13 (24)	137
RT 1	17	16	12	4 (300)	300
RT 2	18	17	12	2 (306)	306
F_{nm}	0	0	0	0 (20)	20
Capacity evacuation centers	670	450	220	650	1,990

$$\min z = 670.3 + 38.2 + 412.3 + 107.10 + 113.3 + 24.13 + 300.4 + 306.2 + 0 = \mathbf{6\,855€}$$

The Index Method takes into account the rates and their sizes and for the same rates, the box with the larger quantity and the lowest price coefficient is preferred. When applying the index method, we transform the original table, taking into account the size of the rates c_{ij} and finding the ratio of requirements of boarding places to the capacity of the evacuation center. Then, we sort the values in the table in ascending order according to the size of the rates c_{ij}, while with the same value of rate c_{ij} we take into account the share of requirements and capacity of the center and gradually add the sorted values to the table (Table 3).

Table 3. Initial solution of the index method.

	LL PS	LL HC	P HC	RT PS	Requirements of boarding places
LL 1	3 (131)	2 (450)	7 (83)	18 (44)	708
LL 2	2 (519)	3	10	21	519
P	7	6	3 (137)	13	137
RT 1	17	16	12	4 (300)	300
RT 2	18	17	12	2 (306)	306
F_{nm}	0 (20)	0	0	0	20
Capacity evacuation centers	670	450	220	650	1990

$$\min z = 131.3 + 450.2 + 83.7 + 44.18 + 519.2 + 137.3 + 300.4 + 306.2 + 20.0 = \mathbf{5\,927€}$$

Vogel's Approximation Method is based on the determination and apply of the row and column penalty R_i and R_j, which represents the difference between the lowest rates in a row or columns. In the row (or column) with the highest penalty, the field with the lowest rate is occupied, the row and column penalty are recalculated again and the procedure is repeated. In case of occurrence of two identical lowest rates, $R_{ij} = 0$. We will expand the table with the input by one row and one column, in which we will add the values of the row and column penalty (Table 4).

Table 4. Transportation problem with row and column penalty.

	LL PS	LL HC	P HC	RT PS	Requirements of boarding places	Row penalty
LL 1	3	2	7	18	708	1
LL 2	2	3	10	21	519	1
P	7	6	3	13	137	3
RT 1	17	16	12	4	300	8
RT 2	18	17	12	2	306	10
F_{nm}	0	0	0	0	20	0
Capacity evacuation centers	670	450	220	650	1990	
Column penalty	2	2	3	2		

We start the solution with a row, resp. column with the highest value of penalty, while we try to fill the capacity of the evacuation center and at the same time distribute as many inhabitants as possible. Then the row, resp. the column is not taken into account, again the penalty for increasing rows and columns are recalculated and the procedure is repeated (Table 5).

Table 5. Initial solution of the Vogel's approximation method.

	LL PS	LL HC	P HC	RT PS	Requirements of boarding places
LL 1	3 (151)	2 (450)	7 (107)	18	708
LL 2	2 (519)	3	10	21	519
P	7	6	3 (113)	13 (24)	137
RT 1	17	16	12	4 (300)	300
RT 2	18	17	12	2 (306)	306
F_{nm}	0	0	0	0 (20)	20
Capacity evacuation centers	670	450	220	650	1990

$$\min z = 151.3 + 450.2 + 107.7 + 519.2 + 113.3 + 24.13 + 300.4 + 306.2 + 0 = \mathbf{5\ 603€}$$

Comparing the solutions apply individual selected methods for the initial solution of the transportation problem (The North West Corner Method, The Index Method and The Vogel's Approximation Method), we found out that the optimal solution came out applying the Vogel's approximation method (Table 6). The least optimal solution is applying The North West Corner Method, because this method does not take into account the rates c_{ij}.

Table 6. Comparation of transport costs according to initial solutions.

Applying method	minz
The North West Corner Method	6,855€
The Index Method	5,927€
The Vogel's Approximation Method	**5,603€**

4.2 Verifying Optimality of the Initial Solution

To verify, if the best of the initial solutions obtained by means of the Vogel's approximation method is the optimal solution, apply the potential method with matrix of differences. The matrix of differences determines the optimality criterion of the solution. For the matrix of differences D was applied formula (2):

$$C - C^{*} = D, \tag{2}$$

where C is a matrix of rates and C^{*} represents a matrix of indirect rates. To determine the matrix of indirect rates, it is necessary to determine row and column numbers, where the sum of the row and column number equals the respective rate ($r_i + s_j = c_{ij}$). The determination of the matrix of indirect rates was based on Table 5 from the initial solution of The Vogel's Approximation Method. The resulting matrix of differences D obtained according to formula (2) will have the form (3):

$$\begin{pmatrix} 3 & 2 & 7 & 18 \\ 2 & 3 & 10 & 21 \\ 7 & 6 & 3 & 13 \\ 17 & 16 & 12 & 4 \\ 18 & 17 & 12 & 2 \\ 0 & 0 & 0 & 0 \end{pmatrix} - \begin{pmatrix} 3 & 2 & 7 & 17 \\ 2 & 1 & 6 & 16 \\ -1 & -2 & 3 & 13 \\ -10 & -11 & -6 & 4 \\ -12 & -13 & -8 & 2 \\ -14 & -15 & -10 & 0 \end{pmatrix} = \begin{pmatrix} 0 & 0 & 0 & 1 \\ 0 & 2 & 4 & 5 \\ 8 & 8 & 0 & 0 \\ 27 & 27 & 18 & 0 \\ 30 & 30 & 20 & 0 \\ 14 & 15 & 10 & 0 \end{pmatrix}. \tag{3}$$

The elements of resulting matrix of differences D are larger or equal than zero $(d_{ij} \geq 0)$, the optimality criterion is met. The initial solution determined by The Vogel's Approximation Method can be considered as an optimal solution.

4.3 Verifying Optimality of the Initial Solution by Software Tools

To verify our results, we used a freeware software tool at web site www.easycalculation.com (Fig. 5). The applied software tool offers the solution of classic transportation problems and at the same time optimizes the initial solution by the Optimal solution modification method (MODI). This software tool confirmed the assumption that the initial solution by The Vogel's Approximation Method is the optimal solution.

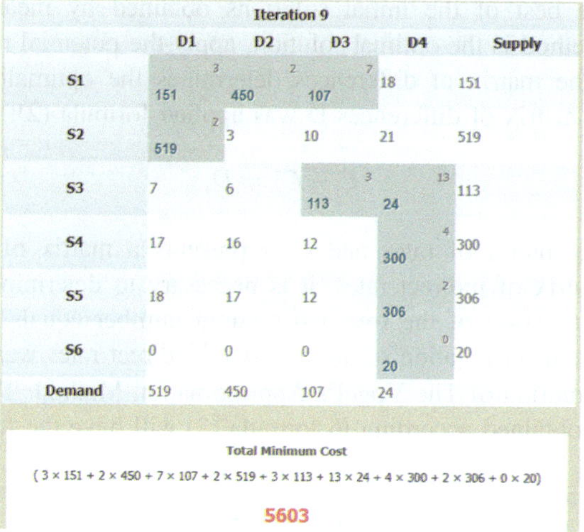

Iteration 9

	D1	D2	D3	D4	Supply
S1	3 151	2 450	7 107	18	151
S2	2 519	3	10	21	519
S3	7	6	3 113	13 24	113
S4	17	16	12	4 300	300
S5	18	17	12	2 306	306
S6	0	0	0	0 20	20
Demand	519	450	107	24	

Total Minimum Cost

(3 × 151 + 2 × 450 + 7 × 107 + 2 × 519 + 3 × 113 + 13 × 24 + 4 × 300 + 2 × 306 + 0 × 20)

5603

Fig. 5. Solution of the transportation problem by software tool [10].

5 Interpretation of Results

From transportation problem, it was necessary to determine the optimal distribution plan of the inhabitants affected by the flood from the municipalities of Lietavská Lúčka, Porúbka and Rajecké Teplice to the evacuation centers located in these three municipalities. We obtained the initial solution applying three methods: The North West Corner Method, The Index Method and The Vogel's Approximation Method. Through the optimization test, we found that the solution by The Vogel's Approximation Method is optimal, and then we determine the optimal distribution plan (Fig. 6).

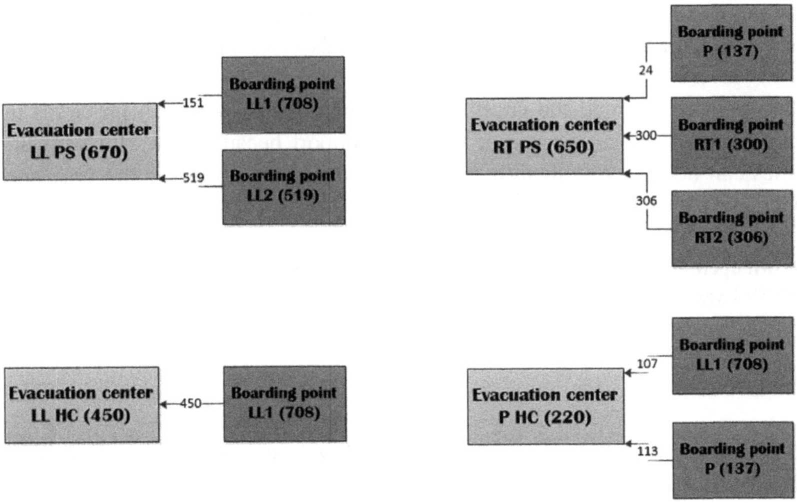

Fig. 6. Optimal distribution plan.

The results of the transportation problem of area evacuation follows that 670 inhabitants from Lietavská Lúčka should be transported to the evacuation center LL PS (151 inhabitants from the boarding point LL1 and all requirements - 519 inhabitants from boarding point LL2). The evacuation center LL HC will be filled with 450 inhabitants from the boarding point LL1. The village of Porúbka has an evacuation center with a capacity of 220 people, which will be filled by 113 inhabitants of Porúbka and 107 inhabitants of Lietavská Lúčka. Inhabitants from two boarding places in Rajecké Teplice (RT1 and RT2) and 24 inhabitants of Porúbka will be transported to the evacuation center RT PS with a capacity of 650 people.

6 Conclusion

The methods of operational analysis use mathematics as a tool to solve problems that may arise in management. The results of the evaluation of mathematical calculations determine the decisions, evaluate their applications, they significantly affect the lives of inhabitants in their protection from natural disasters or industrial accidents that may cause a crisis situation with the need for evacuation.

The methods of operational analysis are the basis for decision support in order to find the optimal solution to a problem. The optimal route in the graphs is used in the area evacuation for searching and planning evacuation routes, which include the transport of evacuees or the distribution and import of basic foods and products. Using network analysis methods, it is possible to determine the sequence of individual evacuation activities, the expected duration of individual activities and the total time of evacuation [11]. For transportation problem that are part of linear programming, it is important optimizing the transfer of evacuees from boarding points to evacuation centers based on the capacity of vehicles, but also free places in evacuation centers [12].

The article presented the possibility of applying transportation problems in crisis management as well as the use of software support because powerful software technologies can help us significantly in this case mainly by reducing the time needed to prepare and prevent crisis events.

Acknowledgements. "This article was supported by the scientific research project VEGA 1/0159/19 Evaluation of the level of resilience of key elements of land transport infrastructure."

References

1. Seidl, M., Tomek, M., Vičar, D.: Evakuácia osôb, zvierat a vecí. EDIS – vydavateľstvo ŽU, Žilina (2014)
2. Abdelgawad, H., Abdulahai, B.: Emergency evacuation planning as a network design problem: a critical review. In: Transportation Letters: In International Journal of Transportation Research (2009). https://www.researchgate.net/publication/247898894_Emergency_evacuation_planning_as_a_network_design_problem_A_critical_review
3. Ballay, M., Macurová, L., Kohut, P., Copiak, M.: Development of road safety status and the evaluation criterion causes of specific traffic accidents. In: Transport Means, pp. 765–770 (2018)
4. Coutinho-Rodrigues, J., Sousa, N., Natividade-Jesus, E.: Design of evacuation plans for densely urbanized city centres. In: Municipal Engineer. Institution of Civil Engineers (2015)
5. Liu, X., Lim, S.: Integration of spatial analysis and an agent-based model into evacuation management for shelter assignment and routing. J. Spat. Sci. **61**(2), 283–298 (2016). https://doi.org/10.1080/14498596.2016.1147393
6. Janáček, J., Sibila, M.: Optimal evacuation plan design with IP-solver. In: Communications – Scientific Letters of the University of Žilina, pp. 29–35 (2009). http://komunikacie.uniza.sk/index.php/communications/article/view/1003/967
7. Jablonský, J.: Operační výskum. Professional publishing, Praha (2007)
8. Distribučné problémy – príklady. http://www.fsi.uniza.sk/ktvi/leitner/2_predmety/OA/Semester/EX05_Distribucne%20problemy%20upr.pdf. Accessed 20 Mar 2021
9. Mapy povodňového ohrozenia a mapy povodňového rizika vodných tokov Slovenska. http://mpompr.svp.sk/okres.php?id=42. Accessed 20 Apr 2021
10. Minimum Transportation Cost Calculator/VAM Calculator. https://www.easycalculation.com/operations-research/minimum-transportation-vogel-approximation-method.php. Accessed 30 Apr 2021
11. Sventeková, E., Lusková, M., Dvořák, Z.: Use of network analysis in conditions of critical infrastructure risk management. In: WMSCI 2016 – 20th World Multi-Conference on Systemics, Cybernetics and Informatics, pp. 247–250, Orlando, United States (2016)
12. Mihoková Jakubčeková, J., Tomek, M.: Optimization of evacuation route in case of disturbance of a part of transport network during the initiation of an extraordinary event. In: Innovation Management and Education Excellence Through Vision 2020, pp. 2197–2203, Norristown (2020)
13. Ballay, M., Sventeková, E., Urbancová, Z., Monoši, M.: Application analyses of state of evolution – ETA on selected an extraordinary events. In: 13th International Scientific Conference on Sustainable, Modern and Safe Transport, pp. 1244–1251 (2019)

14. Janáček, J., Kvet, M.: Semi-fair design of emergency service system with failing centers. CEJOR **25**(3), 665–677 (2016). https://doi.org/10.1007/s10100-016-0456-5
15. Kvet, M.: Computational study of radial approach to public service system design with generalized utility. In: Digital Technologies: The 10th International Conference, Žilina, Slovakia (2014)
16. Máca, J., Leitner, B.: Operačná analýza I, FŠI ŽU, Žilina (1998)
17. Monosi, M., Ballay, M.: Use of telematics road transport systems in Slovak republic and abroad. In: International Conference on Crisis Management and Solution of the Crisis situations, pp. 214–222, Uherské Hradište, Czech Republic (2015)
18. Pekarčíková, M., Fiľo, M., Trebuňa, P.: Metódy riešenia distribučných úloh. In: XIII. Medzinárodná vedecká konferencia Manažérstvo životného prostredia, Bratislava (2013)
19. Sventeková, E., Svetlík, J.: Permeable performance testing of limiting road section. In: 19th International Scientific Conference on Transport Means, pp. 543–546 Kaunas, Lithuania (2015)
20. Šimková, V.: Dopravné úlohy. https://spu.fem.uniag.sk/cvicenia/ksov/fandel/Operacny_vyskum–optimalne_programovanie/10%20-%20dopravne%20ulohy/Literatura/Simkova-Dopravne_ulohy.pdf. Accessed 26 Nov 2020

Supply Chain Synchronization Through Deep Reinforcement Learning

Ilya Jackson[1,2(✉)]

[1] Center for Transportation and Logistics, Massachusetts Institute of Technology, Cambridge, MA, USA
Ilyajack@mit.edu
[2] Transport and Telecommunication Institute, Riga, Latvia
Jackson.i@tsi.lv

Abstract. Synchronized supply chains can mitigate a cascading rise-and-fall inventory dynamic and prevent cycles of over and under-production. This paper demonstrated that a deep reinforcement learning agent could only perform adaptive coordination along the whole supply chain if end-to-end information transparency is ensured. Operational and strategic disruptions caused by the COVID-19 pandemic and the post-pandemic recovery can become a necessary kick-starter for required changes in information transparency and global coordination. This paper explores the capabilities of deep reinforcement learning agents to synchronize commodity flows and support operational continuity in the stochastic and nonstationary environment if end-to-end visibility is provided. The paper concludes that the proposed solution can perform adaptive control in complex systems and have potential in supply chain management and logistics. Among discovered benefits, it is essential to highlight that the proximal policy optimization is universal, task unspecific, and does not require prior knowledge about the system.

Keywords: Supply chain · Reinforcement learning · Deep RL · PPO

1 Introduction

We live in an era of rapidly accumulating data volumes and processing power. This tendency notably favors algorithms capable of leveraging the increased computational resources. In recent years, deep learning has been the most highly debated topic in artificial intelligence. However, this success was mainly achieved within the supervised learning paradigm [1]. Nevertheless, the potential of deep learning extends classical supervised learning. Deep reinforcement learning (RL) is the paradigm of choice for applying deep learning to non-supervised scenarios. Deep RL is a machine learning technique that has gathered significant interest in the last years. The main reason for this success is that Deep RL can be used to solve many problems that cannot be solved with other machine learning techniques, such as supervised learning.

Deep RL has already demonstrated remarkable performance in the fields of unmanned aerial vehicles [2], road traffic navigation [3], autonomous vehicles [4], and robotics [5]. However, what really served as the inspiration for this work is the ability

O. Prentkovskis et al. (Eds.): TRANSBALTICA 2021, LNITI, pp. 490–498, 2022.
https://doi.org/10.1007/978-3-030-94774-3_48

of systems based on deep RL to beat professional human players in games distinguished by long-term planning horizons, information asymmetry, partial observability, incomplete information, and high dimensionality of possible state and action spaces. The remarkable success of AlphaZero in mastering chess, shogi and Go through self-play [6] can face criticism and be depreciated by pointing out the relative simplicity of rules and determinism. However, it appeared that deep RL is capable of mastering video games that partially capture and reflect the complexity of the real world. Notable examples include Dota 2 [7], StarCraft 2 [8], and Minecraft [9]. Given these facts, a prominent research question arises: "Can the principles behind deep RL be applied to control problems in the context of supply chain management?".

In the past years, supply chain practitioners were mainly concerned about operational components within the supply chain, including production, inventory, and transportation. However, the focus has the tendency to shift toward the management of information flows. One of the most critical drivers of this tendency is growing system complexity. Nowadays, supply chains incorporate many interconnected business processes performed by independent economic agents, extremely complex in their own way. Adaptive control in such systems requires robust handling of these complex interactions. This objective can be achieved by transforming the supply chain into a synchronized system akin to gears and cogs in a mechanical clock [10].

Moreover, synchronized supply chains can avoid a cascading rise-and-fall inventory dynamic widely known as the "bullwhip effect" [11] and mitigate ripple effects caused by operational failures [12]. Indeed, a deep RL agent can perform adaptive coordination along the whole supply chain only if end-to-end informational transparency is provided. Strategic and operational disruptions during the pandemic can become a kick-starter for necessary changes in supply chain visibility and global coordination [13]. This paper aims to demonstrate how deep RL can synchronize demand and supply and support business continuity in the face of uncertainty if end-to-end visibility is provided.

2 Related Work

First and foremost, it is worth highlighting that the numerical experiment presented further in this paper is based on the recent paper that provides a detailed technical description of both simulation environment and deep RL agent [10].

Additionally, it is important to emphasize [14], which demonstrated the efficiency of the deep RL approach based on Q-learning. The method was applied to dynamic inventory control in a multi-echelon supply chain model. The supply chain simulation is distinguished by nonstationary Poisson demand. Barat et al. approached a closed-loop demand-replenishment synchronization problem by applying the Advantage Actor-Critic algorithm [15]. A deep RL agent built upon the Q-learning algorithm was capable of mitigating and partially preventing the "bullwhip effect" was designed in the recent paper [16]. Besides, it is especial to highlight the OR-Gym [17], an open-source library that contains various operations research (OR) problems in the form of RL environments, including the inventory management environment used in this study.

3 Methodology

3.1 Supply Chain Environment

The supply chain environment (SCE) was developed by Hubbs et al. [17] based on the seminal work [18]. The setting is based on a single-product multi-stage supply chain model. Several critical assumptions are made. Economic agents incorporated into the supply chains can be represented as economic entities (or stages) involved in production and distribution $M = \{0, 1, \ldots m_{end}\}$. Entity 0 stand for a retailer that fulfills arising customer's demand. Entity m_{end} stands for a raw material supplier. Entities from 1 to $m_{end} - 1$ represent middlemen involved in the product lifecycle, for example, wholesalers and manufacturers (see Fig. 1). One unit from the previous entity is transformed into one unit at the following entity until the final product is obtained. Replenishment lead times between entities are constant, measured in days, and include both manufacturing and transportation times. Both production and inventory capacities are limited for all the entities. Raw material supply makes an exception and is assumed to be infinite. During the simulation at each time step $t \in T$, the following sequence of events take place:

1. All the entities except the raw material pool place orders downstream the supply chain.
2. Replenishment orders are satisfied according to production capacities and inventory on hand.
3. Daily demand is satisfied according to the inventory on hand at entities 0 (retailer).
4. Unfulfilled demand is lost (backlogging is not allowed).
5. Holding costs are charged for each unit of excess inventory [10].

The following equations prescribe the dynamics of the SCE:

$$I_{t+1}^{m} = I_{t}^{m} + Q_{t-Lm}^{m} - \zeta_{t}^{m}, \quad \forall m \in M, \forall t \in T, \tag{1}$$

$$V_{t+1}^{m} = V_{t}^{m} - Q_{t-Lm}^{m} + Q_{t}^{m}, \quad \forall m \in M, \forall t \in T, \tag{2}$$

$$Q_{t}^{m} = \min\left(c^{m+1}, I_{t}^{m+1}, \hat{Q}_{t}^{m}\right), \quad \forall m \in M \backslash \{m_{end}\}, \forall t \in T, \tag{3}$$

$$\zeta_{t}^{m} = \begin{cases} Q_{t}^{m-1}, \; if\; m > 0 \\ \min\left(I_{t}^{0} + Q_{t-Lm}^{0}, D_{t}\right), \; if\; m = 0 \end{cases}, \quad \forall m \in M, \forall t \in T, \tag{4}$$

$$U_{t}^{m} = \hat{Q}_{t}^{m-1} - S_{t}^{m}, \quad \forall m \in M, \forall t \in T, \tag{5}$$

$$P_{t}^{m} = p^{m}\zeta_{t}^{m} - r^{m}Q_{t}^{m} - k^{m}U_{t}^{m} - h^{m}I_{t+1}^{m}, \quad \forall m \in M, \forall t \in T, \tag{6}$$

where at the beginning of each period $t \in T$ for each entity $m \in M$, I stands for the inventory level. V stands for the commodities that are ordered and on the way (pipeline inventory). Q corresponds to the accepted reorder quantity and $\widehat{Q}$ to the requested reorder quantity. L stands for replenishment lead time between entities. Demand D is a discrete random variable under Poisson distribution. The sales ζ at each period equal to the customer demand fulfilled by the retailer at entity 0 and the accepted reorder quantities from the succeeding entities from 1 to m_{end}. U denotes unfulfilled demand and unfulfilled reorder requests. The profit P equals sales revenue minus procurement costs, the penalty for unfulfilled demand, and inventory holding costs. p, r, k, and h stand for the unit sales price, unit procurement cost, unit penalty for unsatisfied demand, and unit inventory holding costs. If production capacities and available inventory are sufficient (do not exceed capacity constraints c), Q would be equal to $\widehat{Q}$. However, if capacities are insufficient, it imposes an upper bound on the reorder quantity that can be accepted. Since it is assumed that the inventory of raw materials at entity m_{end} is infinite, the term $I_t^{m_{end}}$ can be ignored. The recent paper describes this dynamic in more detail [10].

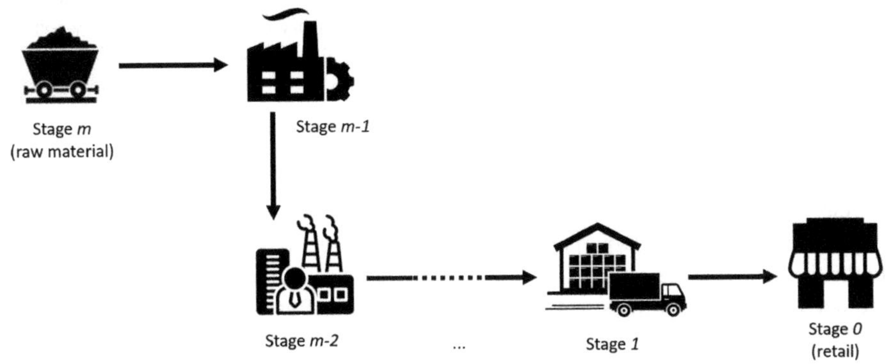

Fig. 1. Multi-echelon supply chain with m entities (stages).

3.2 Reinforcement Learning

SCE can be naturally formulated as the Markov Decision Process (MDP). The MDP is a formalized framework for goal-directed through interaction with a stochastic simulated environment. An RL or deep RL agent interacts with an environment over multiple time steps (see Fig. 2).

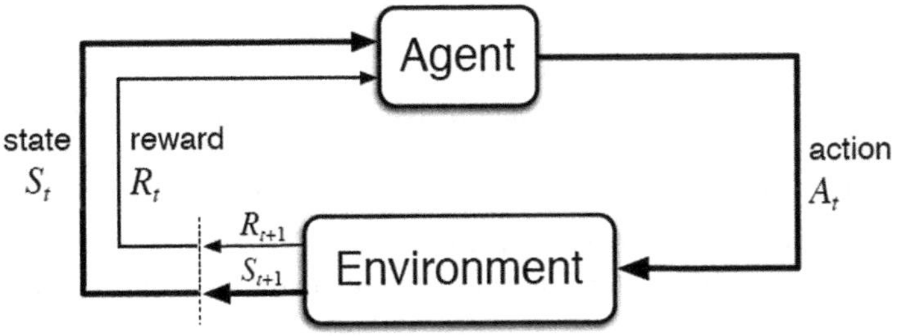

Fig. 2. Agent-environment relations in a Markov decision process [19].

The RL agent and simulated environment interact at each instance of a sequence of discrete-time steps $t \in T$. At time step t, the agent receives a representation of the environmental state $S_t \in S$ and numerical reward $R_t \in R$ in response to made action $a_t \in A$. After that, the agent moves to a new state S_{t+1}. The sequence of states, actions, and rewards can be considered as a trajectory S_0, a_0, R_1, S_1, a_1, R_2, S_2, a_2, R_3, and so on. At each time step, the agent performs several (or multiple) trials. The agent observes both S_t and R_t performing actions in accordance with policy π. In such settings, the agent is assigned a straightforward goal to learn a policy that maps states to actions with the highest cumulative reward over a time horizon [20].

In the context of SCE, the agent must decide how many goods to order for each entity m at each time step t. The action a_t^m is a value that corresponds to order size at each entity across the supply chain. The state S_t is a vector that includes the inventory levels for each entity and the previously taken actions for each of the *max(L)* time steps. Therefore, the RL agent attempts to synchronize the supply chain in such a way that the total profit over the time horizon is maximized (Eq. 6).

3.3 Proximal Policy Optimization Algorithm

Proximal Policy Optimization (PPO) is used to train the RL agent. PPO is originally developed by the OpenAI team and characterized by generality, straightforward implementation, simple hyperparameters tuning, and relatively low sample complexity. In this context, sample complexity refers to the minimal size of the obtained dataset that the deep RL agent requires to learn a target function. Besides, an extra advantage of PPO is the capability to produce decent results with default parameters or minimal hyperparameter tuning [21].

PPO is an actor-critic approach to deep RL. It uses two deep artificial neural networks, *actor* and *critic*. The actor generates actions at each time step. The critic, on its turn, predicts the corresponding rewards. The difference between the critic's predictions and the actual reward received from the environment is reflected in the loss function $L(\theta)$, where θ stands for the vector of hyperparameters (Eq. 7). This loss function constitutes the pivot distinguishing feature of PPO and demonstrated stable learning across various benchmarks.

$$L(\theta) = \min\left[\frac{\pi_{k-1}}{\pi_k}, clip\left(\frac{\pi_{k-1}}{\pi_k}, 1-\varepsilon, 1+\varepsilon\right)\right]\hat{A}_t, \tag{7}$$

where π_{k-1} and π_k stand for the previous policy and the new policy. k is the number of updates to the policy since the moment of initialization. The function *clip(.)* imposes the constraint of the form $1 - \varepsilon \leq \pi_{k-1}/\pi_k \leq 1 + \varepsilon$. ε is a tunable hyperparameter that ensures gradual policy updates. The advantage estimation of the state is the sum of the discounted prediction errors over T time steps $\widehat{A}_t = \sum \gamma^{T-t+1}\delta_t$, where δ_t is the prediction error of the critic artificial neural network, and γ stands for the discount rate.

4 Numerical Experiment

4.1 Implementation and Configurations

SCE is implemented using OR-gym, an open-source collection of problems related to logistics and operations research formulated as RL environments. A numeric experiment is conducted with a 5-stage supply chain, and the simulation lasts for 30 periods (modeling days). Table 1 contains the parameters of SCE used in the experiment.

Table 1. Parameters values for numeric experiment with SCE.

Parameter	Entity 0 (retail)	Entity 1 (wholesale)	Entity 2 (semi-finished)	Entity 3 (finished)	Entity 4 (raw)
Average demand (D)	20	–	–	–	–
Initial inventory (I_0)	50	50	50	50	∞
Unit sales price (p)	2.5	2.0	1.5	1.0	0.75
Unit replenishment cost (r)	1.5	1.0	0.75	0.5	0.5
Unit lost-sales fee (k)	0.1	0.075	0.05	0.025	0.025
Unit holding cost (h)	0.15	0.1	0.05	0.05	–
Maximum production capacity (c)	–	120	120	120	120
Replenishment lead time (L)	2	3	3	5	–

PPO is implemented in parallel using Ray, the framework capable of performing actor-based computations controlled by a distributed scheduler and single execution engine [22]. In the numerical experiment, a multilayered perceptron with two hidden layers and 256 neurons at each layer is used for both actor and critic artificial neural networks. Exponential linear unit (ELU) is used as an activation function, ε equals to 0.3, and the learning rate is 10^{-5}. Experiments are conducted in a reproducible manner using Google Colaboratory. Therefore, results can be reproduced and verified by anyone [23].

4.2 Results

In 50,000 training episodes, the PPO agent could derive a policy that leads to an average reward of 377.69 abstract monetary units with a standard deviation of 21.2. The learning procedure is demonstrated in Fig. 3.

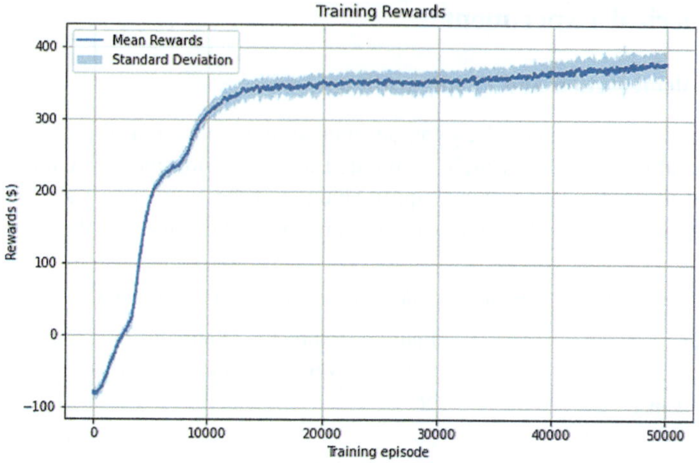

Fig. 3. Learning curve. The performance of the PPO agent increases during training episodes.

Figure 4 illustrates the inventory dynamics at all the entities under the control of the trained PPO agent. During the set of experiments, unusual behavior was observed. The PPO agent terminates ordering additional inventory closer to the end of the time horizon. The potential explanation for this phenomenon is the fact that the simulation time is limited by 30 days, and the learning agent predicts that there will not be enough time to sell all the products, and holding costs associated with excessive inventory will become an unprofitable burden. Indeed, such a problem can be tackled by modifying the SCE, for example, by adding a "cool down" period to the initial time horizon. The drive of the RL agent to exploit vulnerabilities and imperfections of virtual environments is unfortunately common and especially well-known in competitive cybersport. Therefore, this phenomenon deserves specific attention and primary considerations in real-world applications.

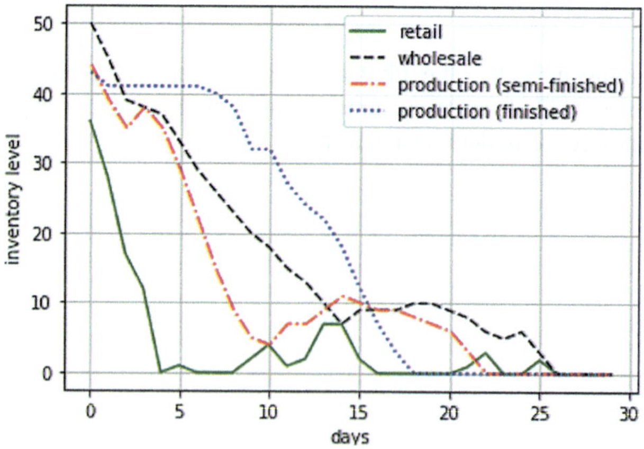

Fig. 4. Inventory dynamics under the PPO agent's control.

5 Conclusions

PPO algorithm can perform adaptive control of complex systems and has potential in real-world application in supply chain management and logistics. The discovered benefits include generality, universality (the algorithm is task unspecific), low sample complexity, and simple hyperparameter tuning. In addition, its already implemented in parallel within the contemporary frameworks. It is worth keeping in mind the RL agents entirely rely on virtual environments. Thus, it is essential to avoid oversimplification in simulation design. Ideally, the virtual environment must capture all the pivotal relations between entities of a real-world supply chain or any other business-driven system. The tendency of RL agents to exploit vulnerabilities and imperfections of simulation mechanics deserve primary consideration in development, implementation, and deployment.

References

1. Agrawal, T.: Hyperparameter Optimization in Machine Learning. Apress, California (2021)
2. Ng, A., et al.: Autonomous inverted helicopter flight via reinforcement learning. Exp. Rob. **9**, 363–372 (2006)
3. Fridman, L., Terwilliger, J., Jenik, B.: Deeptraffic: Crowdsourced Hyperparameter Tuning of Deep Reinforcement Learning Systems for Multi-agent Dense Traffic Navigation. arXiv preprint, arXiv:1801.02805 (2018)
4. Isele, D., Rahimi, R., Cosgun, A., Subramanian, K., Fujimura, K.: Navigating occluded intersections with autonomous vehicles using deep reinforcement learning. In: Proceedings of the 2018 IEEE International Conference on Robotics and Automation (ICRA), pp. 2034–2039 (2018)
5. Gu, S., Holly, E., Lillicrap, T., Levine, S.: Deep reinforcement learning for robotic manipulation with asynchronous off-policy updates. In: Proceedings of the IEEE International Conference on Robotics and Automation (ICRA 2017), pp. 3389–3396 (2017)
6. Silver, D., et al.: A general reinforcement learning algorithm that masters chess, shogi, and go through self-play. Science **362**, 1140–1144 (2018)
7. Berner, C., et al.: Dota 2 with Large Scale Deep Reinforcement Learning. arXiv preprint, arXiv:1912.06680 (2019)
8. Vinyals, O., et al.: Grandmaster level in StarCraft II using multi-agent reinforcement learning. Nature **575**, 350–354 (2019)
9. Guss, W.H., et al.: The MineRL Competition on Sample Efficient Reinforcement Learning using Human Priors. arXiv preprint, arXiv:1904.10079 (2019)
10. Kegenbekov, Z., Jackson, I.: Adaptive supply chain: demand-supply synchronization using deep reinforcement learning. Algorithms **14**(8), 240 (2021)
11. Lee, H.L., Padmanabhan, V., Whang, S.: Information distortion in a supply chain: the bullwhip effect. Manage. Sci. **43**(4), 546–558 (1997)
12. Li, Y., Chen, K., Collignon, S., Ivanov, D.: Ripple effect in the supply chain network: forward and backward disruption propagation, network health and firm vulnerability. Eur. J. Oper. Res. **291**(3), 1117–1131 (2021)
13. Ivanov, D., Dolgui, A.: A digital supply chain twin for managing the disruption risks and resilience in the era of Industry 4.0. Prod. Planning Control **32**, 1–14 (2020)

14. Mortazavi, A., Arshadi, K.A., Azimi, P.: Designing of an intelligent self-adaptive model for supply chain ordering management system. Eng. Appl. Artif. Intell. **37**, 207–220 (2015)
15. Barat, S., Khadilkar, H., Meisheri, H., Kulkarni, V., Baniwal, V., Kumar, P., Gajrani, M.: Actor based simulation for closed loop control of supply chain using reinforcement learning. In: Proceedings of the 18th International Conference on Autonomous Agents and Multi Agent Systems, pp. 1802–1804 (2019)
16. Oroojlooyjadid, A., Nazari, M., Snyder, L.V., Takac, M.: A deep Q-network for the beer game: deep reinforcement learning for inventory optimization. Manuf. Serv. Oper. Manage. 1–20 (2021)
17. Hubbs, C., Perez, H.D., Sarwar, O., Sahinidis, N.V., Grossmann, I.E., Wassick, J.M.: OR-Gym: A Reinforcement Learning Library for Operations Research Problem. arXiv preprint, arXiv:2008.06319 (2020)
18. Glasserman, P., Tayur, S.: Sensitivity analysis for base-stock levels in multiechelon production-inventory systems. Manage. Sci. **41**(2), 263–281 (1995)
19. Sutton, R.S., Barto, A.G.: Reinforcement Learning: An Introduction. MIT press, Cambridge (2018)
20. Bellman, R.: A Markovian decision process. J. Math. Mech. **6**(5), 679–684 (1957)
21. Schulman, J., Wolski, F., Dhariwal, P., Radford, A., Klimov, O.: Proximal Policy Optimization Algorithms. arXiv preprint, arXiv:1707.06347 (2017)
22. Moritz, P., et al.: Ray: a distributed framework for emerging {AI} applications. In: Proceedings of the 13th {USENIX} Symposium on Operating Systems Design and Implementation ({OSDI} 18), pp. 561–577 (2018)
23. Executable Colab Notebook: Google Colaboratory. https://colab.research.google.com/drive/1iEpfmkGDHOdn5iWZSXSMAUVuNbnY4l5b?usp=sharing. Accessed 02 Oct 2021

Possibilities of Contrailer Transportation Development in Lithuania

Aldona Jarašūnienė and Enrika Tarasevičiūtė(✉)

Faculty of Transport Engineering, Department of Logistics and Transport Management, Vilnius Gediminas Technical University, Plytinės Street 27, 10105 Vilnius, Lithuania
enrika.tarasevicíute@stud.vgtu.lt

Abstract. The article analyses the importance of the development of contrailer transportation, assesses the current situation of such transport in Lithuania, presents a SWOT analysis of contrailer transportation in Lithuania, analyses the technologies of contrailer transportation applied in Europe, calculates the criteria for expert evaluation and determines the importance of the requirements. The study's primary goal is to discover the main problems in developing contrailer transport in Lithuania and provide suggestions. Based on the qualitative study results, the most crucial criterion is the lack of specialized rolling stock. The least essential measure is the reduction of the carrying capacity, of the rolling stock, when transporting contrailers. The identified problem areas show that more detailed studies on technological, technical and organisational elements are needed to apply to haul contrailer trains in Lithuania.

Keywords: Contrailer transport · Intermodal transport · Market analysis · Cargo-Beamer · Mega-Swing · Running Highway · Modalohr · Lo-Lo system

1 Introduction

Relevance of the Study: Despite the global pandemic, freight flows only increase, resulting in heavy road traffic, congestion and correspondingly longer transport times. In the current context, rail transport is an excellent alternative to road transport. Rail transport is not only safer and cheaper but also less harmful to the environment. Unfortunately, door-to-door rail transportation is not feasible, so communication with other transportation providers and innovative information technologies is critical to improving rail operations' efficiency. One of the most efficient ways of transporting goods, in many European countries and America, is contrailers [1]. A contrailer is a two-axle or three-axle semi-trailer with a covered or open body, transported on unique platforms [2]. Based on the best practices of foreign countries (Austria, Switzerland), it can be argued that contrailer transport is efficient. However, this method has not yet been applied in Lithuania, due to several obstacles.

Research Problem: Launching a new service is a complex process because of the lack of research, the need for significant initial capital, and the unpredictability of consumer response to the innovation. As contrailer transport is successfully implemented in many

O. Prentkovskis et al. (Eds.): TRANSBALTICA 2021, LNITI, pp. 499–508, 2022.
https://doi.org/10.1007/978-3-030-94774-3_49

countries, it is essential to analyse the above vehicle, identify problem areas and make suggestions based on research results and foreign best practices. In Lithuania contrailer transport does not take place yet because there are several obstacles. The main problems are the lack of a legal framework and unclear pricing policy. Here is also the problem that we do not yet have a narrow-gauge railway and that, due to a lack of suitable rolling stock and special equipment, it is not possible to convert contrailer transport trains from narrow gauge to broad gauge.

2 The Concept and Importance of Intermodal Transport For the Development of Contrailers Transport

To ensure efficient transport of goods, two or even more modes of transport are often used. The freight itself is not reloaded when the mode of transport is changed [3]. According to Jarzemskis [4], intermodal transport units are trailers, semi-trailers and swap trailers. Intermodal transport offers advantages for all participants in the transport process [5]. Intermodal transport is developing rapidly and successfully almost worldwide. There are obstacles such as intermodal growth which requires dedicated road-rail terminals. Since such large-scale terminals require considerable investment, they are scarce, so the combination of these modes is best suited for long distances. By increasing effectiveness of logistic activity and providing logistic services that are less harmful to the environment, proper infrastructure development can partially contribute to resolving the aforementioned issues [6]. Modern transportation is one of the environmental problems that pollute the environment and harm human health. It is the reason why "green logistics" is increasingly being developed. Intermodality is a way to transport goods at a lower cost in a shorter time and partially solves the environmental problem. In the absence of successful intermodal transport in Lithuania and abroad, it is necessary to create an economic basis to apply the principles of green logistics properly. All companies desire and need to expand their business, make a profit, use their resources and innovate. Road transport is one of the most popular modes of transporting goods and people. Consequently, it is recognised as one of the primary sources of chemical, mechanical and noise pollution in the environment. There are various sources of toxic substances in cars, such as exhaust gases and fuel vapours. The largest share of environmental pollution, from road traffic, is the exhaust gases from internal combustion engines.

Chmelnickaja [7] identifies the main factors affecting the level of pollution: the use of old cars; poor road infrastructure, irrational use of traffic lights, leading to congestion; low public interest in greener transport logistics processes; lack of national regulations to promote the use of green logistics and others.

2.1 Assessment of the Current Situation of Contrailers Freight in Lithuania

In Lithuania is a problem that many carriers prefer road transport, which leaves the advantages of rail unused. The main benefits of rail transport are that it can transport long distances and heavy loads, it does not pollute the environment and it is the only mode of transportation that can carry almost all types of goods. Jarzemskis argues that rail is indispensable in certain situations. One of the most efficient modes of transport for combining road and rail transport is contrailer transport. A contrailer transport trailer is a two-axle or three-axle semitrailer with a covered or open body, transported on unique platforms [2]. Trials with the transport of contrailers have already been conducted in Lithuania. The first test was carried out with the project launch of the Viking container train, which has been operating successfully since then. At that time, an attempt was made to transport contrailers from Odessa to Draugystes station. Unfortunately, the attempt was unsuccessful, as the transport itself required a significant investment and was much more expensive than road transport. Although some time has passed since the first attempt, the situation has not changed much. Experts say that this is due to the low road taxes in Lithuania. Moreover, particular rolling stock is needed to organise the transport of contrailers, which requires high investments. It is difficult to predict how quickly this method will make financial sense in Lithuania.

2.2 SWOT Analysis of Contrailers Transport in Lithuania

A SWOT analysis has been carried out based on scientific sources and an analysis of Lithuania's attempts to transport contrailer vessels (Table 1).

Table 1. SWOT analysis of contrailer shipping in Lithuania [8–11].

Strengths	Weaknesses
• Not affected by climate conditions • No congestion; • Drivers work in shifts, so there is no stoppage; • Long-distance transport possible; • Environmentally friendly transport	• Unclear tariff policy; • High levels of investment • Carriers' lack of confidence in the new service; • Lack of terminals and other infrastructure elements; • Lack of specialised rolling stock; • Low utilisation of rolling stock carrying capacity
Opportunities	**Threats**
• Rail and road transport interactions; • Rail and terminal development; • Using global experience to develop the rolling stock-terminal system; • Traffic safety; • Pollution reduction; • Ease of transport; • Relatively low investment for terminals; • Lower cost of transport services; • Timetabled transport; • Shorter customs procedures	• Arranging for the transport of the driver of the tractor unit (if there is a need for carriage together); • Lack of terminals and other infrastructure elements; • Interest from hauliers; • Lack of specialised rolling stock; • Low utilisation of rolling stock carrying capacity

The SWOT analysis shows that the introduction of contrailers to ship in Lithuania requires both technical, technological and organisational elements. On the technical side, it is necessary to upgrade terminals to receive contrailers for transhipment properly and modify/procure particular rolling stock to transport semi-trailers. It would ensure a wide range of freight forwarding services and facilitate the daily routine of freight transportation. According to Shapkin semi-trailer terminals, unlike freight stations can have their track and be connected to contrailer switches [12].

In summary, the potential for contrailerised transport in Lithuania is relatively high. Therefore, it is necessary to consider the potential threats and existing weaknesses and seek to establish this mode of transportation as one of the most environmentally friendly and safest modes of transport shortly.

2.3 Contrailer Transport Technology is Used in Europe

In many European countries, not only are trailers successfully transported by contrailers but specific technologies have been developed.

Table 2. Contrailer transport technologies in Europe.

Country	Title	Special features	Benefits	Disadvantages
Germany	Cargo-Beamer	Specialised platforms and rail terminals	Use of standard size wheels, versatility, high loading/unloading capacity	High cost of platforms and terminals, operational difficulties, the precise location of semi-trailer unloading
Switzer-land	Mega-Swing	Specialised platforms and multi-purpose terminals	Use of standard size wheels, easy to operate, no fixed unloading point, versatile, high loading/unloading capacity	High cost of platforms
Austria	Running Highway	Specialised platforms and multi-purpose terminals	Relatively inexpensive, easy to operate, no fixed unloading point	Not versatile, requires a passenger wagon for drivers, small wheels, long loading/unloading times
France	Modalohr	Specialised platforms and rail terminals	Use of standard size wheels, high loading/unloading capacity	High cost of platforms and terminals, operational difficulties to unload semi-trailers precisely
ES	Lo-Lo system	Universal platforms and universal terminals	Relatively inexpensive, easy to operate, no fixed unloading point, versatile	Unescorted transport, special equipment required for loading/unloading

Source: compiled by the authors based on [10]

The main reason for the development of this transport technology is the reduction of production costs, in the handling of transit cargo flows and, at the same time, the possibility of improving the safety of goods. Germany is a leader in contrailer transport in Europe for obvious reasons such as its size, developed economy, access to the main ports in Scandinavia and Baltic Seas with access to Scandinavia, the Baltic States and Russia. These sea routes are also used to transport a large amount of cargo, the system Ro-Ro, which is ideal for onward transport by rail [3]. Since conventional platforms with a width of 1435 mm are not suitable for transporting semi-trailers, designing appropriate platforms for this purpose was necessary[13]. Different technical solutions have been applied in other countries (Table 2).

3 Methodology

The experts evaluate the main problem factors affecting the development of contrailer traffic in Lithuania. Each expert evaluates each criterion, in the order of importance during the survey. Balezentis [14] argues that expert evaluations require specific expertise and experience. In the empirical part, two tests are conducted: the coefficient of knowledge is determined and the significance coefficient is evaluated.

An expert is a person who has a relatively high level of expertise and reliable information that helps to solve specific problems [15]. The first step of the study is selecting experts, according to the following selection criteria: education, position held, professional experience, and all respondents' involvement in a project related to the development of contrailer shipping.

The expert survey provides for a structured questionnaire, assessing the importance of each of the problematic factors in the development of contrailer shipping in Lithuania, according to the given criteria: unclear tariff policy; project requiring significant investments; lack of terminals and other infrastructure elements; lack of specialised rolling stock; carriers' lack of confidence in the new service.

This study aims to substantiate the scientific literature and assess the main problems encountered in the development of contrailer transport in Lithuania. After analysing the evaluations obtained by the experts, a range is constructed and the answers are expressed as coefficients. The respondents were briefed before the assessment of the research being carried out and the purposes for which the information would be used. The confidentiality of the information obtained is also ensured. The rating scale is based on a review of the scientific literature, and the responses are processed in Microsoft Excel.

3.1 Expertise Quotient

After the experts have evaluated the criteria, a competence factor is calculated. In all quantitative studies, it is necessary to remove outliers because outliers distort the final result. The first step of the survey is to determine if there are any experts whose answers differ from the majority. The following formulas are used:

$$K_i^0 = \frac{1}{m}, j = 1, \ldots, m, \tag{1}$$

where m = number of experts.

The following steps are used to calculate the competence factor:

$$X_j^t = \sum_{i=1}^{m} K_i^{t-1} \cdot x_{ij}, j = 1, \ldots, n, \tag{2}$$

$$\lambda^t = \sum_{j=1}^{n} \sum_{i=1}^{m} x_j^t \cdot x_{ij}, \tag{3}$$

where λ – lambda.

$$K_i^t = \frac{1}{\lambda^t} \cdot \sum_{j=1}^{n} x_j^t \cdot x_{ij}, \sum_{i=1}^{m} K_i^t = 1, \tag{4}$$

$$\overline{K}_i^t - 1.96s \leq K_i^t \leq \overline{K}_i^t + 1.96s. \tag{5}$$

If an investigation reveals that there are differences between experts, then we must remove those differences. Only after the outliers have been removed, can further research be targeted.

3.2 Significance of the Criteria

The experts surveyed are asked to rate the importance of the criteria on a scale of 1 to 100. The survey reveals which problem criterion has the most significant impact on contrailer traffic in Lithuania. To calculate the importance of the criteria, it is first necessary to obtain the assessment of each expert on the requirements. In the second step, the evaluations of the requirements (in sum) are added. In the third step, the importance of the criteria is calculated according to the formula:

$$\text{Significance of criteria} = \frac{Total}{800}. \tag{6}$$

These steps reveal which criterion is the most and least significant.

4 Empirical Part

4.1 Analysis of the Expert Competence Quotient

Expertise coefficients are coefficients that measure whether there are outliers among experts, i.e., whether there are experts whose answers differ from the majority. For all experts to be competent, the coefficient of expert competence must fall within the range [0.0935; 0.1565] (Table 3).

From the coefficients in the table, it can be assumed that all the selected experts are competent, since the coefficient of competence for all of them is in the range [0.0935; 0.1565]. Therefore, all further tests are performed without any changes.

Table 3. Expertise quotient.

Expert	Coefficient
1	0.1373
2	0.0938
3	0.1322
4	0.1454
5	0.1276
6	0.1251
7	0.1277
8	0.1108

4.2 Analysis of the Significance of the Criteria

Based on the calculations, it became clear which of the problematic criteria is the most significant and the least influential, in developing contrailer transport in Lithuania (Table 4).

Table 4. Calculation of the significance of the selected criteria.

	Experts								Total	Significance of criteria
	1	2	3	4	5	6	7	8		
1. Unclear tariff policy	0	20	10	5	10	15	17	20	97	0.12
2. High-investment project	20	20	20	20	20	25	20	25	170	0.21
3. Lack of terminals and other infrastructure	35	20	20	20	30	20	20	20	185	0.23
4. Lack of specialised rolling stock	35	10	40	50	30	30	35	20	250	0.31
5. Low utilisation of rolling stock carrying capacity	0	10	5	0	5	0	3	10	33	0.04
6. Carriers' lack of confidence in the new service	10	20	5	5	5	10	5	5	65	0.08

The calculation, of the significance, of the criteria shows how the experts scored each of the requirements. Table 5 indicates the ranking of the criteria in order of importance:

Table 5. Determining the significance of the criteria.

Criteria	Rank in order of importance
Unclear tariff policy	4
High-investment project	3
Lack of terminals and other infrastructure	2
Lack of specialised rolling stock	1
Low utilisation of rolling stock carrying capacity	6
Lack of confidence of hauliers in the new service	5

Overall, the lack of specialized rolling stock is a significant obstacle to contrailer transport in Lithuania. The lack of terminals and other infrastructure elements, in Lithuania, is equally essential. According to the experts, the least important criteria is that the carrying capacity of rolling stock is reduced when semi-trailers are transported on railway platforms.

5 The Contrailers Development Model

The country's well-developed infrastructure allows it to make economical use of the transport available. Transport is how the country's trade, both internationally and domestically, is conducted. The development of contrailer vehicles in Lithuania would increase transit traffic and imports/exports, thus, generating significant added value for the national economy. A model for the development of contrailer shipping is presented below (Fig. 1). The model presented for the development of contrailer transport includes both external and internal factors. The model identifies the following key factors: society, public transport policy, specialized rolling stock and terminals, and other infrastructure elements. These factors generate the added value of contrailer transport to the economic growth of the country.

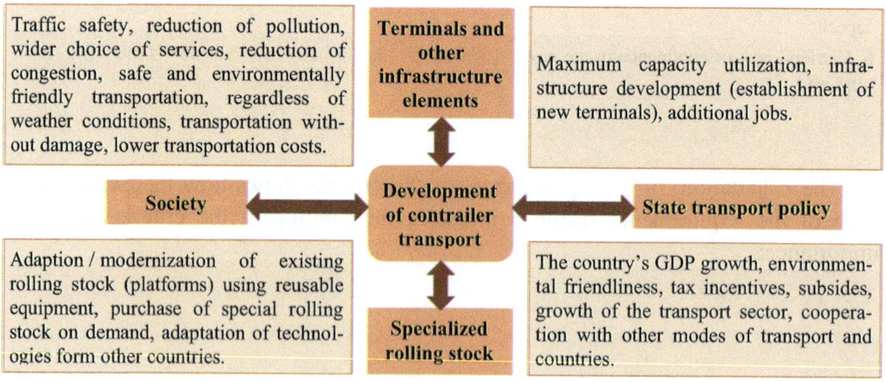

Fig. 1. The contrailers development model.

Based on best practices in other countries, investments in rolling stock and new terminals pay off, create new jobs, make transport more accessible, cheaper, and reduce negative environmental impacts. This increases a country's mobility and creates the opportunity to establish new relations with other countries. The development of contrailer transport increases the range of services that are an excellent alternative to road transport, making transport both efficient and environmentally friendly, all while improving the country's GDP.

6 Conclusion

1. The concept of intermodality has been examined and it can be argued that intermodal transport is an excellent opportunity to combine several modes of transport, thus ensuring safer, cheaper and more efficient transport.
2. The scientific analysis of the literature has shown that contrailerised transport in Lithuania would offer several opportunities, such as simplification of transport, reduction of pollution, use of global experience in the formation of a rolling stock-terminal system and many others. However, to implement this mode of transport in Lithuania, it is important to take into account the potential threats and weaknesses.
3. After calculating the competence coefficient of the experts, it was found that all the experts were competent, and therefore further research was carried out without any changes. After a substantial explanation of the criteria analysis, there are shortcomings of specialized rolling stock in the transportation of the main duration development trails in Lithuania. The shortcomings of terminals and other infrastructure elements are no less significant.
4. The model identifies the following key factors: society, public transport policy, specialised rolling stock and terminals and other infrastructure elements. These factors generate the added value of contrailerised transport for the country's economic growth. Based on best practices in other countries, investments in rolling stock and new terminals not only pay for themselves, but also create new jobs, make transport simpler, have a lower cost and reduce negative environmental impacts.
5. The development of contrailer ships increase the mobility of the country and offers the opportunity to establish new relations with other countries. The development of contrailerised transport increases the range of services, which is an excellent alternative to road transport.

References

1. Vasilis Vasiliauskas, A., Kabashkin, I.: Analysis of indicators measuring performance of rail-road terminals. In: Proceedings of 10th International Conference. Transport Means, pp. 93–96 (2006)
2. Kuznetsova, A.I., Postovalov, A.I.: Vyyavlenie potentsiala energosberezheniya v promyshlennykh otraslyakh, v sfere zhkkh i vnutrigorodskoy transportnoy sisteme. [In the case of energy-saving energy products, in the case of energy-saving and energy-saving products]. Transportnoe delo Rossii. [Transport business in Russia] **5**, 171–173 (2014) In Russian

3. Lur'ye, V.: Pochemu buksuet kontreyler? [Why is the contrailer skidding?]. Journal TIR (TransInfoRoad) 42–45. In Russian (2014)
4. Jaržemskis, A., Jaržemskis, V.: Krovininis transportas: vadovėlis. [Freight transport: textbook]. Vilnius: Technika. In Lithuanian (2014)
5. Kvaraciejūtė, J.: Transporto politika Lietuvai įstojus į Europos Sąjungą. Magistro baigiamasis darbas. [Transport policy after Lithuania's accession to the European Union. Master's thesis]. Vilnius. (2011). C:/Users/13780/Downloads/1892205.pdf
6. Vasilis Vasiliauskas, A., Vasilienė-Vasiliauskienė, V., Čižiūnienė, K.: Solutions toward faster development of green warehouses in Lithuania. In: Transformation of International Economic Relations: Modern Challenges, Risks, Opportunities and Prospects, pp. 230–239 (2017)
7. Chmelnickaja, Z.B.: Use of technologies of "green" logistics in greening of motor transport (2017)
8. Hlopov, K.: Piggyback traffic foreign experience and development directions in Russia. Rossijskij vneshnejekonomicheskij vestnik [Russian export reporter] **9**, 101–108 (2011)
9. Kuz'min, D.: Organizatsiya regional'noy seti kontreylernykh terminalov: dis. kand. tekhn. nauk. Moskva: MIIT. [Organization of a regional network of piggyback terminals: dis. Cand. tech. Sciences]. In Russian (2015)
10. Fedorina, A.V., Cyganov, A.V., Pikalov, V.A.: Problemy razvitija kontrejlernyh perevozok v Rossii [Piggyback traffic development problems in Russia]. Sb. nauch. tr. Studentov [Proceeding] Molodjozh'. Nauka. Budushhee [Youth. Science. Future] 14, (2014)
11. Sładkowski, A., Pamuła, T.: Intelligent transportation systems - problems and perspectives, vol. 32. Springer, Studies in Systems Decision and Control, Switzerland (2016)
12. Shapkin, A.S.: Vybor tekhniko-tekhnologicheskikh parametrov sistemy kontreylernykh perevozok na zheleznodorozhnykh napravleniyakh seti: dis. … kand. tekhn. nauk/. [The choice of technical and technological parameters of the piggyback transportation system on the railway lines of the network: dis. ... Cand. tech. Sciences]. Moskva: MIIT, 154 c. In Russian (2005)
13. Kontseptsiya organizatsii kontreylernykh perevozok na "prostranstve 1520". [The concept of organizing piggyback transportation in the 1520 area]. Moskva: OAO RZhD (2011)
14. Baležentis, A., Žalimitaitė, M.: Ekspermentinių vertinimų taikymas inovacijų plėtros veiksnių analizėje. Lietuvos inovatyvių įmonių vertinimas. [Management theory and studies for rural business and infrastructure development]. Vilnius: Mykolo Riomerio universitetas. http://mts.asu.lt/mtsrbid/article/viewFile/269/298. In Lithuanian (2011)
15. Tidinis, R.: Socialinių mokslų tyrimų metodologija. [Social science research methodology]. Lithuanian University of Law, Vilnius (2003)

Application of Quality Criteria in the Development of Partial Load Transportation

Aldona Jarašūnienė and Evija Savickė(✉)

Faculty of Transport Engineering, Department of Logistics and Transport Management, Vilnius Gediminas Technical University (VILNIUS TECH), Plytinės Street 27, 10105 Vilnius, Lithuania
evija.savicke@stud.vgtu.lt

Abstract. The article examines the importance of the quality of partial load transportation by road transport under contemporary transport market conditions. Quality criteria and specifics of partial load transportation are discussed. It is thus emphasized that companies should prioritize adequate quality management, consequantly it is neccessary to consider such aspects as organization of service processes, as well as the perception of services from the client point of view. Based on the analysis of scientific literature, the problem areas of transport service quality and partial load quality criteria are highlighted. Methods to solve particular problems are presented.

Keywords: Quality management · Partial cargo · Cargo consolidation · Quality criteria

1 Introduction

Relevance of the Research: The ever-increasing competitiveness among transport companies encourages to focus more on service quality. Quality becomes one of the major factors determining client choice. Not only does improvement of the quality satisfy the client, but also it increases client base and company's profit. Companies engaged in partial load transportation by international routes encounter complex organizational issues pertaining to transportation, thus a little less attention is being paid to the improvement of services. Clients who choose this mode of transportation often expect not only the organization alone, but also a wide range of services necessary to transport their partial load. There is a constant demand to observe client expectations and adjust quality managent methods in accordance with quality criteria as clients have a direct impact on transportation services and different perceptions of quality service exist.

Research Problem: Realizing the value of quality, most companies try to obtain the ISO 9001 quality certification. However, this seems to be the only effort to improve their service quality. Quality is often measured in terms of the number of orders and profits earned. It is very difficult to assess the quality of partial load transportation.

O. Prentkovskis et al. (Eds.): TRANSBALTICA 2021, LNITI, pp. 509–519, 2022.
https://doi.org/10.1007/978-3-030-94774-3_50

Companies that deploy advanced technologies in their operations and have an understanding on how to determine quality assessment criteria, can manage processes associated with quality improvement more easily and in turn may encourage their business partners to contribute to overall quality improvement.

2 Theoretical Aspects of Partial Loads

2.1 The Concept of Partial Loads and Quality Management

Partial loads are cargoes of different sizes and weights that are loaded into a single vehicle to minimize transportation costs. Cargoes of various volumes that are loaded from different consignors and delivered to logistics terminals in which the selection and distribution processes take place depending on final location can be ascribed to this category. Clausen et al. [1], Dondo [2] claim that transportation of partial loads is the best method of transportation to process the highest number of orders where cargoes from consignor to the consignee are transported by the means of logistics centers. The partial load must meet the following requirements: the weight of the cargo should be less than a maximum permissible weight of the vehicle; the volume of one unit of the vehicle must be greater than the volume of the cargo. Fineeva [3] claims that the service of transporting partial loads becomes an important part of the businesses as it satisfies demands of diverse companies aiming to optimize transportation costs whilst delivering cargoes of small volumes. Such method of transportation is not only safe and convenient, but also economically beneficial to companies. Clients do not need to pay for unused volume of the vehicle and transport companies do no longer need to carry out transportation by half empty vehicles. The structure of partial loads encompasses load lines established by freight forwarders. The shipment process not only involves consignor and the consignee. This process is rather connected by local, regional, and international logistics centers that are in the countries of consignor and the consignee. The activities of such centers are monitored by freight forwarders controlling the transportation of partial loads.

Zvereva [5] identifies the peculiarities of transportation quality assessment:

- Transportation service has no quantitative evaluation criteria due to its intangible nature.
- Transportation service is provided during its production.
- Perception of transportation service from the client point of view.
- Direct customer influence on the provision of transportation service.
- The impact of personal contact and environment.
- Non-existent feedback from service users.
- No unified assessment system for transportation services.

Oruch [6] argues that similarly to other services, the ones that are provided by transport companies do not have a unified assessment system due to their diversity, however, it is possible to determine quality indicators of transportation services considering specific spectrum of services provided.

Daihovskij et al. [7] state that quality indicators of partial loads are as follows: minimal delivery times, adherence to departure schedules and arrival of cargo on time, ensuring the safety of cargo during transportation process, ensuring the transportation of larger cargo without dividing it into smaller shipments.

After analysing the demands of clients in transport sector, Fashiev et al. [8] identify the following quality criteria:

- Service flexibility;
- Ability to aid with the preparation of necessary documentation;
- Reliable transportation;
- Fast order fulfilment;
- Ability to provide additional services;
- Ensuring cargo safety up to delivery location;
- Cargo delivery at the recipient door;
- Possibility to repack the cargo;
- Having the necessary loading equipment at terminals;
- Provision of the required price;
- Possibility to offer the services of customs brokers.

2.2 Problem Aspects of Partial Load Quality Assessment

Oruch [6] states that assessing the quality of service is much harder than the quality of goods as it pertains to major service characteristics.

Researchers distinguish the following aspects among other specific characteristics of the services:

- *Intangibility*;
- *Instability*;
- *Dependence on the Service Consumer*.

Artistov [9] highlight major factors influencing the quality of transportation services:

- Delivery delays;
- Low level of cargo safety during transportation;
- Shortcomings of integrated service system;
- The need for cooperation with different transport companies;
- Employee reluctance to work with cargo demanding non-standard solutions;
- Cargo tracking delays;
- Problems pertaining to customs formalities;
- Poor English language skills;
- No employee training system;
- Discrepancy between transportation quality and price.

Among the aforementioned quality-related problems, Sprogyte et al. [10] add up such factors as inability to present the required documentation on time and diverse unforeseen risks that transport companies may encounter during transportation process or prior to organizing their service provision.

2.3 Quality Management of Partial Load Transportation

Quality, as one of the most important indicators of competitive advantage, does not happen suddenly, thus it requires constant control. Owing to quality management, companies not only implement their business goals successfully, but also retain their employees and long-term business partners. All employees participate in quality management process – from managers to regular employees. Managers not only develop quality management methods, but also constantly revise its efficiency, improve efficiency, and provide their employees with necessary tools to contribute to quality management. As the quality in partial load business directly depend on honest work on behalf of all members in the process, transport companies must introduce them with quality requirements and involve their partners into their quality management system. Quality management system not only helps to determine the source of client dissatisfaction, but also to eliminate them so that they do not recur. Constant monitoring of the provided service, installing modern measures into transportation service development, employee training and information sharing are vital to manage quality properly and correspond to customer expectations. Quality management is not only significant to company providing transportation service, but also to clients, partners, employees, and managers.

Selivestrov et al. [11] claim that quality management is necessary for successful activities in each company as service improvement and development directly depend on it. Quality management is more than monitoring service quality. These are operative actions performed by a company aspiring to make its service attractive to clients.

Kupriyanova [12] and Selivestrov et al. [11] identifies the following functions of quality management:

- Quality planning;
- Organization of operations pertaining to quality;
- Employee motivation to perform active work whilst improving service quality;
- Quality control and assessment;
- Establishment and organization of diverse and effective quality methods.

Quality management process is highly influenced by quality control, which helps to identify any deviations from quality requirements and standards. Kupriyanova [12] states that quality management encompass determination of client demands, assessment of the provided quality, establishment, realization, and organization of different methods with an aim to develop service quality.

3 Criteria Affecting Company's Clients and Business Partners Research

A great deal of attention is given to analyzing the concept and value of quality itself. A close cooperation with partners form different logistics terminals in various countries is an integral part in carrying out the transportation of partial loads by road transport on international routes. A company that is concerned with developing its quality, finds it important to make sure that its clients quality perception is in line with theirs.

Transshipment of partial loads takes place not only in logistics terminals of national significance, but also in terminals of different countries. In order to define the quality criteria of diverse logistics companies, ten managers of transport companies who communicate with major clients and partners at least twice per quarter, were interviewed. To maintain the accuracy and reliability of expert assessment, the expert group is recommended to be comprised of at least five experts. Under certain conditions the number of experts may well reach several dozen people, and the minimal recommended group size is three experts. Most researchers believe that an optimal group number is from eight to ten experts.

The expert assessment is based on the premise that a decision can only be reached after assessing the consistency of the expert overall opinion. Considering quality criteria that are imposed by clients and quality criteria aimed at business partners, it is necessary to assess the consistency of the expert overall opinion by calculating Kendall concordance coefficients W [13]. Expert assessments are considered contradictory when $W \to 0$, and $W \to 1$ if these are similar. The consistency of expert opinion is calculated in accordance with the fourth and fifth question provided for expert companies. The following is an example of calculation of the concordance coefficient.

Experts were given a questionnaire with a request to answer general and more specific questions, with an aim to assess quality criteria that affect their clients perception of quality and also to assess their own quality criteria. Expert responses were listed in Table 1 and evaluations of each expert eight criteria were calculated.

Table 1. Criteria assessment into rank model.

Expert no.	Criterion encryption model							
	K1	K2	K3	K4	K5	K6	K7	K8
E1	4	6	2	1	7	3	5	8
E2	6	5	2	1	7	3	5	8
E3	4	5	2	3	8	1	6	7
E4	5	7	1	2	8	3	4	6
E5	5	6	3	1	8	2	4	7
E6	7	5	1	2	8	3	4	6
E7	4	6	2	3	8	1	5	7
E8	3	7	4	6	7	4	7	8
E9	8	4	6	7	8	4	8	6
E10	1	8	3	4	6	1	6	8
$\sum_{i=1}^{n} B_{ij} = B_j$	47	59	26	30	75	25	53	71

Each expert response was turned into ranks. First expert E1 K1 criteria assessment − 4 was turned into rank – 5. This was done with all criteria assessments, as shown in Table 2.

By the means of the formulas, the sum of the ranks and averages were calculated. Also, a correlation between estimates and ranks was established. Relevance coefficient sum is equal to the sum of rank average – 33.4. The difference of constant measure and rank sum for each criterion, as well as the square of this difference is calculated. As there are no associated ranks, Kendall concordance coefficient is calculated: W = 0.6729.

Kendall concordance coefficient is higher than 0.5, thus, a conclusion can be drawn that expert opinions are consistent.

As the number of criteria in assessing quality is higher than 7, i.e., $m > 7$, according to the formula, the weight of concordance coefficient is calculated on the basis of χ^2(chi-kvadrat) Pearson criterion.

$\chi^2 = 47.1$ or more than χ^2kr, which is equal to 20.092, thus the assessment of the respondents is considered to be consistent.

To calculate the smallest value of concordance coefficient $Wmin = 0.2870 < < 0.6729$, it is possible to conclude that expert opinions are consistent.

Table 2. Criterion opinion consistent model.

Expert no.	Criterion encryption model							
E1	5	3	7	8	2	6	4	1
E2	3	4	7	8	2	6	5	1
E3	5	4	7	6	1	8	3	2
E4	4	2	8	7	1	6	5	3
E5	4	3	6	8	1	7	5	2
E6	2	4	8	7	1	6	5	3
E7	5	3	7	6	1	8	4	2
E8	6	2	5	3	2	5	2	1
E9	1	5	3	2	1	5	1	3
E10	8	1	6	5	3	8	3	1
$\sum_{n=1}^{n} R_{ij} = R_{ij}$	43	31	64	60	15	65	37	19
$\overline{R}_j = \frac{\sum_{i=1}^{n} R_{ij}}{n}$	4.3	3.1	6.4	6	1.5	6.5	3.7	1.9
$\sum_{i=1}^{n} R_{ij} - \frac{1}{2}n(m+1)$	−2	−14	19	15	−30	20	−8	−26
$\left[\sum_{i=1}^{n} R_{ij} - \frac{1}{2}n(m+1)\right]^2$	4	196	361	225	900	400	64	676

After completing the calculations, it is possible to claim, that expert opinions are consistent in regard to eight criteria affecting client perceptions on quality, thus their opinions are generalized.

Further on, criteria importance indicators are calculated (Table 3).

Table 3. Criteria importance indicators.

Indicator mark	Criteria encryption symbol								Control number
	K1	K2	K3	K4	K5	K6	K7	K8	
$\bar{q} = \frac{\bar{R}_j}{\sum_{j=1}^{m} R_j}$	0.1287	0.0928	0.1916	0.1796	0.0449	0.1946	0.1108	0.0569	1
$d_j = 1 - \bar{q}_j = 1 - \frac{\bar{R}_j}{\sum_{j=1}^{m} R_j}$	0.8713	0.9072	0.8084	0.8204	0.9551	0.8054	0.8892	0.9431	7
$Q_j = \frac{d_j}{\sum_{j=1}^{m} d_j} = \frac{d_j}{m-1}$	0.1245	0.1296	0.1155	0.1172	0.1364	0.1151	0.1270	0.1347	1
$\overline{Q}_j = \frac{\sum_{i=1}^{n} B_{ij}}{\sum_{i=1}^{n} \sum_{i=1}^{m} B_{ij}}$	0.1407	0.1766	0.0778	0.0898	0.2246	0.0749	0.1587	0.2126	1
Factors by importance	5	3	7	6	1	8	4	2	

Q_j indicator enables to determine not only how one criterion is more important than the other, but also how many times does one criterion is more important than the other. Criteria distribution is presented in a diagram (Fig. 1).

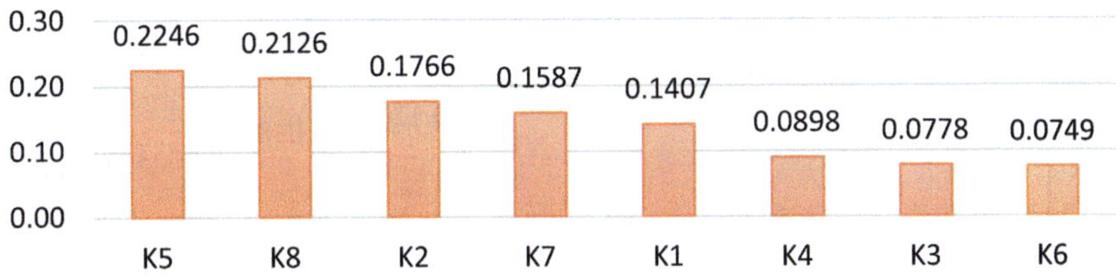

Fig. 1. Criteria affecting company's clients quality perception.

The most important criteria affecting company's clients quality perception are as follows: K5 – ensuring cargo safety, K8 – prompt information sharing, K2 – ability to make decisions and provide suggestions, K7 – order execution within a specified timeframe, K1 – service flexibility, K4 – ability to transport cargo "at the door", K3 – Possibility to offer the services of customs brokers, K6 – possibility to provide additional services.

Ten experts were given criteria that affect the quality assessment and perception of company's clients who use the service of partial load transportation on international routes. The most important factors were K5 and K8 or ensuring cargo safety and prompt information sharing, less important ones – K3 and K6 or possibility to offer the services of customs brokers and possibility to provide additional services, such as cargo insurance or labelling.

Kendall concordance coefficient calculation methodology was applied in the analysis of quality criteria aimed at business partners and the result obtained: $W = 0.792$.

This indicator shows that opinions of all respondents are consistent (Fig. 2).

In accordance with the importance, it is possible to distinguish the following criteria aimed at company's partners working with partial loads: I – application of advanced technologies, F – employee professionalism, B – prompt servicing of the new order placed, A – working time flexibility, G – all warehousing and handling services for cargo processing, D – work performed on time, E – providing solutions to problems, C – fast transportation to international logistics terminal, H – ability to aid with the preparation of necessary documentation.

After expert assessment of these criteria on companies that frequently use their logistics partners services that help them organize transportation of partial loads on international routes, it was concluded, that the most important criteria are I and F, or application of advanced technologies and employee professionalism, and the less important are C and H or fast transportation to international logistics terminal and ability to aid with the preparation of necessary documentation.

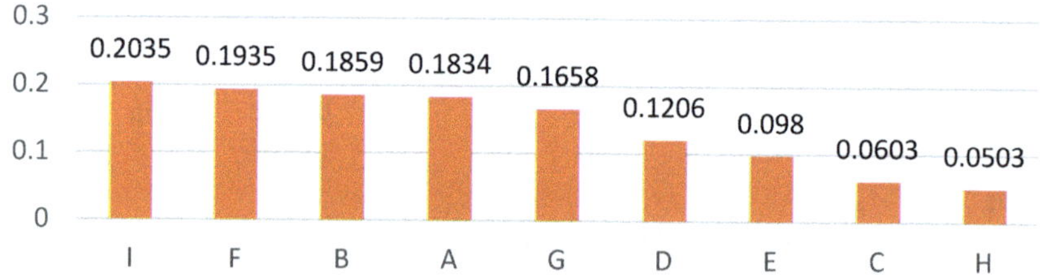

Fig. 2. Quality criteria aimed at business partners.

4 Suggestions to Solving Problems

After completing the expert assessment, it was clear that opinions of company's managers are consistent, as shown by quality indicators applicable to client companies. Company's clients do value such quality criteria as safe transportation and prompt information sharing. Additionally, transport companies do consider such factors as the ability to use advanced technologies and employee professionalism as important (Fig. 3).

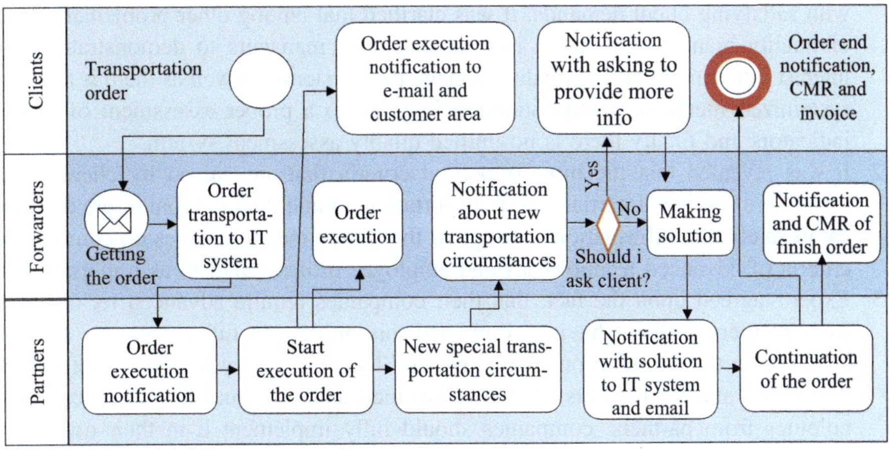

Fig. 3. Advanced system application in transport company (made by authors).

One of the most important criteria that is valued by clients is a prompt information sharing. This criterion can be satisfied by several methods – to provide information by standard measures – through company's employees which makes a whole process to slow or to make use of the advanced technologies. Often demanding advanced technologies from their partners as one of the most important quality criteria, companies need to start implementing those technologies within their own companies. By constant investment and employee encouragement to use advanced technologies, companies will be able to clearly express their requirements to business partners whose advanced technologies may contribute to establishing the best quality service provision to clients. Accurate information sharing by the best possible way in the shortest time is only possible when advanced technologies will operate in agreement with the ones in hired companies, as well as with partners and clients. Below is an example of the advanced system application showing just how this system can meet one of the most important quality criteria.

By implementing systems with business partners, companies may obtain information on cargo and inform their clients on their status, as well provide suggestions to solve arising problems at a certain stage of the order execution. Relevant information thus may be transmitted in real time, which in turn helps the client to observe the order execution progress and, if necessary, to adjust order requirements. Such system will not only enable business partners to be in the highest demand, but also will help to solve the problem of lengthy payment process. Witnessing just how quickly all orders are processed, transport companies will be prepared to make timely payment for ordered services.

5 Conclusions

1. The scientific literature analysis has shown that one of the most important indicators to increase company's competitiveness is the level of service quality, which is established by adequate quality management and service quality is directly linked

with satisfying client demands. It was clarified that among other problematic areas of quality management, there is a tendency for managers to demonstrate a low interest into functioning of quality management systems, as well as there is a lack of systemized methods, a little attention is given to a proper assessment of quality indicators and finally there is no unified quality assessment system.

2. It was revealed that the most important criteria that are valued by client whilst selecting a service of partial load transportation on international routes, are ensuring cargo safety and information sharing. At the same time, companies do consider the criteria of advanced technologies and employee professionalism as important.
3. Experts agreed upon the fact, that their companies require advanced technologies from partners, without having completed this process to full extent. An accurate information transmission on a real time and location is very important not only to companies and their clients, thus it is assumed that by demanding advanced technologies from partners, companies should fully implement it in their own environment to better address the quality criteria and improve service quality in the whole logistics process.
4. To analyse the quality criteria aimed at business partners, the calculation methodology of Kendall concordance coefficient was applied and the following result was obtained: $W = 0.792$ i.e., expert opinions are consistent.
5. To analyse the quality criteria by clients to transport companies, the calculation methodology of Kendall concordance coefficient was applied, and the following result was obtained: $W = 0.6729$ i.e., expert opinions are consistent.
6. By implementing advanced technologies, which involve company's clients and partners, transport companies not also will be able to remain competitive, but also provide a quality service. Clients will always receive relevant information on time and make transactions for services faster.

References

1. Clausen, U., Goedicke, I., Mest, L., Wohlgemuth, S.: Combining simulation and optimization to improve LTL traffic. Procedia. Soc. Behav. Sci. **48**, 1993–2002 (2012)
2. Dondo, R.: Cargo consolidation and distribution through a ter-minals-network. A branch-and-price approach. In: Conference: 44th Argentine Conference on Informatics (44 JAIIO), At: Rosario, September 2015 (2015)
3. Финеева, М.: Организация перевозки сборных грузов в РФ, в системе компании ООО «Русталогистик». [Fineeva, M.]. (2016). https://cyberleninka.ru/article/n/organizatsiya-perevozki-sbornyh-gruzov-v-rf-v-sisteme-kompanii-ooo-rustalogistik/viewer
4. Brabander, C., Braun, M.: Bringing economies of integration into the costing of groupage freight. J. Revenue Pricing Manage. **19**, 366–386 (2020)
5. Зверева, А.: Особенности Оценки Качества Транспортных Услуг. [Zvereva, A. 2013]. Актуальные проблемы авиации и космонавтики, 177–178 (2013)
6. Oruch, T.: Research of service quality customer services: problems and methodology. Mod. Res. Soc. Probl. **9**(29), 757 (2013)
7. Дайховский, С.В., Кивал, Н.Г.: Организация перевозки сборных грузов. [Daihovskij, S. V.; Kival, N.G. Organization or partian cargoes]. Вологдинские чтения 64–65 (2007)

8. Фасхиев, Х., Целищев, В.: Оценка Качества Транспортных Услуг. [Fashiev, Ch.; Celishev, V.]. Менеджмент Качества **03**(47), 226–235 (2019)
9. Аристов, А.: Современные проблемы повышения качества грузовых автомобильных перевозок в Российской Федерации. [Artistov, A. 2012]. Вестник Ленинградского государственного униеститета им. А. С. Пушкина 89–100 (2012)
10. Sprogytė, I., Zinkevičiūtė, V.: Transporto įmonės kokybės va-dybos sistemos projektavimas, įvertinant vartotojų poreikius. Verslas XXI amžiuje (2014)
11. Селивестров, А., Постнов, В., Уткин, Д., Семидотченко, А., Николаева, К.: Стандартизация системы управления качеством. [Selivestrov, A; Postnov, V.; Semidotchenko, A.; Nikolaeva, K.]. Материалы IV науч.конференции «Экономическая наука и право», проходившей в г. Чита, апрель 2018, pp. 24–26. Чита: ООО «Молодой учёный» (2018)
12. Kupriyanova, L.: Quality management as a factor of sustainable business. World New Econ. **4**, 89–100 (2015)
13. Kendall, M.G.: Rank Correlation Methods, 4th edn. Charles Griffin, London (1975)

Proposal for the Improvement of Security Measures in the Slovak Republic During the Exit and Transfer of Sports Fans from Railway Stations

Michal Szatmári(✉) and Patrik Lahuta

Faculty of Security Engineering, University of Žilina,
Univerzitna 1, 010 26 Žilina, Slovakia
{michal.szatmari,patrik.lahuta}@uniza.sk

Abstract. Organizing sporting events brings with it some of risks. One of the potential risks is the transfer of fans to sporting events. Fans often use railway station to move. It is at the point of their exit at the railway station that a process begins in which the event organizers, local and state security forces, the railway police and other participating subjects begin to be responsible for the security situation in a diversified manner. Football and hockey matches have long been one of the riskiest events in the Slovak Republic. Despite the currently epidemic situation, which restricts the movement of people between countries, districts, but also the overall very limited sometimes restricted participation of spectators in sporting events, it is necessary to explore ways to improve the situation when the society will operate as in previous years. In addition to material losses during the transfer and departure of fans to events or returning to the railway station, there is also a risk of health damage to the health and lives of disinterested people.

Keywords: Transfer fans · Railway stations · Security measures

1 Introduction

In the Slovak Republic, it is necessary to state Act no. 1/2014 Coll. on the organization of public sports events. The law directly defines the conditions of a risky sports event where there are three basic conditions:

- A football or a hockey match between the two highest leagues in the adult category or the last four rounds of the cup in football and hockey in the adult category,
- an event involving 4,000 or more spectators,
- an event in which more than 90% of the occupancy of a sports facility with a capacity of 2,000 or more spectators is expected [1].

To understand this complex problem of moving a fan or group of people with help to train stations, it is necessary to realize that the stadium or hall where the sporting event is to take place does not have to be within walking distance of the train station. In

O. Prentkovskis et al. (Eds.): TRANSBALTICA 2021, LNITI, pp. 520–529, 2022.
https://doi.org/10.1007/978-3-030-94774-3_51

the article, however, in order to specify the actions and measures, the walking distance from the railway station to football meetings will be considered.

2 Justification of the Solved Problem

When organizing large or international sporting events, there are many unstable factors that cause high risks when organizing sporting competitions. In order for a sporting event to take place successfully, it is necessary to anticipate the unstable factors associated with transporting fans to and from the venue in order to reduce the impact and threat of these factors [2].

The seriousness and spread of the undesirable phenomenon of spectator violence, which first began at football events in England and gradually spread to other European countries in the 1970s and 1980s, was the reason for it to be on the Council's premises in 1985 [4]. European Convention (19 August 1985; Strasbourg) concluded a basic framework of organizational and security measures for the organization of sporting events European Convention CETS 120 on Violence and Detention of Spectators at Sporting Events, especially at football matches. The member States of the Council of Europe which have acceded to the Convention are committed to cooperating and involving all actors in the elimination of spectator violence. These entities include the police, sports associations, the state, football clubs, municipalities, travel agencies, carriers, fan clubs. According to stakeholders, the only way to tackle this problem adequately is to tackle it together so that the development and culture of sport can progress [3, 5].

These can be not only other spectators present at the sporting event, but also people outside this place. These are situations where violently behaving fans move after the event, e.g. from the stadium towards the railway station and on the way they commit various riots, while the subject of their attack may also be persons accidentally occurring in the area and their property. During the transfer, they use pyrotechnics, attack other people in front of shopping malls, restaurants, throw bottles from the windows while riding the train, and damage the trains and railway stations themselves [6].

This is why the issue of spectator violence should still be of interest to the general professional public, because it is the frustrated social with violent potential, often intoxicated by alcohol, that poses the greatest threat to sporting, but often also cultural, events [7].

The need to solve the problem is proved, among other things, by a study of the football environment from the UEFA World Cup 2016. These were mainly nicknamed "ultra" supporters. It discusses psychological trauma, violence, dissatisfaction and other factors of the participants. Up to 11 patients who required surgery under general anesthesia were registered, with up to 3 of them participating in the accident not at the stadium but during transport to the stadium [8].

Based on previous experience, a FIFA handbook has been issued with the conditions that individual events must meet if they want to perform under its brand or participate in competitions organized by FIFOU. The stadiums used during FIFA events are divided into five different circuits as shown in Fig. 1, the transfer of fans thus concerns the public zone, which pointed to the fact that the public zone, often from the railway station towards the stadium, is at the same time the largest and riskiest [9].

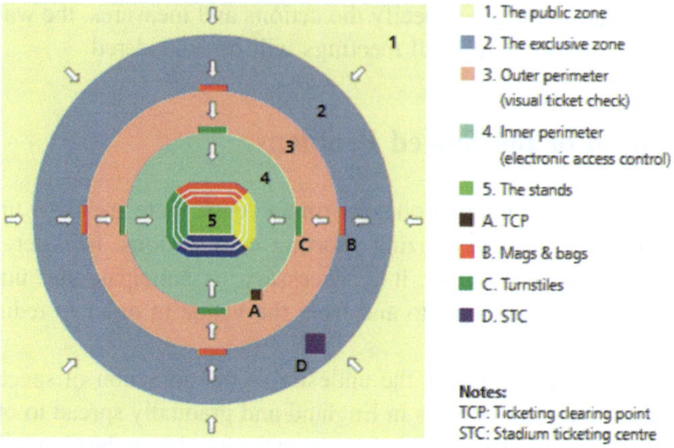

Fig. 1. Zones according to the Fifa manual.

2.1 Approaches of the Police Force in the Transfer of Sports Fans

As far as the police are concerned, in recent years a new approach to solving problems in mass events, such as sports events, has been applied worldwide, namely **the low profile policing** approach. It consists mainly of non-provocative monitoring and a friendly relationship at the fan-police level.

Low profile policing means the presence of the police in public since the arrival of fans at the railway station in such numbers and equipment that its presence does not arouse negative emotions and does not declare the expectation of a clash. While law enforcement units are out of visual contact with the public, but in an appropriate place in terms of the possibility of adequate and prompt intervention [10].

The study involved the police force and academia, which showed a real reduction in conflicts in football stadiums, in the transfer of fans, in the streets towards transport terminals and other places with fans at the UEFA 2004 (Union Europe'enne de Football Association) in Portugal showed the world that it is also possible to use low profile policing for a massive international event [11].

The opposite of low profile policing is **hard profile policing**, this approach evokes from the very name the deployment of equipped law enforcement units with helmets, helmets, heavy equipment and the necessary accessories for the possibility of intervention. Hard profile policing and its advantages/disadvantages for society have been addressed by a number of authors and it is still applied in practice, especially at risky events [12–14].

For a better visual presentation of what the approaches of the low profile policing and hard profile policing police forces look like in practice, we present Fig. 2.

Fig. 2. Hard profile policing vs low profile policing.

2.2 The Main Police Forces in the Slovak Republic in the Transfer of Fans from Railway Stations

Police specialists in spectator violence from the ranks of the criminal police, also called spotters, move among the fans in civilian clothes. They use a low profile monitoring approach and their main task is to monitor risky behavior and risky fans. They accompany risky fans during the transfers from the place of entry into the vehicle to the place of the stadium, but their activity is mainly of a preventive nature. Preventive nature due to the fact that with the help of personal knowledge of the fans, they can not only consult and guide certain activities in advance, but also estimate their behavior and the whole development of the situation. Their further role is to cooperate with other police forces and inform them about the development of the situation not only in the event itself, but also in the preparation of the entire security measure [15].

The strongest emphasis on police spotters is in The prevention of violence in sport. They cannot be individual, and accompanying persons must never act as police informers or "spotters" [16].

2.3 Anti-conflict Teams of the Police Force

The Slovak Republic decided to set up a PZ anti-conflict team in 2014. The main task of the anti-conflict team is to prevent tense and conflicting actions of people participating in various public events, especially sports events, and to move fans. They are mainly used when an offensive exchange of views and attitudes between participants is expected [17].

2.4 Law Enforcement Unit of the Police Force

The action of the law enforcement unit in eliminating the mass disturbance of public order with elements of spectator violence has in the vast majority of cases a repressive form of eliminating the conflict situation. In its operation, it thus directly interferes with the fundamental human rights and freedoms of persons operating within the crowd in the event of a mass violation of public order. These include - the principles of legality,

legality, opportunity, proportionality, subsidiarity, prevention, speed and determination, vigilance and vigilance, instruction, cooperation, cultivated service and professional observance [18]. Order units are characterized and are the most common form in the Slovak Republic of hard profile policing.

2.5 Railway Police of the Police Force

For larger matches in the Slovak Republic, a special train connection is often ordered by a fan club, while the Railway Police has the main task of accompanying these fans throughout the train ride to the entrance to the railway station and it is no less important to check whether only fans really get on the train. The situation is different, but if the fans travel by regular train service. In this case, it seeks to select fans from ordinary passengers most appropriately by moving fans only in certain wagons accompanied by the railway police [19].

The railway police use a hard profile monitoring approach because they are already training in training or crowding out a crowd of fans, securing the perpetrator or selecting problem "leaders" of hooligan groups. If the fan club has not purchased a separate train connection, they must actively intervene in the distribution of wagons and the placement of fans in the safest possible variant from uninterested persons traveling on the same connection. The Railway Police is the first and at the same time the most important component of the Police Force of the Slovak Republic, which comes into contact with fans moving by rail and from the place of exit of the railway station to the place of the sports meeting. Nevertheless, during the interventions against aggressive fans, they also cooperate with law enforcement units when getting off at the railway station.

The solution to the number of offenses and the corrective actions of individual components is shown in Fig. 3.

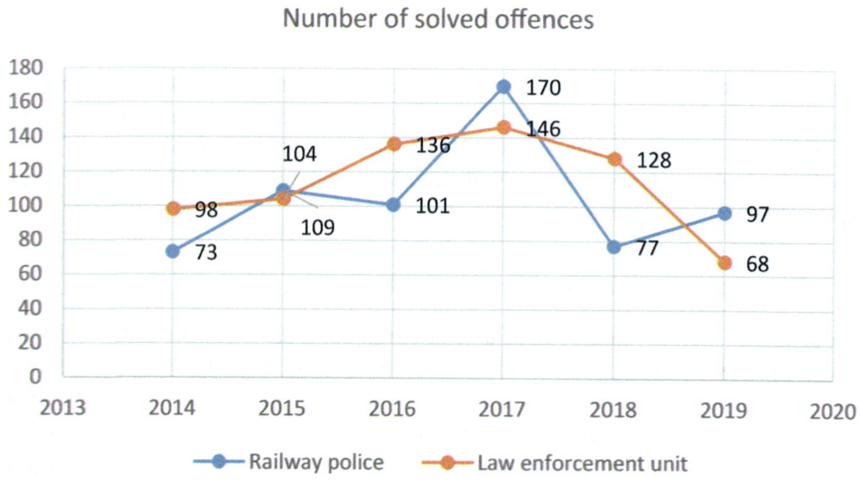

Fig. 3. Number of solved offenses of the railway police and the riot police for the period 2014–2019.

Through the acquired practical and theoretical experience through professional training, they should prevent the emergence of aggressive manifestations at demonstrations or sporting events. There are police officers from various fields on this team. In Slovakia, an anti-conflict team is used mainly when moving fans to the venue from the railway station, because it is the most used way of transporting the main fan club, whether in hockey or football matches [18]. The anti-conflict team uses low profile policing as well as spotters.

3 Identified Problem Areas

After communication with five organizers of sports events, a member of the state police, a member of the railway police, consultations at the Faculty of Safety Engineering in Zilina, the following main causes of stagnation in the transfer of sports fans from the railway station and the railway were chosen as possible reasons for not improving the situation and organizing safety measures:

1. Sports fans have not become accustomed to running anti-conflict teams on the way from railway stations to the venue of sports events even after 5 years, and conflicts have not stopped declining significantly, but also have not escalated significantly against anti-social or illegal activities, so their effectiveness is difficult to statistically prove.
2. Insufficient communication between individual police forces and event organizers as well as with official fan bases (information on the possible location of clashes of hostile basic fans, information on the transfer from the railway station to sports events, confirmed number of fans in member bases, etc.).
3. Weak technical and organizational measures of units of the police force, railway police, equipment of railway stations, retraining of units to manage stressful situations, etc.
4. Weak entry control of the railway police against sports fans on pre-booked fan trains and the impossibility to perform a similar control if they enter classic civilian railway sets and exit at railway stations under the influence of alcohol and other substances.
5. Excessive use of hard profile monitoring, which motivates or even provokes possible perpetrators of anti-social and illegal activities.

4 Proposals for Measures to Improve the Current Situation

Based on interviews and communication with individual entites of the police forces and common identified problems, we made a logical flowchart with a methodological procedure and with a brief description of the main problems.

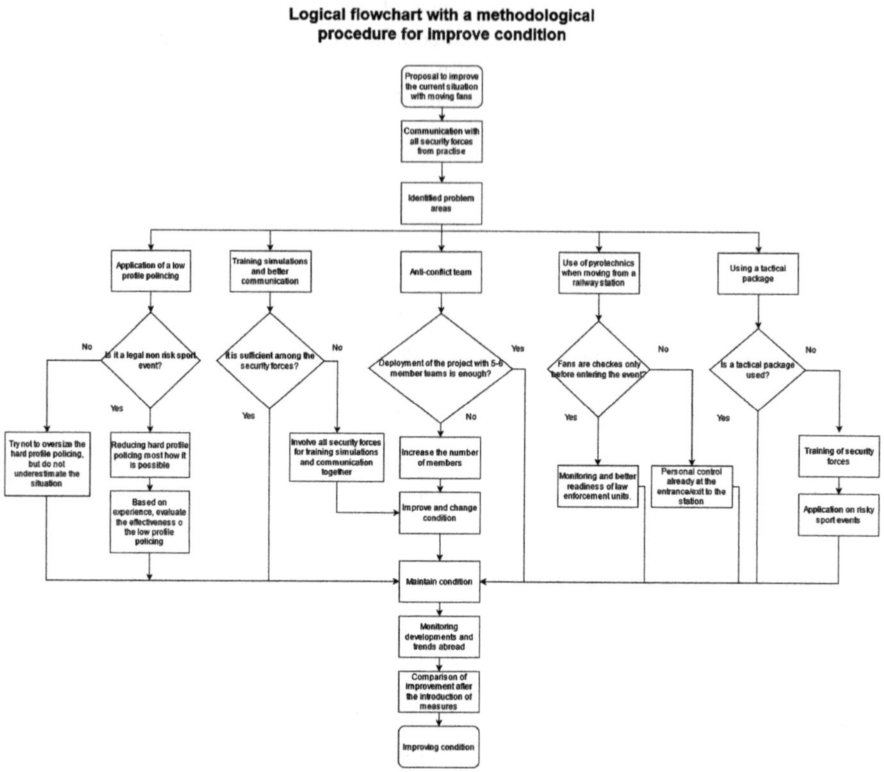

1. Anti-conflict teams in the Slovak Republic mostly work in small groups of 5 to 6 members. Such a small group does not have the opportunity to prevent a large crowd and the escalating negative phenomenon of riots and physical attacks by fans on innocent people and property located between the hall/stadium and the train station. For this reason, and on the basis of consultations, we think that the number of members and the overall deployment should be decided not only by the commander of security measures on the basis of a proposal from the coordinator of the Regional Directorate of the Police Force. Mainly to the organizers of the sporting event and other entities involved in monitoring and supervising the peaceful and calm course of the event directly at the sporting event and moving to and from the place of the railway station. Since even six years after the launch of the project of anti-conflict teams in the Slovak Republic, it is not possible to consider the approach of deploying anti-conflict teams, which they use, i.e. hard policing for improving development. It would certainly be possible to test the deployment of low profile policing in the form of spotters or small groups of police officers in several domestic risky sports, as in other countries abroad, which are not identifiable as clearly different from fans by visible phosphating vests.

2. Insufficient communication between entities that work together to maintain a smooth transition of fans from railway stations and to railway stations could be improved by simulated exercises, such as riot control units, but the problem identified is that in training simulations other units such as the railway police and anti-conflict teams are not involved, because mainly these three units of the Police Force represent state forces and a safe situation for peaceful fans and disinterested people in the transfer of fans. Similarly, organizers of sporting events and representatives of official fan bases are not involved in these simulated exercises.
3. Weak and insufficient organizational and technical equipment of railway stations means an insufficient camera system, incomplete marking or separation of the route of arrival and departure of sports fans from disinterested civilians. Weak measures in the form of the hard profile policing approach is certainly important, especially at events of fan bases where in the past there have been significant clashes of fans when moving from railway stations. This factor was also realized by the Police Corps of the Slovak Republic, which began to use organizational security measures in the form of the use of tactical separation by barriers. During the analysis and communication with the Police Force, the author Geršicová [20] found that it is still not used at all or at least applies the commonly used method of transferring fans from railway stations from abroad in the form of a tactical package (see Fig. 4). The tactical package consists of a hard profile policing approach and the principle is that it is a closed transfer, no one can enter or exit it. Members of the intervention group, the railway police, the riot unit walk around the perimeter of the package and, if necessary, disperse the crowd heading for the package. This method is, however, necessary to simulate and practice what is not done at regular intervals, and many new police officers have not had similar training although they are involved in the package as needed.

Fig. 4. Application of the security measures - barriers and the tactical package [18].

4. In the event that fans travel to a sports meeting with regular traffic, their personal control is impossible and will take place before entering the stadium. This condition is just dangerous when ascending the railway station and moves to the stadium when pyrotechnics can be used. By communicating with the practice, we came to the conclusion that the prevention of similar acts is possible only by an increased

safety situation directly at the railway stations or in a thorough monitoring of the route of the march to the stadium and prepared emergency units. The railway police do not have the legislative authority to carry out personal checks on ordinary passengers in conventional trains and the passenger does not seriously endanger other passengers.

5. In the Slovak Republic, the approach to sports fans in the form of hard profile monitoring is firmly entrenched. Abroad, however, several studies and examples have shown that the apparent deployment of a large number of police officers and visible robust security measures can lead to the fact that even initially non-conflicting sports fans may feel offended or restricted by the crowd psychosis. In the event of incidents during transfers from railway stations or directly at sports venues, official fan bases are also expressed in a very negative and fully condemning manner on a regular basis as sports clubs. A similar trend, especially on the part of the fan base, essentially greatly eliminated the previous trend that killing or destroying property when moving to a stadium is a degree of heroism and superiority and has pushed similar people to the brink. Also for this reason, after consultations and literature studies, we think that it would be appropriate to apply a low profile monitoring approach to those legally non-risk events and slowly increase the culture of sports fans more and more.

5 Conclusion

The main goal of the article was to suggest improvements in security measures in the Slovak Republic during the ascent and transfer of sports fans from railway stations. The authors declared this goal on the basis of communication with practice for the real factual situation in the Slovak Republic and two internationally recognized approaches of police forces in the form of hard profile monitoring and low profile monitoring. The main problems identified were identified in consultation with practice and measures were proposed that could be applied to improve the safety situation when moving fans as well as the railway station itself or in its vicinity. In conclusion, it is necessary to state the fact that there is no methodology, law, decree, standard or any regulation in the Slovak Republic, which would stipulate the minimum security of railway stations against any anthropogenic and deliberate threat, which is also presented in the article sports fans. It is also a problem for academia to determine whether railway stations should be considered as the protection of soft targets or rather as critical infrastructure.

References

1. ACT of 4 December 2013 on the organization of public sports events and on the amendment of certain laws. https://www.slov-lex.sk/. Accessed 01 Jan 2016
2. Zhang, H.: Risk early warning safety model for sports events based on back propagation neural network machine learning. Saf. Sci. **118**, 332–336 (2019)
3. Leitner, B., Môcová, L., Hromada, M.: A new approach to identification of critical elements in railway infrastructure. Procedia Eng. **187**, 143–149 (2017)

4. ETS No.120 European Convention on Spectator Violence and Misbehaviour at Sports Events and in particular at Football Matches
5. Ballay, M., Sventekova, E., Urbancova, Z., Monosi, M.: Application analyses of state of evolution - ETA on selected extraordinary events. In: 13th International Scientific Conference on Sustainable, Modern and Safe Transport (TRANSCOM), pp. 1244–1251 (2019)
6. Romanet, I.: Hooligans and maxillofacial trauma. In: The Surgeon Should Be Warned, Bratislava, p. 17 (2019)
7. Odlerová, M.: The current status and the possibilities of elimination of spectator violence. In: Academy of the Police Chamber in Bratislava, Bratislava, pp. 28–30 (2020)
8. Horváthová, M.: Psychological context of spectator violence. In: The Social Struggle Against Spectator Violence, Bratislava, p. 90 (2014)
9. Federation Internationale de Football Association. http://www.resource.fifa.com. Accessed 20 Dec 2019
10. Dvořák, Z., Leitner, B., Milata, I., Novák, L., Soušek, R.: Theoretical background and software support for creation of railway transport model in crisis situations. In: WMSCI 2010, vol. 3, pp. 343–347. International Institute of Informatics and Systemics (2010)
11. Clifford, S., Adang, O., Livingstone, A., Schreiber, M.: Tackling football hooliganism. Psychol. Public Law **14**, 115–141 (2008)
12. Morgan, S.L., Pally, J.A.: Analysis of recorded crime incidents and arrests in Baltimore City. In: Cities Initiative at Johns Hopkins University, pp. 246–252 (2016)
13. Desmond, M., Parachritos, A., Kirk, D.S.: Police violence and citizen crime reporting in the black community. Am. Sociol. Rev. **81**, 857–876 (2016)
14. Pyrooz, D.C., Decker, S.H., Wolfe, S.E., Shjarback, J.A.: Was there a Ferguson effect on crime rates in large U.S. cities. J. Crim. Justice **46**, 1–8 (2016)
15. Comeron, M.: The prevention of violence in sport Manuel Comeron Hooliganism Prevention City of Liège. Council of Europe Publishing, Strasbourg (2003)
16. Baťková, L.: Spectator violence in the Czech Republic: Statistics and trends. http://ochab.ezin.cz/Oa-B_2015_B/2015_B_01_batkova.pdf. Accessed 04 Mar 2021
17. Kaczor, S.: The role of anti-conflict teams in the process of eliminating mass disturbance of public order, pp. 376–385 (2018)
18. Cigánik, Ľ., Lôffler, B.: Order units. Police College, p. 146 (2001)
19. Durech, P., Sventeková, E., Ballay, M.: The process of rebuilding organizations Interaction of damaged emergencies with regard to the growth of the security population. In: 21st International Scientific Conference on Vehicles, pp. 967–972 (2017)
20. Geršičová, R.: Footbal hooligans in Trnava. https://is.muni.cz/th/kw62c/BP_Gersicova_Romana.pdf. Accessed 05 Mar 2021

Adjustment of Waste Disposal with the Use of Modern Information Technology

Kevin Škerlič, Robert Muha, and Sebastjan Škerlič(✉)

Faculty of Maritime Studies and Transportation, University of Ljubljana,
Pot pomorščakov 4, 6320 Portorož, Slovenia
sebastjan.skerlic@fpp.uni-lj.si

Abstract. Modern information technology solutions enable the optimization of work processes at every operational level within an enterprise. New waste recycling requirements will have a significant impact on the environment both now and in the future. The problem is that the new requirements result in higher costs, which can be reduced through improved waste collection and disposal processes. The article presents an analysis of the state of waste collection in the Municipality of Izola and proposes improvements in the field of waste collection using modern information technologies, based on the findings of the analysis. Research clearly shows that modern technological solutions in logistics can contribute to more efficient waste disposal and sustainable urban development.

Keywords: Waste · Waste disposal · Information technology · RFID · Internet of Things

1 Introduction

Waste is one of the most important environmental, social and economic topics in today's economy. A great quantity of waste is generated as a product of industrialization and the consumer-oriented society. On the one hand, waste represents a business opportunity for certain individuals and, on the other hand, a major environmental issue. Environmental protection and proper waste separation have become very important in today's society. Public awareness about waste management has significantly increased in the last ten years, as users are aware of the problem of excessive amounts of waste and the potential consequences of this waste accumulating directly in the environment. Public utility companies keep users regularly informed about proper waste separation, which has also significantly contributed to the reduction of the number of illegal dumping sites for all types of waste.

Public utility companies prioritize effective logistics, which ensures competitiveness and easier waste collection planning. These companies operate a fleet of specialized waste collection and disposal vehicles. Without an efficient fleet of waste management vehicles, utility companies would not be able to carry out their activities. These companies rely on a concept of logistics consisting of the optimal organization of transport routes to ensure the most efficient collection of waste at the lowest possible cost. The logistics activities of public utility companies include the daily routing of

O. Prentkovskis et al. (Eds.): TRANSBALTICA 2021, LNITI, pp. 530–541, 2022.
https://doi.org/10.1007/978-3-030-94774-3_52

vehicles, identifying optimal transport routes, timing waste collection according to the fill level of waste containers and monitoring fuel consumption.

Modern information and transport technology solutions enable the optimization of work processes at every operational level within an enterprise. New waste recycling requirements will have a significant impact on the environment both now and in the future. New requirements result in higher costs, which can be reduced through the optimizations of waste collection and disposal processes. There are several studies investigating the implementation of efficient waste management systems. Svub et al. [1] highlight the use of RFID in a municipal waste management system. Lee and Wu [2] also developed a waste management system in Hong Kong using RFID technology. Wang et al. [3] conducted a study on the application of IoT (Internet of Things) in the treatment and disposal of household waste. As a new network technology, IoT has significant advantages for its traceability, dynamic characteristics in solving the problems above.

The research problem is therefore focused on proposing a solution for waste disposal in the Municipality of Izola using modern information technologies, which will contribute to the improvement of work processes. The purpose of the research is to demonstrate that modern technological solutions in logistics can contribute to more efficient waste disposal and sustainable urban development.

Following the introductory part, the second section will present modern technological solutions that can effectively aid in optimizing waste disposal. The third section will present the research methods, followed by a presentation of the proposal for improved waste collection in the Municipality of Izola. The last section presents the conclusions of the research.

2 Literature Review: The Use of Modern Information Technology

Many authors who discuss the importance of modern information systems in waste management highlight the advantages of these systems, which enable fast processing of transactions, information management and support the decision-making process. The most promising technologies in terms of providing these advantages are RFID and The Internet of Things (IoT), which as suggested by Wen et al. [4] will operate in an integrated manner in future waste management.

The Internet of Things is a new phenomenon in the field of communication where smart objects communicate with each other and respond to user requests. The IoT provides an integrated framework that ensures interoperability across platforms. One of the most important components of the Internet of Things are wireless sensor networks. Sensor networks play a key role on the lowest level of the Internet of Things [5].

In the Internet of Things (IoT), sensor nodes from various applications collect highly correlated data in their neighborhoods that influence the outcome of any decision in cloud data centres [6, 7]. In these applications, the data is unstructured, intermittent, and somewhat dynamic. Raw data collected by the nodes must be processed locally and analyzed in edge and cloud data centres to optimize the use of available resources. Raw data must be aggregated within the network to reduce its connectivity.

Each node that is unaware of the neighbor's detection range collects data in its own surroundings. The detection range of two or more nodes may overlap, leading to the collection of similar data [8]. Each node must perform local data aggregation to discard multiple copies of the same data. Data aggregation within a network reduces redundancy in the trade-off between data volumes and available resources in edge and cloud data centres [9]. The presence of resource-starved nodes means that the data fusion approach must be simple, robust, and adaptable to the requirements of the application. Data aggregation alone is not enough to optimize the use of available network resources. Field data that is aggregated upward toward cloud data centres should be evenly distributed among edge servers [10].

The advantages of using IoT in waste management are also highlighted by current research. Wang et al. [3] conducted a study on the use of IoT in the treatment and disposal of household waste. The authors outline the current situation and existing problems using a specific location. Wen et al. [4] in address the design, implementation, and evaluation of the Internet of Things (IoT) using sensor-based technology to improve food waste management in restaurants in Suzhou, China. This IoT-based system covers the collection, transport and final disposal of food waste. Tamakloe and Rosca [11] (2020) focus on the use of smart systems and the Internet of Things (IoT) with a view to ensuring an efficient and effective approach to waste management. This project designed and manufactured a prototype of a solar-powered, self-compacting smart bin with a server-side monitoring application. The prototype smart bin is capable of monitoring internal rubbish levels, compacting, and freeing up approximately 25% of the space with each compaction. The bin also monitors total weight and is capable of sending all this information to a secure server-side application. The accompanying web application monitors the state of each smart bin and proposes optimal routes for pick up.

An increasing number of organizations around the world are considering the introduction of Radio Frequency Identification (RFID) systems to improve their business and operational processes. RFID is also considered the next step in supply chain management, as it can increase operational efficiency by exchanging information in real time and tracking goods, and can ensure full visibility in the supply chain. A typical RFID system includes transponders (tags) and interrogators (readers): the tags are attached to objects/persons, and readers communicate with the tags via radio signals in their transmitting areas [12, 13]. RFID holds several advantages over conventional identification methods, such as barcoding, as they include the ability to read tags without a field of vision, the ability to read multiple tags simultaneously, and the ability to store and modify information on an RFID tag. In recent years, companies have started to use this technology in many areas of business management and beyond. Potential RFID applications include retail activities; inventory control and logistics; production and configuration management; and authentication, anti-counterfeiting, and security [14]. Tajima [15] highlights the strategic value of RFID in supply chain management and lists five key advantages: reduced shrinkage, reduced material processing, greater data accuracy, faster exception management, and improved information exchange. Because of these features, RFID technology is a potential successor to barcodes and may be able to fully complement or replace them in the future [16, 17].

The advantages of using RFID in waste management are also highlighted by current research. Svub et al. [1] discuss the application of RFID technology to identify the time and place of containers with municipal waste emptying and the benefits of the system based on this principle. Lee and Wu [2] conducted a pilot test of an RFID-based waste management system in Hong Kong. The results showed that this RFID-based waste management system improved the recycling rate. The contribution of the paper was in providing a novel approach by incorporating RFID technology and mobile app waste management technology and raising public awareness about the importance of waste sorting for recycling and waste minimization. Condemi et al. [18] presented an in-depth economic and technical analysis for the enhancement of e-waste hierarchy applied to the Radio Frequency Identification tags. The results showed that passive RFID tags represent a major candidate for improving the e-waste hierarchy enhancement at every level, demonstrating that it is more convenient for the manufacturer to consider ecologically aware design and promote a take-back system for tags in order to take advantages from the solution proposed for the RFID e-waste hierarchy.

3 Methods

The analysis of waste disposal was performed in a public utility company located in the Municipality of Izola, Slovenia. The compilation method was employed to sum up the observations, views, conclusions and results of individual authors who researched the field of modern information solutions. The method of analysis and synthesis was used to show the current situation in waste collection in the Municipality of Izola. This was also the basis for formulating a proposal for improvements in the optimization of waste disposal using modern information solutions.

4 Results

The town of Izola has seen a growth in the number of inhabitants in recent years, as many new residential neighborhoods and houses are being built. As a result, planning and implementing waste collection is becoming increasingly difficult year after year. The waste collection area has become diverse and, above all, unpredictable. In the past, the most critical points were always located in the same spots, which made them easier to manage. Nowadays, every single container could be a critical point and predicting optimal emptying is almost impossible, which is why new steps need to be made with the aid of modern technology to optimize waste collection and ensure that the town's streets are always clean and tidy, and maintain customer satisfaction.

It is important to keep in mind that the separate waste collection rate will not increase in the coming years without the introduction of new measures and approaches. In order to achieve better results, a qualitative leap in the field of waste collection is warranted. The leap should be systematic and involve the entire Ecology and Waste Management Working Unit. It should include improvements and upgrades to the existing waste collection and disposal infrastructure, the construction of new waste collection and disposal infrastructure, the modernization of collection methods, as well as the organization and transport of waste, in order to ensure a more pleasant living environment in the long run.

The first measure or improvement consists in installing underground ecological islands, which would represent a step forward in waste management, an upgrade of the separate waste collection system and a way to modernize the method of waste collection and transport. Underground ecological islands would replace single waste containers and some smaller ecological islands in the old town centre. As for the town's hinterland, the construction of centralized underground ecological islands would reduce the number of containers and optimize the planned route of waste collection vehicles. The installation of underground ecological islands would also free up public areas, improve the appearance of the town of Izola and increase safety in the city centre. The frequency of waste collection would be reduced, as the ecological island would consist of five different underground reservoirs, which would be up to five times larger than an ordinary container for separate waste collection. The ecological island would consist of underground reservoirs for: mixed municipal waste, volume 5,000 L; mixed packaging, volume 5,000 L; paper and cardboard packaging, volume 5,000 L; biodegradable waste, volume 3,000 L and glass packaging, volume 3,000 L. Figure 1 shows the layout of an underground ecological island located near the fishing pier in Izola.

Fig. 1. Visualization of an underground ecological island.

The construction of ecological islands would result in the centralization of waste disposal, user identification, improved waste collection logistics, and a lower frequency of waste collection, which would also lower costs and improve the appearance of the town of Izola. Due to the larger capacity of the collection point, the introduction of such a system would also reduce traffic, as the number of garbage truck journeys would be significantly lower.

Additionally, a sensor to measure the fill level of each container separately would be installed on the inside of the container or collector in the underground ecological island. Fill-level data of the underground container would then be sent to the company's server, as well as to the garbage truck on a tablet mounted on the dashboard. That would enable the company to empty the containers according to the actual fill-level, instead of relying on a pre-set emptying schedule, thereby facilitating a transition from static to dynamic waste collection. This type of system would result in significant savings in the waste removal process, as the underground reservoir would only be emptied when full.

A reduced number of workers needed to empty the underground ecological island would contribute to additional savings. Emptying this type of collection points requires the use of a special vehicle, which is technologically very advanced and can only be operated by a qualified worker, who also functions as the driver. The vehicle is equipped with all the tools and accessories required to ensure a safe and uninterrupted work process, as it needs to secure the location during the container emptying phase with special barricades to prevent unauthorized persons from accessing the work area. Lateral loading technology is used to lift and empty the underground reservoir. The vehicle is operated via cameras from the cab, which enables the operator to execute the entire container emptying cycle. Due to these changes to the work process, which previously required a team of three workers, a single worker can execute all the tasks.

Another measure used to optimize waste disposal is the upgrade of waste collection and disposal containers with intelligent systems for a more user-friendly management of the entire waste management process. This type of system would be achieved by mounting a container fill level sensor on all waste collection containers, as shown in Fig. 2.

Fig. 2. Waste container equipped with a fill level sensor [19].

The fill level sensor monitors and controls the fill level of the container by means of continuous data transmission via an Internet connection. The sensor informs the control centre of the current level of waste. The control centre is able to monitor the precise fill level of the containers. The head office of the Ecology and Waste Management Working Unit would serve as the control centre. The municipal supervisor in charge of the monitoring and his or her immediate supervisor would have access to the data. In addition to measuring the fill level, real-time container fill level sensors also measure the temperature inside the container, the slope of the container and detect the presence of smoke. Since all the containers would be equipped with sensors, special software could be used for automatic garbage truck route planning, in order to find the optimal route in terms of time and distance driven. All the containers are already equipped with RFDI chips, which show the location and emptying of the container. Combining them with fill level sensors could greatly optimize waste collection and disposal management through the use of modern information technology. We would also be able to cut down on waste disposal costs, as waste would be collected with only two vehicles in the winter, as opposed to the three vehicles currently used. In the summer, however, waste would be collected with three vehicles, as opposed to the four vehicles currently used. Figure 3 shows the real-time traceability of the waste container's fill level utilizing a level sensor.

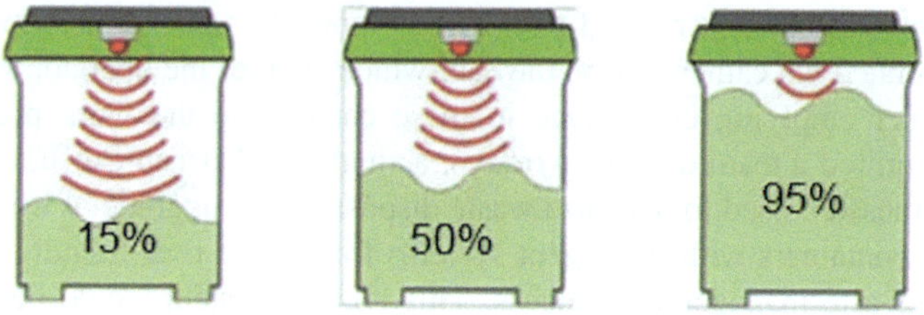

Fig. 3. Real-time tracking of the container fill level utilizing a sensor [20].

In addition to equipping all the containers with waste level sensors and RFDI chips, all garbage vehicles should be equipped with smart tablets. The tablets would be mounted on the garbage vehicle's dashboard as shown in Fig. 4. The tablet would allow the driver to access the daily waste collection plan. Since the container emptying detection system with RFDI chips, the vehicle tracking system and the container fill-level sensors system would be connected, the tablet would allow the driver to see when a container has been emptied and the location of the next container that needs to be emptied. The driver would also be able to monitor all the containers that need to be emptied on a particular day in real time.

Fig. 4. A tablet mounted in the vehicle to monitor all activities and execute the waste collection plan. [21].

The containers that have already been emptied would be shown in green on the screen, while those that have not yet been emptied would appear red, as shown in Fig. 5. The route driven by the vehicle would be shown in yellow, while any sections of the route driven twice on that day would be marked red.

Fig. 5. Emptied containers in a daily plan as shown in the application on the vehicle's tablet and at the control centre in the head office [22].

Communication with the driver would take place via an application installed on the tablet. The vehicle's route and all its activities would be monitored in real time from the head office. The application would also enable communication via text messages and optimized implementation of activities along a precisely delineated route. The system would therefore enable two-way communication between the vehicle and the head office of the Ecology and Waste Management Working Unit, as well as the synchronization of daily plans and routes for the collection of waste containers.

The main improvement would be the establishment of a unified ERP (Enterprise resource planning) system, which would connect all the smart systems used by the Ecology and Waste Management Working Unit in the execution of their waste collection and disposal activities. An ERP system, which is used to manage a company's operations, would enable the company to plan all the work processes of the Unit in an integrated, efficient and transparent manner. All the smart systems currently in use by the Unit or yet to be introduced in the future would be merged into this unified system. The unified system would combine a vehicle tracking system, a map with all pick-up and drop-off locations, a container chipping system using RFDI chips, a container fill-level monitoring system with fill-level sensors, an application to communicate with the vehicle, a dashboard tablet, the head-office control centre, waste collection plans, vehicle employee plans and all other activities used to manage and monitor waste collection and disposal.

A typical workday using the interconnected systems would run smoothly and be divided into the following steps or phases: The workers assigned to the vehicle would start their workday by boarding the vehicle. When the driver starts the truck, the tablet mounted on the dashboard automatically turns on. As all containers would be equipped with waste level sensors, the containers that need to be emptied would be displayed on the vehicle's tablet. The system determines which containers need to be emptied based on the data provided by the sensors located on the containers (Fig. 6).

Fig. 6. Smart waste collection [23].

At precisely 6.00 am, when a typical workday starts, the system automatically calculates the optimal container collecting route for each vehicle for that day. The tablet displays the route for the driver to follow and the locations of each container to be collected. Once emptied, the containers would instantly turn green in the application. Since the system would be connected to the control centre, all this data would be easily monitored from the head office and would no longer need to be managed directly from the vehicle's location. Additionally, the vehicle's planned route would be continuously displayed and, if needed, adjusted in real time, as new instructions would be sent from the system on a regular basis, should any additional containers become full and need emptying. This type of system would enable the company to drastically reduce the number of kilometres driven, which would otherwise be wasted on emptying containers that are already empty or partially empty.

The system would also eliminate waste removal at the request of the clients (businesses), as the sensors would detect when a container needs emptying. The system would also record the number of containers emptied for a particular client, which means that the driver would no longer have to manually fill out the documentation for each client separately. Since the sensors located inside the containers would detect the exact volume of waste in each container, the system would also know when the vehicle needs to be emptied at the Collection Centre and plan the optimal route to the Centre with as few empty kilometres as possible (Fig. 7).

Fig. 7. Transfer of container fill-level data to the vehicle [24].

5 Conclusions

A review and analysis of the use of modern information solutions to optimize transport have revealed ample opportunities for optimization, improving the company's operations and reducing costs. Drawing on the finding of the analysis of the state of waste collection in the Municipality of Izola, this paper proposes improvements in the field of waste collection using modern information technologies. The first step towards visible improvements consists in installing a smart underground ecological island, which would be equipped with a fill-level sensor that would send the data from the collection point directly to the control centre located at the head office and to the waste collection and disposal vehicle. An underground ecological island would increase the volume of the ecological island, improve the town's appearance, free up public areas and, last but not least, reduce the frequency of waste collection. Another measure that would optimize waste collection and disposal with the use of modern information technologies is upgrading waste collection containers by installing fill-level sensors in each and every container across town. The containers would be equipped with RFDI chips and fill-level sensors and connected into a unified system. In addition, all garbage trucks would be equipped with a computer displaying the planned route and the containers already collected.

The main improvement that would contribute to more efficient waste collection and to the sustainable development in the town of Izola and its surroundings would be the introduction of a unified system connecting all the smart systems managed by the Ecology and Waste Management Working Unit. The unified system would have to combine the vehicle tracking system, a map showing the locations of the containers, a container chipping system using RFDI chips, a container fill-level monitoring system with fill-level sensors, a communication application, employee plans and all other activities used by the Unit in the course of its work. The sensors in the containers would send the fill-level data to the garbage truck's tablet and automatically display the containers that need to be emptied. Since all the systems would be connected to each other, the vehicle would be able to receive the shortest possible route, and would only have to collect the containers that are full and actually need emptying, resulting in optimal vehicle utilization. Introducing this type of smart system would not only contribute to the town's sustainable development but would also reduce the total time required for waste collection and disposal, which would, in turn, contribute to the improvement of work processes.

The proposals are consistent with related studies and proposals highlighted by Lee and Wu [2], who demonstrated on the example of the city of Hong Kong that RFID and data mining technique can help provide a sustainable waste management by analysing the waste disposal habit. Svub et al. [1] demonstrated that in the long run, such a system will enable automatic identification and data capture of the municipal waste containers emptying process. Such information will bring added value both to the customer, who can be sure that he is paying for the real services, as well as to the service providers, allowing them to easily and reliably prove the contractual service of the collection and treatment of waste.

Although the research focuses on the proposal for the application of modern information technology in waste disposal, it is limited to the local area. However, it serves as a springboard for a sustainable solution for waste collection in the Municipality of Izola. The proposals leave open additional possibilities for the use of modern information technology in the future, in particular through the application of smart systems and the Internet of Things (IoT) to ensure an efficient and effective approach to waste management.

References

1. Svub, J., Stasa, P., Benes, F., Unucka, J.: RFID application in municipal waste management system. Inzynieria Mineralna – J. Polish Miner. Eng. Soc. (1), 71 (2017)
2. Lee, C.K.M., Wu, T.: Design and development waste management system in Hong Kong. In: 2014 IEEE International Conference on Industrial Engineering and Engineering Management, Selangor, Malaysia. IEEE (2014)
3. Wang, J.-Y., Cao, Y., Yu, G.-P., Yuan, M.-Z.: Research on application of IOT in domestic waste treatment and disposal. In: Proceeding of the 11th World Congress on Intelligent Control and Automation, Shenyang, China. IEEE (2014)
4. Wen, Z., et al.: Design, implementation, and evaluation of an Internet of Things (IoT) network system for restaurant food waste management. Waste Manag. **73**, 26–38 (2018)
5. Heidari, E., Motameni, H., Movaghar, A.: A meta-heuristic clustering method to reduce energy consumption in Internet of Things. Int. J. Nonlinear Anal. Appl. **12**(1), 45–58 (2021)
6. Shen, Y., Zhang, T., Wang, Y., Wang, H., Jiang, X.: Microthings: a generic IoT architecture for flexible data aggregation and scalable service cooperation. IEEE Commun. Mag. **55**(9), 86–93 (2017)
7. Khan, F., Jan, M.A., Rehman, A.U., Mastorakis, S., Alazab, M., Watters, P.: A secured and intelligent communication scheme for iIoT-enabled pervasive edge computing. IEEE Trans. Ind. Inf. **17**, 5128–5137 (2020)
8. Ding, W., Jing, X., Yan, Z., Yang, L.T.: A survey on data fusion in internet of things: towards secure and privacy-preserving fusion. Inf. Fusion **51**, 129–144 (2019)

9. Alam, F., Mehmood, R., Katib, I., Albogami, N.N., Albeshri, A.: Data fusion and IoT for smart ubiquitous environments: a survey. IEEE Access **5**, 9533–9554 (2017)
10. Jan, M.A., et al.: An AI-enabled lightweight data fusion and load optimization approach for Internet of Things. Future Gener. Comput. Syst. **122**, 40–51 (2021)
11. Tamakloe, C.N., Rosca, E.V.: Smart systems and the Internet of Things (IOT) for waste management. In: IEEE International Conference on Computational Intelligence and Virtual Environments for Measurement Systems and Applications, Tunis, Tunisia (2020)
12. Xiao, Y., Yu, S., Wu, K., Ni, Q., Janecek, C., Nordstad, J.: Radio frequency identification: technologies, applications, and research issues. Wirel. Commun. Mob. Comput. **7**(4), 457–472 (2007)
13. Ngai, E.W.T., To, C.K.M., Moon, K.K.L., Chan, L.K., Yeung, P.K.W., Lee, M.C.M.: RFID systems implementation: a comprehensive framework and a case study. Int. J. Prod. Res. **48** (9), 2583–2612 (2010)
14. Gaukler, G.M., Seifert, R.W.: Applications of RFID in supply chains. In: Jung, H., Jeong, B., Chen, F.F. (eds.) Trends in Supply Chain Design and Management. SSAM, pp. 29–48. Springer, London (2007). https://doi.org/10.1007/978-1-84628-607-0_2
15. Tajima, M.: Strategic value of RFID in supply chain management. J. Purch. Supply Manag. **13**(4), 261–273 (2007)
16. Xiao, Y., Yu, S., Wu, K.Q., Janecek, C., Nordstad, J.: Radio frequency identification: technologies, applications, and research issues. Wirel. Commun. Mob. Comput. **7**(4), 457–472 (2007)
17. Jun, H.B., Shin, J.H., Kim, Y.S., Kiritsis, D., Xirouchakis, P.: A framework for RFID applications in product lifecycle management. Int. J. Comput. Integr. Manuf. **22**(7), 595–615 (2009)
18. Condemi, A., Cucchiella, F., Schettini, D.: Circular economy and E-waste: an opportunity from RFID TAGs. Appl. Sci. **9**(16), 3422 (2019)
19. ECUBELABS. https://www.ecubelabs.com/wp-content/uploads/2018/01/fill-level-sensor-in-waste-bin-1.png. Accessed 01 Apr 2021
20. WENGLOR. https://www.wenglor.com/en/Ultrasonic-Sensors/. Accessed 05 Apr 2021
21. CDNASSETS. https://cdnassets.hw.net/8e/dd/d887a6c14ef3b2ba65d73879275e/ssv10-in-cab.jpg. Accessed 05 Apr 2021
22. VTS. www.vts.si/sledenje, edited by the author. Accessed 06 Apr 2021
23. CDN. https://cdn.openpr.com/R/8/R81784815_g.jpg. Accessed 06 Apr 2021
24. TELELINK. https://telelink-city.com/wp-content/uploads/2020/05/Waste_management.png. Accessed 06 Apr 2021

Proposal of an Algorithm for Evaluation of Wet Gap Crossing Using Geoprocessing Tool

Martin Sedláček(✉), Filip Dohnal, and Ota Rolenec

University of Defence, Kounicova 65, 60200 Brno, Czech Republic
{martin.sedlacek,filip.dohnal,ota.rolenec}@unob.cz

Abstract. The article deals with the area of engineer and geographical support within the operation of troops and their planning process. This is one of the tasks in terms of measures to ensure the mobility of own troops. The issue of the article is aimed primarily at identifying the geospatial data source and the criteria that affect wet gap crossing by troops and at the proposal of an algorithm that can be used for the development of a geoprocessing tool of wet gap crossing. The contribution of the article is a comparison of the limit values of the vehicles used in wet gap crossing with the natural environment in the operational area and the proposal of an algorithm in the form of flowcharts usable for a geoprocessing tool of wet gap crossing. The significance of the article lies in the proposal of the development of the information system of command and control and its engineer subsystem as an element of C4ISTAR systems.

Keywords: Algorithm · Digital landscape model · Digital terrain model · Engineer support · Geoprocessing tool · Geographical support · Hydrology · Wet gap crossing

1 Introduction

Wet gap crossing can be defined as an activity to which individual types of troops contribute within the operational area. Evaluation and selection of a suitable area to cross the wet gap is a basic prerequisite for the actual execution by the troops. Wet gaps are characterized by elements of the natural environment, but can also be affected by the activities of the enemy or own troops [1, 2].

The staff (HQ) of Task Force is responsible for a correct understanding of the task [3] and evaluation of the situation in the operational area. The staff must collect data and information on all aspects of the operation [4] so that they can be properly analyzed in a common operational picture [5–7]. Data and information from the operational area can be collected in the databases of individual command and control subsystems [8]. The assessment of a suitable area of wet gaps should be assessed within the engineer command and control subsystem, which is linked to the information system of command and control (IS C2).

O. Prentkovskis et al. (Eds.): TRANSBALTICA 2021, LNITI, pp. 542–551, 2022.
https://doi.org/10.1007/978-3-030-94774-3_53

Suitable tools are databases with geographical and hydrometeorological data [9, 10]. By modifying these databases and taking into account the tactical-technical limitations (TTD) of selected vehicles, a proposal of algorithm has been set for a geoprocessing tool, which enables the engineer staff officer to recommend proposals of areas with wet gaps crossing within the planning process of the Task Force. A geographic information system (GIS) was used to create the geoprocessing tool. This tool takes into account the variant solution of the staff planning process and, based on accurate information, can speed up and refine this process.

2 Source Geographic Database

Digital Landscape Model 25 (DMU 25) contains such layers with object types and corresponding attributes that become criteria when taking into account the operational area with wet gap crossing. However, DMU 25 is not complete for the purposes of application software of wet gap crossing (APV PVP) development in the form of a geoprocessing tool and some data are missing or are out of date or represent a statistically calculated value, which becomes a shortcoming. However, in comparison with other databases, DMU 25 can be assessed as the most complete database that can be used for proposal of algorithm and subsequent development of APV PVP. Furthermore, these shortcomings are also reflected and balanced in the creation of the proposal of algorithm and the development of APV PVP.

For the proposal of the algorithm for evaluation of wet gap crossing and development of APV PVP only some layers from the DMU 25 database are used (FLOW CHARACTERISTICS, FLOW CENTERLINES and BRIDGE). The point layer FLOW CHARACTERISTICS contains information about the water flow on its specific cross-section. The information is sparsely spatially distributed and qualitatively expresses the long-term value of the hydrological quantity on the given profile without other related information. For this reason, it was decided to assign information from the point layer to the line layer FLOW CENTERLINES so that the line representing the course of the flow centerline was divided into 50m sections and these sections were assigned information from point layer based on the spatial relationship. In the event that a permanent bridge (SMPK) leads through a watercourse, the information about the given bridge was assigned from the BRIDGE layer to the line layer with a specific 50m section. These modifications resulted in the creation of a single line layer, in which its geometry represents the course of the center of the watercourse and qualitative data are assigned to each 50 m section based on the spatial relationship.

Digital Terrain Model 5 (DMR 5) contains data that create prerequisites for a comprehensive understanding and evaluation of the operational area in the proposal of the algorithm and the development of APV PVP. It is an elevation model of the relief, the spatial accuracy of which cannot be obtained from the available data contained in DMU 25 (contour relief model) [11]. For the proposal of the algorithm and development of APV PVP in the form of a geoprocessing tool, DMR 5 was used to determine the bank slope of watercourses. For each 50 m section, the average slope on the left and right bank was calculated from the raster model DMR 5.

3 Identification of Criteria for Proposal of Algorithm

Based on the selected input models DMU 25 and DMR 5, decisive criteria are identified that influence the evaluation of the area to cross the wet gaps. These criteria are also based on the possibilities of establishing crossing sites as basic ways of wet gap crossing by the Army of the Czech Republic (ACR) in conducting tactical activities of troops. The individual crossing sites are:

- permanent bridges (SMPK);
- fords (BRP);
- bridges (MOP);
- raft sites/vessels (PLP);
- ferries (PRP);
- underwater transport (ride of tanks under water), (POP).

With regard to the identified crossing sites, it is also necessary to determine the anticipated vehicles that these crossing sites will use or establish. Based on the organizational structure mechanized brigades and engineer regiment of Czech army forces, ten vehicles are identified, which are expected to be used in the performance of activities in the operational area with wet gap crossing and their representation within the Land Forces (LF) are the most important. These are:

- bridge vehicle (AM-50);
- pontoon bridge set (PMS);
- amphibious transporter (PTS-10);
- bridge tank (MT-55);
- bridge carrier (PM-55);
- infantry fighting vehicle (BVP-2);
- tank (T-72M4CZ);
- wheeled armoured vehicle (KBVP Pandur II);
- Tatra truck (T-810 6x6);
- Tatra truck (T-815 8x8).

Additional vehicles can be added to the APV PVP that would expand the database and could also refine the planning and decision-making process of the Task Force staff. These are, for example, the vehicles of NATO countries in the framework of international operations [12]. Selected vehicles are limited in the establishment or use of individual crossing sites. They are influenced by the geographical and hydrological conditions of the operational area [13], but also by the limit values of the vehicles. These are:

- depth of the wet gap;
- fording depth;
- limit speed of flow;
- limit width of the wet gap;
- weight of the vehicle;

- width of the vehicle;
- height of the vehicle;
- climbability;
- floating ability.

Table 1 gives an overview of selected objects and their attributes depending on the crossing site. Table 1 also sets out the binding of attributes to TTD resources and also explains attribute abbreviations, further it contains a consideration of the DMR 5 database for selected crossing sites and the link to TTD vehicles.

Table 1. Assignment of objects and attributes of DMU 25 and DMR 5 to individual crossing site.

Object	**Attribute**	**Crossing site**					
		Permanent bridges	Fords	Bridges	Ferries	Raft sites/vessels	Underwater transport
BRIDGE, BRIDGING	LC5	weight of the vehicle					
	SHC	the width of the vehicle					
	OHC	the height of the vehicle					
	EXS	object status					
FLOW CHARACTERISTICS	HDP 1-12*		fording	the depth of the wet gap	the depth of the wet gap		the depth of the wet gap
	WVA		limit speed of flow	limit speed of flow	limit speed of flow	limit speed of flow	limit speed of flow
	WD3			limit width of the wet gap – bridge length			
DMR 5			climbability	climbability at work	climbability, climbability at work	climbability	climbability

* depth modifications (HDP) for individual calendar months (1-12) [14].
LC5 – load capacity, SHC – width, OHC – height, EXS – object status, HDP – depth, WVA – speed of flow, WD3 – width of wet gap.

4 Proposal of the Algorithm

The proposal of the algorithm is based on the evaluation of DMU 25 data, modification of DMU 25 from points to line elements, DMR 5 and own database of selected vehicles. It is compiled for the development of APV PVP and thus also for the user himself and his ability to understand the user environment. Flowcharts have become the basic element of algorithm proposal. This graphical representation of the individual steps allows a deterministic approach to monitor the interrelationships and conditions between the input data and the data entered from the user.

The proposal of the algorithm for the development of APV PVP and the understanding of the user environment are divided into two submodels. Figure 1 shows the string of inputs, outputs and the assigned PVP_Submodel1. In the first step, the user has the option to set the display of raster equivalents of topographic (RETM) or military maps (REVM). In addition, the user has the option to set the orthophoto as a map product and to have a layer with permanent bridges displayed. The user then runs PVP_Submodel1.

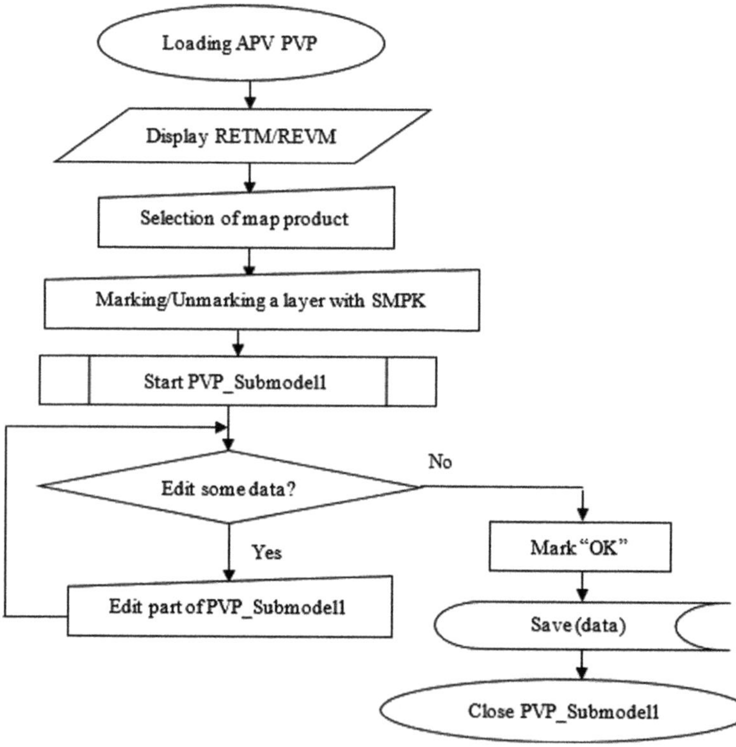

Fig. 1. Flowchart - settings of the map product, SMPK layer and PVP_Submodel1.

The performed Flowchart PVP_Submodel1 (Fig. 2) reduces the sections on wet gaps for calculations and thus the necessary data. A new database is created corresponding only to the defined operational area. The user has the option to generate and save coordinates in the Universal Transverse Mercator coordinate system (UTM) from the specified areas in the map product.

The determination of the polygon is set so that the connectors do not cross. Thus, the user can select points on the underlying map in any order. The determination of the polygon for both the limited access area and the minefields must always include the wet gap which is in the database, so that the line of wet gap together with the adjacent areas (banks) is excluded from the calculations.

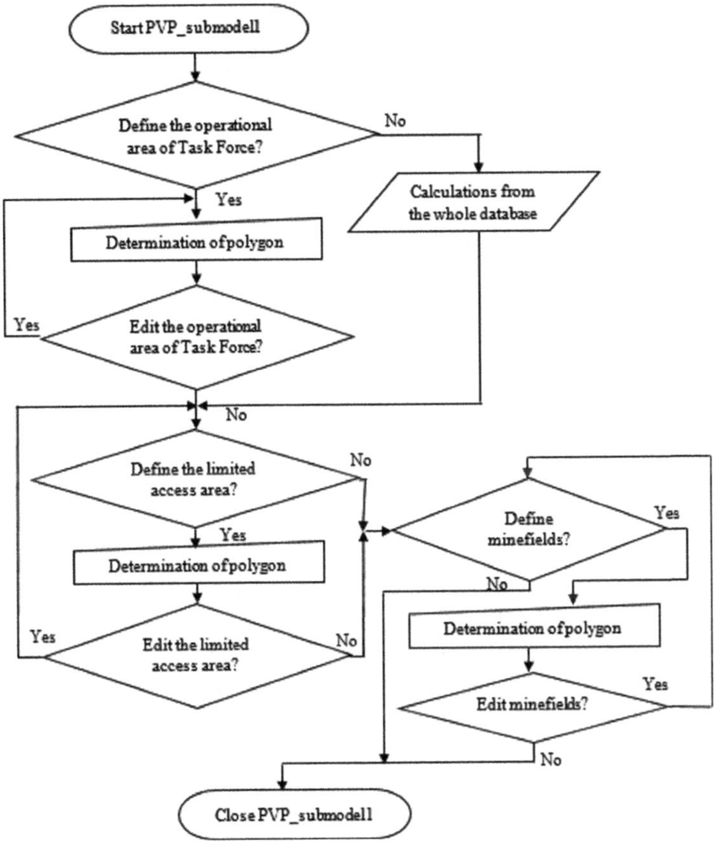

Fig. 2. Flowchart PVP_Submodel1.

Figure 3 shows all the inputs that affect the calculations in PVP_Submodel2. These inputs are the determination of the destroyed permanent bridges, the determination of the rivers of interest to the user, the determination of the calendar month for which the calculated depth is set, the selection of the vehicles and their number (related to bridge vehicles).

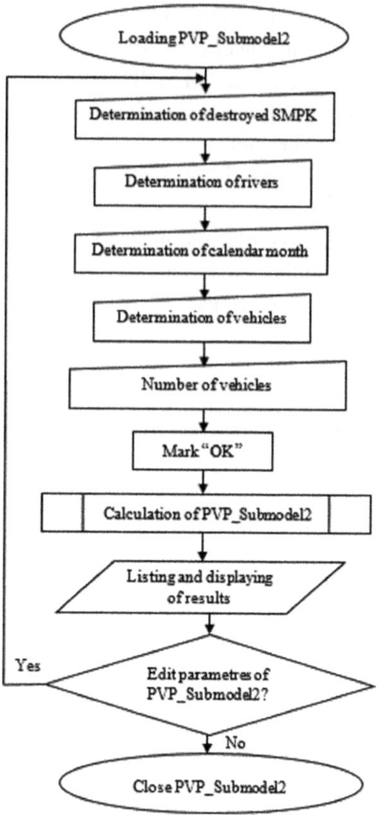

Fig. 3. Flowchart PVP_Submodel2.

Flowcharts can also become an input model and backbone for the further developed APV PVP or a basis for the development of a separate application software (APV) with the possibility of implementation in the IS C2. Flowcharts can also be used as a basis for development other APVs related to the tasks of engineer support for the activities of troops, in which there would be basic databases DMU 25 and DMR 5. That should be [15]:

- engineer reconnaissance of roads and objects on them;
- roads maintenance;
- setting up detours;
- culverts maintenance;
- bridges maintenance;
- emplace the explosive obstacle on the roads;
- maintenance of military bases of all kinds;
- maintenance of existing airfields, airfield damage repair, building, repair and maintenance of logistic infrastructure;
- engineer reconnaissance for water extraction;

- building of water treatment and distribution sites;
- building, repair and maintenance of railways.

Only two rivers (Morava and Becva) are shown in the flowchart due to their inclusion in the demo version of the APV PVP. These are rivers that have significant parameters of depth, width, speed of flow and bank slope. In terms of the scope of the PVP_Submodel2 calculation flowchart, only part of this flowchart is illustratively shown in Fig. 4 for permanent bridges. The evaluation of the other crossing site is similar to the evaluation of the SMPK.

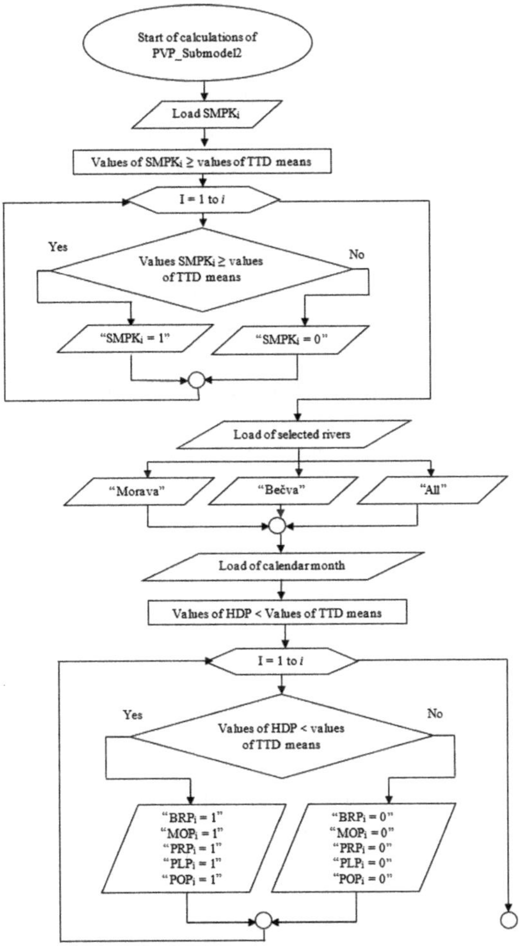

Fig. 4. Flowchart – Selected part of calculations of PVP_Submodel2.

PVP_submodel2 evaluates whether a given crossing site can be used by at least one specified vehicle and also evaluates the individual vehicles to the individual crossing sites. The user is thus able to submit proposals for individual crossing sites as part of the planning process at the Task Force HQ. The user always has the possibility to find out the reason why a particular crossing site is not proposed by the geoprocessing tool by studying the individual values that enter the process as a condition to the limit TTD of the vehicles. The values are clearly listed in the attribute table. However, the ability to recognize an exceeded value depends on the user's knowledge of the TTD vehicles. In addition, the user has the opportunity to evaluate the extent of the exceeded values by his own judgement and to decide on his own, based on experience and knowledge, about a possible tolerance interval, i.e., the condition is fulfilled but with a limitation.

5 Conclusion

The classification of criteria for proposal of the algorithm with evaluation of wet gap crossing is determined with the links between the own vehicles database and their limit TTD with respect to the DMU 25 and DMR 5 databases.

The proposal of the algorithm is based on the description and display of flowcharts. The same proposal is based on the settings of the map product, the SMPK layer, the PVP_Submodel1, PVP_Submodel2 and its calculation. The PVP_Submodel1 defines the operational area, limited access area and minefields. This reduces the data in terms of the complexity of the duration of the calculations. Within the PVP_Submodel2, the options for determining the destroyed SMPK, the selection of rivers, the calendar month, the number and selection of vehicles to be covered by the calculations are displayed and described.

The implementation algorithm and APV PVP in the planning process of the staff Task Force is primarily proposed for the 2nd phase (task analysis and intelligence preparation of the battlefield), but the use of APV PVP can be implemented in all phases of planning process. The main keys of APV PVP in this process will be engineer officers of staff Task Force, usually the intelligence, planning and operational groups or departments. As part of the planning process, APV PVP is proposed to be included on the same level as another application of engineer subsystem command and control. A propose of an algorithm can speed up the second phase of the planning process and also makes outputs more accurate.

Although the applicability of the algorithm and the related APV PVP is addressed only on the territory of the Czech Republic, development within NATO is possible, but depends on the availability of suitable data that would be converted according to the transformation key into the current form of the algorithm.

References

1. Hošková-Mayerová, Š., Talhofer, V., Otřísal, P., Rybanský, M.: Influence of weights of geographical factors on the results of multicriteria analysis in solving spatial analyses. ISPRS Int. J. Geo-Inf. **9**(8), 489 (2020). ISSN 2220-9964. IF 2.239

2. Hubáček, M., et al.: Modelling of geographic and meteorological effects on vehicle movement in the open terrain. In: Central Europe Area in View of Current Geography, pp. 149–159. Masaryk University, Brno (2016). ISBN 978-80-210-8313-4
3. Wesselényi, J., Kompan, J.: Use of methods for decision-making support in anti-terrorism operations. Studia nad Autorytaryzmem i Totalitaryzmem **39**(4), 19–26 (2017). https://doi.org/10.19195/2300-7249.39.4.2
4. Hrnčiar, M.: Tactical variables – a tool for mission analysis. In: The 25th International Conference. The Knowledge-Based Organization: Conference Proceedings - Management and Military Sciences, pp. 86–90. Nicolae Bălcescu Land Forces Academy, Sibiu, Romania (2019). https://doi.org/10.2478/kbo-2019-0014. ISBN 978-973-153-355-1
5. Sedláček, M., Zelený, J.: Requirements for engineer information with wet gap crossing. Vojenské rozhledy **28**(60), 44–62 (2019)
6. Rolenec, O., Šilinger, K., Žižka, P., Palasiewicz, T.: Supporting the decision-making process in the planning and controlling of engineer task teams to support mobility in a combat operation. Int. J. Educ. Inf. Technol. **2019**(13), 33–40 (2019)
7. Lucia, A., Feng, Y.: Global terrain methods. Comput. Chem. Eng. **26**(4–5), 529–546 (2002)
8. Dawid, W., Pokonieczny, K.: Visualisation of a military topographic spatial database with use of GIS servers. In: 2019 International Conference on Military Technologies (ICMT), pp. 1–8 (2019). https://doi.org/10.1109/MILTECHS.2019.8870118
9. Heštera, H., Pahernik, M.: Physical-geographic factors of terrain trafficability of military vehicles according to western world methodologies. Hrvatski geografski glasnik/Croatian Geogr. Bull. **80**(2), 5–31 (2021). https://doi.org/10.21861/HGG.2018.80.02.01. ISSN 13315854. https://hrcak.srce.hr/213787
10. Stock, A., Tolvanen, H., Kalliola, R.: Crossing natural and data set boundaries: coastal terrain modelling in the South-West Finnish Archipelago. Int. J. Geogr. Inf. Sci. **24**(9), 1435–1452 (2010)
11. Hubáček, M., Kovařík, V., Kratochvíl, V.: Analysis of influence of terrain relief roughness on DEM accuracy generated from LIDAR in the Czech Republic Territory. Int. Arch. Photogramm. Remote Sens. Spatial Inf. Sci. **XLI-B4**, 25–30 (2016). Praha: ISPRS, p. 25–30. ISSN 1682-1750
12. Cibulová, K., Rolenec, O., Garba, V.: A selection of mobility support engineering devices of NATO armies usable in the Czech armed forces combat operations. In: Proceedings of International Conference of Military Technologies Brno, Brno, p. 8870016. Institute of Electrical and Electronics Engineers Inc. (2019)
13. Cibulová, K., Sobotka, J.: Utilization of perspective materials for negotiation of watercourses. In: Transport Means 2019, pp. 655–659. Kaunas University of Technology, Palanga (2019)
14. Sedláček, M., Dohnal, F.: Proposal of optimization of depth values in wet gap crossing military operations. Adv. Mil. Technol. **16**(1), 35–47 (2021)
15. Sedláček, M., Dohnal, F.: Possibilities of using geographic products in tasks of military engineering. In: Challenges to National Defence in Contemporary Geopolitical situation, pp. 145–155. General Jonas Žemaitis Military Academy, Vilnius, Lithuania (2020)

Classification of Transport and Logistics Enterprises in Ukraine According to the Level of Innovation Potential

Olena Komchatnykh(✉), Iryna Klymenko, and Olena Levishchenko

National Transport University, Omelianovycha-Pavlenka Street 1,
Kyiv 01010, Ukraine
komchatnykh@ntu.edu.ua

Abstract. The paper presents classification according to the level of innovative potential of Ukrainian transport and logistics enterprises specializing in freight transportation with the use of hierarchical cluster analysis on the basis of STATISTICA software. Hierarchical cluster analysis was performed on the basis of the structural components' calculated values of the enterprise innovation potential (production and technological, financial, personnel, scientific and technical, organizational and managerial and marketing) by comparing the actual values of indicators that characterize them with the criteria of comparison established for them. The criteria of comparison were the normative values of indicators from the economic and financial literature, industry average indicators, the most optimal enterprises' indicators in the field of transport logistics, and combined values based on these indicators. According to the results of hierarchical cluster analysis, it is determined that Ukrainian transport and logistics enterprises can be divided into four groups: with high, medium, low and critical level of innovation potential. Innovatively active transport and logistics enterprises are characterized by a high and medium level of innovation potential, innovatively inactive ones are characterized by a low and critical level of innovation potential. The common features and prospects of each enterprise's development level are determined. The principle of using hierarchical cluster analysis is quite universal and can be used to classify the enterprises of other industries, provided that the composition of indicators and their criteria values are adjusted specifically to any of the activities.

Keywords: Innovative potential · Transport and logistics enterprise · Classification · Hierarchical cluster analysis · STATISTICA

1 Introduction

In modern economic conditions, transport and logistics enterprises' survival is largely based on their innovative potential. However, the available innovation potential is used by participants in the freight market only by 7–10%. Without having time to respond to market needs and without using the available potential in accordance with the economic development innovative directions, enterprises face the threat of losing their competitiveness.

O. Prentkovskis et al. (Eds.): TRANSBALTICA 2021, LNITI, pp. 552–560, 2022.
https://doi.org/10.1007/978-3-030-94774-3_54

For modern transport and logistics enterprises, whose activities are based on scientific and technical achievements embodied in the techniques and technologies of motor transport services, rapid and efficient implementation of innovations is connected with sufficient resources, which is only possible using the reasonable assessment of the results and searching for the innovative development resources and reserves. The development of the enterprise's innovative potential contributes to the strengthening of its adaptive capabilities, increasing the ability to make the right management decisions, increasing the variability of the used methods and technologies. In these circumstances, the need to find an objective method of assessing the innovation potential and make up an effective management system for innovative development of the enterprise which becomes especially important.

The object of study is Ukrainian transport and logistics enterprises. The subject of study is theoretical and methodological provisions for improving the management of transport and logistics enterprises' innovative potential. The information base of the research includes processed and generalized results of 25 Ukrainian transport and logistics enterprises' activity. The statistical reports "Survey of innovative activity of the enterprise during the period 2016–2018" and financial statements of these enterprises at the end of 2018, which corresponds to the period of the last CIS in Ukraine were used as materials. Mathematical calculations were performed using the computer program Statistica.

The authors suggest that the implementation of this study results will increase the scientific management level of transport and logistics enterprises' innovative potential.

2 Literature Review

For a long period of time, scientists had not been paying much attention to the study of innovations in services, instead they were studying only the innovations of industrial enterprises. Service innovation was considered only as an activity that complements production processes. However, with the development of services all around the world and increasing society's needs in the products of this area, more and more scientists are beginning to pay attention to the factors of its effective functioning, including innovation. Among the scientists who have devoted their work to the study of this issue, the greatest achievements belong to such scientists as X. Vence, A. Trigo [1], I.Miles [2], J. Sundbo [3], F. Gallouj [4], R Barras [5], K. Pavitt [6], B.S. Tether [7].

Nowadays, there are not many works dedicated to innovations in the field of transport and logistics services. According to the study of S.M. Wagner [8] at the beginning of 2008 only 6 scientific works were devoted to transport and logistics innovation, so it can be confirmed that this topic had finally attracted the scientists' attention only in the last decade. One of the most thorough studies in this direction is the work [9], in which authors used mathematical modeling and binary regression models to analyze the statistics of freight forwarding and logistics companies in Turkey. They proved that the following factors had a positive impact on innovation: openness of external sources of information, efficient use of one's own sources of information, research costs and development (including internal), cooperation in the development of logistics network, use of financial support, number of employees as well as the market size. A significant contribution of E. Przybylska should be noted. In her works [10–12] she investigates the specifics of innovations on the example of Polish freight forwarding and logistics enterprises. However, despite the presence of a certain amount of scientific research, the issue of formation, evaluation and development of the innovative potential of transport and logistics enterprises has not yet been given enough attention.

3 Results

Enterprise's innovative potential assessment lies in monitoring the level of its constituent elements development for further comparison with existing ideas about their level. Innovative potential is an organic component of general economic potential, it interacts with other potentials of the enterprise (production and technological, financial, scientific and technical, organizational and managerial, personnel and marketing) and is formed as a result of this interaction. On the one hand, innovation potential is a part of each of the enterprise‘s overall potential elements, on the other hand, each of the components of the overall potential of the enterprise provides a level of innovation potential development [13]. Determination of the enterprise innovation potential level does not require the full value of production, personnel or scientific and technical potential, but only that part of them that is directly involved in its formation. Therefore, the composition of innovation potential should include the components of those potentials which most fully meet the requirements and reasonably determine its level, but not the potentials in relevant areas only.

Properly defined level of the enterprise's available innovation potential will allow not only to assess the real innovation potential of the enterprise and reasonably approach the choice of the future innovation strategy, but also to avoid financial investments in those projects that the company is eventually unable to implement. On this basis, objective knowledge of its own innovation capabilities allows an enterprise to avoid irrational costs connected with the development and implementation of inefficient innovation projects. Due to the low awareness of other participants in the trucking market, this task is quite difficult for the company to complete. For this problem to be solved, the authors suggest using cluster analysis tools.

Cluster analysis is a tool for classifying complex, not obviously interconnected objects by their common features. In this case, the cluster is understood as a real subset of the studied objects, which are characterized by a certain homogeneity. The main objectives of cluster analysis are [14]: to determine the homogeneity of objects, in case it cannot be established using common methods; formation of new subsets of objects as carriers of certain characteristics; meaningful interpretation of these subsets role in the transformation of the surrounding socio-economic reality.

To conduct a cluster analysis, a sample of 25 transport and logistics enterprises specializing in trucking was formed. To ensure the representativeness of the sample, the enterprises that differ in size (large, medium and small), occupy different market positions (leaders, contenders, followers, representatives of the “niche”), have different geography of freight (within their region, all over the Ukraine, international transportation) were selected. Among the selected enterprises there are both innovation-active enterprises. And the ones which did not engage in innovation activities during the analyzed enterprise, that is where innovation-inactive. All selected enterprises have been operating in the Ukrainian market for more than 10 years, have their own fleet and are engaged in transport and logistics activities.

As the next step, the matrix that describes the characteristics of the studied observations significance sample was constructed (see Table 1). In this context, each observation is recorded as a separate line consisting of the values of the used characteristics, and each characteristic in the matrix is represented by a separate column that consists of this characteristic's values for all objects of the sample: X_1 – the value of production and technological component of innovation potential; X_2 – the value of the financial component of innovation potential; X_3 – the value of the personnel component of innovation potential; X_4 – the value of the scientific and technical component of innovation potential; X_5 – the value of the organizational and managerial component of innovation potential; X_6 – the value of the marketing component of innovation potential. The calculation of the integrated values of innovation potential components was carried out according to the data at the end of 2018, which corresponds to the period of the last CIS-survey of these enterprises, according to the formula:

$$X_j = \frac{\sum_{i=1}^{m} a_i}{m}, \tag{1}$$

where a_i – the point value of the i_{th} indicator, m – the number of indicators that characterize the j_{th} component of innovation potential.

The indicators selected to identify the innovation potential components characterize the provision of transport and logistics enterprise with the necessary material and financial resources for the innovative development, intellectual capital and available innovative results due to the constituent components of the enterprise's general economic potential: production and technological (coefficient of the new equipment's development, coefficient of the equipment's suitability, coefficient of innovations' introduction, share of expenses for equipment and technologies' acquisition); financial (autonomy ratio, financial stability ratio, absolute liquidity ratio, coverage ratio); personnel (share of employees with higher education, share of employees involved in R&D, share of employees who have been trained and retrained, the share of training costs and staff training for innovation); scientific and technical (intellectual property ratio, coefficient of the intellectual property suitability, share of R&D costs); (organizational and managerial – the size of the enterprise, the growth index of the innovation costs, the share of innovation costs in the enterprise's income); marketing (market share of the enterprise in the group of the nearest competitors, the level of cooperation with suppliers, the level of consumer loyalty). In order to bring the obtained values to a comparable form, they were standardized by assigning score values. For this purpose, the obtained data were compared to the established comparison criteria. The criteria of comparison were the normative values of indicators from the economic and financial literature [15–18], industry average indicators, the most successful indicators of competing enterprises and the values combined on their basis.

Table 1. Initial data for cluster analysis.

Enterprise	X_1	X_2	X_3	X_4	X_5	X_6
1	0.75	0.85	0.93	0.66	0.76	1
2	0.78	1	0.66	0.63	0.75	1
3	0.63	0.98	0.65	0.65	0.74	0.79
4	0.73	0.95	0.83	0.4	0.49	0.62
5	0.69	0.74	0.64	0.6	0.78	0.75
6	0.5	1	0.78	0.46	0.46	0.57
7	0.06	1	0.4	0	0.05	0.34
8	0.01	0.19	0.38	0	0.06	0.48
9	0.08	0.87	0.31	0.15	0.08	0.43
10	0.13	0.75	0.5	0	0.3	0.63
11	0.5	0.01	0.27	0	0.09	0.45
12	0.22	0.61	0.5	0	0.03	0.43
13	0.28	0.19	0.5	0	0.05	0.46
14	0.41	1	1	0.33	0.45	0.58
15	0.14	0.32	0.5	0	0.02	0.33
16	0.16	0.5	0.46	0	0.02	0.35
17	0.42	1	0.67	0.49	0.38	0.61
18	0.25	0.52	0.46	0.06	0.02	0.57
19	0.43	0.68	0.43	0.37	0.52	0.65
20	0.22	0.58	0.5	0	0.02	0.6
21	0.25	0.56	0.5	0	0.01	0.67
22	0.65	1	0.69	0.58	0.74	0.94
23	0.28	1	0,5	0	0.01	0.67
24	0.11	0.08	0.39	0	0.07	0.37
25	0.67	0.93	0.71	0.64	0.8	0.71

Since integrated values of the studied enterprises' innovation potential components had significant differences as a result, the method of full communication was used as a method of combining observations into clusters, according to which the distance *d(Y, Z)* between clusters *Z* and *Y* was determined by the formula below:

$$d(Y, Z) = max\, d(i, k), \tag{2}$$

where *d(i, k)* – the distance between the i_{th} object of the cluster *Y* and the k_{th} object of the cluster *Z*.

This method of cluster analysis was implemented using STATISTICA software. The result of the cluster analysis is presented in the form of a horizontal tree dendrogram. It is considered that using more sophisticated aggregation methods, such as Ward's method, which uses statistical distances between clusters, gives better results.

However, when using Ward's method, enterprises were divided into clusters in a similar way, which confirmed the previously gained results. The analysis of the horizontal dendrogram begins on the left – moving to the right in the figure, you can see how objects begin to merge and form clusters. The horizontal axis of the dendrogram characterizes the distance of the formed associations. According to the conducted hierarchical analysis, four clusters were formed in the sample of surveyed enterprises (see Fig. 1). The first two clusters included all enterprises that were engaged in innovation activities during the analyzed period, that is were innovation was being active. The third and fourth clusters included enterprises that did not show innovation activity in the analyzed period.

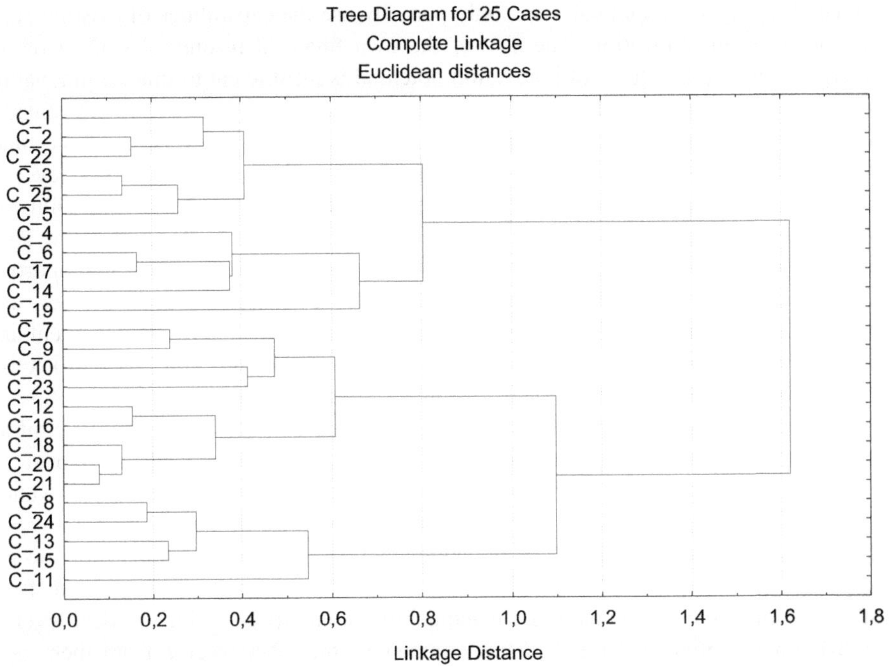

Fig. 1. Dendrogram of hierarchical cluster analysis of transport and logistics enterprises according to the of innovation potential structural components development level at the end of 2018.

All enterprises in the group with a high level of innovation potential have a fairly high level of development for all of its components. Depending on the level of development of innovation potential components, these companies are the most relevant to the innovation leadership and simulation innovation strategies.

As for the companies that are grouped by an average level of innovation potential, they all demonstrate a stable financial position and major investments in the innovative development. Due to the relatively small volume of their activities, enterprises grouped by an average level of innovation potential have less free financial resources and are mostly engaged in rolling stock renewal, and not paying enough attention to modern IT technologies and scientific research. In enterprises with an average level of innovation potential, research and development is conducted at a low level, and they themselves do not take leading positions in the market, instead, they usually adher to the protective or opportunistic innovation strategy.

The problems of most of the enterprises with a low level of innovation potential are primarily related to their management. The problem is the management in such companies usually ignores the issue of innovation development. These companies use outdated rolling stock and software, only partially use the capabilities of existing staff, do not invest in innovation. The size of the own financial resources of most of the enterprises with a low level of innovation potential is insufficient to finance innovation in the required amount, and management, focusing on current issues, does not consider it possible to attract external financial resources for these purposes. At the same time, the management of these enterprises ignores the fact that with the current level of competition in the market, their further survival depends entirely on the ability to innovate. In the case of a decision to intensify innovation, companies with a low level of innovation potential can use the traditional innovation strategy aimed at supporting the requirements for service quality by gradually updating the rolling stock and improving customer service. However, in the long run, with effective management and improvement of financial condition, such companies should turn to a protective innovation strategy to maintain their positions.

Enterprises with a critical level of innovation potential have a very low level of all components' development, and the difficult financial situation leads to almost no opportunities for innovative development. Enterprises with a critical level of innovation potential should review all internal factors that may contribute to the innovative development of these enterprises. If the company does not have any opportunity to get out of the crisis on its own, one should reconsider even the factors that govern its form of ownership. Thus, in the case of the acquisition of enterprises with a critical level of innovation potential by large logistics companies, they can receive from them new technologies or a new type of transport services, that is to begin using a dependent innovation strategy.

As the conducted hierarchical cluster analysis of transport and logistics enterprises showed, innovation-inactive enterprises are characterized by low and critical level of innovation potential, and innovation-active enterprises have medium and high level of innovation potential, only cases of high and medium level of innovation potential for innovation-active enterprises and low and critical for innovation-inactive enterprises were considered (see Table 2).

Table 2. Alternatives to innovative development of transport and logistics companies.

		The level of innovation potential			
		High	Medium	Low	Critical
Recommendations for increasing the level of innovation potential		Carrying out organizational and managerial measures to eliminate weaknesses and effectively implement innovations in the activity	Strengthening the scientific and technical component, primarily through the introduction of modern IT technologies for automation of logistics activities and communications	Activation of innovative activity due to deep transformation of technics and the equipment for work with freights at all stages of a supply chain	It is worth reviewing all the internal factors that can contribute to innovative development, even those that regulate the form of enterprise's ownership
Alternative innovation strategies					
Innovative activity of the enterprise	High	Innovative leadership strategy	Opportunistic innovation strategy	–	–
	Medium	Simulation innovation strategy	Protective innovation strategy	–	–
	Low	–	–	Traditional innovation strategy	Dependent innovation strategy

4 Conclusions

Using the cluster analysis tools, the classification of transport logistics enterprises was carried out and the existence of four levels of innovation potential for enterprises in this field of activity - high, medium, low and critical - was determined. On the basis of the available innovative potential of the enterprise it is possible to estimate enterprise's ability to innovative activity and to define strategic directions of its future innovative development. Innovative enterprises are characterized by a high level of innovation potential and the desire for its formation and development. Such enterprises have significant competitive advantages, as their inherent susceptibility to innovation allows them to timely protect their activities from external threats by staying ahead of them and constantly finding new solutions for future development. The conducted cluster analysis of transport and logistics enterprises according to the level of development of their innovation potential allowed to identify common features and develop recommendations for further innovative development for enterprises of each cluster. With the level of innovation potential being increased, so does the number of opportunities for development available to the company. Enterprises with a low and critical level of innovation potential have the least opportunities, as the main obstacle for such enterprises is the lack of financial resources. The principle of using cluster analysis is quite universal and can be applied to enterprises in other industries, provided that the composition of indicators and their regulatory values are adjusted in accordance with the certain activity specifics.

References

1. Vence, X., Trigo, A.: Diversity of innovation patterns in services. Serv. Ind. J. **1**(12), 1635–1657 (2009)
2. Miles, I.: Services Innovation: A Reconfiguration of Innovation Studies. PREST, University of Manchester, Manchester (2001)
3. Sundbo, J.: Management of innovation in services. Serv. Ind. J. **3**, 432–455 (1997)
4. Gallouj, F.: Innovating in reverse: services and the reverse product cycle. Eur. J. Innov. Manag. **1**(3), 123–138 (1998)
5. Barras, R.: Towards a theory of innovation in services. Res. Policy **15**(4), 161–173 (1986)
6. Pavitt, K.: Sectoral patterns of technical change: towards taxonomy and a theory. Res. Policy **13**(6), 343–373 (1984)
7. Tether, B.S.: Do services innovate (differently)? Insights from the European Innobarometer Survey. Ind. Innov. **12**(2), 153–184 (2005)
8. Wagner, S.M.: Innovation management in the German transportation industry. J. Bus. Logist. **29**(2), 215–231 (2008)
9. Burmaoglu, S., Şeşen, H., Kazançoğlu, Y.: Determinants of logistic sector innovation creating common value nodes in supply chain. LAÜ Sosyal Bilimler Dergisi. **6**(2), 37–58 (2015)
10. Przybylska, E.: Innowacyjność branży TSL [Innovation in TSL Sector]. Zeszyty Naukowe Politechniki Częstochowskiej **24**(2), 235–245 (2016)
11. Przybylska, E.: Potencjalne źródła innowacji w branży TSL [Potential sources of innovation in the TSL industry]. Zeszyty Naukowe Politechniki Śląskiej. Seria Organizacja I Zarządzanie **101**, 401–410 (2017)
12. Przybylska, E.: Typologia innowacji w branży TSL [Typology of innovation in the TSL industry]. Zeszyty Naukowe Politechniki Śląskiej. Seria Organizacja I Zarządzanie **99**, 399–409 (2016)
13. Kozak, L.S. (ed.): Perspektyvy ekonomichnoho rozvytku ta pidpryiemnytstva v umovakh hlobalizatsii [Prospects for economic development and entrepreneurship in the context of globalization]. NTU, Kyiv (2021)
14. Simchera, V.M.: Metody mnogomernogo analiza statisticheskih dannyh [Methods for multivariate analysis of statistical data]. Finansy i statistika, Moscow (2008)
15. Kalchenko, O.M., Androsenko, Ya.S.: Metodychni pidkhody do otsiniuvannia finansovoi bezpeky pidpryiemstva [Methodical approaches to assessing the financial security of the enterprise]. Probl. Prospects Econ. Manag. **2**, 244–250 (2015)
16. Lynenko, A.V.: Otsiniuvannia efektyvnosti vidtvorennia nematerialnykh aktyviv [Evaluating the effectiveness of the reproduction of intangible assets]. Probl. Theory Methodol. Account. Control Anal. **3**(21), 199–204 (2011)
17. Lysevych, S.H., Velikanova, V.K.: Upravlinnia finansovoiu stiikistiu ta faktory, shcho vplyvaiut na nei [Management of financial stability and factors influencing it]. Scientific works of Poltava State Agrarian Academy. Econ. Sci. Ser. **1**(6), 211–215 (2013). T. 2
18. Budrina, Ye.V. (ed.): Ekonomika otrasli. Avtotransport [Economy of the industry. Road transport]. Izdatelstvo Yurait, Moscow (2018)

Rating of Railway Development in Europe Country's

Gediminas Vaičiūnas(✉)

Vilnius Gediminas Technical University,
Saulėtekio al. 11, 10223 Vilnius, Lithuania
gediminas.vaiciunas@vilniustech.lt

Abstract. The authors of the article suggest a way to compare the development of countries' rail systems. This is calculating the geometric mean from the normalized values from the ratio of the length of the railways to the country indicators. This approach is illustrated by the example of European railways. The results of such an assessment show that the most developed railways are not in the countries with the longest railway network, but in those countries where there are many railways compared to other indicators.

Keywords: Railway development · Development indicators · Railway network

1 Introduction and Literature Review

As relations between countries develop, there is often a need to compare the development of the rail network between countries. For example, the construction of a new railway line across several countries raises the question of which countries will find this route more important and which less so [11, 12]. As well as the assessment of railway development is relevant for the representation of countries in various organizations and institutions. Planning the development of the railways also requires an objective comparison of the development of the railway system between the countries. Innovative new railway operators also enter international connections – the result of market liberalization [10]. When studying the development of railways in a particular country, it are usually considered through the prism of issues such as tariff policy for transport, technical and environmental requirements for rolling stock and infrastructure, the impact of competition in road transport on railways, state support for railways, regional social development and the need for new lines. This are confirmed by several examples from the literature below. Studies show that modern railway lines have a greater impact on less economically developed regions than on more developed ones [5]. It is also necessary to assess the historical context of the development and integration of railway systems [6]. The literature deals with research on the relationship between the performance of road and rail freight transport and transport infrastructure in EU countries [4]. Is examines the relationship between public investment in infrastructure in India and regional incomes and whether the character of this relationship is affected both by the nature of investment and the level of socio-economic development of the different regions [7]. The use of railway infrastructure, whether freight or passengers, shall be

O. Prentkovskis et al. (Eds.): TRANSBALTICA 2021, LNITI, pp. 561–567, 2022.
https://doi.org/10.1007/978-3-030-94774-3_55

subject to a charge. Depending on the infrastructure and economic characteristics of the country, this charge is different. There are also differences in the methodology used to calculate this charge, which can be read separately in the literature [8]. Rail transport rates tend to vary from country to country [15], the requirements for the railway line itself also differ, and the peculiarities of nature, climate and geography differ [1]. The railway lines are influenced by the cultural and political diversity of the countries it crosses. The studies shall take into account the level and nature of urbanization of the areas [2]. Different countries have different options for providing space for infrastructure facilities [13]. Particularly peculiar are countries located on regional borders such as Turkey or Azerbaijan (on the border between Europe and Asia). In countries such as these, the priority given to rail transport as a rule has recently been recognized. More developed road transport. The peculiarities of rail transport in such countries are analyzed in the literature [1].

Lithuanian research studies decided that a study should be started from the indicators that are no complicated, yet cover a broader range of issues. The indicator representing the social aspect could be the number of residents in a country, the indicator representing the economic aspect would be the country's gross domestic product, the geographic one would be area of the country and the indicator of the impact on the infrastructure could be the aggregate length of existing railway lines in the country [14]. Researchers have already done several studies on this principle. Researchers analysed the distribution of significance of the railway line Rail Baltica (constructed across Poland, Lithuania, Latvia and Estonia) and of the container train Viking route (going across Lithuania, Belarus and Ukraine) by countries [12]. The aim of this study is to assess the development of the European railway system in terms of economic, social and geographical indicators: the country's gross domestic product, population, and area. Multi-criteria methods are used for this purpose [9]. Research methodology developed based on these methods.

2 Methodology of Investigation

The simplest of the applied methods is the sum of the places of all criteria [3]. The complex criterion for each "*j*" object is determined according to the formula:

$$V_i = \sum\nolimits_{i=1}^{m} m_{ij}, \tag{1}$$

where m_{ij} is the location of the "*i*" criterion for the "*j*" alternative ($1 < m_{ij} < m$). The value of the best criterion *V* is the smallest. However, this method has drawbacks due to its simplicity. When evaluating the criteria (comparing their values), their order according to the alternatives is determined, however, the proportions are not evaluated. For example, whether the values differ by 5% or 5 times is not evaluated by this method.

More accurate is the SAW method (Simple Additive Weighting). Each member of the same alternative is multiplied by its significance and added to the other members of the alternative:

$$S_j = \sum_{i=1}^{m} w_i \cdot r_{i,j}, \tag{2}$$

where S_j is the value of the multi-criteria evaluation of the j-th alternative, w_i is the weight of the i-th criterion; r_{ij} is the normalized value of the i-th criterion for the j-th alternative. The best alternative is $S_j = S_{max}$.

The authors of the article suggest a way to compare the development of countries' rail systems. This is calculating the geometric mean from the normalized values from the ratio of the length of the railways to the others country indicators. First, the values of the indicators are collected and systematized. This study assesses the length of the rail network, the countries'gross domestic product, the country's area, and population. The next step is to calculate the ratio of the length of the rail network to other indicators in the country:

$$R_{ij} = \frac{L_i}{I_{ij}}, \tag{3}$$

where L_i – the length of the rail network in the "i" country, km. I_{ij} – other ("j") indicator in the country "i".

Calculating this ratio based on different indicators gives different units of measurement, which complicates the possibility of comparing these values with each other. This problem is easily solved by the normalization method. For comparison, the proportion of the indicator value for that object is calculated from the sum of the indicators for all objects:

$$N_{ij} = \frac{R_{ij}}{\sum_{1}^{j=k} R_{ij}}, \tag{4}$$

where R_{ij} – the ratio of the length of the rail network to "i" indicator in the country, k – number of country's.

For each country, as many N_i values as the criteria are considered (indicators excluding the length of the rail network). As these are normalized indicators derived from relative indicators, it is recommended to summarize them on the principle of geometric mean (although other ways of generalization are possible):

$$N_i = \prod_{1}^{i=n} N_{ij}, \tag{5}$$

where n – number of indicators (excluding the length of the rail network).

Then the countries are evaluated according to the value of the indicator N, the higher N is – the higher the development of the country's railways.

3 Research Process and Results

The length of the railway network is generally taken as an indicator of the development of the railway in the country [16, 17]. Often such an approach is perfectly acceptable. In this case, the countries with the most developed railways are shown in Fig. 1.

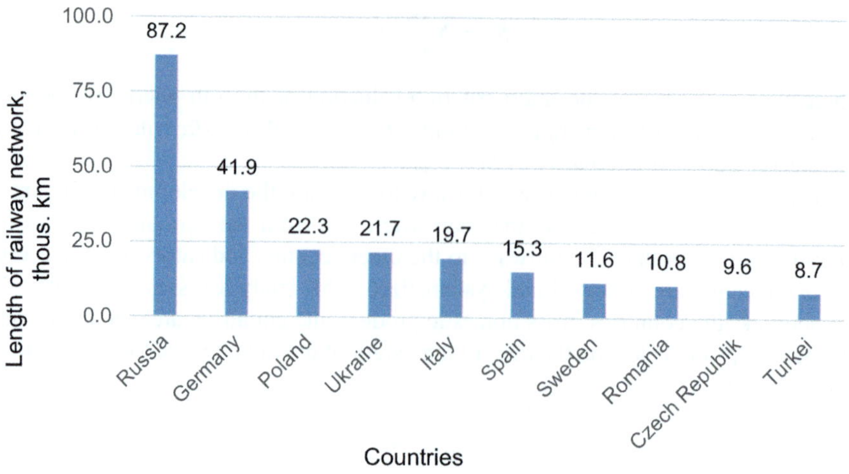

Fig. 1. The countries with the most developed railways.

The length of the Russian railway network is more than twice the length of the German railway network and four times that of Poland (Although Russia is not only in Europe but also in Asia, most of its railways are in Europe). However, the length of the railways is not the only indicator of a country's development. Which means that the length of the rail network can also be considered in relation to other indicators. The size of a country is well defined by its gross domestic product, area, population. Therefore, the length of the rail network can be estimated as a ratio to the above indicators (Table 1).

Table 1. Country development indicators.

No.	Country	Length of railway network, thous. km	GDP, trln. USD	Area, mln. km^2	Population, mln. people
1	Russia	87.2	2.2	1,710.00	140.70
2	Germany	41.9	2.9	0.36	8.14
3	Poland	22.3	0.667	0.31	38.65
4	Ukraine	21.7	0.337	0.60	42.02
5	Italy	19.7	1.8	0.30	60.00
6	Spain	15.3	1.4	0.50	45.72
7	Sweden	11.6	0.349	0.45	10.14
8	Romania	10.8	0.271	0.24	18.83
9	Czech Republic	9.6	0.663	0.08	10.58
10	Turkey	8.7	0.9065	0.78	84.76

Ratio of the length of the railway network to the indicators of country are shown in Table 2.

Table 2. Ratio of the length of the railway network to the indicators of country.

No.	Country	Ratio of length of railway network to GDP, thous. km/trn USD	Ratio of the length of the railway network to the area of the country thous. km/million km^2	Ratio of the length of the railway network to the population, thous. km/million
1	Russia	39.6	0.1	0.62
2	Germany	14.4	117.3	5.14
3	Poland	33.5	71.4	0.58
4	Ukraine	64.3	35.9	0.52
5	Italy	11.0	65.5	0.33
6	Spain	10.9	30.3	0.33
7	Sweden	33.3	25.9	1.15
8	Romania	39.8	45.4	0.57
9	Czech Republic	14.5	122.0	0.91
10	Turkey	9.6	11.1	0.10

Normalized ratio of the length of the railway network to the indicators of country are shown in Table 3.

Table 3. Normalized ratio of the length of the railway network to the indicators of country.

No.	Country	Normalized ratio of length of railway network to GDP	Normalized ratio of the length of the railway network to the area of the country	Normalized ratio of the length of the railway network to the population	Geometric mean of normalized ratios
1	Russia	0.146	0.000	0.060	0.010
2	Germany	0.053	0.255	0.502	0.190
3	Poland	0.123	0.155	0.056	0.103
4	Ukraine	0.237	0.078	0.050	0.098
5	Italy	0.040	0.142	0.032	0.057
6	Spain	0.040	0.066	0.033	0.044
7	Sweden	0.123	0.056	0.112	0.092
8	Romania	0.147	0.099	0.056	0.093
9	Czech Republic	0.054	0.265	0.089	0.108
10	Turkey	0.035	0.024	0.010	0.020

The Geometric mean of normalized ratios of Countries is shown in Fig. 2.

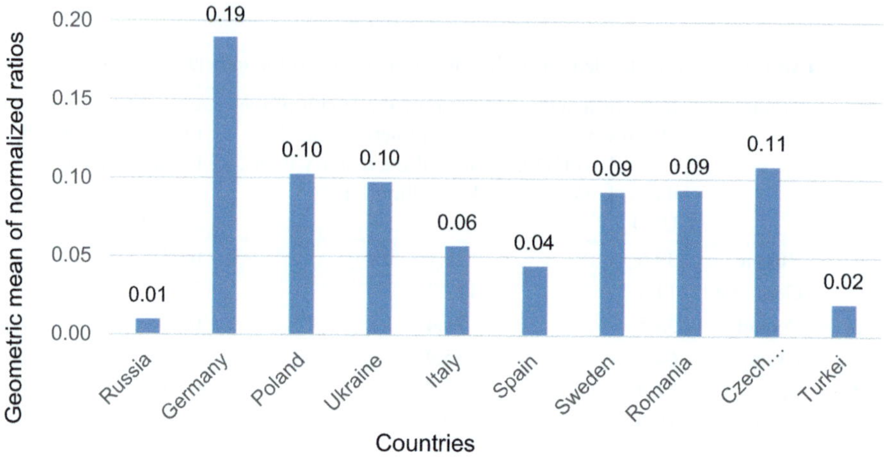

Fig. 2. Geometric mean of normalized ratios of Countries.

The results of assessment show that the most developed railways are not in the countries with the longest railway network, but in those countries where there are many railways compared to other indicators. For example, the German railway network is almost twice as short as the Russian railway network, but due to its high density and high economic development indicators, Germany leads in the aggregate railway development indicator.

4 Conclusions

1. One from ways to compare the development of countries' rail systems is calculate the geometric mean from the normalized values from the ratio of the length of the railways to the country indicators.
2. The biggest geometric mean from the normalized values from the ratio of the length of the railways to the country indicators are for Germany, Czech Republic, Poland and Ukraine.
3. According to the indicator in question, German railways are about twice as developed as the railways of other leading countries. Due to its high rail density and high economic and social performance, Germany leads the way in terms of overall railway development.
4. In countries such as Russia, Turkey or Spain, the significance of the indicator in question is relatively small, as the countries have large and large significance for other indicators.
5. The results of assessment show that the most developed railways are not in the countries with the longest railway network, but in those countries where there are many railways compared to other indicators.

References

1. Babalik-Sutcliffe, E.: Pro-rail policies in Turkey: a policy shift? Transp. Rev. **27**, 485–498 (2007)
2. Chen, C., Loukaitou-Sideris, A., Ureña, J., Vickerman, R.: Spatial short and long-term implications and planning challenges of high-speed rail: a literature review framework for the special issue. Eur. Plan. Stud. **27**, 415–433 (2019)
3. Ginevičius, R., Podvezko, V., Ginevičienė, V.B.: A complex evaluation of the work productivity of university teaching staff. In: Seminar Proceedings of 8th Baltic Region Seminar on Engineering Education, pp. 44–46. Kaunas University of Technology, Kaunas (2004)
4. Gnap, J., Varjan, P., Durana, P., Kostrzewski, M.: Research on relationship between freight transport and transport infrastructure in selected European countries. Transp. Probl. **14**(3), 63–74 (2019)
5. Liang, Y., Zhou, K., Li, X., Zhou, Z., Sun, W., Zeng, J.: Effectiveness of high-speed railway on regional economic growth for less developed areas. J. Transp. Geogr. **82**, 102621 (2020)
6. Martí-Henneberg, J.: European integration and national models for railway networks (1840–2010). J. Transp. Geogr. **26**, 126–138 (2013)
7. Ravi Kumar, T.: The impact of regional infrastructure investment in India. Reg. Stud. **36**(2), 194–200 (2002)
8. Sánchez-Borràs, M., López-Pita, A.: Rail infrastructure charging systems for high-speed lines in Europe. Transp. Rev. **31**, 49–68 (2011)
9. Sivilevičius, H., Maskeliūnaitė, L.: The model assessing the impact of price and provided services on the quality of the trip by train: MCDM approach. Econ. Manag. **22**(2), 51–67 (2019)
10. Seidenglanz, D., Taczanowski, J., Król, M., Hornak, M., Nigrin, T.: Quo vadis, international long-distance railway services? Evidence from Central Europe. J. Transp. Geogr. **92**, 102998 (2021)
11. Vaičiūnas, G.: Assessment of railway lines: an efficiency rating analysis for Baltic countries. Transp. Probl. **13**(1), 49–58 (2018)
12. Vaičiūnas, G., Steišūnas, S.: Investigation of priority directions of rail baltica extension from Warsaw. In: Procedia "Transbaltica 2017", pp. 40–45. Vilnius Gediminas Technical University, Vilnius (2017)
13. Wang, K., Li, G., Liu, H.: Porter effect test for construction land reduction. Land Use Policy **103**, 105310 (2021)
14. Yu, M., Fan, W.: Fan Accessibility impact of future high speed rail corridor on the piedmont Atlantic megaregion. J. Transp. Geogr. **73**, 1–12 (2018)
15. Zakeri, J.A.: Investigation on railway track maintenance in sandy-dry areas. Struct. Infrastruct. Eng. **8**(2), 135–140 (2012)
16. Central Intelligence Agency. The World Factbook. https://www.cia.gov/library/publications/the-world-factbook/rankorder/2121rank.htm. Accessed 10 Jan 2021
17. World population. https://countrymeters.info/en/World. Accessed 10 Jan 2021

Research of Geographical Information Systems of Graded Transport Flow Networks of Ukraine

Ievgeniia Ugnenko[1(✉)], Elena Uzhviieva[1], Nataliia Sorochuk[1], Yevhen Korostelov[1], and Gintas Viselga[2]

[1] Department of Research and Design of Communication Paths, Geodesy and Land Management, Faculty of Construction, Ukrainian State University of Railway Transport, Kharkiv, Ukraine
kostya_90_@ukr.net

[2] Department of Mechanics and Materials Engineering, Faculty of Mechanics, Vilnius Gediminas Technical University, Vilnius, Lithuania
gintas.viselga@vilniustech.lt

Abstract. The study presents the local geometric characteristics of graded networks and describes an innovative method of presenting graded networks – the tuple method – by means of the local characteristics organized by special tuples. The authors suggested the method for presenting organization networks, which fully complies with the requirements for coherence, algorithmic simplicity, clarity and efficiency. The choice of a motorway on the prospective residential area developments requires such a layout which satisfies the planning requirements for urban streets in terms of the best performance factors of the automobile transportation and traffic safety. Each road layout is determined by various conditions, such as the historical street network, production, geomorphologic, ecological, hydrological, and other factors which impact the formation of a design road network. These data can be used for forecasting growth opportunities and the final layout of an urban territory. The authors also suggested the method of ecological and energy analysis in order to determine the radius (remoteness) of designed bypass roads around the urban areas of Group 1 according to the classification approved. This method is called the simultaneous access method. The travel time ratio between the given points (along the bypass roads and urban streets) was taken as an indicator for the ecological and energy substantiation of the bypass road radius. The research is based on the method of simultaneous access which has been used for the ecological and energy substantiation of the designed bypass roads for linear residential areas (Groups II and III).

Keywords: Geological information systems · Graded networks · Tuple method · Roadway · Transport flow · Ecological and energy method · Simultaneous access method · Bypass road radius · Residential area · Linear

O. Prentkovskis et al. (Eds.): TRANSBALTICA 2021, LNITI, pp. 568–577, 2022.
https://doi.org/10.1007/978-3-030-94774-3_56

1 Introduction

The network and graph are classical concepts of discrete mathematics [1, 2]. They are traditionally used for modelling and research into the network's objects, such as transport networks (railway, automobile, etc.), organization structures, networks of business and legal interaction of economic entities, and information exchange between the objects [3, 4]. Studies [5, 6] give the concept of a graded network and explain that most networks in practice are graded. The graded network G is a network with the homomorphism $\varphi : G \to \langle\{1, 2, ..., N\}; \leq\rangle$ to the initial section of a natural row $\{1, 2, ..., N\}$, ordered in a natural way. Meaningfully, the mapping φ assigns to the elements (entities) of the network G a natural number, priority or rank, which indicate the significance and importance of an entity in the network in terms of its production functions.

Let us assume that we have the graded network G and its graduation $\varphi : G \to \langle\{1, 2, ..., m\}; \leq\rangle$. The graduation φ naturally divides the set of the nodes of the network G into levels (equivalent classes) $G = G_1 \cup G_2 \cup ... \cup G_m$ of nodes of the same rank. Here $G_j = \varphi^{-1}(j)$, which implies that the level G_j contains all the nodes of the network G with the rank j.

The levels $G_1, G_2, ..., G_m$ can be presented as stacked horizontal plates; each plate contains all the nodes with the rank equal to the plate number [7]. Here the edges of the network G can connect the nodes of the level G_j with the nodes of the levels G_{j-1}, G_j and G_{j+1}, but they cannot skip a level (e.g. connect the nodes of the levels G_{j-1} and G_{j+1}) because it contradicts the definition of the graded network.

The presentation of networks through adjacency matrices (or flow matrices) is a standard widely used method to assign and store networks in digital form [1]. Suppose that we have the network $G = \langle\{P_1, P_2, ..., P_N\}; E\rangle$, the nodes of which are numbered with consecutive natural numbers (here k- number of the node P_k, N – the number of nodes in the network G, E – the number of sections (edges)). The adjacency matrix (g_{ij}) of the network G has the dimensions $N \times N$ and it is determined as follows:

$$g_{ij} \begin{cases} 1, \text{ if there is an edge between the nodes } P_i \text{ and } P_j \\ 0, \text{ if there is no edge between the nodes } P_i \text{ and } P_j \end{cases}$$

As it is known, any network is determined by its adjacency matrix unambiguously and accurately to an isomorphism [1]. It explains a widespread use of adjacent matrices for storage, analysis and computer processing of information on organization networks and other network structures. However, the method of presenting networks with adjacency matrices has its drawbacks. First of all, this matrix method is excessive (especially with a priori information about the network organization and its diagram). The bulk memory in adjacency matrices are occupied by the same numbers, the matrices are large, therefore, the work of various standard algorithm of networks analysis (particularly, exhaustive algorithms, such as a Hamilton cycle or finding the shortest way) is difficult and requires much more time.

The redundancy of the method of presenting networks with adjacency matrices is explained by its universal nature and possibility to be applied in any situation. It stands to reason that for specific networks, such as graded ones, there is a simpler, clearer and more efficient way of presentation, which, nevertheless, has all the advantages and useful characteristics of the matrix representation. Moreover, regarding this method it is reasonable to impose requirements for simplicity and high speed of standard algorithms for the network analysis.

2 Application of the Method of Graded Networks to Substantiate Bypass Roads for the Residential Area of Ukraine

The research by the authors of the article presents the innovative method of representing organization networks, which fully conforms to the requirements for an ambiguity, algorithmic simplicity, visual clarity and efficiency.

Let us assume the graded network $G = \langle G_1 \cup G_2 \cup \ldots \cup G_m; E \rangle$.

Let us define two characteristics for each node $P \in G_j \subseteq G$ on the level G_j: $R(P)$ – the number of edges from the node P to the nodes of the same rank, i.e., to the nodes of the level G_j; $S(P)$ – the number of edges from the node P to the nodes of the next rank, i.e. to the nodes of the level G_{j+1}.

Let us randomly index the nodes of each level G_j with natural numbers, i.e. denote each node $P \in G_j$, $\varphi(P) = j$ with the natural number $i = i(P) \in \{1, 2, \ldots |G_j|\}$– the number of this node. Here $|G_j|$ is the number of nodes at the level G_j. Naturally, let us assume that the indexing $i : G_j \to \{1, 2, \ldots |G_j|\}$ is a bioactive mapping; the numbers of different nodes cannot coincide: $i(P) \neq i(Q)$.

After assigning the numbers to the nodes of each level of the network G, let us associate each node P of the network G with the following tuple $K(P)$ of natural numbers:

$$P \mapsto K(P) = \begin{pmatrix} j = \phi(P),\ i = i(P); \\ r_1, r_2, \ldots r_{R(P)}; n_1, n_2, \ldots n_{S(P)} \end{pmatrix}. \tag{1}$$

In the tuple $K(P)$ the first two numbers are the rank $\varphi(P)$ of the node P and the number $i(P)$ of the node P on its level G_j in the accepted numeration. Let us call the first two numbers j, i of the tuple $K(P)$ the prefix part. The numbers $r_1, r_2, \ldots r_{R(P)}$ are the intent of the nodes on the level G_j adjacent to the node P on the level G_j (all these nodes have the rank equal to the rank of the node P). Assume that if the node P does not link to the nodes of its level G_j, in the tuple $K(P)$ instead of a group of the numbers $r_1, r_2, \ldots r_{R(P)}$ is zero. The numbers $n_1, n_2, \ldots n_{S(P)}$ are the intent of the number of the nodes on the level G_{j+1}, adjacent to the node P, i.e. the numbers of the nodes lying on the next level G_{j+1} and linked to the node P (all such nodes have the rank $\varphi(P) + 1$). And assume that if the node P does not link to the nodes on the next level G_{j+1}, the tuple $K(P)$ has zero instead of $n_1, n_2, \ldots n_{S(P)}$. Let us call the numbers $(r_1, r_2, \ldots r_{R(P)}; n_1, n_2, \ldots n_{S(P)})$ the suffix part of the tuple $K(P)$.

By associating each node of the network G with a corresponding tuple, we obtain the tuple set $K(G)$ in the form of (1) for all nodes of the network G.

According to the tuple set $K(G)$ let us build the tuple table $T(G)$ which is organized the following way: the j-th line of the tuple table $T(G)$ has all tuples of the nodes of the level G_j in the increasing number order, i.e. the prefix part of each tuple $K(P)$ has the line number and the column number of the table cell with the given tuple $K(P)$. It should be noted that on completion of the table the prefix part of each tuple $K(P)$ can be removed, as it is easily restored according to the location of the tuple $K(P)$ in the table $T(G)$ [8]. An example of the graded network G and the corresponding tuple table $T(G)$ is given in Fig. 1.

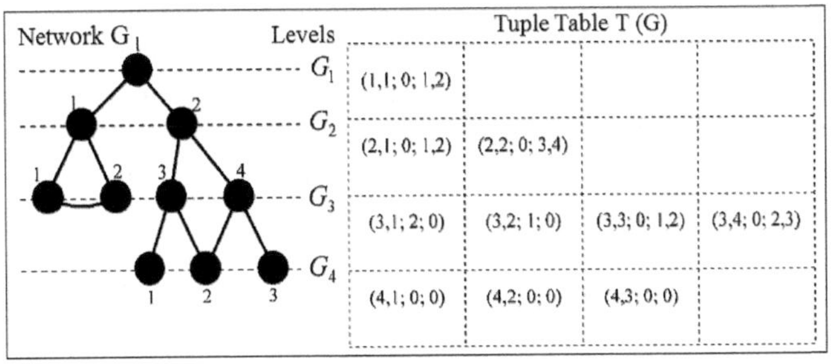

Fig. 1. Example of the graded network G and the corresponding table.

Let us take the tuple table $T(1)$. If for the graded network G we can index the nodes on each level so that the given taple table T is the tuple table of this network G, i.e. $T = T(G)$, let us say that the network G satisfies the tuple table T, or, which is the same, that the tuple T is fulfilled on the network G.

Let us say that two graded networks G and H are isomorphic if there is the bijection $\psi : G \rightarrow H$ is such that for any two nodes $g_1, g_2 \in G$ of the network G and their corresponding nodes $\psi(g_1)$, $\psi(g_2)$ in the network H the following is fulfilled:

a) the nodes g_1 and $\psi(g_1)$ are of the same rank;

b) the nodes g_1, g_2 are connected by an edge in the network G only when their images $\psi(g_1)$, $\psi(g_2)$ are connected by an edge in the network H.

According to the authors of the article the process of decision taking is as follows: to introduce the initial data, to assign a number/dislocation variant (according to the sector or the nearest exit from the region) of graded networks, and to calculate the transport expenditures for motor vehicles. The analysis of the results obtained is used for an intermediate decision on the study of the transport flows and substantiation of the bypass roads for the residential areas [9]. The calculation is repeated for each assigned variant; results are used for making the final decision on efficiency of a variant of bypass roads for the residential areas. The best variant is chosen according to the requirement of the graded network G. The variants are calculated simultaneously by motor vehicles.

3 Substantiation of the Bypass Road Construction for the Linear Residential Areas

The road designing for the territories of existing and promising urban development of Ukraine is characterized by a probable conversion of the roadways into the urban streets with the appropriate road traffic and urban growth characteristics due to expansion of the territory. Therefore, the road planning on the residential areas should comply with the planning requirements for the street network which fulfills the best transport and forwarding requirements for road transportation and traffic safety [10].

Taking into account the prospective growth of the residential areas it is reasonable to classify them by growth characteristics. The trends of development and the outline of prospective urban area will be useful for setting the territory on which the planning requirements should be met for main road sections [11] and choosing a certain road network layout. One of the most important issues in designing a roadway network on the territory of prospective urban development is to substantiate the construction of bypass roads [12]. The tailor-designed and constructed bypass roads can direct the transit motor vehicles out of the residential area and service part of the urban traffic. The classification of urban areas is given in Table 1. The authors suggested the method of ecological and energy substantiation of the radius of the bypass roads designed around the residential areas of Group 1 according to the classification presented in Table 1 [12]. This method is called the simultaneous access method. The travel time ratio for the given points was taken as an indicator for the ecological and energy substantiation of the bypass road radius while comparing a travel by the bypass road and the urban street. The research is based on the method of simultaneous access for the ecological and energy substantiation of the bypass roads designed for the linear residential areas (Groups II and III).

Table 1. Classification of urban areas.

Groups of urban areas	Character of urban growth	Typical forms of population distribution (subgroups)
I	Urban areas of free growth	A. Single compact urban area B. Metropolitan area
II	Urban areas of partial growth	A. Single compact urban area B. Linear residential areas
III	Urban areas of restricted growth	A. Single compact urban area B. Linear residential areas

In the simplest case these bypass roads are two parallel roadways along the residential area looped at both ends. The procedure of choosing the bypass road outlay around such urban areas should include two stages: for the roads entering the urban area from the longitudinal sides and for the roads entering the urban area from the transverse sides. If a road enters from the end section of the urban area and if the network of transverse streets is dense (Fig. 2), the bypass road will be more convenient if the travel from the point *A* (junction of the main section to the bypass road) to the point D (in the urban territory) along the route $ABCD$ (along the bypass road) takes less time than along the route AED (along the entry road), i.e.

$$\frac{R+a}{V_k}+\frac{R-b}{V_{\mathrm{av}}}\leq\frac{a+b}{V_{\mathrm{av}}}, \tag{2}$$

where R – distance from the entry road to the bypass road section, m; a – distance from the transverse bypass road section to the point E, m; b – distance from the entry road to the point D, m; V_{av} – average speed of traffic flow, km/h; V_k – traffic speed during peak hours, km/h.

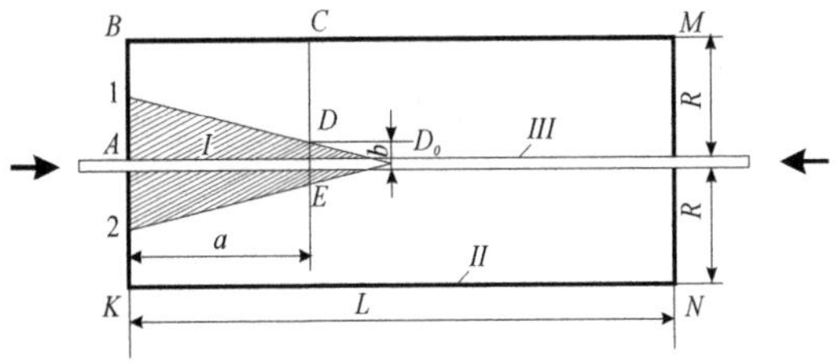

Fig. 2. Configuration of the urban territory in the zone of transport impact from the entry road: I – simultaneous access line; II – bypass road; III – longitudinal entry road.

Solution to inequality (2) can be used for finding the position of the simultaneous access line for motor vehicles directing to the urban area along the bypass road. The change of inequality (2) gives $2bV_k \geq RV_{\mathrm{av}}+aV_{\mathrm{av}}+RV_k-aV_k$, from which:

$$b=\left(\frac{R+a}{V_k}\right)+\frac{V_{\mathrm{av}}}{V_k}\leq\frac{R-a}{2}. \tag{3}$$

For borderline cases at $a=0$,

$$b\geq\frac{R}{2}\left(\frac{V_{\mathrm{av}}}{V_k}+1\right) \tag{4}$$

at $b=0$,

$$a=\frac{R}{\left(1-\frac{V_{\mathrm{av}}}{V_k}\right)}\left(\frac{V_{\mathrm{av}}}{V_k}+1\right) \tag{5}$$

The function $b=f(a)$, expressed by equality (3), is shown graphically in Fig. 2 for motor vehicles entering the urban area along one of the entry roads.

In the course of the design work, the authors of the article determined that the diagram demonstrates the urban area distribution by the separate zones serviced by a road. The urban area surrounded by the simultaneous access lines (shadowed) can be serviced only by a longitudinal entry road, the other territory – by a bypass road. It should be taken into account that the zone is changeable and depends on the end sections of the urban area which are becoming closer, and the transverse sections of the bypass road which are becoming more remote. At the same time the zone serviced by the bypass road will enlarge. If it is impossible to develop the end area, the urban area serviced by the bypass road will not change. It can be used for location of various administrative and residential buildings, which naturally do not attract a lot of motor vehicles [12]. With two entry roads at the end sides the alignment of the similar although differently oriented diagrams allows distinguishing several transport impact zones on the urban area (Fig. 3).

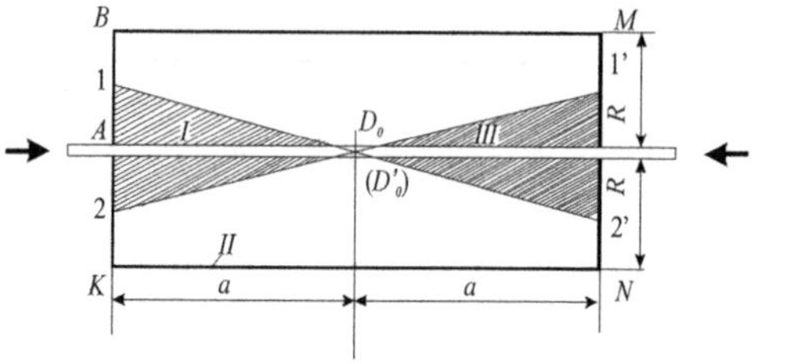

Fig. 3. Location of the urban territory included in the transport impact zones of two entry roads at an urban territory length of 2*a*.

Figure 3 demonstrates an expansion of the area serviced by the bypass roads during a further increase of the urban area in length. If the length of an urban area exceeds 2*a*, the territory between the points D_0 and D_0' will be serviced by the bypass road along the whole width.

The urban areas serviced separately by the bypass and entry roads can be taken as the criterion of efficient construction of bypass roads for the linear urban areas.

With road entries adjoining the urban area in the transverse direction (Fig. 4), the simultaneous access condition for the cars moving along the entry road *BCDE* and the bypass road *BLKOE* can be written as the equation [12]:

$$\frac{2A - a + 2R}{V_k} + \frac{R - b}{V_{\text{av}}} = \frac{R + b + a}{V_{\text{av}}}, \tag{6}$$

where *A*– distance from the entry road to the transverse section of the bypass road; *R* – distance from the longitudinal section of the bypass road to the longitudinal entry road; and *a* and *b* – coordinates of the point on the simultaneous access line.

After solving Eq. (6), we will obtain the formula for determination of the coordinates of points on the simultaneous access line for cars.

The distance *b* is defined by the formula [12]:

$$b = (A + R)\frac{V_{\text{av}}}{V_k} - \frac{a}{2}\left(1 + \frac{V_{\text{av}}}{V_k}\right). \tag{7}$$

The entry roads adjoining to the urban area in the transverse direction (Fig. 4). For borderline cases at $b = 0$,

$$a_{b=0} = \frac{2(A+R)\frac{V_{av}}{V_k}}{1+\frac{V_{av}}{V_k}}. \tag{8}$$

At $b = R$,

$$R = (A+R)\frac{V_{av}}{V_k} - \frac{a}{2}\left(1+\frac{V_{av}}{V_k}\right),\ a_{b=R} = \frac{2\left[(A+R)\frac{V_{av}}{V_k} - R\right]}{1+\frac{V_{av}}{V_k}}. \tag{9}$$

At a = 0 and $b = R$, from formula (7),

$$R = (A+R)\frac{V_{av}}{V_k},$$

from which:

$$A = R\left(1-\frac{V_{av}}{V_k}\right)\frac{V_k}{V_{av}}. \tag{10}$$

Therefore, at such a layout of the transverse entry roads from the end section of the bypass road the travel is convenient both across the urban area and along the bypass road. Under the condition $A > R\left(1-\frac{V_{av}}{V_k}\right)\frac{V_k}{V_{av}}$ the travel by car to the opposite part of the urban area along the bypass road is not convenient.

As it is seen in Fig. 4 if there are transverse entry roads only from one side of the urban area, most urban area (not shadowed) is serviced by the bypass road; it particularly refers to the urban zones located at the junction of the entry road.

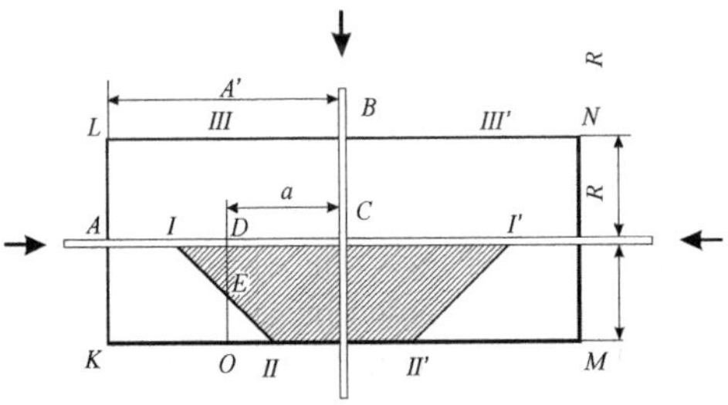

Fig. 4. Configuration of the urban territory located in the impact zone of the transverse entry road: I – entry road; II – bypass road; III – transverse entry road.

Thus, it should be noted, that the bypass roads around the linear urban areas can completely direct transit motor vehicles beyond the urban territory. The construction of bypass roads around the linear residential areas can be regarded as the most efficient way for entry roads along the urban areas [12]. Such bypass roads can service all transit motor vehicles and part of the urban traffic moving to the opposite urban districts. If the transverse entry roads adjoin from two sides, the urban zone serviced by the bypass road decreases and occupies only the end parts of the urban territory. With several transverse and longitudinal entry roads being in service at the same time, this zone decreases even more intensively (Fig. 5).

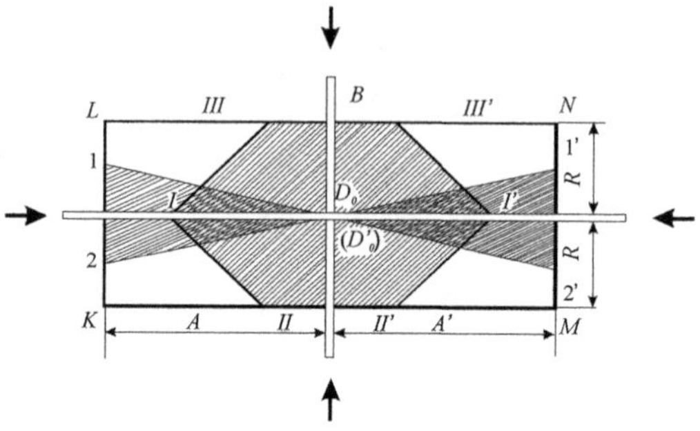

Fig. 5. Combination of the urban territories located in the transport impact zones of the longitudinal and transverse entry roads.

The bypass roads around the linear residential areas can be of great value together with the transverse entry roads located at one side only. Here, the urban territory located at the transverse entry roads is completely serviced by the bypass road.

4 Conclusions

A change in the indexing of the nodes on the level *Gj* of a graded organization network causes a simple replacement of the tuples in the *j*-th line of the table *T*(*G*) and appropriate replacement of the second parts of the suffixes of the tuples in the (j – 1) line, if any. It is clear that such a change of the tuple table *T*(*G*) is a simple effective process which allows building the tuple table of the graded network *G* by changing the numeration of the nodes.

The bypass roads around the linear residential areas can completely direct transit motor vehicle beyond the residential areas. The construction of the bypass roads around the linear residential areas can be regarded most efficient for the entry roadways running along the residential areas. Such a bypass road can service all transit motor vehicles and part of the urban traffic moving to opposite urban districts.

If the transverse entry roadways adjoin from two sides, the urban zone serviced by the bypass road considerably decreases and occupies only the ends of the urban territory. With several transverse and longitudinal entry roadways being in service at the same time, this zone decreases even more.

The bypass roads around the linear residential areas can be of great value even with the transverse entry roads at one side only. Here, the urban territory located on the side of the transverse entry roads is completely serviced by the bypass road.

References

1. Емеличев, В.А., Мельников, О.И., Сарванов, В.И., Тышкевич, Р.И.: Лекции по теории графов. Наука, М., Гл. ред. физ.-мат. лит. (1990). 384 с.
2. Филлипс, Д., Гарсиа-Диас, А.: Методы анализа сетей. Мир, М. (1984). 496 с.
3. Сай, В.М.: Планетарные структуры управления на железнодорожном транспорте. Транспорт: наука, техника, управление. ВИНИТИ РАН, М., no. 4, pp. 8–11 (2002)
4. Сай, В.М.: Планетарная структура: корпоративные варианты. Мир транспорта (1), 96–102 (2003)
5. Сай, В.М., Сизый, С.В.: Организационные структуры как мультиоператорные сети. Задачи прочности и устойчивости. Транспорт Урала **2**(21), 5–9 (2009)
6. Сизый, С.В.: Модель образования и распада организационных сетей в экономико-правовом пространстве. ВИНИТИ РАН Транспорт наука техника управление (3), 16–26 (2010)
7. Сизый, С.В.: Сетевая поддержка предприятий в градуированных организационных сетях. Вестник УрГУПС. Екатеринбург: УрГУПС, no. 1, pp. 33–45 (2010)
8. Кейслер, Г., Чэн, Ч.Ч.: Теория моделей. Мир, М. (1977). 614 с.
9. Угненко, Є.Б., Ужвієва, О.М.: Матеріали Х Міжнародної науково-технічної конференції «АВІА-2011» (м. Київ, 19–21 квітня 2011 р.). НАУ, Київ, vol. 3, pp. 24.59–24.62 (2011)
10. Бабков, В.Ф.: Ландшафтное проектирование автомобильных дорог: [учеб.]. Транспорт. М. (1980). 192 с.
11. Проектування автомобільних доріг: [підруч. для студ. вищ. навч. закл.]/Білятинський О. А. [та ін.]. Вища школа, К. (1997). 517с.
12. Угненко, Є.Б., Ужвієва, О.М.: Удосконалення методу обґрунтування будівництва обходів населених пунктів з урахуванням екологічних показників Монографія. ХНАДУ, Харків (2014). 136 с.

Review of Engineering Research Methods for the Formation of a Digital Model of the Area with the Determination of the Accuracy and Compliance

Ievgeniia Ugnenko[1(✉)], Anna Shevchenko[1], Oleksander Shevchenko[2], and Gintas Viselga[3]

[1] Ukrainian State University of Railway Transport, Feierbakh Square 7, Kharkiv 61050, Ukraine
[2] Kharkiv Design Institute "Teploelektroproekt-soyuz", Shevchenko Street, 233B, Kharkiv 61000, Ukraine
[3] Vilnius Gediminas Technical University, J. Basanavicius Street, 28, Vilnius, Lithuania
gintas.viselga@vilniustech.lt

Abstract. The article considers the main methods of engineering surveys (leveling by diameters, tacheometric survey, laser survey, survey by laser profilers, GPS-survey) survey of the area with the main advantages and disadvantages. Processing of the received data at system automated designing as a result of topographic and geodetic works is a digital model of district (which consists of an array of points) of a design zone. During geodetic survey or its in-house processing, the points are given certain attributes necessary for adequate modeling of surfaces, situations and correct implementation of all subsequent design procedures. Since in real projects of complex structures, the engineer deals with arrays of tens and hundreds of thousands of points, it is natural that modern computer-aided design systems must have advanced means of visual editing. For modeling of surfaces there are different types of structures: horizontally, on structural lines, on diameters to a route or the main course, statistical. The problem of constructing a surface by the method of triangulation is one of the basic ones in computational geometry. It includes many others related to surface modeling and solving spatial problems in machine graphics, computer-aided design systems and geographic information systems. Also, a study of the dependence of the height of the terrain and the percentage of error of the obtained digital terrain model. Therefore, based on the results of calculations, it can be concluded that the materials of the survey follow the instructions for topographic surveying for the use of these works in further design.

Keywords: Engineering surveys · Digital terrain model · Three-dimensional terrain model · Error and mean error · Laser imaging · Triagulation · Point "clouds" · Error percentage · Laser-location image

O. Prentkovskis et al. (Eds.): TRANSBALTICA 2021, LNITI, pp. 578–588, 2022.
https://doi.org/10.1007/978-3-030-94774-3_57

1 Methods of Engineering and Geodetic Surveys

Engineering and geodetic surveys for the construction (reconstruction) of roads should provide topographic and geodetic materials and data on the situation and terrain, existing buildings and structures, planning elements necessary for a comprehensive assessment of natural and man-made conditions of the construction site [1].

The composition of engineering and geodetic surveys for the construction of roads includes [3, 14]:

- collection and processing of materials of engineering researches of the last years, topographic and geodetic, cartographic, aerial photography and other materials and data;
- reconnaissance survey of the territory;
- creation (development) of geodetic reference networks, including geodetic networks of special purpose for construction;
- creation of plan-height survey geodetic networks;
- topographic (ground, aerial photographic, stereophotogrammetric) shooting;
- updating of engineering-topographic and cadastral plans in graphic, digital, photographic and other forms;
- engineering and hydrographic works;
- in-house tracing and preliminary selection of competitive route options for field work and surveys;
- field tracing;
- survey of existing railway and highways, assembly of longitudinal and transverse profiles, intersections of power lines, communication lines, radio facilities and main pipelines.

Regardless of the method of geodetic surveys of communication routes, it is important that the density of survey points was high and uniform in the longitudinal and transverse directions. Thus, you can get the most adequate reflection of the existing surface.

The main factors that must be taken into account when choosing a search technology are the speed of work and the reliability of the results.

Leveling by diameters technology of geodetic surveys, based on the leveling of diameters along the project route, is the most traditional and brought from the technology of surveys of new roads of the period 50–60 - ies of the twentieth century. Its essence is that the route is traced along the axis of the road, the route is fixed, the picket is broken and, with a given step, the leveling of diameters normal (perpendicular) to the axis of the designed road is performed. This technology is extremely simple, which requires the use of the simplest geodetic tools (theodolites, levels, rails, tape measures), which ensures its survivability even today. However, this technology has a number of shortcomings that do not allow to consider it as a base for geodetic surveys for the design of reconstructions and repairs of roads:

- first, tracing (in the case of reconstruction and repair) in the field, and even on the basis of traditional tracing elements, does not allow to perform this procedure quite well, i.e., with the maximum approximation of the designed route in relation to the existing one;

- secondly, at the same time, it is assumed that the designed route repeats the outlines of the existing one. On this basis, perform further search procedures;
- thirdly, changes in design decisions along the route at the stage of in-house works are no longer possible. This fact is characteristic in general for geodetic surveys of paths "picket" method.

Tacheometric survey provides the required accuracy of measurements, but, at the same time, quite time consuming, especially in conditions of high traffic congestion of the designed road. Currently, the most common type of geodetic measurements in surveys for the design of roads is tacheometric survey.

Among the currently existing variety of laser geodetic techniques, the most effective application for the search for paths, laser finishers.

The computer laser system allows to determine with geodetic accuracy the marks of cross-section points with a step of 10 cm. The device is equipped with a special trolley with a built-in path meter and has electronic segments with a matrix scheme of photodiodes. A separately available emitter generates a beam in the visible spectrum, which, falling on a certain segment and photodiode of the device, causes the operation of the corresponding circuit of the electronic circuit and is written to RAM. The frequency of the registered points is regulated and makes 100–300 points on diameter, causing display of an actual surface in the form of dense sequence of points [4].

After converting the received information into a digital model of the computer-aided design system, it is possible to start the design process on the basis of complete information on the outlines of the existing surface of the repaired (modernized) roads.

Laser profilography is the most productive. During one shift you can shoot 100–150 km. However, due to the fact that laser profilographs are usually a vehicle attachment, the accuracy of such measurements is not high. This is due to the fact that the measurement error is made by the action of the sprung suspension of the car.

The operation of the laser scanner is based on the measurement of the inclined range D from the measurement source (laser) to the ground object (linear), which is an obstacle to the propagation of the laser beam. Such an obstacle will cause the appearance of a reflected pulse, which will be registered by the receiver, and the delay time from the moment of emission of the probe pulse to the registration of the reflected pulse can determine the range D.

At the same time determine the coordinates of the spatial position of the carrier X, Y, Z through the use of satellite navigation system, as well as the angles of orientation of the probe beam. Knowledge of these six parameters of external orientation allows you to mathematically move to the coordinates of the point that caused the display. The main result of the laser locator is to obtain a laser-location image or "cloud" of laser points (see Fig. 1). An important detail - the laser-location image is always discrete [5].

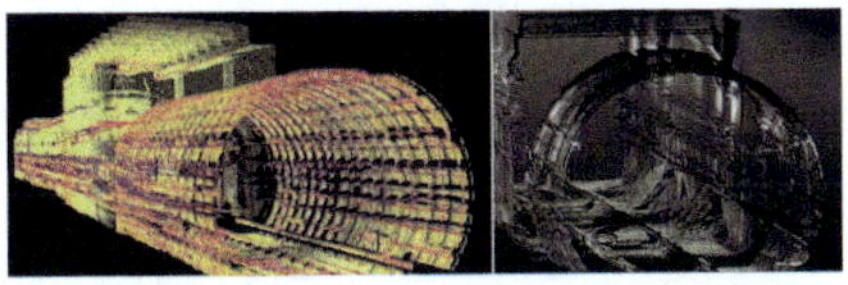

Fig. 1. The results of laser scanner work and the underground subway tunnel.

GPS photography (satellite navigation systems). This type of survey, in recent years, is widely used in the study of communication routes. However, due to the fact that the device ("kinematics" mode) is installed on the car (sprung part), the accuracy of such measurements remains low. In "static" and "stop" modes, GPS is a decent alternative to tacheometric shooting. A significant disadvantage of this method is that in a closed area (forested, building) GPS readings can give failures. This can be avoided by joint use of satellite and gyroscopic systems.

1.1 Digital and Mathematical Modeling of the Terrain

In systemic automated design, the result of topographic and geodetic works is a digital model (array of points) of the design area [12, 13]. During geodetic survey or its in-house processing, the points are given certain attributes necessary for adequate modeling of surfaces, situations and correct implementation of all subsequent design procedures [4].

Thus, to build the terrain and the situation of the design area, the following initial data are formed:

- points that have a name (Name), coordinates (x, y, z), codes of symbols, codes of their accessories to lines and contours [4, 7, 10];
- structural lines, along which there is a violation of the smoothness of the surface (cliff lines, watersheds, thalwegs, rivers, lakes, man-made structures, etc.);
- situational lines and contours - data about the area, such as the location of forests, rivers, lakes, roads, houses, etc.

The points are usually displayed graphically in the form of an array of points with the signature of their height position (see Fig. 2) [5, 7]. A point can carry information about point objects (column, tree, geodetic sign, etc.). The designations of these objects are selected from the corresponding libraries of symbols and placed, as a rule, in a separate display layer.

Since in real projects of complex structures, the engineer deals with arrays of tens and hundreds of thousands of points, it is natural that modern computer-aided design systems must have advanced means of visual editing. In point editing mode, a context menu is usually provided. The composition of context menu commands depends on the number and type of selected points [6].

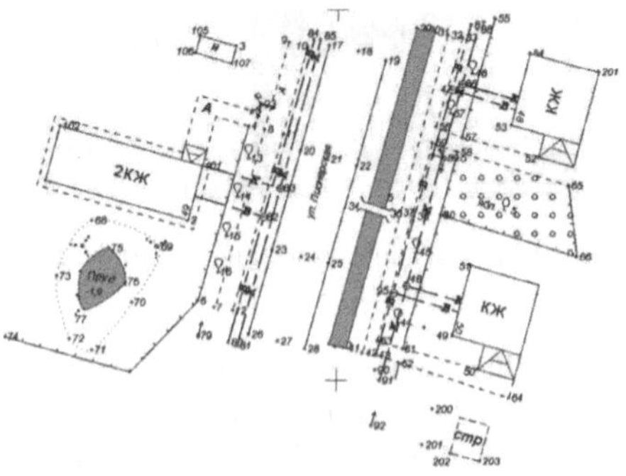

Fig. 2. Initial data for the construction of relief and situation.

Structural lines are built strictly on relief points and in essence they are three-dimensional broken lines (see Fig. 2). It should be noted that the structural lines may not intersect, but may be adjacent to each other. Structural lines can have a significant effect on the contours of surfaces. Situational lines and contours can pass both on relief points, and on situational (do not have z coordinate). The condition for the passage of these lines and contours through the points is not strict. They are also not prohibited from crossing and overlapping.

Both situational and structural lines have color, thickness, and style attributes. Outlines have fill or texture fill attributes (symbols).

In the case of filling contours, one layer (for example, the situation) can close another (triangulation). The sequence of overlapping layers is determined by the designer and can be changed during the design process, if necessary.

1.2 Delaunay Triangulation and Methods of Its Editing

For modeling of surfaces there are different types of structures: horizontally, on structural lines, on diameters to a route or the main course, statistical [11].

The array of points for regular models can be represented as follows:

$$F, m, n, X0, Y0, Z11, \ldots, Z1m, \ldots, Znm, \tag{1}$$

where F – is the grid step; m – is the number of points horizontally; n – is the number of vertical points; $X0$, $Y0$ – coordinates of the starting point of the grid; $Z11, \ldots, Z1m, \ldots, Znm$ – marks of points in grid nodes.

Thus, to unambiguously represent a regular grid of dimension $m \times n$, it is necessary to store all $m \times n + 5$ numbers. However, an objective representation of the surface with a given accuracy requires a high density of points, which is associated with a significant multitasking work on the preparation of the source information. In addition, given the limited speed of computers and data arrays, you have to choose between the accuracy of the representation (cell size) and the size of the processed surface [6].

For irregular models, the array of points is described by the sequence:

$$\Sigma Xi, Yi, Zi, Ti, Ri, Li, \quad (2)$$

where *Xi, Yi, Zi* – are the coordinates of the i-th point (array $i = 1, \ldots, k$); *Ti, Ri, Li* – respectively, the affiliation of the i-th point to the *Ti*-triangle, the connection of the *i*-th point with *Rii Li* – points in the triangle.

The dimension of the irregular grid is 6k, which is almost 6 times larger than the dimension of the regular grid, but, at the same time, to adequately reflect the surface requires a significantly smaller number of points.

The problem of constructing a surface by the method of triangulation is one of the basic ones in computational geometry. It includes many others related to surface modeling and solving spatial problems in machine graphics, computer-aided design systems and geographic information systems [8].

It is important that the flip operation can be used not only as part of the algorithm for constructing the optimal triangulation of the surface, but also as part of the tools of dialog editing, which allows you to get the desired properties for the designer of this surface [5, 9].

The choice of a particular triangulation algorithm significantly affects the efficiency of the entire design system, as well as the reliability of the results. The issue of choosing a data structure to represent triangulation is also important.

There are 3 main types of objects in triangulation:

- nodes (points, vertices);
- ribs (segments);
- triangles.

In the process of performing specific design tasks, it is necessary to perform computational operations with these objects. Here are just some of the possible operations:

- triangle → nodes: required to obtain the altitude mark of the design point located inside a particular triangle. To do this, set the coordinates of the nodes of this triangle, and then in the equation of the plane passing through these three nodes, the coordinates x, y of the design point are substituted.
- node → ribs: required for the analysis of drainage on the considered surface. A list of all adjacent ribs is set at the node, each of which is analyzed for the possibility of "overflow" of water.
- triangle → triangle: required to construct isolines of the considered surface. The list of triangles adjacent to it is established on a triangle and on them the geometrical place of points with the set height mark is calculated.

1.3 Relief Analysis (Surfaces)

One of the basic tasks of analysis of triangulation surfaces is the construction of sections: vertical (profiles) and horizontal (isolines).

In the problem of constructing profiles on the surface, a straight line, broken line or route is set, along which you want to build a section (see Fig. 3a). For this purpose it is necessary to pass along this line (broken, route), calculating consistently points of its intersection with edges of triangulation. When designing linear objects, this task is extremely important, as many longitudinal and transverse profiles are built along the designed routes. The result of this work will be a vertical section (see Fig. 3b). When entering data from geological surveys, this information will be displayed on all sections with the ability to determine the volume of each layer.

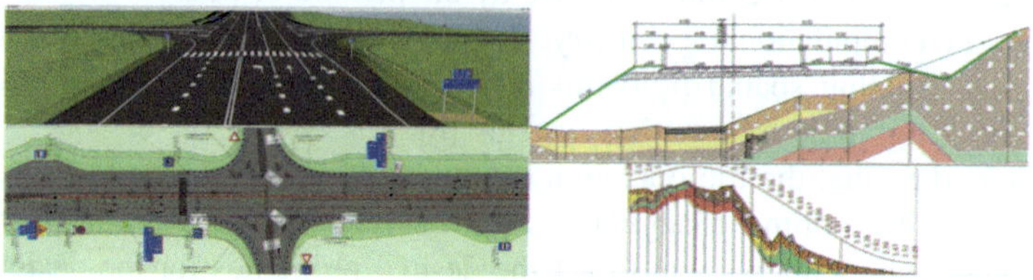

a) b)

Fig. 3. Surface with track (a) and 3D visualization and profile with cross section along track (b).

Isolines are the lines of intersection of horizontal planes of level h with the triangulation surface. In the practice of manual design, isolines were practically the main symbols for displaying and reading the relief of topographic plans. It was used to determine watershed and thalwega lines, catchment area and slope exposures, etc. Longitudinal and transverse profiles were built on the basis of isolines, using interpolation and extrapolation methods.

In the automated design of the isoline is only one way to visually interpret the surface. They are not computational elements of the models, but rather serve as an aid for "engineering reading" of surfaces in traditionally formed concepts. But even in this capacity, isolines still remain in demand in design and cartographic activities.

The algorithm for constructing isolines consists of two steps:

- denote each triangle of triangulation along which the isoline passes, ie the condition is fulfilled:

$$min(z1, z2, z3) < max(z1, z2, z3), \tag{3}$$

where *z1…3* – is the height of its three vertices;

- for each such triangle, we trace such an isoline in both directions from this triangle until one end reaches the other or the boundary of the triangulation.

The main disadvantages of isoline construction algorithms are sharp fractures and their strong oscillation. This is due to the non-uniformity of the obtained isoline line points and the linear interpolation method used.

The simplest and clearest way to smooth out isolines is to use high-degree polynomials. However, in this case, the intersection of isolines of different levels is possible, which is a sign of inadequate surface ratio.

Another way of smoothing is that non-overlapping corridors are built for all isolines. And then within them isolines in the form of broken minimum lengths or smooth Bezier curves are already under construction.

A more obvious reflection of the surface, compared with isolines, is its reflection by isocontours, which by the nature of construction are close to isolines. The representation of the surface by isocontours in the color spectrum will be especially obvious, which allows you to quickly detect low (elevated) places on the surface, thalwegs, watersheds, saddles.

Isocontours between levels h1 on h2 are the geometric location of points on the surface having height h (h1, h2).

You can also increase the visibility due to the fact that different elements of the road (slopes, curbs, roadway) will be painted in different colors.

Of practical interest is the construction of gradients on the triangulation surface of each triangle. The gradient is usually indicated by an arrow in the center of the triangle. The tip of the arrow indicates the direction of the greatest slope in this triangle. The magnitude of the arrow is normalized by the magnitude of this slope. Thus, it is possible to track the directions of the largest water flow, as well as visually capture degenerate triangulation triangles.

For the design of roads, as well as for master plans, one of the purposes of design is to ensure the drainage of water from the projected facility. Therefore, the task of visualization of stagnant water zones is also one of the most popular tasks. The algorithm for constructing such zones is as follows:

- among all the vertices of the triangulation are those that are lower than all adjcent vertices (two vertices are adjacent if they are connected by an edge). Thus, all the found peaks are "bowls" - places where the zones of stagnation of the water will gather;
- fill each bowl with water and find areas of stagnant water. To do this, move from the vertex in the center of each bowl in different directions along the edges to other vertices until some edge is bent, that is, until we find the top - the inflection point of the surface. These inflection points have the property that if the puddle overflows, water will flow out of the stagnant water zone through this point. Among all the inflection points, we choose the one that has a lower height, and on this basis we can determine the capacity of the stagnant water zone;
- if small puddles are formed, they will be checked for merging.

From the standpoint of road safety assessment, the task of constructing zones and lines of visibility is also relevant. The essence of the task is that according to the given position of the observer (the driver of the moving car) determine which parts of the surface he can see and which - no. Most often, this problem is solved by a simplified algorithm:

- build rays coming from the point of the observer in different directions;
- build surface profiles along these rays;

– the floating horizon method forms visible and invisible parts of these profiles.

Above we have considered only some of the problems of surface analysis and given general descriptions of algorithms for solving these problems. In design practice, the list of such problems is extremely wide and each of them can be solved using an algorithm: it all depends on the nature of the problem, the required accuracy of calculations, the form of presentation and scope of the results.

Digital terrain model is the basis for automating the design of objects related to the transformation of the terrain that is modeled (adding new or reconstruction of existing transport, communications, vertical terrain planning, etc.). Given the accuracy of the construction of the digital terrain model and economic considerations, it is necessary to calculate the optimal height of photography and the distance between the planned and high-altitude identification marks. To do this, we use the formulas known in photogrammetry [3]. The calculation of the height of the photograph and the distance between the height starting points is carried out according to the formula.

$$m_z = 0,089\frac{H}{b}m_q\sqrt{n^3 + 154n + 56}, \tag{4}$$

where m_z – is the root mean square error of determining the marks of points from the spatial route phototriangulation; b – is the value of the basis on the aerial photograph; m_q – is the root mean square error of the measurement of transverse parallaxes; n – is the number of bases between the height points of support.

The number of bases between the planned identification marks is calculated by:

$$m_{x,y} = 0,2474\frac{m}{V}m_q\sqrt{n^3 + 11n + 35}, \tag{5}$$

where $m_{x,y}$ – is the root mean square error of determining the planned coordinates of points for the construction of spatial route networks of phototriangulation; m – denominator of the scale of aerial photography; V – increase (for the analytical method of processing $V = 3 - 4$); n – is the number of bases between the planned identification marks.

We assume the root mean square error of the points is m = 0.05 m and was calculated by measuring the differences on the control and basic tack by the formula

$$m = \sqrt{\frac{\{\upsilon^2\}}{2 \cdot 5381}}, \tag{6}$$

where υ – is the difference between the values of the measurements of the main and control tack; *5381* – the number of differences.

The allowable root mean square error is calculated according to (7):

$$m_0 = \mathrm{p}(\%) \cdot Z, \tag{7}$$

where p (%) – is the allowable mean square error as a percentage for the category of relief complexity adopted by the project for this type of work; Z – is the average height/depth of the river in the area of works.

Based on the ratio of error differences, the criterion of tolerance of residual systematic error was confirmed (8):

$$|[\upsilon]| \leq [|\upsilon|]. \tag{8}$$

After performing research for the average sample of hilly terrain data, and calculating the error of each of the attempts, the results of which are presented in Fig. 4.

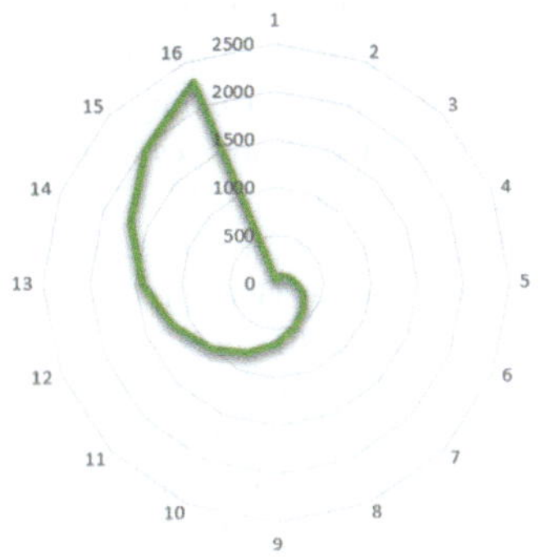

Fig. 4. Graph dependence on the height of the survey (m) and the error of the digital model (%).

2 Conclusions

The article considers the main methods of engineering surveys of the area with the main advantages and disadvantages of each of them. Processing of the received data at system automated designing as a result of topographic and geodetic works is a digital model of district (which consists of an array of points) of a design zone. Since in real projects of complex structures, the engineer deals with arrays of tens and hundreds of thousands of points, it is natural that modern computer-aided design systems must have advanced means of visual editing. For modeling of surfaces there are different types of structures: horizontally, on structural lines, on diameters to a route or the main course, statistical. The problem of constructing a surface by the method of triangulation is one of the basic ones in computational geometry. It includes many others related to surface modeling and solving spatial problems in machine graphics, computer-aided design systems and geographic information systems.

The results of the dependence of the height of the survey and the percentage of error were obtained from the performed research. Therefore, based on the results of calculations, it can be concluded that the materials of the instruction [2] are suitable for use in this further design.

References

1. Pidlipna, M.: The use of GIS technologies for land zoning. Young Sci. **2**(17), 8–10 (2015)
2. SCST-2.04-02-98 Instruction on topographic surveying in the scales of 1: 5,000, 1: 2,000, 1: 1,000 and 1: 500, Moscow (1982)
3. Li, Z., Zhu, Q., Gold, C.: Digital Terrain Modeling: Principles and Methodology. CRC Press, Boca Raton (2005)
4. Sorochuk, N., Timchenko, O., Sorochuk, Y.: Geodetic data processing and digital terrain modeling in the state cadastre with the help of the software complex "CREDO". In: Modern Technologies of Land Management, Cadastre and Land Management: VI All-Ukrainian Scientific-Practical Conference, pp. 6–7. Kyiv, NAU (2020)
5. Ugnenko, E., Viselga, G., Timchenko, O., Uzhvieva, E., Sorochuk, N.: Land cadastral works as a basis for maintaining the state real estate cadastre. In: Science - Education, Production, Economy: 18th International Conference, Republic of Belarus, Minsk (2020)
6. Ugnenko, E., Timchenko, O., Sorochuk, N.: Application of the software complex "CREDO" for processing of geodetic data and digital modeling of the area. In: Modern Technologies of Land Management, Cadastre and Land Resources Management. Proceedings of the V All-Ukrainian Scientific-Practical Conference, pp. 131–132. Kyiv, NAU (2019)
7. Oksanen, J.: Digital elevation model errorin terrain analysis. Academic Dissertation in Geography, Publications of the Finnish Geodetic Institute, No. 134. Helsinki University Press, Helsinki, KIRKKONUMMI (2006). 59 p.
8. Ugnenko, E., Shevchenko, A., Matviienko, O., Maliavin, A., Viselga, G., Turla, V.: Analysis of existing train directions and international transport corridors of Ukraine. In: TRANSBALTICA XI: Transportation Science and Technology, pp. 622–632 (2019)
9. Shevchenko, A., Orel, E., Manuylenko, V.: Prospects of GIS technologies in the survey of Ukrainian railways. In: Prospects for the Institutional Development of Land Relations in Ukraine: Proceedings of the All-Ukrainian Conference, Poltava, pp.150–152 (2019)
10. Droj, G.: Improving the Accuracy of Digital Terrain Models. Studia Univ. Babes Bolyai, Informatica, Volume LIII, Number 1, pp. 65–72 (2008)
11. Wilson, J.P.: Digital terrain modeling. Geomorphology **137**, 107–121 (2012)
12. Hengl, T., Reuter, H.I.: Geomorphometry: Concepts, Software, Applications (2009)
13. Zhou, Q., Lees, B.G., Tang, G.-A.: Advances in Digital Terrain Analysis. LNGC. Springer, Heidelberg (2008). https://doi.org/10.1007/978-3-540-77800-4
14. Hengl, T., Gruber, S., Shrestha, D.P.: Reduction of errors in digital terrain parameters used in soil-landscape modeling. Int. J. Appl. Earth Obs. Geoinf. **5**, 97–112 (2004)

Estimation of Technological Development of Transport Enterprises

Kristina Vaičiūtė(✉), Gintautas Bureika, and Darius Bazaras

Vilniaus Gediminas Technical University, Saulėtekio Avenue 11, 10223 Vilnius, Lithuania
{kristina.vaiciute,gintautas.bureika, darius.bazaras}@vilniustech.lt

Abstract. The article examines the problems of technological development of transport companies and the use of this development as one of the tools for the competitiveness of transport companies and ensuring the quality of transport companies' transportation services. Modern Information Technology (IT) of transport companies improves the management of information flows and enables better service of the company's customers or passengers. The technological development of transport companies is inseparable from the availability and use of information technologies and telecommunications. Based on the analysis of literature sources, an expert evaluation questionnaire with the concepts of factors – explanations were formed and prepared, and a list of experts to be interviewed was compiled. During the structural analysis of the technological development of enterprises, the components of the technological development process of the transport enterprise and the factors determining the technological development of the transport enterprise are identified. An expert evaluation questionnaire is developed according to the identified quality criteria of the impact of transport technologies on freight or passenger transport. After processing the data of the expert survey, the criteria are arranged and the results of the research are presented. Conclusions and suggestions are presented at the end.

Keywords: Transport company · Technological development · Multi-criteria evaluation method

1 Introduction

The constantly increasing flows of information transmitted by transport companies challenge the need for innovative technologies and communication systems needed by transport companies. These systems quickly transfer an unchanged flow of information from one unit to another. To ensure a better quality of transport service, it is necessary for transport companies to constantly invest in the quality of technological infrastructure [1]. Zhou [2], in researching the technological development of enterprises, found that the efficiency and effectiveness of technological development should be given priority [2]. The technological development of the transport company is related to the availability of information technologies and telecommunications and their use for the transportation service [3]. According to Mnif [4], the technological development of

O. Prentkovskis et al. (Eds.): TRANSBALTICA 2021, LNITI, pp. 589–598, 2022.
https://doi.org/10.1007/978-3-030-94774-3_58

a transport company symbolizes the constant ability of the transport company to improve cheaper and faster.

Given the growth of the global economy and the impact of economic growth on the development of transport technology infrastructure, factors remain to be considered and studied [5]. One of them is the management of components of technological development processes in the transport sector. The main functions of information technology development processes in transport companies: data modelling; database design and creation; database indexing; data entry and updating; implementation and design of the search procedure; adapting communication networks to access databases; realization of various other functions [6, 7]. Technological development is essential for the exchange of information in transport companies between service providers, information must be provided quickly and efficiently. Transmission of information of a transport company by technological means, efficient exchange of information between different modes of transport performing the service, good technological information management is mandatory [8]. The application of innovative technologies in a road transport company covers various directions and innovative technologies are the main ones that influence other activities performed in the company [9]. Expanding the technological infrastructure of a transport company requires discovering a new promising area [10]. Håkansson and Waluszewski [11] explored the technological development of transport companies as an ongoing process that allows new directions to be taken in collaboration between road transport companies. The possibility of technological development processes of transport companies is related to the transport company's ability to obtain certain information through equipment [12, 13]. Thus, in summary, it can be stated that the main criteria for the study of the components of the technological development processes of the main transport company are: the design of the database and its creation; data modelling; data entry and updating; database indexing realization of search procedure; adapting communication networks to access databases; realization of various other functions.

The object of this research is the technological development processes of transport companies. In studying the impact of technological development process components on the quality assurance of services provided by the company, the authors applied research methodologies: analysis of scientific literature, grouping, AHP method and ranking of indicators.

2 Methodology of Research of Technological Development Processes of Enterprises

Quantitative research proposed by scientists Kardelis, Tidikis [14, 15] and its method – a questionnaire – were chosen to evaluate the interaction between transport company management improvement and technological development. Based on the analysis of the literature, an expert evaluation questionnaire with the concepts of factors – explanations were formed and prepared, and a list of experts to be interviewed was compiled. One of the most important characteristics of experts is competence, therefore the experts were required to have competence and experience in the field of study. When compiling the questionnaires, a lot of attention was paid to the formulation of the questions, as the reliability of the survey results largely depends on it. The questions are formulated clearly, ambiguity is avoided. The questionnaire included both closed and open-ended questions for the experts.

The data obtained from the expert survey are processed. Processing is required to obtain aggregated data and new information contained in the expert evaluation questionnaire. Based on the processing results, a possible solution to the problem is formulated. The consistency of the opinions of more than two experts can be quantified by the value of the concordance coefficient. The concordance coefficient indicates the level of coherence of the expert group if the number of experts is greater than two. Expert assessments obtained from the completed questionnaires are listed in the table.

The expert group n quantifies m objects. Estimates form a matrix of n rows and m columns [16]. Evaluations can be – indicator units, unit parts, percentages, in a ten-point system. The ranking of expert indicators is suitable for calculating the concordance coefficient. The ranking is a procedure in which the most important indicator has given a rank (*R*) equal to one, the second indicator a second rank, and the last indicator a rank m (m is the number of benchmarks). The average of the sum of ranks is calculated [17–19]:

$$\sum_{i=1}^{n} R_{ij} = \frac{1}{2} n(m+1). \tag{1}$$

Given the indicators of expert evaluation, the consistency of their opinions is determined by calculating the concordance coefficient of Kendal ranks. If *S* (variance) is the real sum of the squares calculated according to formula (1), then the concordance coefficient (2), in the absence of related ranks, is defined by the ratio of the resulting S to the corresponding maximum S_{max} (2):

$$W = \frac{12S}{n^2 m(m^2 - 1)} = \frac{12S}{n^2(m^3 - m)}. \tag{2}$$

The questionnaire chosen for the data collection method consisted of 4 questions.

3 Results of the Research of Technological Development of Transport Companies

To achieve the aim of the research, the factors determining the efficiency of technological development of transport companies were investigated. The obtained data are processed and the concordance coefficient of Kendal ranks was estimated. The study involved 8 experts, all of whom have management experience in land transport companies (Decision Makers) from 3 to 10 years. Experts agreed (65.4%) that transport workers (respondents) lack the fast transmission of information using the existing information system.

The experts were asked to assess the main factors determining the efficiency of technological development of transport companies:

(a) planning and organization of work;
(b) improvement of the transport system of customer service quality;
(c) sales order management;
(d) the introduction of technological measures to ensure the rapid, reliable and high-quality transmission of information;
(e) use of IT for intermodal freight transport;
(f) implementation of measures to implement freight flow management;
(g) establishment of an IT system shared database to provide operational traffic information to all subscribers.

The distribution of the factors determining the technological development of the transport company according to the above-mentioned criteria (a–g) was evaluated by experts with points from 1 to 7: 1 being the most important, 7 being the least important. The distribution of ranking is shown in Fig. 1.

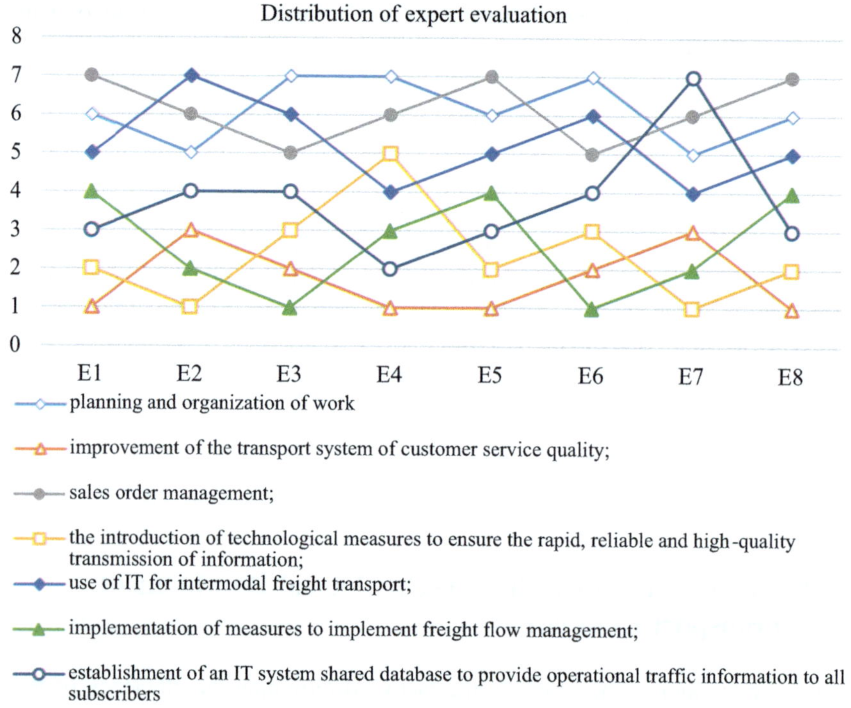

Fig. 1. Distribution of ranks of factors determining the technological development of a transport company (compiled by the Authors).

The data of the analysis and calculation of the distribution of the ranks of the eight experts were listed in Table 1.

Table 1. Ranking table of the importance of the factors determining the technological development of the transport company (compiled by the Authors).

Ranking formulas	Factor designation symbol (m = 7)						
	a	b	c	d	e	f	g
$\sum_{i=1}^{n} R_{ij}$	49	14	49	19	42	21	30
$\bar{R}_j = \frac{\sum_{i=1}^{n} R_{ij}}{n}$	6.125	1.75	6.125	2.375	5.25	2.625	3.75
$\sum_{i=1}^{n} R_{ij} - \frac{1}{2}n(m+1)$	17	−18	17	−13	10	−11	-2
$\left[\sum_{i=1}^{n} R_{ij} - \frac{1}{2}n(m+1)\right]^2$	289	324	289	169	100	121	4

The concordance coefficient W was calculated according to formula (3) when there are no associated ranks.

$$W = \frac{12S}{n^2(m^3 - m)} = \frac{12 \times 1296}{8^2(7^3 - 7)} = 0.7232. \tag{3}$$

The impact of technological developments is important factors determining the main technological development of a transport company is 7 the number $m > 7$. The weight of the concordance coefficient χ^2 is then calculated according to formula (4) and a random value is obtained.

$$\chi^2 = n(m-1)W = \frac{12S}{nm(m+1)} = \frac{12 \times 1296}{8 \times 7(7+1)} = 34.714. \tag{4}$$

χ^2 calculated value was 34.714 it was higher than the critical value of (12.5916) that is why the opinion of the respondents is considered to be compatible, and the average ranks show the general opinion of the experts:

$$W_{\min} = \frac{\chi^2_{\nu,\alpha}}{n(m-1)} = \frac{12.5916}{8(7-1)} = 0.2623 < 0.7232. \tag{5}$$

The minimum value of the concordance coefficient $W_{\min}$ was calculated according to formula (5), $W_{\min} = 0.2623 < 0.7232$, so that the opinions of all the 8 respondents on the 7 main criteria determining the importance of the transport company's technological development, which are important, are still considered harmonized.

The indicators of the importance of the factors determining the technological development of the transport company are calculated – Q_j. The obtained data are presented in Table 2.

Table 2. The main indicators of the importance of the technological development of a transport company Q_j (compiled by the Authors).

Indicator marker	Factor designation symbol (m = 8)							Sum
	a	b	c	d	e	f	g	
q_j	0.219	0.063	0.219	0.085	0.188	0.094	0.134	1.00
d_j	0.781	0.938	0.781	0.915	0.813	0.906	0.866	6.00
Q_j	0.130	0.156	0.130	0.153	0.135	0.151	0.144	1.00
Q_j'	0.067	0.223	0.067	0.201	0.098	0.192	0.152	1.00
Factor layout	7	1	6	2	5	3	4	

According to expert assessments and calculations, the rankings of the importance of the factors determining the efficiency of technological development of transport companies are:

1. Improvement of the transport system of customer service quality;
2. Fast, reliable and high-quality information transmission;
3. Rational management of cargo flows;
4. Operational information on travel duration, costs, etc. available to all IT system subscribers.

The experts assessed the main criteria of the components of the technological development process of the transport company:

(a) data modelling (database design processes, analysis);
(b) the design and construction of the database;
(c) indexing of databases;
(d) data entry and updating;
(e) implementation and design of the search procedure;
(f) communications network application databases to achieve;
(g) realization of various other functions.

The distribution of the components of the technological development process of transport companies according to the above-mentioned criteria (a–g) was evaluated by experts from 1 to 7: 1 – most important, 7 – least important. The distribution of ranking is shown in Fig. 2.

The data of the analysis and calculation of the distribution of the rankings of the questionnaires of the eight expert questionnaires were listed in Table 3.

The concordance coefficient W was calculated according to formula (6) when there are no associated ranks.

$$W = \frac{12S}{n^2(m^3 - m)} = \frac{12 \times 478}{8^2(7^3 - 7)} = 0.2702. \tag{6}$$

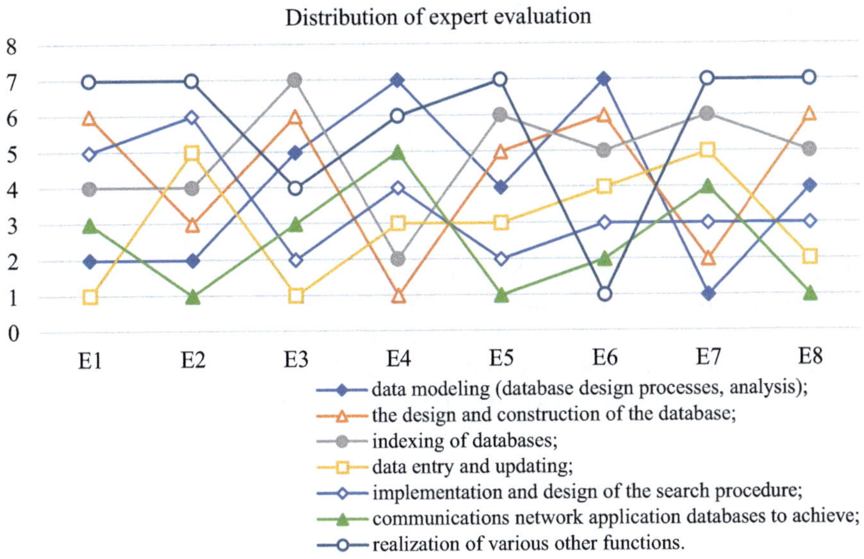

Fig. 2. Distribution of ranks of components of the technological development process of transport companies (compiled by the Authors).

Table 3. Ranking table of the main components of the technological development process of a transport company (compiled by the Authors).

Ranking formulas	Factor designation symbol (m = 7)						
	a	b	c	d	e	f	g
$\sum_{i=1}^{n} R_{ij}$	32	35	39	24	28	20	46
$\overline{R}_j = \frac{\sum_{i=1}^{n} R_{ij}}{n}$	4	4.375	4.875	3	3.5	2.5	5.75
$\sum_{i=1}^{n} R_{ij} - \frac{1}{2}n(m+1)$	0	3	7	−8	−4	−12	14
$\left[\sum_{i=1}^{n} R_{ij} - \frac{1}{2}n(m+1)\right]^2$	0	9	49	64	16	144	196

The number of important main components of the transport company's technological development process (m) is 7, i.e. $m > 7$. Then the weight of the concordance coefficient χ^2 is calculated according to formula (7) and a random quantity is obtained.

$$\chi^2 = n(m-1)W = \frac{12S}{nm(m+1)} = \frac{12 \times 478}{8 \times 7(7+1)} = 12.804. \tag{7}$$

χ^2 the calculated value of 12.804 was higher than the critical value (equal to 12.5916), therefore the opinion of the respondents is considered to be consistent, and the average ranks show the general opinion of the experts.

$$W_{\min} = \frac{\chi^2_{v,\alpha}}{n(m-1)} = \frac{12.5916}{8(7-1)} = 0.2623 < 0.2702. \quad (8)$$

The lowest value of the concordance $W_{\min}$ coefficient $W_{\min} = 0.2623 < 0.2702$ was calculated according to formula (8). Consequently, the views of all 8 respondents on the 7 key components of the transport company's technological development process, which are important, are still considered harmonized.

The indicators of the components of the technological development process of the transport company are calculated – Q_j. The obtained data are presented in Table 4.

Table 4. The main indicators of the timportant components of the technological development process of the transport company.

Indicator marker	Factor encryption symbol (m = 7)							Sum
	a	b	c	d	e	f	g	
q_j	0.143	0.156	0.174	0.107	0.125	0.089	0.205	1.00
d_j	0.857	0.844	0.826	0.893	0.875	0.911	0.795	6.00
Q_j	0.143	0.141	0.138	0.149	0.146	0.152	0.132	1.00
Q_j'	0.143	0.129	0.112	0.179	0.161	0.196	0.080	1.00
Factor layout	4	5	6	2	3	1	7	

Based on expert assessments and calculations, the order of importance of the main components of the transport company's technological development process is as follows:

(f) adaptation of communication networks to access databases;
(d) data entry and updating;
(e) implementation and design of the search procedure;
(a) data modelling (database design processes, analysis).

The analysis of the research results revealed that the components of the technological development process of transport companies are related to the application of communication networks in improving the quality of service in the transport system.

4 Conclusions

After processing the results of the survey, it was found that the majority of employees (65.4%) of transport companies lack information on fast data transmission using the existing information system. The research revealed that the order of importance of the factors determining the efficiency of technological development of transport companies is: 1) improved quality of service in the transport system; 2) fast, reliable and high-quality information transmission; 3) rational management of cargo flows; 4) operational information on travel duration, costs, etc. available to all IT system subscribers.

It was stated that the order of the indicators of the components of the technological development process of transport companies in terms of importance is as follows: 1) adaptation of communication networks to access databases; 2) data entry and updating; 3) implementation and design of the search procedure; 4) data modelling.

It can be said that this study reaffirms the importance of information and data processing for the company's operations, its technological process and the quality of customer service. There have been many discussions lately about how data arrays can be processed and how their potential can be exploited in the technological process. The study confirmed the need and importance of providing databases and access to them. This proves that it is not good enough when data is only stored in databases, but access to and use of it is insufficient. In practice, in such cases, it can be argued that there is a "data ballast" available that only takes up space in the data loggers.

Another important thing is updating the data. This aspect is especially important when analyzing the operational data received and generated in the company's activities. Operational data that emerges "here and now" requires their rapid and efficient updating – otherwise, instead of providing support and assistance in the management of the technological process, they can become the basis for erroneous decisions. On the other hand, it is important to create arrays of this type of data that would allow other actions such as correlation, analysis, and forecasting. Appropriate technological possibilities to work with data allow to perform other actions – to model, plan, evaluate, visualize and search for regularities.

References

1. Munim, Z.H., Schramm, H.J.: The impacts of port infrastructure and logistics performance on economic growth: the mediating role of seaborne trade. J. Ship. Trade **3**, 1–19 (2018)
2. Zhou, T., Tan, R., Sedlin, T.: Planning modes for major transportation infrastructure projects (MTIPs): comparing China and Germany. Sustainability **10**(10), 1–23 (2018). https://www.mdpi.com/2071-1050/10/10/3401
3. Rakauskienė, G., Tamošiūnienė, R.: Šalies konkurencingumo pokyčio optimizavimas. Verslo Sistemos ir Ekonomika **3**(2), 167–176 (2013)
4. Mnif, S.: Bilateral Relationship between Technological Changes and Income Inequality in Developing Countries (2016)
5. Wang, L., Xue, X., Zhao, Z., Wang, Z.: The impacts of transportation infrastructure on sustainable development: emerging trends and challenges. Int. J. Environ. Res. Public Health **15**, 1172 (2018). https://www.ncbi.nlm.nih.gov/pubmed/29874785
6. Jarašūnienė, A., Batarlienė, N., Vaičiūtė, K.: Application and management of information technologies. In: Transbaltica 2015: Proccedings of the 9th International Scientific Conference, 7–8 May 2015, pp. 336–343 (2015). Procedia Engineering
7. Zijlstra, H.: Cost savings through better exchange of information. IT Board-Room **4**, 82–87 (2012)
8. Stevanović, M., Stevanovic, K.: New directions in the design of railways stations. Građevinar **66**(8), 739–747 (2014). https://doi.org/10.14256/JCE.975.2013

9. Drejeris, R., Oželienė, D.: New approach to the technological aspect of corporate sustainable development. Bus. Theory Pract. **20**, 363–371 (2019). https://doi.org/10.3846/btp.2019.34
10. Bolodurina, M., Mishurova, A.: Intellectualization of transport and logistics infrastructure agents network interaction through adaptive information and communication technologies introduction. In: Proceedings of the 7th Scientific Conference on Information Technologies for Intelligent Decision Making Support (ITIDS 2019), vol. 166, pp. 8–14. Atlantis Press (2019). https://doi.org/10.2991/itids-19.2019.2
11. Håkansson, H., Waluszewski, A.: Managing Technological Development. Routledge, London and New York (2002). https://doi.org/10.4324/9780203217535
12. Lovelace'as, R., Parkinas, J., Cohenas, T.: Atviros prieigos transporto modeliai: tvaraus transporto planavimo sverto taškas. Transporto politika **97**, 47–54 (2020)
13. Zhao, J.L., Fan, S., Yan, J.: Overview of business innovations and research opportunities in blockchain and introduction to the special issue. Financ. Innov. **2**(1), 1–7 (2016). https://doi.org/10.1186/s40854-016-0049-2
14. Kardelis, K.: Mokslinių tyrimų metodologija ir metodai. Mokslo ir enciklopedijų leidybos centras, Vilnius (1997)
15. Tidikis, R.: Socialinių mokslų tyrimų metodologija. Lietuvos teisės universiteto Leidybos centras, Vilnius (2003)
16. Sivilevičius, H.: Application of expert evaluation method to determine the importance of operating asphalt mixing plant quality criteria and rank correlation. Baltic J. Road Bridge Eng. **6**(1), 48–58 (2011). https://doi.org/10.3846/bjrbe.2011.07
17. Podvezko, V.: Agreement of expert estimates. Technol. Econ. Dev. Econ. **9**(2), 159–172 (2005)
18. Bureika, G., Gaidamauskas, E., Kupinas, J., Bogdevičius, M., Steišūnas, S.: Modelling the assessment of traffic risk at level crossings of Lithuanian Railways. Transport **32**(3), 282–290 (2017)
19. Vaičiūtė, K., Bureika, G.: Research on transport enterprise technological development directions. Mokslas – Lietuvos Ateitis/Science – Future of Lithuania **12** (2020). https://doi.org/10.3846/mla.2020.11790

A Study of the Impact of Social Responsibility on the Technological Development of a Transport Company

Sigita Pagirienė, Kristina Vaičiūtė(✉), and Darius Bazaras

Vilnius Gediminas Technical University, Saulėtekio Avenue 11, 10223 Vilnius, Lithuania
{sigita.pagiriene,kristina.vaiciute, darius.bazaras}@vilniustech.lt

Abstract. The growing public need to see the application of social responsibility in the activities of transport companies has a significant impact on cooperation between enterprises and provokes transport companies to engage in socially responsible activities. Therefore, the application of social responsibility in the management of a transport company becomes the basis for organizing the activities of the company. In order for a transport company to successfully expand its transport activities and remain competitive in the transport market, it must supplement its activities with various criteria of social responsibility. The article examines the problems of the application of the social responsibility of the transport company in the activities and the use of the technological development option as one of the tools for ensuring the competitiveness of transport companies. The structural analysis of social responsibility determines the components of social responsibility of the transport company and the factors determining technological development, such as the relationship between social responsibility and technological development. On the basis of a literature analysis, an expert evaluation questionnaire has been developed and prepared. An expert evaluation questionnaire shall be established. After processing the data of the expert survey, the criteria are arranged and the results of the study are presented.

Keywords: Social responsibility · Transport company · Technological development · Multi-criteria method

1 Introduction

The growing expectations of society and cooperating companies increase the pressure on transport companies to act in accordance with social responsibility norms and take on additional responsibilities. Michalski, Montes-Botella and Figiel [1] emphasize that social responsibility (CSR) is a public interest requirement for the operation of companies and not an expression of the altruism of transport companies themselves. From the point of view of the authors, it is important to assess the factors that encourage transport companies to implement CSRs.

O. Prentkovskis et al. (Eds.): TRANSBALTICA 2021, LNITI, pp. 599–609, 2022.
https://doi.org/10.1007/978-3-030-94774-3_59

The aim of the research is to investigate the effectiveness of social responsibility of transport companies for technological development.

Research methodology:

- analysis of scientific literature;
- grouping and synthesis;
- ranking of indicators.

The introduction of social responsibility by transport companies under Zenisek [2] the main factor defining the primary goal of a transport company includes gaining a benefit. Išoraitė [3] states that the level of attention focused on the CSR of transport companies in different periods is focused on the pursuit of material and intangible benefits of transport companies, the rational use of available resources. In order to ensure CSR indicators, it is necessary for transport companies to constantly invest in the quality of technological infrastructure [4]. According to Mnif [5] the technological development of a transport company symbolizes the constant ability of a transport company to develop cheaper and faster which allows to strengthen the application of CSR in the activities of a transport company.

2 Study on the Effectiveness of Social Responsibility of Transport Companies for Technological Development

Factors of social responsibility in the field of transport are related to the basic principles of operation of transport companies as business entities. Two different approaches can be distinguished here: economic and social. Činčalová and Hedija [6] states that the goal of all businesses is profit, therefore it is based on the provision that the final benefit of social responsibility to the company must be reflected in the financial results [6]. According to Piecyk and Björklund [7], in the 21st century it is not enough to associate the success of a company's activities only with financial indicators. The benefits of social responsibility can be justified by social, ethical, environmental performance [7]. Different conclusions of Kavecké and Paužuolienė [8] presuppose that the link between CSR and financial results (e.g. profitability) is indirect and complex. In addition it should be emphasized that profits are not only financial benefits for shareholders, but also the basis for further investments in the improvement of transportation processes and technological development. Haski-Leventhal [9] states that the more financially successful (profitable) a company is, the higher the expectations for it are raised by society, as evidenced by the Carroll [10] model, which treats economic responsibility as the basis for CSR. Tüzün Rad and Gülmez [11] clarify that the additional costs for the technological development of transport companies are generated by investments in aid to the local community, improvement of workers' qualifications, purchase of environmentally friendly materials, inventory upgrades, application of alternative energy sources, etc. Environmental initiatives of a socially responsible company directly contribute to the reduction of operating costs through technological development. However, companies operating in the market and interested in the long-term perspective tend to cooperate with socially responsible companies, as they understand that investments focused on the social responsibility of transport companies have higher returns in the future. Banabakova [12] associates social responsibility in transport with urban, mass events transport processes, transportation of people with disabilities, adoption of strategic decisions in the activities of transport companies into technological development, which helps the environment and as an aid to the local environment. Socially oriented responsibility in transport is like a mechanism that usually aims to optimize the allocation of resources on the basis of not only economic but also social criteria [12]. For transport companies investment in the application of innovative technologies in a transport company includes various directions and innovative technologies are the main ones that influence other activities performed in the company and at the same time CSR [13]. Technological development in the transport company is linked to the transport company's access to certain information through equipment in order to save natural resources [14]. Given that transport is one of the largest sectors with high-energy consumption, recent attention has been paid to the environment and the reduction of environmental pollution and quality management. Transport companies are gradually introducing ISO standards to demonstrate the quality of the services or products offered [15]. Ye Li points to the planning of a freight route using information technology (hereinafter IT) as an effective tool to reduce CO_2 emissions [16].

In conclusion, the criteria for the application of technological development in a transport undertaking for the development of social responsibility measures are: application of technological development to the improvement of the transportation service, improvement of the qualification of employees, technological development in order to save natural resources or increase productivity and realization of various other functions.

3 Research Methodology

Transport companies selected scientists to assess the interaction between CSRs and technological development [17, 18] the proposed quantitative study and its method is a questionnaire survey. On the basis of literary analysis, an expert evaluation questionnaire has been formed and prepared. The questions are worded clearly, multi-prominence has been avoided when it is not clear exactly what is being asked. The experts also asked closed and open-ended questions in the questionnaire. A list of experts to be interviewed has been drawn up. One of the most important characteristics of experts is competence, which has led to the requirements for experts with regard to competence and experience in the field of research.

Following an expert survey, the data obtained are processed due to the need to obtain aggregated data and new information contained in the expert evaluation questionnaire. Based on the results of processing, a solution to the problem is formulated. The compatibility of the opinions of more than two experts may be quantified by the value of the concordance coefficient [19]. The concordance coefficient indicates the degree of compatibility of the expert group if the number of experts exceeds two. Expert evaluations obtained from completed questionnaires shall be drawn up in a table.

The expert group n quantifiable m objects. Estimates form a matrix of n-rows and m columns [19]. For evaluations can be – in indicator units, unit parts, percentages, ten-point system. The ranking of expert indicators is suitable for calculating the concordance coefficient. Ranking is the procedure where the most important indicator is assigned a grade (R) equal to one, the second indicator a second grade, the last indicator a rank m (m is the number of benchmarks). Calculates the average of grade amounts [20]:

$$\sum_{i=1}^{n} R_{ij} = \frac{1}{2}n(m+1). \tag{1}$$

With expert evaluation indicators, the compatibility of their opinions shall be determined by calculating the Kendal grade concordance factor. If S (variance) is the real sum of squares calculated according to the formula (1), then the concordance factor (2) in the absence of the associated grades is defined by the ratio of the resulting S to the corresponding maximum $S_{\max}$ (2):

$$W = \frac{12S}{n^2m(m^2-1)} = \frac{12S}{n^2(m^3-m)}. \tag{2}$$

4 Results of the Study on the Effectiveness of Social Responsibility of Transport Companies for Technological Development

The study involved 8 experts, all with managerial experience in transport from 3 to 10 years.

Experts noted that 75% have ISO 14001 standard installed, 12% have ISO 9001 and 13% ISO 26000 standards.

The experts were asked to assess the main factors determining the effectiveness of the social responsibility measures of the transport undertaking for technological development: a) training of arbuors on "responsible logistics"; b) the most optimal cargo transportation route and conditions for IT technologies, making economic use of vehicles; c) looking for the most optimal delivery option (as soon as possible and at lower cost); d) the renewal of the transport park to nature by clean means;e) the use of liquefied natural gas in poor freight operations; f) the recycling of waste from ece-usable or defective products using innovative technologies; g) technological development with a view to saving natural resources or increasing productivity.

The distribution of the ranch is shown in Fig. 1.

The data for the analysis and calculation of the breakdown of the rankings of expert questionnaires were listed in Table 1.

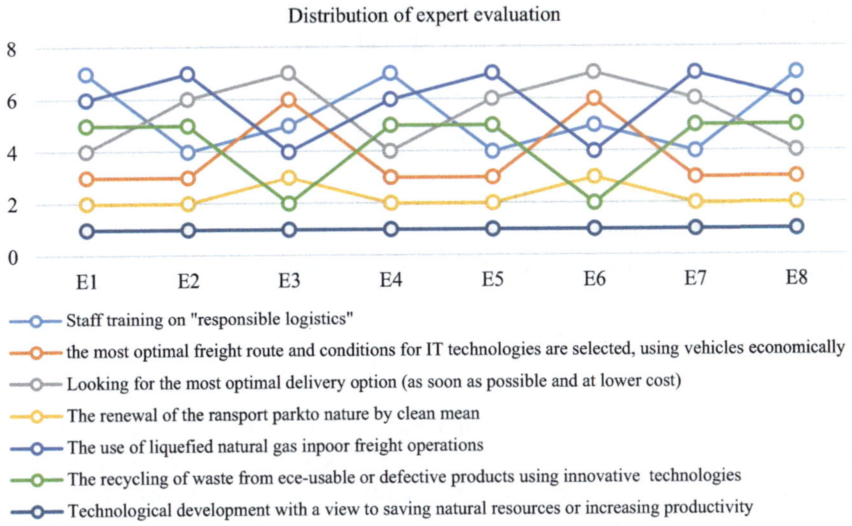

Fig. 1. Ranking of the main factors determining the effectiveness of the social responsibility measures of the transport undertaking for technological development (compiled by the authors).

Table 1. Table on the relevance of the main factors determining the effectiveness of the social responsibility measures of the transport undertaking for technological development (compiled by the authors).

Respondent No.	Factor encryption symbol (m = 7)						
	a	b	c	d	e	f	g
$\sum_{i=1}^{n} R_{ij}$	43	30	44	18	47	34	8
$\overline{R}_j = \frac{\sum_{i=1}^{n} R_{ij}}{n}$	5.375	3.75	5.5	2.25	5.875	4.25	1
$\sum_{i=1}^{n} R_{ij} - \frac{1}{2}n(m+1)$	11	–2	12	–14	15	2	–24
$\left[\sum_{i=1}^{n} R_{ij} - \frac{1}{2}n(m+1)\right]^2$	121	4	144	196	225	4	576

Calculated concordance factor W according to formula (3) in the absence of associated grades.

$$W = \frac{12S}{n^2(m^3 - m)} = \frac{12 \times 1270}{8^2(7^3 - 7)} = 0.7087. \tag{3}$$

The number of important factors determining the effectiveness of the main, most necessary, measures of social responsibility of the transport company in technological development (m) is 7, there are $m > 7$. The weight of the concordance coefficient χ^2 is then calculated according to the formula (4) and a random value is obtained.

$$\chi^2 = n(m-1)W = \frac{12S}{nm(m+1)} = \frac{12 \times 1270}{8 \times 7(7+1)} = 34.018. \quad (4)$$

χ^2 the estimated value of 34.018 is higher than the critical value (equal to 12.5916), which is why the opinion of the responders is considered to be balanced and the average grades reflect the general opinion of the experts.

$$W_{\min} = \frac{\chi^2_{v,\alpha}}{n(m-1)} = \frac{12.5916}{8(7-1)} = 0.2623 < 0.7087. \quad (5)$$

According to formula (5), the minimum value of the $W_{\min}$ coefficient of concordance was calculated, which states $W_{\min} = 0.2623 < 0.7087$ that the opinions of all 8 respondents on the 7 main criteria of relevance of the factors determining the effectiveness of the social responsibility measures of the transport undertaking for technological development, which are important, are still considered to be harmonized.

The main indicators of the importance of the factors determining the effectiveness of the social responsibility measures of the transport undertaking for technological development are calculated – Q_j. The data obtained are shown in Table 2.

Table 2. Key indicators Q_j (compiled by the authors) of the main drivers of the effectiveness of the social responsibility measures of the transport undertaking in technological development.

Indicator label	Factor encryption symbol (m = 8)*							Amount
	a	b	c	d	e	f	g	
q_j	0.192	0.134	0.196	0.080	0.210	0.152	0.036	1.00
D_j	0.808	0.866	0.804	0.920	0.790	0.848	0.964	6.00
Q_j	0.135	0.144	0.134	0.153	0.132	0.141	0.161	1.00
Q_j'	0.094	0.152	0.089	0.205	0.076	0.134	0.250	1.00
Arrangement of factors	5.000	3.000	6.000	2.000	7.000	4.000	1.000	

According to expert evaluations and calculations, the rankings of the factors determining the effectiveness of the main social responsibility measures of the transport undertaking for technological development are:

1. Technological development with a view to saving natural resources or increasing labour productivity;
2. Renewal of the transport fleet by clean means for nature;
3. The most optimal cargo transportation route and conditions are selected by IT technologies, making economical use of vehicles;
4. Recycling of waste from discarded or defective products with innovative technologies.

The experts were asked to assess the main criteria for the effective components of the social responsibility measures of the transport undertaking in the context of technological development:

a) support/assistance to the local community,
b) improvement of products/services,
c) investment of funds for the upskilling of employees (courses, trainings, etc.),
d) support for socially supported areas (culture, sport, education, etc.),
e) awareness of the enterprise in publicizing its social activities,
f) worker safety/satisfaction with working conditions,
g) implementation/updating of information systems to reduce costs.

The distribution of the ranch is shown in Fig. 2.

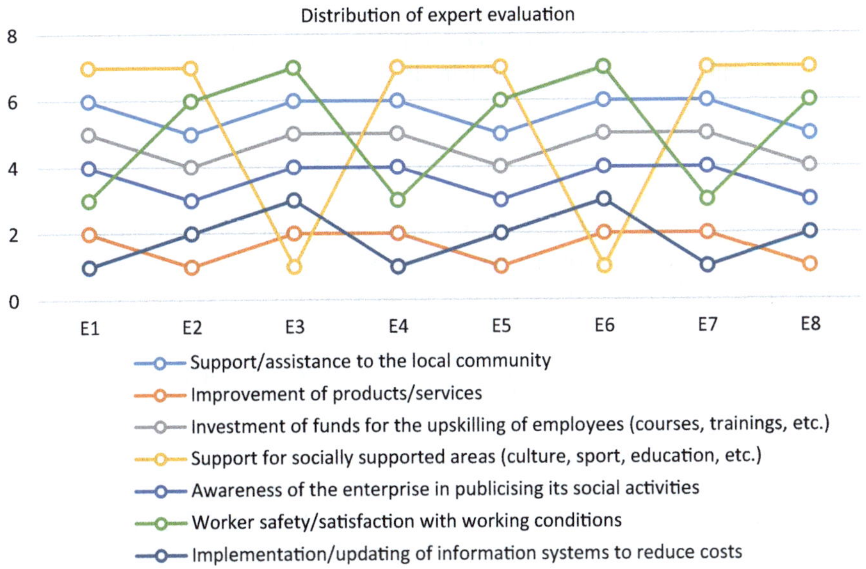

Fig. 2. Ranking of the main components of the social responsibility measures of the transport company in the course of technological development (compiled by the authors).

The analysis and calculation of the breakdown of the rankings of expert questionnaires shall be recorded in Table 3.

Calculated concordance factor W according to the (6) formula in the absence of associated grades.

$$W = \frac{12S}{n^2(m^3 - m)} = \frac{12 \times 1078}{8^2(7^3 - 7)} = 0.601. \tag{6}$$

The number of important components (m) of the main social responsibility measures of the transport company in the context of technological development is 7, i.e.

Table 3. Table of the main components of the transport undertaking's social responsibility measures in the context of technological development (compiled by authors).

Respondent no.	Factor encryption symbol (m = 7)*						
	a	b	c	d	e	f	g
$\sum_{i=1}^{n} R_{ij}$	45	13	37	44	29	41	15
$\overline{R}_j = \frac{\sum_{i=1}^{n} R_{ij}}{n}$	5.625	1.625	4.625	5.5	3.625	5.125	1.875
$\sum_{i=1}^{n} R_{ij} - \frac{1}{2}n(m+1)$	13	–19	5	12	–3	9	–17
$\left[\sum_{i=1}^{n} R_{ij} - \frac{1}{2}n(m+1)\right]^2$	169	361	25	144	9	81	289

$m > 7$. The weight of the concordance coefficient χ^2 is then calculated according to formula (7) and a random value is obtained:

$$\chi^2 = n(m-1)W = \frac{12S}{nm(m+1)} = \frac{12 \times 1078}{8 \times 7(7+1)} = 28.875. \tag{7}$$

χ^2 estimated value 28.875 is greater than the critical value (equal to 12.5916), which is why the opinion of the responders is considered to be balanced, and the average grades reflect the overall opinion of the experts.

$$W_{min} = \frac{\chi^2_{v,\alpha}}{n(m-1)} = \frac{12.5916}{8(7-1)} = 0.2623 < 0.601. \tag{8}$$

According to the (8) formula, the minimum value of the $W_{\min}$ coefficient of concordance was calculated, which states $W_{\min} = 0.2623 < 0.601$, that the opinions of all 8 respondents on the 7 main components of the social responsibility measure of the transport undertaking in the context of techno-development, which are important, are still considered to be harmonized.

The indicators of the components of the social responsibility measures of the transport company in the course of technological development are calculated – Q_j. The data obtained are presented in Table 4.

According to expert evaluations and calculations the order of importance indicators of the effective components of the main means of social responsibility of the transport company in the course of technological development is as follows:

1. Implementation/updating of information systems for cost reduction;
2. Improvement of products/services;
3. The awareness of the enterprise in publicizing the social activities carried out;
4. Funds for the improvement of the qualification of employees (courses, trainings, etc.).

Table 4. Indicators of the importance of the effective components of the main means of social responsibility of the transport company in the course of technological development – Q_j.

Indicator label	Factor encryption symbol ($m = 8$)*							Amount
	a	b	c	d	e	f	g	
q_j	0.201	0.058	0.165	0.196	0.129	0.183	0.067	1.00
d_j	0.799	0.942	0.835	0.804	0.871	0.817	0.933	6.00
Q_j	0.133	0.157	0.139	0.134	0.145	0.136	0.156	1.00
Q_j'	0.085	0.228	0.121	0.089	0.156	0.103	0.219	1.00
Arrangement of factors	7.000	2.000	4.000	6.000	3.000	5.000	1.000	

In conclusion, when analyzing the application of social responsibility measures of transport companies in the company's activities, there is a connection with the technological development applied in the company.

5 Conclusions

The study found that of the transport companies surveyed, the majority were in place (75%) ISO 14001 standard.

Studies have shown that the rankings of the 4 main factors determining the effectiveness of the social responsibility measures of the transport company for technological development were located: 1) technological development in order to save natural resources or increase the efficiency of work; 2) renewal of the transport fleet by clean means for nature; 3) the most optimal cargo transportation route and conditions are selected using IT technologies, making economical use of vehicles; 4) recycling of waste from discarded or defective products using innovative technologies.

The research found that the order of importance indicators of the 4 main components of the social responsibility measures of the transport company in the course of technological development is as follows: 1) implementation/updating of information systems for cost reduction; 2) improvement of products/services; 3) raising the awareness of the company in social activities; 4) investment of funds in the upskilling of employees.

At the same time, it should be noted that there is a certain trend that can be analyzed in various ways. It is an investment in information systems and in raising the qualification of employees at the same time. Companies strive to retain existing skilled workers, but at the same time facilitate working conditions and minimize costs. It is likely that with the development of the idea of corporate and corporate social responsibility, employee training will be linked to this topic as well. One of the most anticipated social responsibility activities is support and assistance to the local community. This allows us to conclude that business sufficiently understands the forms of social responsibility that are relevant in the business environment. This reaffirms the assumption that a business-friendly attitude towards the local community is relevant to business.

References

1. Michalski, M., Botella, J.L.M., Figiel, A.: Corporate social responsibility in supply chain management: a new model approach. Int. J. Logist. Syst. Manag. **30**(4), 477–502 (2018)
2. Zenisek, T.: Corporate social responsibility: a conceptualization based on organizational literature. Acad. Manag. Rev. **4**(3), 359–368 (1979). https://doi.org/10.2307/257192
3. Išoraitė, M.: Alaus gamybos įmonių socialinės atsakomybės tyrimas. [Beer production enteprises corporate social responsibility research in colleges]. Bus. Syst. Econ. **3**(2), 248–265 (2013). (in Lithuanian). https://doi.org/10.13165/VSE-13-3-2-10
4. Munim, Z.H., Schramm, H.-J.: The impacts of port infrastructure and logistics performance on economic growth: the mediating role of seaborne trade. J. Ship. Trade **3**(1), 1–19 (2018). https://doi.org/10.1186/s41072-018-0027-0
5. Mnif, S.: Bilateral relationship between technological changes and income ine-quality in developing countries. Atlantic Rev. Econ. **1** (2016). http://hdl.handle.net/10419/191976
6. Činčalová, S., Hedija, V.: Firm characteristics and corporate social responsibility: the case of Czech transportation and storage industry. Sustainability **12**(5) (2020). https://doi.org/10.3390/su12051992
7. Piecyk, M.I., Björklund, M.: Logistics service providers and corporate social responsibility: sustainability reporting in the logistics industry. Int. J. Phys. Distrib. Logist. Manag. **45**(5), 459–485 (2015). https://doi.org/10.1108/IJPDLM-08-2013-0228
8. Kavecke, I., Paužuolienė, J.: Transporto įmonių aplinkosaugos tendencijos ir galimybė pereiti prie žaliosios logistikos. [Environmental trends of transport companies and the possibility of moving to green logistics]. Reg. Formation Dev. Stud. **33**(1), 17–28 (2021). (in Lithuanian)
9. Haski-Leventhal, D.: Strategic corporate social responsibility: tools & theories for responsible management. SAGE Publications, London (2018). https://uk.sagepub.com/en-gb/eur/strategic-corporate-social-responsibility/book255155
10. Carroll, A.B.: Carroll's pyramid of CSR: taking another look. Int. J. Corp. Soc. Responsib. **1**(1), 1–8 (2016). https://doi.org/10.1186/s40991-016-0004-6
11. Tüzün Rad, S., Gülmez, Y.S.: Green logistics for sustainability. Int. J. Manag. Econ. Bus. **13**(3), 603–614 (2017)
12. Banabakova, V.: Role of logistics services for improving social policies. Knowl. Int. J. **41**(1), 55–62 (2020). https://ikm.mk/ojs/index.php/KIJ/article/view/4235
13. Drejeris, R., Oželienė, D.: New approach to the technological aspect of corporate sustainable development. Bus. Theory Pract. **20**, 363–371 (2019). https://doi.org/10.3846/btp.2019.34
14. Lovelace, R., Parkin, J., Cohen, T.: Open access transport models: a leverage point in sustainable transport planning. Transp. Policy **97**, 47–54 (2020). https://doi.org/10.1016/j.tranpol.2020.06.015
15. Gargasas, A., Samuolaitis, M., Mūgienė, I.: Quality management systems in logistics. Manag. Theory Stud. Rural Bus. Infrastruct. Dev. **41**(2), 290–304 (2019). https://doi.org/10.15544/mts.2019.24
16. Li, Y., Yu, Y.: The use of freight apps in road freight transport for CO_2 reduction. Eur. Transp. Res. Rev. **9**(3), 1–13 (2017). https://doi.org/10.1007/s12544-017-0251-y
17. Kardelis, K.: Mokslinių tyrimų metodologija ir metodai [Research methodology and methods]. Mokslo ir enciklopedijų leidybos centras, Vilnius (2016). (in Lithuanian)

18. Tidikis, R.: Socialinių mokslų tyrimų metodologija. [Methodology of Social Sciences Research]. Publishing Centre of the Law University of Lithuania, Vilnius (2003). https://repository.mruni.eu/handle/007/15459MRU
19. Sivilevičius, H.: Application of expert evaluation method to determine the importance of operating asphalt mixing plant quality criteria and rank correlation. Baltic J. Road Bridge Eng. **6**(1), 48–58 (2011). https://doi.org/10.3846/bjrbe.2011.07
20. Podvezko, V.: Agreement of expert estimates. Technol. Econ. Dev. Econ. **11**(2), 101–107 (2005)

A Study of Technological Development (Automation) of Land Transport Companies in Supply Chain

Dovydas Vaičius, Nijolė Batarlienė, and Kristina Vaičiūtė(✉)

Vilnius Gediminas Technical University, Plytinės 27, 10105 Vilnius, Lithuania
dovydas.vaicius@stud.vgtu.lt,
{nijolė.batarlienė,kristina.vaiciute}@vilniustech.lt

Abstract. The supply chain management of a land transport companies is an integral part of the activities of organizations. Improperly organized supply chain activities cause significant losses that can make a company uncompetitive. One of the most important factors in the supply chain is technology. The article examines the supply chain processes of technological development (automation) of land transport companies as a tool to ensure the competitiveness of transport companies and the quality of their services. An expert study was performed and the results were presented. Expert assessments of the main criteria for the creation and development of the information system of the transport system in land transport companies were calculated using the AHP methodology.

Keywords: Land transport companies · Technological development (automation) · Supply chain

1 Introduction

In the world today, everything happens globally, and many people in their daily lives do not do without transportation services like delivery, transportation and the like of ordered or purchased goods. Increased loads highlight the importance of freight transport, which affects the proper management of freight transport. In order to improve the smoothness and efficiency of cargo transportation, activities must be properly planned, organized and controlled, as a result of which investment in technological development is needed, which would help automate transportation processes and speed up work productivity.

The supply chain covers not only the manufacturers and suppliers, but also the carriers, storage service providers and other intermediaries, as well as the ordering of the supply (supply chain starts from the customer order and the needs expressed in it). Entities in the stages are connected to the streams of materials, information and finances on the ground floor, but supply in the mains is not entirely obligatory. The structure of the supply chain will depend on the needs of the customer and the actions of each stage entity. As far as clearing is concerned, only one entity is responsible for a separate supply chain stage. Manufacturers can obtain materials from several suppliers, and have their product delivered to several caterpillars, and so on. Therefore, the supply

O. Prentkovskis et al. (Eds.): TRANSBALTICA 2021, LNITI, pp. 610–620, 2022.
https://doi.org/10.1007/978-3-030-94774-3_60

chain should be treated as a network, as it should be considered as a supply chain structure, it is appropriate to consider the term supply networks.

Supply chain management (abbreviated as SCM) is a similar concept, but it is also possible to include suppliers and customers, as well as equivalent chain components.

Supply chain – is the totality of entities directly and indirectly involved in the implementation of the needs expressed by the customer.

A typical supply chain includes a large number of participants: a customer, a retailer, a major artist, a manufacturer, a raw material supplier [1–3].

The structure of the supply chain will depend on the demand of the customer and the actions of each stage. As far as clearing is concerned, only one entity is responsible for a separate supply chain stage. Manufacturers can obtain materials from several suppliers, and have their product delivered to several caterpillars, and so on. Therefore, the supply circuit should be treated as a network [4].

Object of research – technological development of a transport companies (automation).

Research methodology – analysis of scientific literature, collection of statistical data, multi-criteria evaluation, AHP method.

The aim of the article is to conduct a study of the technological development (automation) of the land transport companies in the supply chain and, using the AHP method, to identify the main factors that have the greatest impact on technological development.

2 Technological Development of Transport Companies (Automation)

The aim of science and technology is to enable companies and individuals to use technology more efficiently, as this reduces costs and increases productivity. The development of new technologies paves the way for the production of new cheaper goods and the accumulation of capital, and in this respect for greater international competitiveness of individual countries, as well as better quality of research institutions and contributing to the development of cultural and political society [5]. Technological change is an opportunity to change the market environment by reducing costs. As a result, technological change may increase the speed with which products are placed on the market [6]. Faure [7] sees technology as what is learned and acquired knowledge or technical skills as the best way to do everything. Mnif, Vaičiūtė and Bureika [8, 9] technological development means the constant ability of a companies to develop its activities better, faster and cheaper. Technological progress is the way forward in the development of humanity towards equality between the levels of population and nations [10].

Technological development is the introduction of new strategies and practices to increase the efficiency of a product or process. Therefore, TS mechanisms are necessary to ensure the strategy of supply chain members [11]. On one hand, the additional benefits should be clearly defined rather than simply attributed to the total profits increase of manufacturing enterprises. In addition, benefits should be known by their partners and be distributed at a reasonable distribution coefficient [12]. This is when organizations come up with or improve a product or process to get a better reward for the same amount of work. In addition, it is considered an important key for companies to survive all the changes faced by the markets [13]. In addition, technological change is bringing innovation, and over the last twenty years it has become a very important topic due to the fact that it is vital and significant for organizations to survive [6]. As well, technological change makes it possible to produce the same amount of production at lower cost and can benefit [10, 12]. Technological development has become a central theme of all issues, as development in recent centuries to the present day has been driven by technology, including the economic, labor, political and cultural dimensions of new product development [8].

One of the most important elements for the efficient use of transport is the provision of information [14, 15]. The main goal of the creation and development of the information infrastructure of the transport system is to facilitate and accelerate the movement of material and information flows in the transport system by computerizing and automating the work of the elements of the transport system regulating this movement:

- to implement technological (information, methodological) measures to ensure fast, reliable and high-quality information transfer both in the transport system and its interaction with the outside world;
- to improve the customer service of the transport system by systematizing and computerizing customer service procedures;
- to create databases of the branches of the transport system and a national database of common use of the transport system, allowing to obtain operative, reliable and detailed information on economic and commercial activities of interest to all their subscribers;
- to implement measures implementing the interaction of the elements of the transport system with other elements of the domestic and foreign information infrastructure (banks, tax, statistical, customs systems, etc.).

3 Research of Technological Development (Automation) of Land Transport Companies in the Supply Chain

3.1 Experts Research Methodology

Based on the AHP methodology, the expert group n quantifies *m* objects. Estimates form a matrix of *n* rows and *m* columns. Estimates form a matrix of *n* rows and *m* columns [16]. Evaluations can be − indicator units, unit parts, percentages, in a ten-point system. The ranking of expert indicators is suitable for calculating the concordance coefficient. Ranking is a procedure in which the most important indicator is given a rank of *R*, equal to one, the second indicator is given a second rank, and the last indicator is given a rank of *m* (*m* is the number of benchmarks). The average of the sum of the ranks is calculated [17]:

$$\sum_{i=1}^{n} R_{ij} = \frac{1}{2} n(m+1), \tag{1}$$

$$S = \sum_{j=1}^{m} \left(R_j - \overline{R}\right)^2, \tag{2}$$

Having expert evaluation indicators, the consistency of their opinions is determined by calculating the concordance coefficient of the Kendall ranks. If S is a real sum of the square value calculated by Eq. (2), the concordance coefficient W is described (when there are no related ranks) by the ratio of the calculated S value (3):

$$W = \frac{12S}{n^2 m(m^2 - 1)} = \frac{12S}{n^2(m^3 - m)}. \tag{3}$$

The random value is distributed according to χ^2:

$$\chi^2 = n(m-1)W = \frac{12S}{nm(m+1)}, \tag{4}$$

with the degree of freedom $v = m - 1$. Based on the selected confidence level α (which is assumed to be 0.05 or 0.01), the critical value $\chi^2{}_{v,\alpha}$ is found from the table of χ^2 distribution with the degree of freedom $v = m - 1$. If the value of χ^2 calculated by Eq. (4) is larger than $\chi^2{}_{v,\alpha}$, it shows that the experts' estimates are consistent.

The smallest value of the concordance coefficient $W_{\min}$ can be estimated by applying Eq. (5):

$$W_{\min} = \frac{\chi^2_{v,\alpha}}{n(m-1)}, \tag{5}$$

where n is expert opinions; m is the number of comparative criteria that indicates the quality of an object under analysis with the selected levels of significance α and degree of freedom $v = m - 1$.

The study involved 8 experts, all of whom have management experience in land transport for 3 to 10 years.

3.2 Results of the Study

The experts were asked to evaluate the main criteria for the creation and development of the information technology infrastructure of the transport system in the land transport companies in order of importance: 1 most important, 7 least important. The distribution of ranking is presented in Fig. 1.

The data of the analysis and calculation of the distribution of the rankings of the questionnaires of the eight expert questionnaires were listed in Table 1.

The main criteria for the creation and development of the transport system information infrastructure:

a – to create databases of branches of the transport system;

b – to improve the transport system by systematizing and computerizing customer service procedures;

c – to improve the transport system by systematizing and computerizing customer service procedures;

d – to implement technological (information, methodological) measures to ensure fast, reliable and high-quality information transfer;

e – create an automated supply chain management;

f – to implement measures realizing the interaction of the elements of the transport system with other elements of the domestic and foreign information;

g – to create a national transport system sharing database, which will provide operative information to all their subscribers.

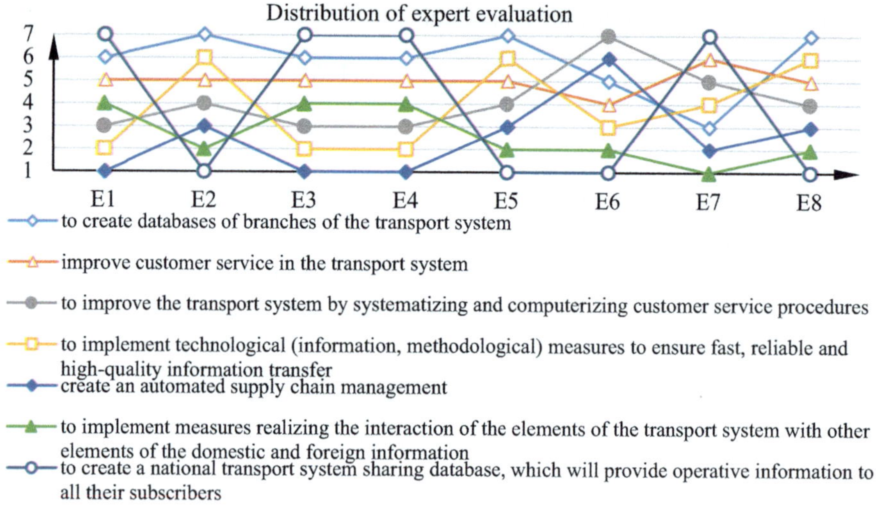

Fig. 1. Distribution of expert ranks (compiled by the authors).

The concordance coefficient W was calculated according to Eq. (3) when there are no associated ranks.

$$W = \frac{12S}{n^2(m^3 - m)} = \frac{12 \times 556}{8^2(7^3 - 7)} = 0.3102.$$

The number (m) of important major information system infrastructure developments and developments (automation) in the land transport companies is 7, i.e., $m > 7$. Then the weight of the concordance coefficient χ^2 is calculated according to formula (4) and a random quantity is obtained.

Table 1. Ranking table of the main criteria for the importance of the creation and development (automation) of the information technology infrastructure of the transport system in a land transport companies (compiled by the authors).

Expert Que. No.	Factor encryption symbol (m = 7)						
	a	b	c	d	e	f	g
$\sum_{i=1}^{n} R_{ij}$	47	40	33	31	20	21	32
$\bar{R}_j = \frac{\sum_{i=1}^{n} R_{ij}}{n}$	5.875	5	4.125	3.875	2.5	2.625	4
$\sum_{i=1}^{n} R_{ij} - \frac{1}{2}n(m+1)$	15	8	1	−1	−12	−11	0
$\left[\sum_{i=1}^{n} R_{ij} - \frac{1}{2}n(m+1)\right]^2$	225	64	1	1	144	121	0

$$\chi^2 = n(m-1)W = \frac{12S}{nm(m+1)} = \frac{12 \times 556}{8 \times 7(7+1)} = 14.893.$$

χ^2 the estimated value of 14.893 was higher than the critical value (equal to 12.5916), which makes the respondents' opinion considered consistent, and the average ranks reflect the overall opinion of the experts.

$$W_{\min} = \frac{\chi^2_{v,\alpha}}{n(m-1)} = \frac{12.5916}{8(7-1)} = 0.2623 < 0.3103.$$

The minimum concordance was calculated according to Eq. (5) $W_{\min}$ coefficient value where positive $W_{\min} = 0.2623 < 0.3103$, that the opinions of all 8 respondents on the 7 main criteria for the creation and development of the information system infrastructure of the transport system in the land transport companies, which are important for land transport companies in the impact of information technology development (automation) on freight transport in the supply chain, are considered harmonized.

Indicators of the importance of the impact of technological development on freight transport in logistics, which are important for road transport companies, are calculated – Q_j. The obtained data are presented in Table 2.

Table 2. Indicators of the importance of the main creation and development of the information infrastructure of the transport system in the company's Q_j (composed by the authors).

Indicator marker	Factor encryption symbol (m = 7)							Sum
	a	b	c	d	e	f	g	
q_j	0.2098	0.1786	0.1473	0.1384	0.0893	0.0938	0.1429	1.0000
d_j	0.7902	0.8214	0.8527	0.8616	0.9107	0.9063	0.8571	6.0000
Q_j	0.1317	0.1369	0.1421	0.1436	0.1518	0.1510	0.1429	1.0000
Q_j'	0.0759	0.1071	0.1384	0.1473	0.1964	0.1920	0.1429	1.0000
Order of factors by importance	7	6	5	3	1	2		

Based on expert assessments and calculations, the main criteria for the creation and development (automation) of the transport system information technology infrastructure in the land transport companies are as follows:

1. Create an automated supply chain management system (e);
2. To implement measures realizing the interaction of the elements of the transport system with other elements of the domestic and foreign information infrastructure (f);
3. To implement technological (information, methodological) measures to ensure fast, reliable and high-quality information transmission (d);
4. Establish a national transport system shared database to provide operational information to all their subscribers (g).

Less important criteria are:

c – to improve the transport system by systematizing and computerizing customer service procedures;
b – improve customer service in the transport system;
a – to create databases of branches of the transport system.

Experts were also asked to assess (in order of importance: 1 most important, 7 least important) what needs to be addressed in order to ensure quality organization of the transport companies in the supply chain.

The distribution of ranking is presented in Fig. 2.

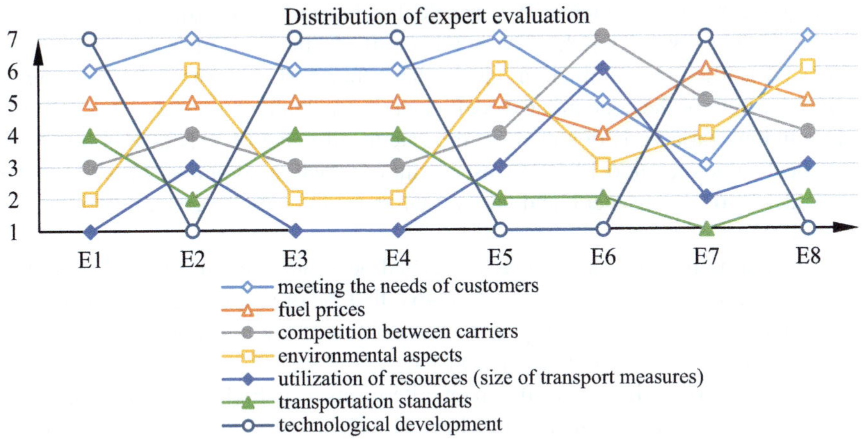

Fig. 2. Distribution of expert ranks (compiled by the authors).

The data of the analysis and calculation of the distribution of the rankings of the questionnaires of the eight expert questionnaires were summarized in Table 3.

Table 3. Ranking table of the main criteria for ensuring the quality organization of the supply chain in the supply chain (compiled by the authors).

Expert Que. No	Factor encryption symbol (m = 7)						
	a	b	c	d	e	f	g
$\sum_{i=1}^{n} R_{ij}$	25	47	36	25	29	46	16
$\overline{R}_j = \frac{\sum_{i=1}^{n} R_{ij}}{n}$	3.125	5.875	4.5	3.125	3.625	5.75	2
$\sum_{i=1}^{n} R_{ij} - \frac{1}{2}n(m+1)$	−7	15	4	−7	−3	14	−16
$\left[\sum_{i=1}^{n} R_{ij} - \frac{1}{2}n(m+1)\right]^2$	49	225	16	49	9	196	256

The main criteria that ensure the quality organization of the transport companies in the supply chain are the following:

a – meeting the needs of customers;
b – fuel prices;
c – competition between carriers;
d – environmental aspects;
e – utilization of resources (size of transport measures);
f – transportation standards;
g – technological development.

The concordance coefficient W was calculated according to Eq. (3) when there are no related ranks. $W = 0.44642$. The number (m) of important main information technology infrastructure (automation) in the land transport companies is 7, i.e. $m > 7$. Then the weight of the concordance coefficient χ^2 is calculated according to formula (3) and a random variable is obtained. χ^2 the calculated value of 21.428 was higher than the critical value (equal to 12.5916), as a result of which the opinion of the respondents is considered to be consistent, and the average ranks indicate the general opinion of the experts.

The minimum concordance was calculated according to Eq. (5) $W_{\min}$ coefficient value where positive $W_{\min} = 0.2623 < 0.4464$, that the opinions of all 8 respondents on the 7 main criteria ensuring the quality organization of the transport companies in the supply chain are still considered harmonized.

The importance indicators of the main criteria ensuring high-quality organization of the transport companies in the supply chain are calculated – Q_j. The obtained data are presented in Table 4.

Table 4. Key indicators of importance – Q_j, ensuring high-quality organization of the transport companies in the supply chain.

Indicator marker	Factor encryption symbol (m = 7)							Sum
	a	b	c	d	e	f	g	
q_j	0.1116	0.2098	0.1607	0.1116	0.1295	0.2054	0.0714	1.000
d_j	0.8884	0.7902	0.8393	0.8884	0.8705	0.7946	0.9286	6.000
Q_j	0.1481	0.1317	0.1399	0.1481	0.1451	0.1324	0.1548	1.000
Q_j'	0.1741	0.0759	0.1250	0.1741	0.1563	0.0804	0.2143	1.000
Factor arrangement	2	7	5	3	4	6	1	

Based on expert assessments and calculations, the main list of priorities that ensure the quality organization of the transport companies in the supply chain should be arranged in the following order and the 4 main ones are presented:

1. Technological development (g);
2. Meeting the needs of service customers (a);
3. Environmental aspects (d);
4. Utilization of resources (size of transport measures) (e).

Less important criteria are:

b – fuel prices;
c – competition between carriers;
f – transportation standards.

4 Conclusions

The analysis of the results of the study according to the 4 main criteria singled out by the experts revealed the need to: create an automated supply chain management system (e); to implement measures realizing the interaction of the elements of the transport system with other elements of the domestic and foreign information infrastructure (f); to implement technological (informational, methodological) measures to ensure fast, reliable and high-quality information transmission (d); to establish a national transport system shared database to provide operational information to all their subscribers (g).

Based on expert assessments and calculations, the main list of priorities that ensure the quality organization of the transport companies in the supply chain should be arranged in the following order and the 4 main ones are presented: technological development (g); meeting the needs of service customers (a); environmental aspects (d); utilization of resources (size of transport measures) (e).

References

1. Židonis, Ž.: Verslo logistika. Vilniaus vadybos kolegija, Vilnius (2002)
2. Garalis, A.: Logistika. Bendrieji pagrindai. Šiaulių universiteto leidykla, Šiauliai (2003)
3. Palšaitis, R.: Šiuolaikinė logistika. Vadovėlis. Technika, Vilnius (2010)
4. Vasiliauskas, A.V., Zinkevičiūtė, V.: Gamybos logistika. Gamybos vadyba. SMK, Klaipėda (2013)
5. Çalışkan, H.K.: Technological change and economic growth. Procedia – Soc. Behav. Sci. **195**(3), 649–654 (2015). https://doi.org/10.1016/j.sbspro.2015.06.174
6. Al-Khafaji, A.J., Al-Ameri, T., Abeed, Q.: Oil and gas play and prospect assessments of Babel, Diwania and Karbala Governorates (Middle Euphrates Region), Iraq. Arab. J. Geosci. **7**(8), 3147–3161 (2013). https://doi.org/10.1007/s12517-013-0980-8
7. Faure, A.P.: Money Creation: Role of Bank Liquidity, 1–7 (2013). https://doi.org/10.2139/ssrn.2256310
8. Mnif, S.: Bilateral relationship between technological changes and income inequality in developing countries. Atlantic Rev. Econ. **1** (2016)
9. Vaičiūtė, K., Bureika, G.: Research on transport enterprise technological development directions. Mokslas – Lietuvos Ateitis/Science – Future of Lithuania **12** (2020). https://doi.org/10.3846/mla.2020.11790
10. Haile, G.A., Srour, I., Vivarelli, M.: The impact of globalization and technology transfer on manufacturing employment and skills in Ethiopia. IZA Discussion Paper No. 7820, pp. 1–25 (2013)
11. Zhong, F.M., Zeng, W., Zhou, Z.B.: Mechanism design in a supply chain with ambiguity in private information. J. Ind. Manag. Optim. **16**(1), 261–287 (2020). https://doi.org/10.3934/jimo.2018151
12. Acciaro, M., Renken, K., El Khadiri, N.: Technological change and logistics development in European ports. In: Carpenter, A., Lozano, R. (eds.) European Port Cities in Transition. SS, pp. 73–88. Springer, Cham (2020). https://doi.org/10.1007/978-3-030-36464-9_5

13. Masad, Q.A.: Factors driving technological change and its impact on human resources diversity management practices: test of the mediator role for strategic renewal (Vol. Thesis Submitted in Partial Fulfillment of the Requirements for the Degree of Master in Business). Amman - Jordan: Administration Department of Business Administration Middle East University (2016)
14. Wang, X., Guo, H., Wang, X.: Supply chain contract mechanism under bilateral information asymmetry. Comput. Ind. Eng. **113**, 356–368 (2017)
15. Attaran, M.: Digital technology enablers and their implications for supply chain management. Supply Chain Forum Int. J. **21**, 158–172 (2020). https://doi.org/10.1080/16258312.2020.1751568
16. Podvezko, V.: Agreement of expert estimates. Technol. Econ. Dev. Econ. **11**(2), 101–107 (2005)
17. Sivilevičius, H.: Application of expert evaluation method to determine the importance of operating asphalt neixing plant quality criteria and rank correlation. Baltic J. Road Bridge Eng. **6**(1), 48–58 (2011). https://doi.org/10.3846/bjrbe.2011.07

Organization of Transport Provision of Export of Grain Cargo Under the Conditions of Stochastic Nature of Receipt of Service Requirements

Oleksandr Gryshchuk[1], Anatoliy Petryk[2(✉)], Arkadiy Kozlov[2], Olena Bura[2], and Maryna Tyshkevych[2]

[1] Faculty of Management, Logistics and Tourism, National Transport University, Kyiv, Ukraine
gryshchuk@ntu.edu.ua

[2] Faculty of Transport and Information Technologies, National Transport University, Kyiv, Ukraine

Abstract. The article simulates the organization of transport support for grain exports under the condition of modeling the phased maintenance of requirements, identifies the main characteristics of queuing systems using the technology of pre-accumulation of grain cargo in port and port elevators. The possibility of solving the problem of rational transport service is substantiated, when the incoming application of the flow of requirements of the incoming flow is gradually served by several applications of another flow. It is proved that the main components of transport systems can be the development of methods of transport services and justification of rational interaction of different types of transport for the transportation of grain cargo for export purposes in integrated production systems. As a result, transport processes in the systems of transport service of export cargo flows in the work were considered as the interconnected work of economic entities, means of one or different modes of transport, taking into account the impact of the environment.

Keywords: Transport service · Export cargo flows · Transport systems · Step-by-step maintenance of requirements · Transport nodes

1 Introduction

The creation of transport systems in agro-industrial production requires the development of scientific bases for reliable transport services and the solution of a number of practical problems. Such a set of measures to ensure agro-industrial production with efficient operation of transport systems involves determining the patterns of formation of cargo flows and the necessary components of the transport process, development of methods of transport services, justification of rational interaction of different modes of transport for integrated cargo delivery systems [1]. If it is necessary to provide transport support for export cargo flows in the daily work of transport nodes, there is a whole class of tasks, when the incoming application of the flow of requirements is gradually

O. Prentkovskis et al. (Eds.): TRANSBALTICA 2021, LNITI, pp. 621–631, 2022.
https://doi.org/10.1007/978-3-030-94774-3_61

served by several applications of another flow [2]. Typical examples of such systems are the creation of a consolidated consignment of goods in specialized terminals, the maintenance of railway cars by cars, the delivery of containers by wholesale consignment cars, and so on. In this case, for these queuing systems, the set of requirements is considered as an extraordinary Poisson flow [3].

Analysis of the results of previous studies of transportation processes in agro-industrial production systems shows that in the vast majority of scientific papers the process of transport service of export cargo flows is considered separately from the conditions of origin and formation of cargo-forming arrays and characteristics of material flows [4]. In addition, in the process of practical application of scientific recommendations is not always taken into account the influence of random factors on the characteristics of transport services, and in most cases the initial parameters of material flow do not always respond adequately to the impact of transport [5, 6]. In this case, the transport processes in the systems of export orientation should be considered as the interconnected work of economic entities, means of one or different modes of transport, taking into account the impact of the environment [7].

2 Feature of Infrastructural Support of Export Deliveries of Grain Cargoes in Transport Knots

The peculiarity of the transport of export supplies on the example of grain cargo indicates the need to model this process in transport hubs as open systems [8]. That is, the sources of service requirements and supply of existing infrastructure elements are not limited [9]. The structure of such a queuing system (QS) is reduced to a model of a single-channel system, in which the first stream is an incoming stream of service requests, and the second stream is a stream of limited offers, numerical values which are characterized by technical capabilities of certain devices [10]. Therefore, the features of transport services for grain exporters (traders) are closely related to the need to determine the performance of production activities of such economic structures (Fig. 1).

Assume that the probability density distributions of the time intervals between the receipt of applications and the time intervals between the ends of services have the form

$$f(t) = \lambda e^{-\lambda t}, \quad t \geq 0; \tag{1}$$

$$g(x) = \frac{r\mu(r\mu x)^{r-1}}{(r-1)!} e^{-r\mu x}, \quad x \geq 0$$

where λ, μ – the intensity of the requirements of the 1st and 2nd streams, respectively; r – the number of stages of servicing the requirements of the 1st stream by the applications of the 2nd stream.

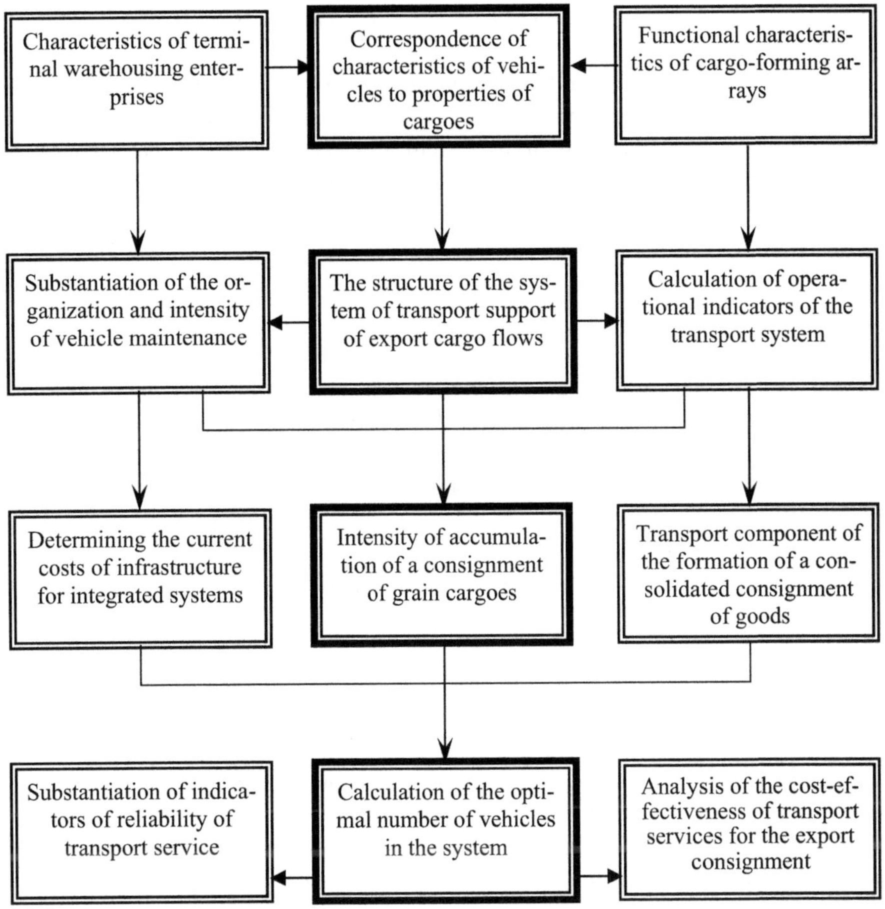

Fig. 1. Calculation of indicators of transport support for grain exports in the formation of export consignments of grain.

Since the analysis of such a system, in addition to calculating the number of applications in the QS, you must also determine the number of stages through which to serve the service application, each application that is in the service queue should be described by the number of stages *r* required to complete service. Therefore, the state of the QS at any time it is advisable to describe the total number of stages of service, which may be all the applications that are currently in the system, until the complete completion of service [11].

If the current state of the system is characterized by the presence of *k* applications, and the incoming application is in the *i-th* stage of service $(i = 1, \ldots, r)$, then the number of stages through which must pass all applications in the QS is equal to:

$$j = (k-1)r + (r-i+1) = kr - i + 1. \quad (2)$$

To create mathematical models of transport support for grain exports, we introduce the notation: $P_k(t)$ – the probability of k applications in the service system; $P_j(t)$ – the probability that all applications in the system must pass j stages of maintenance. The relationship between the number of applications and the number of stages is determined by the mathematical dependence:

$$P_k(t) = \sum_{j=(k-1)r+1}^{kr} P_j(t), \quad k = 1, 2, \ldots \quad (3)$$

The task of determining the numerical indicators of the functioning of such a system is to calculate the probabilities $P_k(t)$ and $P_j(t)$ for the stationary mode of operation of the QS. The probabilities of system transitions from one state to another in a short period of time Δt up to an infinitesimal value of $0(\Delta t)$ of the order higher than Δt will look like:

$$\begin{aligned}
&P(k \xrightarrow{\Delta t} k) = e^{-\lambda \Delta t} \cdot e^{-r\mu \Delta t} = (1 - \lambda \Delta t + \ldots)(1 - r\mu \Delta t + \ldots) = 1 - (\lambda + r\mu)\Delta t + 0(\Delta t);\\
&P(k-1 \xrightarrow{\Delta t} k) = (1 - e^{-\lambda \Delta t}) \cdot e^{-r\mu \Delta t} = (\lambda \Delta t + \ldots)(1 - r\mu \Delta t + \ldots) = \lambda \Delta t + 0(\Delta t);\\
&P(k+1 \xrightarrow{\Delta t} k) = e^{-\lambda \Delta t} \cdot (1 - e^{-r\mu \Delta t}) = (1 - \lambda \Delta t + \ldots) \cdot r\mu \Delta t = r\mu \Delta t
\end{aligned} \quad (4)$$

Using the probability of transition of the system at time Δt, the probabilities of the states of the system at time $t + \Delta t$ are determined using the method of compiling Kolmogorov differential equations. When using these mathematical dependences, conditions are imposed on the states of the system, which are that the probabilities of states with negative indices are zero. Taking into account that the densities of the time interval distribution between the events of the input and output streams are given by the exponential distribution and the Erlang distribution, respectively, the following expressions were obtained [12]:

- the probability that the system will not receive a new request during Δt and it is not affected by the maintenance process

$$\begin{aligned} e^{-\lambda \Delta t} &= 1 - \lambda \Delta t + 0(\Delta t);\\ e^{-r\mu \Delta t} &= 1 - r\mu \Delta t + 0(\Delta t) \end{aligned} \quad (5)$$

- the probability that the system will not receive a new request during Δt and it is not affected by the maintenance process

$$\begin{aligned} 1 - e^{-\lambda \Delta t} &= 1 - (1 - \lambda \Delta t + 0(\Delta t)) = \lambda \Delta t + 0(\Delta t);\\ 1 - e^{-r\mu \Delta t} &= 1 - (1 - r\mu \Delta t + 0(\Delta t)) = r\mu \Delta t + 0(\Delta t) \end{aligned} \quad (6)$$

Using these relations for the probabilities of states at time $t + \Delta t$, mathematical dependences are obtained:

$$\begin{aligned}
&P_0(t+\Delta t) = P_0(t)(1-\lambda\Delta t) + P_1(t)r\mu\Delta t + 0(\Delta t);\\
&P_j(t+\Delta t) = P_j(t)[1-\lambda\Delta t - r\mu\Delta t] + P_{j-1}(t)\lambda\Delta t + P_{j+1}(t)r\mu\Delta t + 0(\Delta t)\\
&j = 1, 2\ldots
\end{aligned} \tag{7}$$

Given the division of the right and left parts of the relations by Δt and the transition to the boundaries at $\Delta t \to 0$, the system of differential equations will have the form

$$\begin{aligned}
&P_0'(t) = -\lambda P_0(t) + r\mu P_1(t);\\
&P_j'(t) = -(\lambda + r\mu)P_j(t) + \lambda P_{j-r}(t) + r\mu P_{j+1}(t);\\
&j = 1, 2
\end{aligned} \tag{8}$$

Using the basic provisions of the theory of optimization of material resources, it becomes possible to minimize logistics costs in servicing transit and export grain cargo flows.

3 Application of the Method of Step-by-Step Servicing of Requirements for Long-Term Planning of Export Transportation of Grain Cargoes

Consider the operation of the transport system in a steady state. Since the input and output flows of the QS are given by the exponential distribution and the Erlang distribution, respectively, this system is Markov and therefore there are stationary probabilities. Since in this case $P_j(t) = 0$; $(j = 0,\ 1)$, then the composite system of differential equations is transformed into a system of algebraic equations:

$$\begin{aligned}
&\lambda P_0 = r\mu P_1;\\
&(\lambda + r\mu)P_j = \lambda P_{j-z} + r\mu P_{j+1}, \quad j = 1, 2.
\end{aligned} \tag{9}$$

To solve this system of equations, we use the method of generating functions, which are determined by equality:

$$F(z) = \sum_{j=0}^{\infty} P_j z^j. \tag{10}$$

Given that $F(1) = 1$, the constant P_0 can be calculated by the Popital rule, provided that each j-th equation is multiplied by z^j, we obtain the dependence:

$$\sum_{j=1}^{\infty} (\lambda + r\mu)P_j z^j = \sum_{j=1}^{\infty} \lambda P_{j-r} z^j + \sum_{j=1}^{\infty} r\mu P_{j+1} z^j. \tag{11}$$

In this system, the intensity of applications is equal to λ, and the average service time is $1/\mu$ regardless of r. Given the selection of $F(z)$, we obtain:

$$(\lambda + r\mu)[F(z) - P_0] = \lambda z^r F(z) + \frac{r\mu}{z}[F(z) - P_0 - P_1 z]. \quad (12)$$

Using the proposed equality for further simplification, we obtain:

$$F(z) = \frac{r\mu P_0[1 - (\frac{1}{z})]}{\lambda + r\mu - \lambda z^r - (\frac{r\mu}{z})}. \quad (13)$$

The constant P_0 can be calculated as:

$$F(1) = 1 = \frac{r\mu P_0}{r\mu - \lambda r}. \quad (14)$$

Therefore, the utilization factor of the system is equal to $\rho = \lambda/\mu$. For the case $r = 1$, which corresponds to a single-channel QS, the generating function will look like:

$$F(z) = \frac{\mu(1-\rho)(1-z)}{\mu + \lambda z^2 - (\lambda + \mu)z} = \frac{(1-\rho)(1-z)}{1 + \rho z^2 - (1+\rho)z}. \quad (15)$$

The denominator of this fraction is decomposed into prime factors in the form $(1 - z)\ (1 - \rho z)$. Then we obtain the inverse transformation according to the tables:

$$P_k = (1-\rho)\rho^k, \quad \mathrm{k} = 0, 1, 2, \quad (16)$$

In the case $r = 1$ it is obvious that $p_k = P_k$, and Eq. (12) gives the distribution of the number of applications in a single-channel QS. For $r > 1$, the standard method of inverting the transformation $F(z)$ is to decompose this fine-rational function into simple fractions and invert each of them separately. Before decomposing into simple fractions, you need to find $(r + 1)$ the roots of the polynomial, which is in the denominator. The analysis shows that one of them is equal to one. Therefore, the denominator can be written in the form $(1 - z)\ [r\mu - \lambda\ (z + z^2 + \ldots + z^r)]$. Moreover, the last r roots (denoted by $z_1, z_2, \ldots, z_r$) are the roots of the polynomial, which is in square brackets. Having determined these roots, we can write the denominator in the form $r\mu(1 - z)\ (1 - z/z1)\ldots(1 - z/zr)$:

$$P_j = (1-\rho)\sum_{i=1}^{r} A_i (z_i)^{-1}, \quad j = 1, 2, \ldots \quad (17)$$

where

$$A_i = \prod_{n=1,n\neq i}^{r} \frac{1}{(1 - z/z_i)}. \tag{18}$$

Thus, for a given service system, the distribution of the number of service steps is a weighted sum of geometric distributions [13].

Taking into account the stated theoretical provisions, it becomes possible to solve the issue of long-term planning in the organization of transport support of export transportation of grain and coordination of power parameters of transhipment mechanisms and vehicles in the general structure of the transportation process. Given the random nature of the numerical values of overload capacity $f(x)$ in the case of an unstable number of vehicles, it is advisable to determine the change in the allowable limits of these parameters.

4 The Results of Mathematical Modeling of the Organization of Transport Service of Grain Cargo Flows in Transport Nodes

An urgent issue in the process of forming a joint consignment of grain cargo for export in the transport hub is to determine the rational organization of delivery and accumulation of these goods. The main economic criterion in this case is the total cost of moving goods in the transport system. The example of the formation of combined export consignments of grain cargoes simulates the situation when to ensure a given level of reliability of transport services in addition to their own n and involved vehicles y use additional capacity of the production and terminal system of the sea trade port [14].

In this case, the operating costs include both the costs associated with the current transportation of goods B_1 by cars, and the costs of pre-storage of goods in the terminal premises B_2 and loading and unloading. The additional involvement of vehicles is characterized by the costs usually associated with a reduction in load capacity D_1, and the cost of rail transport for the delivery of grain to the elevator D_2. Losses of the transport system from downtime C of vehicles in the absence of demand for transport services and possible losses E associated with the lack of necessary material flows in the system were determined by conventional methods. The problem was considered under the assumption that the capabilities of the reloading installation in the port are characterized by a continuous random variable with a distribution density $f(x)$. That is, the demand for grain cargoes, and accordingly the vehicles for their delivery, are determined by the mathematical expectation m_x and the standard deviation σ_x of the random variable x. These characteristics can be determined by experimental studies and field observations.

Given the uncertainty of the proposals of additional vehicles u and the capabilities of the terminal economy v for the accumulation of grain, it was assumed that the exponential distribution of these values with density:

$$\begin{aligned} g(u) &= \lambda e^{-\lambda u} \\ h(v) &= \mu e^{-\mu v}, \end{aligned} \tag{19}$$

where λ, μ – distribution parameters.

Taking into account these theoretical provisions, the function of total costs of the production system on the example of the formation of a joint grain lot will be determined by the number of required vehicles n and depend on the probability distributions of random variables of demand for them x, spare cars u and available elevator storage v in general should be characterized by mathematical dependence:

$$g(n,x,u,v) = \begin{cases} B_1 x + C(n-x) & npu \quad 0 \le x \le n \\ B_1 x + D_1(x-n) & npu \quad n \le x \le n+u \\ B_1 n + (B_1+D_1)u + (B_2+D_2)(x-n-u) & npu \quad x+u \le x \le n+u+v \\ B_1 z + (B_1+D_1)u + (B_2+D_2)v + E(x-n-u) & npu \quad x \ge n+u+v \end{cases} \tag{20}$$

The problem of determining the rational organization of the transport process was considered under the condition of equality of indicators $n = n_O$, when the function of specific costs $B^{(n)}$ has a minimum value:

$$B^{(n)} = \frac{B(n)}{n} \min \quad B(n). \tag{21}$$

The calculation of logistics costs per 1 ton of grain cargo is performed under the condition of preliminary accumulation in the port elevator with subsequent delivery to the sea trade port at a distance of 50 km. In this case, the bulk of the grain is delivered to the port by road, and the loading of the vessel is performed in combination with the involvement of the previous accumulation of grain directly in the port elevator.

According to the calculations, the change in the specific costs of transport and handling work is characterized by an ambiguous dependence on the intensity $f(x)$ of handling work.

The minimum value of B_m under the condition α = 75 tons/hour is explained by the absence in this case of involved road transport, and hence the additional costs associated with its use. With increasing productivity of the reloading node within $n \le x \le n + u + v$ additional operating costs of the involved vehicles $(B_1 + D_1)u$ and the use of elevator and storage capacity $(B_2 + D_2)v$ port to some extent increase the resulting B_m. However, under the condition x = 120… 150 tons/hour, the rate of increase of the total costs is smaller in comparison with the increase of the quantity of processed cargo. Therefore, in the specified range of transhipment capacity, the total cost of B_m tends to decrease. The size of the export consignment of grain cargoes is closely related to the cost of downtime. In this case, with insufficient capacity of maintenance mechanisms, the cost of transport and handling works increases significantly (Fig. 2).

Thus, mathematical modelling of the impact of the organization of the formation of an export batch of grain cargo in seaports suggests that if it is impossible to ensure intensive supply of the required amount of grain by road, a certain amount of cargo should be pre-accumulated in port elevators.

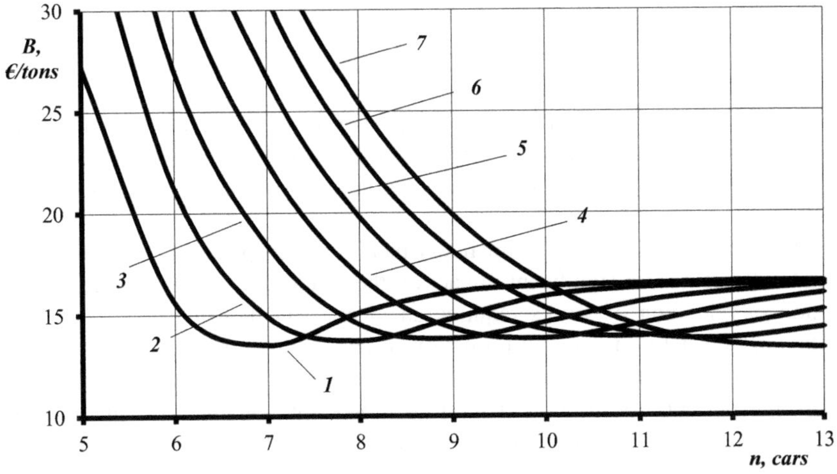

Fig. 2. Dependence of total costs on the available number of vehicles at the capacity of the reloading unit, tons/hour: 1 – 140; 2 – 160; 3 – 180; 4 – 200; 5 – 220; 6 – 240; 7 – 260.

According to the analysis performed using the mathematical dependence (20), there is a tendency of interdependence of indicators *f(t)* and *r*. Calculations performed using the criterion of minimization of total costs *B(n)*, make it possible to determine the possible limits of change of the influencing indicators of the capacity of transhipment vehicles and their transport. Thus, under the condition of preliminary delivery of grain by rail to the port elevator at a distance of l = 500 kms and its current supply by road trains with a capacity of q = 20 tons at a distance of l = 45 kms, the change in total costs confirms the previous theoretical assumption rolling stock. Moreover, with the increase in the capacity of reloading vehicles, the lack of cars is becoming more and more noticeable. The increase in the numerical value of $B^{(n)}$ in such cases is due to the significant impact of costs associated with downtime of the transport system.

In addition, with the change of transhipment capacity in the port with the appropriate transport support, the optimal level of total costs changes ambiguously. The decrease in the numerical value of $B^{(n)}$ at relatively small values of *f(x)*, namely m_x = 80 tons/hour, the impact of costs associated with the possible additional use of vehicles and the previous accumulation of cargo *D* in the port elevator is insignificant. As the numerical value of transhipment capacities m_x increases, due to the random nature of the quantity *f (x)*, the increase in costs $B^{(n)}$ is explained by the change in the numerical value of the mathematical expressions $\int_0^\infty \left\{ \int_0^\infty \left[\int_{n+v}^{n+u+v} (x-n-u)f(x)dx \right] g(u)du \right\} h(v)dv$ and $\int_0^\infty v \left\{ \int_0^\infty \left[\int_{n+u+v}^{\infty} f(x)dx \right] g(u)du \right\} h(v)dv$. A further increase in f(x) reduces the costs

$E\int_{0}^{\infty}\left\{\int_{0}^{\infty}\left[\int_{n+u+v}^{\infty}(x-n-u-v)f(x)dx\right]g(u)du\right\}h(v)dv$ associated with ship downtime.

Thus, mathematical modelling of the impact of the organization of the formation of an export batch of grain cargo in seaports suggests that if it is impossible to ensure intensive supply of the required amount of grain by road, a certain amount of cargo should be pre-accumulated in port elevators.

Under such conditions, even relatively increased transport costs for grain delivery by rail will not have a significant impact on the resulting B_m performance. The organization of reloading of works requires an increase in the intensity of their implementation, especially in cases of formation of large ship parties.

An urgent issue of long-term planning in the organization of transport support for export transportation of grain is the coordination of the parameters of the capacity of transhipment mechanisms and vehicles in the overall structure of the transportation process. Given the random nature of the numerical values of the overload capacity $f(x)$ in the case of an unstable number of vehicles, it is advisable to determine the change in the allowable limits of these parameters. Thus, to optimize the total costs $B^{(n)}$ along with a comprehensive pre-planning of the process of movement and accumulation of goods, it is important to ensure the stable operation of the transport system.

5 Conclusions

According to the results of the characteristics of cargo-forming arrays in systems with extraordinary input flow under the condition of batch receipt of requirements, the probabilities of queuing system states are obtained and operational characteristics are calculated, which allow to take into account changes in numerical values of cargo-forming arrays in agro-industrial cargo flows. The process of formation of cargo flows in transport systems at step-by-step process of service of requirements is modelled, the basic characteristics of systems of queuing at use of the extended technology of preliminary accumulation of grain cargoes in stationary and temporary objects of transport infrastructure are defined. Models of formation of consolidated consignments of grain cargoes for queuing system with two input streams of requirements are developed and the main characteristics of transport system at batch receipt of requirements on condition of formation of the specified requirements in limited sources of applications are defined.

References

1. Danchuk, V., Bakulich, O., Svatko, V.: Identifying optimal location and necessary quantity of warehouses in logistic system using a radiation therapy method. Transport **34**(2), 175–186 (2019). https://doi.org/10.3846/transport.2019.8546
2. Taji, T., Tanigawa, S.-I., Kamiyama, N., Katoh, N., Takizawa, A.: Finding an optimal location of line facility using evolutionary algorithm and integer program. J. Comput. Sci. Technol. **2**(3), 362–370 (2008). https://doi.org/10.1299/jcst.2.362

3. Vorkut, T., Volynets, L., Bilonog, O., Sopotsko, O., Levchenko, I.: The model to optimize deliveries of perishable food products in supply chains. Eastern-Eur. J. Enterp. Technol. **5**(3–101), 43–50 (2019). https://doi.org/10.15587/1729-4061.2019.177903
4. Prokudin, G., Chupaylenko, O., Dudnik, O., Dudnik, A., Omarov, D.: Improvement of the methods for determining optimal characteristics of transportation networks. Eastern-Eur. J. Enterp. Technol. **6**(3(84)), 54–61 (2016). https://doi.org/10.15587/1729-4061.2016.85211
5. Sonmez, A.D., Lim, G.J.: A decomposition approach for facility location and relocation problem with uncertain number of future facilities. Eur. J. Oper. Res. **218**(2), 327–338 (2012). https://doi.org/10.1016/j.ejor.2011.10.028
6. Sharai, S., Oliskevych, M., Roi, M.: Development of the procedure for simulation modeling of interrelated transport processes on the main road network. Eastern-Eur. J. Enterp. Technol. **5/3**(101), 70–83 (2019). https://doi.org/10.15587/1729-4061.2019.179042
7. Prokudin, G., Chupaylenko, O., Dudnik, O., Prokudin, O., Dudnik, A., Svatko, V.: Application of information technologies for the optimization of itinerary when delivering cargo by automobile transport. Eastern-Eur. J. Enterp. Technol. **2**(3(92)), 51–59 (2018). https://doi.org/10.15587/1729-4061.2018.128907
8. Melo, M.T., Nickel, S., Saldanha da Gama, F.: Dynamic multi-commodity capacitated facility location: a mathematical modelling framework for strategic supply chain planning. Comput. Oper. Res. **33**(1), 181–208 (2006). https://doi.org/10.1016/j.cor.2004.07.005
9. Crainic, T.G., Perboli, G., Rosano, M.: Simulation of intermodal freight transportation systems: a taxonomy. Eur. J. Oper. Res. **270**(2), 401–418 (2018). https://doi.org/10.1016/j.ejor.2017.11.061
10. Beloglazov, A., Banerjee, D., Hartman, A., Buyya, R.: Improving productivity in design and development of information technology (IT) service delivery simulation models. J. Serv. Res. **18**(1), 75–89 (2014). https://doi.org/10.1177/1094670514541002
11. Apfelstädt, A., Dashkovskiy, S., Nieberding, B.: Modeling, optimization and solving strategies for matching problems in cooperative full truckload networks. IFAC-PapersOnLine 49(2), 18–23 (2016). https://doi.org/10.1016/j.ifacol.2016.03.004
12. Harris, I., Wang, Y., Wang, H.: ICT in multimodal transport and technological trends: unleashing potential for the future. Int. J. Prod. Econ. **159**, 88–103 (2015). https://doi.org/10.1016/j.ijpe.2014.09.005
13. Shin, S., Roh, H.-S., Hur, S.: Characteristics analysis of freight mode choice model according to the introduction of a new freight transport system. Sustainability **11**(4), 1209 (2019). https://doi.org/10.3390/su11041209
14. Ritzinger, U., Puchinger, J., Hartl, R.F.: A survey on dynamic and stochastic vehicle routing problems. Int. J. Prod. Res. **54**(1), 215–231 (2015). https://doi.org/10.1080/00207543.2015.1043403

Improvement of the Urban Transport System by Developing the Platform "Park and Ride" in Vilnius City

Darius Bazaras, Aldona Jarašūnienė(✉), and Martynas Norkūnas

Faculty of Transport Engineering, Department of Logistics and Transport Management, Vilnius Gediminas Technical University, Plytinės Street 27, 10105 Vilnius, Lithuania
aldona.jarasuniene@vilniustech.lt

Abstract. The article emphasizes that in the current time it is difficult to imagine a car-free world, many cities are paying more and more attention to environmentally friendly transport policy. The biggest challenge implementing the policy is to reduce the use of cars in densely populated city areas with the highest traffic flows. In order to make cities more environmentally friendly to the environment and humans, there is a need to ensure continuous investment in the renewal, improvement and development of transport infrastructure, as well as to raise people's awareness about the problems caused by cars and their impact on the environment and human health. Having in mind people's daily transport needs, it is important to restructure mobility trends radically, by encouraging people to use public transport, cycle or walk or use "Park and ride" services. Since free and unrestricted movement is very important in today's world, it is essential to guarantee the necessary mobility while limiting the use of cars in cities. The article presents the measures for solving the problems and analyzing the development of the platform "Park and ride".

Keywords: Urban transport system · Transport · The platform "Park and ride" · Economic viability

1 Introduction

The main reason for people's constant desire to move is the wide territorial location of objects of human interest. With the constant increase in passenger and freight traffic in cities, it is important to ensure the possibility of uninterrupted movement. The larger the urban area, the more difficult it is to maintain and regulate the entire transport system. Avoiding transport problems in the city, a properly designed, well-used and constantly renewing urban transport system. The urban transport system consists of three main elements, that is, vehicles, transport infrastructure, passengers or freight. In case of failure of any of these elements, the whole system becomes inefficient. The main task of the urban transport system is to ensure proper, efficient, high-quality movement of passengers and freight throughout the city, taking into account all the needs of the participants of the system. A person chooses a trip vehicle taking into account time, price, comfort and safety factors. In most cases it is a public transport:

O. Prentkovskis et al. (Eds.): TRANSBALTICA 2021, LNITI, pp. 632–647, 2022.
https://doi.org/10.1007/978-3-030-94774-3_62

underground, tram, bus, or a personal transport: cars, bikes, scooters, multimodal services are becoming relevant too.

2 The Characteristic of Vilnius City Transport System

Juškevičius, Burinskienė, Paliulis and Gaučė [1] state that the last decade the urbanization process has been considered as an unmanageable phenomenon due to a constant people's migration from villages to cities. According to Juškevičius [2], each city dweller has certain different communication needs which are determined by a person's lifestyle.

Juškevičius and Valeika [3] note that city residents and guests are considered to be the main users of the system and they determine the number of passengers and the load of the streets' networks. According to Juškevičius and others [1], the urban transport system has changed radically in recent years. Fewer and fewer people are using public transport services showing increased preference of their own cars. Vytautas and Andrius Jaržemskiai [4] claim that an own car has become a comfortable and convenient guarantee of travel. Currently, about seven hundred million cars are registered worldwide. Private transport is expected to grow to one billion by 2030.

Banister [5] claims that urban transport problems are due to the growth and expansion of cities. By 2050, seventy percent of the world's population is expected to live in cities. Increased demand for transport and road traffic have caused serious congestion, delays, accidents and environmental problems. Cities' transport emission makes up a quarter of the whole CO_2 that is released by transport and about sixty-nine percent of all accidents occur here [6]. According to Barauskas [7], the problems of the transport system arise when the transport infrastructure can no longer provide the growing needs of consumers. Banister [5] highlights the seven main problems of urban connectivity, for which city dwellers and guests show the greatest dissatisfaction. These include falling demand for public transport, heavy traffic congestion, high accident rates, pollution of the urban environment, risks and noise from vehicles to people, difficulties in pedestrian and cyclist traffic and parking difficulties.

The population density of Vilnius city is much lower than in similar European cities [8]. This is due to the rather widely spread territorial location of Vilnius city between work, recreation, entertainment and residence and consequently longer journeys. In 2019 about 1.6 million tourists visited Vilnius, who spent more than one day in the city. Nor should we forget the people coming here to work from other towns or cities close to Vilnius [8–10].

Public transport operates in order to create suitable conditions for the city's residents and guests to travel freely within the city. The survey of Vilnius population conducted in 2019 reveals that Vilnius public transport statistically is mainly used by center city residents, schoolchildren, people living alone, unemployed, pensioners or persons with lower incomes [11, 12]. In 2020, people mostly used the city's public transport services on weekdays. Around 535,000 trips are made every working day. The most loaded routes are those that connect the central part of the city with residential districts. These routes make about 16% of all daily journeys [13]. Although the level of satisfaction of Vilnius city residents with public transport services is increasing every year. Nevertheless, in 2020, only 25% of the townspeople used public transport services in Vilnius. The number of trips made by public transport in Vilnius decreases every year. This is influenced by the rapidly increasing level of motoring in the city.

Table 1. Distribution of trips made by Vilnius city residents according of travel (compiled by the authors based on the Vilnius sustainable mobility plan).

Way of traveling	Modelic distribution of trips, %	
	2016	2020
Public transport	24.3	25.0
Bike	0.7	1.0
Walking	29.5	30.0
Own car	45.0	43.0
Public car	0.5	1.0

It can be noted that for many years the most popular and commonly used vehicle for traveling around Vilnius city are cars (Table 1). In 2020, about 43% of the city's population preferred personal cars for their trips [14]. At the end of 2019, according to the Statistics Department in Lithuania, 335,107 passenger cars were registered in Vilnius County [15]. Analyzing the data provided by the SE "Regitra", a trend in the growth of the number of passenger cars is significant.

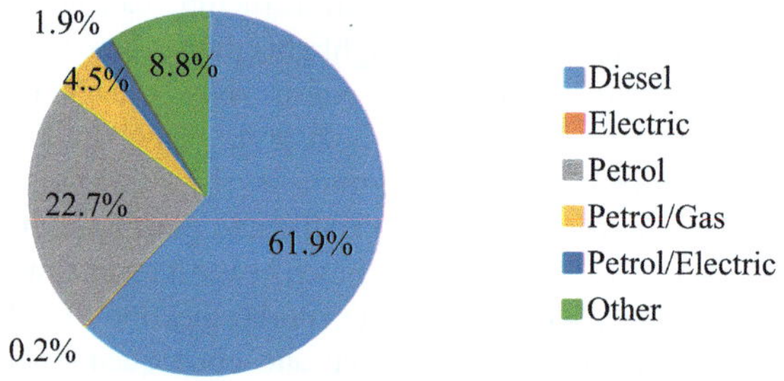

Fig. 1. Distribution of passenger cars in Vilnius by type of fuel used (compiled by the author, based on the data of SE "Regitra").

Analyzing the distribution of passenger cars by the type of fuel run in Vilnius, it can be seen in the chart that in Vilnius dominate diesel-powered vehicles, which pollute the environment quite heavily. In 2020, just over 42,000 of them were registered. This number makes about two-thirds of all vehicles registered in the city. Although electric and hybrid cars are becoming more popular every year, only a little over 2% of them are currently registered (Fig. 1). According to SE "Regitra" after the introduction of the motor vehicle pollution tax in Lithuania, the registration of vehicles run by less polluting fuels has slightly increased, however diesel-powered cars are still popular [16].

About 67% of Vilnius residents living farther from the city center make their daily trips in their own cars. This is mainly influenced by the fact that about 50% of all jobs are located in the central part of the city and its surroundings. As Vilnius city is not densely populated, the distance from residential districts to the central part of the city is quite large. The city center is the destination of the daily trips of the majority of the population. The relatively high level of automation and the low dispersion of jobs in the city often lead to congestion, which causes additional problems such as an increased risk of accidents, higher than normal noise and pollution levels, and so on [14].

2.1 Problems of Vilnius City Transport System and Reasons for Their Occurrence

Traffic flows in the city are greatly influenced by seasonality, school holidays, religious and public holidays, weather conditions and more.

The largest flows are recorded during the morning and evening peaks on weekdays. During 2019, the residents of Vilnius spent about 167 h or in other words a week in traffic jams [17]. Congestion is not the only consequence of a high level of automation. Approximately 60–70% of urban emissions are released by heavy and passenger cars. Maximum concentration of pollutants. Comparing the major cities of Lithuania, it is Vilnius that records the highest average annual values of pollutants [18]. The problem of noise in the city can not be omitted. Studies have shown that about 65 thousand of Vilnius residents live close to high-intensity streets. 78,000 people experienced an excess of noise during the day [14]. It is also worth mentioning the problem of traffic accidents in Vilnius city. Most accidents in Vilnius, as well as many other transport-related problems, are caused by heavy traffic flows and heavy traffic. The biggest accident rate in Vilnius city is recorded in autumn and spring, when traffic flows are the highest and the lewerest rate of accidents is in summer [19].

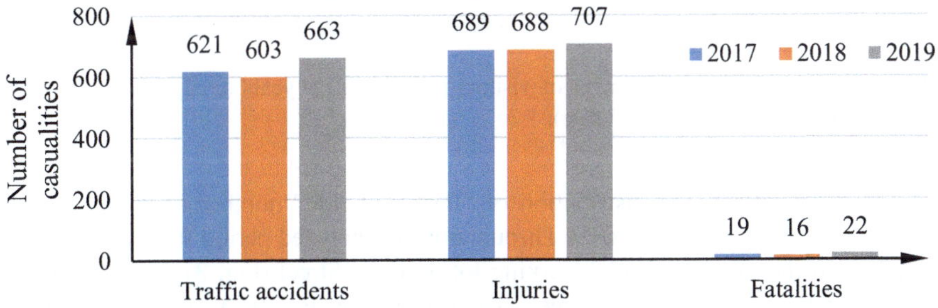

Fig. 2. The number of accidents and victims in Vilnius city in 2017–2019 (compiled by the authors, based on the data of the SE "Lithuanian road administration").

It is noticed that the number of accidents occurring in Vilnius city during the year is really high (Fig. 2). The annual damage in Vilnius due to injuries and fatalities in traffic accidents ranges from 40 million euro to 50 million euro [19].

The three most common types of accidents recorded in Lithuanian cities are car collisions involving another car, car collisions involving pedestrians and car collisions involving cyclists [19].

3 "Park and Ride" Parking Lots Characteristic

One of the ways to tackle congestion or other problems caused by heavy traffic flows in Vilnius city were chosen "Park and ride" parking lots so favourable abroad. In 2017, for the first time in Lithuania and in Vilnius were arranged the parking lots "Park and ride" that connected private and public transport. In 2017, three parking lots of this type were arranged in the main places of Vilnius connecting the city center and residential districts and provided parking space for about 250 cars. In 2019 another "Park and ride" parking was arranged for almost 100 cars on Savanorių Avenue [20, 21].

Since July 2017, when the first parking lots of this type were opened in the city until the beginning of 2018, more than 1.7 thousand people used this service, so a decision was made to reduce the price of this service four times from 2 euros to 50 euro cents. The reduced cost of the service has attracted a larger number of its users [21]. In 2018, about 15 thousand people used this service, in 2019 – about 33 thousand people, the number of the users doubled compared to the previous year.

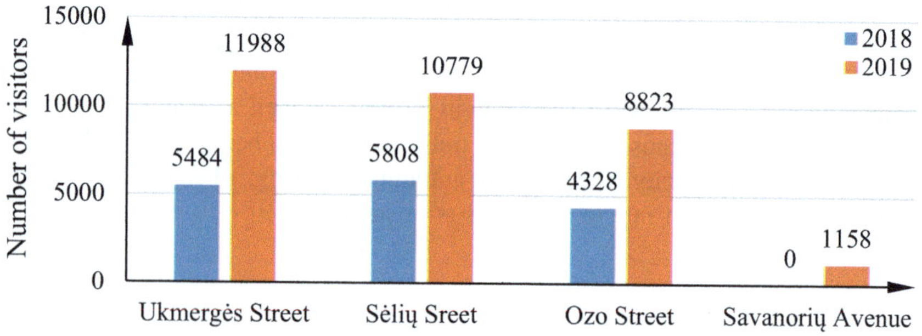

Fig. 3. "Park and ride" the number of visitors in 2018–2019 years (compiled by the authors based on SE "Transport service" data).

The chart reflects the distribution of "Park and ride" number of visitors in all 4 parking spaces in 2018 and 2019. Throughout the analyzed period the least number of people used the "Park and ride" parking lot on Ozas Street (Fig. 3). In 2018 the most popular parking was on Sėlių Street, which was used by about 10 thousand people. However, the research revealed that this car parking is mostly used by people working in the neighbourhood area as a cheap parking space, rather than by people traveling towards the central part of the city. Meanwhile, in 2019 the largest number of visitors was recorded on Ukmergės Street arranged "Park and ride" site, making almost 12 thousand [22].

Each car parking should be analyzed separately, taking into account their location, public transport service and so on. The results of the analysis of the existing "Park and ride" parking lots in Vilnius are presented in the second table. Assessment values are (+) – meet the criteria, (–) – do not meet the criteria, (+/–) – partially meet the criteria. For more information see Table 2.

Table 2. Assessment of "Park and ride" parking's arranged in Vilnius city (compiled by the authors, based on the carried out analysis).

Assessment aspect/name of the parking	246 Ukmergės Street	14 Ozas Street	62 Sėlių Street	124 Savanorių Avenue
Appropriate location of the parking	+/–	–	–	+/–
Frequency of public transport towards city center (+/– every 10 min.)	+	+	–	+/–
Priority of public transport on streets (number of A lane)	+	+/–	–	+/–
Availability of the parking	–	+/–	+/–	+
The cost of the service	+	+	+	+

Summarizing the whole current situation, it can be stated that all four "Park and ride" parking lots in Vilnius are located in the central part of the city. "Park and ride" parking sites arranged on Ukmergės Street and Savanorių Avenue serve the purpose the most appropriately. The success of car parking's that connect private and public transport significantly depends on the quality of the public transport network. Three of the four "Park and ride" parking's are located near the stops of the public transport, where a bus/trolleybus runs towards the central part of the city approximately every 10 min. However, the high demand for public transport on these stops causes buses/trolleybuses at peak times to be overcrowded. Automatically traveling by public transport becomes uncomfortable and unattractive for a person. The main advantage of using "Park and ride" parking service is lower cost than traveling to the city center by own car. Thus, having assessed the whole current situation, it can be concluded that the main criteria were not fully evaluated for arranging "Park and ride" parking lots in Vilnius city.

3.1 The Research Methodology

In order to analyze and examine a client's and user's needs, expectations and satisfaction, various research methods are applied. The research includes a questionnaire with questions submitted on the internet space. The main purpose of the survey is to identify and name possible disadvantages and problems of "Park and ride" parking lots from the perspective of the users of this service. The main task of the research is to assess the attitude of the residents of Vilnius city and its surroundings toward the service of "Park and ride" operating in Vilnius city. Additional tasks arising from the main purpose of the survey are: to determine the respondents' ways of traveling in the city and their reasons, to assess the informativeness about the location of the sites among the survey respondents and evaluate different aspects of the "Park and ride" (price, locations, etc.) and their influence on the choice of respondents to use or not this service.

The survey is divided into three main groups of questions which suit and enable to achieve the named tasks of the survey. The first group of the questions from the first to the sixth, is created to identify the travel habits of respondents in Vilnius city. The second group consists of questions from the seventh to the fifteenth. This group is created to clarify the respondents' opinion about the "Park and ride" parking network in Vilnius. The third group of the survey consists of questions from the sixteenth to eighteenth and identifies the demographic data of respondents, gender, age and social status.

In order to evaluate the quality of the provided service properly, it is very important to select the right respondents participating in the survey. Therefore, assessing and analyzing the quality of the service provided by "Park and ride" in Vilnius city, respondents have to be chosen according to the importance of this service for them. The survey is carried out during the period of two months.

In order to find out the reliability of the conducted survey were taken into consideration some possible mistakes in the process of calculation. The formula (1) shows how the reliability of the executive sample is calculated [23]:

$$n = \frac{z^2 \times s^2}{\Delta^2}, \tag{1}$$

where z – the coefficient derived from the Stuart Distribution Table, which is selected based on the reliability we seek and want to obtain; s – the average square deviation of the sample; n – the number of cases in the sample group; Δ the optional overall average.

To calculate the reliability of the survey the most optimal mistake was chosen = 7 and population 522,368. The population of this survey was chosen in 2020 according to the number of working people in the capital region [24]. After calculating the reliability of the survey, it was found that at least one hundred and ninety-six respondents should be interviewed in order to reflect the opinion of the interviewees. In conclusion, the aim of an online survey is to evaluate the quality of the service provided by "Park and ride" in Vilnius city. The results will be used to prepare the service quality assessment as well as to identify problematic and to be improved aspects.

A forecasting method will also be used to predict the economic viability of the "Park and ride" parking network development and its renewal project. Evaluation of the economic viability of the project will be carried out in three main stages. The first stage is the collection of the necessary data on the occupancy of the currently operating "Park and ride" parking lots in Vilnius. The economic profitability of the "Park and ride" parking lots project will be estimated on the basis of 2017–2019 statistics. The statistics of 2020–2021 will not be included in the forecast due to the Covid-19 pandemic that hit the world in 2020 and the universal quarantine introduced in Lithuania. The second step is filtering of the collected data which will allow us to select necessary and accurate data needed for precise estimation. In the article the economic payback will be estimated on the basis of the number of available parking spaces per day expressed as a percentage. This indicator shows how the number of "Park and ride" visitors changed in the relevant years. The last third stage is the preparation and evaluation of the forecast. The economic payback will be calculated by forecasting how the number of "Park and ride" visitors will change over the years and the revenue received from the users of this service, as well as by estimating the project implementation and parking maintenance costs.

4 The Results of the Research of the Services Provided by the Parking Lots "Park and Ride" in Vilnius

In order to identify the travel habits of Vilnius residents and its guests and the reasons for it, as well as to identify possible disadvantages of "Park and ride" parking lots, informativeness about these parking lots and different aspects of them (price, location, etc.) and how they influence the consumers' choice to use or not to use this service, a survey of the residents of Vilnius city and Vilnius surroundings was conducted. Below is presented the summary of the survey results.

Total number of interviewed 184 people. Among them 108 are women, that's about 59%, and 76 are men, that's about 41%. The majority of respondents belonged to the age group from 21 to 30 years, which makes up 43% of all respondents. Nearly 27% of respondents belonged to the age group of 31 to 40 years and 19% of respondents were older than 41 years old. The vast majority of respondents, 125, almost 68%, indicated that their social status is currently "employed".

Most of the surveyed people, about 53%, live in the western part of the city, about 23% of respondents live in Vilnius district and other cities / towns, another 16% live in the northern part of the city, the remaining part of respondents, almost 9% live in the southern part or in the center (for more details see chart 4).

Answering the survey questions, 71% of the respondents indicated that they mostly choose personal cars for their daily trips in the city. The results of the survey show that about 45% of respondents choose this type of travel in the city because of convenience, shorter travel time or lower costs.

When asked to comment on how often they use their own cars to travel to the central part of Vilnius, almost 22% of respondents said that they travel to this part of the city 5 or more times a week. Another nearly 16% said they travel once/twice a week. When asked to comment on where they most often park their cars in the central part of the city, the majority, almost 38%, indicated that it was on paid parking's.

Summarizing the first and second section of questions, it can be stated that the profile of users who tend to use the "Park and ride" parking service make men and women aged from 21 to 40, currently working or studying, mainly living in the western and northern part of the city or in Vilnius district or other cities and towns located near Vilnius, mostly traveling in the city by cars.

Evaluating combined city trips, almost 60% of respondents rated them positively, as long as it helps them reach their destination faster. Combined travel is also viewed positively because it promotes sustainable movement, reduces environmental damage (Fig. 4). So it means that there is a demand for combined parking among the surveyed people (Fig. 5).

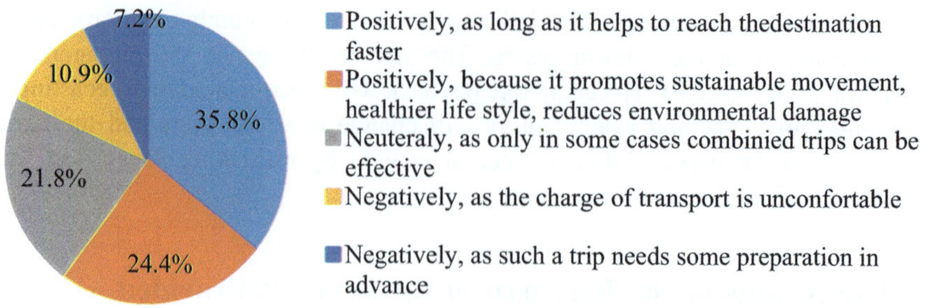

Fig. 4. Evaluation of combined trips made in the city (compiled by the authors, based on the survey results).

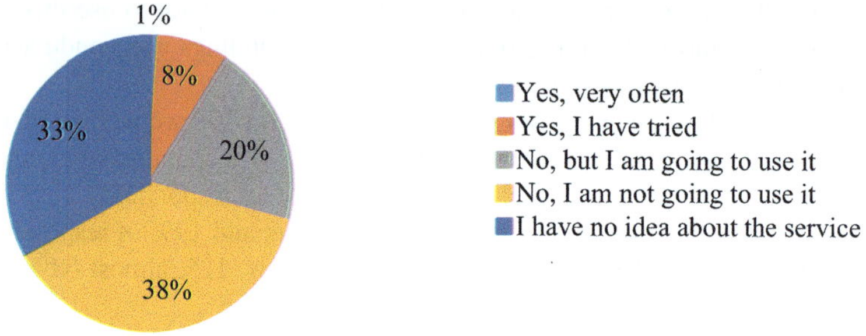

Fig. 5. Popularity of "Park and ride" parking lots (compiled by the author based on the survey results).

However, only a little more than 8% of the respondents have tested the "Park and ride" parking lots in Vilnius. The majority of respondents, that is about 58%, have not tried this service at all. However, about 20% of the respondents are planning to try this service in the future. When asked to comment on why they do not use the "Park and ride" parking service, the majority, almost 23%, indicated that they do not use the service because it does not help to reduce travel time. Another 15% indicated that they do not use this service due to the inconvenient network of "Park and ride" parking lots. When asked to rate the suitability of the "Park and ride" parking's on 1–5 point scale, 6% rated it with two points (inappropriate) and another 13% rated it with 3 points (neither appropriate or inappropriate). Other reasons why respondents do not use this service are: overcrowded city public transport, uncomfortable public transport, underdeveloped public transport network in the city and so on.

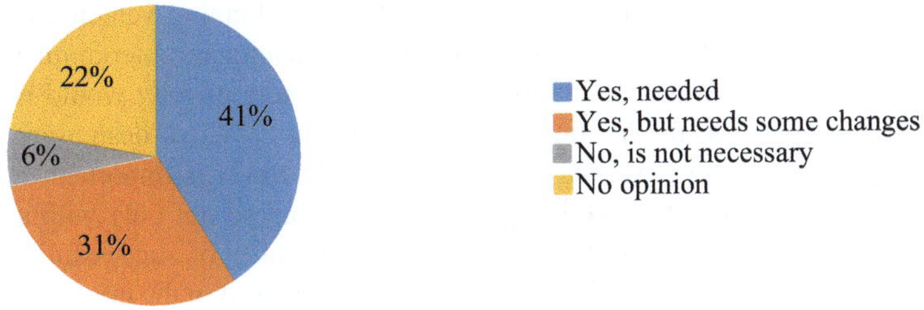

Fig. 6. Respondents' opinion on the need for a "Park and ride" car parking network in Vilnius city (compiled by the authors based on the survey results).

Almost 72% of respondents think that the "Park and ride" parking network is needed in the city. 31% of respondents state that the network of "Park and ride" parking lots currently operating in Vilnius requires some changes, but in general, it is necessary (for more details, see Fig. 6). When asked to comment on what they think should be changed to make the service more accessible, most respondents indicated the need to upgrade the public transport fleet, to increase the number of car parks and also to change the location of existing car parks. Summarizing the results of the survey, it can be stated that the majority of respondents positively evaluated the combined way of traveling in the city. It can be concluded that the demand for such parking lots does exist, but the problem is that the majority of respondents do not know or do not use the project of specific combined parking lots "Park and ride" operating in Vilnius city. However, most of the interviewed people believe that the "Park and ride" parking network in Vilnius is still necessary, but it requires certain changes. After some changes, this service would become much more popular among Vilnius residents and city guests.

4.1 Problematic Areas of Vilnius City Transport System

Analyzing the Vilnius city transport system, some of the problematic areas appeared. The problematic areas of the Vilnius city transport system are the following: large number of diesel-powered cars in the city; high level of motorization in the city; long journeys by public transport in the city; the price of the public transport service does not correspond to the quality of the service; lack of combined parking lots.

4.2 Suggestion of the Development and Renovation of "Park and Ride" Parking Network

The main everyday objects of human‘s needs are located in the central part of Vilnius. The constant growth of Vilnius population and the relatively rapid development of the suburbs lead to a large number of daily trips and growing level of motorization. In order to solve the problems caused by high levels of motorization, are raised parking prices, changed road markings and directions, Vilnius residents and city guests are encouraged to use constantly renewal public transport services, are being created new and reconstructed existing bicycle paths, developed "Park and ride" parking network and other. The Old town is the busiest area of the city. Banning vehicle traffic in this part of the city, where there are many job places, educational, entertainment, service and cultural facilities, would create a CO_2-free zone where priority would be given to pedestrians, cyclists or public transport, but not cars. When designing new and reconstructing the existing cycling paths in the Old Town, it is important to take into account that the layout of the cycling paths would be linked not only within this area, but also to the cycling paths outside this area. The number of successfully operating platforms of renting and sharing bikes could also be increased in the city. It is important that public transport routes in this urban area would be designed so that pedestrians and cyclists could reach them without much difficulty. Moreover, the public transport network should be accessible to people with disabilities. The quality of public transport is a prerequisite for restricting vehicle traffic in the Old Town. Banning car traffic in the Old Town would lead to free parking spaces, which are located in this part of the city. Most of them could be converted into bicycle storage areas adapted to the storage and maintenance of different types of bicycles and electric or simple scooters. Other car parks would continue to serve their main purpose but would be better adapted to electric cars with the necessary infrastructure. In order to ensure more convenient access to the Old Town for the guests and residents of Vilnius, the “Park and ride” parking ring will be developed on the outskirts of the city. Restricting car traffic in the Old Town and developing the “Park and ride” car parking network, it should be located and installed in places that are easily accessible to both citizens and city guests, ensure good access to the existing public transport network, and provide fast, convenient and cheap travel by public transport. In this way the service would become more popular and acceptable to the public. It would also reduce the transit transport running through the city. A public survey shows that the majority of respondents are in favor of a combined mode of travel, as long as it allows them to reach their destination faster. Having analyzed the composition of Vilnius city public transport network and human‘s needs objects, current mobility situation (distribution of traffic flows, most intense traffic crossroads and etc.) impact of noise, air pollution and accidents caused by transport and currently operating “Park and ride” parking network in Vilnius, the following suggestions would not only increase the demand for the “Park and ride” service but also reduce the high level of motorization in the city as well as the side effects caused by transport.

Two of the four “Park and ride” parking lots currently operating in Vilnius do not meet the set criteria and requirements. They are located on Ozo and Sėlių Streets. These parking lots will be completely dismantled and converted into ordinary parking’s. The new “Park and ride” parking lots will be installed in more suitable places of the city, that is the parking lot from Sėlių Street will be moved closer to the edge of the western part and the parking lot from Ozas Street will be moved closer to the edge of the northern part of the city. In both the western and northern parts of the city, the new “Park and ride” parking lots will be installed on the land belonging to Vilnius City Municipality.

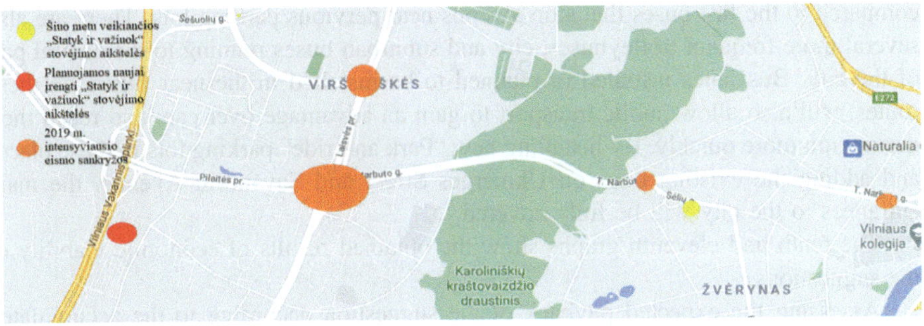

Fig. 7. The plan of the most intense traffic crossroads in 2019 and "Park and ride" parking lots in the western part of the city (compiled by the authors, based on the data of Vilnius City Municipality).

Figure 7 simulates how the situation would change if the current "Park and ride" parking lot on Sėlių Street was turned into a simple parking and a new one would be installed closer to the western edge of the city.

Figure 8 simulates how the situation would change if the current "Park and ride" parking lot on Ozo Street was turned into a simple parking lot and a new one would be installed closer to the northern edge of the city.

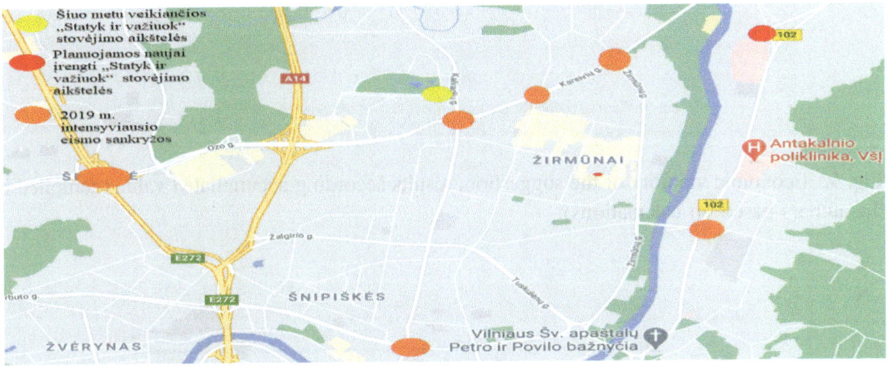

Fig. 8. The plan of the most intense traffic crossroads in 2019 and "Park and ride" parking lots in the northern part of the city (compiled by the authors, based on the data of Vilnius City Municipality.

It can be seen from the eighth and ninth pictures that the locations of the newly planned "Park and ride" parking lots meet the set criteria and requirements much better than the previous ones. The parking lots installed in these places will allow the user to avoid city congestion, as they will be accessible before reaching the busiest streets and intersections in the city. There are also public transport stops (400–500 m on foot) for the 4G fast city bus going to the central part of the city, close to the newly built "Park and ride" car parks. The buses on this route are less crowded during the peak hours compared to the fast buses that stop at stops near pervious parking lots. There are also several more frequent trolleybuses, city and suburban buses running to the central part of the city. Bus lanes installed or planned to be installed in the near future on these routes, will also allow public transport to gain an advantage over cars and reach their destination more quickly. By installing new "Park and ride" parking lots in these places and adding the existing ones on Ukmergės Street and Savanorių Avenue, the main entrances to the city will be fully covered.

The tenth and eleventh graphs show the obtained results of economic viability of the suggestions.

Assessing the expected payback of the suggestion according to the accumulated value, it can be seen from the tenth graph that the funds invested in the implementation of the "Park and ride" parking lots development and renovation will pay off in six years, that is in 2026 (Fig. 9).

However, it is also important to assess the payback of "Park and ride" parking spaces in terms of operating costs.

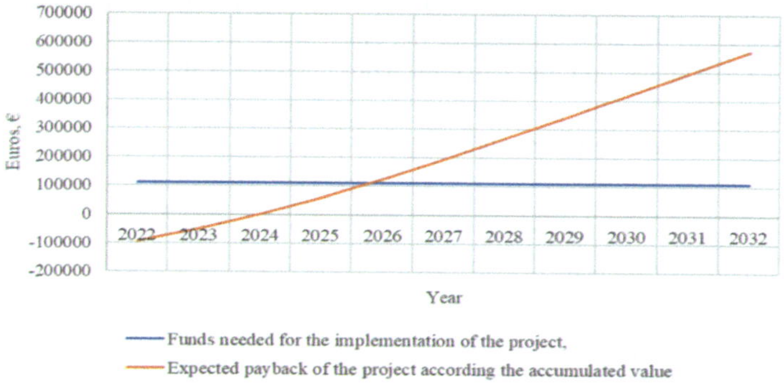

Fig. 9. Economic viability of the suggestion, results according accumulated value (compiled by the authors based on calculations).

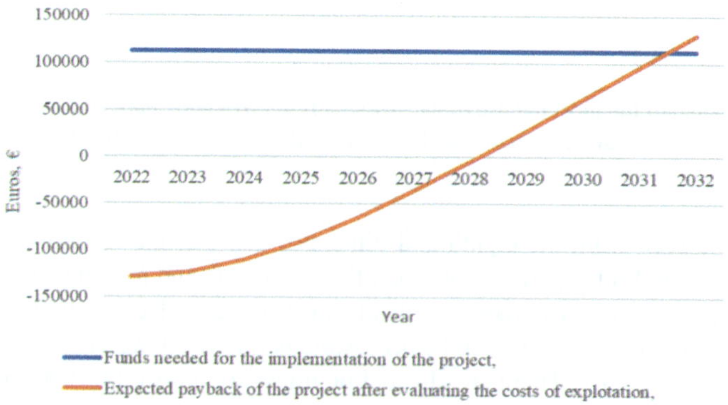

Fig. 10. Economic viability of the suggestion, after estimating operating costs (compiled by the authors, based on calculations).

According to the eleventh graph, it can be concluded that after the implementation of the "Park and ride" parking lots' renovation and development, this service will become profitable in about nine years, that is, in 2031 (Fig. 10).

5 Conclusions

The relocation of the "Park and ride" parking lots from the locations that do not meet the basic criteria and requirements to more suitable locations and increasing the number of parking lots on the "Park and ride" parking's, the number of visitors should double and reach about 70,000 per year. However, the "Park and ride" car parking network development and renewal project alone is not enough to solve the city's current problems caused by high traffic flows, high levels of motorization and its side effects. Banning the traffic in Vilnius Old Town and implementing the project of development and renewal of the "Park and ride" parking network, positive changes will not be noticeable immediately, but the biggest changes should be noticeable in the Old Town. Banning car traffic in the Old Town would create more space for bike paths. This would increase the popularity of bicycles in the city, which is still not very high for some reasons. Also, banned car traffic in the Old Town would mean safer streets for pedestrians. In the Old Town will be created an environment that takes into account the different needs of people, with a special focus on the most vulnerable groups (the elderly, people with physical or sensory disabilities, people using walking aids or wheelchairs) that meet people's expectations. The space of the Old Town will become safe and attractive where an active street life and pedestrian culture will flourish.

It is expected that the introduction of certain restrictions on the movement of cars in the Old Town and this part of the city, creating suitable conditions for comfortable walking, cycling, electric or simple scooters, renovating and expanding the "Park and ride" parking network, with time being will change the distribution of travels in the city. It means a decrease of car trips and an increase of travels by public transport, walking or cycling. The planned distribution of trips should look like this.

- Of all daily trips in the city, 30% will be made by personal cars;
- Of all daily trips in the city, 28% will be made on foot;
- Of all daily trips in the city, 7% will be made by bicycle, electric or simple scooters, and other non-motorized vehicles;
- 30% of all daily trips in the city will be made by public transport;
- Of all daily trips in the city, 5% will be made by other road transport (public shared cars, taxis, etc.).

Such a model of change in the distribution of travel in the city would reduce the traffic congestion, which is quite often caused by high levels of motorization. Reducing congestion in the city would also reduce the number of accidents and the undesirable side effects of vehicles, such as the level of emissions and additional noise from vehicles. Assessing the ecological viability of the project, it is estimated that 3,202,168 kg or 1,630,195 m^3 of CO_2 less emissions will be emitted into the environment in a decade than at present.

References

1. Juškevičius, P., Burinskiene, M., Paliulis, G.M., Gaučė, K.: Urbanistika: procesai, problemos, planavimas, plėtra. Technika, Vilnius (2013). https://doi.org/10.3846/1447-S
2. Juškevičius, P.: Miestų planavimas. Technika, Vilnius (2003). https://doi.org/10.3846/533-S
3. Juškevičius, P., Valeika, V.: Lietuvos miestų susisiekimo sistemos. Technika, Vilnius (2019). https://doi.org/10.20334/2019-041-M
4. Jeržemskis, A., Jaržemskis, V.: Keleivinis Transportas. Technika, Vilnius (2017)
5. Banister, D.: Great cities and their traffic: Michael Thomson revisited. Built Environ. **41**(3), 435–446 (2015). https://doi.org/10.2148/benv.41.3.435
6. World health organization (2020). https://www.who.int/phe/health_topics/outdoorair/databases/en/
7. Barauskas, A.: Miesto aplinkkelių poveikio transporto srautų pasiskirstymui gatvių tinkle vertinimas ir modeliavimas [Assessment and modelling of urban bypasses' influence on transport flows distribution in street network]. Technika, Vilnius (2019). https://doi.org/10.20334/2019-036-M
8. Vilniaus miesto bei Vilniaus rajono gyventojų skaičius 2021 metų pradžioje. Retrieved from Rodiklių duomenų bazė - Oficialiosios statistikos portalas (2020)
9. Labiausiai apgyvendinti Vilniaus miesto mikrorajonai 2020 metais. Retrieved from "Microsoft Power BI" (2020)
10. Turistų skaičius Lietuvoje pagal apgyvendinimo įstaigas (2020). Retrieved from Rodiklių duomenų bazė - Oficialiosios statistikos portalas: https://osp.stat.gov.lt/statistiniu-rodikliu-analize/
11. Vilniaus viešojo transporto tinklas (2020). Retrieved from SĮ "Susisiekimo paslaugos"/ Viešasis transportas: Vežėjai: http://www.vilniustransport.lt
12. Viešuoju transportų pravežtu keleivių skaičius pagal viešojo transporto rūšis (2020). Retrieved from VT 2020 ataskaita.pdf: http://www.vilnius.transport
13. Viešojo transporto keleivių srautų pasiskirstymas pagal paros metą ir maršrutą (2020). Retrieved from Vilniaus miesto savivaldybė. Vilniaus miesto savivaldybė - Vilniaus viešuoju transportu – daugiau kaip pusė milijono kelionių kasdien, populiariausi – greitieji autobusai: http://www.vilnius.lt
14. Vilniaus miesto darnaus judumo planas.pdf (2020). Retrieved from Vilniaus miesto savivaldybės darnaus judumo planas: http://www.vilnius.lt
15. Lengvųjų automobilių skaičius Vilniaus apskrityje 2019 metais (2020). Retrieved from Rodiklių duomenų bazė - Oficialiosios statistikos portalas: http://www.statgov.lt
16. Lengvųjų automobilių Vilniuje pasiskirstymas pagal naudojamą degalų rūšį (2020). Retrieved from "Regitra": https://www.regitra.lt/lt/atviri-duomenys/
17. Eismo spūsčių tyrimas Vilniuje (2020). Retrieved from Vilnius traffic report, TomTom Traffic Index
18. Vilniaus miesto savivaldybės aplinkos oro kokybės valdymo programa 2020–2025 m (2020). Retrieved from Microsoft Word - Vilniaus OKVP.docx: http:www.vilnius.lt
19. Eismo įvykių statistika Lietuvoje 2019 metais (2020). Retrieved from eismo_ivykiu_statistika_2016–2019.pdf: http://www.lrv.lt
20. "Statyk ir važiuok" stovėjimo aikštelių Vilniuje įrengimas (2020). Retrieved from 2017 m. veiklos ataskaita.pdf: http://www.vilniustransport.lt
21. "Statyk ir važiuok" stovėjimo aikštelės Savanorių prospekte įrengimas (2020). Retrieved from Vilniaus miesto savivaldybė - Savanorių prospekte atidaryta "Statyk ir važiuok" aikštelė: http://www.vilnius.lt

22. "Statyk ir važiuok" stovėjimo aikštelių lankytojų skaičius 2017–2019 metai (2020). Retrieved from 2019 m. veiklos ataskaita.pdf: http://www.vilniustransport.lt
23. Grigorjeva, T., Andriušytė, A.: Mokslinių tyrimų pagrindai. Technika, Vilnius (2015). https://doi.org/10.3846/1535-S
24. Lietuvos statistikos depertamentas (2020). Retrieved from https://osp.stat.gov.lt/en: https://ivpk.lrv.lt/lt/naujienos/auga-elektroniniu-budu-teikiamu-paslaugu-skaicius

Impact of COVID-19 on Shaping Comprehensive Services Within Intermodal Terminals

Ludmiła Filina-Dawidowicz[1(✉)] and Mariusz Kostrzewski[2]

[1] West Pomeranian University of Technology, al. Piastów 41, 71-065 Szczecin, Poland
ludmila.filina@zut.edu.pl

[2] Warsaw University of Technology, ul. Koszykowa 75, 00-662 Warsaw, Poland
mariusz.kostrzewski@pw.edu.pl

Abstract. COVID-19 pandemic significantly influenced transport and logistics sector's operation, including intermodal terminals functioning. Pandemic appearance evoked the operators of such terminals to face various challenges related to decision-making processes. One of the challenges was dealing with comprehensive services offer shaping. The article aims to analyse the influence of COVID-19 on shaping the comprehensive services offer at intermodal terminals during the first four month of the pandemic. The questionnaire was a tool applied to collect and investigate the opinions of intermodal terminals representatives on analysed issue. The case study of terminals located in Poland was considered. On the basis of the conducted research, it was stated that terminals operators' representatives generally perceived negatively the impact of pandemic on terminals operation during its initial phase. The areas mostly influenced by the pandemic included organization of terminals operation and transhipment volume. Among the criteria, which influenced decisions connected to shaping of terminal comprehensive services offer and affected by the pandemic, the costs and profits related to service provision, and the demand for services were mentioned.

Keywords: Intermodal terminal · Comprehensive service · COVID-19 · Logistics · Services shaping

1 Introduction

COVID-19 pandemic evoked serious challenges that transport, and logistics sector have faced. These challenges have dealt i.a. with uncertainty of decisions and their consequences due to the market changes. Such challenges were also faced by intermodal terminals that provide comprehensive services for operations on containers, swap-bodies, trailers and semitrailers transhipment.

Intermodal terminals, including rail-road terminals and those located in seaports (container and ferry terminals), play significant role in operation of transport and supply chains. Their services offer may give their clients a possibility to receive all

O. Prentkovskis et al. (Eds.): TRANSBALTICA 2021, LNITI, pp. 648–657, 2022.
https://doi.org/10.1007/978-3-030-94774-3_63

necessary favours within comprehensive services offer in one place, meeting expectations related to time, quality and price. COVID-19 pandemic induced that intermodal terminals had to adjust their activity to occurred situation.

The article aims to analyse the influence of COVID-19 on shaping the comprehensive services offer at intermodal terminals during the first four month of the pandemic. The case study of intermodal terminals located in Poland was analysed. The criteria influencing comprehensive services shaping have been determined. The questionnaire tool was applied to collect and investigate viewpoint of intermodal terminals' representatives of the COVID-19 impact on the terminals' operation. On the basis of achieved data analysis, the appropriate conclusions have been drawn.

2 Literature Analysis

The authors applied the following code to the one of acclaimed scientific databases, namely Scopus. This code was as follows: "COVID-19" AND "intermodal terminal*". The results, obtained on 26 May 2021, were surprisingly modest in number comparing to numerous research publications assigned to the pandemic period. They found one editorial [5], one conference paper [12] and three articles [2, 16, 19]. As the mentioned works do not directly focus on comprehensive services within intermodal terminals it is worth digging deeper the scientific literature. Therefore, apart from mentioned publications, it was decided to make a desk research independently of any databases, relying on the authors own scientific intuition.

Belhadi et al. [2] underlined that the recent global disruption evoked by the COVID- 19 outbreak gives rise to the new ways of adaptation, especially that the pandemic disrupted a state of equilibrium in complex and intertwined supply chains. They paid attention to application of agent-based system in the context of Artificial Intelligence (AI) technology in intermodal terminal functionality, which may be especially used in the mentioned disruption. Muravev et al. [13] claimed an important aspect of services in intermodal terminals. It was highlighted that proactive delivery of services from a port management organization to its stakeholders is desirable as a process in the strategic planning of intermodal terminals (or actually any other logistics facility's types). This statement was mentioned unbound of COVID-19 era. Belhadi et al. [2] and Munien and Telukdarie [12] focused on supply chain resilience modelling in logistics and industry areas of companies' interest.

Russell et al. [16] undertook the research that was found strongly linked to selected research area. They considered the logistics triad of transport carriers, port operators, and logistics service providers in the context of current pandemic challenges. The representants of this triad were selected due to the fact that their interactions and service capacities constitute the overall capacity of the logistics system – the authors of current paper add here that shaping a comprehensive service also constitute the mentioned capacity. Russell et al. [16] expressed their opinion that growth in containerized trade reveals increasing in demand for port services and growing operational challenges for ports, which perfectly stands align getting comprehensive service in shape. This statement corresponds in their opinion, quoted after Behdani et al. [1], not only for physical handling of containers traffic yet also it assures high-quality service which relates to actual logistics and supply chain strategies. Any strategies can be verified with simulation methods as they are applied for what-if analyses. As the abovementioned Muravev et al. [13] applied simulation methods in their research, also Ivanov [8] presented the results of a simulation study enabling both short-term and long-term impacts of epidemic outbreaks on supply chains.

COVID-19 pandemic has led to widespread interest and re-examination in various areas of the economy, society, systems development and accompanying services. Yu et al. [18] were interested in the aspect of medical waste management due to the exponentially increasing medical waste materializing within a very short period. The authors developed multi-objective multi-period mixed integer program for reverse logistics network design facing the challenges of a pandemic. Their aim was to determine the deserved locations of temporary facilities and the transportation strategies for effective management not only for waste management, yet for services as well to make them comprehensive ones.

Notteboom et al. [14] compared the current pandemic with the 2008–2009 financial crisis. The outcomes of this paper were indirectly focused on services in intermodal terminals especially that the authors considered how facilities such as ports and the shipping industry fit within complex supply chains and the cargo composition handled by ports.

Zhang et al. [20] analysed the current pandemic-based challenges of the shipping and port industry and its impact on ports, containers, tankers, dry bulk cargoes, cruise ships, etc. The authors assessed that the current recovery is a slow process (which is not surprising as pandemic is continually concerning all over the globe), and the negative impact of COVID-19 occurs continuously on the shipping and port industry. Therefore, other researchers consider these aspects as follows.

Donnan et al. [4] analysed the current situation of ports and the provided services. The authors described optional disruptive scenarios of emergent and future conditions in logistics facilities of that kind, including hybrid scenarios involving the actual pandemic.

Depellegrin et al. [3] investigated decreasing in maritime fishing and goods transportation as well as reduction in passenger maritime traffic.

Interesting viewpoint was presented in Luke and Rodrigue (2008). In general, when pandemic was mentioned in the literature, the lockdowns impact on economic situation was considered. Nevertheless, the authors mentioned that maritime, rail, air, and trucking industries relies on sparse numbers of specialized employees who can be hardly replaced in case of death, illness or voluntary absenteeism. On the other hand, many service providers had no other option but to cease operation [7].

Gray [6] analysed transportation services themselves as well as new demands for transportation services in agricultural supply chains. The author observed and mentioned (after [17]) that intermodal containerized movement of some products was disrupted due to the lack of empty containers in North America. It might be connected to other branches of global economy as well. In similar manner as Gray [6], Loske [10] considered food retail in Germany. The author has observed it causes a conflict of interest between transportation companies and food retail logistics. Meanwhile, Munawar et al. [11] analysed transport sector in Australia, and Pluciński et al. [15] in Poland and Ukraine.

The literature review revealed that the topic of intermodal terminals in the era of COVID-19 is still a sparse one. It is worth then considering the impact of pandemic on intermodal terminal especially in the context of services shaping provided by facilities.

3 Methodology

In order to conduct the research, the following methodology was applied (Fig. 1). On the basis of literature analysis and discussions conducted with selected representatives of intermodal terminals, the questionnaire was developed. This questionnaire, despite questions related to identification of the respondent's type and terminal location, included the research questions. The respondents were asked to answer the following research questions:

1. What is/was the impact of COVID-19 pandemic on the terminals service activity?
2. What areas of terminal functioning were impacted by COVID-19 pandemic?
3. What criteria were taken into account while providing the comprehensive service at the intermodal terminal before the pandemic?
4. How has the COVID-19 pandemic affected the services provided so far in terms of given assessment criteria?
5. What other criteria were taken into account when shaping the service offer at the terminal?
6. Will the changes in the terminal operation be maintained after COVID-19 cessation?

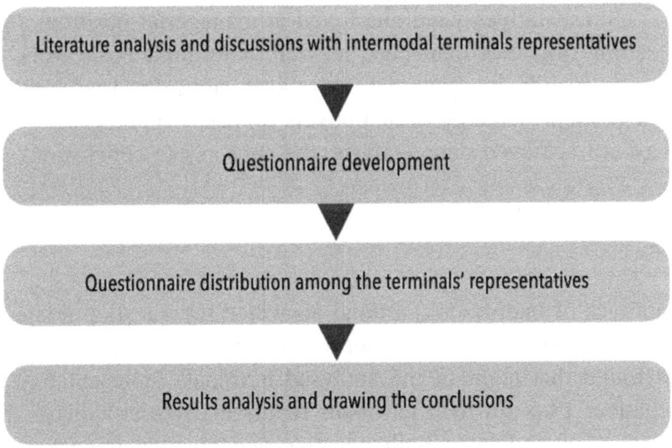

Fig. 1. The research methodology scheme.

Thirteen criteria were selected for assessment of terminals degree of compliance while making a decision on adding the service to terminal's offer and changes of criteria perception due to pandemic impact. The following criteria were analysed:

1. Costs related to the provision of the service (cr1),
2. Profits related to the provision of the service (cr2),
3. Demand for service (cr3),
4. Reliability of the service (cr4),
5. Risk of failure to undertake service (cr5),
6. Risk of cargo quality loss (cr6),
7. Duration of service launch (cr7),
8. Duration of service provision (cr8),
9. Importance of the service to the customer (cr9),
10. Feasibility of innovative solutions implementation (cr10),
11. Compatibility of the service with other services (cr11),
12. Flexibility of service (cr12),
13. Possibility to undertake the service in outsourcing (cr13).

The research was carried out in June 2020 and covered first four months after start of the pandemic. The developed questionnaire was sent to representatives of 32 intermodal terminals located in Poland. Among the analysed terminals there were rail-road intermodal terminals, maritime container terminals and ferry terminals. The questionnaire was sent by email, moreover, phone calls were made additionally to inform the terminals' representatives about conducted data collection.

The opinions of 10 terminals were obtained and analysed in detail. These terminals are located in the following voivodeships of Poland: Greater Poland (Wielkopolskie), Masovian (Mazowieckie), Lodz (Łódzkie), Pomeranian (Pomorskie), Silesian (Śląskie), West Pomeranian (Zachodniopomorskie). Four of mentioned terminals are situated within seaports facilities (three of them are container terminals and one – ferry terminal) and other six terminals are rail-road terminals. 60% of respondents declared more than 10 years of professional experience. 90% of interviewee were men. All of the terminals' representatives were employed at managerial positions (there were i.a. operational, selling, logistics managers, as well as terminal's director). The analysed terminals provide mainly international operations, only one representative indicated that activity of the represented terminal covers domestic market.

On the basis of achieved data analysis, the appropriate conclusions were drawn.

4 Results

The representatives of intermodal terminal answered the question related to impact of COVID-19 pandemic on the terminals service activities developed from March to June 2020. It was found that eight of the analysed terminals representatives assessed the impact as negative. Two rail-road terminals representatives articulated different viewpoint: one respondent indicated both negative and positive impact, and one person mentioned that the pandemic had not significantly influenced the terminal's activity.

Respondents also pointed particular areas of terminal functioning that were influenced by COVID-19 (Fig. 2). The assessments were given in Likert scale [9] from 1 to 6, where 1 – the impact is insignificant, 6 – very significant impact. It was found that initial phase of the pandemic mainly influenced organization of Polish intermodal terminals operation, as well as transhipments volume. There were some challenges related to cooperation with clients and changes in cargo storage time and cooperation with services provider. However, it should be noted that pandemic made rather small impact on the availability of services provided so far at the terminals.

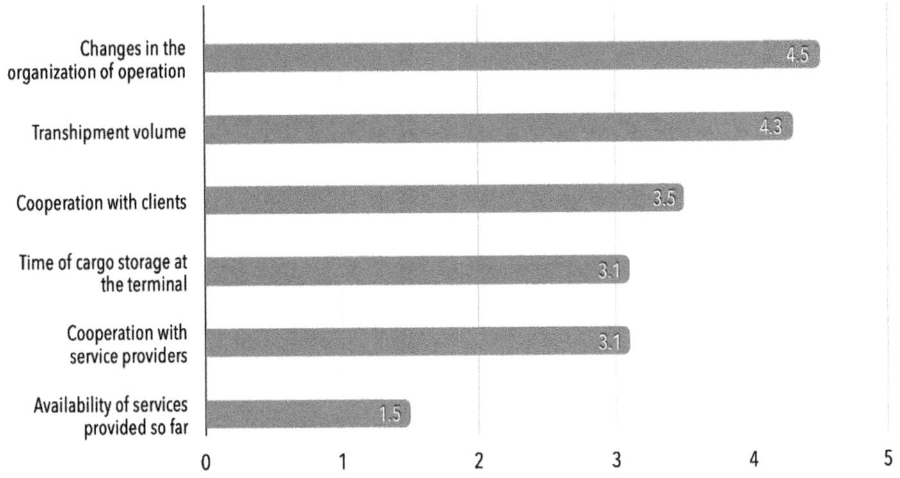

Fig. 2. The areas of terminals operation influenced by the pandemic (mean values).

Among other areas that were influenced by the pandemic and required changes provided into previous functionality of terminal, the respondents mentioned: introducing longer time for changing employees between shifts, the need to store both full and empty containers longer than usual, remote documents transfer, restrictions on the movement of drivers within the terminal building, lack of face-to-face meetings with potential suppliers, high costs of epidemiological protection, etc.

Terminals' representatives were also asked to assess selected criteria deciding on the comprehensive service shaping at the intermodal terminal. Respondents separately evaluated the degree of compliance with the selected criteria in deciding whether to add services to terminal's offer before the pandemic, as well as COVID-19 pandemic impact on the services provided so far in terms of given assessment criteria. Respondents rated individual criteria from 1 to 6, where 1 – insignificant impact, 6 – very significant impact. The mean value of each criterion assessment indicated by terminals' representatives was calculated. The achieved assessments rates are presented in Fig. 3 and Table 1.

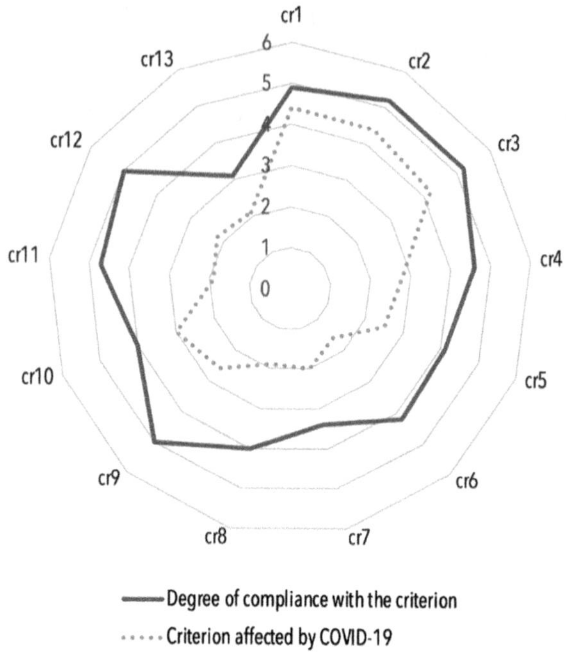

Fig. 3. Criteria that are taken into account while comprehensive service shaping at intermodal terminal (mean values).

Table 1. Ranking of the assessment criteria (mean values).

Degree of compliance with the criteria in deciding whether to add services to terminal's offer		Pandemic impact on the services provided so far in terms of a given assessment criterion	
Criterion	Mean value	Criterion	Mean value
cr2	5.2	cr1	4.4
cr3	5.2	cr2	4.4
cr9	5.0	cr3	4.2
cr12	5.0	cr10	3.0
cr1	4.9	cr4	2.8
cr11	4.7	cr9	2.6
cr4	4.6	cr5	2.5
cr6	4.2	cr12	2.2
cr5	4.1	cr13	2.1
cr10	4.0	cr7	2.0
cr8	4.0	cr8	1.9
cr7	3.4	cr11	1.9
cr13	3.1	cr6	1.6

It should be noted that intermodal terminals located in Poland among the criteria taken into account while adding the services to the offer take into account profits related to the provision of the service, as well as demand for service. High degree of compliance is also observed within such criteria as importance of the service to the customer, flexibility of service and costs related to the provision of the service. The smallest degree of compliance was given to duration of service launch and possibility to undertake the service in outsourcing.

The COVID-19 pandemic impact on the services provided so far was faced within the following assessment criteria: costs related to the provision of the service, profits related to the provision of the service and demand for service. Less affected criteria included: duration of service provision, compatibility of the service with other services and risk of cargo quality loss. It could be stated that cost- and demand-related criteria were pointed as the most sensitive and significant for comprehensive service shaping for intermodal terminals. The respondents' perception of these aspects' importance was also affected by the pandemic.

Among other criteria taken into account when shaping the service offer at terminal, their representatives mentioned: loads' mass, standardization, efficiency of operations, impact on terminal basic functions, geographic criterion, customer specificity, suppliers' prices, competitive position of other terminals. It was found that terminals expect to receive unified comprehensive service offer from cooperating enterprises (e.g. rail transport companies) to provide the safe and effective delivery of cargo within integrated transport chains. Intermodal terminals representatives also highlighted that they prefer adding services related to their core business, and do not take into account temporary trends or seasonal demand.

The last question was to investigate the opinion whether the changes in the terminal operation are going to be maintained in future. The opinions of terminals representatives were varied. Six respondents believed that terminals would return to normal operation and pandemic will not have continuous effect on terminals' activities (there were both maritime and rail-road terminals). However, two opinions expressed the viewpoint that pandemic will have long-term effect. Two representatives of intermodal terminals had difficulties in unambiguous determination of the possible impact.

5 Conclusion

On the basis of conducted research, it may be concluded that the COVID-19 pandemic made significant impact on intermodal terminals operation, including comprehensive services shaping. This impact was generally assessed as negative. Pandemic forced changes in organization of intermodal terminals operation and influenced transhipments volume.

The ranking of compliance degree with assessment criteria deciding whether to add services to terminal's offer was achieved, as well as the ranking of assessment criteria related to services provided so far impacted by pandemic was compiled. It was stated that cost- and demand-related criteria were mostly affected by the pandemic in relation to comprehensive service shaping for intermodal terminals.

The attention should be paid to the fact that intermodal terminals, as key nodes of international integrated transport chains, operate in the conditions of the economic environment and their operation is influenced by market situation, including supply and demand for cargo and co-related services. Therefore, achieved results should be referred to the conditions prevailing in Poland. It should be also noted that the achieved results are limited to the selected case study analysis – intermodal terminals located in Poland. Moreover, the small number of responses received from intermodal terminals representatives may influence the results. Therefore, it is worth continuing the research and analysing the changes in terminals representatives' viewpoint on comprehensive services offer shaping for example a year after pandemic started.

The achieved results may be of interest both for intermodals terminals and their clientele. There results show the areas of terminals operation that were impacted by pandemic and influenced comprehensive services offer shaping.

Acknowledgements. The authors would like to gratefully acknowledge certain transhipment terminals' representatives who agreed to complete the survey questionnaire.

References

1. Behdani, B., Wiegmans, B., Roso, V., Haralambides, H.: Port-hinterland transport and logistics: emerging trends and frontier research. Marit. Econ. Logistics **22**, 1–25 (2020)
2. Belhadi, A., Mani, V., Kamble, S.S., Khan, S.A.R., Verma, S.: Artificial intelligence-driven innovation for enhancing supply chain resilience and performance under the effect of supply chain dynamism: an empirical investigation. Ann. Oper. Res. (2021). https://doi.org/10.1007/s10479-021-03956-x
3. Depellegrin, D., Bastianini, M., Fadini, A., Menegon, S.: The effects of COVID-19 induced lockdown measures on maritime settings of a coastal region. Sci. Total Environ. **740**, 140123 (2020)
4. Donnan, R.C., et al.: Enterprise resilience of maritime container ports to pandemic and other emergent conditions. In: 2020 Systems and Information Engineering Design Symposium (SIEDS 2020), pp. 1–6. IEEE, Piscataway, NJ, USA (2020)
5. Frazzon, E.M., Freitag, M., Ivanov, D.: Intelligent methods and systems for decision-making support: toward digital supply chain twins. Int. J. Inf. Manage. **57**, 102281 (2021)
6. Gray, R.S.: Agriculture, transportation, and the COVID-19 crisis. Can. J. Agric. Econ. **68**(2), 239–243 (2020)
7. Hossain, M.: The effect of the Covid-19 on sharing economy activities. J. Cleaner Prod. **280** (1), 124782 (2021)
8. Ivanov, D.: Predicting the impacts of epidemic outbreaks on global supply chains: a simulation-based analysis on the coronavirus outbreak (COVID-19/SARS-CoV-2) case. Transp. Res. Part E: Logistics Transp. Rev. **136**, 101922 (2020)
9. Joshi, A., Kale, S., Chandel, S.K., Pal, D.K.: Likert scale: explored and explained. British J. Appl. Sci. Technol. **7**, 396–403 (2015)
10. Loske, D.: The impact of COVID-19 on transport volume and freight capacity dynamics: an empirical analysis in german food retail logistics. Transp. Res. Interdisc. Perspect. **6**, 100165 (2020)

11. Munawar, H.S., Khan, S.I., Qadir, Z., Kouzani, A.Z., Parvez Mahmud, M.A.: Insight into the impact of COVID-19 on Australian transportation sector: an economic and community-based perspective. Sustainability **13**(3), 1–24 (2021)
12. Munien, I., Telukdarie, A.: COVID-19 supply chain resilience modelling for the dairy industry. Procedia Comput. Sci. **180**, 591–599 (2021)
13. Muravev, D., Hu, H., Rakhmangulov, A., Mishkurov, P.: Multi-agent optimization of the intermodal terminal main parameters by using AnyLogic simulation platform: case study on the NingboZhoushan port. Int. J. Inf. Manage. **57**, 102133 (2021)
14. Notteboom, T., Thanos, P., Rodrigue, J.P.: Disruptions and resilience in global container shipping and ports: the COVID-19 pandemic versus the 2008–2009 financial crisis. Marit. Econ. Logistics **23**, 179–210 (2021)
15. Pluciński, M., Filina-Dawidowicz, L., Mańkowska, M., Savelieva, I., Shemayev, V.: Impact of initial phase of COVID-19 pandemic on operations of stevedore companies in seaports: a case study of Poland and Ukraine. In: Soliman, K.E. (ed.), Proceedings of the 36th International Business Information Management Association (IBIMA), pp. 9881–9890. International Business Information Management Association, USA (2020)
16. Russell, D., Ruamsook, K., Roso, V.: Managing supply chain uncertainty by building flexibility in container port capacity: a logistics triad perspective and the COVID-19 case. Marit. Econ. Logistics (2020). https://doi.org/10.1057/s41278-020-00168-1
17. Shih, W.: COVID-19 and global supply chains: Watch out for bullwhip effects. Forbes. https://www.forbes.com/sites/willyshih/2020/02/21/covid-19-and-global-supply-chains-watch-out-for-bullwhip-effects/#6928db537195. Accessed 09 Jul 2011
18. Yu, H., Sun, X., Solvang, W.D., Zhao, X.: Reverse logistics network design for effective management of medical waste in epidemic outbreaks: Insights from the coronavirus disease 2019 (COVID-19) outbreak in Wuhan (China). Int. J. Environ. Res. Public Health **17**(5), 1770 (2020)
19. Zhang, W., Zhang, X., Tian, X., Sun, F.: Economic policy uncertainty nexus with corporate risk-taking: The role of state ownership and corruption expenditure. Pac. Basin Finan. J. **65**, 101496 (2021)
20. Zhang, Y.-F., Gong, J.-W., Yin, M.: Influences and response measures of COVID-19 epidemic on shipping and port industry in China. Jiaotong Yunshu Gongcheng Xuebao/J. Traffic Transp. Eng. **20**(3), 159–167 (2020)

Modern Ecological Approach to Air Transportation Management

Viktoriia Ivannikova(✉), Maryna Boldyrieva, and Valentyna Konovalyuk

National Aviation University, Kyiv 03058, Ukraine

Abstract. The paper analyzes the specifics of the impact of air transport services on the environment and provides methodology to solve emerging problems. This article is dedicated to the research of a relevance to upgrade the fleet for running cost-effective business. The main negative ecological influences of the aviation industry operation are noise, electromagnetic radiation, dangerous chemical substances emissions including carbonic acid, harm to animals, plant organisms and adjacent territories, waste production leading to climatic fluctuations, nature destruction and health problems, economic collapses. In addition, aviation sector itself will be affected by its disastrous actions – the main appearing problem can be loss of optimality in carrying out air transport service providers' activity. The performed researches allowed to find out new ways of air transport disastrous impact shortage. According to the research results, there were determined positive direction for the environment modifications in the airline industry activity – updated assessment method in the form of mathematical model of aircraft fleet usage, which will help to decrease the emissions of old aircraft. Mathematical model of aircraft replacement in the most optimal period of time allowing achieve ecologically safer and cost-effective carriage has been developed.

Keywords: Environment · Air transport ecological effects · Aviation sector · Fleet renewal · Noise · Air pollution · Aviation impact

1 Current Ecological State in Aviation Industry Worldwide

1.1 Global Impact

Nowadays the world is fighting against the great climatic catastrophe, which has resulted in each part of the Earth in multiple ways. Transportation no less than other human industries facilitates such damaging effects. Share of air carriages in the nature contamination has become a significant issue requiring attention and immediate actions as demand on it is still in steady growth.

According to the results of studies on the adverse impact of aviation on the environment [1], the following categories of factors exist in the vicinity of airports:

O. Prentkovskis et al. (Eds.): TRANSBALTICA 2021, LNITI, pp. 658–670, 2022.
https://doi.org/10.1007/978-3-030-94774-3_64

Chemical impact determinants:

- emission of harmful substances by aircraft engines and their effect on the ozone layer of the atmosphere;
- thermal pollution.

Physical impact indicators:

- noise during the operation of aircraft;
- sound shock;
- electromagnetic radiation;
- polluted runoff from the airport territory;
- waste from premises, warehouses and other facilities.

Table 1 presents the sources of the physical impact of civil aviation on the environment and the population [2].

Table 1. Sources of physical impact of civil aviation on the environment and the population.

Aircraft noise	Electromagnetic radiation	Thermal effects	Ionizing radiation	Vibration
Aircraft engines, auxiliary power units and wing mechanization elements	Radio equipment, magnetron ovens, radiotelephones, power transmission lines, transformer stations, power plants, computers, video monitors	Aircraft and special vehicle engines, boiler rooms	Tank level sensors on aircraft, devices with luminous dials, installations for radiation control at customs, electronic computers	Aircraft

To clarify and measure entire influence of emissions produced by air transportation sector on climate, the Intergovernmental Panel on Climate Change (IPCC) has determined that the total environmental effects of aerial operations performance is 2–4 times higher in comparison with its straight carbon dioxide emitting amount. This is evaluated as radiation exposure.

According to the IPCC research, among the climatic changes caused by human activity the aviation share takes approximately 3.5%, consisting of both CO_2 and non-CO_2 impacts. By its scientists were also developed forecasts of this indicator quantity in 2050. The potential value shows growth of aviation input up to 5% if any actions are not provided to manage these emissions, but there is also version of 15% portion. Besides, in case of substantial greenhouse gas emissions decrease in another sectors, the share of air transport industry considering residual emitting sources will grow as well [3].

Quantity of technogenic carbon (i.e. appeared after human activity) is shown in permanent increase since 2015. That is caused by the growth of emitting this product in coal, oil and gas combustions. According to the research of WMO, 2019 appeared to be one of the hottest years in the history, as well as for the last ten-year period. In the Fig. 1 it is shown increased amount of CO_2, CH_4 and N_2O density [4]. The lines colored in blue show quantity of these substances per month across the globe and in red ones – average value of emissions per month globally for the five year periods.

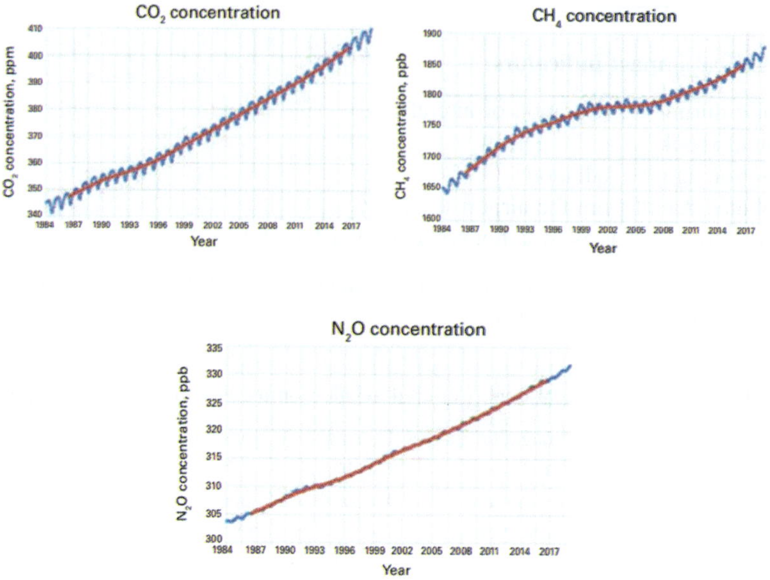

Fig. 1. Worldwide mean amount of concentrations of CO_2 in ppm (parts per million), CH_4 in ppb (parts per billion) and N_2O in ppb.

There were also detected the world areas, where this kind of pollution is concentrated the most (Fig. 2). Yellow points in the model show the emissions from the aircraft usage. The network for the methane is analogous [5].

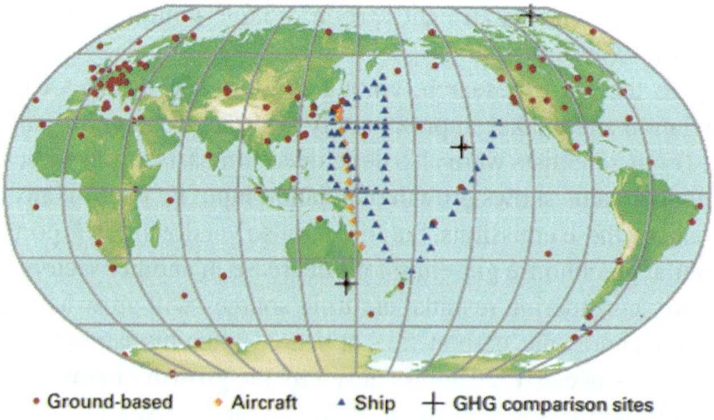

Fig. 2. The GAW (Global Atmosphere Watch) global network for carbon dioxide in the last decade (2010–2020).

Reviewing the received data about the speed of permanent CO_2 values expending in the 2015–2018 period in WMO's Global Climate 2015–2019 report, the revealed amount equals 18% higher than that over 2011–2015 (see Table 2).

Table 2. Concentration and growth rates of CO_2 (ppm), CH_4 and N_2O (ppb) in 2011–2015 and 2015–2018, their percentage in relation to period before 1750.

	Concentration			*Growth rate*		
	2015–2018	*2011–2015*	*2015–2018 % to pre-industrial*	*2015–2018*	*2011–2015*	*% change*
CO_2	404.2	395.5	145	2.5	2.2	+18%
CH_4	1856.0	1826.4	257	9.0	7.2	+21%
N_2O	329.6	326.2	122	0.95	1.0	−5%

After researchers' estimation there were received figures demonstrating the emission of specific substances by three hundreds of air vessels designed for transcontinental flights on the daily basis: 3.7 tons of carbon monoxide, 2 tons of hydrocarbon compounds and 1.7 tons of nitrogen oxides. Emissions of harmful substances in the airport area depend on the take-off and landing cycle for various types of aircraft [1, 6].

Despite the fact the scientist forecasted decrease of greenhouse gas emissions at least by 6% in 2021, which was forced by the COVID-19 pandemic restrictions and prohibitions in performing of air transportation services and economic downturn as well, there is no evidence of keeping such enhancement steady and constant. It may be called only environmental respite, so the climatic conditions continue the transformation process as the economic crisis has been resumed so the issues connected with greater emissions will be seen again not so far [7].

The graph below (Fig. 3) shows the share of all transport modes in the emitting of carbon dioxide and other harmful contents. According to the data of UK government, air carriages have the second position after road transport in amount of substances polluting air by one passenger per one kilometer traveled [6].

The data plotted in the next graph (Fig. 4) comprise the calculations of aviation impact on air pollution with carbon. The emissions included in this figure comprise those produced during passenger, cargo transportation, as well as military missions. The figures are taken from the Global Carbon Project, an organization aiming to measure the greenhouse gases and the reason generating them [8].

The red line shows aviation sector, which is an originator of 1.04 billion tons of carbonic acid.

As one of the main drawbacks must be also mentioned aviation noise. Noise caused by air transportation processes is an important factor in the negative attitude of the population to aviation in the territories adjacent to the airport. It affects a relatively large number of people living in the surrounding area, as well as airport employees and passengers. Aircraft noise has a negative impact on people's health (most often it is hearing impairment, stress conditions, problems associated with concentration) [5, 9].

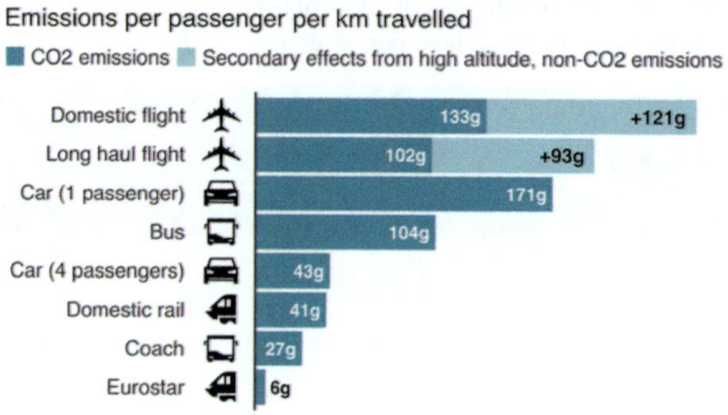

Fig. 3. The amount of emissions exuded by different modes of transport.

The main causes of aircraft noise are disturbances of air and gas flows created by the operation of aircraft engines, among which jet engines have the worst noise indicators. At the airport, the noises of take-off and landing moving along taxiways are joined by intense noises during the preparation of aircraft for departure as well as noises that occur at special sites during engine tests.

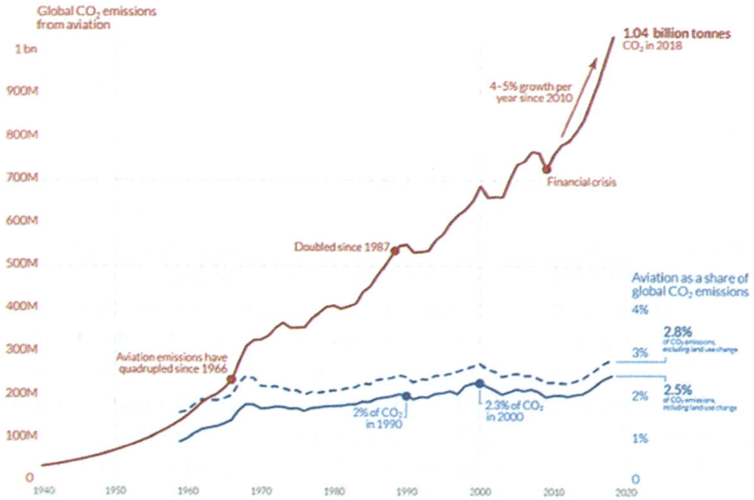

Fig. 4. Global carbon dioxide emissions from aviation.

All ecosystems are affected by the harmful effects of aircraft noise: humans, animals and plant organisms. The expected growth of air traffic means that potentially more people risk being exposed to aircraft noise in future. In fact, air traffic is forecasted to increase in the next years, as a result of an increasing demand for air travel, as shown in Fig. 5 from the European Aviation Environmental report (EAER) 2019.

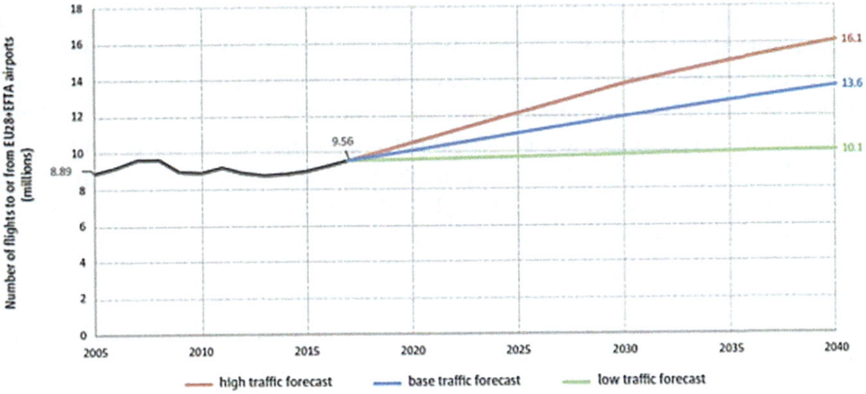

Fig. 5. Expected increase of air traffic for 3 different scenarios.

According to the EAE report, in 2017 3.2 million people were exposed to L_{den} (day-evening-night sound pressure level) 25 levels higher than 45 dB and 1.4 million people to L_{night} (night-time sound pressure level) 26 levels above 40 dB around the 47 major European airports. According to WHO (World Health Organization), these levels cause symptoms of high annoyance and sleep disturbance. For the same airports, it was also estimated that 1 million people were exposed every day to more than 50 aircraft noise events above 70 dB, when harmless noise does not exceed 30 dB.

Noise pollution around airports and air routes covers millions of square kilometers of territory, including residential areas. When the airport is close to the city the social problem of aviation noise, which interferes with the sleep and rest of the population and negatively affects health is acute [10].

2 Updated Technology in Provision of Ecologically-Friendlier Air Transportation

2.1 Mathematical Modeling in the Protection of Environment

Considering appearing trends and their popularity growth in the form of new vessels design, form and composition that are environmentally-friendly, the airlines should be aware of necessity to replace their fleet in not so far perspective. Another way to start being ecologically responsible is by reusage of aircraft parts and replacement of components that produce noise, emissions and other negative environmental influences [11].

Protection of the environment has become one of the parameters indicating the development, modernity and profitability of the airlines and airports as well. Ecological negative influences of the air transportation are controlled not only in the sphere of legislation in the form of regulations and requirements, climate sciences and ethics but also addressing to the taxation measures.

A number of foreign countries have introduced a tax on noise pollution caused by aviation. In France, the aircraft noise tax consists of two payments: the first of them is determined by multiplying the landing fee at the airport by five different coefficients, ranging from 0.9 (for "quiet" aircraft) and up to 1.2 (for "noisy" aircraft), which then goes to the airport budget; the second payment is from 0 to 20% of the amount of the landing fee and goes to a special fund, the money from which are used to finance measures for sound insulation of residential areas adjacent to the airfield area. In Denmark and Japan, such a tax depends on the mass of the aircraft and on the strength of the sound during take-off and landing [12]. One of the examples what charges are set in the European countries is shown in the Table 3 [13, 14].

Table 3. Estimated average aviation fuel tax necessary to cover main externality costs (noise and air pollution) (GBP per liter of aviation fuel).

Airplane Type	Noise Tax	Air Pollution Tax	Total Tax
A310	0.01	0.05	0.06
A340	0.01	0.06	0.07
BAe146	0.00	0.05	0.05
B737-100	0.06	0.04	0.10
B737-400	0.00	0.05	0.05
B747-400	0.01	0.06	0.07
B757	0.01	0.05	0.06
B767-300	0.01	0.05	0.06
B777	0.00	0.06	0.06
F100	0.00	0.05	0.05
MD82	0.01	0.05	0.06
Average	0.01	0.05	0.06

Whether it is an airline or airport, the company may be obliged to pay for contamination of harmful to nature substance, such as carbon dioxide and other chemicals produced while flight performance or other different operations. According to aviation experts, the absolute size of the damage caused by pollution is much greater than the costs required to solve them. As a result, the reason for replacement the aircraft with new, environmentally-neutral or even –friendly vessel turns to be pertinent and demanded in every case: it is necessary action for the present ecological situation globally and the future of the planet in one respect, and will save the costs of the company as the taxation on emissions is high enough to make the activity unprofitable.

The difficulties and challenges of aircraft updates are investments and risks of air carriers to fail with new approaches. To maintain these concerns and investigate how to become environmentally beneficial in a cost-effective manner, there is proposed a mathematical model described below. It will help the airline decide whether to spend money to replace an old aircraft component (for example, emitting carbon engine), or whether the best option is to attract a new or even different type over a certain period of time [15]. The task is based on a well-known equipment replacement model with a certain modification.

To determine the optimal service life of the aircraft by dynamic programming there were indicated the main elements of the model:

- *stage i* – serial number of the year of decision making (i = 1, 2,…n);
- possible decisions at the beginning of *i–stage* are alternatives – whether to continue operating an old aircraft or replace it with a new model;
- state at *i-stage* is life cycle t (age) of aircraft before *i-th year.*

In calculation will be considered the next data:

- *r(t)* – revenue per operation of *t-aged* aircraft per one period;
- *e(t)* – taxes an airline is obliged to pay for the emissions with the operation of *t-aged* aircraft per one period;
- *c(t)* – operational expenses of *t-year* aircraft during one period (replacement of spare parts, maintenance, airport fees, fuel, staff and crew fees, etc.);
- *s(t)* – salvage of *t-year* vessel;
- *l(i)* – costs of new aircraft [16].

For the analysis conduction there was considered a model of the feasibility of replacing a 3-year-old Airbus 320 aircraft with an upgraded vessel within 21 year (n = 21). At the beginning of each stage (every 3 years) there is made a decision on its replacement or further maintenance considering the earned profit and losses. There is assumed the average inflation rate for three-year period is equal to 2.1%. The initial data are shown in the Table 4.

Table 4. Initial data for the Airbus 320.

Aircraft age, t	Revenue, r(t)	Operating expenses, c(t)	Emissions charge, e(t)	Salvage value, s(t)	Aircraft cost, l(i)
0	96	12	0.42	–	80
3	95.9	13	0.4242	60	81.7
6	94	14	0.4284	57	83.4
9	92	15	0.4326	52	85
12	88	17	0.4368	48	86.7
15	84	18	0.441	43	88.4
18	79	20	0.4452	38	90.1

The conducted calculations show that it is more profitable to replace the aircraft on the 12th year of its operation (see Table 5 and Fig. 6).

Table 5. Results of calculations.

Stage 6	Time	Keep r(t)-c(t)-e(t)+s(t+3)	Replace s(3)-l(i) +r(0)-c(0)-e(0)+s(t)	$F_6(t)$	Decision
	3	95.9-13-0.4242+60=142.5	60-81.7+96-12-0.42+60=121.8	142.5	Keep
	6	94-14-0.4284+57=136.6	57-83.4+96-12-0.42+60=117.18	136.6	Keep
	9	92-15-0.4326+52=128.6	52-85+96-12-0.42+60=110.58	128.6	Keep
	12	88-17-0.4368+48=118.6	48-86.7+96-12-0.42+60=104.8	118.6	Keep
	15	84-18-0.441+43=108.5	43-88.4+96-12-0.42+60=98.2	108.5	Keep
	18	79-20-0.4452+38=96.5	38-90.1+96-12-0.42+60=91.5	96.5	**Keep**

Stage 5	Time	Keep r(t)-c(t)-e(t)+ F_6(t+3)	Replace s(t)-l(i)+r(0)-c(0)-e(0)+ F_6(3)	$F_5(t)$	Decision
	3	95,9-13-0.4242+136.6=219.07	60-81.7+96-12-0.42+142,5=204.4	219.07	**Keep**
	6	94-14-0.4284+128.6=208.2	57-83.4+96-12-0.42+142,5=199.7	208.2	Keep
	9	92-15-0.4326+118.6=195.2	52-85+96-12-0.42+142,5=193.08	195.2	Keep
	12	88-17-0.4368+108.5=179.06	48-86.7+96-12-0.42+142,5=187.4	187.4	Replace
	15	84-18-0.441+96.5=162.05	43-88.4+96-12-0.42+142,5=180.7	180.7	Replace

Stage 4	Time	Keep r(t)-c(t)-e(t)+ F_5(t +3)	Replace s(t)-l(i)+r(0)-c(0)-e(0)+ F_5(3)	$F_4(t)$	Decision
	3	95,9-13-0.4242+208.2=282.5	60-81.7+96-12-0.42+219,07=280.9	282.5	Keep
	6	94-14-0.4284+195.2=274.8	57-83.4+96-12-0.42+219,07=276.3	276.3	Replace
	9	92-15-0.4326+187.4=263.9	52-85+96-12-0.42+219,07=269.7	269.7	Replace
	12	88-17-0.4368+180.7=251.3	48-86.7+96-12-0.42+219,07=263.9	263.9	**Replace**

Stage 3	Time	Keep r(t)-c(t)-e(t)+ F_4(t +3)	Replace s(t)-l(i)+r(0)-c(0)-e(t)+ F_4(3)	$F_3(t)$	Decision
	3	95,9-13-0.4242+276.3=358.7	60-81.7+96-12-0.42+282.5=344.4	358.7	Keep
	6	94-14-0.4284+269.7=349.3	57-83.4+96-12-0.42+282.5=339.7	349.3	Keep
	9	92-15-0.4326+263.9=340.5	52-85+96-12-0.42+282.5=333.08	340.5	**Keep**

Stage 2	Time	Keep r(t)-c(t)-e(t)+ F_3(t +3)	Replace s(t)-l(i)+r(0)-c(0)-e(0)+ F_3(1)	$F_2(t)$	Decision
	3	95,9-13-0.4242+349.3=431.7	60-81.7+96-12-0.42+358.7=420.6	431.7	Keep
	6	94-14-0.4284+340.5=420.07	57-83.4+96-12-0.42+358.7=415.8	420.07	**Keep**

Stage 1	Time	Keep r(t)-c(t)-e(t)+ F_2(t +3)	Replace s(t)-l(i)+r(0)-c(0)-e(0)+ F_2(1)	$F_1(t)$	Decision
	3	95,9-13-0.4242+420.07=502.5	60-81.7+96-12-0.42+431.7=493.6	502.5	**Keep**

Thus, in terms of ecology it is also more relevant to address to the new vessel consuming less fuel and emitting less carbonic acid and other harmful particles in the air.

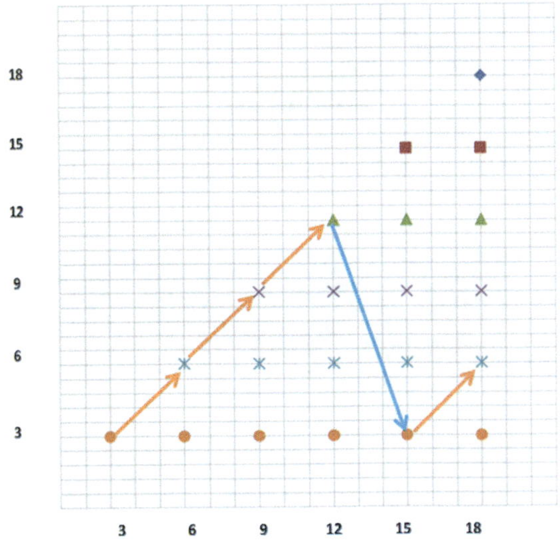

Fig. 6. Scheme of decision-making for the replacement of Airbus 320 (orange lines represent the decision "to keep current aircraft" and blue – "replace with new vessel").

Nowadays, strong market players such as NASA, Lockheed Martin, Airbus and Boeing design new aircraft concepts: they create a hybrid of an airplane and a helicopter, install solar panels on the roof and wings to generate energy, make a glass floor or ceiling to lighten the weight. These innovative transport modes are an example of possible aircraft configuration that will replace an old aircraft and become a solution in optimization of air transportation management sphere [17, 18].

It is possible to increase the fuel efficiency of the aircraft by optimizing the flight dynamics and the distribution of the maximum take-off weight. Developments in the field of variable wing geometry are aimed at solving these problems that will be demonstrated in the form of aircraft with variable wing geometry, named in other materials as sweepback.

Adapting the wing design to the flight conditions (at high speed, a large wing sweep is effective, at low speeds is less) allows to increase the ratio of lift to drag of the wings by 15–20%, which helps to reduce fuel consumption, improve aerodynamics and in such a way decrease the noise.

One of the most promising developments in this area is led by FlexSys Inc., commissioned by NASA. Its technology of transformable flaps has already passed flight tests on the basis of the Gulfstream III aircraft, showing high efficiency (Fig. 7–8). Currently, the turnover of the civil segment of the market for aircraft with variable wing geometry is estimated at $135 billion [19, 20].

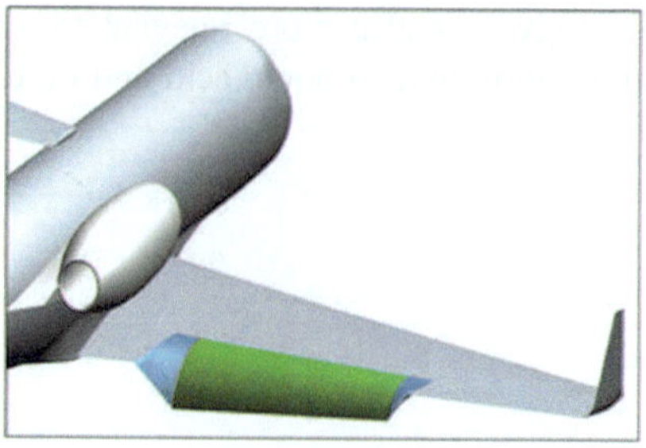

Fig. 7. FlexSys ACTE flap technology bridges gaps in wing for a seamless surface.

Fig. 8. Modified Gulfstream III aircraft tests for the ACTE flexible-flap research project.

3 Conclusions

There was carried out analysis of the specifics of environmental management and research of aviation sector air pollution, as well as were considered possible means to reduce emissions and mechanisms for achieving rational environmental management in the field of ecology protection from the air transport influence. According to data received after the analysis, it was found that the air transportation pollution taxation is one of the most effective and applicable measures that can appear in every country to force the aviation industry parties to pay attention to the environmental issues and modern technology that would be useful for management and decrease of airport and airline emissions and nature contamination. There was proposed for the first time the mathematical model that can be a tool to assist the aviation service providers in implementing of new technologies aimed to replace ecologically ineffective fleet with environmentally neutral aircraft and perform it at the appropriate time and in a cost-effective manner. The future research work will be focused on the deeper investigation of updated aircraft that will appear in not so far perspective.

References

1. Влияние авиации на окружающую среду и меры по ослаблению негативного воздействия / Иванова А. Р. // Труды Гидрометцентра России. Вып. **365**(С), 5–14 (2017)
2. Коробова, О.С., Филиппова, Д.В.: Воздействие объектов гражданской авиации на окружающую среду на примере международного аэропорта "Шереметьево". Горный информационно-аналитический бюллетень. №. **5** С, 299–305 (2017)
3. Wuebbles, D.J., Gupta, M.L., Malcolm, K.W.: Evaluating the Impacts of Aviation on Climate Change. Eos **88**(3), 157–168 (2007)
4. Global Climate in 2015–2019: Climate change accelerates. World Meteorological Organization (WMO), p. 21 (2019)
5. Aviation Noise Impacts. White Paper. State of the Science: Aviation Noise Impacts, pp. 44–66 (2019)
6. Юрчук, А. П.: Влияние авиации на окружающую среду и меры по ослаблению негативного воздействия / А. П. Юрчук. — Текст : непосредственный // Молодой ученый. № 8 (350). — С. 198–201 (2021)
7. Comi, A., Savchenko, L.: Last-mile delivering: analysis of environment-friendly transport. Sustain. Cities Soc. **74**, 103213 (2021)
8. Greenhouse gas reporting: conversion factors 2019. Department for Business, Energy & Industrial Strategy (DEFRA), June 2019. https://www.gov.uk/government/publications/greenhouse-gas-reporting-conversion-factors-2019
9. European Aviation Environmental Report 2019. European Aviation Safety Agency. https://www.easa.europa.eu/eaer/topics/sustainable-aviation-fuels
10. Декалин, А. А., Нечаева, О. А.: Воздействие авиационного шума на экологию. Журнал: E-Scio (2019). https://cyberleninka.ru/article/n/vozdeystvie-aviatsionnogo-shuma-na-ekologiyu
11. IATA. Best Industry Practices for Aircraft Decommissioning (BIPAD). https://www.iata.org/contentassets/ffbed17ac843465aad778867cb23c45c/bipad.pdf
12. Sameh, M.M., Scavuzzi, J.: Environmental Sustainability Measures for Airports. Occasional paper series VII, Centre for Research in Air and Space Law, McGill University (2016)
13. Mantecchini, L.: Airport noise charges and local communities: application to regional airports. J. Eng. Sci. Technol. **11**(11), 1518–1527 (2016)
14. Keen, M., Strand, J.: Indirect Taxes on International Aviation. WP/06/124, International Monetary Fund (2006)
15. Болдирєва, М. О., Тішин, І. А.: Проблематика заміни літаків авіакомпанією. Матеріали Міжнародної науково-практичної конференції «Проблеми організації авіаційних, мультимодальних перевезень та застосування авіації в галузях економіки» 21 листопада (2019)
16. Таха, Х. А.: Введение в исследование операций. М.: Вильямс, 912 с. (2005). http://bookre.org/reader?file=445915&pg=1
17. Yudin, O., Ziubina, R., Buchyk, S., Matviichuk-Yudina, O., Suprun, O., Ivannikova, V.: Development of methods for identification of information-controlling signals of unmanned aircraft complex operator. Eastern-Eur. J. Enterp. Technol. **2/9**(104), 56–72 (2020)
18. Savchenko, V., Laptiev, O., Kolos, O., Lisnevskyi, R., Ivannikova, V., Ablazov, I.: Hidden transmitter localization accuracy model based on multi-position range measurement. In: 2020 IEEE 2nd International Conference on Advanced Trends in Information Theory, IEEE ATIT 2020, Proceedings, pp. 246–249 (2020)

19. NASA. Green Aviation Project Tests Shape Changing Wing Flaps (2019). https://www.nasa.gov/image-feature/green-aviation-project-tests-shape-changing-wing-flaps
20. NASA. Past Project: Adaptive Compliant Trailing Edge Flight Experiment (2017). https://www.nasa.gov/centers/armstrong/research/ACTE/index.html

Possibilities of Subsidized Public Transport in Vilnius

Žaneta Česonienė and Nijolė Batarlienė(✉)

Vilnius Gediminas Technical University, Plytines 27, 10105 Vilnius, Lithuania
zaneta.cesoniene@stud.vgtu.lt,
nijole.batarliene@vilniustech.lt

Abstract. This article describes the possibilities of state-subsidized public transport in Vilnius. An expert survey showed how important it is for cities to have such a model of public transport as is needed in new districts and suburban areas of Vilnius. The study collected information on the socio-economic factors of the population that affect public passenger transport, the mobility of the population, as well as the impact of private vehicles on the city. There are many obstacles to implementing a fully subsidized public transport model; legal, technological and organizational.

Keywords: Public transport · Mobility · Transportation · Transportation areas

1 Introduction

Public transport has always been important to the city's residents. The changing market is also constantly changing citizens' attitudes towards urban public transport. It is very important for cities to have good quality public transport services. City municipalities often oversee urban public transport so that hauliers provide this service in a comprehensive and high-quality manner. Public transport occupies a particularly important place in the world's major cities, as it allows competition between other cities. Public transport allows cities to attract more residents and guests. The services provided by public transport can be directly linked to the satisfaction of city residents, guests and create added value. For these fundamental reasons, public interest in public transport is only growing.

Public authorities are increasingly encouraging residents to use public transport services, and this is becoming a priority in urban policy-making. Addressing these issues requires a comprehensive integrated approach by the city municipality and authorities, which could balance the benefits for the urban population and public transport services. Congestion problems would be partially solved, population movement in the city would be ensured with minimal costs and investments. With a long-term perspective, subsidized public transport would expand to the nearest suburban areas.

Attractive and fully subsidized public transport would attract more passengers, partially reduce congestion on city streets and encourage more visitors to arrive. Using the good experience of other EU countries, it is possible to install fully subsidized public transport in major Lithuanian cities. However, there is a problem - not many

O. Prentkovskis et al. (Eds.): TRANSBALTICA 2021, LNITI, pp. 671–681, 2022.
https://doi.org/10.1007/978-3-030-94774-3_65

studies have been conducted in Lithuania that could show the possibility to implement a fully subsidized public transport model and show its usefulness.

In recent years fare-free public transport (FFPT) [1] found itself at the centre of attention of various groups, such as economists, transport engineers and local authorities, as well as those responsible for the organisation of urban transport.

Although fare-free public transport (FFPT) services, such as limited campaigns and evasion of ticketing by special groups or special services, predominate, there is little evidence of the consequences of introducing a full-fledged FFPT. The case of Tallinn, Estonia, offers a comprehensive experiment that provides a unique opportunity to investigate the effects of FFPT.

2 Theoretical Aspects of the Public Transport System

When addressing the issue of urban public transport, it is important to pay attention to the quality of passenger transport. After Lithuania joined the European Union, it was necessary to reorganize all urban public transport for the second time after regaining independence. The need for passengers described in the scientific literature is the departure from one geographical point to another when the need to travel arises. This is, for example, the need to go to school, kindergarten, work, or another institution. According to the authors A. Jaržemskis and V. Jaržemskis [2], the needs and the level of their satisfaction depend on the cultural characteristics and economic development of the country. The geography of transport is of great importance. Factors determining the need for social services: aging population, disability, emigration, social risk and unemployment [3]. The nominal freedom of urbanistic principles does not correlate with the problems of urbanistics or with the problems emerging in the market [4].

There is a problem that we do not have fully subsidized public transport in larger Lithuanian cities. The main principles of sustainable transport organization [5] would be:

- the right of access for every person in all spaces and in all activities;
- socially just system: non-discriminatory for people of different ages, the disabled, the rural population, etc.;
- a transport system that functions without compromising the requirements of health, safety and quality of life;
- promotion of personal responsibility;
- transport system and physical and functional structure of the city;
- one integrated planning object;
- efficient use of resources;
- balance of financial and economic resources.

The number of cities where attempts to introduce FFPT, diverse in their scope, is currently estimated to exceed 100, most of which are in Europe, especially France and Poland [6]. The cities with FFPT vary in size. In France, for example, FFPT was introduced in some 30 towns and cities, most with a population between 10,000 and 110,000 residents [7]. In Germany, on the other hand, FFPT tests were completed in two cities, each with a population of approximately 15,000 residents. Outside Europe,

FFPT was/is operational in, amongst others, the USA (39 cities) and Brazil (15 cities). One of the Sustainability 2020, 12, 6390 4 of 19 largest experiments concerning FFPT was carried out in the city of Chengdu, China – a city of nearly 15 million residents [8]. In summary, it can be said that FFPT is introduced both to help partially solve actual temporary problems (for example due to flooding or road reconstruction) as well as for strategic reasons.

The example of Stockholm policy shows that the public supports the congestion charging system [9, 10], the use of bicycles in the UK [11]. These studies conclude that when people experience the effects of the system and benefit themselves, the public appreciates and becomes more appreciative of this policy intervention.

Author Kębłowski [12, 13] argues that free society is a policy form of transport (FFPT) is almost complete 100 different municipalities around the world. It is also stated that FFPT shows medium potential in terms of tackling car congestion or improving urban, and very high potential in terms of improving mobility of under-privileged group, across the urban territory [14].

The implementation of free public transport alone has not in itself had a far-reaching impact on environmental policy, nor has it necessarily reduced traffic-related CO_2 emissions, at least in the short term. The examples of Tallinn, Hasselt and Templin show that the increase in passenger numbers is due to the fact that many people who used to be cycling or walking for a short time were now "lured" to use public transport [15].

Key facts about the European Union's transport sector [16–19] provide key guidelines for the transport system as a whole and for public transport [20, 21]. Efficient transport services and infrastructure are essential to reap the economic benefits of all EU regions, to support the internal market and growth, and to promote economic, territorial and social cohesion [22, 23]. As a key player, transport is also closely linked to policies such as the environment, job creation and growth, competition, social policy and digitalization [24, 25].

The main law regulating transport in Lithuania is the Law on the Fundamentals of Transport Activities of the Republic of Lithuania (1991, No. 215-0; 0911010ISTA00I-1863).

Urban passenger transport is licensed in accordance with the procedure established by the Government. In addition to licenses, this business requires permits issued by the Lithuanian Transport Safety Administration under the Ministry of Transport and Communications authorized by the Ministry of Transport and Communications of the Republic of Lithuania. The Ministry of Transport and Communications of the Republic of Lithuania performs the issuance, temporary suspension, revocation and extension of permits. The legislation affects city companies that provide passenger transport services to the city in one way or another. Organizing these services is not easy. Many different governmental organizations are involved in this system of transport services.

3 Possibilities of Introducing Subsidized Public Transport on the Example of Foreign Countries

3.1 Public Transport Organization Models

There are three basic organizational models in public transport that are in place in the European Union. According to Jaržemskis and Jaržemskiene [26], the European Commission commissioned a study to find out the models of passenger transport organization in the community. The research company investigated the models of passenger transport organization in 43 cities of the European Union and summarized the results [26]. According to certain criteria, the company has identified and distinguished three main models for the organization of passenger transport activities: the so-called German, Scandinavian and British. All these models of passenger transport organization have their advantages and disadvantages.

The German model is a non-competitive model, the essence of which is that there is no competition between carriers, unless the general carrier lacks buses and it decides to lease additional contracts from other private companies. This model is applied in a city like Berlin.

The Scandinavian model is a model of regulated competition, which is characterized by the organization and control of passenger transport activities by the institution authorized by the municipality - the public transport agency. This model is used in many European cities.

The British model is a model of regulated competition, it is considered to be one of the first models of passenger transport organization. This model sets out minimum rules and controls. This model is applied in Manchester and many UK cities except London [26].

The main purpose of public transport is to meet the needs of passengers in a variety of means of transport and to provide the widest possible choice [27]. The improvement of passenger transport services has been driven by the UNE - EN 13816 certificate, which sets out specific requirements for public transport. It is a special standard for public passenger transport that is recognized by members of the industry. This certificate helps ensure that organizations monitor, accumulate, measure, and determine key parameters for continuous service improvement (time, availability, security, etc.).

3.2 Following the Example of the City of Tallinn

Public transport in Tallinn became free in 2013. But only for city dwellers. As stated in his speech in 2018, former Mayor Taavi Aasas of the FFPT program when public transport is fully subsidized is considered a success, despite the increase in the number of cars in recent years. According to surveys in 2018, about half of passengers started using public transport more often when it became free. Unlike in Tallinn, 11 Estonian counties that have canceled ticket prices are open to all, regardless of where they register, including tourists. This project is financed from the state budget.

According to researchers Yusak O. Susilo and Triin Reimal [13] this study is based on a large-scale annual survey commissioned by the City of Tallinn. The survey is based on live interviews with a random sample of 1,500 households to examine the impact of FFPT on individual travel habits before and after policy implementation. The analysis takes into account the effects of modal shift and whether they are driven by travel generation or travel change, travel attitudes and satisfaction, as well as impacts on equity, employment prospects, and travel destination choice. Given that FFPT policies are designed to increase the mobility and accessibility of local people, it is essential to analyze and understand how passengers take public transport directions and how these directions affect their daily activities and travel patterns. Although the theoretical discussion attempted to draw conclusions about the overall impact of FFPT based on changes in individual decision-making, empirical analysis has so far attempted to identify changes in the travel pattern at the individual level based on aggregate changes.

Initiating the FFPT policy, the City of Tallinn has announced that lost ticket revenue will be covered by an increase in municipal income tax. Annual revenue from ticket sales in 2012 It amounted to 12 million euros. In Estonia, the share of income tax is determined by the municipality where the person is registered. Some people who have migrated to Tallinn do not change their registration and thus continue to pay income tax to their city of origin. This is especially common among students and people who migrate from the countryside and feel dependent on their place of origin and therefore want to support it financially. Although the exact number of the unregistered population of Tallinn is unknown, municipal officials estimate it at around 25,000 to 30,000. November And in 2013 In November, in the annual municipal survey, before and after the implementation of the FFPT. Data were collected among more than 1,500 Tallinn residents in each period, randomly selected. Summary statistics for both samples are shown. In general, the composition of the sample is similar in both years and represents the urban population. The increase in the registered population of Tallinn in the sample corresponds to the increase in the 10,000 newly registered population in 2013. January - November It is therefore considered that in 2013 At the end, ~ 25,000 residents of Tallinn were registered elsewhere.

Smart technologies must be used to implement fully subsidized public transport [14, 25]. In the first period, these would be smart devices to which the passenger should add a smart passenger card. Thanks to this card and the device, the operator will be able to see at the current time how passengers move, at what time the passengers travel the most and on what routes. By collecting information, the city government will be able to improve passenger service and improve the functionality of urban transport. The collection of this information would help to implement one of the latest technologies, namely facial recognition devices. The passenger will then not necessarily have to mark the card, thus saving time. The passenger would have the opportunity to free his hands. This innovation is called a bus cloud card. Opening the face recognition function requires the user to provide ID card information and photos for authentication and the bank card is tied to the balance replenishment system.

4 Evaluation Study of Passenger Transport on Fully Subsidized Public Transport

4.1 The Expert‘s Survey

An empirical study was conducted on the designed model of subsidized public transport, during which competent experts were invited. These are individuals who have a high level of education and have worked in the field of transport for many years. 10 experts participated, who practically all agreed that the introduction of fully subsidized public transport in Vilnius would reduce the number of private cars in the city. The results of the peer review are analyzed according to the normal peer review procedure.

During the qualitative research, a research on the quality of the fully state-subsidized public transport service was carried out, meeting the needs of passengers. Subjects were judged primarily on the need for a transportation service. The need for a fully subsidized public transport in Vilnius was mainly emphasized. Theoretically, the possibilities of fully subsidized Vilnius city public transport in the city - as an alternative to personal cars.

The main reason for the existence of the public transport service is the satisfaction of the needs of the city residents and guests. Fully subsidized public transport is a preventive measure for cities to solve the problems caused by congestion, pollution and the possibility for residents to move freely in the city.

One of the experts stressed that fully subsidized public transport will not force residents to give up private cars, as a private car is a more convenient vehicle for traveling in the city.

However, several experts (86%) agreed that with the introduction of fully subsidized public transport, residents will be inclined to give up private cars due to difficult parking conditions in the city center.

Also, one of the experts said that as long as cities exist, public transport will live in one form or another. The same expert says that, of course, if the service is subsidized and the passenger does not have to think about a valid or invalid ticket every time he gets on public transport, it will be much more convenient.

One of the experts mentioned that the emergence of fully subsidized public transport will lead to congestion at public transport stops, as the number of people wishing to use these services may increase.

According to one expert, passengers who have previously used bicycles or scooters will start using more subsidized transport.

The majority of experts (84%) said that fully subsidized public transport would not work in terms of quality. Because even if there is an increase in passengers, it will reduce the space in the vehicle itself. In this case, the municipality will be forced to buy additional vehicles.

A review of the experts' responses suggests that the number of passengers will increase as those who previously avoided public transport because they were too expensive will start to travel. The responses showed that congestion will not be reduced because residents with a private car will not want to give up comfort while traveling, as the quality of travel by public transport may suffer. Experts differ. Some experts thought that fully subsidized public transport would be detrimental to the city and the quality of travel would suffer. Others initially agreed that the system was good, but would not reduce congestion.

One of the experts agreed that a fully subsidized public transport would pay off because he had lived in Tallinn for several years. However, he mentioned that in order to force the population to use fully subsidized public transport, it is necessary to attract the Vilnius City Traffic Safety Department, educational institutions and through social dissemination, such as: media, billboards and television.

4.2 Research of the Needs of Vilnius City Residents

The survey showed that transport problems in Vilnius are very sensitive, although two thirds of them have adapted to their place of residence. Over the past decade, a third of the population has moved to new homes or flats, while about 8 percent have moved to Vilnius from other places. One third of the population would like to relocate.

One third of the population gives priority to cultural and entertainment facilities. Assessing other factors, residents often stress that urban transport is sufficiently satisfactory. When choosing a place to live, people give priority to the safety, environment and accessibility of public transport. However, Vilnius is not satisfied with the condition of many streets, squares and courtyards. More than two-thirds of the residents of Naujininkai, Naujamiestis and Senamiestis complain of heavy traffic and air pollution in their districts. All residents would like more parking spaces to be built. Vilnius residents are especially concerned about the problems of public transport. More than half of the population regularly rides buses and trolleybuses; 42% use cars, while a quarter of the population travel by taxi. People rely mostly on public transport to travel to work or study in institutions (34%), while cars are in second place (21%).

Overall, 45% of respondents rely on public transport when traveling to work and 40% use it when visiting Vilnius. A survey conducted in Vilnius revealed the influence of some factors on the specific transport device chosen by passengers. Use of cars due to inability to reach public transport route (22%), distance to bus (trolleybus-bus) stop (20%), long interval in transport schedule (20%), non-regular schedule (3%), travel time and conditions (9%), vehicle changes and incompatibilities (9%), noise (9%), vibration (1%), environment (3%), congestion (12%) and jerky movement (1%). Most respondents are not satisfied with the price of public transport services (75%).

Figure 1 shows the revenue of Vilnius City Municipality company "Susisiekimo paslaugos" ("Communications Services") in 2015–2019.

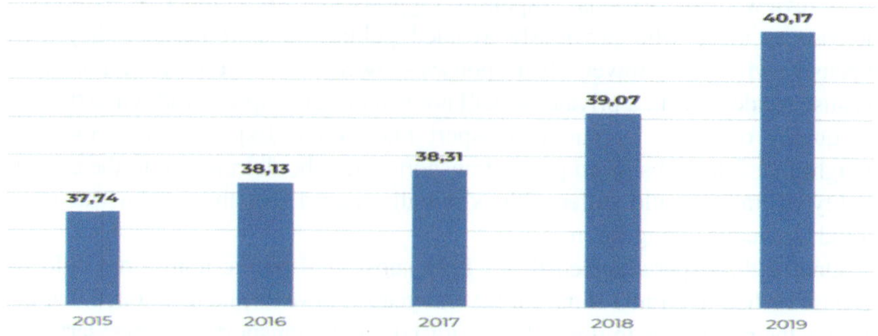

Fig. 1. Revenues of Vilnius City Municipality company "Susisiekimo paslaugos" ("Communications Services") in 2015–2019 in millions of euros [28].

Ticket variants in public transport in Vilnius [28] have two directions: subsidized and non-subsidized. A variety of tickets are offered, which are often very confusing for passengers (Fig. 2).

Fig. 2. Tickets issued by public transport for 2018–2019 have been prepared according to the data of the Public Transport Service.

In 2018, with the renewal of the public transport fleet, almost half of all public transport vehicles currently have automatic passenger counting equipment. Every day, about 80 thousand people receive equipment installed on buses and trolleybuses records (number of passengers embarking and disembarking at stops). This equipment allows to quickly receive passenger flow data from all routes of the Vilnius public transport network and ensures the possibility to carry out passenger flow research without additional natural observations. According to the data of the passenger flow survey, it was found that in 2019, 535 thousand scheduled trips are made per working day, which is about 4% more than in 2018.

Implementing fully subsidized transport services under free market conditions is commercially unprofitable in the first phase and organizations may be reluctant to provide them, but in the long run, this becomes a win-win system.

5 Conclusions

An analysis of the scientific literature reveals a number of key aspects related to passenger services. There are specific requirements for passenger transport services: speed, reliability, frequency, vehicles used that are suitable for modern society. The modern passenger also pays great attention to comfort and safety. Of course, the cost of services is important to the inhabitants of all cities, but, as we know, city municipalities often provide partial grants to such companies.

Improving the public transport system is a complex task. The main factors with the greatest impact must be: sustainable and efficient development of fully subsidized public transport, compact urban and urban sprawl, increase of population density in existing areas and balancing of industrial areas, housing and jobs, development of mixed construction, regulation of start time.

Fully subsidized public transport, following the example of the city of Tallinn, would allow the city's people to move freely and the above-mentioned problems would not also benefit the city of Vilnius and the surrounding areas.

The next step in this study will be to further investigate individual models of public transport travel by performing analysis of various situations before and after the trip. This will be done to determine the role of individual characteristics such as the purpose of the trip.

References

1. Amos, P.: Public and Private Sector Roles in the Supply of Transport Infrastructure and Services. The International Bank for Reconstruction and Development/The World Bank (2004)
2. Jaržemskis, A., Jaržemskis, V.: Keleivinis transportas. Vilnius, Technika (2017)
3. Verkulevičiūtė-Kriukienė, D., Nomeikienė, I.: Demographic and social changes in suburbs of Klaipėda in 2010–2017 / Regional formation and development studies. Nr. **3**(29), 131–143 (2019)
4. Juškevičius, P.: Urbanizacija ir urbanistikos interpretacijos. Acta Academiae Artium Vilnensis, t. **71**, 13–21 (2013)
5. Juškevičius, P.: Lietuvos miestų sistemos raida ir jos ateities perspektyvos. Acta Academiae Artium Vilnensis, t. **76**, 11–34 (2015)
6. Rybus, G.: Should Public Transport Be Free? More Cities Say, Why Not? https://www.nytimes.com/2020/01/14/us/free-public-transit.html. Accessed 05 Dec 2020
7. Wood, R.: Public Transport in France: Can You Get by without a Car? https://www.completefrance.com/living-in-france/getting-by-in-france-without-a-car-1-6540261. Accessed 20 Apr 2020

8. Kebłowski, W.: Fare-free public transport in Chengdu: riding for free in a Chinese metropolis. In: Jason, P., Dellheim, J. (eds.) Free Public Transit: And Why We Don't Pay to Ride Elevators, Black Rose Books: Montreal, QC, Canada, pp. 215–221 (2018)
9. Eliasson, J.: The role of attitude structures, direct experience and reframing for the success of congestion charging. Transp. Res. Part A **67**, 81–95 (2014)
10. Börjesson, M., Hamilton, C., Näsman, P., Papaix, C.: Factors driving public support for road congestion reduction policies: congestion charging, free public transport and more roads in Stockholm. Helsinki and Lyon. Transp. Res. Part A **78**, 452–462 (2015)
11. Chatterjee, K., Sherwin, H., Jain, J., Christensen, J., Marsh, S.: A conceptual model to explain turning points in travel behaviour: application to bicycle use. Transp. Res. Rec. **2322**, 82–90 (2013)
12. Kębłowski, W.: Eight contradictions in tallinn's free public transport. In: Dellheim, J., Prince, J. (eds.) Free Public Transport And Why We Don't Pay To Ride Elevators, Montreal: Black Rosen Books, 1st edition, pp. 102–107 (2017)
13. Cats O., Susilo Y. O., Reimal, T.: The prospects of fare-free public transport: evidence from Tallinn. Transportation **44**, 1086–1104 (2017). https://doi.org/10.1007/s11116-016-9695-5
14. Kębłowski, W.: More than just riding without a ticket? Exploring the geography of fare-free public transport. (Cosmopolis Working Paper). Vrije Universiteit Brussel, Cosmopolis, currently under review for publication in Transportation, Brussels (2017)
15. Dellheim, J., Prince, J.: Free Public Transit And Why We Don't Pay To Ride Elevators, Black Rose Books, 2nd edition, Montreal, p. 250 (2018)
16. European Commission Communication. Keep Europe moving – sustainable mobility for our Continent https://ec.europa.eu/transport/themes/strategies/2006_keep_europe_moving_en
17. European Commission Communication. A sustainable future for transport: Towards an integrated, technology-led and user https://ec.europa.eu/transport/themes/strategies/consultations/2009_09_30_future_of_transport_en
18. European Commission Communication. A European vision for Passengers: Communication on Passenger Rights in all transport modes. Brussels, 19.12.2011 COM, 898 final (2011)
19. European Commission Communication. Developing the citizens' network. https://eur-lex.europa.eu/legal-content/EN/TXT/?uri=celex:51998AR0436
20. Guidelines for Developing and implementing a Sustainable Urban Mobility Plan. European Commission Communicate (2nd edition) (2019)
21. European Commission, Directorate-General for Mobility and Transport: Attitudes of Europeans towards urban mobility. Special Eurobarometer 406. http://ec.europa.eu/public_opinion/archives/ebs/ebs_406_en.pdf. Accessed 15 Dec 2020
22. Cichosz, M., Goldsby, T.J., Knemeyer, A.M., Taylor, D.F.: Innovation in logistics outsourcing relationship – in the search of customer satisfaction. LogForum **13**(2), 209–219 (2017)
23. Litman, T.: Transportation Cost and benefit analysis techniques, estimates and implications. http://www.vtpi.org/tca/. Accessed 14 Dec 2020
24. Vural, C., Tuna, O.: The prioritisation of service dimensions in logistics centres: a fuzzy quality function deployment methodology. Int. J. Logistics **19**(3), 159–180 (2016) https://doi.org/10.1080/1367556.2015.1008438
25. Rossum, J.E.: 7 innovative technologies transforming the logistics industry. https://www.bizjournals.com/bizjournals/how-to/technology/2016/09/7-technologies-transforming-logistics-industry.html. Accessed 05 Dec 2020
26. Jaržemskis, A., Jaržemskiene, I.: Evolution of traveler experience quality perception in European level policy documents and the case study for Siauliai. Transp. Telecommun. J. **18**(3) (2017). https://doi.org/10.1515/ttj-2017-0019

27. Grzelec, K., Jagiełło, A.: The effects of the selective enlargement of fare-free public transport. Sustainability **12**, 6390 (2020). https://doi.org/10.3390/su12166390
28. Reports of Vilnius City Municipality company Susisiekimo paslaugos (Communications Services) for 2015–2019. http://www.vilniustransport.lt/. Accessed 20 May 2021

Prospects of Neuromarketing Application in Communication Activities of Logistics Enterprises

Lina Shenderivska[1], Mykhailo Krystopchuk[2(✉)], Viktoriia Nykonchuk[2], Anna Kniazevych[3], and Vira Shketa[1]

[1] Igor Sikorsky Kyiv Polytechnic Institute, 37, Peremohy Prosp, Kyiv 03056, Ukraine

[2] National University of Water and Environmental Engineering, 11, Soborna Street, Rivne 33028, Ukraine
{m.ie.krystopchuk,v.m.nykonchuk}@nuwm.edu.ua

[3] Academician Stepan Demianchuk International University of Economics and Humanities, Stepan Demyanchuk Street 4, Rivne 33027, Ukraine

Abstract. Neuromarketing has a wide range of applications: to build an effective communication policy with consumers, investors and other components of the business ecosystem. Compliance with ethics of neuroscience, data protection, helps to increase consumer confidence, more efficient, more personalized satisfaction of needs. Companies need to educate consumers about the relevance of neuromarketing. Motivating factors are: better, more complete and comprehensive satisfaction of current and anticipation of future needs to improve the quality of life, and this requires constant research.

The advantage of neuromarketing is that it allows to detect reactions that occur under the influence of new factors, in particular in a Covid-19 pandemic.

Content analysis of leading in Ukraine logistics companies websites showed a comprehensive positioning of companies, paying significant attention to quality standards, environmental safety, corporate social responsibility. In order to achieve the desired emotions, response of consumers more effectively, logistics companies need to improve their websites in the following areas: to place relevant information about services in two versions – concise and more detailed; to ensure the localization of websites, in particular, to follow the usual sites navigation for a particular country. It is expedient to use chatbots more actively for acceleration and convenience of communications. It is advisable to diversify visualization tools that will increase the effectiveness of content perception.

Keywords: Content analysis of websites · Information communications · Logistics companies · Neuromarketing tools · Sensors · Social media platforms

1 Introduction

Globalization, digitalization of the economy, increasing competition, intensification of communication processes, the need for a better understanding of dynamic consumer needs in the context of continuous improvement of product supply – all this requires

O. Prentkovskis et al. (Eds.): TRANSBALTICA 2021, LNITI, pp. 682–693, 2022.
https://doi.org/10.1007/978-3-030-94774-3_66

the search for new effective marketing tools. The application of the concept of neuromarketing allows a better understanding of consumer behavior, as a consequence of more rewarding interaction with them.

At the turn of the current millennium, a scientific revolution took place: the system of concepts changed from "man-rational" to "man-emotional". If before the buyer could compare several available products in a short period of time and choose the one he needed personally, then during the information overload people became physically impossible to process and rationally evaluate information about all alternatives that fall into their "field of view". That is why providing a list of technical capabilities and benefits of the product has become ineffective. He was replaced by the use of techniques to influence the emotional state of a potential consumer. This was the beginning of the neuromarketing development.

Neuromarketing should be used in various economy sectors. But it is especially acute now in conditions when Covid-19 pandemic has caused new consumer behavior factors that need to be thoroughly researched, and has shown the critical importance of logistics to ensure the society viability. In a pandemic, the ability to quickly adapt logistics processes to current consumer demands has become a crucial factor in the continued operation of enterprises. The development of logistics services has made it possible to compensate for losses in other types of economic activity.

2 Analysis of Recent Research

Effective development of the company requires constant study of consumer needs, search and application of effective marketing tools. A large number of scientific publications are devoted to these issues. The study [1] proved that the efficiency of logistics companies is significantly influenced by market segmentation and positioning, and also cross-functional customer focus.

The scientist [2] empirically examines the factors that affect firms' innovation activities in the context of basic functions major supply chain functions: production, delivery and support systems. The study showed innovation efforts across supply chain functions should prioritize strategic resources, first of all, labor that will create competitive advantages.

Marketing is also used to enhance security in enterprises. Using the example of logistics operations in Taiwan ports, researchers [3] conclude that companies need to pay more attention to security marketing, as it has a positive impact on the security climate and employee attitudes to security. Supervisor's safety commitment also has a positive effect on safety climate.

Any company operates in conditions of uncertainty, which causes the risks of its operation, which are the probability of loss of resources, markets, untapped opportunities. Among the factors of uncertainty in supply chains, such as, fluctuations in supply and demand, bias in planning and forecasting, scarcity and loss of resources, inconsistency of the information system with the goals and objectives of flow processes in supply chains; political, natural factors, Lutsenko I.S. [4] names human factor. With the help of neuromarketing tools, it is possible to manage the human factor more effectively.

Neuromarketing is a complex study of the brain in order to determine certain patterns of its response to advertising, goods, price, branding and more. Neuromarketing is a flexible and to some extent universal method of finding out about customer preferences and brand loyalty [5]. The following tools are most often used for research:

1. Quantitative electroencephalography (EEG) - measuring brain activity by recording electrical signals. Neurobiologists claim that every thought and emotion has such a signal [5].
2. Functional magnetic resonance imaging (fMRI) - measurement of blood flow to different parts of the brain caused by neural activity [5].
3. Eye-tracking (oculography) - finding out what a person's gaze is aimed at and how long he is looking at it. For example, when watching a promotional video, it is important to track whether a potential consumer notices the advertised product or brand; how often and/or how long a product or brand is in a person's field of vision; whether there are stimuli that distract from viewing [6]. Eye-tracking confirms that the human face is always in the center of attention, a person subconsciously perceives even a schematic facial image [7].
4. Analysis of micromimics - recording instantaneous changes and movements of facial muscles, through which we unconsciously transmit our emotions and feelings.
5. Analysis of heart rate and pressure [8].

Neuromarketing is used to build thermal maps of websites based on data about the movement of the user's eyes and the reactions that occur as a result of navigating the site. Thus, it is possible to form an optimal scenario for using the site, which improves the conversion of the site and increases the level of trust in this product or service [9].

The findings of investigation of consumer reactions to narrative advertising videos are presented in [10]. Encephalography was used to study the reactions. The study found that special attention should be paid to the fit between story object and story-receiver when creating consumer narratives.

3 Research Methodology

Neuromarketing involves the use of a number of tools to measure the true consumers' response. In this study, authors were guided by direct customer surveys of logistics companies and analysis of corporate websites.

Consumer surveys provided an objective view of their inquiries and expectations about the content of logistics companies' websites. Websites are chosen as objects of analysis because they allow the use of neuromarketing tools (eye-tracking, forums, etc.), and at the same time are the main means of communication of logistics companies with consumers.

The statistical sample of logistics companies for content analysis of websites was conducted by investigation of the main array. It includes the 10 largest logistics companies operating in Ukraine, their total share in the market of logistics services in Ukraine exceeds 70%. The statistical sample of customers of the 10 largest logistics companies is formed by random selection and is 120 respondents, which is 20% of consumers of the surveyed logistics companies.

Taking into account that sales neuromarketing includes a sensory approach, the authors of the article also analyzed the logos of the 10 largest logistics companies in Ukraine.

4 Problem Setup

The application of neuromarketing has a number of difficulties: the presentation of research results in clinical terminology is difficult to perceive by employees (marketers) unfamiliar with this terminology; some neuromarketing tools are expensive, which limits their use in small business; neuromarketing research needs to be repeated from time to time in the conditions of high rates of scientific and technical progress, fast change of consumers' preferences. Sounds, smells, touches, appearance: all this affects customers in different ways. The task of neuromarketing is to effectively use knowledge about sensors, to set the desired direction, to motivate people to select for certain factors.

5 Research Results

With regard to practical marketing methods aimed not at research but at commercial activities, it should be noted the importance of taking into account the impact on the consumers' sensory system. The sensory system, or sense organs, is a way to create the desired associations in the consumer's brain about goods, services, brands, and so on. This is how the unconscious "anchoring" of certain events and situations takes place.

It would seem that the main thing is what product, service, etc. the store offers, and its characteristics. However, after conducting a survey of the logistics services customers, we learned that: 91% – confirm the opinion that the possibility of more detailed acquaintance with each service of the company, including the price affects their desire to continue browsing the site; 87% of consumers pay attention to the convenience of the website menu, the simplicity of the order form; modern site design is important for 69% of respondents; for 64%, the location of the company (its representative offices) on GoogleMaps is important. Therefore, the right influence on the sensory system allows to increase consumer interest in the product. According to the concept of neuromarketing, effective influence is possible if consumer reactions are measured using special devices, methods, and not just established by consumer surveys, because consumers cannot distinguish the subconscious, which also affects decision-making.

Thus, neuromarketing can be compared to a "lie detector" because it provides truthful answers that cannot always be obtained from consumer surveys, also because of their unwillingness to be honest. That is why the Delphi method was developed, where it is possible to express opinions anonymously; e-surveys are being replaced by anonymous offline surveys to alleviate the fear of leaving a digital footprint. It is advisable for respondents to create a relaxed atmosphere to get sincere reactions.

For example, there is a practice: a focus group is invited to discuss a certain topic, then an advertisement is launched, and at this time the correspondent leaves the room, if the advertisement is of interest, the group will watch it. In this way, you can understand whether advertising will really be effective. This allows us to conclude that neuromarketing is not only the use of special measuring devices to assess consumer response, but also the creation of conditions for obtaining sincere reactions/responses of respondents. Neuromarketing combines marketing, psychology, neurosciences [11].

Here is another example: a social experiment was conducted in Australia, participants were asked to evaluate two video ads of one product of the same brand. The main difference between the videos was that the first video was shot at a higher speed, is energetic, and the second - slower, more atmospheric and plot. The first "fast" video received negative reviews from the participants of the experiment, and the second "slow" – positive. Nominally, it was advisable to choose the first option for broadcasting. But physiological studies that correspond to the concept of neuromarketing, namely: electroencephalogram, analysis of evoked potentials showed that the first "high-speed" "viral" video is better remembered and became an incentive to buy goods in the future. The classical method has shown its inefficiency [12].

Another study related to the assessment of visual perception revealed the following relationship: for expensive branded products, the value of color perception for the decision to purchase increases. At the same time, the impact of color on the consumer in the group of goods "mass demand" is minimal [13]. Thus, when designing speciality goods, expensive goods, special attention should be paid to the design of goods, packaging, advertising. In general, the psychology of color is important in the marketing activities of the enterprise. For example, black in certain groups of people causes negative associations. But modern fashion for black design of websites, packaging etc. changes the initially negative attitude of consumers to a positive one.

Guided by a set of features (categories) presented in [14], such as providing information, features of action, features of interaction, features of lateral linkages, features of creative expression, we made a content analysis of leading in Ukraine logistics companies websites (Table 1). The leading logistics companies in Ukraine are: **Küehne + Nagel (founded in Switherland),** DSV Logistics (Denmark), **FM Logistics** Ukraine (France), Raben (Netherlands), **Ekol Ukraine** (Turkey), Zammler (Ukraine), **Pakline Logistics** (Ukraine), UVK (Ukraine), Logistics-Plus(Ukraine), DB Shenker Ukraine(Austria) [15].

Table 1. Content features of leading in Ukraine logistics companies websites.

Set of features	Frequency
Features of Providing information	
Convenience menu	9
Podcast Video or Audio Reports	9
News Articles	8
Features of Action	
Urgent Action Alerts	6
Plans International, National, Local Actions	6
Calendar of Events	2
Features of Interaction	
Support Services or Advice	10
Staff Contact Information (Profiles)	5
Participatory Forums	1
Features of Lateral Linkages	
Links to Social Media	9
Links to Mainstream News	8
Links to Google Maps	8
Features of Creative Expression	
Visual Art	9
Corporate Colors:	
Blue	7
Red	4
Black	3
Grey	1

Source: developed by the authors on the basis of [14].

As the content analysis showed, logistics companies on their sites pay the most attention to the category of "providing information" about the company's services, company history, the business philosophy. At the same time, information about services is not always conveniently presented, it would be worth giving concise relevant information and link to more detailed information. In addition, site navigation may not be typical of Ukraine when a horizontal menu and/or a vertical menu on the left is offered. Therefore, logistics companies when developing websites need to take into account the localization, namely, the features of sites for a particular country. Also, not all investigated companies pay enough attention to the coverage of company news, industry, frequency of updates.

To improve the "action" category, more attention should be paid to informing consumers about the company's plans, promotions, price changes, etc.

Analysis of the "interaction" category showed that companies provide consumers with the opportunity to fill out online forms to obtain advice. Landline numbers predominate among the contact details, but it would also be advisable to provide mobile phone numbers as well as the name of the contact person (this is not always the case on websites). Chatbots should be used more widely to speed up communications.

The category "linkages" is not fully used, the sites mostly contain links to social media platforms.

Sales neuromarketing includes sensory perception, which is the most advanced in the analysis category – "creative expression". Logistic companies efficiently use visual art for the activation of visual comprehension. The logos colors of the analyzed logistics companies correspond to colors selection of the largest global brands: blue – the color 33% of the largest global brands, red – 29%, black, grey, silver – 28%; also correspond to colors prevalent in logistics sphere (red and blue) [16].

Sometimes deviating from the norms allows you to more effectively meet the customers' needs. For example, the straying from the traditional monochrome in car design has led to the introduction on the market of cars painted in several colors (mostly two-tone); understanding people's interest in other people's lives led to the emergence of reality-show, blogosphere. Awareness of people's dependence on social networks stimulates the development of social media marketing and taking into account in the communication policy of enterprises the main trends, topics, message formats of social networks and messengers.

Social networks influence consumer behavior. Given the fact that social media users imitate the behavior of influencers, the placement of advertising by opinion leaders is becoming increasingly popular. And the key selection criterion is not the influencer with the largest audience, but the influencer who covers the target audience of the company [17]. When maintaining social networks, sites, messengers, it is necessary to thoroughly analyze the comments of consumers/readers, who will tell you what needs to be improved in the management of content, business, which topics should be covered. It is also necessary to monitor the Internet resources of competitors, the reactions of consumers to them. At generalization of results of researches it is necessary to segment consumers on the basis of the revealed laws in reactions, it will allow to carry out effectively positioning, to develop more effective marketing strategy.

According to the study [18], it was determined that most organizations and consumers among social media (Facebook, Instagram, Twitter, Google+, Linkedin, YouTube) prefer Facebook. But there are different priorities for consumers to use different social networks (Table 2). For each purpose, there is not a complete list of social media platforms, but only those that received the highest percentage of respondent responses.

Table 2. Distribution of consumer purposes for the use of social media platforms.

Purpose	Consumer preferences regarding the purposes of using social media platforms	Recommendations for the use of social media platforms by organizations
For professional purposes	Google + (40%), Linkedin (36.5%), Twitter (6.8%)	To post company and industry news, create and distribute questionnaires, conduct research
Finding a job	Linkedin (31.8%), Google + (5.2%), Facebook (2.2%)	To search for employees through Linkedin

(continued)

Table 2. (*continued*)

To follow people and organizations of interest to consumer	Twitter (27.4%), YouTube (17.0%), Linkedin (16.5%), Instagram (14.2%), Facebook (11.8%)	To post company and industry news, organize flash mobs
To keep up with what is happening in the world	Twitter (28.8%), Google + (16.5%), YouTube (15.6%)	To post interesting content for consumers, industry and the world news
Communicating with other social media users	Facebook (17.4%), Instagram (11.7%), Google + (6.1%)	Position as a social media for communication with the company, where quick responces are given
For entertainment	YouTube (51.4%), Instagram (17.5%), Facebook (15%), Twitter (12.3%)	To place entertaining content, to communicate with consumers casually, without formalism, as far as the specifics of business, the philosophy of the company allow; to place advertisements with influencers
Sharing opinions with others, through personal posts or comments on others' posts	Twitter (11%), Facebook (9%), Instagram (8.1%)	To distribute questionnaires, promote the product – to get feedback
To know what my friends/acquaintances are doing	Instagram (14.3%), Facebook (12.6%)	To place advertisements with influencers
To meet new people	Instagram (6.8%), Facebook (6.1%), Twitter (5.5%)	To place advertisements, including influencers; promote businesses for which personalities are important, such as the publishing business
To post photos/videos of moments in my life	Instagram (14.6%), Facebook (9.4%), YouTube (2.5%)	To place advertisements with influencers, organize flash mobs

Source: developed by the authors on the basis of [18]

Organizations need to fill each social network with content that meets the priority purposes of consumers for each social network.

It is advisable to use neuromarketing research to build effective business communications in social networks [18]. But the use of neuromarketing tools, including eye-tracking, face coding, voice recognition is perceived negatively by consumers as a company's distrust of its customers. When using neuromarketing tools, companies must take into account the reluctance of consumers to authorize the collection of data on their behavior. This allows us to conclude that the company needs to explain to consumers the reasons for the use of neuromarketing and the benefits that consumers will receive, in particular, providing the content they need, understanding consumer desires, rearrange page visual to make interesting items more visible, finding out the consumer interest areas, choosing the right tone etc. [18]. For example, Facebook Virtual and Augmented Reality Lab has developed a bracelet for the benefit of consumers, to improve human-computer interaction, which reads neural impulses on a person's wrist and uses them to enter information. It is not necessary to make the movement itself. The bracelet is equipped with vibration for realistic interaction with objects in augmented reality, as well as with artificial intelligence. The future task is to teach the system to read information about a person in order to anticipate requests and with the least effort on the part of the person to satisfy them. The Facebook Virtual and Augmented Reality Lab is guided by the norms of neuroethics: seminars are organized to identify and mitigate the possible negative harm of the product to society [19].

In the communication activities of enterprises, including logistics enterprises, advertising plays a substantial role. Modern advertising is becoming more socially oriented, when in the advertising appeal a small part of the content is devoted directly to the brand, and more – to issues related to people. This allows you to show the brand on the other hand – to demonstrate the attitude to the basic values of life. The main messages of social orientation advertising include: "Help others", "Love yourself as you are, do not necessarily meet the standards of beauty", "You have the freedom to do what you like", "You have the strength to achieve the goals, try and strive!", "Maintain strong family relations" and others. The message "The strength of the company in its employees and customers" is most clearly expressed in the promotion of logistics companies' services through their websites. That is, the center of social advertising is the focus on human feelings, aspirations, overcoming fears, anxieties, insecurities, and confession of business value not as a system for earning money, but to meet the needs of consumers and employees. This is what finds a response in all people, and not only in the target audience. Thus, social advertising performs not only the function of brand promotion, but also to support people, solidarity with them, mentoring. This shows the social responsibility of the enterprise. Businesses need to take into account that when making a purchase decision, consumers take into account the holistic perception of the brand, the company's image:

- implementation of a comprehensive approach by the company: fair-minded relation "price-quality", consulting support (including at the design/ordering stage), additional services (remote ordering, delivery, informing about discounts, offering new products, etc.), after-sales service, storage level of service even when buying inexpensive goods or services;
- ecological safety;
- quality of advertising appeals;
- attitude to competitors (hostile, neutral, consolidation of efforts or implementation of joint projects, favorable);
- implementation or participation in socially important projects, position on socially important topics, attitude to employees, business philosophy.

The modern concept of green economy [20] causes a more meticulous attitude of consumers to the environmental friendliness of business. The study [21] proved that environmental warehousing and logistics optimization, social values and ethics have a positive impact on supply chain sustainability, which improves economic performance.

6 Conclusions

The effectiveness of neuromarketing is that it avoids subjectivity. Inasmuch as, the basis of neuromarketing tools is the measurement of consumer reactions. The use of a neuromarketing tools set is costly. But the use of relatively cheap neuromarketing tools, such as eyetracking, as well as integration with Big Data Analysis, creates the potential for wider use of neuromarketing by companies of all sizes.

The peculiarity of the activities of logistics companies is that their first acquaintance with the consumer is mainly through the website. Further communications also take place mainly through the website, mobile application, e-mail, social media pages. Therefore, the quality of web content is critical for attracting customers. To better understand consumer behavior, logistics companies should pay more attention to the frequency of video updates, and analysis of consumer reactions through eye tracking; to study the behavior of consumers on the website, on corporate pages in social networks to get answers to the questions: what information is most often viewed, for how long, at what time. All this knowledge will contribute to the creation of web content, choosing its location, relevant to the real needs of consumers.

The results of content analysis of leading in Ukraine logistics companies revealed a number of shortcomings that could be eliminated by managing neuromarketing tools. Including in particular, the use of neuromarketing tools would optimize the location of customer support information by logistics companies. At present, it is not always easy and fast to find information about customer support on logistics companies websites, which can negatively affect consumer loyalty.

According to the results of the analysis, currently only 5 out of 10 logistics companies post employee profiles. At the same time, eye tracking proves that the human face is always get priority. Therefore, it is necessary more actively to place profiles of employees with whom customers will communicate directly. This would enhance consumer emotional involvement.

The analyzed logistics companies do not pay enough attention to an important neuromarketing tool – forums – only 1 company out of 10 offers a participatory forum. The relevance of this tool is that the forums are characterized by a relaxed atmosphere of communication, as a result, companies can track the real reactions, to identify current market needs.

Thus, neuromarketing is a promising area of marketing research and tools that studies the reaction of consumers to sensory and emotional stimuli. Adherence to the ethics of neuromarketing will increase the level of trust in it, will help expand the range of products, services that will upgrade the quality of life.

References

1. Panayides, P.M.: Marketing in Asia-Pacific logistics companies: a discriminant analysis between marketing orientation and performance. Asia Pac. J. Mark. Logist. **16**, 42–68 (2004). https://doi.org/10.1108/13555850410765122
2. Karaman Kabadurmus, F.N.: Antecedents to supply chain innovation. Int. J. Logist. Manage. **31**(1), 145–171 (2020). https://doi.org/10.1108/IJLM-04-2019-0096
3. Lin, C.-C., Chang, C.-H.: Evaluating employee's perception toward the promotion of safety marketing at ports. Int. J. Shipping Transp. Logist. **13**(3/4), 275–299 (2021). https://doi.org/10.1504/IJSTL.2021.113985
4. Lutsenko, I.S.: Unrecognizability of supply chains: causes, stages, consequences. In: Proceedings of the 2nd International Scientific and Practical Conference "Business, Innovation, Management: Problems and Prospects", Kyiv, Ukraine, 22 April 2021, pp. 268–269 (2021). http://confmanagement.kpi.ua/proc/article/view/231033. Accessed 13 Jul 2021
5. Akulych, M.: Emotional Branding and Neuromarketing. Publishing Solutions, Moscow (2017)
6. Luneva, E.A., Skobelkina, N.G.: Eye tracking in the system of modern neuromarketing technologies. Siberian Trade Econ. J. **3**, 50–53 (2016). https://cyberleninka.ru/article/n/aytreking-v-sisteme-sovremennyh-tehnologiy-neyromarketinga. Accessed 05 Jun 2021
7. Neuromarketing: How it really works. https://www.sostav.ru/publication/nejromarketing-kak-eto-rabotaet-na-samom-dele-23553.html. Accessed 09 Jul 2021
8. Lewis, D.: Neuromarketing in action. How to get into the brain of the buyer. Translation from English: Matskovskaya, M. Mann, Ivanov and Ferber, Moscow (2015)
9. Zhmai, A.V.: Formation and development of the neuromarketing's concept in modern conditions. Mark. Econ. Modern Manage. Theor. Pract. **17**(1/38), 157–170 (2018). https://doi.org/10.18524/2413-9998/2018/1(38).135569
10. Gordon, R., Ciorciari, J., van Laer, T.: Using EEG to examine the role of attention, working memory, emotion, and imagination in narrative transportation. Eur. J. Mark. **52**(1/2), 92–117 (2018). https://doi.org/10.1108/EJM-12-2016-0881
11. Mansor, A.I., Isa, S.M.: Fundamentals of neuromarketing: what is it all about? Neurosci. Res. Notes **3**(4), 22–28 (2020). https://doi.org/10.31117/neuroscirn.v3i4.58
12. Chervyakova, E.A., Vagin, D.Y., Seredina, V., Yashkova, E.V.: Neuromarketing: a modern approach in classical marketing. Innov. Econ. Prospects Dev. Improv. **2**(36), 437–442 (2019). https://cyberleninka.ru/article/n/neyromarketing-sovremennyy-podhod-v-klassicheskom-marketinge. Accessed 05 Sep 2021
13. Gimaletdinov, K.V., Gimaletdinov, W.K.: Some aspects of neuromarketing. Sci. Educ. J. Students Teachers StudNet **2**, 64–70 (2020). https://cyberleninka.ru/article/n/nekotorye-aspekty-primeneniya-neyromarketinga. Accessed 16 Jun 2021
14. Stein, L.: Social movement web use in theory and practice: a content analysis of US movement websites. New Media Soc. **11**(5), 749–771 (2016). https://citeseerx.ist.psu.edu/viewdoc/download?doi=10.1.1.983.1164&rep=rep1&type=pdf. Accessed 09 Jul 2021
15. Top 10 largest logistics companies in Ukraine. https://mintrans.news/logistics/top-10-naybilshikh-logistichnikh-kompaniy-ukraini. Accessed 02 Jul 2021
16. Logo color: how to choose the right one. https://azbyka.com.ua/kak-vybrat-pravilnyj-tsvet-dlya-vashego-logotipa. Accessed 10 Jun 2021
17. Zabrodina, V.V., Shenderivska, L.P.: Tendencies in advertising industry. In: Proceedings of the 2nd International Scientific and Practical Conference Business, Innovation, Management: Problems and Prospects, Kyiv, Ukraine, 22 April 2021, pp. 72–73 (2021). http://confmanagement.kpi.ua/proc/article/view/230468. Accessed 08 Jul 2021

18. Constantinescu, M., Orindaru, A., Pachitanu, A., Rosca, L., Caescu, S.-C., Orzan, M.C.: Attitude evaluation on using the neuromarketing approach in social media: matching company's purposes and consumer's benefits for sustainable business growth. Sustainability **11**, 1–21 (2019). https://doi.org/10.3390/su11247094. Accessed 22 Jun 2021
19. Inside Facebook reality labs: wrist-based interaction for the next computing platform (2021). https://tech.fb.com/inside-facebook-reality-labs-wrist-based-interaction-for-the-next-computing-platform. Accessed 11 Jul 2021
20. Agyabeng-Mensah, Y., Ahenkorah, E., Afum, E., Dacosta, E., Tian, Z.: Green warehousing, logistics optimization, social values and ethics and economic performance: the role of supply chain sustainability. Int. J. Logist. Manage. **31**(3), 549–574 (2020). https://doi.org/10.1108/IJLM-10-2019-0275
21. Kniazevych, A., Olikhovskyi, V., Olikhovska, M.: Clustering of the economy as a means of developing an innovation infrastructure. Baltic J. Econ. Stud. **7**(3), 134–139 (2021). https://doi.org/10.30525/2256-0742/2021-7-3-134-139

Station Capacity Analysis of a Metro Line with Discrete Event Simulation

Mehmet Sinan Yıldırım[1] and Metin Mutlu Aydın[2](✉)

[1] Manisa Celal Bayar University, 45000 Manisa, Turkey
[2] Ondokuz Mayıs University, 55270 Samsun, Turkey

Abstract. This paper demonstrates the utilization of discrete event simulation for the capacity assessment of an existing metro line using performance metrics of train utilization and passenger waiting queues at the stations. The metro line was modelled with using Arena Simulation model blocks of queues and train routing delays and the simulation model was executed with using the hourly passenger arrival schedules, an origin-destination matrix scenario and variable train time headways. The results indicated the significant deviations of the waiting passenger numbers prior to train boarding with failed train boarding resulted from system congestion. The study indicated that the train time headways can influence the system equilibrium and significant congestions are especially prominent for the intermediate stations with high passenger traffic. The characteristics of the O-D matrix was also a significant contributor to the individual station congestion since the train capacity is highly occupied with the passengers of the popular stations.

Keywords: Station capacity discrete event · Metro line · Train · Simulation

1 Introduction

Public metro systems are important transport infrastructures for the cities and their capacities are important servicing the passenger needs. Especially for the metro systems, the line capacity must be enough for the limiting the passenger delays and decreasing the station congestion problems. With the increase of the urban population, traffic congestion and delays are prominent problems of the city life resulting economic and productivity losses. Apart from the roadway transport, the public transport and especially urban commuter train and metro services are very important. Compared with the road transport, these transport modes can serve a comfortable and safe trip for the passengers and also help to decrease the road traffic congestion and air pollution. Beside their advantages, an efficiently operated public train service should target the optimum point of sustaining the comfort and service quality perception of the passengers while cutting the operation costs with minimizing the train trips.

In literature, simulation techniques were extensively utilized for evaluating the capacity and performance of the rail freight and passenger transport systems. The railway microsimulation models can be classified considering their scopes

O. Prentkovskis et al. (Eds.): TRANSBALTICA 2021, LNITI, pp. 694–704, 2022.
https://doi.org/10.1007/978-3-030-94774-3_67

(microscopic, mesoscopic or hybrid) and modelling approaches (general simulation software or railway simulation packages). Especially with the increasing computer power, microscopic methods are turned to be more advanced tools for detailed modelling of train or passenger entities and representing the real system with great details [1].

In previous studies, simulation models are extensively used for train scheduling and time tabling. Some of the studies considered the railway freight transport. In a study [2] used agent-based microsimulation model for scheduling the shuttle freight trains in a congested railway network. In the study, [3] also considered the scheduling of additional freight trains in a constraint railway network. In another study [4] developed a simulation model for analyzing the utilization levels of a railway line sections to investigate the influence of the freight trains on the passenger trains. [5] also used microsimulation for rescheduling of the trains in congested railway networks with combining multi-objective optimization and simulation. [6] used an event-based simulation model for analyzing the railway freight movement in urban environment and the railway utilization levels were examined. All simulation-optimization methods were also used for minimizing train travel times delays [7], calibrating train speed profiles [8], and for improving train schedules [9]. Additional studies also included the train scheduling problem for the metro lines considering energy-efficiency and dynamic passenger demand [10, 11] and train rescheduling in dense-railway networks [5]. Especially for the train scheduling and rolling-stock planning studies, Linear Programming [12], Heuristics [13] and Stochastic Models [14, 15] were also used.

In current studies, there exist very small number of studies considering the microscopic simulation of metro lines and analyzing the performance and reliability of the metro lines. The objective of this paper is to build a discrete-event simulation model for a metro line and using the model for analyzing the performance and capacity of an existing metro system. The simulation model was used for a case study of the İzmir metro system and the line capacity of the system was investigated.

2 Methodology

In this section, the details of the simulation modeling for the metro line were discussed. The metro simulation model (MSM) was constructed with using the Arena 14.0 simulation software and discrete event simulation (DES) approach. The principal role of the MSM was to synchronously simulate the passenger movements, queueing and transfers at the stations and the movements of the metro trains using consecutive delay blocks. MSM includes model animation for trains and passenger movements for model validation phase. MSM employed Create, Dispose, Delay, Queue, Signal and Release blocks for the model construction. The properties and characteristics of the MSM blocks were shown in Table 1.

Table 1. MSM blocks and their characteristics.

Block Name	Definition	Purpose of Usage
Create	Generates arriving entities to a process model	Train and passenger generation
Dispose	Ending point for the model entities	Disposing trains and passengers
Delay	Delays entities by a duration	Train routing between stations
Hold	Holds entities in a queue for a signal	Passenger waiting in stations and trains
Signal	Send signal to a hold block for releasing entities in queue	Passenger train boarding and leaving
Queue	Models the waiting space for entities	Passenger waiting in stations and trains
Search	Search the entities within a queue	Selecting the passengers leaving the train
Remove	Remove the entities from the queue	Modelling train exit process
Assign	Assigns the variables and entity attributes	Assign variables and attributes for passengers and trains

Several assumptions were made for developing the MSM to simplify the modelling phase of the real-word system. The assumption related with the train and passenger behaviors were as follows:

- Hourly passenger arrival rate was constant in each hour. Such as, the arrival rate was not changed if the train was close to the station,
- After a train arrived at a station, passengers exit, and the new passengers get into the train,
- The passenger waiting queue was operated with First-Come-First-Service (FCFS) priority rule,
- If maximum passenger capacity of a train was reached, passengers waited for the next train and they didn't leave the station,
- Passenger entered and leaved the trains within a station dwell time (T_D),
- Poisson distribution was used for modelling passenger arrivals,
- The percentages of the passenger for origin and destination stations were constant.

The assumptions regarding the train operations were as follows:

- No train failures,
- Trains are dispatched from both stations after a specific dwell time was reached, no train resources are defined.

For each metro station and a single line direction, a create block was used for passenger generation with an arrival schedule. Each arrival schedule was constructed with hour intervals which the passenger arrival rates were the input parameters. The arrival time of the passengers, the destination stations were assigned to each passenger with using the assign block. The destination stations were determined with using a discrete probability distribution using the observed passenger origin destination (OD) matrix for each station. The passengers were taken to a hold block attached with a Queue Block for simulating the passenger waiting process at the station.

The trains were generated with a single Create block with constant time headways. The generated train was taken to a delay block for simulating the train routing delay. After the train arrived to a station, consecutive Signal and Search blocks were used for unloading and loading the passengers to the train. For the first and last stations only passenger boarding and leaving functions were triggered. The layout of the MSM was shown in Fig. 1.

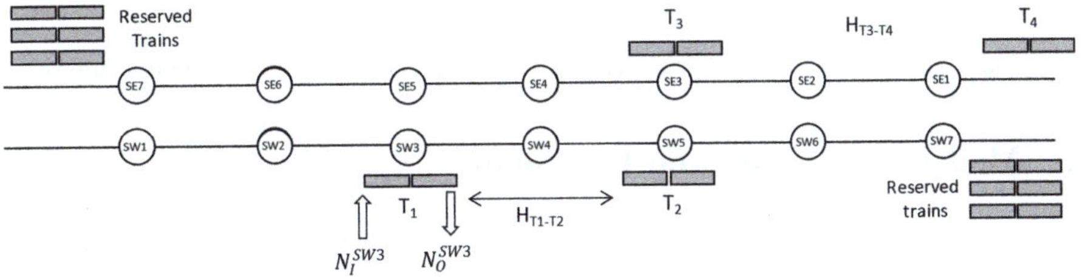

Fig. 1. MSM layout and stations.

MSM was used for recording the statistics of passenger transport capacity, maximum number of passengers' waiting in the stations and train utilization ratio. The model blocks used for simulating the train routing and passenger boarding processes were shown in Fig. 2. MSM blocks used for passenger generation, station waiting and train routing were also given in Fig. 3.

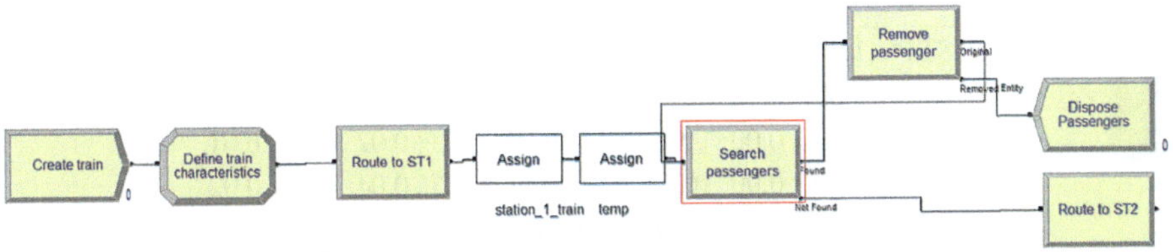

Fig. 2. Modelling blocks for train routing and passenger boarding.

The performance of the MSM was evaluated using two performance metrics such as, the number of passengers waiting at the i-th station (N_P^i) and the number of passengers unable to board the train for i-th station N_P^i. These metrics were used for comparing the system capacity, station congestion and station utilization levels.

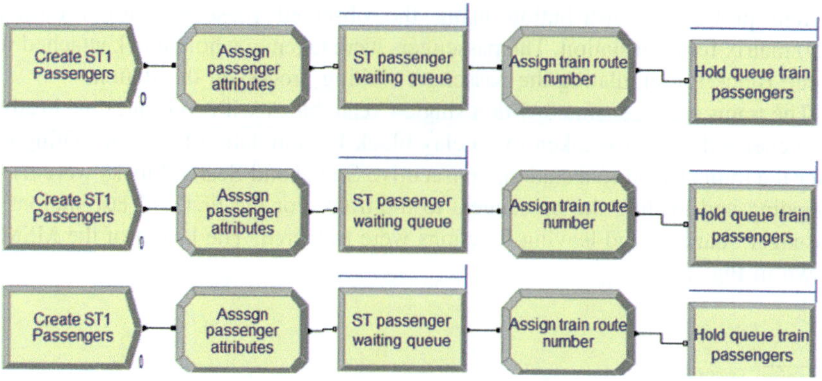

Fig. 3. Simulation blocks for simulating the passenger arrival and station operations.

3 Model Verification and Execution

Model verification is used for testing the reliability of the model construct and used for guaranteeing that the model executes in a proper way. Additional to the verification stage, model validation also implemented for testing the fitness of the model comparing the underlying real-world system. For the model verification, the passenger arrival scenarios were developed for all stations. Each station was named as ST-1 to ST-10 for a single direction. The model was firstly executed with using the passenger OD probability matrix including the probability of the passenger trips between two associated stations. The OD probability matrix is given in Table 2.

Table 2. Passenger OD probability matrix.

	ST1[1]	ST2	ST3	ST4	ST5	ST6	ST7	ST8	ST9
ST1[2]	1.00	0.56	0.58	0.50	0.40	0.39	0.33	0.31	0.30
ST2	0.00	0.44	0.34	0.26	0.23	0.19	0.16	0.13	0.11
ST3	0.00	0.00	0.08	0.12	0.09	0.03	0.02	0.02	0.01
ST4	0.00	0.00	0.00	0.13	0.11	0.05	0.04	0.03	0.02
ST5	0.00	0.00	0.00	0.00	0.18	0.14	0.18	0.17	0.15
ST6	0.00	0.00	0.00	0.00	0.00	0.21	0.18	0.17	0.16
ST7	0.00	0.00	0.00	0.00	0.00	0.00	0.09	0.09	0.08
ST8	0.00	0.00	0.00	0.00	0.00	0.00	0.00	0.08	0.07
ST9	0.00	0.00	0.00	0.00	0.00	0.00	0.00	0.00	0.10
ST10	0.00	0.00	0.00	0.00	0.00	0.00	0.00	0.00	0.00

[1]: Origin station.
[2]: Destination station.

Passenger arrivals for each station during daily (18-h operation period) were developed according a hypothetical scenario considering the morning and evening rush hours for a single line direction (From ST1 to ST2). The daily passenger arrival data with hourly intervals are given in Table 3 considering passenger flow per hour.

Table 3. Hourly passenger arrivals for 18-h operation period.

Hours	ST1	ST2	ST3	ST4	ST5	ST6	ST7	ST8	ST9
6:00–7:00	160	135	76	453	53	112	95	53	317
7:00–8:00	714	653	481	1,939	337	500	457	337	1,357
8:00–9:00	1,174	1,073	930	2,549	651	822	751	651	1,784
9–10 am	615	535	402	1,356	281	431	375	281	949
10–11 am	483	455	325	1,279	228	338	319	228	895
11–12 am	510	516	362	1,482	253	357	361	253	1,037
12–1 pm	577	573	466	1,768	326	404	401	326	1,238
1–2 pm	677	831	448	1,861	314	474	582	314	1,303
2–3 pm	749	687	440	2,218	308	524	481	308	1,553
3–4 pm	746	718	498	2,477	349	522	503	349	1,734
4–5 pm	696	747	514	2,774	360	487	523	360	1,942
5–6 pm	892	740	620	2,576	434	624	518	434	1,803
6–7 pm	2,535	1,012	746	2,226	522	1,775	708	522	1,558
7–8 pm	1,035	671	438	1,693	307	725	470	307	1,185
8–9 pm	450	354	260	1,213	182	315	248	182	849
9–10 pm	274	174	140	920	98	192	122	98	644
10–11 pm	180	126	106	755	74	126	88	74	529
11–12 pm	109	58	66	490	46	76	41	46	343

MSM validation was completed with investigating the model animation with reduced animation speed and 5 min train arrival scenario (HT = 5). The simulation model includes the passenger arrivals with using the exponential distribution function of with a mean of values in Table 3. During this period, an extreme case analysis was performed to check if the model constantly runs with high passenger arrival times. The model verification phase included system variability checks using stochastic variables and constants. The train capacity (CT) was assumed 1,000 people with 3 cars per train according to the length limits of the station boarding platform. CT can also be decreased if the passenger comfort is the principal design factor, but this condition was not considered in this study. The validation results of the MSM considering the total generated passengers and simulation output were shown in Table 4 for the averages of 100 model replications.

Table 4. Comparison of the MSM outputs with the input data.

	ST1	ST2	ST3	ST4	ST5	ST6	ST7	ST8	ST9
Data	34560	8635	3517	12503	14100	1517	8962	19644	28108
Simulation	31652	8225	3474	11507	12938	1502	8800	19375	27753
CI*	±1582	±401	±473	±575	±646	±95	±440	±781	±1287

CI*: 85% confidence interval for 100 model replications.

4 Results and Discussion

After the MSM validation, model was executed for different H_T values and the number of passengers at each station before the train boarding N_P^i was recorded for the averages of 100 model replications. The variation of N_P^i with the train trips N_T were shown in Fig. 4, Fig. 5 and Fig. 6 for different H_T values.

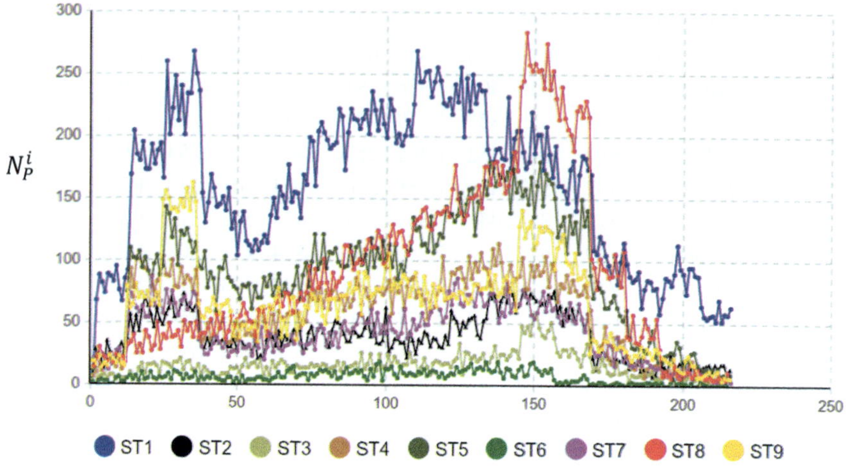

Fig. 4. Variation of the N_P^i with the train trips (H_T = 6 min.).

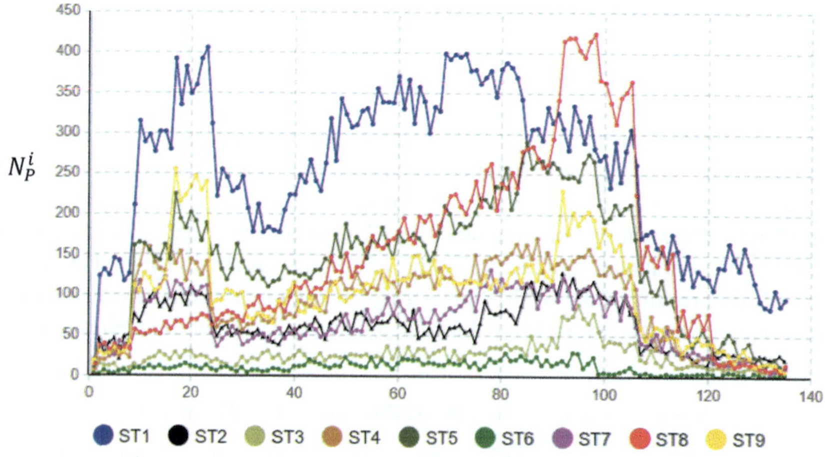

Fig. 5. Variation of the N_P^i with the train trips (H_T = 8 min.).

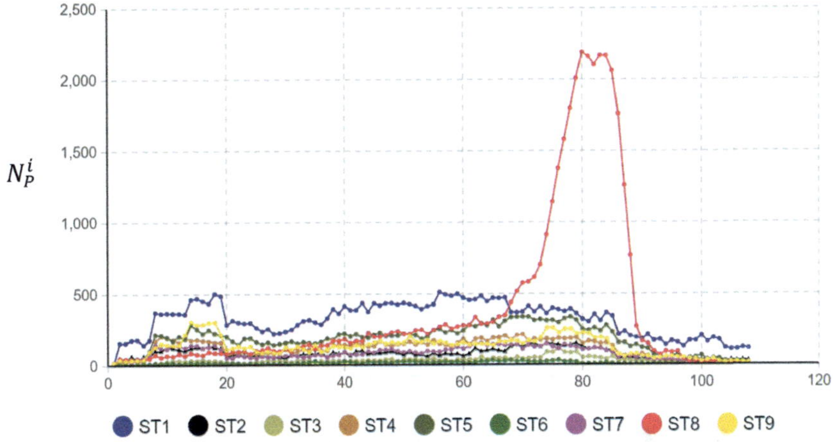

Fig. 6. Variation of the N_P^i with the train trips (H_T = 10 min.).

From the figures, it can be said that, with the morning and evening rush hours, N_P values were increased for all stations. The most extreme increased were observed for ST1 and ST8 stations. The congestion was also significant for the ST8 station since, N_P value continuously increased up to 285 passengers at the evening rush hours according to the Fig. 4. The station congestion was more significant for the Fig. 6 where extreme passenger numbers were observed for ST8 (up to 2,218 passengers) that nearly increased by 7.2 times of the average of the remaining stations. The congestion at the ST8 was also accompanied by the emergence of failed passenger boarding. The station

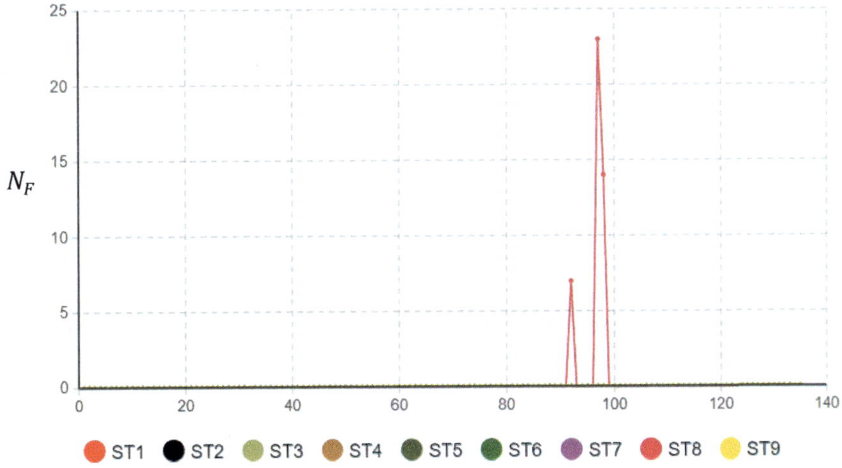

Fig. 7. Variation of NR with the (H_T = 8 min).

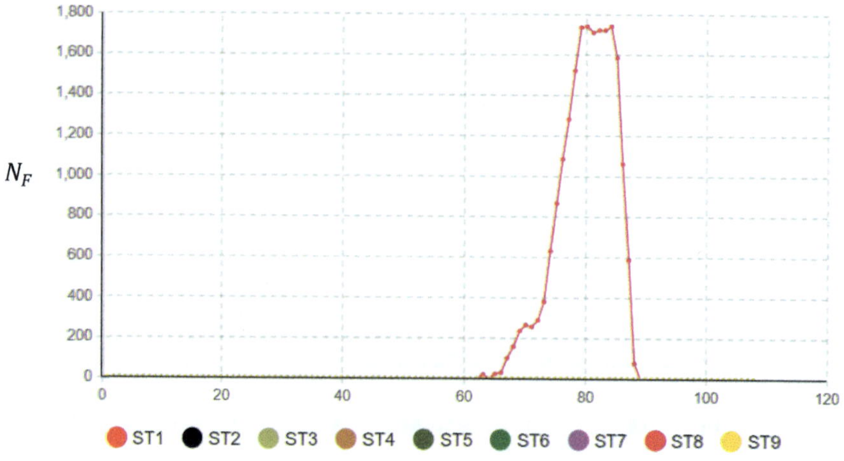

Fig. 8. Variation of NR with the (H_T = 10 min).

congestion was also investigated with using the number of failed passenger boarding (N_f) for all stations with respect to completed train trips. The variation of N_f values were shown in Fig. 7 and Fig. 8 for $H_T = 8$ min and $H_T = 10$ min since for $H_T = 5$ min, no failed passenger boarding was observed for all stations.

The figures indicated that, with the increasing train interarrivals, the number of failed boarding at the ST8 was increased up to 23 passengers for H_T = 8 min. resulted from the insufficient train capacity. This resulted that, the passengers waiting at the ST8 couldn't board the arrived trains. This congestion was extreme for the less frequent train service scenario of H_T = 10 min. since there were vast numbers of passengers failed to board the train at the evening rush hour. This phenomenon was unfortunately resulted from the O-D probability matrix since a significant number of passengers destination was ST9 and the trains are traveled with a full capacity. This resulted that, the intermediate ST8 station could not be serviced and there were heavy congestion problems for this specific station. The second part of the study includes the determination of the optimum train headway for decreasing the number of failed boarding attempts for all stations. This study includes the separation of the train operational hours into sub-operation hours with different train time headways for decreasing the station congestion and minimizing the number of failed passenger boarding. 6 times intervals were considered with 3-h length and the model was executed for 100 replications for observing the averages of the N_f values with associated confidence intervals. The simulation models were executed for different headway values and total N_f values were shown in Table 5.

Table 5. Variation of N_f with operation time interval hours and H_t.

H_t(min)	Operation Time Intervals (hours)					
	6:00/9:00	9:00/12:00	12:00/15:00	15:00/18:00	18:00/21:00	21:00/24:00
5	0	0	0	0	0	0
6	21.4 (±3.1) *	0	0	76.7 (±3.4)	0	0
7	124.1 (±13.4)	0	0	418.2 (±21.4)	0	0
8	459.7 (±28.6)	0	21.2 (±2.7)	2654.4 (±65.7)	54.2 (±6.4)	0
9	1355.6 (±68.2)	0	69.6 (±7.1)	4554.1 (±157.1)	224.7 (±22.7)	0
10	4515.4(±102.1)	16.7 (±2.6)	5.8 (±0.3)	12584.9 (±418.5)	987.4 (±57.6)	0

* average of 100 model replication with 85% of confidence interval

Table 5 indicates that, train interarrival times can be adjusted for minimizing the total trip numbers and also decreasing the number of failed passenger boarding. The optimum train interarrival times for a single operation day was found as 5, 9, 7, 5, 7 and 10 min representing the case with no failed passenger boarding. Additionally, it can be suggested that, the train intervals can also furtherly increase between 9pm and 12 pm since the more relaxed time headway can also be utilized while keeping no failed passenger boarding.

5 Conclusion and Suggestion

In this study, a discrete-event microsimulation model was constructed and executed for investigating the station congestion problem for a specific metro service line. The model was constructed with using Arena simulation software with several model blocks, model was validated and executed for an example passenger arrival scenario and O-D probability matrix. The study firstly concluded that, microsimulation modelling can be used with the passenger travel and OD probability matrix data for simulating a public metro line and the system bottle-necks can be traced triggering the station congestion and trip delays. The conclusion of the study are as follows.

- The results indicated that, with increasing train headways, the number of waiting passengers at the stations significantly increased for the ST8 station. This localized bottleneck was resulted from the passenger trip behaviors.
- OD probability matrix is an important factor for triggering congestion at intermediate stations. Especially anon-homogeneous OD probability matrix results excessive train congestion at several intermediate station with lower passenger demand hence, passenger of this particular station is not able to board the overcrowded train.
- For decreasing the failed boarding, the optimized train headways can be used. The model execution with the scenario data indicated that, $H_t = 5$ can be used for passenger rush hours (6:00/9:00 and 15:00/18:00) for 3 h operation intervals.

In the light of the results of this study, future studies are suggested for decreasing the effects of the non-homogeneous O-D probability matrix and passenger trip characteristics with scheduling additional intermediate train trips or routing the passenger for an alternative mode of transportation. Future study is also planned to be extended to a real-life scenario and the model can be extended for simulating the bidirectional metro line considering the passenger comfort index and perception toward the train crowdedness. Moreover, the model can be coupled with an extensive optimization study for approximating the most feasible train headways while minimizing the total realized train trips.

References

1. Motraghi, A., Marinov, M.V.: Analysis of urban freight by rail using event based simulation. Simul. Model. Pract. Theory **25**, 73–89 (2012)
2. Yıldırım, M.S., Karaşahin, M., Gökkuş, Ü.: Scheduling of the shuttle freight train services for dry ports using multimethod simulation-optimization approach. Int. J. Civil Eng. **19**(1), 67–83 (2021)
3. Michal, G., Huynh, N., Shukla, N., Munoz, A., Barthelemy, J.: RailNet: a simulation model for operational planning of rail freight. Transp. Res. Procedia **25**, 461–473 (2017)
4. Singhania, V., Marinov, M.: An event-based simulation model for analysing the utilization levels of a railway line in urban area. PROMET-Traffic Transp. **29**(5), 521–528 (2017)
5. Altazin, E., Dauzère-Pérès, S., Ramond, F., Tréfond, S.: Rescheduling through stop-skipping in dense railway systems. Transp. Res. Part C Emerg. Technol. **79**, 73–84 (2017)
6. Potti, P., Marinov, M., Sweeney, E.: A simulation study on the potential of moving urban freight by a cross-city railway line. Sustainability **11**(21), 60–88 (2019)
7. Högdahl, J., Bohlin, M., Fröidh, O.: A combined simulation-optimization approach for minimizing travel time and delays in railway timetables. Transp. Res. Part B Methodological **126**, 192–212 (2019)
8. Bešinović, N., Quaglietta, E., Goverde, R.M.: A simulation-based optimization approach for the calibration of dynamic train speed profiles. J. Rail Transp. Plan. Manage. **3**(4), 126–136 (2013)
9. Pouryousef, H., Lautala, P., White, T.: Railroad capacity tools and methodologies in the US and Europe. J. Mod. Transp. **23**(1), 30–42 (2015)
10. Wang, Y., D'Ariano, A., Yin, J., Meng, L., Tang, T., Ning, B.: Passenger demand oriented train scheduling and rolling stock circulation planning for an urban rail transit line. Transp. Res. Part B Methodological **118**, 193–227 (2018)
11. Yin, J., Yang, L., Tang, T., Gao, Z., Ran, B.: Dynamic passenger demand oriented metro train scheduling with energy-efficiency and waiting time minimization: mixed-integer linear programming approaches. Transp. Res. Part B Methodological **97**, 182–213 (2017)
12. Corman, F., D'Ariano, A., Marra, A.D., Pacciarelli, D., Samà, M.: Integrating train scheduling and delay management in real-time railway traffic control. Transp. Res. Part E: Logist. Transp. Rev. **105**, 213–239 (2017)
13. Mu, S., Dessouky, M.: Scheduling freight trains traveling on complex networks. Transp. Res. Part B Methodological **45**(7), 1103–1123 (2011)
14. Kroon, L., Maróti, G., Helmrich, M.R., Vromans, M., Dekker, R.: Stochastic improvement of cyclic railway timetables. Transp. Res. Part B Methodological **42**(6), 553–570 (2008)
15. Hassannayebi, E., Zegordi, S.H., Amin-Naseri, M.R., Yaghini, M.: Train timetabling at rapid rail transit lines: a robust multi-objective stochastic programming approach. Oper. Res. Int. J. **17**(2), 435–477 (2016). https://doi.org/10.1007/s12351-016-0232-2

The Potential of Using Micromobility to Connect to Urban Rail in an Integrated Passenger Transport System

Borna Abramović(✉), Kristijan Tuđa, and Denis Šipuš

Faculty of Transport and Traffic Sciences, University of Zagreb, Vukelićeva 4, 10000 Zagreb, Croatia
{borna.abramovic,denis.sipus}@fpz.unizg.hr

Abstract. There is a challenge of adequate planning and functioning the first and last mile problem in urban mobility. When this issue is not solved according to passenger expectations, they will often start to switch from public passenger system to cars. This is a stumbling block in the attempt to achieve a better modal split in favour of public transport. For the last few years, the new emerging technology has been micromobility. Micromobility is a new and feasible solution that encourages people to use an integrated public transport system more often. There is a useful link between micromobility and the rest of the integrated passenger transport system, especially considering a tariff policy. This paper outlines the results of a field research consisting of a survey that aimed to establish the wishes for the potential use of micromobility. The research was carried out in and around the Croatian capital – Zagreb. Research results represent the first step of implementing micromobility in an integrated passenger transport system.

Keywords: Integrated passenger transport · Public transport · Survey · Quality of public transport system · Micromobility · First mile · Last mile

1 Introduction

Integrated passenger transport is a system that merges two or more transport modes into one system. Integrated passenger transport as a system utilises the advantages of each transport mode and at the same time turns the downsides of every mode into the advantages of the other methods of transport [1]. The backbone of the system is the railway because of its ecological, energy, and infrastructural benefits. Other public transport modes are then confluent to the railway, thus utilising their advantages [2].

From the citizen perspective, mobility is one of the most important factors when valuing the quality of life. Therefore, traffic planning faces the task of successfully solving mobility challenges, ensuring the highest savings for the user [3]. The points of integration, that is, points of changing transport modes, have a substantial impact on the success of an integrated passenger transport system [4]. Also, the first-mile and last-mile represent an important issue to be solved in the transport process, especially when

O. Prentkovskis et al. (Eds.): TRANSBALTICA 2021, LNITI, pp. 705–716, 2022.
https://doi.org/10.1007/978-3-030-94774-3_68

passengers use an integrated passenger transport system. Different approaches for the first-mile and last-mile must promote sustainable, efficient, and socially appropriate solutions [5].

Even though today's mobility relies heavily on cars, the public transport system offers some feasible and better solutions. The critical point of the journey in the public transport system is the first-mile and last-mile. There is a necessity for conducting surveys and, afterwards, detailed analyses with solutions that are most feasible for most users [6]. One of the methods to identify urban public transport users' needs and requests and improve service quality is Quality Function Deployment (QFD) [7]. Also, very important aspects of public transport accessibility become an important attribute of modern society's social and economic activity. The shift to the sustainable paradigm of current cities' development has made public transport the basic mean for population travels execution [8]. Important aspects of micromobility are integration with other transport modes with a focus on public transport. Urban mobility strives for a more sustainable transport system. Therefore, the introduction of micromobility could potentially accelerate this transition. One of the most positive changes could improve accessibility and lead to potential modal shifts away from cars. Of course, these changes need to be incorporated into the legal framework [9]. These reasons push a micromobility in the centre of planning and implementing a user-oriented integrated passenger transport system. This paper aims to conduct a survey and analyse the current situation and propose an optimal solution for first-mile and last-mile challenges in the integrated passenger transport system.

2 Micromobility

Micromobility is an undefined term, or nowadays broadly defined term, associated with the rapidly evolving "light" vehicles that are increasingly appearing on the streets around the world. Micromobility was popularised by Horace Dediu, an American industry analyst and investor. It is related to bicycle, scooter, and moped sharing services. According to Dediu, the term "micro" refers to the use of vehicles that are usually lighter than 500 kg and offer short-distance travel that can be fun, cheap, and enjoyable [10].

The research team Abduljabbar, Liyanage and Dia [11] reviewed the literature on micromobility using an established systematic literature review approach. This paper analyses 328 publications using a software tool that helps identify the changing scientific landscape of the literature by analysing and visualising its bibliometric networks. The results of the paper show a consistent increase in recent research covering the sustainability aspects of micromobility. In particular, research reflects its importance as a low-carbon and transformative mode of urban transport. The co-citation analysis helped to categorise the literature into four main research themes covering: (1) benefits, (2) technology, (3) policy, and (4) behavioural mode choice.

Policy recommendations for the introduction of micromobility devices have been detailed explain in research [12]. They introduced four pillars: (1) strengthening infrastructure and policy capacity, (2) inclusive, participatory policy making, (3) government accountability to resolve short-term and long-term impacts, and (4) adoption of different governance strategies. They also point out that one of the most important actions is to address key risks in the areas of safety, liability, privacy, and cybersecurity.

There is also a definition provided by the International Transport Forum (ITF), which is politically autonomous and administratively integrated within the Organisation for Economic Co-operation and Development (OECD). Micromobility is personal transport using devices and vehicles weighing up to 350 kg and whose power supply, if any, is gradually reduced and cut off at a given speed limit which is no higher than 45 km/h. Micromobility includes the use of exclusively human-powered vehicles, such as bicycles, skates, skateboards, and kick-scooters [13].

The European Union relatively quickly recognised the potential for developing the micromobility market by adopting Regulation (EU) No 168/2013 [14]. It introduced a new vehicle type category L, which comprises powered two-, three- and four-wheel vehicles. Those vehicles include powered cycles, two- and three-wheel mopeds, two and three-wheel motorcycles, motorcycles with side-cars, light and heavy on-road quads, and light and heavy quadri-mobiles.

Certain types of micro-vehicles may be classified in category L1e vehicle (light two-wheel powered vehicle), subcategorised into:

- L1e-A vehicle (powered cycle) and
- L1e-B vehicle (two-wheel moped).

L1e-A category includes electric bicycles with an auxiliary drive that develop a maximum speed of up to 25 km/h and electric power between 250–1,000 W. L1e-B category includes any two-wheeled vehicle with a maximum speed of more than 25 km/h and less than 45 km/h, whose power does not exceed 4,000 W. This category also includes pedelec bicycles with gears. However, most pedelecs with gears have a power of 500–750 W.

Other micro-vehicles that do not fall into the L1e category are:

- human-powered vehicles, such as bicycles, skates, and kick scooters,
- pedelecs, defined as bicycles with pedal assistance of speeds up to 25 km/h and with an auxiliary electric motor having a maximum continuous rated power of up to 250 watts, and
- self-balancing vehicles and vehicles not equipped with a seat (i.e., standing scooters).

Other countries in the world want standardised vehicles used in micromobility, but different backgrounds and views for the future are complicated. Nevertheless, ITF OECD has made an advantageous proposal for micromobility definition and classification. For classification purposes, they used two variables: (1) maximum speed and (2) weight. This approach is very useful because micromobility vehicles are polymorphic devices that do not share a common form factor. According to traction energy and maximum speed, two main groups are (A) unpowered or powered up to 25 km/h and (B) powered with a maximum speed between 25–45 km/h. Each group is divided according to the maximum allowed weight: (1) <35 kg and (2) 35–350 kg. A detailed scheme of classification is outlined in Table 1.

Table 1. Proposed micromobility definition and classification according to ITF OECD.

Type A	Type B	Type C	Type D
Unpowered or powered (<25 km/h)		Powered with a maximum speed (25–45 km/h)	
< 35 kg	35–350 kg	< 35 kg	35–350 kg
Bicycle	Mobility scooter	Powered two-wheeler	Moped
E-bike		E-bike	
E-scooter		Moped	
Electric unicycle		Electric unicycle	
One wheel			

Different micromobility solutions can very successfully solve the first-mile and last-mile problems. These innovative solutions for a relatively wide range of users can be crucial to divert users from personal cars to public transport. Therefore, the standardisation of different micromobility vehicles and suitable legal solutions for their inclusion in everyday traffic would be practically unified. In some countries and regions, there is a tendency to ban the use of micromobility vehicles. Still, such moves are short-sighted and contribute to the even worse use of integrated public passenger transport.

In our research, we shall concentrate on e-scooters. For the last few years, e-scooters have become the most developing vehicles in the frame of micromobility.

3 E-Scooters

An e-scooter is a stand-up or seated scooter propelled by an electric motor. There are two ways to use an e-scooter; the first way is to own one. The price of an e-scooter is reasonable, making it accessible to finance new vehicles. Some users experience important issues of using them in everyday commute. One of the disadvantages is the need for parking, a charging station, and time-to-time maintenance. Thus, disadvantages have been recognised in the business model called shared mobility.

Sharing economy represents a new business model in which individuals now experience a twofold function: they are both consumers and act as service providers. This non-traditional way of doing business has its advantages, such as affordable prices and flexible working hours. However, it brings quite a few disadvantages as well, such as unstable income, and less protection and benefits for workers. Sharing economy is already present in many sectors: accommodation, transportation, household services and professional and financial services. The total market value of sharing economy is estimated to be 250 billion US dollars, with estimations that it may grow to 335 billion euros by 2025 [15].

Today, many companies deal with e-scooters. At the world level, the following are currently the most popular: (1) Lime, (2) Bird, (3) Spin, and (4) Tier. In Croatia, there are currently two companies: (1) Bolt and (2) Dash City. We can easily classify them into two groups: (1) docking system and (2) free floating system.

When describing these two possibilities of sharing an e-scooter, it should be emphasised that neither is wrong. Companies most often opt for one of the methods after an agreement with the local community. However, it is a relatively new means of transport that aims to fit and complement the existing transport network. Both methods have their advantages and disadvantages which are shown in Table 2.

Table 2. Advantages and disadvantages of E-scooter with different stations configurations.

Possibilities	Docking system	Free floating system
Flexibility	–	+
Charging workforce	+	–
Initial investment	–	+
Security against theft	+	–
Marketing space (opportunity)	+	–
Creating supply/demand	–	+

4 Survey Methodology and Analysis

A six-step research methodology was established: (1) purpose of survey, (2) identification of target population and area, (3) selection of survey method, (4) survey questionnaire, (5) analysis of results, and (6) possible solution.

The purpose of the survey was to introduce e-scooters as an alternative means of transport for the first-mile and last mile within the suburban railway line in the city of Zagreb. Zagreb is the capital and the largest city of the Republic of Croatia (European Union). It is in north-western Croatia along the Sava River and lies on the southern slopes of Mount Medvednica. In the last official census of 2011, the city's population was 790,017 and on the surface area of 641,355 km^2. In 2018, the GDP per capita was 22,695 €, which is 85.0% higher than the average Croatian GDP per capita [16].

The target population is daily commuters. Due to the main objective of the research, we decided to conduct the survey in an online form, using Google Forms. The survey was distributed exclusively on Facebook, specifically to commuter groups. We chose this channel to target respondents because this group of users is almost ready to use alternative solutions for the first and last mile as micromobility, especially e-scooters. The title of the survey was "Electric Scooter Survey" and had a total of 12 questions. The survey remained open for one week. All questions were single choice. The average duration of the survey was 2 min, and the average response time was 20 s per question. A total of 249 respondents completed the survey in full. Figure 1 shows the block diagram of the methodology used.

There were four demographic questions – on sex, age structure, average wage level, and profession. Of the total number of respondents, the number of female respondents was 62.7%, while 37.3% were male. In the population, 52.9% are women and 47.1% are men, which shows that slightly more women participated in the survey than in relation to the population.

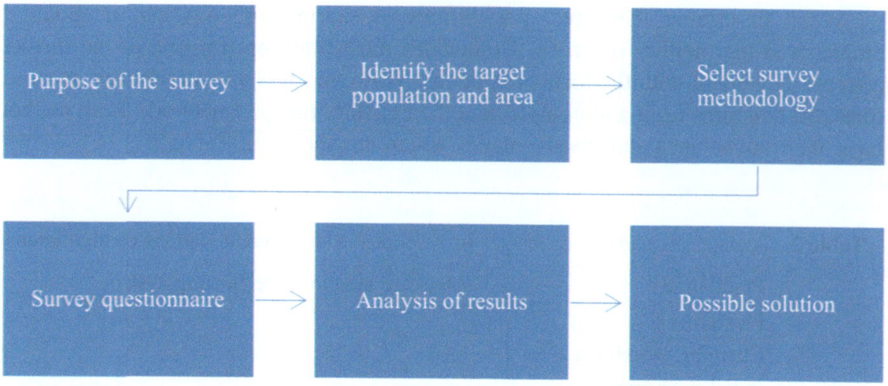

Fig. 1. Block diagram of survey methodology.

The question on age structure was a single-choice scale consisting of the following items: less than 18 years, between 18 and 24, between 25 and 34, between 35 and 44, between 45 and 54, between 55 and 64, and more than 65 years of age. Interestingly, there were no respondents younger than 18, 72 respondents (28.9%) were aged 18–24, 75 respondents (30.1%) were aged 25–34, 62 respondents (24.9%) 35–44, 30 (12.0%) aged 45–54, 8 respondents (3.1%) aged 55–64, and only 2 (1.0%) were more than 65 years old. The working contingent participated fully in the survey, which was expected since they are commuters.

The question on average wage level was a single choice on a scale of less than 1,000 kunas (around 133 €), between 1,000 and 5,500 kunas (133 €–733 €), between 5,500 and 7,500 kunas (733 €–1,000 €), and more than 7,500 kunas (1,000 €). Two respondents (0.8%) earned less than 1,000 kunas (around 133 €), 114 respondents (45.8%) earned between 1,000 and 5,500 kunas (133 €–733 €), 84 respondents (33.7%) between 5,500 and 7,500 kunas (733 €–1,000 €) and 49 (19.7%) more than 7,500 kunas (1,000 €). The average net salary was 7,956 kunas (1,061 €), but the median salary was 7,100 kunas (947 €), suggesting that the survey is about commuters who wish to reduce transport costs to/from work.

The question on the profession was a single choice with offered responses: (1) high school students, (2) university students, (3) working person, (4) unemployed person, and (5) pensioners. Five (2.1%) respondents were high school students, 62 (24.7%) respondents were university students, 164 (66.0%) respondents were employed, 15 (6.1%) respondents were unemployed persons and only 3 (1.1%) were pensioners. The structure of the questions on occupation is completely consistent with the population, which confirms that the relevant groups of people participated in the survey.

The second part of the survey was concentrated on different aspects of habits in the mobility process. This part had a total of 8 questions. The goal was to the determine the potential for introducing E-scooters as an alternative means of transport for the first-mile and last-mile.

We have divided the City of Zagreb into four geographical regions: (1) city centre, (2) western part of the city, (3) eastern part of the city, and (4) the suburbs (outside of the city). According to the results, 18 (7.2%) respondents lived in the city centre, 23 (9.2%) in the western part of the city, 177 (71.1%) in the eastern part of the city, and 31 (12.5%) lived in the suburbs. Given that most respondents (71.1%) came from the eastern part of the city, it can be concluded that in this part, these respondents are most interested in the possible use of e-scooters.

When then ask the respondents: "How often do you use rail for your commuting?". There were 22 (8.9%) everyday commuters, 12 commuted a few times in a month (4.8%), 66 a few times in year (26.5%), and 149 (59.8%) said that they do not use rail for commuting. We then asked this largest group (who do not use rail) to list the main reasons why they do not use the rail for commuting. This question was a single choice with the following choices: (1) great walking distance from the origin (home) for nearest rail stop, (2) great walking distance from the destination (work, school, university, etc.) to the nearest rail stop, (3) high price of rail ticket than tram/bus, and (4) poor traffic connection to the nearest stop. Fifty-three (21.3%) respondents have the much walking distance from their origin to the nearest rail stop, 44 (17.7%) respondents have the much walking distance to their destination to the nearest rail stop, 29 (11.6%) said that ticket price is too high in comparison to tram/bus ticket, and 123 (49.4%) claimed that poor traffic connection makes it impossible for them to use rail stop.

Interestingly, for 39.0% of respondents, the walking distance is too great from/to origin/destination, and another 49.4% have poor traffic connection to reach nearest rail stop. This means that a total of 88.4% cannot use the railway because of practical reasons. The first group or walking group included respondents who are most interested in the possible use of e-scooters.

An interesting fact for using the rail is the distance from origin/destination to nearest rail stop. We divided the distance category into four intervals: (1) under 500 m, (2) 500–1,000, (3) 1,000–1,500, and more of 5,000 m. The first interval can be considered a strictly walking interval. The second one is semi-walking semi-other transport modes. The third one is strictly for using different transport modes, including the micromobility options. The fourth one is strictly for using public transport (such as bus, tram, etc.) and car. The exact distribution of distance is listed in Fig. 2. Also, we can conclude that respondents from the last group (>5,000 m) will not try to use micromobility, that is, e-scooters. Other groups tend to use micromobility, namely, e-scooters, so the focus for future proposals of the system must keep in mind thus persons.

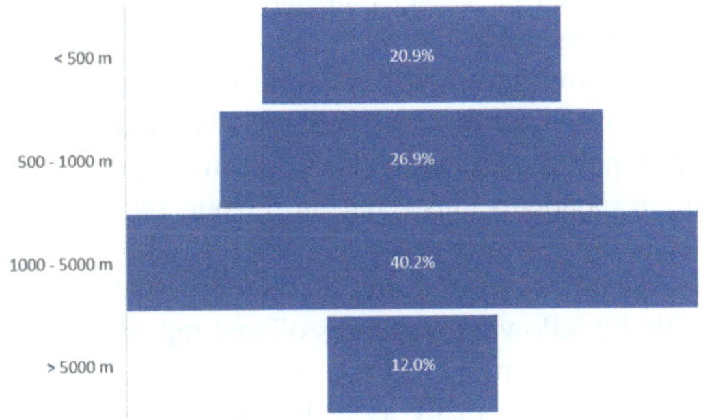

Fig. 2. Distance to nearest rail stop.

The next question was: "Which mode of transport do you use to arrive at the rail stop?". It was a single-choice question with the following choices offered: (1) bus, (2) tram, (3) personal car, (4) bike, and (5) taxi. The exact distribution of transport modes that users use to come to the rail stop is illustrated in Fig. 3. Respondents that said that they use bikes are unlikely to switch to e-scooters. The most challenging category for are respondents that use a personal car, since they use cars as a symbol of freedom.

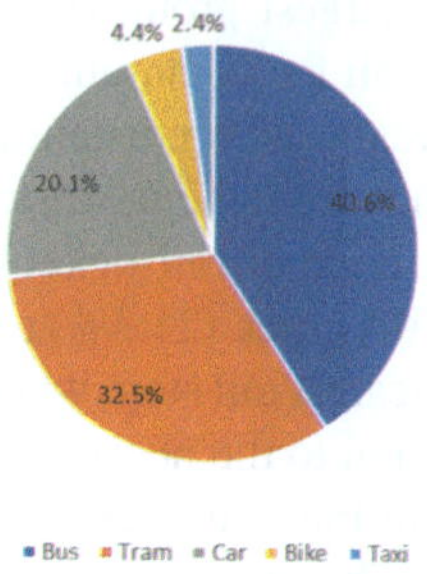

Fig. 3. Mode type used to reach nearest rail stop.

A critical aspect of using the rail for everyday commuting is the possibility of using public transport at to/from rail stop. We investigated how often commuters lose their connection because of the uncoordinated timetables of rail and public transport. In the City of Zagreb, the public transport company is Zagrebački električni tramvaj (ZET). This question was a single choice with the following choices offered: (1) every day, (2) more times per week, (3) once per week, (4) monthly, and (5) never. For 24.9% of respondents, the problem of uncoordinated timetables is a daily struggle, 23.3% of respondents claimed it happened more times per week, 6.8% of respondents encountered uncoordinated timetables once per week, 11.3% of respondents uncoordinated timetables happened monthly, and 33.7% of respondents uncoordinated timetables have never happened. In practice, coordinating timetables is quite an exhausting process, especially when there is no central authority. Therefore, this is an opportunity to ensure coordination and improve the quality of service. Furthermore, there is a window of opportunity to engage micromobility vehicles. Timetable coordination is one of the milestones for the integrated passenger transport system. Figure 4 shows a radar diagram for uncoordinated timetables between rail and public transport.

In the survey, the above-given answers were related to the question on the willingness to use E-scooters in case that they help to avoid traffic jams, waiting for transfer between different transport modes, and today's essential aspects that are a sustainability eco solution. The respondents answered in favour of using e-scooters with an incredibly high percentage of 80.3 percent.

Of course, the crucial aspect of using e-scooters is the price. This was a single-choice question with the following responses offered regarding price intervals: (1) less than 2 kunas per km (0.27 €/km), (2) 2 kunas per km (0.27 €/km), and (3) more than 2 kunas per km (0.27 €/km). As many as 63.9% of the respondents stated they were

Fig. 4. Uncoordinated timetables.

willing to pay the e-scooters if it was less than 2 kunas per km (0.27 €/km), 30.9% of respondents expressed their willingness to pay 2 kunas per km (0.27 €/km), and only 5.2% respondents would pay the e-scooters more than 2 kunas per km (0.27 €/km). All in all, two thirds of the survey participants would pay less than 2 kunas.

5 A Possible Solution for Introducing E-Scooters

Introducing e-scooters in the transport offer can go in the two directions: (1) as a stand-alone independent system and (2) as part of the integrated passenger transport system.

The former is a fully stand-alone independent system. A user registers with personal data on the e-scooters app. After this process, or better said, by activating an account, users can start activating, driving, and deactivating E-scoters. At the end of the ride, depending on the used service, the payment is automatically made through a bank account, and the confirmation of payment is received by e-mail. If the user wants to use another mode of transport on the same trip, he/she buys a ticket separately.

In the latter system, e-scooters are a part of the integrated passenger transport system. The method of using e-scooters is the same as in the first system, but the method of charging for the ride is different. The user buys one ticket (daily/weekly/monthly/annual), which is valid for the whole system.

Another important decision to be made is which system to use for e-scooters: (1) a docking system or (2) a free-floating system. Regardless of the chosen system, each rail stop within its boundaries should provide parking for e-scooters, making it a place of integration of different transport modes. The user can from that parking lot use the e-scooter service within the 24/7/365 sharing model. The ride can be allowed in a specific gravitational area. Therefore, based on the period of use, we can easily determine passenger flows, i.e., demand, and therefore increase/decrease the fleet of e-scooters, i.e. supply in a given area. The option with a free-floating system would be more optimal due to the expensive construction of the lot and because the service, in that case, is more flexible. The disadvantage of such a system is battery-charging because, in the docking system, the battery is charged automatically once the user leaves an E-scooter at the dock. It is crucial to monitor the battery level of individual e-scooter as this ensures service reliability. Although the standard charge for driving an e-scooter is typically expressed in the amount of money per minutes, we proposed that the charge be made in the amount of money per km. The reason for this is user safety –users should drive faster to pay a lower price.

6 Conclusion

For the integrated passenger transport system to function properly, harmonised timetables in different modes of transport are needed. This is often impossible to achieve, especially since different modes of transport are involved in the system. Because of that, there is a niche for micromobility. In our research, we focused on the advantages and disadvantages of micromobility as a part of an integrated passenger transport system with an emphasis on e-scooters. We followed this up by conducting a survey in the city of Zagreb. Finally, we proposed a possible solution for e-scooters.

The advantages of railways such as (1) speed, (2) mass capacity, and (3) infrastructure (that it does not share with other modes of transport) must come to the fore. However, for the railway to increase the number of users, it must improve the connection from/to the station/stop. In other words, it must establish cooperation with other transport modes that also include micromobility.

Micromobility is an emerging technology that is entering quite successful to the transport offer. There are two different systems: (1) docking system and (2) free floating system. Each has some advantages and disadvantages, and the system needs to be picked depending on local conditions. There is also a different approach to how to include micromobility in the system. The approaches are (1) a stand-alone independent system and (2) part of the integrated passenger transport system. For sure, sustainable transport system and practical usage from users side the integration within an integrated passenger transport system is the most reasonable solution. Although the standard charge for driving an E-scooter is typically expressed in the amount of money per min, we proposed that the charge be made in the amount of money per km. The reason for this is increased user safety. Today we need mobility to be flexible and available for all types of users. And of course, in the first-mile and last-mile problem, the cost and time are very important factors driving usage of different transport modes.

For the transport service to be considered successful, it must meet the following criteria: (1) speed, (2) safety, (3) accuracy, (4) reliability, and (5) economy. A sure combination of rail and E-scooter can undoubtedly achieve this all criteria without any problems.

In our research, we looked at the advantages and disadvantages of micromobility, focusing on e-scooters. Then we did a survey, which we conducted in the city of Zagreb. After that, we proposed a possible solution for e-scooters. Of all our respondents, 90.7% students and workers participated in our survey, representing daily commuters. Interestingly, for 39.0% of the respondents, the walk from/to the origin/destination to the nearest rail stop was too long, and another 49.4% have poor transport accessibility to the nearest rail stop. This means that a total of 88.4% cannot use the rail stops for practical reasons. The first group or the walking group are the respondents most interested in the use of e-scooters. Respondents with a rail stop between 500–1,000 m away amounted to 26.9% and most of this group is inclined to use e-scooters. But interestingly, as many as 40.2% of the respondents are within the distance of 1,000–5,000 m. This group has a significant potential to switch to e-scooters. Regarding the desire to pay for the service, 63.9% of the respondents indicated that they would pay less than 2 kunas per km (0.27 €/km).

After the introduction of the e-scooter system, future research should examine the satisfaction with the new service. Specifically in this kind of research, the focus should be on user groups and the willingness to pay for the service. According to this research, the service level should be improved and satisfaction surveys should be conducted cyclically, every 3–6 months. Furthermore, one of the directions for future research is to include the e-scooter service in the tariff systems of integrated passenger transport, so that all methods of public transport would be united in one system that would be financially acceptable to its users.

References

1. Stopka, O., Bartuška, L., Kampf, R.: Passengers evaluation of the integrated transport systems. Naše More **62**(SI), 153–157 (2015)
2. Abramović, B., Šipuš, D., Jurešić., D.: Organisation of Integrated Passenger Transport on the Zagreb – Velika Gorica Route. Transp. Res. Procedia **44**, 342–347 (2020)
3. Šipuš, D., Abramović, B.: Tariffing in integrated passenger transport systems: a literature review. PROMET – Traffic Transp. **30**(6), 745–751 (2018)
4. D'Orso, G., Migliore, M.: A GIS-based methodology to estimate the potential demand of an integrated transport system. In: Gervasi, O., et al. (eds.) ICCSA 2017. LNCS, vol. 10407, pp. 525–540. Springer, Cham (2017). https://doi.org/10.1007/978-3-319-62401-3_38
5. Bruzzone, F., Cavallaro, F., Nocera, S.: The integration of passenger and freight transport for first-last mile operations. Transp. Policy **100**, 31–48 (2021)
6. Tuđa, K.: Alternative Transport Means to Support Railway within the Integrated Passenger Transport System. University of Zagreb, Faculty of Transport and Traffic Sciences (2020)
7. Bajčetić, S., Tica, S., Živanović, P., Milovanović, B., Đorojević, A.: Analysis of public transport users' satisfaction using quality function deployment: Belgrade case study. Transport **33**(3), 609–618 (2018)
8. Rossolov, A., Naumov, V., Popova, N., Vakulenko, E., Levchenko, O.: Estimation of transport accessibility in case of rational transport hub location. Transport **36**(1), 1–12 (2021)
9. Oeschger G., Carroll P., Caulfield B.: Micromobility and public transport integration: the current state of knowledge. Transp. Res. Part D Transp. Environ. **89**(1), 102628 (2020)
10. Horace Dediu. https://micromobility.io/blog/2019/2/23/the-micromobility-definition. Accessed 26 Jul 2021
11. Abduljabbar, R.L., Liyanage, S., Dia, H.: The role of micro-mobility in shaping sustainable cities: a systematic literature review. Transp. Res. Part D Transp. Environ. **92**(1), 102734 (2021)
12. Pande, D., Taeihagh, A.: The governance conundrum of powered micromobility devices: an in-depth case study from Singapore. Sustainability **13**(11), 6202 (2021)
13. ITF OECD. https://www.itf-oecd.org/sites/default/files/docs/safe-micromobility_1.pdf. Accessed 26 Jul 2021
14. Regulation (EU) No 168/2013. https://eur-lex.europa.eu/legal-content/EN/TXT/PDF/?uri=CELEX:32013R0168&from=EN. Accessed 26 Jul 2021

15. Brozović, V., Hrastić, F., Meštrović, L., Bilić, H., Naletina, D., Petljak, K.: Ekonomija dijeljenja: Poslovni model budućnosti? Zbornik Ekonomskog Fakulteta u Zagrebu **17**(1), 33–61 (2019)
16. City of Zagreb. www.zagreb.hr. Accessed 23 Aug 2021

Assessment of Environmental Risks of the Gas Transportation Process by Main Pipelines

Oleg Mandryk[1], Liubomyr Poberezhny[1], Pavlo Maruschak[2(✉)], Liubov Poberezhna[1], Oksana Maniuk[1], and Mykhailo Maniuk[1]

[1] Ivano-Frankivsk National Technical University of Oil and Gas, Ivano-Frankivsk, Ukraine
[2] Ternopil Ivan Puluj National Technical University, Ternopil, Ukraine

Abstract. The problem of ensuring technological reliability and environmental friendliness of the natural gas transportation process is considered along with the main approaches to risk assessment and management at industrial facilities. To increase the trouble-free operation of the gas transmission system, a comprehensive risk management system has been proposed. Investigations into the ecological condition of the ground layer of air are underway in Bogorodchany district. Maps representing the distribution of harmful matter concentrations, coefficients of concentrations, coefficients of ecological danger, and the map of total contaminating index are used in the investigations. General chart of ecological risk is represented here. In particular, the objects for transporting gas are analyzed. The method for calculating the area detonation and the explosive cloud distribution is considered during damage of the main gas pipeline. Area detonating and distance of mixture explosive cloud distribution are calculated at different wind speeds.

Keywords: Gas transportation systems · Main pipelines · Ecological risk

1 Introduction

Ukraine's gas transportation system is the second largest in Europe after Russia. It consists of a network of gas pipelines with a total length of more than 38.5 thousand km (pipe diameter 1,020–1,420 mm), 73 compressor stations and 13 underground gas storages with an active gas volume of 32 billion m. The inlet system capacity is 290 billion m^3 of gas per year, and output - 176 billion m^3 per year.

The pipeline system of Ukraine has received great development in the Ivano-Frankivsk region. In particular, large main gas pipelines "Soyuz", "Progress", "Urengoy - Pomary - Uzhgorod", "Ananyiv - Chernivtsi - Bohorodchany", "Torzhok - Dolyna" and others have been laid and transported gas on the territory of the region. The above-mentioned gas pipelines have a high level of wear, so there is a high risk of accidents. One of the important gas transportation hubs in Prykarpattia is Bohorodchany. Nowadays, the oil and gas industry give great attention to providing the

O. Prentkovskis et al. (Eds.): TRANSBALTICA 2021, LNITI, pp. 717–725, 2022.
https://doi.org/10.1007/978-3-030-94774-3_69

reliability and safety during the construction and operation of main pipelines. Despite this fact, the emergency situations associated with gas transportation by pipelines occur very often [1–3]. In general, such accidents include high material costs for the oil and gas pipeline's operating companies and significant damage to the environment, people, and property in the vicinity of the damaged pipelines. Therefore, decreasing the probability of emergency situations with gas mains is one of the immediate issues in this field. The problem is especially relevant for pipelines, which are located and operated in special climatic conditions. The main factors, which contribute to the emergency occurrence, include using significant amounts of flammable and explosive substances in operating procedure; causing abrasion of equipment and pipelines; operating procedure under pressure; harsh climatic conditions and others.

Pre-accident analysis is an important aspect of ecological risk-management of gas transport. The main causes of gas pipelines accidents are pipeline corrosion (including internal and external), construction defects, mechanical damage (including excavation work), diversions, unauthorized junctions, third party activity, operational imperfection, manufacturing defects of pipes and equipment, natural disasters [4, 5]. During the transportation, the loss of piping integrity may occur, which would lead to release of hazardous substances into environment. The main result of such accident is the transported gas leakage, which creates toxic pollution of the environment. Moreover, as a result of gas leakage evaporation, the cloud of explosive mixture is formed. The presence of ignition source within the cloud may lead to its inflammation and explosion. The highest risk of accidents on pipelines is connected with longitudinal damages. They can occur during the formation of corrosive gas pocket or guillotine ruptures both on the base part of pipe and in the zone of welding joints. Therefore, the loss of piping integrity is considered as the most frequent initiating event, which causes the hazardous substance's release into environment [6, 7]. The loss of piping integrity mode influences the types of leakage. If discharge area is small, relatively longtime outflow through a hole is observed. On the other hand, if the pipeline integrity is damaged significantly, large volumes of hazardous substance go into environment immediately.

2 Methodology

Phenomenological approach is based on determining the possibility of emergencies using the analysis results that represent conditions associated with the natural laws. This approach is the easiest to use and gives reliable results, if operating processes have sufficient margin to limiting levels. However, it is unreliable, when an abrupt change of substances or systems occurs. This approach is preferable when comparing the safety margins of different potentially dangerous objects and is useless in analysis of emergencies, which depends on reliability of different parts of the object or its safety facilities. Figure 1 shows the conceptual frame layout of the dynamic risk assessment.

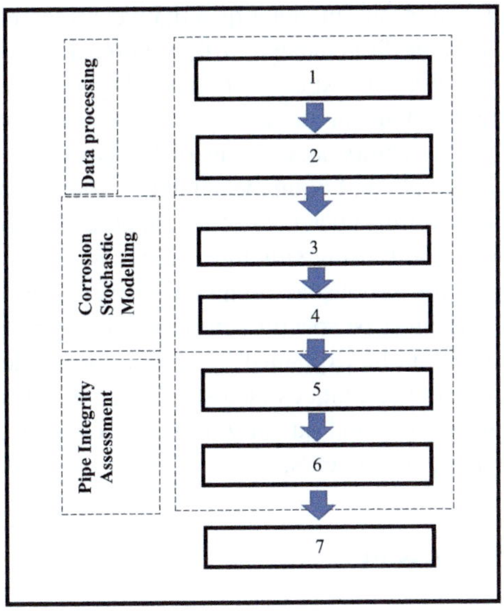

Fig. 1. Conceptual layout of dynamic risk assessment: 1 – data collection; 2 – data tidying process; 3 – degradation process selection; 4 – mixed Lévy process parameters estimation; 5 – mean time to failure; 6 – identification of leak-prone segments; 7 – maintenance support decision-making.

Deterministic approach provides for an analysis of the sequence of the accident development from the initial event through a series of suggested failures. Emergency process is studied and predicted using mathematical modeling, construction of simulation models and complex calculations. Disadvantages of this approach include the gaps between some important events in accident development; difficulty in adequate mathematical models construction; complexity and high cost of experimental studies to validate computer models, Fig. 2.

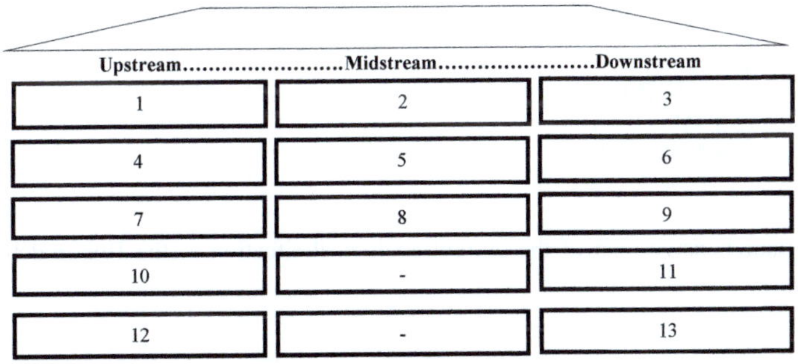

Fig. 2. Predictive analytics steps to be carried out in pipeline system: 1 – asset maintenance; 2 – transportation optimisation; 3 – demand optimisation; 4 – optimisation in exploration; 5 – pipeline risk assessment; 6 – trade optimisation; 7 – production optimisation; 8 – storage optomasation; 9 – price optimisation; 10 – risk assessment; 11 – market analysis; 12 – optimisation in drilling; 13 – marketing effectiveness.

Expert approach is based on obtaining quantitative risk assessments by processing experts' opinions. The main disadvantage of this approach is the necessity to analyze the objectivity and reliability of experts' opinions. Figure 2 describes the predictive analysis sequences to be done in oil and gas pipeline industry. The probabilistic approach involves the probability assessment of the accident. This approach allows analyzing the chain of events and equipment failures and estimating the total probability of accident. The main disadvantage of this approach is the lack of statistics on equipment failure. In addition, using simplified design schemes reduces the reliability of the risk assessment for serious accidents. However, this approach is currently considered as one of the most perspective ones. In this paper, probabilistic approach was used. Its application can be explained by capacity to model all possible pipeline accident scenarios. Different techniques (statistical data, event tree analysis, fault tree analysis, etc.) allow to identify and quantify all scenarios, as well as to determine the protective actions if emergency occurs.

3 Results and Discussion

The analysis of the ecological condition of the surface air layer of Bohorodchany district was carried out on the basis of the constructed maps of distribution of concentrations of harmful substances, concentration coefficients, coefficients of ecological danger and the map of the total pollution indicator, Fig. 3.

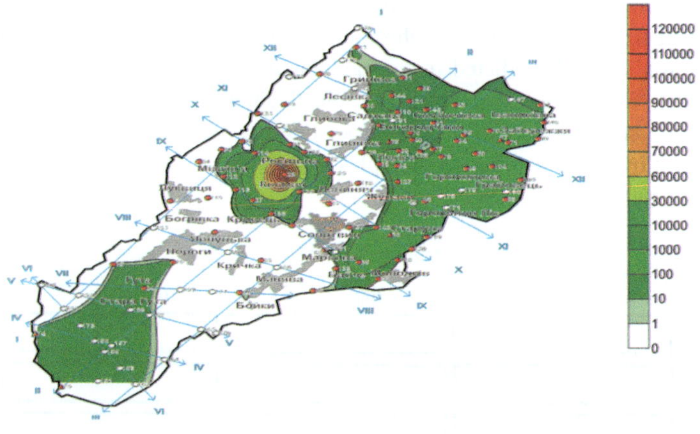

Fig. 3. Comprehensive map of the current ecological situation of the Bohorodchany node (in Ukrainian).

Analyzing the state of atmospheric air in general, we can draw next conclusions:

- the percentage of oxygen in the air of the main gas pipelines decreased, in some places to 20.175%. This decrease is most noticeable in the western part of the study area characterized by the densest network of gas pipelines. The highest oxygen

content in the air was found in the southwestern and northeastern parts, because they are somewhat distant from the pipelines;
- the content of carbon monoxide is the highest in places with the lowest concentration of oxygen. In the same areas, a slightly higher content of carbon dioxide was found, and its highest concentration was in the west near the village of Mezhigorye;
- NOx and SOx have a mutually reinforcing effect;
- the most polluted air is found along the main gas pipelines, especially in the western part of the node, where the densest network of their laying. The boundaries of areas with anomalous values almost coincide, so the snow cover is most polluted in the following places: in the western part near the village of Lovagy, in the south - in the village of Starunya and its environs, in the east – the village of Lypivka. Particularly significant air pollution was found between the villages of Bohorodchany and Burial on the territory of main gas pipelines management.

In the snow cover, we found patterns of distribution of such heavy metals as copper, zinc, lead. Areas with concentrations other than zero were found in rugged terrain in the west in the village of Lovagy - this also explains the pollution of heavy metals at this point - the result of cross-border transfers. In addition to heavy metals found in the snow cover from the air, the distribution of other compounds was analyzed: carbon dioxide, carbon monoxide, oxides of nitrogen and sulfur, dust, gasoline, acetone and oxygen. The areas of distribution of these pollutants coincide with the wind rose for the study area. The main regularity of the research was also revealed - high concentrations of pollutants were found in areas with a dense network of roads, boiler houses, and near natural gas transportation facilities.

It is known that natural gas also has a negative effect on human life, as it has narcotic and toxic effects. As part of the problem we are considering, the influence of the gas pollution level due to the destruction of the main gas pipeline is of some interest. Calculations showed that the maximum amount of air pollution does not exceed 250–300 m. This distance can increase depending on the pipeline diameter and the gas jet flow direction during its depressurization. In almost half of the cases, the gas ignites at the site of damage, then the scale of air pollution decreases.

Therefore, pipeline transport negatively affects the environment of its laying. Therefore, the main task of environmental protection is to reduce the man-made impact during the construction and operation of gas pipelines, and, on the other hand, it is necessary to reduce the negative impact of natural factors on their safety and reliability. To solve this problem, we use the methods of environmental risk analysis [3, 6].

The main factors of environmental risk of accidents on gas pipelines can be divided into two groups - man-made and natural. Man-made defects include pipes, installation errors, non-compliance with the profile and plan of the trench for laying the pipeline, various damages during earthworks, etc. To reduce such factors, it is necessary to strictly adhere to technological discipline and timely conduct control diagnostic measures. Natural factors include corrosion of various types and mechanical action of soils. The intensity of these processes is characterized by regional features of the territory. The most important of them are soil temperature and humidity, which depend on the morphological characteristics of the terrain.

The linear part of gas pipelines is a potentially dangerous object and has a large energy potential that can adversely affect the environment. Therefore, it is important to solve the problem of zoning within the route and forecasting the parameters of dangerous areas for the population and territories [3]. When choosing methods for assessing the risk of accidents on main gas pipelines, it is necessary to pay attention to the cartographic model. In addition to the ability of quickly predicting, simulation allows developing recommendations for reducing the risk and eliminating the consequences of emergencies [6].

Consider the method of calculating the detonation zone and the distance of the cloud of explosive mixture during an accident on the main gas pipeline. It is necessary to take into account that accidents during depressurization of gas pipelines are accompanied by the following processes: gas leakage before shut-off valves, shut-off valves closing, gas leakage from the emergency section of the gas pipeline, which is cut off by valves. In places of damage, there is a gas leak under high pressure into the environment (Fig. 4).

Fig. 4. Liquidation site of the accident that occurred on the Dolyna - Bilche Volytsia gas main.

Due to the depressurization of the gas pipeline, methane rises into the atmosphere, and other gases and their mixtures are deposited in the surface layer, mixing with air gases form a cloud of vybunon-hazardous mixture. According to statistics, almost 80% of accidents are accompanied by fires and explosions. Fire and explosion hazards occur when the concentration of methane in the air exceeds 5% by volume. Explosive combustion during accidents on gas pipelines occurs in one of two modes - deflection or detonation. During operational forecasting, it is assumed that the process develops in the detonation mode [8].

The distance to which the cloud of explosive mixture extends in the wind direction is determined by the empirical formula [8]:

$$L = 25\sqrt{M/V}, \tag{1}$$

where M is the mass second gas consumption, kg/s; 25 is the coefficient of proportionality; V is the wind speed, m/s.

The boundary of the detonation zone with radius r_0, resulting from the gas leakage in the case of pipeline depressurization, is determined by the formula:

$$r_0 = 12.5\sqrt{M/V}. \tag{2}$$

The mass second flow rate of gas M from the pipeline for the critical leakage regime, when its main parameters (flow rate and leakage rate) depend only on the parameters of the depressurized pipeline, is determined by the formula:

$$M = \psi \cdot F \cdot \mu\sqrt{P_r/V_r}, \tag{3}$$

where ψ is the coefficient that takes into account the gas flow rate from the leak state (for the sound leakage rate $\psi = 0.7$); F is the leakage hole area, which is taken to be equal to the cross-sectional area of the pipeline, m^2; μ is the gas consumption factor, which takes into account the hole shape (for calculations it is taken equal to 0.8); Pr is the gas pressure in the pipeline, Pa.

V_r is the specific volume of transported gas, which is determined from the equation:

$$V_r = R_0\frac{T}{P_r}, \tag{4}$$

where T is the temperature of the transported gas, K.

R_0 is the specific gas constant, which is determined by the partial gas content q_k and molar masses of the components of the mixture from the ratio:

$$R_0 = 8314\sum_{t\to 1}^{n} q_k/m_k, \tag{5}$$

where 8314 is the universal gas constant, J/(kmol • K); m_k is the molar mass of components, kg/kmol; n is the number of components.

When predicting the consequences of accidents on the gas pipeline [9], the detonation zone is taken into account along with the wind direction, assuming that the detonation zone boundary extends from the pipeline in the wind direction to a distance of $2r_0$. In the case of early prediction, the detonation zone spreads in the form of bands along the pipeline with a width of $2r_0$ on each side. This is because the cloud of explosive mixture can spread in any direction from the pipeline, depending on the wind direction.

For calculations, the temperature of the° transporting gas is 40 °C. The content of ordinary gas in the absence of data can be taken in the ratio: methane - 90%, ethane - 4%, propane - 2%, H-butane - 2%, isopentane - 2%.

The area of possible spread of the explosive mixture on the topographic maps is indicated in the form of a circle, semicircle or sector. The center of the circle, semicircle or sector coincides with the pipeline depressurization site (chosen arbitrarily):

1. At a predicted wind speed of 0.5 m/s, the detonation zone has the form of a circle with radius ro.
2. At the predicted wind speed of 0.6–1.0 m/s, the detonation zone has the form of a semicircle. The section angle is equal to 180°, the bisector of the angle coincides with the cloud distribution axis and is oriented in the wind direction.
3. At the predicted wind speed of more than 1.0 m/s, the detonation zone has the form of a sector with radius $2r_0$. At a wind speed of 1.1–2.0 m/s the angle of the sector is 90°, and at a speed of more than 2 m/s–45°.

According to the above method, the detonation zone and the range of the explosive mixture cloud at different wind speeds were calculated.

The results obtained:

- at 0.4 m/s: L = 4,325 m; r_0 = 2,162.5 m;
- at 0.8 m/s: L = 3,075 m; r_0 = 1,537.5 m;
- at 3 m/s: L = 1,581; r_0 = 787.5 m.

4 Conclusions

The main reasons for the gas pipeline failure are depreciation and metal corrosion, which are caused primarily by local climate conditions. The ecological damage is the main adverse effect for such accidents.

Using cartographic and model methods to assess the impact of the detonation zone and the range of the explosive mixture cloud during accidents on main gas pipelines allows for predicting and helps developing recommendations for reducing the risk of dangerous situations. This, in turn, will reduce the impact of negative impacts on the environment, infrastructure, as well as prevent human losses.

References

1. Peekema, R.M.: Causes of natural gas pipeline explosive ruptures. J. Pipeline Syst. Eng. Pract. **4**(1), 74–80 (2013)
2. Jo, Y.D., Ahn, B.J.: Analysis of hazard areas associated with high-pressure natural-gas pipelines. J. Loss Prev. Process Ind. **15**(3), 179–188 (2002)

3. Mandryk, O., Vytyaz, O., Poberezhny, L., Mykhailiuk, Y.: Increase of the technogenic and ecological safety of the natural gas transportation due to displacement of explosive mixtures with nitrogen. Arch. Mater. Sci. Eng. **106**(1), 17–27 (2020)
4. Grudz, V., et al.: Optimal gas transport management taking into account reliability factor. Manage. Syst. Prod. Eng. **28**(3) 202–208 (2020)
5. Maruschak, P., Bishchak, R., Prentkovskis, O., Poberezhnyi, L., Danyliuk, I., Garbinčius, G.: Peculiarities of the static and dynamic failure mechanism of long-term exploited gas pipeline steel. Adv. Mech. Eng. **8**(4), 1687814016641565 (2016). https://doi.org/10.1177/1687814016641565
6. Yavorskyi, A.V., Karpash, M.O., Zhovtulia, L.Y., Poberezhny, L.Y., Maruschak, P.O.: Safe operation of engineering structures in the oil and gas industry. J. Nat. Gas Sci. Eng. **46**, 289–295 (2017)
7. Cruz, A.M., Krausmann, E.: Hazardous-materials releases from offshore oil and gas facilities and emergency response following Hurricanes Katrina and Rita. J. Loss Prev. Process Ind. **22** (1), 59–65 (2009)
8. Hoff, A.M.: An experimental study of the ignition of natural gas in a simulated pipeline rupture. Combust. Flame 51–55 (1993)
9. Lytvynenko, I., Maruschak, P., Prentkovskis, O., Sorochak, A.: Modelling kinetics of dynamic crack propagation in a gas mains pipe as cyclic random process. In: Kabashkin, I., Yatskiv, I., Prentkovskis, O. (eds.) RelStat 2017. LNNS, vol. 36, pp. 262–269. Springer, Cham (2018). https://doi.org/10.1007/978-3-319-74454-4_25

Accessibility of Regional Public Transport

Rasa Ušpalytė-Vitkūnienė[1] and Justina Ranceva[2(✉)]

[1] Research Institute of Territorial Planning,
Vilnius Gediminas Technical University, Vilnius, Lithuania
rasa.uspalyte@vilniustech.lt
[2] Road Department, Vilnius Gediminas Technical University, Vilnius, Lithuania
justina.ranceva@vilniustech.lt

Abstract. Passengers living in remote areas should prioritise public transport over private cars, which increase congestion, parking and environmental problems. In order to achieve this goal, public transport must be accessible to every inhabitant of the region. The development of a regional public transport system must provide for balanced access to regional public transport stations/stops and the frequency of public transport routes to satisfy the needs of passengers. This article aims to analyse the serious public transport accessibility situation in Lithuanian regions and to develop a methodology for the setting regions where public transport is completely inefficient and where alternatives to public transport should be explored.

Keywords: Accessibility · Bus stops · Frequency of routes · Regional public transport

1 Introduction

Settlements in rural areas and around cities continue to expand rapidly [1, 2]. Passengers living in remote areas mostly use private cars, increasing congestion, parking, and environmental problems [3]. According to the International Transport Forum forecasts, demand for non-urban travel, which is defined by passenger kilometres, will grow faster than in urban areas [4]. Forecasts appear that for 2015–2050 CO_2 emissions from regional passenger transport will be double [5].

Public transport is an advanced, efficient and sustainable mode of transport [6]. Public passenger transport must be accessible to all interested transport participants, regardless of gender, age, nationality, etc. [7]. Public transport is becoming an important eco-friendly alternative for the country's citizens, those who cannot buy/drive a car [8], whereas most of the population consists of children, adolescents, seniors, the disabled, women on parental leave, etc. [9]. Accordingly, this problem can lead to social exclusion in the outermost settlements. Social exclusion is defined as the process by which people don't have the opportunity to participate in the economic, political and social life of a community due to reduced access to services, lack of mobility in society and the environment [10]. In this case, the rural population begins to decline [9].

Public transport in rural areas is underdeveloped due to the small number of consumers and therefore does not meet the needs of passengers: it is inaccessible,

O. Prentkovskis et al. (Eds.): TRANSBALTICA 2021, LNITI, pp. 726–737, 2022.
https://doi.org/10.1007/978-3-030-94774-3_70

unpredictable, high in price and congested [11]. In this case, the car is the only practical option in remote settlements [12]. According to official calculations of the use of a private car for official purposes, the use of a car is cheaper only for short distances provided that more than one person is transported [9].

In small and very small populated rural areas (<200 people), there are few social services, and for this reason, residents are forced to travel: go to school, see doctors, go shopping, go to cultural events and so on [9]. Therefore, good accessibility of public transport also improves the accessibility of other services [13]. Finally, Johnson et al. [14] propose to consider the accessibility of the public transport system as a vital parameter for employment.

Public transport faces major challenges in rural and suburban areas because low population density makes it difficult to provide high-quality services [15]. The public transport services in rural and suburban areas are the worst due to the prevalence of dispersed travel patterns and limited overall travel demand, making it difficult to provide high-quality public transport services [16]. The availability of public transport is an important indicator of the quality of the passenger public transport system. Public transport is the basis of mobility due to its capacity, economy, spatial rationality and ecological acceptability. Key indicators of public transport services are [17]:

- availability;
- travel time;
- departure frequency;
- price;
- safety.

It is important to mention that public transport can become more attractive by providing door-to-door service [18]. Demand Responsive Transport (DRT) models, widely used worldwide, can help people get where they need to go, responsibly plan daily routes around destinations, and not following inflexible schedules [19]. The Community Transport Association [20] defines DRT as a passenger-oriented public transport service model with flexible routes and smaller capacity vehicles travelling between pick-up and drop-off points according to the needs of passengers. In Wales, this service was launched in 2009 [19]. Two successful examples of DRTs are Bwcabus and Grass Routes. Daily routes and stops are dynamically planned according to user needs, using mapping, routing and GPS technologies. Pre-registration is required, and it is free to call an emergency number. Most passengers are picked up at their local bus stops, but people with reduced mobility or living in very remote areas can be picked up from the door of their homes. The cost of the trip is calculated based on the travelled distance [19]. These days, making a request has become more accessible: the user sends the request with a mobile app indicating their departure (usually the user's current location), place of arrival (destination), and the desired time of arrival. The on-demand public transport system responds by sending the consumer an offer indicating which bus stop he has to walk to and which bus he will pick up, as well as the estimated time of arrival. This system avoids unnecessary empty kilometres [21]. Bwcabus and Grass Routes show that DRT can be successfully implemented in rural areas where regular buses are not viable or where planned routes are not convenient [19]. DRT model allows trips that could not previously be done without a car [22], and going only where needed can help reduce costs and emissions [20, 23].

This article aims to analyse how serious the public transport accessibility situation in Lithuanians regions is and to develop a methodology for the setting regions where public transport is completely inefficient and where alternatives to public transport should be explored (PT on demand, multi-mobility, carpooling, etc.)

2 Accessibility Indicators for Regional Public Transport

Michniak [24] defines regional spatial accessibility as the ease of reaching a region from other cities or regions. All residents must have access to public transport.

It is known from previous studies that accessibility determines the demand for public transport between rural settlements. Accessibility is a very important factor as it promotes mobility and social inclusion. This is the social aspect of mobility [25].

The general characteristics of local travel differ from regional travel, which is dominated by longer distances, lower service frequencies, fewer stops, and so on [26]. The results of the Trembošová et al. [7] study showed that there is a specific limit on travel distance, time, and number of transfers, which makes people choose to travel in their own car.

This chapter examines regional public transport accessibility (distance to stops) and route frequencies.

2.1 Distance to Regional Public Transport Stops (Availability)

The integration of regional public transport networks is an important factor in ensuring the accessibility of the system. Developing a regional public transport system must provide for balanced access to regional public transport stations/stops. According to the relevant regulations, the distance between public transport stations in different areas varies: in the city is dominated a distance of 300–500 m. In contrast, in the suburbs, the walking distance is much longer [27]. Bruzzone et al. [28] examined the difference between the attitudes of urban and peripheral residents towards bus use. Respondents living in suburban areas said that they would like to use less their cars and use bus services, but do not do due to shortcomings in the public transport system. They stressed the need to walk more than 1 km to reach the nearest bus station and complained about the long waiting times at the stations. Respondents also noted low bus frequencies, inconsistencies in bus schedules and opening hours, and inefficient bus routes.

Akbari et al. [29] define accessibility to a regional public transport stop within a radius of 800 m. The distance of 400–800 m can go in about 5–10 min on foot. Many studies follow this walking distance, but other studies show that many passengers do not object to going further for health reasons [26]. For example, 90% of Amsterdam residents travelling on regional public buses on foot walk distances of 1,200–1,500 m [30]. Perth suburban population in Australia trips to stops are even longer, as a large number of passengers walk to stations / stops as much as 2–3 km [31]. The pedestrian speed is generally considered to be 1 m/s, which corresponds to 3.6 km/h [32].

Based on the collected data, Plazinić & Jović [25] found that the acceptable distance to the bus stop in the regions is up to 700 m, and for cyclists from 700 to

2,945 m. Meanwhile, Stentzel et al. [33] indicated that the maximum distance to public transport stops 1,000 m and in case of transfer need – 250 m distance between bus stops.

The Integrated Transport System of the South-Moravian Region (IDS JMK) has adopted a distance standard that the reach of the nearest stop must be no more than 2 km from each area of the region of residence (depending on local conditions) [9] (see Fig. 1).

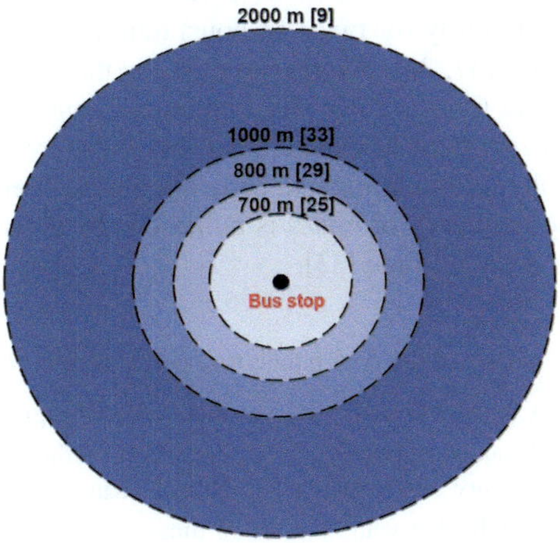

Fig. 1. Walk distance to the bus stop in the regions.

The better the settlement is accessible, the more people get used to the possibility of having a public transport stop nearby. The people living in more remote rural areas prefer a longer distance to bus stops than those living in more accessible rural areas [25].

According to Wong et al. [34] model results, the travel of the elderly people would increase by reducing the walking time to the public transport stop and the waiting time for the bus. The availability of bus stops, reflected in pedestrian time, influences the willingness of elderly people to travel. Therefore, better access to public transport services can increase their willingness to travel.

2.2 Frequency of Regional Public Transport Routes

As Kwan [35] explains, the criterion of proximity alone is not sufficient to assess the accessibility of public transport. Even if a person lives near a bus stop, accessibility also depends on other factors, including the frequency of service provision. The report of the Commission for Integrated Transport [36] calls for a new approach to rural transport, based on the fact that the quality of public transport varies between major cities and suburbs or rural areas and in remote areas, public transport services are infrequent, end early in the evening and not provided on weekends. For example, in Sisak-Moslavina County in Croatia, the mobility of the population is insufficient, as buses run only every 3 h on the suburban route Sisak-Blinjski Kut [17].

The Czech South Moravian Region expected public transport coverage throughout the territory and quality service, which would satisfy the maximum requirements of the population. The Integrated Transport System of the South-Moravian Region (IDS JMK) has published a requirement that each rural settlement is served by public transport at least 6 times a day on weekdays and at least 3 times a day on weekends public holidays. It is stated that the frequency of 6 times a working day applies only to the outermost regions, while rural areas closer to major roads need to be serviced more frequently [9].

Weinland in Switzerland is known as a region with a high quality of public transport services, with hourly (or twice an hour) train and bus services on weekdays from 6 AM to midnight [37, 38]. Another example is the Lü village (62 inhabitants), which is served by buses 6 times a day and 2 additional routes in the summer [37] (see Table 1). Buses run every hour in remote areas, but there are exceptions when routes are not running (usually mid-morning and early afternoon). Very often, in the suburbs and remote areas, the frequency of routes is influenced by the schedule of school trips of students and their vacation time [33].

Table 1. Frequency of regional public transport routes in foreign countries.

Reference	Country/Region	Frequency of regional public transport
Šipuš and Abramović 2017	Sisak-Moslavina County in Croatia	every 3 h on the suburban route Sisak-Blinjski Kut
Šťastná and Vaishar 2017	South Moravian Region in the Czech Republic	6 times a day on weekdays and at least 3 times a day on weekends and public holidays
Peterson 2014; Petersen and Mees 2010	Weinland in Switzerland	1 time or 2 per hour on weekdays from 6 a. m to midnight
Peterson 2014	Lü village in Switzerland	6 times a day and 2 additional routes in the summer

An efficient public transport system contributes to sustainable economic growth and the development of a healthy social area, which in turn ensures the security of people, taking into account the mobility of goods and services for todays and future needs. Research has been carried out on the Malaysian peninsula, providing an empirical analysis of the level of service (LOS) of fixed-route urban and rural bus services. During analysis, the frequency of routes >60 min was unattractive for all passengers, and the frequency of 21–30 min was unattractive for some passengers [39].

The results of an experiment conducted by Bos [40] showed that public transport users most value travel time and frequency. A direct public transport route without intermediate stops with a frequency of 30 min is particularly valued. It was concluded that public transport users prefer the 30 min frequency of routes the most, and the least – every hour.

In many cases, due to the low frequency of scheduled transport, incoming buses remain overcrowded and is unable to accommodate all passengers waiting at stations. In this case, passengers are forced to wait a very long time for the next bus [41]. Afukaar et al. [42], in a study of travel comfort, interviewed passengers who specified that bus congestion causes very high public transport discomfort. However, respondents still indicated that congestion is more appropriate for them than waiting an hour or two for the next transport.

The frequency of public transport services, which is reflected in waiting times, also influences the willingness of older people to travel [34]. Szeto et al. [43] research show that older people usually travel by public transport between 10 and 11 AM to avoid the crowd. However, during this time interval, the frequency of most public transport routes is reduced, which again leads to longer waiting times. Especially if there are no pavilions at the bus stops, waiting for the buses causes additional inconveniences (snow in winter, rain, or a hot summer day).

Accessibility of public transport, better routes management and the development of integrated public transport traffic systems would contribute to changing emigration trends in rural areas and thus improve the quality of life in the regions [17]. Passengers must be satisfied with the public transport service that should become an attractive alternative to the private car [44].

3 Used Methodology

During the research, the accessibility of public transport in Lithuanian regions was examined. According to accepted practice, a public transport stop in the regions must be no more than one kilometre from the urbanised areas. Therefore, during the study, the territories of Lithuanian regional municipalities were covered by taking a distance of 1 km from the stops.

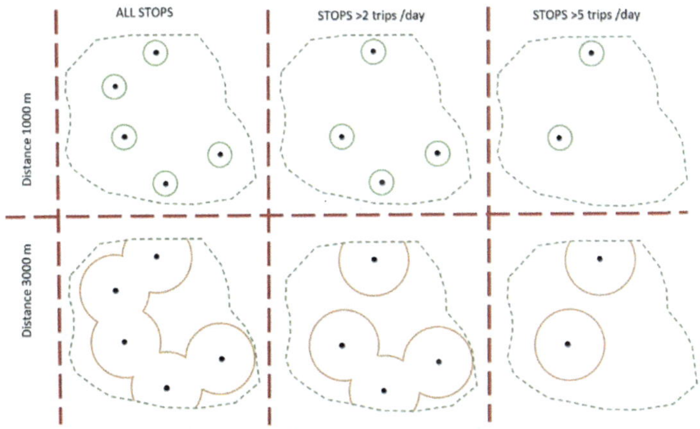

Fig. 2. Methodology used for analysis.

The convenience of the population to use public transport according to the number of trips served per stop per day has also been taken into account. According to this indicator, the convenience of the population to use public transport was divided into three steps (see in Fig. 2). In the first step, the accessibility area of 1 km was taken from all stops, thus obtaining overall accessibility. Stops that are served by only 1–2 trips per day are then rejected, with residents not being able to choose even the time of day when they want to make their trip. Moreover, the third step is to give residents the option to travel. During this step, have been left stops that are served over five trips a day. In such circumstances, the resident can already choose when he needs the trip the most.

The second large sample of the study was analysed by taking the maximum possible distance from the stop when residents still choose public transport – 3 kms. Areas covered at a distance of 3 kms from the stops were also analysed in 3 steps: all stops over two trips per day and stops over five trips per day.

Cities municipalities were not analysed.

4 Public Transport Sccessibility in Lithuanian Regions

Analysing the accessibility of public transport by one kilometre in Lithuanian municipalities, the analysis showed (see in Fig. 3A) that none of the regional municipalities has a public transport accessibility indicator higher than 50%. However, the focus should be on municipalities, which, even without additional conditions, have a public transport accessibility rate of less than 20%.

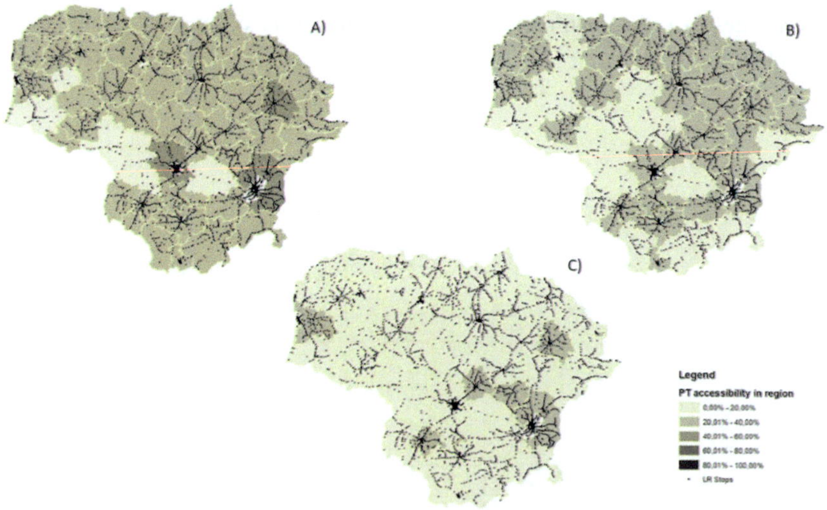

Fig. 3. Public transport stops 1 kms accessibility in percents in Lithuanian municipalities. A) from all stops; B) from the stops with more than two trips per day; C) from the stops with more than five trips per day.

There are 7 such municipalities in Lithuania. Most in central and western Lithuania. It should be noted that in Šakiai district, only 8.1% is accessible at a distance of one kilometre from public transport stops. Adding the condition that the population of the regions must be able to choose at least the time of day, the share of municipalities with less than 20% access to public transport has increased to 31.67% (see Fig. 3B). Under these conditions of analysis, 5 municipalities had less than 10% access area of public transport. The analysis further assesses the comfortable conditions for the inhabitants of the regions, when they do not need to apply their travel targets to public transport schedules (leaving stops with only 5 or more trips per day). Under these conditions, only 15% of regional municipalities remain, with public transport access covering more than 20% of the area per kilometre (see Fig. 3C).

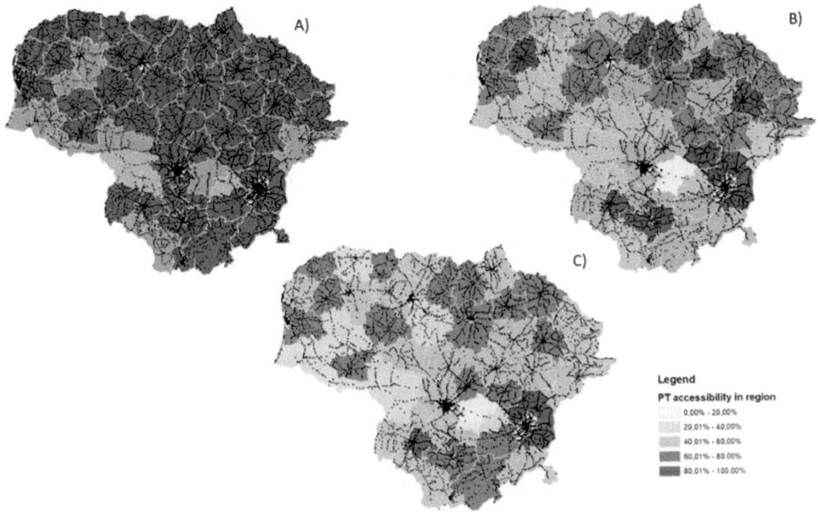

Fig. 4. Public transport stops 3 kms accessibility in percent's in Lithuanian municipalities. A) from all stops; B) from the stops with more than two trips per day; C) from the stops with more than 5 trips per day.

First, the conditions of maximum accessibility when residents want to use public transport and can walk a greater distance to reach a stop with five or more trips to reach their destination when they need it was analysed (Fig. 4). Under these conditions, many municipalities already have sufficient coverage of accessible areas. But there is still one, Kaišiadorys municipality, where less than 20% of the whole municipality (13.13%) is reached even at a distance of 3 kms. Case of concern the accessibility of 8 more municipalities with 3 km of stops served by 5 or more trips ranges from 20% to 40% of the area. In the second step, the number of municipalities with 3 km of public transport reachable areas has already decreased to 2 (Kaišiadorys - 22.85% and Šakiai - 36.32%). 6 municipalities have also been identified, under the analysed conditions of which the accessibility of public transport covers 40–60% of the whole territory. This shows the bad situation in these regions, where there is no social service, which must ensure the entire Lithuania population's opportunity to make a trip for domestic, work, scientific or cultural purposes. In these regions, residents are forced to choose a car to make their trips, and without it, they become dependent on other people and on their income to book a trip in other available means of transport. The last step, after analysing the coverage of the accessibility of all Lithuanian stops at a distance of 3 kms, showed that the biggest problems with the public transport service, which must be solved immediately, are in Šakiai, Elektrėnai and Kazlų Rūda municipalities, where up to 60% of territory can be reached in 3 km from public transport stops.

Further decisions in the problematic municipalities highlighted in Step 2 and Step 3 should be made immediately by changing the routes of public transport services, their frequency or the nature of the public transport service itself. It is recommended to analyse the population density in the identified problematic municipalities, excluding forests and water areas. In municipalities with a higher population density, decisions should focus on changing public transport routes and increasing frequency. In municipalities with a low population density (<20 in/km excluding forests and water areas), the change in the nature of the public transport service should be analysed by introducing an on-demand service or a shuttle service. This would reduce the need for public transport funding without covering the low occupation public transport run cost and providing a higher level of service to the population.

5 Conclusions

1. The literature analysis concludes that the availability of bus stops, reflected in pedestrian time, influences the willingness of elderly people to travel. Therefore, better access to public transport services can increase their willingness to travel.
2. Researchers around the world have differently indicated the distance at which a public transport stop should be reached for the population of not high-density region's, but many have found that public transport accessibility should be planned at a distance of 1 km. The maximum distance recommended by scientists, when the population still using public transport, was 3 kms.
3. The Integrated Transport System of the South-Moravian Region, after the deep analysis, has published a requirement that each rural settlement should be served by public transport at least 6 times a day on weekdays and at least 3 times a day on weekends and public holidays. It is stated that the frequency of 6 times a working day applies only to the outermost regions, while rural areas closer to major roads need to be serviced more frequently
4. Analysing the accessibility of public transport by one kilometre in Lithuanian municipalities showed that none of the regional municipalities has a public transport accessibility indicator higher than 50%. However, the focus should be on seven municipalities in central and western Lithuania, which, even without additional conditions, have a public transport accessibility rate of less than 20%.
5. The analysis of the comfortable conditions for the inhabitants of the regions, when they do not need to apply their travel targets to public transport schedules (leaving stops with only 5 or more trips per day) show that under these conditions, only 15%

of regional municipalities remain, with public transport access covering more than 20% of the area per kilometre.

6. Case of concern the accessibility of 8 municipalities where within accessibility of 3 km from stops served by two or more trips range to 40% of the area. In these regions, residents are forced to choose a car to make their trips, and without it, they become dependent on other people and on their income to book a trip in other available means of transport.
7. Further decisions in the problematic municipalities should be made immediately. Based on the population density, it is recommended to focus on changing public transport routes and increasing frequency or on changing the nature of the public transport service by introducing an on-demand service or a shuttle service.

References

1. EEA, European Environment Agency: Urban Sprawl in Europe (EEA Report No 11/2016) (2016). https://doi.org/10.2800/143470
2. WHO & UN, World Health Organization & United Nations: Hidden Cities: Unmasking and Overcoming Health Inequities in Urban Settings (2010)
3. UITP, International Association of Public Transport: Demographic Changes and Challenges for Regional Transport (Core brief). Author, Brussels (2013)
4. ITF, International Transport Forum: ITF Transport Outlook 2017. OECD Publishing, Paris (2017). https://doi.org/10.1787/9789282108000-en
5. Chen, G., Benezech, V.: Non-urban Passenger Travel Demand and CO_2 Emissions: A Global Perspective (No. 18-03905) (2018)
6. Ribeiro, P., Fonseca, F., Santos, P.: Sustainability assessment of a bus system in a mid-sized municipality. J. Environ. Planning Manage. **63**(2), 236–256 (2020)
7. Trembošová, M., Dubcová, A., Nagyová, Ľ, Cagáňová, D.: Chosen aspects of a spatially functional accessibility by public transport: the case of Trnava self-governing region (Slovakia). Acta Logistica **7**(2), 121–130 (2020)
8. Szilva, I., Caganova, D., Bawa, M., Pechanova, L., Hornakova, N.: Knowledge management perception in industrial enterprises within the CEE region. In: Longo, A., Zappatore, M., Villari, M., Rana, O., Bruneo, D., Ranjan, R., Fazio, M., Massonet, P. (eds.) IISSC/CN4IoT -2017. LNICSSITE, vol. 189, pp. 66–75. Springer, Cham (2018). https://doi.org/10.1007/978-3-319-67636-4_8
9. Šťastná, M., Vaishar, A.: The relationship between public transport and the progressive development of rural areas. Land Use Policy **67**, 107–114 (2017)
10. Kenyon, S., Lyons, G., Rafferty, J.: Transport and social exclusion: investigating the possibility of promoting inclusion through virtual mobility. J. Transp. Geogr. **10**(3), 207–219 (2002)
11. Hine, J.: Good policies and practices on rural transport in Africa: planning infrastructure and services (2014)
12. Květoň, V., Chmelík, J., Vondráčková, P., Marada, M.: Developments in the public transport serviceability of rural settlements with examples from various types of micro-regions. AUC Geographica **47**(1), 51–63 (2017)
13. Abreha, D.A.: Analysing Public Transport Performance Using Efficiency Measures and Spatial Analysis: The Case of Addis Ababa, Ethiopia. ITC (2007)

14. Johnson, D., Ercolani, M., Mackie, P.: Econometric analysis of the link between public transport accessibility and employment. Transp. Policy **60**, 1–9 (2017)
15. Petersen, T.: Watching the Swiss: a network approach to rural and exurban public transport. Transp. Policy **52**, 175–185 (2016)
16. White, P.: Public Transport: Its Planning, Management and Operation, 5th edn. Routledge, London (2009)
17. Šipuš, D., Abramović, B.: The possibility of using public transport in rural area. Procedia Eng. **192**, 788–793 (2017)
18. Yatskiv, I., Budilovich, E., Gromule, V.: Accessibility to Riga public transport services for transit passengers. Procedia Eng. **187**, 82–88 (2017)
19. Goodwin-Hawkins, B.: Demand Responsive Transport in Rural Areas (2020)
20. Community Transport Association. The Future of Demand Responsive Transport (2017)
21. Melis, L., Sörensen, K.: The on-demand bus routing problem: a large neighborhood search heuristic for a dial-a-ride problem with bus station assignment (No. 2020005) (2020)
22. Tracks. The Future of Rural Bus Services in the UK. Campaign for Better Transport (2018)
23. Interreg Europe. Demand-responsive Transport. A Policy Brief from the Policy Learning Platform on Low-carbon economy (2018)
24. Michniak, D.: Vplyv dostupnosti na rozvoj cestovného ruchu vo vybraných regiónoch na Slovensku. Geographia Cassoviensis **4**(1), 114–117 (2010)
25. Plazinić, B.R., Jović, J.: Mobility and transport potential of elderly in differently accessible rural areas. J. Transp. Geogr. **68**, 169–180 (2018)
26. Hansson, J., Pettersson, F., Svensson, H., Wretstrand, A.: Preferences in regional public transport: a literature review. Eur. Transp. Res. Rev. **11**(1), 1–16 (2019). https://doi.org/10.1186/s12544-019-0374-4
27. Sun, C., Chen, X., Zhang, H.M., Huang, Z.: An evaluation method of urban public transport facilities resource supply based on accessibility. J. Adv. Transp. **2018**, 1-11 (2018)
28. Bruzzone, F., Scorrano, M., Nocera, S.: The combination of e-bike-sharing and demand-responsive transport systems in rural areas: a case study of Velenje. Res. Transp. Bus. Manage. **40**, 100570 (2020)
29. Akbari, S., Mahmoud, M.S., Shalaby, A., Habib, K.M.N.: Empirical models of transit demand with walk access/egress for planning transit oriented developments around commuter rail stations in the Greater Toronto and Hamilton Area. J. Transp. Geogr. **68** (C), 1–8 (2018)
30. Brand, J., Hoogendoorn, S., van Oort, N., Schalkwijk, B.: Modelling multimodal transit networks integration of bus networks with walking and cycling. In: 2017 5th IEEE international conference on models and Technologies for Intelligent Transportation Systems (MT-ITS), pp. 750–755. IEEE (2017, June)
31. Ker, I., Ginn, S.: Myths and realities in walkable catchments: the case of walking and transit. Road Transp. Res. **12**(2), 69 (2003)
32. Abedi, N., Bhaskar, A., Chung, E., Miska, M.: Assessment of antenna characteristic effects on pedestrian and cyclists travel-time estimation based on Bluetooth and WiFi MAC addresses. Transp. Res. Part C Emerg. Technol. **60**, 124–141 (2015)
33. Stentzel, U., Piegsa, J., Fredrich, D., Hoffmann, W., van den Berg, N.: Accessibility of general practitioners and selected specialist physicians by car and by public transport in a rural region of Germany. BMC Health Serv. Res. **16**(1), 1–10 (2016)
34. Wong, R.C.P., Szeto, W.Y., Yang, L., Li, Y.C., Wong, S.C.: Public transport policy measures for improving elderly mobility. Transp. Policy **63**, 73–79 (2018)
35. Kwan, M.P.: Beyond space (as we knew it): Toward temporally integrated geographies of segregation, health, and accessibility: space–time integration in geography and GIScience. Ann. Assoc. Am. Geogr. **103**(5), 1078–1086 (2013)

36. Commission for Integrated Transport: A New Approach to Rural Public Transport. HMSO, London (2008)
37. Peterson, T.: Public Transport Beyond the Fringe. The Public City: Essays in Honour of Paul Mees Melbourne University Press, Melbourne (2014)
38. Petersen, T., Mees, P.: A case of good practice: the Swiss 'network' approach to semi-rural public transport. In: World Congress on Transport Research, 12th, 2010, Lisbon, Portugal (No. 02049) (2010)
39. Ponrahono, Z., Bachok, S., Ibrahim, M., Osman, M.M.: Assessing disparities in the urban-rural service: a case of public bus services in Peninsular Malaysia. In: McLellan, B. (ed.) Sustainable Future for Human Security, pp. 189–201. Springer, Singapore (2018). https://doi.org/10.1007/978-981-10-5433-4_13
40. Bos, P.: Self-driving bus to improve accessibility of rural areas in the Netherlands. People mover as a first-and last mile solution (2017)
41. Abramović, B., Šipuš, D., Jurešić, D.: A preparatory survey in integrated passenger transport planning: a case study. Transp. Res. Procedia **53**, 16–22 (2021)
42. Afukaar, F., Damsere-Derry, J., Peters, K., Starkey, P.: Rural transport services indicators: using a new mixed-methods methodology to inform policy in Ghana. Transp. Res. Interdisc. Perspect. **3**, 100074 (2019)
43. Szeto, W.Y., Yang, L., Wong, R.C.P., Li, Y.C., Wong, S.C.: Spatio-temporal travel characteristics of the elderly in an ageing society. Travel Behav. Soc. **9**, 10–20 (2017)
44. Lunke, E.B.: Commuters' satisfaction with public transport. J. Transp. Health **16**, 100842 (2020)

Intelligent Vehicles and Infrastructure

Resilience as a Feature of Intelligent Railway Infrastructure

Zdeněk Dvořák(✉) and Katarína Hoterová

Faculty of Security Engineering, University of Zilina,
Ul. 1. maja 32, 010 26 Zilina, Slovak Republic
{zdenek.dvorak,katarina.hoterova}@uniza.sk

Abstract. The transport infrastructure of the state is an important part of modern society. The quality of the transport infrastructure is crucial for the functioning of logistics processes and for transport services. Our research into the resilience of major infrastructure is closely linked to the use of information and communication technologies. The article presents the state of art of individual parts of resilience, then presents the proposals of researchers and the results of research solutions focused on railway infrastructure. Resilience is divided into robustness, renewable and adaptability. The research of individual parts of resilience was solved within the RESILIENCE project in the Czech Republic, the CIERA methodology was created. The mentioned methodology was tested in the Slovak Republic with the conclusion that it is very complicated for practical use. For this reason, the methodology was modified. For the purposes of practice, the research was focused on the renewable part and focused on material, finance, personnel and recovery technologies. An application in the excel environment was prepared for practical use, which enables effective evaluation of renewable energy with the participation of experts. The resulting renewable is complemented by real measures focusing on sub-indicators. The aim of the article to create a methodology for the renewability of railway infrastructure facilities has been met.

Keywords: Railway infrastructure · Resilience · Recoverability · SMART

1 Introduction

Research in the field of intelligent vehicles and intelligent infrastructure is carried out in all developed countries of the world. The interest in full automation and a significant increase in transport efficiency and safety is key. The University of Zilina was originally the only transport-oriented university in the former Czechoslovakia. Current challenges and projects for all types of transport have been addressed here for almost 70 years. At present, efforts are mainly devoted to the interconnection of cutting-edge information and communication technologies (ICT) into individual parts of transport systems. We see the most significant shift in management and information systems, where SMART technologies are currently being applied in passenger information, in traffic management, and in monitoring the delivery of consignments. A new area of research today is the deployment of new ICT in vehicles and transport infrastructure

O. Prentkovskis et al. (Eds.): TRANSBALTICA 2021, LNITI, pp. 741–750, 2022.
https://doi.org/10.1007/978-3-030-94774-3_71

with the aim of full automation, which will gradually lead to the creation of autonomous transport systems. Our research is focused on resilience and its parts - robustness, renewable and adaptability. Our research is aimed at creating a detailed methodology that will be a good basis for information support for intelligent railway infrastructure. The aim of the article is to present the current level of knowledge in building a modern railway infrastructure with a high level of resilience. The current result is a tool for measuring resilience, focusing on the field of renewable, which was created in collaboration with academia and experts from practice.

2 Analysis of the Railway Infrastructure System

The state's transport system consists of road, air, water and rail transport. These modes of transport focus on passenger and freight, in-house, national and international transport. Railway infrastructure is considered to be promising worldwide in terms of sustainability. It needs to be modernized to create a dynamic, ecological and economic choice for future rail transport. The current challenge for the countries of Central Europe is to return to public transport at the expense of individual motoring. The absence of transport policy in the last 30 years after the change of the system in the Slovak Republic brought a huge increase in the number of registered cars and on the other hand only minimal maintenance costs and very low expenditures on expansion and modernization of transport infrastructure. The authors focus in the article on railway transport and railway infrastructure.

2.1 State of Art Intelligent Railway Infrastructure

The current railway infrastructure is the backbone of the railway transport system. A prerequisite for the construction of intelligent railway infrastructure is a high-quality high-speed internet connection for all parts and objects of the railway infrastructure. In the case of the Slovak Republic, all major railway lines are run together with high-quality information and communication infrastructure. This creates the preconditions for the development of highly efficient smart solutions, which will become the basis for future autonomous solutions. From the point of view of future tasks, we divided the objects located on the territory of the Slovak Republic into three groups, namely:

- Point objects - objects that are formed by a concentrated and closed unit, are located on a small area (control elements and smaller elements of railway infrastructure), are connected with line objects.
- Line objects - objects that provide transmission, delivery or transport between two physically separate places are an important basic group that forms the connection between point and area objects.
- Area objects - objects that represent the area unit and places of concentration of point and line objects are the most complex objects,
- Moving objects - are railway sets carrying specific types of dangerous goods and passenger trains [1, 2].

Another important view of the railway infrastructure is the possible division of railway infrastructure engineering objects according to types into:

- Railway bridges.
- Railway tunnels.
- Railway stations.
- Intermodal railway terminals [1].

Railway bridges and tunnels are currently one of the most important railway infrastructures. Their damage or failure could lead to the dysfunction of rail transport on the route in question and, consequently, to the complicated provision of an alternative mode of transport [3].

Railway infrastructure on the basis of its characteristics as well as legal norms of critical infrastructure issues (e.g. Act No. 45/2011 on Critical Infrastructure, Council Directive 2008/114/EC on the identification and designation of European Critical Infrastructure and the assessment of the need to improve their protection, and others) is classified as one of the subsectors of the transport sector. Critical infrastructure is a part of national infrastructure, the destruction or disruption of which as a result of an adverse event will affect the life, health and property of the population, but also the economic operation of the state. If it is an element of national critical infrastructure, then to the operation of the state in which the element of critical infrastructure is located, if it is an element of critical infrastructure of an international nature, to the other states concerned. In order to include a building in the system of critical infrastructure, it is necessary to apply national sectoral and cross-sectional criteria, and in the case of the most important buildings also international sectoral and cross-sectional criteria. An important step for the European Commission is the issuance of a new directive on the resilience of critical entities in 2020. This directive, in close connection with the NIS directive, will force Member States to focus on cyber security, which is gradually becoming the most important part of security for future autonomous infrastructures systems.

2.2 Research on the Resilience of Railway Infrastructure

We have been researching railway infrastructure at the University of Zilina for more than 60 years. In the last decade, the attention in our team has been focused on creating a knowledge base for the protection of critical infrastructure. The basis was the project APVV 0471-10 on the protection of critical infrastructure in the transport sector. Among other results, these parameters of railway objects were investigated:

- Railway bridges - bridge height, length of the bridge, load on the bridge, number of bridge pillars, construction material of the bridge, age of the bridge, access to the bridge, number of transported persons and commodities in 24 h, number of tracks across the bridge, estimated bridge renewal time, total economic losses.
- Railway tunnels - tunnel length, number of tracks in the tunnel, age of the tunnel, current safety security of the tunnel, tunnel lining material, access to the tunnel, the amount of rock pressure on the tunnel (overburden), number of transported persons and commodities in 24 h, economic losses.

- Railway stations - number of passengers in 24 h, current security of the station, number of passenger and freight train crossings, importance of the station in terms of international transport, location of the station in terms of the negative effects of adverse events, economic losses.
- Intermodal railway terminals - quantity of transported goods in 24 h, quantity of manipulated intermodal transport units in 24 h, economic losses [4].

In addition to the above technical parameters, the research was focused on determining the resilience of each object. Gradually, the individual parts of robustness, recoverability and adaptability were addressed in detail (Fig. 1).

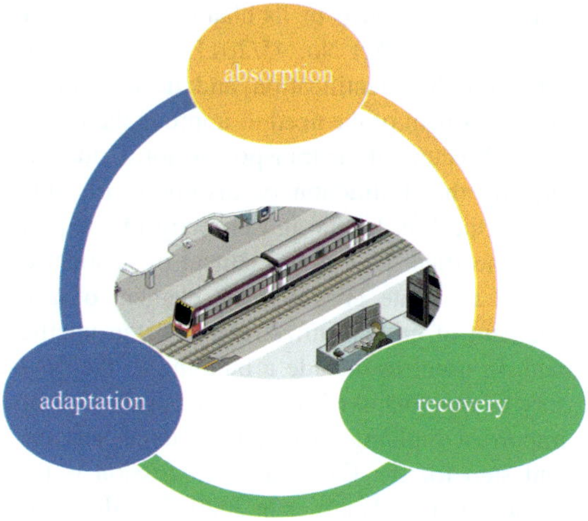

Fig. 1. Resilience and its parts.

In research, resilience was defined as the ability to absorb, adapt and quickly recover from the potential effects of a dangerous event. Robust systems show a smaller decrease in performance during the occurrence of an adverse event and the system will return to the operating level in less time. Within technical resilience, the researchers' attention was focused on the robustness of the system - assessment of crisis preparedness, redundancy, detection capability, response capability and physical resilience. These robustness sections are generally evaluated quantitatively if all available statistics and additional information are available. Otherwise, with the participation of experts, a qualitative assessment of the relevant parts will be made with a final statement, whether e.g. crisis preparedness is excellent, good or insufficient. If an insufficient level of any of the parts of the robustness is found, the whole robustness is evaluated as insufficient.

The level of recoverability is given by the real level of prepared material resources, financial resources, human capacities and the quality of technical and technological processes of renewal. These parts of transport infrastructure recovery were examined in detail in the research of authors in cooperation with the national administrator of railway infrastructure - the company Railways of the Slovak Republic [5].

2.3 Intelligent and Resilient Infrastructure

The railway infrastructure in Europe is constantly evolving. Developed countries are completing the deployment of the ETCS system on all express lines, other countries are renewing their railway infrastructure to introduce higher speeds, and a third group of countries are planning to renew their more than 100-year-old railway infrastructure. European Railway Agency declared the year 2021 as the European Year of rail with the main goal of boosting rail as a sustainable, innovative, interconnected and inter-modal, safe and affordable mode of transport in line with the objectives set out in the European Green Deal and the Sustainable and Smart Mobility Strategy [6].

In developed countries, modern information and communication technologies are currently being developed at a rapid pace. 5G technology is an ideal choice for various sectors, including transport, which requires real-time communication and guaranteed quality of service [7].

In terms of railway traffic management, currently the biggest challenge is to increase the resilience of all parts of railway transport - infrastructure, means of transport, management and information systems, renewal capacities and personnel. The initial phase is the identification of threats - is the creation of a list of threats that can theoretically operate in a given place. Several dozens of these threats can be defined for a given subsystem and a given site. Subsequently, it is necessary to group them into suitable groups that can be addressed together [2]. Within the research of the security environment, security pillars were defined.

New groups of threats should be correlated with the security pillars - physical security, information security, occupational safety and health, fire safety, environmental safety and operational safety [8]. In order to measure the impact of threats within individual security pillars, it is appropriate to create a set of indicators that will address individual groups of security threats [9]. According to the research results, it is recommended to define min. 5 indicators for each group of security threats. A key feature of the proposed indicators is their real measurability.

2.4 Analysis of Scientific Information Sources

Many researches have been devoted to the field of railway transport and its resilience, which has resulted in publications that have been used in our research. As in every area, it can be argued that individual ideas and approaches differ from author to author. Bureika et al. in the publication (Applying analytical hierarchy process to assess traffic safety risk of railway infrastructure) uses the process of analytical hierarchy for the process of assessing the risks of railway infrastructure. Their application of a hierarchical process of risk assessment with a focus on railway infrastructure facilities was used in our study. The model is designed for railway managers to improve strategic transport safety, to set priorities and required measures. The model makes it possible to classify the basic objects of railway infrastructure according to traffic safety, helps to identify weak points and to plan technical and organizational measures necessary to eliminate the occurrence of a non-adverse event [10].

It is important in critical infrastructure protection in general to determine the order of the infrastructure elements of the highest protection priority, as stated by Zhang et al., In the Quantified Approach for the evaluation of critical node and linear elements of railway infrastructure. Node elements include train stations and line elements include railway lines, which are modelled as nodes and connections in the transport network diagram. The aim of the methodology is to better understand the importance and prioritization of the protection of railway infrastructure resources as well as other critical infrastructures [11].

Another publication on the identification of potential elements in rail transport is "A new procedure for identifying the importance of infrastructure elements in the rail transport subsector". It is primarily focused on the problem of assessing the importance of individual elements in the railway infrastructure. It contains the main features of the proposed theoretical approach of the significance of key typological elements of railway infrastructure and the initial proposal of the methodology for the evaluation of such elements [12]. In order to maintain the continuity of railway transport on the most important routes, it is important to identify and evaluate the criticality of individual bridges and provide them with increased protection against adverse events as reported by Dvořák et al., In the publication "Criticality assessment of railway bridges" [13].

The creation of alternative routes is one of the ways to increase the safety of the railway system. The publication "Comparison of railway routes enabling freight transport from the eastern to the western border of Poland" deals with the design of new alternative routes and their comparison was subsequently performed using parameters with the currently used routes. The parameters were, for example, the length of the route, the number of tracks, economic evaluation and others. By calculating the average value of these parameters, it is possible to evaluate the route and compare it with others [14].

In order to maintain continuous rail transport, it is necessary to ensure its resilience. As Rehak states in the publication "Assessment and Strengthening of Organizational Resilience in the Critical Infrastructure System: A Case Study of the Slovak Republic", it is important that both technical and organizational resilience be ensured primarily by critical infrastructure owners and operators [15]. Along with the growing technological trend, there is an increased resilience in traffic management, but little attention is paid to this area. According to Achillopoulou in "Monitoring transport infrastructure exposed to multiple hazards: a resilience plan", digital technologies have the potential to provide rapid and up-to-date resilience assessments for transport infrastructure that is exposed to many natural and man-made hazards and different burdens [16].

3 Recoverability as a Tool to Increase the Safety of Railway Infrastructure

Ensuring the effective operation of any system requires knowing the current state of security. A possible way to assess safety is to assess the resilience of a reference object. Resilience is generally perceived as the ability of a reference object to retain its basic functions both during and after the action of an undesirable phenomenon. The evaluation of the resilience of a reference object is focused on its three parts - robustness, recoverability and adaptability. In the research, we focused on the renewable, which is limited by the indexes, which are shown in the following Fig. 2.

Fig. 2. Recoverability and its indexes.

These four indexes are defined as follows:

- Material resources (O.M.) - availability of the necessary technological for the implementation of repair or replacement of damaged parts of the reference object,
- Financial resources (O.F.) - availability of financial resources/reserves enabling financing of the necessary renewal of a part of the reference object,
- Human capacities (O.P.) - the availability of the staff needed to carry out the refurbishment of the reference object itself with the necessary qualifications,
- Restoration processes (O.PO.) - the availability of the processes needed to quickly restore the reference object and the estimated time needed to restore the reference object [17].

For the needs of assessing the resilience of a specific reference object of railway infrastructure, which is a railway bridge, a methodology for assessing renewable energy was developed as part of the resilience assessment. However, for the purposes of this article, the entire methodology will not be shown. Within the indexes, categories of indicators are created. Each indicator belonging to the index has a verbal description of the indicator and also a point rating in the scale 10,5,0 (Table 1).

Table 1. Indicator of the availability of the required material.

Evaluation of railway bridge recovery				
No	Index	Indicator	Verbal description	Score [10, 5, 0]
O.M.1	Material	Availability of the required material	Assessment of material availability (through a contract for a future contract, placement in a storage, and so on)	10 – the required material is available in a larger capacity and/or is placed in a storage 5 – the required material is available in the exact capacity (without reserve) through a contract and/or is placed in a storage 0 – the required material is not available in the exact capacity or through the contract and is not even placement in a storage

The result of the added values to the individual indicators is a graph that displays the results of the resilience evaluation with reaching the maximum number of points (maximum value) and the current state of resistance (acquired value), the graph is shown in the following figure. Based on this graph, it is possible to find out which index has deficiencies and measures must be taken to increase the current state of resilience of the reference object (FIg. 3).

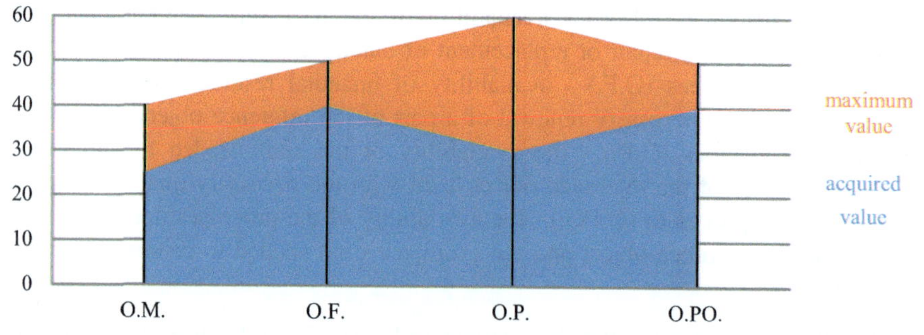

	O.M.	O.F.	O.P.	O.PO.
Acquired value	25	40	30	40
Maximum value	40	50	60	50

Fig. 3. The resulting graph of the level of resilience parts.

As part of our research, we, in cooperation with experts from practice, evaluated several important bridge structures with the help of the above-mentioned methodology and obtained specific results - the O.M. = 62.5%, financial security index O.F. = 80%, O.P. 50%, O.PO. recovery process index. = 80%. The evaluation of the individual indices leads to the conclusion, which is useful for the infrastructure manager, that the level of financial security and the index of the recovery process are at a good level, but the material security index is satisfactory and the staffing index is unsatisfactory. In this particular case, several measures will be recommended to improve the latest indices. The overall renewable of the railway bridge due to these low values is below 68% and this is a sufficient level of renewable [17].

4 Conclusion

In the conclusion, we present a discussion of the results and benefits of research. The methodology for evaluating transport infrastructure recovery is based on the Czech CIERA methodology, which deals with the evaluation of the resilience of critical infrastructure elements. Based on it, an analysis of the recoverability of railway bridge structures was performed. The methodology for evaluating renewable energy is based on indices and indicators for evaluating renewable energy. The developed proposal for the methodology for evaluating the renewables of railway bridge structures was double-commented and improved in cooperation with experts from practice.

The main benefit for the practice is the information support of the management employees of the individual companies that are responsible for the renewal of railway bridges. In the specific conditions of the Slovak Republic, these are the companies ŽSR, EUROVIA Kosice, Doprastav, and others. Respondents, by using the methodology of evaluating the renewable nature of railway bridges, are led to the need for the existence of preventive measures taken to ensure the timely provision of materials, finances, personnel and to ensure an efficient entire restoration process. Also, by extending this methodology into practice, there will be room to realize the complexity of ensuring the renewable nature of railway bridge structures.

Acknowledgement. The publication of this paper was supported by the project IGP 2019/02 Proposal of a methodology for assessing the resilience of infrastructure facilities in the transport sector.

References

1. Luskova, M., Dvorak, Z.: Specifics of identification of critical infrastructure elements in the railway subsector. Crisis management, pp. 332–337 (2016). https://www.fbi.uniza.sk/uploads/Dokumenty/weby/rksarchiv/2016/articles/Luskova_Dvorak.pdf. 02 Feb 2020
2. Rehak, D., Hromada, M., Senovsky, P., Krocova, S., Apeltauer, T., Pidhaniuk, L.: Summary of methods for evaluating the quality and resilience of infrastructure. Final report on the public contract of the Office of the Government of the Czech Republic. Ostrava: Vysoka skola banska - Technical University of Ostrava, Faculty of Safety Engineering. 120 s. (2016)

3. Dvorak, Z., Sousek, R., Sventekova, E., Leitner, B., Cizla, M.: Risk management in railway transport. 1st ed. Pardubice - Jan Perner Institute, o.p.s., 286 s. (2010). ISBN 978-80-86530-71-0
4. Project APVV 0471-10 Critical Infrastructure Protection in Sector Transportation
5. Rehak, D., Hromada, M., Senovsky, P.: Resilience of critical infrastructure: Theory, principles, methods, Association of Fire and Safety Engineering, p. 108. (2019). ISBN: 978-80-7385-224-5
6. European Year of Rail 2021|ERA (europa.eu)
7. Samsung a Deutsche Telecom demonstrated guaranteed latency in the networks 5G| TECHBOX.sk (dennikn.sk)
8. Dvorak, Z., Leitner, B. Rehak, D.: Critical Infrastructure Protection Specifications in the Transport Sector. (Z. Cekerevac, Ed.) MEST J. **7**(2), 31–40 (2019). https://doi.org/10.12709/mest.07.07.02.04
9. Vidrikova, D., Boc, K., Dvorak, Z., Rehak, D.: Critical Infrastructure and Integrated Protection. 1st edit. Ostrava, Czech Republic: The Association of Fire & Safety Engineering, 2017. 172 p. (2017). ISBN 978-80-7385-190-3
10. Bureika, G., Bekintis, G., Liudvinavičius, L., Vaičiūnas, G.: Applying analytic hierarchy process to assess traffic safety risk of railway infrastructure. Maintenance Reliability **15**(4), 376–383 (2013)
11. Zhang, Z., Li, X., Li, H.: A quantitative approach for assessing the critical nodal and linear elements of a railway infrastructure. Int. J. Crit. Infrastruct. Prot. **8**, 3–15 (2015)
12. Leitner, B., Rehak, D., Kersys, R.: The new procedure for identification of infrastructure elements significance in sub-sector railway transport. Communications **20**(2), 41–48 (2018)
13. Dvorak, Z., Luskova, M., Rehak, D., Slivkova, S.: Criticality assessment of railway bridges. TRANSBALTICA XI: Transp. Sci. Technol., 474–483 (2019). https://doi.org/10.1007/978-3-030-38666-5_50
14. Kaminski, W., Sładkowski, A.: Comparison of railway routes enabling freight transportation from the eastern border of poland to the west. TRANSBALTICA XI: Transportation Science and Technology, pp. 171–184 (2019)
15. Rehak, D.: Assessing and strengthening organisational resilience in a critical infrastructure system: Case study of the Slovak Republic, Safety Science, vol. 123 (2020). ISSN 0925-7535
16. Dimitra, V.A., Mitoulis, S.A., Argyroudis, S.A., Wang, Y.: Monitoring of transport infrastructure exposed to multiple hazards: a roadmap for building resilience. Sci. Total Environ. **746** (2020). ISSN 0048-9697
17. Hoterová, K.: Evaluation of static resilience of railway infrastructure objects. Dissertation thesis. Zilina: FBI UU, 127 p. (2021)

High Order Bézier Curves Based Path Planning for Autonomous Movement in Single Lane Roundabouts

Paulius Skačkauskas(✉)

Vilnius Gediminas Technical University, Plytinės 27, 10105 Vilnius, Lithuania
paulius.skackauskas@vilniustech.lt

Abstract. None of the possible advantages of autonomous vehicles are going to become reality if the autonomous vehicles are not going to be able to move properly in various different urban scenarios like lane changing, intersections, roundabouts, etc. One of the first tasks, which must be solved in order to be able to properly autonomously move in different urban scenarios, is the path planning. However, the path planning problems, while taking into consideration different urban scenarios, are analysed not equally and the roundabouts is the scenario, which is still behind. Thus, this work analyses the path planning for roundabouts problem, while applying one of the most popular interpolation based approaches – the Bezier curves. A novel way to select the control points for high order Bezier curves, intended for entering a single lane roundabout, driving within the roundabout and exiting the roundabout, is presented.

Keywords: High order · Bezier curves · Autonomous vehicle · Path planning · Single lane · Roundabout · Autonomous movement

1 Introduction and Literature Review

Currently autonomous ground vehicles is one of the most discussed topics while taking into consideration many different possible application possibilities and advantages. Although the potential of the autonomous ground vehicles is very high, the actual advantages are not proven yet. In the research works only possible future outcomes, regarding the autonomous ground vehicles, are being predicted [1]. In order to ensure that the outcomes of the integration of the autonomous ground vehicles into the transport system are going to be positive, such a key function of autonomous movement as path planning must perform without any doubts about the efficiency and reliability in any complex urban traffic scenarios. According to the [2], among urban environments, roundabouts are considered to be challenging driving scenarios, however, as the roundabouts have demonstrated improvements over unsignalized and signalized intersections regarding safety, traffic flow, access management, etc., it is clear that the autonomous ground vehicles should satisfactorily deal with the problem of path planning at these circular intersections [3]. Nevertheless, despite all of the progress made in the field of intelligent transportation systems, the path planning in roundabouts remains a problematic subject [4]. Thus, this section further gives a

O. Prentkovskis et al. (Eds.): TRANSBALTICA 2021, LNITI, pp. 751–761, 2022.
https://doi.org/10.1007/978-3-030-94774-3_72

concise review of how the path planning problem for the autonomous vehicles movement in roundabouts is being solved in other research works.

In the research works different complexity approaches are being used for the path planning in roundabouts. An interesting approach, mostly based on the clothoid curves, is described in detail by the same authors in [3, 5]. In these two works the authors state that the proposed approach could be used as a global path planner, which is suitable for planning paths in roundabouts and other scenarios. The proposed path planner interpolates the desired configurations of the autonomous vehicle at the roundabout entrance and exit with G^2 curvature continuity as traveling along the circular roadway, however, the authors further noted that in their approach it is important to more properly compute the parameters, which define the geometry of the roundabout [3]. In [6] a parametric based path planner for roundabouts is presented. In this case the proposed path planner achieves a G^1 continuous path, while minimizing curvature steps to increase smoothness and dividing the driving process into three stages: entrance manoeuvre, driving within the roundabout and exit manoeuvre. In the work it is explained that each stage has a specific path generation based on the parametric curves [6]. A hybrid path planner, which consists of the Bezier curves and Model Predictive Control (MPC) approach, is presented in [7]. The authors describe that the developed approach consists of using a nominal trajectory based on a Bezier curves with a linear MPC, in charge of generating a speed reference according to the predictions made with the vehicles involved [7]. The performed experimental tests showed that the development of the hybrid approach was successful – the autonomous vehicle was able to execute the planned path. Another hybrid path planning approach for moving in roundabouts, also based on the numerical optimization of Bezier curves, while using the Bound Optimization BY Quadratic Approximation method, is presented in [8]. From the research works it can be noticed that the Bezier curves approach is actually one of the most popular approaches, applied for the path planning in roundabouts. For example, the Bezier curves control points selection to enter and exit a roundabout is described in [9]. The authors suggested to use two separate 3rd order Bezier curves for entering and exiting the roundabout. To simplify the process it was assumed that the entering lane is straight. Similarly, in [10] the authors also proposed to use the 3rd order Bezier curves to enter and exit the roundabout. It is important to mention that in both of these two works it was assumed that different approaches and not the Bezier curves will be used for the path planning while the autonomous vehicle is going to move through the circular part of the roundabout. The application of the Bezier curves for the path planning in different urban scenarios, including the roundabouts, is also described in [11]. The authors described the Bezier curves control points equations, which define the roundabout entrance and exit control points, while taking into consideration the roundabout centre point, its radius, entrance angle and exit angle. It is worth noting that in the named work [11] the authors used not the 3rd order, but 4th order Bezier curves. Another novel real time application of the 3rd order Bezier curves for the path planning in roundabouts is described in [12]. Specifically, in [12] the 3rd order Bezier curves were used as a core of a nonlinear local optimization method, which used the principle of the Bezier curves' convex hull to generate a safe path. The convex hull property is an important characteristic of designing the control points of the Bezier curves, thus, the authors in [13] developed a hybrid approach, which combines η^3 splines and 7th order Bezier curves. The developed hybrid approach combines the benefits of both curves with closed-form expressions. The main reason why this work [13] is different from other research works, which uses the Bezier curves – in [13] the high order Bezier curves were used. It can be noticed that various research works, which solve the research problem of the path planning in roundabouts, mostly use the low order Bezier curves and, in most of the cases, the Bezier curves are used only for entering and exiting the roundabout. The main reason for it – the high order Bezier curves (5th and higher) can become unstable. However, this disadvantage of the high order Bezier curves can be solved while properly selecting the control points. Thus, the main aim of this work – to apply the high order Bezier curves to all three parts of the roundabout: entrance, driving within the roundabout, and exit.

2 Application of the Bezier Curves

2.1 Basics of the Bezier Curves

Bezier Curve of degree n is a parametric curve composed of the Bernstein basis polynomials of degree n and it can be defined as [6]:

$$P(t) = \sum\nolimits_{i=1}^{n} \mathbf{P}_i B_{i,n}(t) = \sum\nolimits_{i=1}^{n} \mathbf{P}_i \cdot \frac{n!}{i!(n-i)!} t^i (1-t)^{n-i}, \tag{1}$$

where $\mathbf{P}_i$ are the control points which define the Bezier curve in the plane x and y, B_i is the Bernstein polynomial, n is the Berzier curves order, t is the parameter used to interpolate the Bezier curve, $t = [0, 1]$.

The Bezier curves have a number of properties, but three of these properties are very important while using this approach for the path planning in roundabouts [11, 12]:

1. The Bezier curve will lie into the convex hull formed by the control points;
2. The Bezier curves are continuous geometrically and in curvature, and this continuity can be preserved in the joints of different curves;
3. The first point tangent vector (at $t = 0$) will be given by $\overrightarrow{\mathbf{P}_0\mathbf{P}_1}$ and the last point tangent vector (at $t = 1$) will be given by $\overrightarrow{\mathbf{P}_{n-1}\mathbf{P}_n}$.

2.2 Development of the Bezier Curves Based Approach for Path Planning in Single Lane Roundabouts

While taking into consideration the basics and main properties of the Bezier curves, the approach to select the control points for the path planning in single lane roundabouts with the inscribed circle radius of $16\,\text{m} < r_{ins} < 20\,\text{m}$, possible different movement scenarios and different geometrical properties, is being further developed. Respectively, these movement scenarios are considered: right turn, through, left turn, "U" turn. The roundabout scheme for the approach development is given in Fig. 1.

Firstly, the safe distance, required for the autonomous vehicle to safely enter the roundabout is calculated while taking into consideration the pedestrian crossing:

$$S_{min} = (S_{cross} + (v_a \cdot t_d)) \cdot \sin\varphi, \tag{2}$$

where S_{cross} is the distance from the inscribed circle to the pedestrian crossing, v_a is the constant velocity of the autonomous vehicle, t_d is the assumed delay time of the path planning approach computational time, φ is the entrance/exit installation angle.

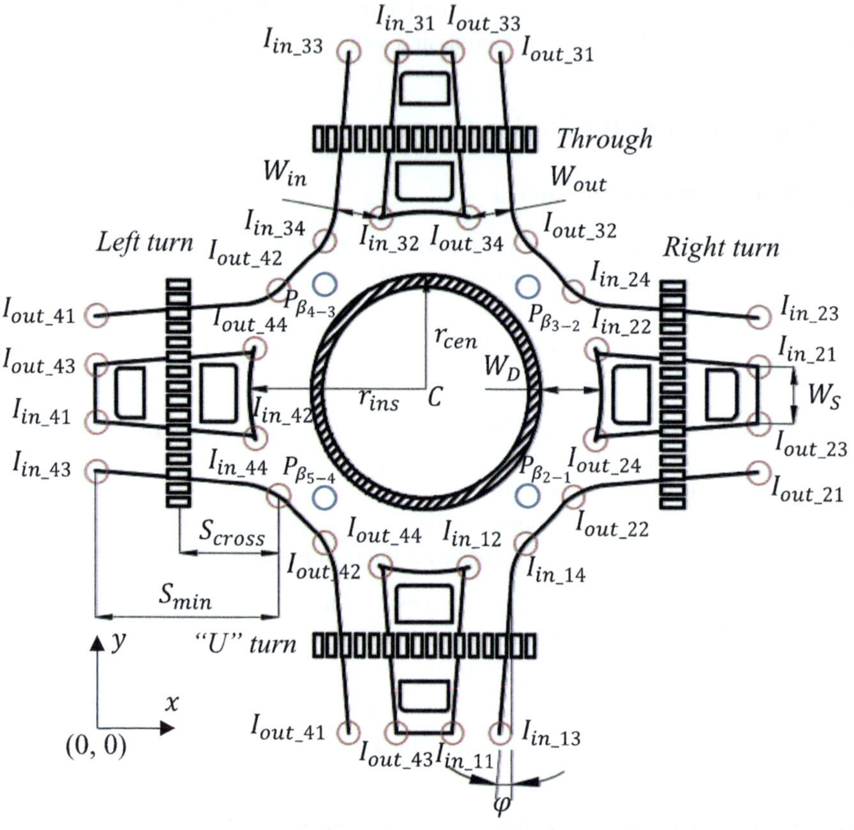

Fig. 1. Roundabout scheme for the Bezier curves based path planning approach development.

Entrance to the Roundabout. While taking into consideration the assumption that, before entering the roundabout, the autonomous vehicle should be moving in a straight line (steering angle is $\delta = 0°$) and the property that the first tangent vector will be given by $\overrightarrow{\mathbf{P}_0\mathbf{P}_1}$, it can be stated that the initial path section of the entrance to the roundabout has to be described by at least three control points: $\mathbf{P}_{in\_0}$, $\mathbf{P}_{in\_1}$, $\mathbf{P}_{in\_2}$. While further considering that the last point tangent vector will be given by $\overrightarrow{\mathbf{P}_{n-1}\mathbf{P}_n}$ and, while assuming that at the moment when the vehicle enters the roundabout and starts to move in circular part, the steering angle also should be $\delta = 0°$, three more control points are required to properly describe the entrance to the roundabout: $\mathbf{P}_{in\_3}$, $\mathbf{P}_{in\_4}$, $\mathbf{P}_{in\_5}$. Thus, in this work, for the entrance manoeuvre path planning it is proposed to use the fitfh order Bezier curve. The proposed equation for the control point selection:

$$\mathbf{P}_{in} = \begin{cases} \mathbf{P}_{in\_0}\left(S_{min} + r_{ins} + \frac{W_S}{2} + \frac{W_{in}}{2}, 0\right) \\ \mathbf{P}_{in\_1}\left(\frac{P_{in\_0(x)} + P_{in\_2(x)}}{2}, \frac{P_{in\_0(y)} + P_{in\_2(y)}}{2}\right) \\ \mathbf{P}_{in\_2}\left(\frac{P_{in\_0(x)}}{2} + \frac{I_{in\_12(x)} + I_{in\_14(x)}}{4}, \frac{P_{in\_0(y)}}{2} + \frac{I_{in\_12(y)} + I_{in\_14(y)}}{4}\right) \\ \mathbf{P}_{in\_3}\left(I_{in\_12(x)}, I_{in\_12(y)}\right) \\ \mathbf{P}_{in\_4}\left(\frac{P_{in\_3(x)} + P_{in\_5(x)}}{2}, \frac{P_{in\_3(y)} + P_{in\_5(y)}}{2}\right) \\ \mathbf{P}_{in\_5}\left(\frac{r_{ins} + r_{cen}}{2} + \cos\beta_{2-1} + C_x, \frac{r_{ins} + r_{cen}}{2} + \sin\beta_{2-1} + C_y\right) \end{cases}, \tag{3}$$

where r_{ins} is the inscribed circle radius, W_S is the width of the splitter island, W_{in} is the width of the entrance, $P_{in\_i(x)}$ is the x coordinate of the control point for the entrance, $P_{in\_i(y)}$ is the y coordinate of the control point for the entrance, $I_{in\_i(x)}$ is the x coordinate of the point, which describes the entrance geometry of the roundabout, $I_{in\_i(y)}$ is the y coordinate of the point, which describes the entrance geometry of the roundabout, r_{cen} is the radius of the central island, β_i is the angle between the point P_{β_i}, which describes the exit from the roundabout, and vertical axis, C_x is the x coordinate of the roundabout centre point,C_y is the y coordinate of the roundabout centre point.

In the given equation (Eq. 3) for the control points, the control points $\mathbf{P}_{in\_0}$, $\mathbf{P}_{in\_1}$, $\mathbf{P}_{in\_2}$ are set in a straight line and the control point $\mathbf{P}_{in\_1}$ is actually a middle point of the vector $\overrightarrow{\mathbf{P}_{in\_0}\mathbf{P}_{in\_2}}$. Similarly, the control points $\mathbf{P}_{in\_3}$, $\mathbf{P}_{in\_4}$, $\mathbf{P}_{in\_5}$ are also set in another straight line, where the control point $\mathbf{P}_{in\_4}$ is a middle point of the vector $\overrightarrow{\mathbf{P}_{in\_3}\mathbf{P}_{in\_5}}$. While setting the control points in the described way it is ensured that the manoeuvre to enter the roundabout is going to begin and end when the steering angle of the autonomous vehicle is $\delta = 0°$.

Right Turn. While considering the right turn manoeuvre it is assumed that the autonomous vehicle actually does not move further through the circular part of the roundabout, i.e., after finishing the entrance manoeuvre the autonomous vehicle immediately starts to perform the exit manoeuvre. Respectively, in this scenario, the control points for exiting the roundabout while performing the right turn manoeuvre are selected according to the same assumptions as in the previously described entrance scenario. Thus, the proposed equation for the control points selection for the right turn path planning:

$$\mathbf{P}_{out} = \begin{cases} \mathbf{P}_{out\_0}\left(P_{in\_5(x)}, P_{in\_5(y)}\right) \\ \mathbf{P}_{out\_1}\left(P_{out\_0(x)} + \left(P_{in\_5(x)} - P_{in\_4(x)}\right), P_{out\_0(y)} + \left(P_{in\_5(y)} - P_{in\_4(y)}\right)\right) \\ \mathbf{P}_{out\_2}\left(P_{out\_1(x)} + \left(P_{in\_4(x)} - P_{in\_3(x)}\right), P_{out\_1(y)} + \left(P_{in\_4(y)} - P_{in\_3(y)}\right)\right) \\ \mathbf{P}_{out\_3}\left(I_{out\_21(x)} - P_{in\_2(y)}, \left(2 \cdot S_{min} + 2 \cdot r_{ins}\right) - P_{in\_2(x)}\right) \\ \mathbf{P}_{out\_4}\left(\frac{P_{out\_3(x)} + P_{out\_5(x)}}{2}, \frac{P_{out\_3(y)} + P_{out\_5(y)}}{2}\right) \\ \mathbf{P}_{out\_5}\left(I_{out\_21(x)}, I_{out\_21(y)} + \frac{W_{out}}{2}\right) \end{cases}, \tag{4}$$

where $P_{out\_i(x)}$ is the x coordinate of the specific control point for the exit, $P_{out\_i(y)}$ is the y coordinate of the specific control point for the exit, $I_{out\_i(x)}$ is the x coordinate of the point, which describes the exit geometry of the roundabout, $I_{out\_i(y)}$ is the y coordinate of the point, which describes the exit geometry of the roundabout, W_{out} is the width of the exit.

While comparing the equations Eq. 3 and Eq. 4, it can be seen that, in order to avoid discontinuities in the planned path, the control points $\mathbf{P}_{in\_5}$ and $\mathbf{P}_{out\_0}$ are coincident. It can be also seen that the control points of the proposed equations are directly related.

Moving Through. In this scenario the planned path is divided into two parts: 1. Moving through the circular part of the roundabout; 2. Exiting the roundabout. While taking into consideration the basics of the Bezier curves and the already described assumptions, for moving through the circular part it is proposed to use the sixth order Bezier curve. The proposed equation for the control point selection, while seeking to perform the path planning in order to move through the circular part of the roundabout:

$$\mathbf{P}_{c} = \begin{cases} \mathbf{P}_{c\_0}\left(P_{in\_5(x)}, P_{in\_5(y)}\right) \\ \mathbf{P}_{c\_1}\left(P_{c\_0(x)} + \frac{P_{c\_2(x)} - P_{c\_0(x)}}{2}, P_{c\_0(y)} + \frac{P_{c\_2(y)} - P_{c\_0(y)}}{2}\right) \\ \mathbf{P}_{c\_2}\left(P_{c\_0(x)} + \frac{\left(0.5 \bullet I_{out\_22(y)} + 0.5 \cdot I_{out\_24(y)}\right) - P_{in\_5(y)}}{\tan\gamma}, \frac{I_{out\_22(y)} + I_{out\_24(y)}}{2}\right) \\ \mathbf{P}_{c\_3}\left(S_{min} + 2 \cdot r_{ins} + 0.2 \cdot W_{D}, S_{min} + r_{ins}\right) \\ \mathbf{P}_{c\_4}\left(P_{c\_6(y)} + \frac{P_{in\_5(x)} - \left(0.5 \cdot I_{in\_22(y)} + 0.5 \cdot I_{in\_24(y)}\right)}{\tan\gamma}, \frac{I_{in\_22(y)} + I_{in\_24(y)}}{2}\right) \\ \mathbf{P}_{c\_5}\left(\frac{P_{c\_4(x)} + P_{c\_6(x)}}{2}, \frac{P_{c\_4(y)} + P_{c\_6(y)}}{2}\right) \\ \mathbf{P}_{c\_6}\left(\frac{r_{ins} + r_{cen}}{2} + \cos\beta_{3-2} + C_{x}, \frac{r_{ins} + r_{cen}}{2} + \sin\beta_{3-2} + C_{y}\right) \end{cases}, \tag{5}$$

where $P_{c\_i(x)}$ is the x coordinate of the specific control point for moving in the circular part of the roundabout, $P_{c\_i(y)}$ is the y coordinate of the specific control point for moving in the circular part of the roundabout, γ is the angle between the horizontal axis and the positioned control points, i.e., the vector $\overrightarrow{\mathbf{P}_{c\_0}\mathbf{P}_{c\_2}}$, W_D is the width of the circulatory roadway.

The proposed equation for the control points selection in order to exit the roundabout:

$$\mathbf{P}_{out} = \begin{cases} \mathbf{P}_{out\_0}\left(P_{c\_6(x)}, P_{c\_6(y)}\right) \\ \mathbf{P}_{out\_1}\left(\frac{P_{out\_0(x)} + P_{out\_2(x)}}{2}, \frac{P_{out\_0(y)} + P_{out\_2(y)}}{2}\right) \\ \mathbf{P}_{out\_2}\left(I_{out\_34(x)}, I_{out\_34(y)}\right) \\ \mathbf{P}_{out\_3}\left(P_{in\_2(x)}, P_{out\_5(y)} - P_{in\_2(y)}\right) \\ \mathbf{P}_{out\_4}\left(P_{in\_1(x)}, P_{out\_5(y)} - P_{in\_1(y)}\right) \\ \mathbf{P}_{out\_5}\left(S_{min} + r_{ins} + \frac{W_S}{2} + \frac{W_{out}}{2}, 2 \cdot S_{min} + 2 \cdot r_{ins}\right) \end{cases} \tag{6}$$

Left Turn. In the left turn scenario, the planned path is also divided into two parts. In order to perform the path planning for the left turn manoeuvre in the roundabout, as the arc length gets bigger, for the circular part path planning it is proposed to use the eight order Bezier curve. The proposed equation for the control point selection:

$$\mathbf{P}_c = \begin{cases} \mathbf{P}_{c\_0}\left(P_{in\_5(x)}, P_{in\_5(y)}\right) \\ \mathbf{P}_{c\_1}\left(P_{c\_0(x)} + \frac{P_{c\_2(x)} - P_{c\_0(x)}}{2}, P_{c\_0(y)} + \frac{P_{c\_2(y)} - P_{c\_0(y)}}{2}\right) \\ \mathbf{P}_{c\_2}\left(P_{c\_0(x)} + \frac{\left(0.5 \cdot I_{out\_22(y)} + 0.5 \cdot I_{out\_24(y)}\right) - P_{in\_5(y)}}{\tan\gamma}, \frac{I_{out\_22(y)} + I_{out\_24(y)}}{2}\right) \\ \mathbf{P}_{c\_3}\left(1.15 \cdot S_{min} + 2 \cdot r_{ins}, S_{min} + r_{ins}\right) \\ \mathbf{P}_{c\_4}\left(1.15 \cdot S_{min} + 2 \cdot r_{ins}, 1.3 \cdot \left(\frac{r_{ins} + r_{cen}}{2} + \sin\beta_{3-2} + C_y\right)\right) \\ \mathbf{P}_{c\_5}\left(S_{min} + r_{ins}, 1.25 \cdot S_{min} + 2 \cdot r_{ins},\right) \\ \mathbf{P}_{c\_6}\left(\frac{I_{in\_32(x)} + I_{in\_34(x)}}{2}, \left(\frac{r_{ins} + r_{cen}}{2} + \sin\beta_{4-3} + C_y\right) + \frac{0.5 \cdot I_{in\_32(x)} + 0.5 \cdot I_{in\_34(x)} - P_{c\_8(x)}}{\tan\gamma}\right) \\ \mathbf{P}_{c\_7}\left(\frac{P_{c\_6(x)} + P_{c\_8(x)}}{2}, \frac{P_{c\_6(y)} + P_{c\_8(y)}}{2}\right) \\ \mathbf{P}_{c\_8}\left(\frac{r_{ins} + r_{cen}}{2} + \cos\beta_{4-3} + C_x, \frac{r_{ins} + r_{cen}}{2} + \sin\beta_{4-3} + C_y\right) \end{cases} \tag{7}$$

The equation for the control points selection in order to exit the roundabout:

$$\mathbf{P}_{out} = \begin{cases} \mathbf{P}_{out\_0}\left(P_{c\_8(x)}, P_{c\_8(y)}\right) \\ \mathbf{P}_{out\_1}\left(\frac{P_{out\_0(x)} + P_{out\_2(x)}}{2}, \frac{P_{out\_0(y)} + P_{out\_2(y)}}{2}\right) \\ \mathbf{P}_{out\_2}\left(I_{out\_44(x)}, I_{out\_44(y)}\right) \\ \mathbf{P}_{out\_3}\left(P_{in\_2(y)}, P_{in\_2(x)}\right) \\ \mathbf{P}_{out\_4}\left(\frac{P_{out\_3(x)} + P_{out\_5(x)}}{2}, \frac{P_{out\_3(y)} + P_{out\_5(y)}}{2}\right) \\ \mathbf{P}_{out\_5}\left(0, I_{out\_41(y)} - \frac{W_{out}}{2}\right) \end{cases} \tag{8}$$

"U" Turn. In this scenario the path planning is also divided. Because during the "U" turn manoeuvre the arc length gets even bigger, for the circular part it is proposed to use the tenth order Bezier curve. The proposed equation for the control point selection:

$$\mathbf{P}_c = \begin{cases} \mathbf{P}_{c\_0}\left(P_{in\_5(x)}, P_{in\_5(y)}\right) \\ \mathbf{P}_{c\_1}\left(P_{c\_0(x)} + \frac{P_{c\_2(x)} - P_{c\_0(x)}}{2}, P_{c\_0(y)} + \frac{P_{c\_2(y)} - P_{c\_0(y)}}{2}\right) \\ \mathbf{P}_{c\_2}\left(P_{c\_0(x)} + \frac{\left(0.5 \cdot I_{out\_22(y)} + 0.5 \cdot I_{out\_24(y)}\right) - P_{in\_5(y)}}{\tan\gamma}, \frac{I_{out\_22(y)} + I_{out\_24(y)}}{2}\right) \\ \mathbf{P}_{c\_3}\left(1.25 \cdot S_{min} + 2 \cdot r_{ins}, S_{min} + r_{ins}\right) \\ \mathbf{P}_{c\_4}\left(1.25 \cdot S_{min} + 2 \cdot r_{ins}, 1.2 \cdot \left(\frac{r_{ins} + r_{cen}}{2} + \sin\beta_{3-2} + C_y\right)\right) \\ \mathbf{P}_{c\_5}\left(S_{min} + r_{ins}, 1.5 \cdot S_{min} + 2 \cdot r_{ins}\right) \\ \mathbf{P}_{c\_6}\left(0.25 \cdot S_{min}, 1.2 \cdot \left(\frac{r_{ins} + r_{cen}}{2} + \sin\beta_{4-3} + C_y\right)\right) \\ \mathbf{P}_{c\_7}\left(0.25 \cdot S_{min}, S_{min} + r_{ins}\right) \\ \mathbf{P}_{c\_8}\left(P_{c\_10(x)} - \frac{\left(0.5 \cdot I_{out\_22(y)} + 0.5 \cdot I_{out\_24(y)}\right) - P_{in\_5(y)}}{\tan\gamma}, \frac{I_{out\_42(y)} + I_{out\_44(y)}}{2}\right) \\ \mathbf{P}_{c\_9}\left(\frac{P_{c\_8(x)} + P_{c\_10(x)}}{2}, \frac{P_{c\_8(y)} + P_{c\_10(y)}}{2}\right) \\ \mathbf{P}_{c\_10}\left(\frac{r_{ins} + r_{cen}}{2} + \cos\beta_{5-4} + C_x, \frac{r_{ins} + r_{cen}}{2} + \sin\beta_{5-4} + C_y\right) \end{cases} \tag{9}$$

The equation for the control points selection in order to exit the roundabout:

$$\mathbf{P}_{out} = \begin{cases} \mathbf{P}_{out\_0}\left(P_{c\_10(x)}, P_{c\_10(y)}\right) \\ \mathbf{P}_{out\_1}\left(\frac{P_{out\_0(x)} + P_{out\_2(x)}}{2}, \frac{P_{out\_0(y)} + P_{out\_2(y)}}{2}\right) \\ \mathbf{P}_{out\_2}\left(I_{out\_44(x)}, I_{out\_44(y)}\right) \\ \mathbf{P}_{out\_3}\left(2 \cdot S_{min} + 2 \cdot r_{ins} - P_{in\_1(x)}, P_{in\_1(y)}\right) \\ \mathbf{P}_{out\_4}\left(2 \cdot S_{min} + 2 \cdot r_{ins} - P_{in\_2(x)}, P_{in\_2(y)}\right) \\ \mathbf{P}_{out\_5}\left(S_{min} + r_{ins} - \frac{W_S}{2} - \frac{W_{in}}{2}, 0\right) \end{cases} \tag{10}$$

Although all of the described path planning stages are related, it must be understood that, at this approach development stage, the proposed equations for the control points selection serve more as a primary tool, required to demonstrate that the high order Bezier curves can be efficiently applied to the path planning.

3 Example of the Proposed Approach Application

The proposed approach was tested while assuming that: the velocity v_a is 8 m/s, r_{ins} is 19 m, r_{cen} is 14.5 m, φ is 5°, W_S is 4 m, W_D is 4.5 m, W_{in} and W_{out} each are 3.5 m, t_d is 0.9 s. The simulations were performed while using the *MATLAB/Simulink* software package. Figures 2, 3, 4 and 5 present the results of the performed simulations.

From the provided results it can be seen that, in all of the described different roundabout scenarios, the application of the high order Bezier curves was successful: the planned paths did not violate the geometric constraints of the roundabout, there were no discontinuities between the planned paths, while taking into consideration the different parts of the roundabout, and the planned paths were feasible for the autonomous vehicle (Figs. 2 and 3, part a). Respectively, it can be also seen that each part of the moving in the roundabouts starts and ends when the steering angle of the autonomous vehicle is $\delta = 0°$ (Figs. 2 and 3, part b).

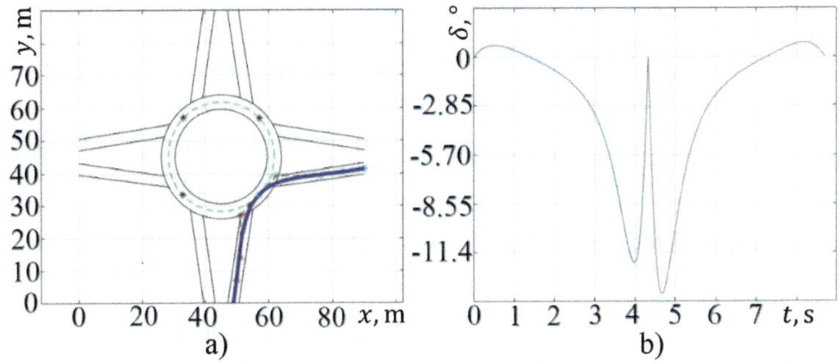

Fig. 2. Example of the right turn scenario: a – planned path; b – steering angle values.

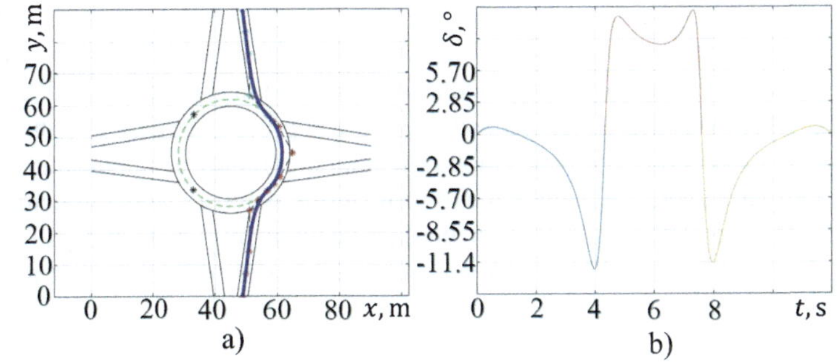

Fig. 3. Example of the moving through scenario: a – planned path; b – steering angle values.

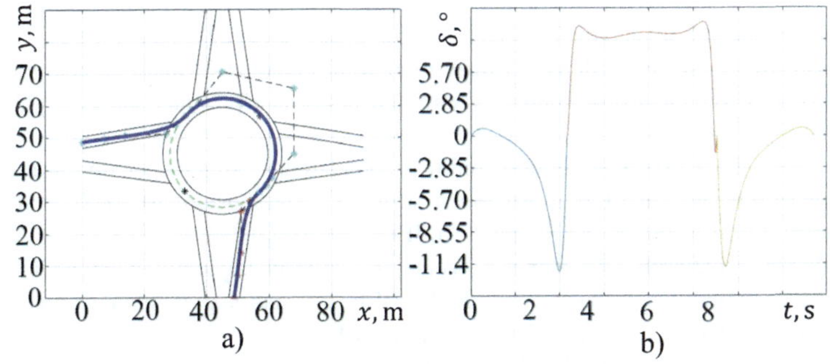

Fig. 4. Example of the left turn scenario: a – planned path; b – steering angle values.

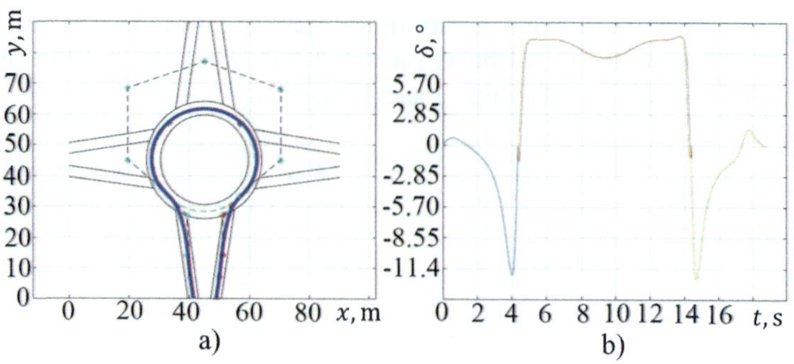

Fig. 5. Example of the "U" turn scenario: a – planned path; b – steering angle values.

Based on the results it can be stated that a successful path planning, while using the high order Bezier curves, is definitely possible and should be further developed.

4 Conclusions

This paper described a possible way to define the control points for the high order Bezier curves, while having roundabouts in mind. It must be explained that, in the current state, the proposed equations and the entire approach are definitely not prepared to employ in autonomous vehicles. A lot of development, improvements and testing are still required. However, the main achievement of this manuscript is the demonstration that the high order Bezier curves can be successfully applied to the path planning. Thus, the future work is to further develop this approach and to apply it in an actual autonomous vehicle.

References

1. Skrickij, V., Šabanovič, E., Žuraulis, V.: Autonomous road vehicles: recent issues and expectations. IET Intel. Transport Syst. **14**(6), 471–479 (2020)
2. Morsali, M., Frisk, E., Aslund, J.: Spatio-temporal planning in multi-vehicle scenarios for autonomous vehicle using support vector machines. IEEE Trans. Intell. Vehicles, 1–11 (2020)
3. Silva, A. R. J., Grassi Jr, V.: Path planning at roundabouts using piecewise linear continuous curvature curves. In: Latin American Robotics Symposium (LARS) and Brazilian Symposium on Robotics (SBR), pp. 1–6. IEEE, Brazil (2017)
4. Hidalgo, C., Lattarulo, R., Perez, J., Asua, A.: Hybrid trajectory planning approach for roundabout merging scenarios. In: 2019 IEEE International Conference on Connected Vehicles and Expo, pp. 1–6. IEEE, Austria (2019)
5. Silva, A. R. J., Grassi Jr, V.: Clothoid-based global path planning for autonomous vehicles in urban scenarios. In: IEEE International Conference on Robotics and Automation (ICRA), pp. 4312–4318. IEEE, Australia (2018)
6. Gonzalez, D., Perez, J., Milanes, V.: Parametric-based path generation for automated vehicles at roundabouts. Expert Syst. Appl. **71**(1), 332–341 (2017)

7. Hidalgo, C., Lattarulo, R., Perez, J., Asua, E.: Hybrid trajectory planning approach for roundabout merging scenarios. In: IEEE International Conference on Connected Vehicles and Expo (ICCVE), pp. 1–6. IEEE, Austria (2019)
8. Lattarulo, R., Gonzalez, L., Perez, J.: Real-time trajectory planning method based on n-order curve optimization. In: 24th International Conference on System Theory, Control and Computing (ICSTCC), pp. 751–756. IEEE, Romania (2020)
9. Perez, J., Godoy, J., Villagra, J., Onieva, E.: Trajectory generator for autonomous vehicles in urban environments. In: IEEE International Conference on Robotics and Automation (ICRA), pp. 409–414. IEEE, Germany (2013)
10. Rastelli, J.P., Penas, M.S.: Fuzzy logic steering control of autonomous vehicles inside roundabouts. Appl. Soft Comput. **35**, 662–669 (2015)
11. Lattarulo, R., Gonzalez, L., Marti, E., Matute, J., Marcano, M., Perez, J.: Urban Motion Planning Framework Based on N-Bézier Curves Considering Comfort and Safety. Journal of Advanced Transportation, 1–13 (2018)
12. Lattarulo, R., Perez, J.: Fast Real-Time Trajectory Planning Method with 3rd-Order Curve Optimization for Automated Vehicles. In: IEEE 23rd International Conference on Intelligent Transportation Systems (ITSC), pp. 1–6, IEEE, Greece (2020)
13. Hsu, T. W., Liu, J. S.: Design of smooth path based on the conversion between η^3 spline and Bezier curve. In: American Control Conference (ACC), pp. 3230–3235, IEEE, USA (2020)

Human - Machine Interaction, Behavioural Sciences

Fatigue Risk Assessment of Cabin Crew for Aviation Transport Safety

Irem Çevik(✉), Kalle Elfvengren, and Ajantha Dahanayake

School of Engineering Science, Lappeenranta-Lahti University of Technology, 53850 Lappeenranta, Finland
irem.cevik@student.lut.fi, {kalle.elfvengren, ajantha.dahanayake}@lut.fi

Abstract. Airlines recognize fatigue as a threat to flight safety. Fatigue risk management (FRM) is crucial to maximizing safety levels. FRM necessitates measuring and assessing fatigue, sleep, and workload to identify risks, and to determine associated risk indicators, risk severity, risk probability, and risk level. This research evaluates the causes of fatigue and their risks in order to develop an effective fatigue risk assessment system for aviation cabin crew, in collaboration with four commercial airlines. The bowtie method is used to assess fatigue causes, risk identification, and management strategies. This study identifies fatigue causes and risk-related components for developing a practical fatigue risk assessment tool for the aviation cabin crew.

Keywords: Aviation transport safety · Fatigue risk assessment · Cabin crew · Fatigue · Bowtie analysis

1 Introduction

Aviation cabin crew members work long hours, overtime, in shifts, and deal with the differences in time zones during long-distance flights, all of which contribute to fatigue. Operational data is typically collected to estimate the safety risks associated with fitness and performance, though metrics and collection methods vary among airlines.

As yet, there is no integrated system to address measuring fatigue for aviation cabin crews. This article focuses on the reliability of subjective and objective measurements using sleep and fatigue data. It presents relationships between sleep parameters, fatigue and wearable device data to measure sleep, and gauges the vigilance of airline workers' fit-to-fly assessments.

This research aims to determine the hazards and risks that threaten flight safety caused by cabin crew fatigue. This study intends to address the following research question: What causes and consequences of risks can constitute a practical fatigue risk assessment tool for the aviation cabin crew?

O. Prentkovskis et al. (Eds.): TRANSBALTICA 2021, LNITI, pp. 765–774, 2022.
https://doi.org/10.1007/978-3-030-94774-3_73

The research examines case studies from four commercial airlines; interviews four airlines to explore their fatigue risk assessment methods, including data collection, risk assessment, and mitigation strategies; and undertakes a bowtie analysis to identify fatigue-related causes and risks for cabin crews.

2 Background

Fatigue is a widespread problem, affecting mental and physical well-being and disrupting business performance [1]. Individuals may face decreased response time, reduced wakefulness, diminished decision-making ability, poor judgment, etc. Fatigue has many causes, e.g. insufficient sleep, shift work, workload, circadian variability, mismanagement. Also, the sleep-wake cycle is affected by fatigue [2]. Adverse effects inevitably cause accidents and other incidents in the airline sector, where risk and safety have always been significant concerns [3]. The modern approach to enterprise risk management is by implementing risk management systems. Safety frameworks and risk management need to be successfully integrated into organizations' safety policies [4].

2.1 Fatigue Risk Management Regulations in Aviation Industry

A broad international consensus has emerged across several shift working sectors. A structured method called the 'Fatigue Risk Management System' (FRMS) is widely seen as the best way to handle and reduce employee fatigue risk [5]. The ICAO (International Civil Aviation Organization) Doc 9966 manual defines several elements for the oversight of fatigue management approaches. The causes and effects of fatigue need to be recognized by regulators, employers, and employees, and one of the primary purposes of air operators' regulations is to restrict workers' exposure to fatigue-related risks [6]. There are several models on which aircraft and air transport operations base their risk and safety management: the Swiss Cheese Model, Fault tree analysis (FTA), Common cause analysis (CCA), bowtie analysis [7, 8].

Bowtie Risk Analysis. Bowtie analysis is a technique for qualitative risk assessment. It offers a means of quickly communicating complex risk situations in a visual format that is easy to understand, illustrating the relationships between the causes of unwished incidents and the potential for failure and harm to escalate. The bowtie shows hazard-specific steps to prevent a trigger event from occurring, and outcome-specific recovery steps ready to limit potential effects should the trigger event occur anyway [9, 10]. The bowtie diagram organizes the "causes" or "threats," "events", and "consequences", as shown in Fig. 1, with "controls" inserted to prevent hazards from turning into consequences [10]. In aviation, safety standards levels are high, and communication and interaction systems imply a shift in opportunity for safety and performance improvement [12, 13].

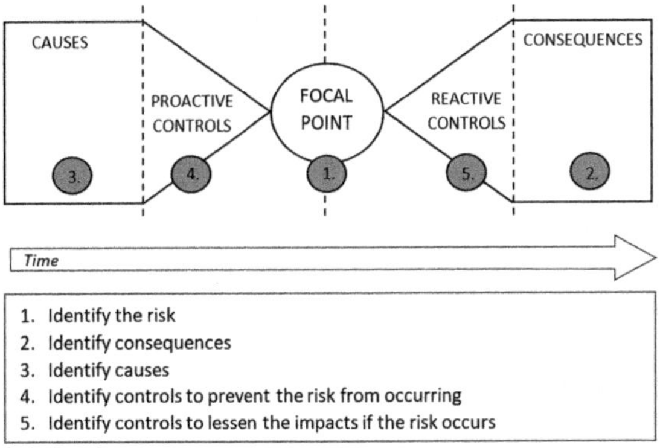

Fig. 1. The bowtie diagram [11].

3 Case Studies

Four commercial airlines took part in the case studies, providing interviews. The airlines' names are not disclosed due to privacy and security agreements.

3.1 Airlines' Risk Assessment Procedures

Operators design and define FRMS policy to achieve their safety targets. An action group usually leads it. All four airlines comply with the Fatigue Risk Management System according to their activity and organizational structure. Table 1 presents the scope of the FRMS procedures of four airlines.

Table 1. FRM processes of 4 airlines.

Processes	Airline 1	Airline 2	Airline 3	Airline 4
FRMS implemented as an incremental part of SMS (Safety Management System)	✓	✓	✓	✓
Fatigue-related hazard identification	✓	✓	✓	✓
FRMS Organization	✓	✓		
Fatigue Working Team in place	✓	✓		✓
Risk Assessment	✓	✓	✓	✓
Risk Mitigation	✓	✓	✓	✓
Control & Re-Assessment			✓	

Data Collection. The estimated level of fatigue risk is the most crucial factor to consider in choosing fatigue assessment steps. Data collection and analysis interventions need resources. Table 2 shows the data collection methods of these four airlines.

Table 2. Data collection methods of 4 airlines.

Data collection	Airline 1	Airline 2	Airline 3	Airline 4
Surveys and audits	✓	✓	✓	✓
Reports such as hazard, safety reports	✓	✓	✓	✓
Flight Data Monitoring (FDM)	✓	✓	✓	✓
Calls of fatigue from cabin crew		✓		
Customized Rostering Module	✓	✓	✓	✓
Investigations related to incidents and accidents	✓	✓	✓	✓
Gap Analysis		✓		
Following trends, studies for FRMS processes		✓	✓	

Airline 1. The collection and review of data is part of the day-to-day FRMS operation, and various information is available for hazard identification and risk assessment. Airline 1 has introduced a monitoring framework for reporting systems, biomathematical models and collects relevant data from multiple sources.

Airline 2. Data collection includes reports, flight schedules, operational data, and resources provided by current scientific studies.

Airline 3. For fatigue risk management, airline employees are tracked and monitored through data obtained from various sources and FRMS policy.

Airline 4. Central to the company culture, fatigue-related reporting plays a vital role in airline 4. Reports notified by the employees are recorded in the database of the safety department.

Airlines' Risk Assessment Techniques. When the regulator and company have identified a fatigue threat, it is essential to determine the risk it presents and to decide whether to minimize it. Safety Management System (SMS) concepts are accompanied by fatigue risk assessment. Operators develop and implement risk evaluation protocols to determine the probability and possible seriousness of fatigue-related incidents, and to determine when associated risks can be minimized.

Airline 1. The methods of fatigue risk assessment are compatible with the methods used in SMS, integrating risk detection, review, assessment, and mitigation, but tailored to assess fatigue risk. Airline 1 determines the potential for harm, damage to facilities, or failure due to a fatigue hazard, and offers risk management guidelines. The risk level is determined by evaluating the flight duty, rest, and operational indicators. The severity level is assessed according to the risk scale.

Airline 2. SMS concepts are followed by fatigue risk assessment. Airline 2 measures the potential for fatigue-related injury, damage to facilities, or failure, and guides management of the risk. Airline 2 takes subjective assessments from reports presented as the first step of the evaluation process.

Airline 3. Airline 3 carries out risk assessment methods under the "Safety Risk Management" part of the Custom in-House Module, including probability and severity.

Airline 4. The safety department carries out a risk assessment. After the hazard is accepted, the safety department evaluates the probability of an unwanted event and the severity of an unwanted event's outcome. Besides, risk exposure is investigated by the group responsible.

Mitigation. When action is deemed required for a specific fatigue risk, controls and mitigations must be defined and enforced.

Airline 1. Airline 1 determines whether to prevent or eliminate the risk due to fatigue based on the data. The responsible staff examines the rest period and working roster. Depending on the research, changes are made to resting periods and accommodation places.

Airline 2. Airline 2 implements fatigue reduction strategies to ensure compliance with safety requirements. The FRM team evaluates and offers solutions to reduce the risk of fatigue. If the safety directorate department approves the proposals, the responsible department informs the relevant units and persons. Scales are used to evaluate the effectiveness of risk reduction recommendations. The processes are monitored continuously by the personnel or team responsible for the risk analysis.

Airline 3. While determining risk reduction methods, the airline primarily tracks rest and working periods. Circadian effects on sleep and waking times are analyzed. Airline 3 tracks the applicability of the measures taken and the results of the application.

Airline 4. Monitors flight crew working hours, flight period tracking, and control according to the Flight Time Limitations (FTL) schedule to reduce the risk level. As a preventive, they aim to minimize sleep loss. The process considers rest days before and after the flight, rest periods depending on the day or night flights, and creating the right sleep environment based on operational factors.

3.2 Bowtie Analysis of Fatigue Risk Assessment

A risk, defined as any threat that an event or action that will adversely affect business and objectives, requires assessing the consequences of a hazard expressed in terms of the predicted probability and severity. Scales are typically applied to determine the effects of human factors and to provide feedback when risk is measured in terms of its probability and severity. The airline establishes the scale and decides what each point on the scale implies [9, 10]. The probability assessment is based on indicators. Table 3 presents a likelihood scale for hazard assessment.

Table 3. The assessment of probability.

Rare	Sometimes	Very probable
- Almost never - Unlikely - May occur only in exceptional circumstances	- Occasionally - Might occur sometimes - Will probably occur at some time	- Frequent - Expected to occur frequently

Table 4 shows the severity scale, which considers loss or harm caused or caused.

Table 4. The assessment of severity.

Minor	Harmful	Severe
- Slight - Negligible - No injuries - The worst possible injury is minor	- Critical - Serious incident - The worst possible injury is serious - Equipment damage	- Catastrophic - Serious injury - Fatal - The worst possible injury is deadly - Equipment destroyed

The software application applied to visualize Fig. 2 and Fig. 3 is BowTieXP supplied by CGE Risk Management Solutions B.V, The Netherlands (www.cgerisk.com/products/bowtiexp/). In Fig. 2, the probability and severity levels are combined into a 'risk category', and represented by colour coding. The risk matrix determines the risk weight, and contributes to the evaluation of safety-related events. The action step is determined according to the risk type.

The primary purpose of hazard identification and risk assessment is to provide and maintain safe working conditions for employees. Therefore, it is essential to identify hazards that can cause injury to personnel, death, material loss, etc. Table 5 details risks related to fatigue.

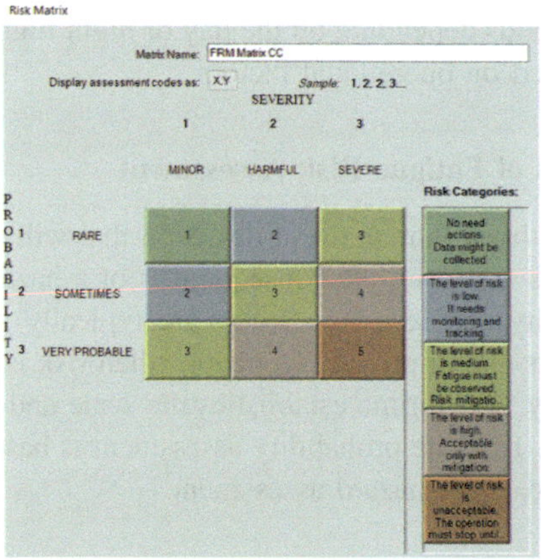

Fig. 2. Risk matrix.

Table 5. Identified risk list.

	Identified risk	Severity	Probability
1.	Serious event in cabin	3	1
2.	Crew incapacitation	3	2
3.	Deviation from standards	2	2
4.	Well-being disruptions	2	2
5.	Sleep disruption	3	3
6.	Disruption of circadian synchronization	3	2
7.	Disability	3	1
8.	Poor performance	3	2
9.	Mood changes	3	2
10.	Reputation	3	2
11.	Financial impact	3	2

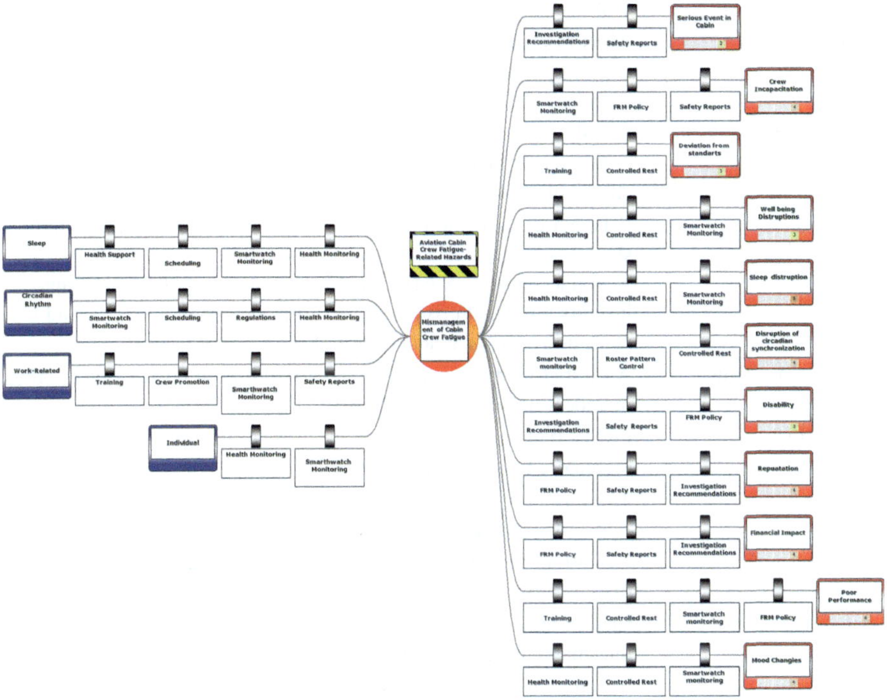

Fig. 3. The bowtie diagram to manage cabin crew fatigue.

Categorizing risks contributes to the development of risk improvement and reduction methods. In Fig. 3, the bowtie begins with identifying hazards necessary to analyze sleep, workload, circadian rhythm, and individual factors that cause fatigue under four main categories. Identifying the risk can be defined as the most critical step in the analysis, as it is where control of the hazard starts. This identification is called the

Top Event [10], where cabin crews and airlines expose the hazard's potential harm. Loss, damage, and undesirable events are the consequences. Barriers are then placed on both sides of the bowtie to prevent unwanted occurrences.

Bowtie provides a visual summary of possible accident and incident scenarios that resulting from the hazard. Threats and situations that may pose a danger are examined in four leading indicators in the study. Indicators and proactive/reactive controls are determined depending on the literature review, company policies, sectoral approaches, individual differences, and technological developments. In the risk assessment, one of the hazard categories is defined as sleep. Sleep deprivation and irregular sleep affect flight safety. Regulating work and rest periods, eating healthy, and monitoring sleep records with smartwatches can help develop an action plan to reduce cabin crew exposure to the risks caused by fatigue. The controls that can minimize the impact of events and brief explanations are given in Table 6.

Table 6. List of reactive controls.

Barriers	Explanation
Health support	Aims to provide health support for cabin crew. Airlines can offer their employees various forms of health support, such as diet, exercise
Scheduling	Scheduling is a flight sequence designed to satisfy operational requirements and manage resources, including cabin crew, effectively [14]
Smartwatch monitoring	Personal data, including body temperature, heart rhythm pattern, sleep tracking, physical exercises, etc., are obtained through wearable sensors. A large amount of data can potentially improve insights into health risks and performance, monitor effects, and take measurements for guidance [12, 13]
Health monitoring	An initial medical evaluation is carried out on all new cabin crew members in the aviation industry [14]
Regulations	FRMS regulations establish a mechanism for operators and regulators to handle fatigue risk [14]
Training	Training programs are designed to ensure that all stakeholders can fulfill their FRMS responsibilities [14]
Crew promotion	Promoting crew to identify in-field hazards and risks leading to the organization to mitigate them [14]
Safety reports	Cabin crews should be encouraged to report events and situations that may endanger flight safety due to fatigue [14]
Investigation recommendations	Reactive mitigation, following an investigation, to avoid recurrence. These recommendations can also be generated as lessons to learn
FRM Policy	Airline and stakeholders must define the elements and scope of the FRMS [14]
Controlled rest	An effective fatigue reduction for flight crews is managed rest on the flight deck [14]
Roster pattern control	Working times can be arranged or improved by creating appropriate rostering. For example, additional steps can be taken during extended or night shifts [14]

The risks that cause fatigue are shaped with the Bowtie model. Before initiating the required preventative activities, the model process attempts to ensure proper preparation. Risk indicators are also used to collect data that will be used to specify how adequate the controls are. This model is created with the concept of constant improvement in mind. It constructs a theoretical demonstration and visualization that needs reliability testing. Once adapting the data from the risk analysis, proactive and reactive controls are re-examined and formulated.

4 Conclusion

This study has identified the main categories of a cabin crew fatigue risk assessment tool as being sleep, workload, circadian rhythm, and individual factors resulting from the bowtie analysis conducted on four airlines. These four categories need to be components of a practical fatigue risk assessment tool for the aviation cabin crew and data collection strategies. The aim is to create an effective evaluation system for cabin crews to manage fatigue, reduce risk, prevent adverse events, and eliminate shortcomings. Identifying four main categories that constitute a reliable cabin crew fatigue risk assessment tool is a novel contribution. It is crucial to define, proactively identify, and apply practical approaches to address flight safety risks. The risk assessment tool can serve as a new framework for the assessment of cabin crew' fatigue.

This paper combines different approaches to arrive at its results. First, the risk assessment strategies of various airlines are discovered through interviews. Afterward, the leading and lagging indicators of fatigue are investigated with a literature review. The bowtie analysis is used to identify fatigue-related hazards and mitigate fatigue-related risks. The assessment reflects the identification of fatigue-related hazards, preventive and corrective actions to ensure that fatigue is at an acceptable level. These results point to a future initiative towards implementing the identified four main fatigue categories as a decision support tool for assessing the cabin crew's fatigue level and fit to fly assessment to assure aviation transport safety.

References

1. Akerstedt, T.: Consensus statement: fatigue and accidents in transport operations. J. Sleep Res. **9**(4), 395 (2000)
2. Lerman, S.E., et al.: American College of Occupational and Environmental Medicine Presidential Task Force on Fatigue Risk Management. J. Occup. Environ. Med. **54**(2), 231–258 (2012)
3. Janić, M.: An assessment of risk and safety in civil aviation. J. Air Transp. Manage. **6**(1), 43–50 (2000)
4. Yazdi, M.: The application of Bow-tie method in hydrogen sulfide risk management using layer of protection analysis (LOPA). J. Fail. Anal. and Preven. **17**, 291–303 (2017)
5. Moore-Ede, M.: Evolution of fatigue risk management systems: The “tipping point” of employee fatigue mitigation. CIRCADIAN® White Paper (2009)
6. Gander, P., et al.: Fatigue risk management: Organizational factors at the regulatory and industry/company level. Accid Anal Prev. **43**(2), 573–590 (2011)

7. Netjasov, J., Janic, M.: A review of research on risk and safety modeling in civil aviation. J. Transp. Manage. **14**(4), 213–220 (2008)
8. Culwick, M.D., Merry, A.F., Clarke, D.M., Taraporewalla, K.J., Gibbs, N.M.: Bow-tie diagrams for risk management in anesthesia. Anaesth Intensive Care. **44**(6), 712–718 (2016)
9. Voicu, I., Panaitescu, F.V., Panaitescu, M., Dumitrescu, L.G., Turof, M.: Risk management with Bowtie diagrams. ModTech International Conference – Modern Technologies in Industrial Engineering VI. Constanta, Romania (2018)
10. de Ruijter, A., Guldenmund, F.: The Bowtie Method: A Review. Safety Science, Elsevier Ltd., vol. 88, pp. 211–218 (2016)
11. RiskInsight. https://www.riskinsightconsulting.com/do-you-use-bowties. Accessed 27 May 2021
12. Valdes, R.A., Gomez, C.V.: Aviation 4.0. More safety through Automation and digitization. In: Brebbia, C.A. (ed.) WIT Transactions on the Built Environment, pp. 225–236. WIT Press (2018)
13. Neis, S.M., Blackstun, M.I.: Feasibility analysis of wearables for use by airline crew. In: 2016 IEEE/AIAA 35th Digital Avionics Systems Conference (DASC); 2016 Sept 25–29, Sacramento, CA, USA. EEE, pp. 1–9 (2016)
14. International Civil Aviation Organization (ICAO). https://www.icao.int/safety/fatigue management/FRMS%20Tools/FRMS%20Implementation%20Guide%20for%20Operators %20July%202011.pdf. Accessed 27 May 2021

Semiquantitative Evaluation of Societal Vulnerability in Case of Long-Term Power Failure in Railway Stations

Michal Szatmári(✉), Mária Lusková, and Bohuš Leitner

Faculty of Security Engineering, University of Žilina,
Univerzitna 1, 010 26 Žilina, Slovakia
{michal.szatmari,maria.luskova,
bohus.leitner}@uniza.sk

Abstract. Vulnerability means a degree of reference object to withstand external or internal threats. The reference object can have human or material character. The paper will address to the reference object like a human character concretely the societal vulnerability due to a power failure in the railway station and the specific event will be described in the form of a case study. Each part of the railway network, whether it be a depot, a station, a terminal or others, is largely dependent on the supply of the electricity network to its functional elements in the system. To assess the vulnerability, it should be quantified using available methods. Adequate measures must be taken to reduce societal vulnerability.

Societal vulnerability can be quantified by social indicators with their special characteristics. In the paper was use socio structural, socio economic and socio urban indicators.

Keywords: Societal vulnerability · Power failure · Railway station

1 Introduction

The whole system of railway infrastructure, not only the operation of individual objects, such as transport hubs, is dependent on a continuous supply of electricity. This is important especially with regard to the reliable and safe operation of important structural objects and most of the provided transport and other support services. Human resources can provide operational staff of the train set, manual sale of tickets, manual management of traffic on sections, as well as control of operational properties and security of the facility, but they cannot provide technical performance of transport of passengers and material by rail without technical means and reservations in the trainset or monitoring and identification of unwanted persons who are not authorized to enter the building or specific rooms and many other activities. In the current era of automation and technological progress, the company is jointly vulnerable to the need for electricity supply, with the area of ensuring the reliable operation of railway infrastructure and its permeable performance being among the most vulnerable in the field of land transport.

O. Prentkovskis et al. (Eds.): TRANSBALTICA 2021, LNITI, pp. 775–785, 2022.
https://doi.org/10.1007/978-3-030-94774-3_74

The concept of vulnerability can be defined and understood from the perspective of several research areas. Experts and authors differ in their definitions of vulnerability. Just a few examples.

- Vulnerability can be understood as one of the characteristics of the environment, which is not uniformly defined, it is assessed on the basis of various facts [1].
- The vulnerability of the developed world is increased in particular by the growing complexity and interconnectedness of modern societies. The economic and technological systems of modern societies represent systems of nodes such as the city center and the connection between them, which can be a railway line or power lines. The interdependence of individual nodes and connections is so great that the failure of one can cause the collapse of the entire system [2].
- Vulnerability generally means the property of any material object, technical device or social entity to lose the ability to perform its natural or established function as a result of external or internal threats of varying nature and intensity [3].
- The vulnerability of the object and the effect of the threat directly or indirectly condition the possibility of greater risk [4].
- Vulnerability is considered to be the result of the interaction between physical (territorial) characteristics and the sensitivity and ability of the socio-economic system to adapt and deal with a particular danger, which we express by a non-dimensional index [5].
- Deficiency, weakness or condition of the analyzed asset, or an entity, object, system, organization, region, or part thereof, that may be severely exploited to exert its adverse effects [6].
- Vulnerability is the ability of a system and its components consisting of elements to accept hazard in the form of loss and damage [7].
- Vulnerability is also represented by those parts of the object, or those elements of the protection system that do not provide the required degree of protection, are weak or easily overcome element, or create favorable conditions for endangering the object, increase the probability of attack and its success [8].

In general, however, vulnerability is not a negative factor. Its negative impact will only be felt if the activated threat exploits a specific vulnerability. This means that the reference object of interest (resources, services, material, people, etc.) must be exposed to an existing or potential threat, which by its negative effect creates a causal chain of cause creating a possible consequence. The consequence of a negative phenomenon can be understood as overcoming the vulnerability itself.

2 Societal Vulnerability of Passengers in Case of Long-Term Power Failure in Railway Station

The term vulnerability originated from the social sciences and was initially used almost exclusively in the social sphere. From the authors' point of view, the article will examine the societal vulnerability of (passengers, carrier, community,…) in the event of a power failure in the reference object of railway infrastructure - passenger-oriented railway station.

In security terminology, vulnerability is most often assessed using methods based on qualitative, quantitative or semi-quantitative assessment. These methods consist of a verbal description, a numerical expression or a combination thereof [9].

Societal vulnerability is a complex issue that needs to be assessed from a variety of perspectives. In general, a correct and objective assessment of vulnerabilities of any type requires finding answers to three key questions (see Fig. 1) [10].

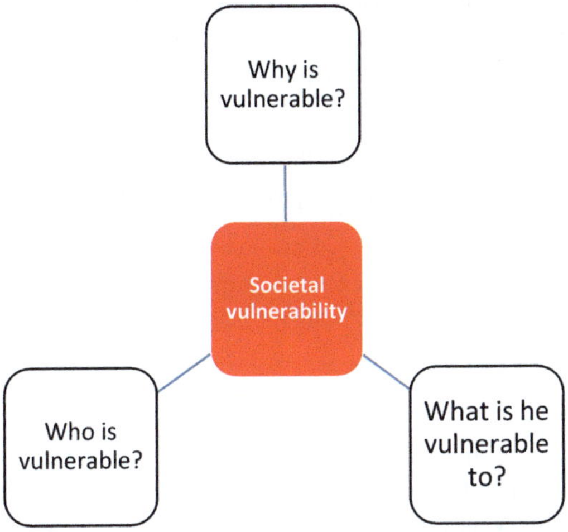

Fig. 1. Three key asks for societal vulnerability.

1. Who / whose vulnerability do we assess, respectively who is vulnerable? The aim is to identify the social actors of the system and their correlation.
2. Why is the object vulnerable? The answer is to identify phenomena and events that can cause social problems or social systems to exist.
3. What is he vulnerable to? The aim is to identify the primary causes causing a predisposition to expose the object to negative phenomena and events [10].

Societal vulnerability is a phenomenon relatively difficult to measure, so to assess the level of societal vulnerability, groups of so-called social indicators. We are by no means able to describe the variability of reality (phenomena and links between elements of the system) that we are examining with one indicator. For this reason, whole groups of indicators are usually used, and the interpretation of the data and dependencies obtained is then based on a combination of data obtained by evaluating the individual indicators. Social indicators are easily identifiable indicators that make it possible to examine or measure phenomena and processes in the societal environment. They represent a complex of parameters created using suitable quantitative methods, providing data on certain characteristics of societal life. They usually have a dynamic character, they change over time, and are the result of interactions between societal environment social actors and reveal some basic features of social reality. In general, the most frequently used indicators are derived from official statistics and are most often used to assess the level of development of society [10].

3 Societal Vulnerability of Passengers in Case of Long-Term Power Failure in Railway Station

The aim of the authors in the implementation of the case study was to predict and evaluation the expected impacts of long-term power outages in the railway station, focusing on the societal vulnerability of the building and its most important systems and services it provides. The number of passengers by rail in the conditions of the Slovak Republic has had an increasing tendency over the last four years (Table 1). For this reason, it is necessary for passenger transport to be as reliable as possible as one of the basic transport services in rail transport.

Table 1. Number of passengers by rail in the Slovak Republic for the period 2016–2019 [11].

Year	Number of passengers million/year
2016	65.61
2017	72.47
2018	73.81
2019	77.36

Disruption of the continuity of electricity supply can come from natural disasters, bad weather, technical failures, human factor, labor conflicts, sabotage, but also terrorism or military interventions. The fault has its starting point in the initiating event, which means that the threat carrier activates a negative scenario acting on the reference object [10].

3.1 Conditions and Limitations Considered in the Case Study

In terms of time, it is assumed that a significant part of the electricity infrastructure has been damaged by an extreme natural event. In view of the extent of the damage to the transmission and distribution network, the resumption of electricity supplies is not expected until several days later, unless an alternative solution is implemented. Scenarios and events that may have a negative effect on the reference object have been identified for the purpose of assessing societal vulnerability (see Table 2).

Table 2. The natural factor causing the event with its scenario.

Incident	Scenario
Earthquake	Electrical equipment loses its stability and the power lines are damaged
Flood	The area in which the railway station is located is flooded
	There will be flooding and short circuits in the server room of the railway station
Massive landslides	The slope loses its long-term stability and falls apart, which damages the energy supply
Extreme wind	The wind will cause considerable damage to the power lines

This means that the bearer of the threat in the case study is a natural factor and the result of a specific scenario will always be a longer-lasting interruption of electricity supply to the railway station building.

In the Table 3 is an overview of considered potential social actors, that mean clarification of who, what societal phenomenon and how they are vulnerable in the event of a long-term power outage at the railway station facilities.

Table 3. Who is vulnerable in the event of a power outage, why is he vulnerable and how.

Who is vulnerable?	How?	Why?
Passengers relying on rail transport	Impossibility to travel to work	Restrictions on the possibility of traveling by other means of transport
Unemployed	Poverty	Education and qualifications
Kids	Stress	Limited transportation options
Seniors	Impossibility to travel to the doctor	Social status
People with disabilities	Crime	Disability

Passengers who rely on rail transport (do not own a car, do not have a driver's license, etc.) in the event of a power failure, for example, do not have another mode of transport to the place of work or move to a greater distance, so their options are limited. During a long outage and the impossibility of transport to work, their job positions would be endangered.

Unemployed people often cannot afford road transport due to their financial side of imminent poverty. They often rely on rail transport. In addition, they may be vulnerable to their lack of education and qualifications, making it very difficult for them to find a job that would lead them to a better economic and social situation. They often have only limited opportunities in the form of gainful employment, such as a seasonal employment agreement or occasional temporary job offers without long-term stable employment. Due to a long-term power outage at the railway station, but they are losing the possibility of their work potential where they could be used.

Children may have a common way of going to school or certain places by rail. In the event of an outage and a necessary change in the usual way, they can be affected by a quick non-standard situation. As children cannot drive a car, they have limited means of transport to the place with the help of another person holding a driving license and a car or an alternative to public transport.

Seniors have a higher incidence of health problems and diseases than the younger generations. Many seniors have no one to transport during the day to a larger city for a medical examination and therefore often use rail transport. A power outage at a railway station and the cessation of the operation of a long-term interval may lead to the cancellation of their original intention to visit medical assistance on a given date. This could cause potential health problems or failure to obtain the necessary medication for their health problems. Their social status as people who are often reliant on the help of doctors makes them socially vulnerable actors.

People with disabilities due to a power outage at a railway station will potentially be forced to use another form of passenger transport. These people usually cannot afford to spend money on a taxi service and must therefore use other passenger transport options. Boarding points, for example for bus transport, do not have to be in close proximity to a railway station and are therefore sometimes forced to pass around the environment in which the magnets of crime are located [12]. Potential attackers may see them as an easy target, as they are predisposed to be unable to defend themselves as a disabled person against a robbery or violent attack [13].

We need to assess societal vulnerability in the event of a power failure at the selected reference station of the railway station, we have set the following three indicators, which contain their specific characteristics (see Fig. 2).

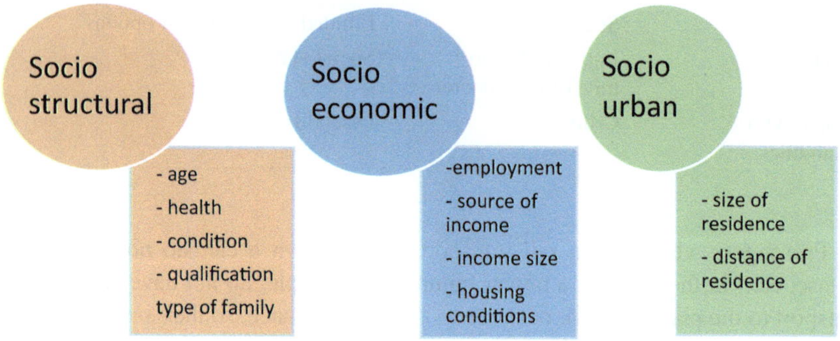

Fig. 2. Social indicators with their special characteristics [10].

To evaluate the level of societal vulnerability was in the article use the Social Vulnerability Index (SVI), which is evaluated by a calculation based on the use of the values of special characteristics of social indicators of selected social actors. As already mentioned, vulnerability can be assessed by various methods, a quantitative point method was chosen for the case study. For the realized point evaluation, the value 1 in Tables 4, 5 and 6 means the minimum level of societal vulnerability and, the highest value from the range of point evaluation represents the maximum level of vulnerability, individual thresholds of indicators with personal special characteristics are also summarized.

Table 4. Socio-structural indicator and special characteristics with point evaluation.

Special characteristics	Point evaluation		
	1	2	3
Age	≤ 29	30–59	60+
Health condition	Healthy	Health restrictions	Half/full invalid
Qualification	No qualifications	High school	University
Type of family	Complete/incomplete	2 members and more	Single

Table 5. Socio-economic indicator and special characteristics with point evaluation.

Special characteristics	Point evalution					
	1	2	3	4	5	6
Employment	State/public administration	Private sector	Business	Working agreement	Brigades	Without job
Source of income	State	Legal person	Self employed	Agreement of service	Part time Job	Social need benefits
Income size	Over 1,500 € gross	1499–1,000 € gross	999–750 € gross	749–500 € gross	On the edge of poverty	Below the subsistence level
Housing conditions	Own house	Own flat	Long term rent	Short term rent	With parents	Hostel

Table 6. Socio-urban indicator and special characteristics with point evaluation.

Special characteristics	Point evalution			
	1	2	3	4
Size of residence	County town	District town	Village	Sparsely populated areas
Distance of residence	to 120 km	to 100 km	to 80 km	to 60 km

To evaluation societal vulnerability, we will use the internationally used Social vulnerability index, which can be expressed by:

$$\mathrm{SVI} = \frac{\sum_{i=1}^{n} p_i}{n}$$

,

where p_i the sum of the point evaluation of the special characteristics of a particular social indicator, n number of indicators [14, 15].

3.2 Application to the Selected Object

To demonstrate the applicability of the above method of evaluation societal vulnerability, the Zvolen railway station was chosen as the object of research in the form of a case study. It is considered that, in view of the extent of the damage to the transmission and distribution network, the resumption of electricity supply is expected only after a few days, unless an alternative solution is implemented.

Quantified values of the social vulnerability index can be assigned to a specific level of societal vulnerability belonging to a given social actor and its quantified characteristics of individual indicators. The social actors listed in Table 3 were evaluated, for which expert estimates of 20 possible variants or groups of the population using rail transport with their relevant social indicators and monitored characteristics. The minimum score for the set indicators is therefore 0.1 point, on the contrary, the maximum possible score is 4.4 points.

Table 7. The scale of societal vulnerability.

SVI	Level of societal vulnerability
0–1.19	Small vulnerability
1.2–2.19	Medium vulnerability
2.2–3.19	High vulnerability
3.2–4.4	Very high vulnerability

An unacceptable threshold of 2.2 or more was set as a critical threshold for societal vulnerability, which, according to the range chosen by the authors, represents a large and very high vulnerability. Exceeding the critical vulnerability threshold means that the negative consequences of a potential natural threat, which would cause a long-term power outage, can seriously affect the individual groups of social actors, whether financially, socially or indirectly to the health side of individual groups of social actors.

3.3 Example of Calculation SVI

A healthy passenger a 54-year-old man who relies on rail transport, who has a high school education, has a 2 members of family. He works in the state, public administration and receives a payment from the state over 1,500 € gross and also the is owner of the flat. He lives in a district town located within 120 km from the place of power outage at the railway station.

We evaluate the socio-structural indicators according to Table 4: 2, 1, 2, 2

We evaluate the socio-economic indicators according to Table 5: 1, 1, 1, 2

We evaluate the socio-urban indicators according to Table 6: 2, 1

We then calculate the evaluated indicators to obtain the value:

$$2 + 1 + 2 + 2 + 1 + 1 + 1 + 2 + 2 + 1 = 15.$$

Number of our social indicators markes as n is 10. In the last step, we can use the values into a formula and find out the resulting value of the societal vulnerability of the actor:

$$SVI = \left(\frac{\sum_{(i=1)}^{n} p_i}{n}\right) \text{ that meanin our example } 15/10 = 1.5.$$

The remaining outcome values of the societal vulnerability of the actors in Table 8 were calculated according to the same presented procedure. The resulting values of the quantified Social vulnerability index in Table 8 could be compared with the threshold values of the SVI range in Table 7.

Table 8. Evaluation of societal vulnerability due to power outages.

Social actor	A	HC	Q	ToF	E	SoI	IS	HS	SoR	DoR	Σ	SVI
Passengers relying on rail transport	A2	HC1	Q2	ToF2	E1	SoL1	IS1	HS2	SoR2	DoR1	15	1,5
	A1	HC2	Q2	ToF3	E6	SoL6	IS5	HS6	SoR1	DoR2	33	3,3
	A2	HC2	Q3	ToF3	E4	SoL4	IS3	HS3	SoR2	DoR4	30	3,0
	A3	HC3	Q1	ToF1	E2	SoL1	IS2	HS1	SoR4	DoR3	21	2,1
Unemployed	A1	HC1	Q2	ToF1	E2	SoL3	IS1	HS2	SoR1	DoR2	16	1,6
	A2	HC2	Q1	ToF3	E5	SoL5	IS3	HS3	SoR2	DoR1	27	2,7
	A3	HC3	Q2	ToF1	E4	SoL2	IS4	HS4	SoR4	DoR4	31	3,1
	A2	HC1	Q3	ToF3	E4	SoL5	IS4	HS5	SoR3	DoR3	33	3,3
Kids	A1	HC1	Q1	ToF1	E6	SoL6	IS6	HS5	SoR1	DoR1	29	2,9
	A1	HC1	Q2	ToF1	E5	SoL5	IS5	HS5	SoR2	DoR3	30	3,0
	A1	HC1	Q1	ToF1	E6	SoL6	IS6	HS5	SoR3	DoR2	33	3,3
	A1	HC2	Q1	ToF1	E6	SoL6	IS6	HS5	SoR4	DoR4	36	3,6
Seniors	A3	HC1	Q2	ToF3	E1	SoL1	IS1	HS1	SoR3	DoR1	17	1,7
	A3	HC3	Q3	ToF3	E6	SoL6	IS4	HS2	SoR2	DoR3	33	3,3
	A3	HC3	Q2	ToF1	E4	SoL5	IS3	HS1	SoR4	DoR2	28	2,8
	A3	HC2	Q1	ToF3	E6	SoL6	IS4	HS5	SoR1	DoR3	34	3,4
People with disabilities	A2	HC2	Q3	ToF3	E2	SoL2	IS1	HS1	SoR2	DoR2	20	2,0
	A2	HC3	Q2	ToF1	E4	SoL3	IS1	HS5	SoR3	DoR3	27	2,7
	A3	HC2	Q2	ToF2	E3	SoL1	IS2	HS2	SoR1	DoR1	19	1,9
	A2	HC3	Q1	ToF1	E6	SoL6	IS5	HS3	SoR4	DoR4	35	3,5

A = Age; HC = Health condition; Q = qualification; ToF = Type of family; E = employment; SoI = Source of income; IS = Income size; HS = Housing conditions, SoR = Size of residence; DoR = Distance of resid.

4 Results

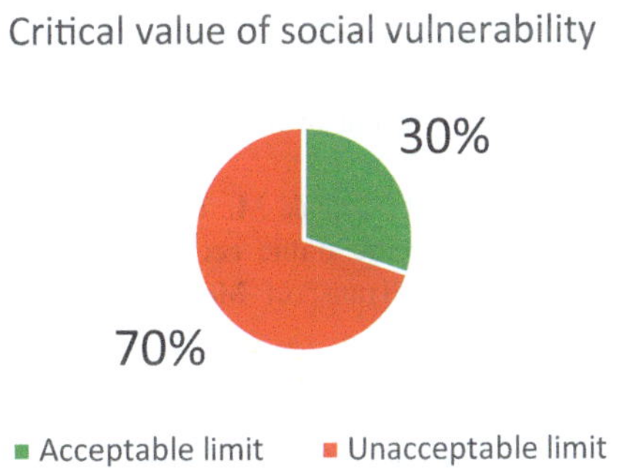

Fig. 3. The resulting ratio of acceptability of societal vulnerability by social actors.

The results of exceeding the critical threshold of societal vulnerability (see Fig. 3) for the selected 20 different types of social actors are exceeded from the total number of 70%, 14 exceeded and at the same time 30%, 6 social actors did not cross the scale. As a result, we have found that the impact of a power outage due to a natural factor can endanger a different population group. According to the results, the most societal vulnerable groups will be children, the seniors and the unemployed, where together up to 10 potential vulnerable social actors from these groups will have difficulty continuing their activities, common transport practices, finances, health, etc.

5 Discussion

From the interpreted results, it is therefore possible to conclude that in the event of the occurrence of a given event, there may be a significant stop or even a decrease in the trend in the number of passengers using rail transport. The state, a public self-governing region or a private investor can support other forms of passenger transport in a given locality and in this way stop the vision and goal of the largest passenger railway carrier ZSSK in the Slovak Republic to transport up to 1 billion people a year by 2030 in competition for a customer using passenger transport services. In addition to the many negative societal impacts of this potential natural threat, which will prevent the supply of electricity and functionality in a certain section of rail transport to society, we can also assume escalating CO_2 emissions in the case of increasing passenger car traffic, which would also endanger the environment.

Acknowledgment. This work has been supported by VEGA grant No. 1/0371/19 named "Societal vulnerability assessment due to failure of important systems and services in electricity sector" and by project GS UNIZA KOR/1043/2020 named "Integration of progressive approaches to design protection system of railway stations".

References

1. Gallina, V., et al.: A review of multi-risk methodologies for natural hazards: Consequences and challenges for a climate change impact assessment. J. Environ. Manage. **168**, 123–132 (2016)
2. Hofreiter, L.: The security environment of the current world. Radim Bačuvčík-VeRBuM, Zlín (2016)
3. Ballay, M., Macurova, L., Kohut, P., Copiak, M.: Development of road safety status and the evaluation criterion causes of specific traffic accidents. In: 22nd International Scientific Conference on Transport Means (Transport Means) 2018, pp. 765–770. Kaunas Univ Technol, Trakai, Lithuania (2018)
4. Erasim, E., Karagiannis, D.: Sicherheit in Informationssystemen. vdf hochschulverlag ag an der eth zürich, Vienna (2002)
5. Paron, P., Baldassarre, G., Shroder, J.F.: Hydro-Meteorological Hazards, Risks, and Disasters. Elsevier, Amsterdam (2018)
6. Cardona, O.D., Carreño M.L.: Updating the indicators of disaster risk and risk management for the Americas. In. Journal of Integrated Disaster Risk Management, pp. 27–47 (2011)
7. Blišťanová, M.: Vulnerability assessment as part of environmental analysis. Metes Proceedings of the International Conference, Bratislava (2017)
8. Murray, A.T., Grubesic, T.H.: Critical Infrastructure Reliability and Vulnerability. Springer, Berlin (2007). https://doi.org/10.1007/978-3-540-68056-7
9. Hofreiter, L.: Object protection management. Žilinská univerzita v Žiline/EDIS, Žilina (2015)

10. Hofreiter, L., Byrtusová, A.: Safety indicators. Radim Bačuvčík-VeRBuM, Zlín (2016)
11. ZSSK: Number of passengers carried. https://www.zssk.sk/vysledky/. Accessed 25 Mar 2021
12. Ivor, J.: Criminology as part of criminal policy. Leges, s. r. o., Praha (2018)
13. Application analyses of state of evolution - ETA on selected extraordinary events
14. Armas, J., Gavriş, A.: Social vulnerability assessment using spatial multi-criteria analysis (SEVI model) and the Social Vulnerability Index (SoVI model) - A case study for Bucharest, Romania. In: Natural Hazards and Earth System Sciences 13(6), pp. 1481–1499 (2013)
15. Spielman, S.E., et al.: Evaluating social vulnerability indicators: criteria and their application to the Social Vulnerability Index. Nat. Hazards **100**(1), 417–436 (2020). https://doi.org/10.1007/s11069-019-03820-z

Smart Workplace: Students' Opinion on Digital Readiness of Educational Institution

Inga Bartusevičienė[1(✉)] and Elena Valionienė[2]

[1] World Maritime University, Fiskehamnsgatan 1, P.O. Box 500, 21118 Malmö, Sweden
ib@wmu.se

[2] Lithuanian Maritime Academy, I. Kanto str. 7, 92123 Klaipėda, LT, Lithuania
e.valioniene@lajm.lt

Abstract. The shift from the traditional physical working places to smart ones by utilizing growing digitalization is actively taking place in transportation industry. The specialists utilizing smart technologies have to be ready to take new opportunities. The discussion of the competences of future professionals is going on and the question of the readiness of educational institution to prepare specialists for the future has to be investigated. The objective of the paper is to present results of the investigation about the students' opinion on digital readiness of educational institution to prepare future professionals being able to work at smart workplace in transportation industry. The paper presents the findings of the exploratory research by applying multi-dimensional model of competences and analysing the quantitative and qualitative data of the students' survey. The research results demonstrate that students understand the importance of digital readiness as a condition to be able to work at smart workplace. However, the higher educational institutions have to design study programmes in a way to pay more attention to the development of digital, soft and especially self-awareness competences to ensure readiness of the graduates to accept the digitization challenges at labour market in transportation industry.

Keywords: Smart workplace · Multi-dimensional model of competences · Readiness of educational institution · Digitalization

1 Introduction

Transportation industry creates working places for a lot of people. Approximately 5 per cent of all workers in the world are employed in transport sector, this corresponds to almost two hundred million jobs. In addition, transportation and warehousing is an industry with many career options and it is characterized with the large number of small and medium companies (96%). The changes in industries, labour market and education boosted intensive transformational processes for the adapting new methods and technologies, so the future workforce should be prepared not only for the use them professionally and socially but also should be ready for new challenges in the context of new smart professions demand in the smart market.

O. Prentkovskis et al. (Eds.): TRANSBALTICA 2021, LNITI, pp. 786–805, 2022.
https://doi.org/10.1007/978-3-030-94774-3_75

The COVID-19 pandemic situation induced shift of traditional physical workplaces to the remote ones enabling employees to work smartly by utilizing growing digitalization of physical objects possibility. The development of the new ways of working by introducing smart workplace and smart technologies is one of the ways to improve workforce efficiency in transportation and warehousing.

However, the specialists utilizing smart workplaces in transportation and warehousing have to be ready for take new opportunities. The competences of the future professionals advocating development of inner-interaction competences such as self-awareness and metacognition, and outer-interaction knowledge and skills, allowing to connect and act with the outer environment by exhibiting boundary-spanning competences such as teamwork, communication, organizational perspective and networks under the continuously changing digital environment conditions. These conditions foster the need to investigate the *smart workplace* that is defined as a workspace that uses growing digitalization of physical objects to deliver new ways of working and improve workforce efficiency. The discussion of the competences of future professionals is going on and the question of the readiness of educational institution to prepare specialists for the future has to be investigated.

2 Competences of Future Professionals in Transportation

The transportation and logistics industry operate in a global network consisting of shippers and service providers of various shapes and sizes where the digitalization processes foster the global supply chain to become smarter. By introducing digital solutions, the transportation and logistics companies can add more value to real-time services. From this point of view transportation business must not only meet the expectations of customers, but also exceed them in ways they have not yet imagined. But for the realization of these challenges to be intelligent the logistics information systems and their development require utilization of smart workforce having a wide collection of competences related to the processes of digitalization. These competences could be considered as digital readiness which could be measured by digital readiness index (DRI).

The analysis of different models describing competences helps to seek the answers to the questions: What kind of professional is best suited to adapt to the continually changing environment and meet future challenges in transportation? What types of competences and how many of them do the future professionals need? How to educate future specialists to be able effectively and efficiently perform in transportation industry and use smart workplace opportunities? For the purpose of recent research, three models of describing competences were analysed in order to develop the Multi-Dimensional model of Competences: T-shaped professional model [1], RHAM (Reference human-centric architecture model) [2], and Tuning methodology model of competences.

2.1 Model of T-Shaped Professional

The concept of the T-shaped professionals (in the beginning named as "T-shaped people", "Hybrid managers") was introduced by McKinsey& Company and further developed by Guest [3] and Brown [4] at the end of the last century. Generally, this concept implies development of different competences as to be presented in letter "T" form. The horizontal axis of the letter "T" represents the breadth of the competences - the ability of a person to combine the competences of the cross-discipline expertise or ability to apply knowledge across domains; and the vertical axis of the letter "T" conveys the meaning of the depth of the competences - comprehensive discipline expertise and knowledge and the ability to apply functional and disciplinary skills [5].

The detailed description of the concept in relation to the education of T-shaped professionals was presented by Gardner and Estry [6]. As presented in Fig. 1, T-shaped model [6] is based on the concepts of deep disciplinary knowledge, deep system knowledge, breadth boundary spanning abilities, self-actualization.

Deep disciplinary knowledge, such as physics, mathematics, history, chemistry, etc. can be considered as essential for the performance according to particular profession. It is unique to every professional and "encompasses the psychomotor and affective abilities critical to a practitioner's success" [6].

Deep system knowledge is related to systems thinking abilities and understanding of intra- and inter- system complexity which embraces different types of processes, services, events (i.e., economic, political, social, etc.); it creates interconnected resources and variables and involves human connections. To understand behaviour patterns of different systems influenced by external and internal challenges self-learning and open minded, flexible and innovative cognitive abilities are needed.

Breadth boundary spanning abilities are related to interdisciplinary/cross-disciplinary professional skills allowing to work within and across complex systems boundaries (i.e., organizational, political, cultural, societal, etc.). Those abilities are called in the literature as boundary spanning, they can generate innovations and new understanding, because possibility to interchange body of knowledge and methods usually used in particular disciplines. They could be developed in interdisciplinary environments where education and training across, within and through multiple contexts could be organized. The boundary-spanning professional is supposed to be able to build sustainable relationships based on partnership and collaboration [7].

ME – self-actualization is related to knowing person's inner self including understanding of personal core values and motivation. The "ME" in T-shaped model consists of the interlocking dimensions of purpose, confidence and awareness and settles knowledge, skills and attitudes characterized in each of the previously described components of the model.

Gardner and Estry [6] highlight the strength of the T-shaped professional as being innovative – open to creative ways of learning; intentional – learn across all disciplines; integrative – able to integrate learning across experiences and connected events. T-shaped professional [6] are better prepared to act in constantly changing environments comparing to currently educated by higher education I-shaped graduates who have deep disciplinary knowledge.

The concept of T-shaped professional [1], highlights the main quality of the future professionals to be "adaptive innovators" who are able to take opportunities in an ever-changing workplace (Fig. 1).

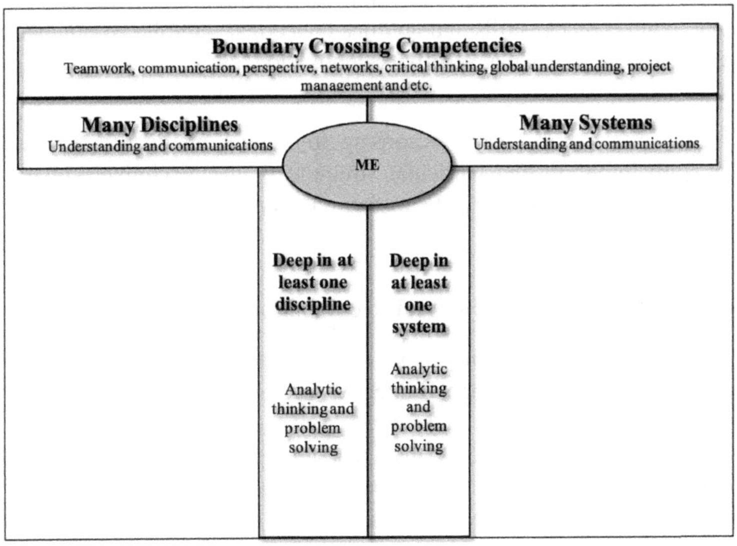

Fig. 1. The T-shaped professional [1].

T-shaped professional concept promotes developing a breadth of knowledge and skills, and exhibiting boundary-spanning competencies such as teamwork, communication, organizational perspective and networks, ability to function innovatively across boundaries between disciplines (i.e., various business activities). In addition, the role of reflection (or meta-cognitive competences) in real working place is also important in the future. Educating students as 'reflective practitioners' enable specialists to analyse own work-based experiences to improve ways of learning and working. That creates supporting environment for encouraging the individual student to develop self-efficacy, analytical, systems and critical thinking skills, and a problem-solving mind-set/approach. The cognitive and self-awareness competences are of high importance to develop on individual level.

2.2 Tuning Methodology Model of Competences

The Tuning methodology as a basic methodology of European Credit Transfer System (ECTS) of higher education, suggests the division of the competences to subject related and transferable/generic. According to Tuning methodology, transferable/generic competences are important for every profession regardless of the field of studies. They can be divided into three main groups [8]:

- *instrumental* competences (competences that function as means or tools for obtaining a given end), such as critical thinking, problem-solving, decision-making, oral communication, writing skills, etc.;
- *interpersonal* competences (different capacities that enable people to interact well with others), such as teamwork, self-motivation, conflict management, etc.;
- *systemic* competences (skills and abilities concerned with the comprehension of an entire set or system), such as creativity, innovation, project management, leadership, etc.

Subject-specific competences, according to Tuning methodology, are related to particular study program and particular structure of the discipline which is directly related to future profession (i.e., chemistry, engineering, geology, history, education sciences, etc.).

2.3 Reference Human-Centric Architecture Model

Sussman [9, 10] highlights importance of educating future professionals for the transportation industry, since for the effective performance in transportation industry the specialists need not only the understanding of the world of transportation systems, institutional structures, technology and its potential – on one hand the new transportation professional will need to perform as part of a team with other specialists, on the other hand, professional expertise in chosen specialty is needed. "The New Transportation *Professional* could contribute substantively to the solution of transportation problems, through detailed knowledge and an enhanced understanding of other professionals' contributions in what is inherently an interdisciplinary field" [9].

Flores et al. [2] have identified five groups of competences of the future professional in RHAM (Reference human-centric architecture model): self-awareness, cognitive, soft, hard and digital, considering self-awareness competences as inner-self interaction, related to internal thinking, and soft, hard and digital - as outer-world interaction competences as associated with external environment through actions and outputs:

- self-awareness competences are related to the understanding of individual consciousness and emotions;
- cognitive competences are about the understanding of physical world by mental processes and interactions with external environment and people;
- soft competences allow the person to communicate, work and collaborate with others and environments;
- hard competences are needed to perform jobs and tasks by using special tools and equipment and applying technical skills;
- digital competences are essential to function in digitally developed society by operating recent technologies and systems if particular settings.

2.4 The Multi-dimensional Model of Competences

The definition of the particular competences to be applied in smart workplace environment is a challenging task. Having in mind the recent demand to shift traditional physical workplaces to the remote ones and need to work smartly by utilizing growing digitalization of physical objects possibility and introducing smart workplace and smart technologies in transportation industry, the authors developed the Multi-Dimensional Model of Competences (Fig. 2) by integrating 5 core groups of competences related to the person and 4 groups of competences related to the industry.

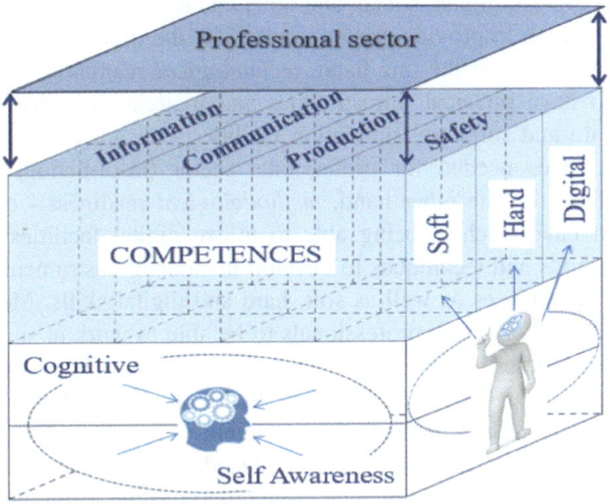

Fig. 2. Multi-dimensional model of competences.

The competences of future professional in transportation industry could be analysed from the perspective of three planes:

- Plane one: two layers of competences (self-awareness and cognitive) are considered mostly as inner individual competences of a person, such as understanding of individual consciousness and emotions; understanding of physical world and self by mental inner processes.
- Plane two: three layers of competences (soft, hard and digital) are considered as connecting inner and outer-world competences - competences allowing the person to connect and interact with the environment (soft, hard and digital) - communication, work and collaboration with others and environments; performing jobs and tasks by using special tools and equipment and applying technical skills; functioning in digitally developed society by operating recent technologies and systems in particular settings.
- Plane three: four layers of competences related to functional groups of activities - characteristics of the environment in this case – transportation industry (information, communication, production and safety); they describe the work area, in which all personal competences have to be realized.

The other set of questions of the investigation was related to the ability and readiness of the higher educational institution to develop particular competences and evaluate the readiness of educational institution for that challenge.

3 Digital Readiness of Educational Institutions to Prepare Students for Working at Smart Workplace

Higher education institutions have to respond to recent changes and be ready to present effective solutions for the development of a future professionals with respective knowledge and practical skills enabling them to work at smart workplaces and adapt to the rapid technological, digital, social, and other changes. The issue of the readiness of educational institution to provide proper services for the digital age could be analysed from two perspectives: on the one hand, *technological* readiness - educational institutions need to have physical infrastructure and material resources ready, such as modern hardware and software, simulation facilities, etc. where students can acquire specific competences needed for professional sector (information, communication, production, safety); on the other hand, *methodological* readiness - educational institutions need to have teachers being able to utilize digital facilities and use proper educational methods and techniques to develop in students instrumental, interpersonal and systemic competences as well as soft, hard and digital skills. Mentioned competences are necessary for future professionals to be able to work at smart workplace in logistic companies based on the functioning of transport intelligent systems.

Assumptions made for the composing the research methodology describe the structure of assessment criteria and allow to identify, that the educational institution also must be ready to educate digitally prepared students for transportation industry and the educational processes should include these criteria as the vision and mission of the organization, resource awareness, the technical infrastructure, software, and also the level of their usability and safe integration with the real-time transportation systems.

4 Methods

The purpose of the investigation was achieved by applying a survey method for the data collection using the originally designed questionnaire and performing qualitative and quantitative analysis of the obtained data using content analysis method and Microsoft Excel Data analysis tools and functions for the required statistical calculations based on the score calculation methodology described below. The questionnaire was designed to obtain the opinions of the respondents regarding their preparedness and the readiness of the educational institution to meet the recent digitalization challenges in developing digital competences and abilities to work with smart working places. The other topic under investigation was related to the perceptions of the students on the understanding of expectations from the labour market and transportation industry in relation to the changes. In the developing of the questionnaire the results of the literature analysis and personal insights of the authors were used.

The questionnaire was designed with Google Forms consisted of 272 questions divided into 6 sections and grouped to meaningful groups:

1. Socio-demographic information (5 questions);
2. The digital readiness of educational institution to prepare specialist to work with smart working places (1st block consisted of 19 questions);

3. The digital readiness of educational institution to prepare digital competences of the future specialists (2nd block consisted of 18 questions);
4. Respondents' perceptions about digital readiness competences (3rd block consisted of 35 questions);
5. Respondents' perceptions about digital readiness to work with smart working places (4th block consisted of 185 questions);
6. Respondents' opinion about the implementation of the idea of the smart working places in transportation (maritime) sector (5th block consisted of 10 questions).

The questionnaire included the open-ended questions, asking the respondents to write their opinion about particular aspects of the research and close ended questions providing 8-points Likert scale responses.

The respondent sample included 87 students of the LMA (Lithuanian Maritime Academy), 26.4% of them were female. The majority of them (88.5%) were full-time students, the others (11.5%) used part-time and session-based modes of studies; 81.6% were state-funded, the others (18.4%) paid for the studies themselves. The median of age of the respondents (21), and the median years of studies (2) indicate that the survey population was relatively young (41% were first year students). This characteristic of the respondents is valuable for the purpose of the research because the young people can have non-biased ideas on the research questions. The survey sample represented all study programs delivered at LMA.

As it was found out in the theoretic description of smart workplace, the competences could be divided into the four functional groups described as the skills required to manage the digital information flow, to communicate, to create products which represent digital content, and to ensure safety ensuring, including the personal and commercial data safety (Table 1).

Table 1. DRI matrix construction for the assessment of digital readiness of institution (technological and methodological) and students (general and specialized).

DRI	Self-Awareness	Cognitive	Hard	Soft	Digital
Information	DRI_{11}	DRI_{21}	DRI_{31}	DRI_{41}	DRI_{51}
Communication	DRI_{12}	DRI_{22}	DRI_{32}	DRI_{42}	DRI_{52}
Production	DRI_{13}	DRI_{23}	DRI_{33}	DRI_{43}	DRI_{53}
Safety	DRI_{14}	DRI_{24}	DRI_{34}	DRI_{44}	DRI_{54}
DRI score visualization	< 1.99	<3.11	<4.23	<5.36	<6.45 <7.6

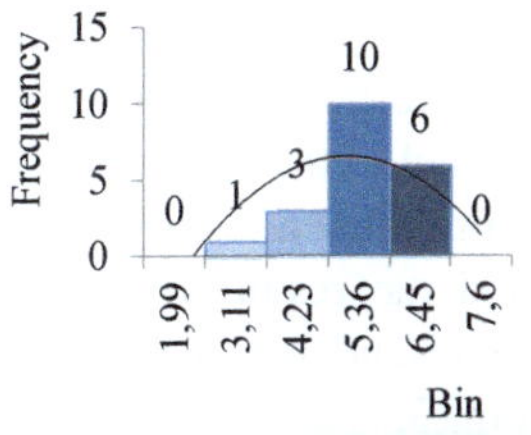

The structure of questionnaire represents the matrix of digital competences which will assess recent strengths and weaknesses of the digital skills and will identify the areas of improving learning content and practice. This is important in global supply chain because transportation industry is developing rapidly and the implementation of the new technologies and software solutions is growing on a daily basis, so the specialists of smart global supply chain need, as well as all professionals working in the maritime transport sector, the fundamental general digital competences which will allow them to adopt faster to dynamically changing workplace technologies and challenges influenced by Industry 4.0.

The construction of this matrix is based on the allocation of all questions to the categories identified by acquisition of groups of attitudes, knowledge and skills divided into four operational groups. Then the DRI was calculated by applying the formula (1) as presented in Table 3. It is important to note, that this calculation methodology was applied for the calculation of four indexes: institutional technological digital readiness, institutional methodological digital readiness, general digital readiness of students and specialized digital readiness to work at smart workplace in maritime industry of LMA students.

The visualization of the distribution of all DRI scores presents ranges of differently developed knowledge and competences by different colours. For the applying the specialized visualization technique all assessed DRI scores in each category were placed into one set which was divided into equal intervals. All DRI were placed by their meanings into the intervals and the frequency of each terminal were calculated for the evaluation of the most probable DRI level identification in the set, but mostly the equal intervals were used for the colouring of matrix elements by the selecting different colours for the different DRI score.

For the assessment of possible ways for the improving of LMA digital preparedness influencing required competences of digital readiness of students, the correlation method was applied and the highly significant ($p < 0.01$) relationships were identified between LMA digital preparedness and LMA students' digital readiness; based on these results the recommendations for the improvement of LMA study programs were formulated.

5 Analysis of Responses to Open-Ended Questions

The responses to open questions were categorized according to Multi-Dimensional Model of the Competences of future professionals. The numbers of mentions of the key word related to each type of the competence as divided into 5 categories were added up and the results of the content analysis summarized in the Table 2. The graphical representation of the outcomes in the form of network diagram is provided in the Fig. 3.

Summary table of content analysis with numbers of mentions according to categories on keywords related to each question and competence is provided in the Annex 1.

The network diagram (Fig. 3), representing the outcomes of the content analysis of the perception of respondents about smart working place in transportation, clearly indicates insufficient understanding of the Self-awareness and Cognitive aspects of the competences of future professionals able to work in smart workplace in transportation industry.

Table 2. The results of the content analysis of the responses to open question.

Categories of the competences	Opinion of the respondents regarding particular aspects in numbers of mentions						
	Q1 Smart working place	Q2 Work expectations	Q3 Needed knowledge	Q4 Missing knowledge	Q5 Communication and teamwork characteristics	Q6 Digital readiness concept	Q7 Needed digital competences
Hard	45	30	19	25	10	42	9
Soft	28	25	16	7	33	10	3
Digital	18	18	54	19	13	23	19
Cognitive	0	0	7	1	0	0	0
Self-Awareness	1	1	2	2	3	1	1

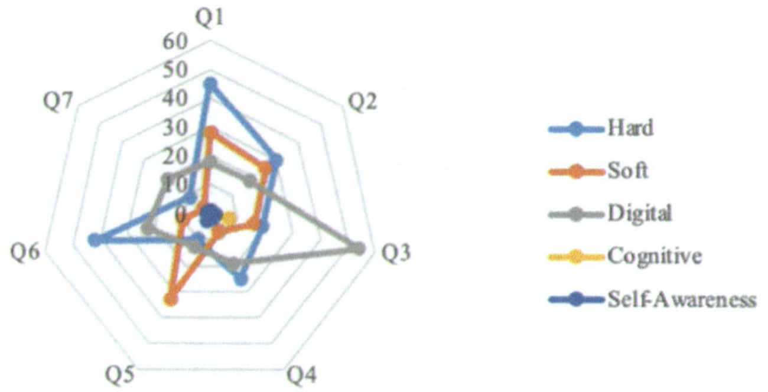

Fig. 3. Network diagram of the perception of respondents about smart working place in transportation.

Based on the findings, it can be said that the respondents underevaluate the inner-interaction competences. The highest value of the Soft aspect was demonstrated by the respondents in respect to imagining the smart workplace (Q1), expectations of the work nature in future (Q2), including the communication and teamwork in transportation industry working from smart working place and with smart team (Q5). The highest importance of the Digital aspect of the competences was mentioned in responses about useful knowledge in 25 years perspectives while working in smart working place in transportation industry (Q3). Interestingly, in understanding of the digital readiness concept (Q6) smart workplace the majority of mentions were related to Hard competences.

6 Statistical Analysis

Based on the findings of quantitative analysis the technological digital readiness of LMA should be increased by the expanding of the digital identity formation and positioning as education institution which is preparing students for working at smart workplace. In comparison with the LMA students' self-awareness, the understanding of the importance of digitalization of maritime transportation industry is higher than their digital competences. This means that students understand and percept the digitization's influence on maritime transportation, but the LMA should adopt its study programmes and identify more clear relationships within these challenges (Fig. 4).

Based on the results of quantitative analysis of the assessment of the digital readiness of LMA it was found out that digital readiness of LMA is high enough as it presented in Fig. 5. Well-developed LMA's material resources are assessed with the highest grade, but they should be used more effectively for the training students to be ready for changes in the maritime transport sector related with the challenges of digitization. The digital competences of LMA students based on their opinion are less

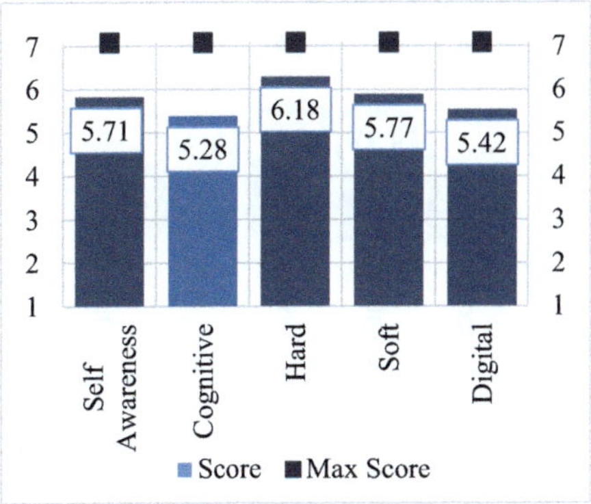

Fig. 4. Digital readiness of LMA: technological aspects of material recourses preparedness.

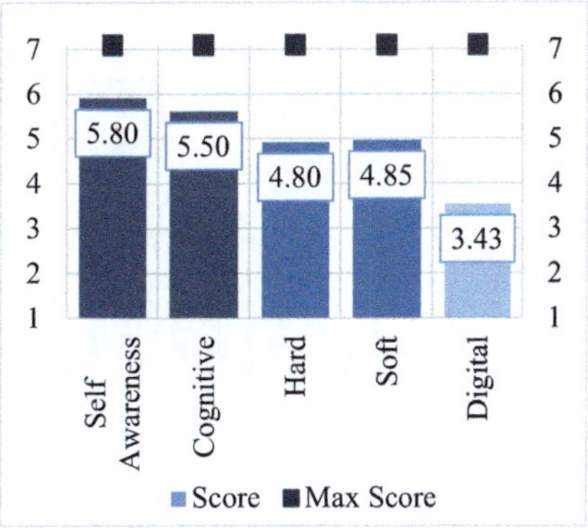

Fig. 5. Digital readiness of LMA: students' digital readiness: overall score.

developed, while the digital preparedness of LMA is not significantly lower than other parameters, the students' digital readiness is very low. As it was found out the self-awareness in the field of digital readiness and knowledge about the digitization is fairly high; however, the level of hard skills, especially in the usability of more software for bigger types of data exploring or software of digitized objects the digital books and personal digital readiness) is low. That implicates the demand to develop wider hard skills required to work more intensively within the data in the study process at LMA.

The difference between LMA technical readiness and low LMA's students' readiness to work at smart workplace could be explained on the basis of results of LMA's methodological readiness score (Fig. 6).

According to the results of LMA it could be resumed that the LMA lecturers apply various methods for the adopting study programmes to the changes happening in the maritime industry and the existence of the left asymmetry enables to resume, that the predominance of assessments was more shifted to the higher grades' (average = 6.3; median = 7, mode = 8; skewness = −1.17 < 0; kurtosis = 0.57 > 0). This fact allows saying that more students participated at the survey assessed methodological readiness of LMA with the bigger grades.

Based on the students' opinion, the strengths of methodological readiness of LMA have two methodological goals: 1) the orientation to train the logical thinking (score = 6.04) and 2) to train the general computer literacy (score = 6.02). But the weaknesses are: low level of training of self-positioning competences (score = 4.7), the competences of conflicts management (score = 4.81), and the competences of working in the intercultural teams with high diversity (score = 4.09).

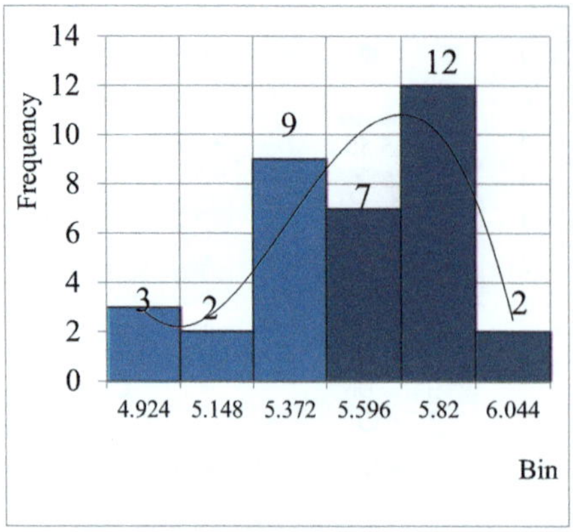

Fig. 6. LMA methodical readiness to educate student's digital readiness competences to work at smart workplace.

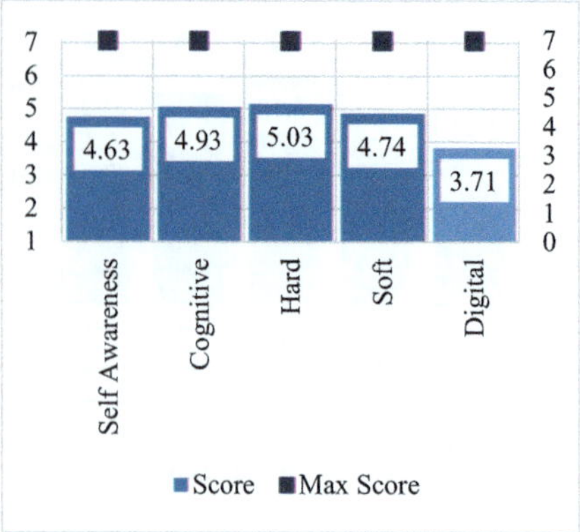

Fig. 7. Q. LMA students' digital readiness to work at the smart place in maritime industry.

By the resuming it could be said that LMA readiness to educate specialists for smart workplace could be increased in the methodological area by paying more attention to promote the creativity of students, by training of competences of self-positioning in labour market, conflicts management, intercultural communication based teamworking.

Based on the correlation analysis with the aim to answer the question which methods would be helpful for the implementation of this recommendations in most effective way it should be noted, that the improving LMA study programmes by methodological principles to promote student's creativity will make positive influence on the general digital readiness of LMA students and also directly on their preparedness to work at smart workplace as it showed in Table 3.

Based on the results of this study, it can be concluded that the methods used by the students of the Academy even if they were poorly assessed by the students but they can increase the digital readiness of students in both perspectives: general and special (professional in the maritime transport sector). However, the link between general and special (professional in the maritime transport sector) training needs to be clarified as well as usage of those interfaces to directly target the application of methods for shaping digital preparedness and developing specific competencies specific to digital preparedness.

Correlation analysis with high level of statistical significance ($p < 0.001$ & $|r|>0.27$) between the LMA students' general digital readiness (results presented in Fig. 5) and their preparedness to work at smart workplace (results presented in Fig. 7) proves that strong influence on the preparedness to work smartly have quiet big amount of general indicators (Fig. 4, 5) which could be identified as the directions for the improvement.

The statistically significant result of the correlation analysis presented in Fig. 8 and Table 3 proves that respondents categories agree that in the short period all workplaces in the maritime industry will become smart (5.4): students who percept that principles of working remotely are based on the *teamworking* (4.6, $r = 0.387$) and on the effective usage of distance *communication* platforms (4.7, $r = 0.408$), also students who understand that the self-positioning content in the digital labour market is very important for the career (4.8, r = 0.317), students who understand their digital competencies as the main tool of their digital identity creation (4.9, r = 0.265) and these students who were interested in the ethics of *communication* of digital era (4.22, r = 0.287) and use advanced search's command in the different search engines (4.49, r = 0.307). It is important to say, that students, who are more engaged to create their own digital content for social media and communication (4.50, r = 0.307) and these students who are more interested in sharing of multi-type material in social media (4.51, r = 0.397), who are using specialized forums for the solving of different type of problems (4.54 r = 0.351, 4.55, r = 0.312) are more attracted to work smartly in the maritime industry (5.10).

Table 3. The LMA's methodological readiness' influence on the student's digital readiness to work at smart workplace ($p < 0.01$).

	SA				C		H		S	D
	4.4	4.6	4.7	4.8	4.17	4.22	4.42	4.44	4.49	4.82
Promotion of creative thinking	–	0.29	0.33	0.41	–	–	–	–	–	0.31
Training Competence of self-motivation	0.40	0.37	0.37	0.38	–	0.46	0.28	0.36	0.30	
Training Competence to work in intercultural teams	0.39	0.39	0.41	0.38	0.30	0.45	–	–	0.28	
Training competences of Conflict management	0.33	0.37	0.34	0.35	–	0.37	–	–	–	
Promotion of creativity actioning	–	0.35	0.34	0.43	–	–	0.29	–		

Analysing these connections, it is possible to see potential relationship between the needs of these students to develop the principles of effective communication by working in teams in international organizations (Table 3).

It is important to note, that students who expect that the smart workplace will influence possibilities to work by flexible and adoptable schedule (5.2) are more aware of the personal big data problems in digital environment (4.13, r = 0.289), also they are more confident with personal data safety regulations, technologies and techniques (4.20, r = 0.297).

The perception of virtual smart workplace in transport sector (5.1) without dependence on location is more confident for the students who use different passwords in each system (4.44, r = 0.248) and change them periodically without the systems' reminders (4.42, r = 0.214). Also it was found out that students who are using laptops not only during lectures and studies but also in another daily activities (4.39.1) are more engaged in globalization and are more independent from the geograpichal location of workplace (5.3, r = 0.327); they also are more confident in the possibilities to work at smart workplace in the maritime transportation industry (5.8, r = 0.364) where all workplaces will become smart in the short forecast (5.4, r = 0.284). It is important to note that globalization of maritime labour market (5.3) is more attractive to the students, who are saying that they can find everything that they need in internet (4.48, r = 0.332) and the students who are using the specialized social networks (such as Flickr, LinkedIn, Blogster, Upwork) for their needs (4.70, r ∈ [0.288;0.314]).

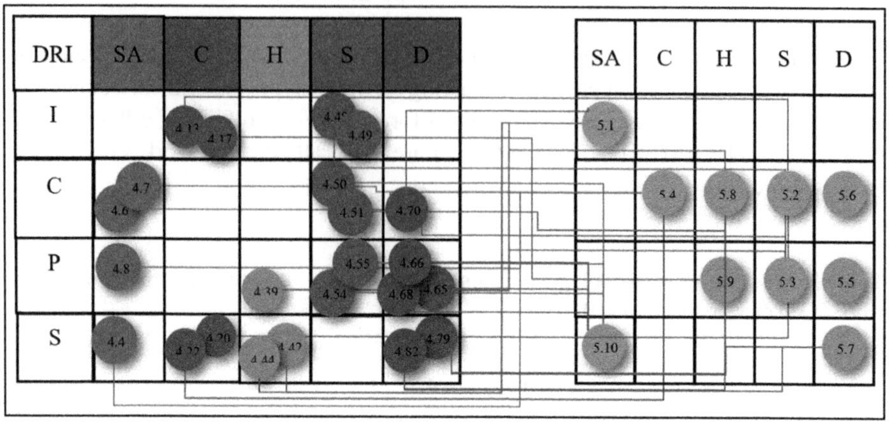

Fig. 8. Students' general digital (4th block) readiness' relationship with the preparedness to work at smart workplace (5th block).

Analysis of results also found out that students prefer to work at smart workplace (5.10) if they understand the principles of machine learning algorithms: students who like the suggestions of search engines (4.66, r = 0.305), automatization of daily routine duties (4.65, r = 0.363), and artificial intelligent solutions (4.68, r = 0.294) are more interested in the training of all required competences for the working at smart workplace (5.10) in maritime industry. Possibly these knowledge and competences of technical data operations for the implementation of required advanced searches, and using different special operations, social media, have positive influence on the decreasing of humans afraid of the newest robotized solutions.

Consequently, the ability to search, use sophisticated search commands, and discover relevant data online may have a stronger formative approach to globalization and automation, helping to avoid the fear of jobs using robotics technologies, which can be seen in the job market when professionals become insecure and see threats in automation processes. On the other hand, the application of more complex and professional operations forms a more systematic approach to the processes taking place in the maritime transport sector. This statement can be supported by research showing that those who use their biometrics for logins (4.79, $r_{5.7} = 0.351$, $r_{5.8} = 0.308$) and two-tier authentication systems (4.80, $r_{5.7} = 0.342$, $r_{5.8} = 0.323$) are more likely to believe that a smart place in the transport sector will emerge when many automated systems (5.7) and indicates that they want to work in a smart job in the maritime transport sector (5.8).

7 Conclusions

The outcomes of the analysis of qualitative data, revealed, that, as it was seen from the responses to open questions, the majority of the respondents highlighted the characteristics of the Smart working place and the work processes in transportation industry by using smart workplace are mostly related to the Hard aspect of the needed competences. It could be noticed, that the second priority was given to the Digital aspect, and Soft aspect also was mentioned in responses to some extent, but, unfortunately, the Cognitive and Self-awareness aspect of the competences did not receive proper attention. To the great surprise of the authors, the competences in cybersecurity and digital hygiene were not mentioned by the respondents at all.

From the theoretical perspective, according to Gardner (2017) future professionals have to be "adaptive innovators", able to take opportunities in an ever-changing workplace and having developed Soft boundary-spanning competences such as teamwork, communication, organizational perspective and networks across boundaries between disciplines (e.g. various business activities). The important value of such a professional is to become 'reflective practitioners' able to facilitate reflective work-based experiences by developing self-efficacy, analytical, systems and critical thinking skills, and a problem-solving mind-set/approach. Those competences are considered as Cognitive and Self-awareness. Our finding show that the respondents understand the value of the Soft competences, but under evaluate the inner-interaction competences, such as self-awareness and cognitive.

The finding of the quantitative analysis demonstrated: the fairly high level of understanding of self-awareness in the field of digital readiness and knowledge about the digitalization by the respondents; low level of hard skills especially in the usability of software for bigger types of data exploring or software of digitized objects the digital books and personal digital readiness) are on the lowest levels in the context of high hard level skills in the field of maritime industry.

The statistical analysis identified that the difference in the digital readiness index between the Academy and students identifies the under-utilized high digital technological readiness of the Academy and this can be observed in the fact that students underestimated the positioning of the Academy as an organization with sufficient resources to develop digital readiness. Students also noted that they lack skills development such as self-positioning in the labor market, conflict management, intercultural communication based teamworking, and they would like to see more emphasis on supporting creativity in a variety of tasks. The study also found that students' awareness of cognitive aspect is high, but they lack general and specific hard, soft, and especially digital competences.

Therefore, based on the results of the correlation analysis, the following key links between general and professional digital competences were identified: readiness to work in a digital workplace in the maritime transport sector could be enhanced by developing hard competencies such as technical means and technology for storing personal data, changing passwords and securing each password. The application of correlation analysis has established strong links between general and specific competencies expressed through complex search engines and the use of specialized communication platforms to solve problems and positioning oneself in the digital space through the creation and sharing of unique digital content in social media and through the development of digital competences such as the use of a wide variety of dedicated communication platforms, knowledge of the principles of search engines and artificial internet, the recognition of lecture plugins, the integrative use of these tools in the resources of the Academy, the effective use of the tools used for personal growth in cyberspace.

The finding can be used in improving the education and training processes by adjusting the LMA study programmes by highlighting:

- value of the cognitive and self-awareness aspects in future jobs, because the environment and the conditions especially in transportation industry will be changing with acceleration;
- orientation of practical tasks to the new tendencies in maritime industry and application of digital innovations in practical training to bigger extend by increasing and soft skills development as dependent on the communicational level and hard skills development as depending on the technological implementational level; the recommendation to improve study programme at LMA is to use more different software, especially which is actual for maritime industry form practical point of view;
- necessity to develop wider hard skills required to work more intensively within the data in the study process at LMA; the recommendation is to use more personal integrated devices in the LMA infrastructure, including simulators, such as mobile phones with specialised apps, personal laptops, expand the integral functionality to connect with personal equipment;
- high importance of the training of digital skills oriented to transportation (in particular – maritime) industry in the fields of communication, production and safety together within the digital skills development; the wider range of possibilities to integrate students in the smart marine business should presented because the DRI for preparedness work smartly at maritime sectors is low. the recommendation is to construct more tasks related to the social networking communication, finding answers in the forums, wide usage the advances search, content creation tasks for the presenting themselves on the labour market and self-positioning, etc.

Additional recommendation to improve study programme at LMA could be exploring the possibility to include more data analysis methodology in different subjects to show the importance of big data for the development of maritime technologies and maritime business; include more data analysis related topics relevant to maritime transportation by identifying more clear relationships within digitalization challenges.

Appendix

Annex 1. The summary of the content analysis with examples.

Questions	Numbers of mentions and examples of responses				
	Hard	Soft	Digital	Cognitive	Self-Awareness
Q1 Smart working place	45 mentions: most recent technical devices and software (25), fully computerized (16), automated (4)	28 mentions: possibility to work remotely (13), work from home (6), allowing to work more effectively (9), comfortable (2)	18 mentions: digitalized (11), interactive (5), AI (2)	0 mention	1 mention: possibility for self-improvement
Q2 Work expectations	30 mentions: the work will be fully computerized (15), monitoring of the working mechanisms and processes (7), properly equipped (4), with recent software (4)	25 mentions: remote work (7), flexible working hours (6), effective processes – saving time (6), less paper work (3), reliable communication (3)	18 mentions: analysis of the data (11), decision making based on the analysis of data (2), automated (5)	0 mentions	1 mention: backpain and eyes problems
Q3 Needed knowledge	19 mentions: practical skills (12), management of robots (5), electronics (2)	16 mentions: economics (3), management (3), logistics (3), psychology (3), communication (2), innovations (2)	54 mentions: computer literacy (21), programming skills (16), analysis of data (7), digital marketing (6), searching of information (4)	7 mentions: knowledge of languages (Russian and English)	2 mentions: ability to learn new things (1), adaptability (1)
Q4 Missing knowledge	25 mentions: ability to work with the technological equipment (14), knowledge and skills of specialty (13)	7 mentions: management (3), management of systems (3), negotiations (1)	19 mentions: programming (10), computer literacy (5), understanding of data (2), analysis of data (1), usage of the newest technologies (1)	1 mention: English language	2 mentions: self-confidence

(*continued*)

Annex 1. *(continued)*

Questions	Numbers of mentions and examples of responses				
	Hard	Soft	Digital	Cognitive	Self-Awareness
Q5 Communication and teamwork characteristics	10 mentions: reliable connection (4), easy access (3), using phones (1), Internet (1), radio communication devices (1)	33 mentions: effectiveness of communication (15), which is clear (4) and helps to make decisions (5), by discussing ideas (3), distributing tasks (3) and based on safe environment (3)	13 mentions: remote communication in the digital/virtual reality (7) with the help of Software which helps to share and correct the information (6)	0 mentions	3 mentions: Oral and written communication
Q6 Digital readiness concept	42 mentions: ability to use technologies (38), work with computers (4)	10 mentions: understanding of systems (3), communication (3), ability to present information (2), effectiveness (2)	23 mentions: ability to work in virtual/digital environment (7), search information (4), digital communication (3), data processing (3), digital ethics (3); digital literacy (3), etc.	0 mentions	1 mention: insufficient physical preparedness
Q7 Needed digital competences	9 mentions: ability to use technologies (5), and equipment (4)	3 mentions: quick adaptability to the changes (1), innovations (1), work standards (1)	19 mentions: programming (7), information technologies (7), ability to work in virtual reality in real time (1), collecting data (2), knowledge of computer systems (2), etc.	0 mentions	1 mention: knowledge about health

References

1. Gardner, P.: Flourishing in the face of constant disruption: cultivating the T-professional or adaptive innovator through WIL. In Bowen, T., Drysdale, M. (eds.) Work-Integrated Learning in the 21st Century: Global Perspectives on the Future. Emerald Publishing Limited, Bingley (2017)
2. Flores, E., Xu, X., Lu, Y.: A Reference human-centric architecture model: a skill-based approach for education of future workforce. Procedia Manuf. **28**(2020), 1094–1101 (2020)
3. Guest, D.: The hunt is on for the renaissance man of computing. The Independent. Accessed 17 Sept 1991
4. Brown, T.: Strategy by design. Fast Company (2005). https://www.fastcom-pany.com/52795/strategy-design
5. Yates, D.J., et al.: Dotting i's and crossing T's: integrating breadth and depth in an undergraduate cybersecurity course. Inf. Syst. Educ. J. **17**(6), 41–62 (2019)
6. Gardner, P., Estry, D.: A primer on the T-professional. Collegiate Employment Research Institute Michigan State University (2017)
7. Williams, P.: The competent boundary spanner. Public Adm. **80**(1), 103–124 (2002). https://doi.org/10.1111/1467-9299.00296
8. Sanchez, A.V., Ruiz, M.P.: Competence-Based Learning: A Proposal for Assessment of Generic Competences. University of Deusto, Bilbao (2008)
9. Sussman, J.: The new transportation faculty: the evolution to engineering systems. Transp. Q. **53**(3), 15–26 (1999)
10. Sussman, J.: Transition in the world of transportation: a system view. Transp. Q. **56**(1), 21–32 (2002)

Relationships Between Road Horizontal Geometry, Driving Behavior and CO_2 Emissions

Gaetano Bosurgi[1], Stellario Marra[1], Orazio Pellegrino[1(✉)], Giuseppe Sollazzo[1], and Massimo Villari[2]

[1] Department of Engineering, University of Messina, Messina, Italy
opellegrino@unime.it

[2] Department of Scienze Matematiche e Informatiche, Scienze Fisiche e Scienze della Terra, University of Messina, Messina, Italy

Abstract. Exhaust gas emission produced by motor vehicles are mainly caused by the elevation features of the road and by the vehicle type. In truth, there is another contribution, probably lower, that can correlate with the driving behavior. This, in turn, is a function of the road geometry that may thus assume a key role not only in terms of functionality and safety, but also in controlling environmental impacts. In this study, the relationships between the longitudinal acceleration, the CO_2 emissions and the geometrical features of the road have been analyzed, to contribute to improving road design quality. The analysis of these potential relationships has been developed through an experimental study in a simulated environment. The results evidenced how the emissions depend on the homogeneity of driving, including variables related to visibility, alignment understanding by the drivers, similarity of adjacent elements. However, some conclusions are interesting because they raise questions regarding the correctness of existing models for operative speed forecasting, available in literature and used to verify the road design homogeneity and consistency. This research may have a positive impact in comparing the road design standards and guidelines of the various countries and, mainly, it may suggest alignment composition procedures based not only on safety and use, but also on environmental impacts.

Keywords: GHG emissions · Road geometry · Driving behavior

1 Introduction

In the last decades, the attention of the scientific research has strongly focused on GHG (Green House Gas) emissions (CO_2, CH_4, N_2O, O_3, etc.), because of the several and remarkable damages caused by the global warming and the consequent climatic changes on the entire Earth. Impact analyses have strongly interested in transportation industries, considering both material production [1, 2], infrastructure construction [3] and vehicle issues [4, 5].

Among the various contributing factors, CO_2 has lower GWP (Global Warming Potential) compared, for instance to CH_4 and N_2O that have, respectively effects 28 and

O. Prentkovskis et al. (Eds.): TRANSBALTICA 2021, LNITI, pp. 806–816, 2022.
https://doi.org/10.1007/978-3-030-94774-3_76

265 times stronger than CO_2. However, CO_2 is almost always assumed as symbolic chemical product for this damage, caused by transport systems too, even because it is directly proportional to the fuel consumption and, thus, easier to be quantified.

Considering the vehicle emissions, generally, the calculation methods for motor vehicles moving along roads available in literature refer to two different procedures [5]:

- measures through dynamometer testing in lab;
- on-board emission measurements.

The first group assumes only mechanical features of the vehicle. According to recent literature evidence, it seems that these measurements underestimate the actual emissions. The second group of measures, instead, determine high costs and are representative of a specific vehicle only. In any case, for this reason, results available in literature are valid only after verification of boundary conditions and hypotheses assumed for the calculations and avoiding generalization to different scenarios [6].

Another scientific thread investigated the influence of road geometry on the emissions. In terms of emissions, in truth, the road geometry has always been considered as a marginal cause that, eventually, assumes importance only when huge flows of trucks move along very high slope gradients. When the territory morphology determines similar conditions, it would be convenient to evaluate mixed transport means. At this regard, Heinold and Meisel [7] identified the limits beyond which it is convenient to select different means of transport. The studies on the road elevation allowed several researchers to find correlations between a certain fixed grade (4%) and the emissions [8]. However, in this case too, the practical utility is limited, since numerous other variables are not involved [9].

Even the horizontal geometry, albeit to a lesser extent, could cause variations in vehicle consumption and, thus, in their emissions since the latter depend on the use of torque or load of the engine. However, while the grade strongly stresses the engine only for limited segments, horizontal inhomogeneities, albeit characterized by less intensity, may have relevance for longer prats of the alignment and, thus, represent a total quantity of emissions particularly remarkable. For example, Park and Rakha [10] evaluated fuel consumption and emissions using a traffic simulation model, after entering some variables into the scenario, such as the speed trend. If the alignment is characterized by huge longitudinal accelerations, the consumptions and the emissions increase proportionally. The homogeneity of the horizontal alignment of the road may be synthetized through the well-known index named Curvature Change Rate (CCR) i.e., the rate of curvature variation of the single curve [11]. Low values of CCR represent long straight segments and flat curves and, thus, high speed almost constant. In these conditions, consumptions and emissions are very limited. Some recent studies have characterized the homogeneity of some roads and, based on some literature-derived models for operative speed forecasting, quantified consumptions and emissions [12–17].

Much of this research concerns the correlation between infrastructure and emissions on such a large scale that the results can only be used for general considerations and actions. In other words, there is no level of detail regarding the individual elements of a road (as straights or curves) and, thus, the possibility of directing designers towards more virtuous choices also from an environmental point of view. For example, the relationship with the horizontal geometry has been poorly investigated and the few available references are based on the correctness of some hypotheses considered by operative speed forecasting models and consistency criteria derived several decades ago.

It is opportune to overpass some limits of the existing scientific literature through the strict calculation of CO_2 emissions. For instance, it is more convenient to consider a different and more specific unit, that may be directly related to the single geometrical elements of the examined road. Consequently, in this paper, the authors propose a procedure for evidencing eventual relationships between horizontal geometry, longitudinal acceleration and CO_2 emissions. These variables have been derived from an experimental framework performed in a simulated environment, avoiding literature correlations. In particular, the emissions have been quantified in g/s instead of g/km, as done in available references.

2 Methods

Due to the limited quantity of references on the relationship between the horizontal geometry and the CO_2 emissions, the experimentation was based on driving tests in a simulated environment, for assuring a more accurate control oc the road context.

The basic hypotheses of the research are:

- the road alignment has not elevation variations; the driving behaviour, then, is influenced by the horizontal geometry only.
- The horizontal alignment is made up of 4 types of transition curves, in compliance with Italian Standards [18], with radii of the residual circular curve equal to 45, 120, 180 and 240 m.
- The straight segments between two curves have values decreasing with the increase of the radii of the adjacent curves; their max and min lengths are equal to 235.66 m (between curves with R = 45 m) and to 20.71 m (between curves with R = 240 m).
- The deviation angles between the straight segments are constant and equal to 50°.
- There is not traffic, to correlate variations of the driving behaviour and, thus emissions, to the horizontal geometry only.
- The output variable is the quantity of CO_2 instantly released, measured in g/s.

The driving simulator at University of Messina

The experimentation has been performed using the driving simulator named SimEASY, produced by AVSimulation, available in the Road Infrastructure Laboratory of University of Messina. This simulator has the following features:

- Three 29″ full HD screens, with a horizontal field of view of 130° and a frequency higher than 50 Hz.
- A steering wheel characterized by a force feedback sensor to simulate the rolling motion of wheels and shocks.
- Sound effects reproduced through several speakers and subwoofers.
- The SCANeR™ studio software, used to design tracks, generate the environmental context and run trials.

- Data collected with a frequency of 10 Hz.
- A family car powered by a 130 hp gas engine, with six manual gears and automatic clutch.

Details of the experimentation

The investigated road is 4.00 km long, with a succession of transition curves made up of entry and exit clothoids with a residual circular arc. In detail, as evidenced in Table 1, 2 curves have a radius equal to 45 m (type named C1), 3 equal to 120 m (C2), 2 equal to 180 m (C3) and, finally, 3 equal to 240 m (C4). These 10 curves, characterized by the same deviation angle equal to 50°, are perfectly in compliance with Italian Standards [18].

Table 1. Main characteristic of the road alignment.

	Radius of the circular curve (m)	Overall length of the transition curve (m)	Number of curves
C1	45	61.27	2
C2	120	145.72	3
C3	180	208.08	2
C4	240	264.44	3

The drivers' sample

The experimentation on the driving simulator involved 5 users, 3 men and 2 women, aging between 21 and 29 years old, with at least 3 years of driving license, without serious accidents, not suffering of pathologies that may negatively influence the driving behaviour.

Emission calculation using the driving simulator

The forecasting calculation of the fuel consumptions and CO_2 emission is hard to be represented through an explicit equation since it depends on the engine mapping and on a series of mechanical and electronical features that vehicle industries maintain confidential. The SCANeR software installed on the driver simulator of University of Messina uses the "Vehicle" modulus to control the vehicle dynamic. In this modulus, there is an advanced management model of the engine component, named CALLAS (Couplé A La Limite d'Adhérence au Sol). This model is used in high-quality simulators for its high level of detail and realistic driving aiming to develop, evaluate and validate vehicles in complex environmental contexts.

In this study, it is important to define a model of vehicle able to process and export information related to the engine load, the fuel consumption and the CO_2 emissions. As mentioned in the Introduction paragraph, it is easier to refer only to CO_2 emissions and not to other elements which, however, will undergo the same proportional variations.

These models are not based on equations, but on a "fuel consumption mapping", measured in g/kWh. This mapping assigns to each engine regime value (rpm) and effective torque, a specific fuel consumption, using appropriate interpolations. The user can customize the mapping in terms of fuel consumption unit (g/kWh o g/s) and/or interpolation type (linear or spline). A simple coefficient allows, then, the analysts to transform the fuel consumptions in CO_2 emissions, as a function of the triggered chemical reactions. In this research context, a “Small Family Car” has been selected as reference vehicle. Then, at each time t_i, the measured values of rpm and torque are used to estimate the fuel consumption at t_i, measured in g/kWh.

3 Results and Discussions

The driving experiments of the 5 users evidenced a common driving behaviour that is analysed in the following. The understanding of this behaviour will improve the next step of this research, represented by a larger experimentation on a sample of at least 20 users, for performing a statistical analysis of the results.

This preliminary analysis regarded the relationship between the single elements of the alignments and two variables that are strictly related, the vehicle longitudinal acceleration and the CO_2 emissions. As evidenced in the following figures, these two functions exhibit very similar trends. Despite this, the authors have decided to provide both graphs to evidence that this similarity and proportionality is not dependent on specific circumstances of the motion or characteristics of the alignment but is globally valid. As anticipated in the previous section, the CO_2 emission is not inferable through an explicit function, but derives from the engine mapping, that is very complex to be numerically described.

Further, for length limitation, only the graphs related to a single user’s driving are provided, but it is underlined that the other drivers provided similar behaviours. However, the conclusions of this research have not absolute validity, but are useful to evidence features of the driving behaviour (conditioned by the horizontal geometry) that may influence CO_2 emission, defining an original contribution compared to literature results.

The first analysis regards the relationship between the longitudinal acceleration and the horizontal geometry. In Fig. 1a, the trend of the longitudinal acceleration has a sharp decrease before of the two C1 curves. In fact, in the straight segment before the first curve, the user decelerated up to -0.6 m/s^2 and, only when he is close to the beginning of the exit clothoid, he restarts the acceleration in view of the long straight before the next curve (about 236 m long). Approximately in the middle of this second straight segment, the user understands that the second curve has a small radius (the same of the previous one) and begins to slow down, keeping this behaviour up to the exit clothoid, where he starts accelerating again. Overall, the small radius curves represent a constraint for the user that drives along the straight segment before the curve and a large part of the same curve in deceleration.

A remarkably different behaviour emerged in another section of the road (Fig. 1b), characterized by curves with radius equal to 120 m (C2). First, the same behaviour of Fig. 1a is noticed on the straight segment. On the contrary, the acceleration begins quickly after the beginning of the entry clothoid. The user, then, drives the entire curve and the first part of the following straight (about 147 m long) in acceleration and, only after having perceived the following curve, at a distance that may guarantee a comfortable deceleration, starts to slow down. In the second and third curves, the graph

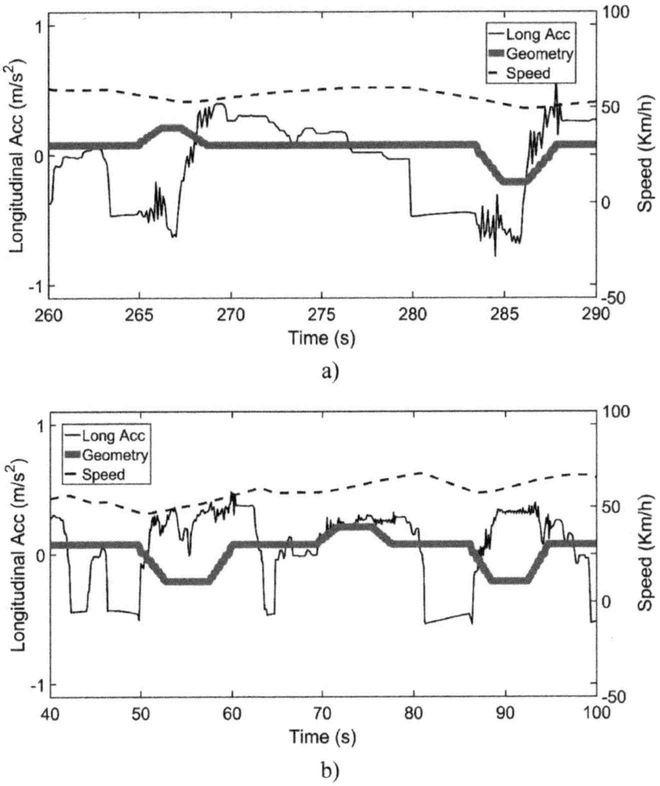

Fig. 1. Relationship between longitudinal acceleration, speed and horizontal geometry: a) a section including two C1 curves; b) a section including three C2 curves. These functions refer to a single driver.

shows the same phenomenon that, obviously, is influenced not only by the radius of the curves, but also by the straight length.

Another different trend of the horizontal acceleration is shown with curves C3 (radius = 180 m), as depicted in Fig. 2a in the time interval between 155 and 180 s. The accelerations are reduced in magnitude because the speed is almost constant in the entire trait. This may be due to an increase of the radius value compared to the previous curve (with a higher speed), but also to a shorter straight (about 81 m long) that prevents further speed increases. In practice, this may represent an ideal condition for the alignment. Similarly, the driving behaviour depicted in Fig. 2b along two C4 curves (radius = 240 m), between 195 and 225 s, is the same of the previous one (Fig. 2a). Only two negative peaks are noticed, indicating a deceleration started at the beginning of the curve, where the user obviously focuses on the mental acquisition of geometry information. This configuration too, with shorter straights than the previous case, favours driving homogeneity and alignment consistency.

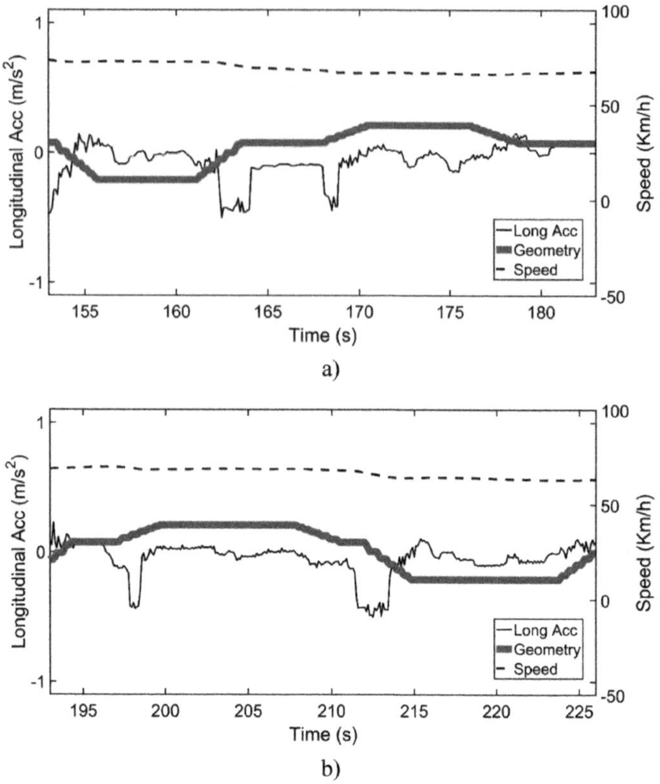

Fig. 2. Relationship between longitudinal acceleration, speed and horizontal geometry: a) a trait including two C3 curves; b) a trait including two C4 curves. These functions refer to a single driver.

Other interesting considerations derive form the analysis of the relationship between horizontal geometry and CO_2 emissions (Fig. 3, Fig. 4), that evidences a trend similar to the horizontal acceleration. In Fig. 3a, the radius of the curves is equal to 45 m and, as expected, in a large part of the curve there is no emission, but it goes up only in the final trait of the exit clothoid, where the drivers increase the speed. This mechanism changes with radius equal to 120 m (Fig. 3b). In this case, the three curves are travelled in acceleration and, thus, with high CO_2 emission values. Along the straight that follows the curve, the user exhibits a dual behaviour. The first trait is travelled in acceleration, while the second, where the user perceives the presence of the following curve, in deceleration. The curves with R = 180 m and R = 240 m (Fig. 4a and 4b) show a similar phenomenon. They are characterized by lower levels of CO_2 emissions (compared to C2 curves) but, mainly, there are not huge variations between max and min values. Even in this case, max and average values of CO_2 emissions strongly depend on the homogeneity of the examined road segment.

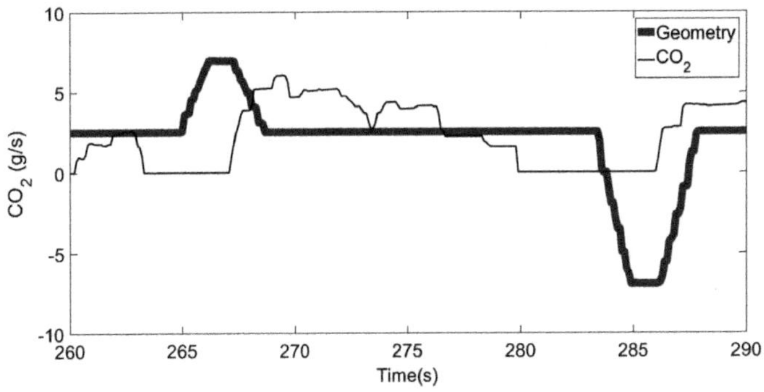

a)

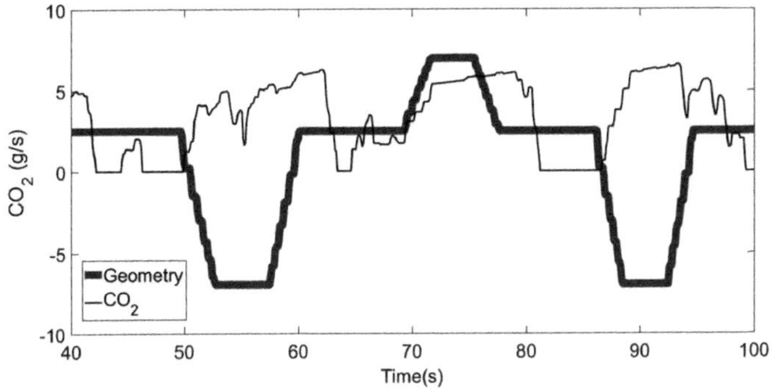

b)

Fig. 3. Relationship between instant CO_2 emission and horizontal geometry: a) trait with C1 curves; b) trait with C2 curves. These functions refer to a single driver.

In summary, Fig. 5 shows some general relationships established between road geometry and CO_2 emissions or longitudinal acceleration. For example, distance A is the short section of the entry clothoid that the user drives before the acceleration and, thus, the emissions begin. It is like the user wants to accurately understand the curve geometry before modifying the vehicle speed, for safety and comfort purposes. Obviously, this distance is function of some variables, such as the driving speed, the curve radius, the deviation angle, the visibility, etc. The evaluation of the segment B is equally interesting. It represents, in practice, the max value of the emissions and it is strongly influenced by the trait homogeneity, in terms of the features of the straight before the curve and the curve itself. Whether the driver feels less the need of acceleration and, thus, consumes less fuel and reduces the emissions, it depends on the speed variation between the straight and the curve. The higher their speed difference, the longer B and, thus, the higher the emissions. Finally, C represents the distance in which the driver decreases his acceleration and, thus, the vehicle emissions. In this case, the influencing variables are the speed of the curve just driven, the length of the following straight and the perception of the curve itself. The combination of these three elements will determine the deceleration of the user.

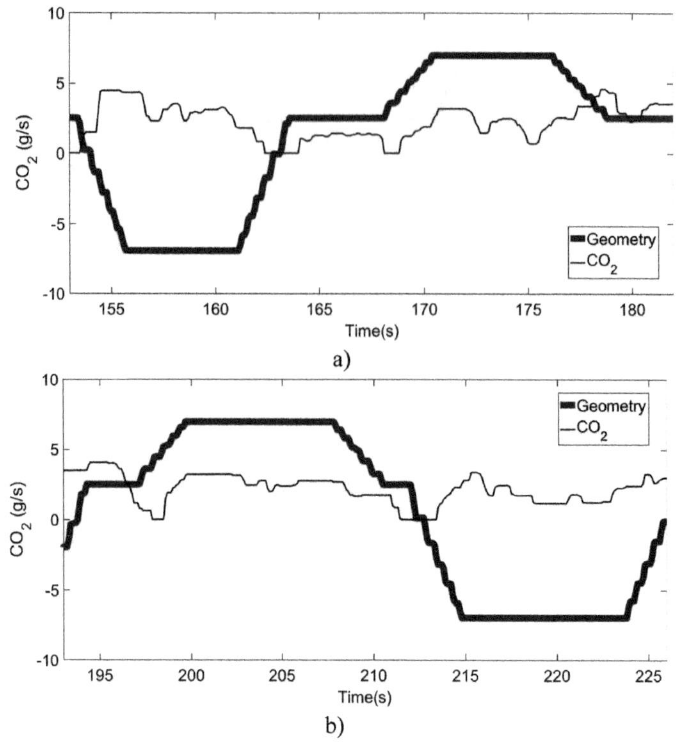

Fig. 4. Relationship between instant CO_2 emission and horizontal geometry: a) trait with C3 curves; b) trait with C4 curves. These functions refer to a single driver.

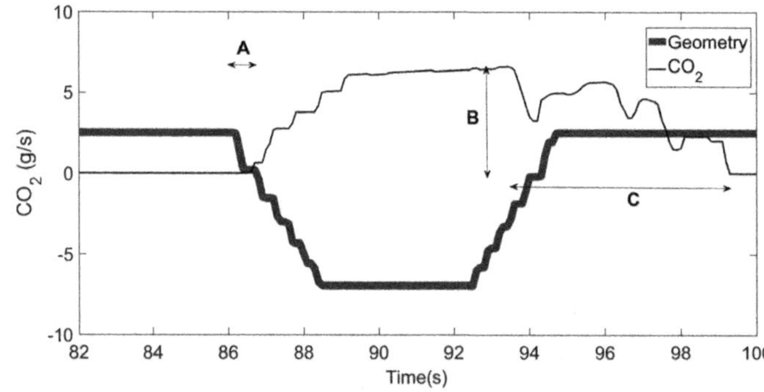

Fig. 5. General relationship between instant CO^2 emission and geometry.

The figures shown above refer to a single driver, but all the others showed similar behaviour. When the experiments are completed with a sample of at least 20–25 users, it will be possible to deduce all the relationships of interest between radius and length of the curves, length of straights and CO_2 emissions, and even to statistically analyse the results.

4 Conclusions

In this first step of the research, the telemetry data regarding longitudinal accelerations and emissions have been derived by a complex driving simulator software using the engine mapping of a reference vehicle. In any case, the results evidence a strong correlation between CO_2 emissions, longitudinal accelerations and, thus, the driving behaviour. These conclusions are, obviously, valid only for the features of this specific road. Obviously, general considerations may be formulated only after the conclusion of the experiments of a sufficient number of users and on real road alignments.

References

1. Franzitta, V., Longo, S., Sollazzo, G., Cellura, M., Celauro, C.: Primary data collection and environmental/energy audit of hot mix asphalt production. Energies **13**, 2045 (2020)
2. Sollazzo, G., Longo, S., Cellura, M., Celauro, C.: Impact analysis using life cycle assessment of asphalt production from primary data. Sustainability **12**, 10171 (2020)
3. Ma, F., Sha, A., Lin, R., Huang, Y., Wang, C.: Greenhouse gas emissions from asphalt pavement construction: a case study in China. Int. J. Environ. Res. Public Health **13**, 351 (2016)
4. Hulail, Z.A., Ayob, A., Omar, W.M.S.B.W.: Carbon footprint of road pavement rehabilitation: case study in Sungai Petani, Kedah. Int. J. Appl. Environ. Sci. **11**, 1285–1302 (2016)
5. Agarwal, A.K., Mustafi, N.N.: Real-world automotive emissions: Monitoring methodologies, and control measures. Renew. Sustain. Energy Reviews **137**, 110624 (2021)
6. Abo-Qudais, S., Abuqdais, H.: Performance evaluation of vehicles emissions prediction models. Clean Technol. Environ. Policy **7**, 279–284 (2005)
7. Heinold, A., Meisel, F.: Emission rates of intermodal rail/road and road-only transportation in Europe: a comprehensive simulation study. Transp. Res. Part D **65**, 421–437 (2018)
8. Ko, M., Lord, D., Zietsman, J.: Environmentally conscious highway design for crest vertical curves. Transp. Res. Record J. Transp. Res. Board **2270**, 96–106 (2012)
9. Boriboonsomsin, K., Barth, M.: Impacts of road grade on fuel consumption and carbon dioxide emissions evidenced by use of advanced navigation systems. Transp. Res. Rec. J. Transp. Res. Board **2139**, 21–30 (2009)
10. Park, S., Rakha, H.: Energy and environmental impacts of roadway grades. Transp. Res. Rec. J. Transp. Res. Board **1987**, 148–160 (2006)
11. Lamm, R., Choueiri, E.M., Hayward, J.C., Paluri, A.: Possible design procedure to promote design consistency in highway geometric design on two-lane rural roads. Transportation Research Record, No. 1195. Transportation Research Board, National Research Council, Washington, D.C., pp. 111–122 (1988)
12. D'Andrea, A., Pellegrino, O.: Application of fuzzy techniques for determining the operating speed based on road geometry. Promet Traffic Transp. **24**(3), 203–214 (2012)
13. D'Andrea, A., Carbone, F., Salviera, S., Pellegrino, O.: The most influential variables in the determination of V85 speed. Procedia Social Behav. Sci. **53**, 1–12 (2012)
14. Llopis-Castelló, D., Camacho-Torregrosa, F.J., García, A.: Analysis of the influence of geometric design consistency on vehicle CO_2 emissions. Transp. Res. Part D **69**, 40–50 (2019)

15. Llopis-Castelló, D., Pérez-Zuriaga, A.M., Camacho-Torregrosa, F.J., García, A.: Impact of horizontal geometric design of two-lane rural roads on vehicle CO2 emissions. Transp. Res. Part D Transp. Environ. **59**, 46–57 (2018)
16. Llopis-Castelló, D., Camacho-Torregrosa, F.J., García, A.: Development of a global inertial consistency model to assess road safety on Spanish two-lane rural roads. Accid. Anal. Prev. **119**, 138–148 (2018)
17. Nobili, F., Bella, F., Llopis-Castelló, D., Camacho-Torregrosa, F.J., García, A.: Environmental effects of road geometric and operational features. Transp. Res. Procedia **37**, 385–392 (2019)
18. MIT. Norme funzionali e geometriche per la costruzione delle strade (in Italian). D.M. no. 6792 of November 5th, 2001. Ministero delle Infrastrutture e dei Trasporti (2001)

Specifics of the Influence of the Perception of the Traffic Situation of Road Users on the Occurrence of an Accident

Michal Ballay[1(✉)], Ľudmila Macurová[2], and Miroslav Rédl[2]

[1] Faculty of Security Engineering, Department of Technical Sciences and Informatics, University of Žilina, Ul. 1. mája 32, 010 26 Zilina, Slovak Republic
michal.ballay@fbi.uniza.sk

[2] Institute of Forensic Research and Education, University of Žilina, Ulica 1. mája 32, 010 26 Žilina, Slovak Republic
ludmila.macurova@uzvv.uniza.sk

Abstract. In order to correctly estimate the moment of recognizing the occurrence of a collision situation, it is essential for the driver to evaluate the method of preventing an accident and to assess the driving technique. In the analysis of an accident, one of the important tasks is to evaluate the driver's ability to perceive the traffic situation according to various aspects. At present, there are various measuring systems of eye movement, which are designed to perform a quality reconstruction of the direction of eye movement of road users as a function of time. In addition to the basic physiology of the perception of road users, the paper deals with the use of the Viewpointsystem® measuring system designed for the analysis of eye movement in the perception of the traffic situation. First, it is characterized by a measuring system, outputs in the human-road-vehicle relationship, as well as its hardware and software. As part of determining the decisive moment for a collision situation, an analysis of the direction of movement of the eyes of the vehicle driver during a dangerous traffic situation is performed while driving on a road section located outside the village.

Keywords: Analysis · Road accident · Collision situation · Road users · Perception · Viewpointsystem

1 Introduction

Road transport is linked to several negative phenomena, including road accidents characterised by extensive material damage, light, or severe consequences for the health of road users, as well as irreplaceable loss of human life. The most common causes of various collision situations are human factor failure, technical failure of the vehicle, unfavourable climatic conditions, unsatisfactory road condition. Drivers are often ruthless, insufficiently foresighted, aggressive and gambling. Pedestrians do not pay attention to road traffic, behave recklessly and may suddenly enter the carriageway. Children and the elderly are particularly at risk. Children behave impulsively, they are unable to estimate distance and time, have a narrowed field of view and insufficiently

O. Prentkovskis et al. (Eds.): TRANSBALTICA 2021, LNITI, pp. 817–824, 2022.
https://doi.org/10.1007/978-3-030-94774-3_77

developed motor coordination. Older people have a reduced ability to see and hear, less ability to concentrate and longer reaction time, leading to more complex adaptation to certain road traffic requirements [6].

By examining the causes and processes of various related conflict situations analysis of traffic accidents deals with road traffic. One of the key parameters is the evaluation of the driver's ability to perceive different traffic situations, all associated stimuli, optical information density, accident and pre-accident event according to different aspects [4]. To perform a quality technical analysis (reconstruction) of the direction of movement of the eyes of drivers and other road users depending on the time, various certified measuring systems consisting of hardware and software are used.

2 Basic Principles of Physiology of Human Perception

Each group of road users has to process a lot of different information, and therefore it is very important to evaluate the extent to which the physiology of perception and the viewing strategy of the participants influence the potential occurrence of a traffic accident. In road traffic, there are basic types of human sensory perception, namely visual, auditory, tactile (receptors in joints, muscles and tendons, for heat and cold), vestibular (maintaining the stability of the human body). Visual perception is especially important when perceiving various information, the share of which is at least 90%. This capability allows also the perception of distant objects at a greater distance in front of them. Perception of information by the rest of the senses (1–10%) is additionally used as a control function [5].

Viewing angle in an adult (pedestrian) with normal vision it is approximately 180–190° in the horizontal direction and 130° in the vertical direction (60° up and 70° down). Visual acuity for pedestrians is 1′ and for drivers of the vehicle is 1.5′–2.0′.

When receiving information, the viewing height of the pedestrian is about 160 cm and of the driver about 115 cm. There is usually a balanced share of concentrated and distributive perception in navigation management. Navigation priorities are a function of object tracking habit and attention, depends on the number of appropriately selected waypoints [9].

In the elderly (with certain health deficits) there are often disadvantages associated with receiving information. These are most often information errors and deficits due to a lack of navigation priorities, longer time required for adaptation and accommodation of the eye (likelihood of visual impairment / visual impairment), disorders caused by the use of prescribed drugs, other symptoms (dizziness, restriction of movement, impaired hearing, etc.). These people mainly use objects in close proximity for navigation, and not in the area of peripheral perception. Because of the fear of falling, they move their eyesight directly into the area of movement. In most cases, they tend to perform inspection visions towards approaching vehicles a few seconds before crossing the road, but after entering the road they perform almost no further inspections [9].

Children begin to perceive when they change their eyesight, leading to a shadow effect and perception with only one eye, as well as a disorder of depth perception. Inexperienced children constantly observe their surroundings, while attention to specific objects lasts only a short time (short-term vision fixation). The behavior of older, more experienced children is already more in line with the strategy of monitoring adults. The proof is a more ordered order of visual fixations and involuntary eye movements. In people with disabilities, the most common visual disturbances can be characterized as visual imaging disorders (blurred/distorted image on the retina), turbidity (decreased visual acuity/decreased contrast perception often combined with increased sensitivity to glare), diseases of the visual system (diseases of the retina, field of vision reduction, blind spots, night blindness, etc.) [9].

3 Measuring System Viewpointsystem®

Viewpointsystem® can be defined as a measuring system of the direction of movement of the drivers' eyes to identify all forms of omitted views to certain locations of the surrounding area, failed (distorted) or insufficient optical information, unnecessary visual fixations and excessive optical information (optically complicated situations as the main cause of the accident, among other circumstances). In particular, the perception-perceptual, physiological, and psychological attributes of perception are investigated. The physiology of human vision may be graphically illustrated in relation to a given place to eliminate the existing risks of road accidents and the associated hazards in accordance with legal and technical standards in the field of road traffic. Viewpointsystem® allows to analyse the behaviour of human vision with accuracy from 15 angular minutes (spatially) or 40 thousandths of a second (time). Very often, with the usual state of the road, the limits of the driver's psychical performance regarding the reception of information are exceeded and this limit is exceeded several times in the area of visual - physiological perception [2, 10].

In the person-road-vehicle relationship, the Viewpointsystem® measuring system may also be used to detect and eliminate safety defects and potential hazards in the event of a road accident and also as part of a (preventive) road safety inspection by the nominated road inspectors [10].

The hardware accessories of the measuring system consist of individual parts, which are eye-tracking glasses and a recording unit. Using sensitive electronic sensors built into glass of glasses, the system can monitor eye movements, motion directions, uncontrolled (saccadic) eye movements, as well as individual blinks. For use in night mode, the glasses are equipped with LEDs, which are built into the frame of the glasses. A camera is built into the front of the glasses frame to make video recording of the scene. All information obtained from the glasses is then recorded in the memory of the recording unit, which can later be transferred to the PC. It is also possible to play individual recordings from the glasses in the recording unit (Fig. 1).

To process individual eye view direction records and to statistically evaluate the necessary parameters, there is a Fact finder software accessory that offers a variety of special visualization functions such as circle, point-fog, navigation, star, saccades, fixation, density, and bubbles (Fig. 2).

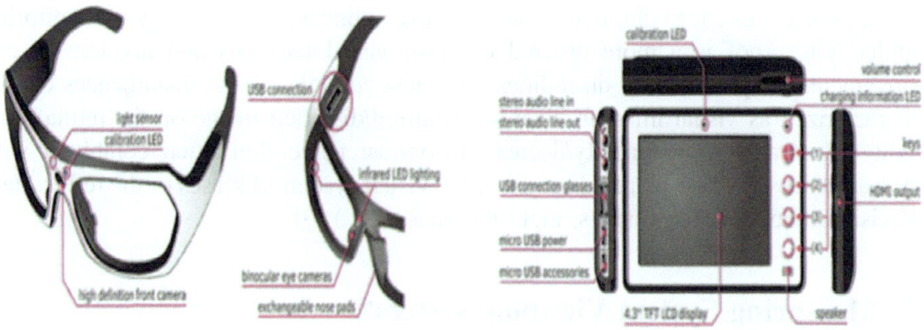

Fig. 1. The hardware accessories of the measuring system [11].

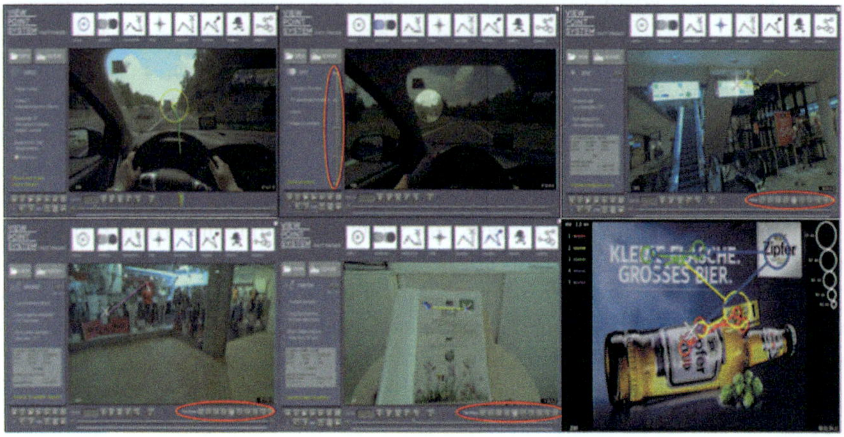

Fig. 2. List of visual functions in Fact finder [11].

Individual types of visualization functions show, for example, a chronological sequence of view and fixation based on the use of variable-size colour circles, visual acuity using two concentric circles, accumulation of several fixation points in a small area of vision of 3°, an overview of significant navigation points and objects with high attractiveness for peripheral detection, an overview of the sequence of individual views, areas of parafoveal vision, etc. [11].

4 Determining the Decisive Moment for a Collision Situation

While driving, the driver pays attention not only to directional driving, but has to also focus on the complex of different events, situations and circumstances. It must evaluate the movement of other road participants (pedestrians, cyclists, motorcyclists, other vehicles), observe traffic signs, traffic lights as well as the data displayed on the dashboard, evaluate the situation in the rear-view mirror, etc. [1].

Test tests were performed to visualize the distribution of the driver's point of view, which were recorded and evaluated by means of the Viewpointsystem measuring system. All points in a certain time space were assessed with respect to their position, in the driver's field of vision. The context of psychological perception has also been clarified and limits of recognition, resp. decisions are in danger. As part of the test tests, an analysis of the direction of eye movement of the vehicle driver during a dangerous traffic situation was performed when driving outside the village. The test subject was a woman, aged 30 years, which had driven more than 10,000 km/year. This person drove a personal motor vehicle on an expressway equipped with 2 + 1 lanes, not far behind a truck with a trailer. To orientate herself in space, the driver used a truck driving in front of her vehicle and, in rare cases, carried out inspection visits to a more remote area on the road, in the space in front of the vehicle (slight clockwise curve) (Fig. 3).

Fig. 3. Orientation of the driver's view using the truck and control view of the driver before performing the prevention [8].

When the widening of the left edge of the right lane to the full width of the left lane has begun in the direction of travel, performed a water control sight and began to precede the truck into the left lane. During the next movement in this lane, the driver noticed a touring car-bus compared to the driving car, which, despite the double solid line, passed into the opposite lane (extremely short latent time from the first opportunity to observe the bus coming from the area of the obscured view to the first fixation on the bus) [8] (Fig. 4).

The driver of the vehicle fixed her gaze on the opposite bus for more than 2 s and examined its driving path with a glance. Subsequently, the driver performed a control view of the right lane from the point of view of its direction of travel and at the same time checked the position of the combination of vehicles with the view. The driver solved the situation by avoiding a maneuver [8] (Fig. 5).

Fig. 4. Fixation of the driver's view of the anti-running bus and examination of its travel path [8].

Fig. 5. Inspection view of the driver before performing the evasive maneuver and view from the driver's position after performing the evasive maneuver [8].

During the entire collision traffic situation, the tested driver did not have a single blink of an eye, which indicates a high degree of concentration of the driver on the traffic situation. As a result of this event, the driver was forced to pursue a critical strategy for monitoring the traffic situation in question, and the directions of view were constantly concentrated in a very narrow defined area [8].

5 Conclusion

Traffic accidents have a stochastic character, and therefore several factors influence their origin, such as unpredictable environmental circumstances, careless drivers, unexpected behavior of road users, bad weather and other difficult conditions when driving a vehicle. An important criterion is represented by specific groups of pedestrians, namely children, adults, the elderly and people with disabilities. There are different visual - physiological differences in the way of perception by sight between specific groups of pedestrians. When analysing the specific conditions of a traffic accident (before the collision situation and during the accident) it is possible to use measuring systems, the so-called analysers of the direction of the eyes' eyes of road users. For this purpose, a "View-points system" is suitable, which can be used to analysis the overall character of the visually perceived scene of a road user and then evaluate the moment of recognition of danger, effective moment of reaction, up to the stage of performing a defensive maneuver. As part of the issue of determining the decisive moment for a collision situation - recognition of danger and the driver's reaction in practice, an analysis of the direction of movement of the test person's eyes (vehicle driver) when driving outside the village was performed. Visualizations of the test person's distribution point views were recorded and evaluated using the View-pointsystem® measurement system. It can be stated that during a collision traffic situation (observation of the bus and recognition of its driving style as a potential danger, until the moment the driver's vehicle passes into the right lane), the driver's vehicle was fully concentrated on the traffic situation. When a forced critical strategy was created, vodka solved the situation by avoiding a maneuver.

Any effective (purposeful) reaction procedure, including the execution of a defensive maneuver, presupposes prior recognition of danger. If this hazard identification is not clearly given, there is a delay in the reaction. The assumption that for all occurrences of traffic events, resp. stimuli, the same set time will be used to react and perform a braking defense maneuver. The total time for an effective (purposeful) reaction action is tied to the moment of the reaction stimulus (danger as a challenge), the self-recognition of the danger and the reaction time [5, 7].

In general, it can be stated that while driving, the choice of waypoints depends to a large extent on the given environmental conditions. The offer of a high number of navigation points while driving causes a high complexity of receiving information, as they affect the length of time the driver fixes the view.

As part of the issue of determining the driver's decisive moment for a collision situation using the Viewpointsystem® measuring system, test tests were also performed during day and night driving in the open air, in a straight line and in a left-hand turn. There are significant differences in the choice of driver navigation points in relation to the viewing distance and the width of the field of view. During the day, navigation points are visible from a distance of about 120 m and at night from a distance of about 50 m from the current position of the driver. In a left-hand bend, the driver is oriented at a distance of approx. 77 m during night driving and at a double distance of approx. 140 m during daytime driving. During the day, the driver is guided mainly by horizontal markings on the road, vegetation at the edge of the road and uses a much larger number of objects for navigation. During the night, the layout of the driver's navigation views is limited to certain objects and the road area. The driver is guided in particular by horizontal traffic signs and reflective directional posts, which are located at greater distances.

The Viewpointsystem® measuring system can also be used in test runs, for example when driving in the city, in populated, densely populated and uninhabited areas, with a focus on recording differences during daytime driving and night driving.

Acknowledgement. The article was created with the support UNIZA grant system number 8010/2020.

References

1. Ballay, M., Macurová, L., Kohút, P., Čopiak, M.: Development of road safety status and the evaluation criterion causes of specific traffic accidents. In: Transport Means - Proceedings of the International Conference, 2018-October, pp. 765–770 (2018)
2. Boratto, L., Manca, M., Lugano, G., Gogola, M.: Characterizing user behavior in journey planning. Computing **102**(5), 1245–1258 (2020). https://doi.org/10.1007/s00607-019-00775-8
3. Čulík, K., Kalašová, A.: Statistical evaluation of BIS-11 and DAQ tools in the field of traffic psychology. Mathematics **9**(4), 433 (2021)
4. Kasanický, G., Kohút, P.: Impact dynamics theory for the analysis and simulation of collisions. In: EDIS, p. 350. University of Žilina, Žilina (2004). ISBN 80-8070-312-4 (2004)
5. Kohút, P., Kasanický, G.: Evaluation of crash tests of the Institute of forensic engineering, University of Zilina, aimed at EES. Komunikacie **15**(2), 124–130 (2013)
6. Kohút, P., Macurová, Ľ., Sventeková, E., Ballay, M.: Safety analysis of pedestrian as the users of road traffic in the Slovak Republic. In: Paper presented at the Transport Means - Proceedings of the International Conference, 2019-October, pp. 446–452 (2019)
7. Ondrus, J., Hockicko, P.: Braking deceleration measurement using the video analysis of motions by SW tracker. Transp. Telecommun. **16**(2), 127–137 (2015)
8. Pfleger, E.: Hazard recognition and reaction in practice - exact time proof by visualization analysis. In: Proceedings – 21st EVU Congress (Annual Congress of the European Association for Accident Research and Analysis), Braşov, Romania (2012)
9. Pfleger, E.: Viewing behaviour as an accident cause – typical risks for elderly people, business people, children, patients with sight disorders. In: Proceedings – 24th EVU Congress (Annual Congress of the European Association for Accident Research and Analysis), Edinburg, United Kingdom (2015)
10. Rábek, V.: Vnímaní a rozhodování účastníků silničního provozu - denní doba. Sborník tuzemských a převzatých cizojazyčných publikací. Univerzita Pardubice, 2014, p. 350 (2014). ISBN 978-80-904944-2-8
11. TPM – Viewpoint Vista Technical Manual (user manual). https://www.docs.oracle.com, Accessed 10 June 2021

Investigation of Criteria of Winter Tires Selection: How the Choice of Experts and Drivers Differs

Justas Bražiūnas(✉), Henrikas Sivilevičius, and Rytis Reinys

Vilnius Gediminas Technical University, Saulėtekio al. 11, 10223 Vilnius, Lithuania
{justas.braziunas,henrikas.sivilevicius}@vilniustech.lt, rytis.reinys@stud.vgtu.lt

Abstract. When analyzing the technical condition of a vehicle, its interaction with the pavement, tires have a special significance, because this interaction takes place through the tires. This paper analyzes the criteria for experts and respondents to select passenger car tires for the winter season. The application of multi-criteria decision making (MCDM) methods to determine the importance of criteria for the choice of tires for the winter season is performed. The study uses the developed questionnaire, the Analytic Hierarchy Process (AHP) method, rank correlations, and ARTIW methods. Descriptions and questionnaires of factors (criteria) for respondents and experts were compiled in the study, surveys of experts and respondents were conducted, as well as paired comparison matrices were used. The study identified what criteria are most important for respondents and experts when choosing passenger car tires for the winter season. After examining the significance of the groups of factors, the experts chose the technical characteristics of the tires as the most important group, and the respondents chose the brand of the tire manufacturer. Examining individual groups and factors it was found that the opinions of Lithuanian drivers and experts on many factors were quite similar, and no significant differences were found in the study.

Keywords: Car tire · Winter tire selection criteria · Expert · Driver · MCDM · ARTIW method · Rank correlation method · AHP method

1 Introduction

The production of automotive vehicles worldwide reached approximately 96 million units in 2016. The total estimated number of vehicles in use in the world is 1.28 billion cars including 335 million commercial vehicles [1]. As a result, the motorization rate increased worldwide. In the developed countries the motorization rate reached over 550 vehicles per 1,000 inhabitants including North America, the European Union (EU), and Japan. Recovery and recycling end-of-life tires incur a significant cost of collection and transportation. Improperly conducted end-of-life collection contributes to excessive vehicle exhaust emissions and high transportation costs [2].

O. Prentkovskis et al. (Eds.): TRANSBALTICA 2021, LNITI, pp. 825–838, 2022.
https://doi.org/10.1007/978-3-030-94774-3_78

In response to the present dynamic market, automobile manufacturing industries are constantly evaluating and improving their manufacturing strategies to stay competitive. It is also evident that they evaluate those strategies considering multiple criteria [3]. In one such case, an automobile project manager must ensure that the selected tire (among the available alternatives) comply with the prospective automobile's performance. The investigation presents an approach for the assessment of alternative automobile tires taking into account multiple criteria, such as tire sustainability, tire-pavement performance, environmental issues, tire purchase cost, and tire pavement grip. Traffic parameters of the pavement transport means are influenced beside the loads to be transported, the travel speeds, the constructive and functional characteristics by the air pressure in the tires which has a direct impact on the tire tread wear, rolling resistance, compaction, soil compaction, and rolling track wear, and finally on the transport operation costs [4].

In winter, snow falls on the pavement surface, which worsens driving conditions [5, 6]. Mulmivuo et al. (2017) compared the safety effects of studded and unstudded winter tires based on fatal road accidents in Finland [7, 8]. The overall effect of tire type on the number of accidents was not significant, although studded tires reduced fatal accidents by 10–15%. Compared to unstudded tires, studded tires reduced accidents significantly only on bald ice in 2005–2012. Drivers using unstudded tires were more experienced and their profession was more frequently related to driving. In addition, the vehicle age was lower for vehicles with unstudded tires. On the other hand, the state of repair was less pertinent for unstudded than for studded tires. These confounding factors offset their effects to some degree.

Knowledge of actual tire-pavement friction potential is most relevant for the safe operation of automated vehicles and for adjusting advanced driver assist systems. Frequently, a vehicle's tire interacts with the pavement surface through an intermediate element. When a layer of water builds between the pavement surface and the rubber tires of the vehicle, the risk of aquaplaning occurs. Aquaplaning may result in considerably reduced friction potential and unstable vehicle behavior for rear-wheel-drive vehicles; thus, awareness of the risk of aquaplaning is crucial [9–11]. To simulate the interaction between the vehicle's wheels and road pavement o lot of studies have been conducted [12–15]. Researchers are examining individual tire tread elements or simply rubber specimens to predict interactions with the pavement surface at an early stage [16].

Road pavement surface macrotexture impacts the wheel's tire friction coefficient [16]. Pavement surface roughness and rutting impact the dynamics of the running vehicle, which influences traffic safety. The development of pavement surface distresses and their impact on the vehicles may be simulated [18]. Rubber deposits change the parameters of the wheel's interaction with the pavement surface [19].

Tire retreading is an economically attractive and environmentally beneficial process for extending the lifetime of used tires [20]. It reduces the negative influence of used tires on the environment as well as natural resources exploitation. According to the latest data, 4.7 million units of used truck tires were retreaded in 2014 in Europe.

Exposure to excessively high noise levels is a relevant health problem in Europe and road traffic noise is the most widespread noise source [21]. The noise level generated on urban roads can be excessive and has a negative impact on the urban environment [22]. Due to the engine work and air resistance in the interaction between the wheel and the pavement surface, noise is generated [23]. This noise level may be measured by using a computer simulation or a natural experiment [24].

Tire failures, including catastrophic tread separation and blowouts, are common events that lead to serious road accidents in Saudi Arabia [25]. Laboratory tests were performed on used and new tires to determine the effect of in-service conditions on tire properties. The laboratory tests included strength, peeling, tensile, elongation, bead unseating, endurance, and high-speed performance tests. Regarding the new tires, all the tested brands passed the internationally adopted tire quality control tests. However, the results indicated some differences between the frequently and infrequently failed tires in terms of some physical properties and the percentage of failed tires during field use.

The aim of the work is to investigate what criteria are most important for drivers and experts when choosing passenger car tires for the winter season. What criteria are important for drivers, what are for the experts? Are they different? Can a driver without special knowledge of tires and the car make the right decision?

2 Methodology of the Research

2.1 Respondents and Experts

Recently, a popular pairwise comparison method called Analytic Hierarchy Process (AHP), which was offered by Saaty (2000) and Saaty et al. (2003) [26, 27], has been widely used. Pairwise comparisons enable to increase the compatibility of evaluations. AHP method enables to identify one level hierarchy indicators' significance in terms of a higher level or hierarchically non-structured significance of indicators. The essence of the method lies in the matrix of pairwise comparison. For the research, respondents and experts living in Vilnius were chosen to be interviewed. A total of 30 individuals, 15 respondents (drivers), and 15 experts each participated in the study (Table 1). The work determines that a respondent is a person who has passed a driving test, drives a vehicle, knows the rules of the road, and has a general knowledge of the operation of the car. Expert - a specialist in the field of transport who has acquired expert knowledge about cars and tires during specialized studies, courses, or training. The survey has been conducted from November 2020 to February 2021.

Table 1. Distribution the number of respondents and experts in the study.

Characteristic	Respondents - 15	Experts - 15
Gender	Woman – 2/Men – 13	Woman – 0/Men – 15
Education	Higher – 15	Prof. Dr. Habil. – 1/Dr. – 1 Higher – 13
Age intervals	Up to 40–11/40 +–4	Up to 40–12/40 +–3

2.2 Criteria and Sub-criteria

All experts and respondents were given questionnaires containing the factors (sub-criteria) divided into 5 groups (criteria) (Fig. 1). These factors were chosen by the authors and the study accepted that these are the most important factors in this area. A three-part questionnaire was used for the survey. In the first part, criteria groups

Group A. Tire Seller Characteristics

- A1. Public awareness from the sources of creating a positive image;
- A2. Subjective opinion and feedback about the seller from individual acquaintances, friends and experts;
- A3. The working time of the tire seller in the market;
- A4. Long-term positive experience when using the services of the same seller more than once;
- A5. Comprehensive ongoing dissemination of information about the seller (online, in specialist magazines, tire shops, radio, television, exhibitions) or on other platforms;
- A6. Smooth fulfillment of the seller's financial and tax obligations.

Group B. Tire Manufacturer's Brand

- B1. Public awareness, popularity in Lithuania and other countries;
- B2. Brand lifespan;
- B3. High marks of the brand tire, confirmed by (ADAC) certified tests;
- B4. Subjective opinion and feedback about the brand from individual acquaintances, friends and experts;
- B5. Long-term experience when choosing a product of the same brand is not the first time;
- B6. A large number of products sold by the brand in Lithuania and other countries;
- B7. A wide range of assortments (tire models) offered by one brand.

Group C. Technical Characteristics of Tire

- C1. Adhesion to wet, snow-covered road surfaces;
- C2. Fuel consumption (rolling resistance);
- C3. Tire noise;
- C4. Tire production week and year;
- C5. Tire compound properties (hard, soft compound);
- C6. Seasonality of the tire (the tire is suitable for summer, winter or all seasons);
- C7. Long tire life (slow tread wear);
- C8. Tread pattern (number of slats, shape and dimensions of blocks, depth and direction of grooves between blocks) and its type (directional, non-directional);
- C9. The length of the road driven by the car is suddenly stopped until it comes to a complete stop;
- C10. Additional design features of the tire (reinforced sidewalls, possibility to drive the tire in case of pressure drop);
- C11. Tire load and speed index (maximum permitted speed and weight maintained as indicated by the tire manufacturer's index).

Group D. Commercial Factors in Tire Sales

- D1. Tire sales low retail price;
- D2. Various payment options (online, in installments, at the point of sale);
- D3. Warranties are provided and compliance with warranty terms;
- D4. Discount programs apply (promotions, loyalty programs);
- D5. A large selection of tires for sale is available;
- D6. Appropriate and easily accessible location of the seller;
- D7. Free acceptance and recovery or regeneration of used tires (rubber reuse);
- D8. Possibility to store tires in the seller's warehouse;
- D9. Various possibilities of product delivery (delivery to the place of residence, workplace, service);
- D10. Communication culture with clients (high-quality, objective and benevolent consultation of competent specialists);
- D11. Aesthetics (appearance, clothing) and orderliness of the premises of the tire seller's employees.

Group E. Factors Increasing the Demand for Tires

- E1. Theft of tires with rims and their intentional damage;
- E2. Continuous and frequent inspection of the conformity of vehicle tires to the season of use and the depth of tread grooves by road police officers;
- E3. Roughness and damage to road and street surfaces damaging tires (punctures, sills, collapses, crushed edges).
- E4. Car accidents involving damage to tires;
- E5. The level of motorization of the country (number of cars per thousand inhabitants) and the number of inhabitants of the country influencing the demand for tires;
- E6. Influence of consumer living standards and increase in purchasing power (tires are replaced before the tread depth limit is reached).

Fig. 1. A description of the criteria and sub-criteria.

A, B, C, D, E, are ranked mutually. In the second part, ranks are assigned to the sub-criteria of the main groups. In the third part, the pairwise comparison matrix was completed, calculated, and checked.

2.3 Assessment of Importance of Criteria and Sub-criteria Using AHP Method

An expert compares all evaluated indicators R_i and R_j $(i,j = 1,2,\ldots n)$, here n – a number of comparative factors (also called criteria, indicators). AHP method is special so that it requires high-level thinking of experts and respondents, therefore just researchers interview several experienced experts [11, 28]. Using experience, competence, knowledge, and intuition, the experts comparing factors with each other (pairwise comparison). Expert has also taken into consideration the significance of the intensity of criteria in terms of each other on a nine-point scale (1-2-3–4-5-6-7-8-9), pointing out how one factor from A-E groups is more important than the other factor from the same group pairwise comparison matrix of all criteria interactions enables to calculate eigenvector ω_i according to the formulas suggested by Podvezko (2009) and Sivilečius (2011) [29, 30]. The significance of the criteria (A-E) was also compared using the same pairwise comparison principle (Fig. 2). A total of 180 pairwise comparison matrices were filled in and counted.

The subjective normalized weight of each criteria or sub-criteria is the eigenvector ω_i, which is calculated from the formula:

$$\omega_i = \frac{\sqrt[n]{\prod_{j=1}^{n} a_{ij}}}{\prod_{i=1}^{n} \sqrt[n]{\prod_{j=1}^{n} a_{ij}}}, \tag{1}$$

where a_{ij} – element of pairwise comparison matrix; $i, j = 1, 2, \ldots, n$ – number of columns and rows of matrix. The following is a pairwise comparison matrix of criteria of one respondent and an explanation of its calculation. The sum of multiplications $\sum_{j=1}^{5} a_{ij}\omega_j$ is divided by the corresponding criterion largest calculated eigen value λ_{max} (Eq. 2), presented in Fig. 2.

$$\lambda_{max} = \frac{1}{n}\sum_{i=1}^{n} \frac{\sum_{j=1}^{n} a_{ij}\omega_j}{\omega_i}. \tag{2}$$

Respondent No 6. Significance of factors influencing the choice of winter tires for passenger cars. 12 Dec 2020.

Main criteria (groups of criteria) (5 criteria)

1. Pairwise comparison matrix $A=[a_{ij}]_{nxn}$

Element interaction code		Rank of criterion	Criterion j=1, 2,….., n					Eigenvector
			A	**B**	**C**	**D**	**E**	ω_i
Criterion i=1, 2,., n	**A**	**4**	**1**	1/5	1/5	1/5	5	0.07091
	B	**1**	5	**1**	3	3	9	0.44852
	C	**2**	5	1/3	**1**	2	8	0.26031
	D	**3**	5	1/3	1/2	**1**	7	0.19208
	E	**5**	1/5	1/9	1/8	1/7	**1**	0.02818
							Sum for verification	1.00000

2. Calculation of eigenvector w_i

A	B	C	D	E
ω''_A	ω''_B	ω''_C	ω''_D	ω''_E
0.04000	405.00000	26.66667	5.83333	0.00040
ω'_A	ω'_B	ω'_C	ω'_D	ω'_E
0.52531	3.32270	1.92839	1.42293	0.20879
		Σ		
7.40811	7.40811	7.40811	7.40811	7.40811
ω_A	ω_B	ω_C	ω_D	ω_E
0.07091	0.44852	0.26031	0.19208	0.02818

3. Highest eigenvalue λ_{max}

		Criterion j = 1, 2, ……, n					Sum
		A	**B**	**C**	**D**	**E**	
Criterion i=1, 2,…,n	**A**	0.07091	0.08970	0.05206	0.03842	0.14092	0.39201
	B	0.35455	0.44852	0.78092	0.57623	0.25366	2.41388
	C	0.35455	0.14951	0.26031	0.38415	0.22548	1.37399
	D	0.35455	0.14951	0.13015	0.19208	0.19729	1.02358
	E	0.01418	0.04984	0.03254	0.02744	0.02818	0.15218

Criterion	$\sum_{i=1}^{5} a_{ij}\,\omega_j$	$\lambda_{max} = \frac{1}{5}\sum_{i=1}^{5}\frac{\sum_{j=1}^{5} a_{ij}\,\omega_j}{\omega_i}$
A	0.39201	5.52837
B	2.41388	5.38187
C	1.37399	5.27835
D	1.02358	5.32899
E	0.15218	5.39941

Arithmetic mean λ_{max}	**5.38340**
Consistency index C.I.	**0.09585**
Consistency ratio C.R.	**0.08635**

Fig. 2. The filling of pairwise comparison matrix and the calculation of eigenvector ω_i, highest eigenvalue λ_{max}, consistency index $C.I.$, consistency ratio $C.R$ of expert No. 6.

The figure shows an example of expert No. 6 calculation results by AHP method. We find the arithmetic mean of these values, which is equal to the largest eigenvalue of matrix λ_{max}. In the calculations of this matrix, it was found that $\lambda_{max} = 5.38$. The closer λ_{max} criteria number n is, the more consistent result is obtained. In our case, the comparison matrix row, or the number of compared criteria (matrix size) $n = 5$ and 5.38 is more than 5. The matrix consistency index $C.I.$ and random index $R.I.$ [26, 29, 30] is presented in Part 3 of Fig. 2. A matrix is considered consistent if the value of ratio $C.R.$ is smaller than 0.1. In our case $0.086 < 0.1$; therefore, the matrix of the presented digital sample is consistent.

2.4 Consistency Options of Respondents and Experts

The next step is to determine the consistency of the assessments of the 15 expert and respondent groups. The idea of the concordance coefficient W is associated with the sum of ranks of each criterion R_j with respect to the estimates of experts and respondents according to the formula suggested by Sivilečius, Maskeliūnaitė (2010) [31]:

$$W = \frac{12S}{n^2(m^3 - m)}, \tag{3}$$

where m – number of criteria or sub-criteria ($j = 1, 2, ..., m$), n – number of experts or respondents ($i = 1, 2, ..., n$).

Also, the average value $\overline{R_j}$ and squares S should be calculated. The level of consistency of respondents' and experts' estimates is shown by the concordance coefficient W. If the estimates of the respondents or experts are consistent, the concordance coefficient W is about 1. When the estimates differ to a great extent, W is about 0. The concordance coefficient W must be greater than its minimum value W_{min}, which depends on the critical *chi-square* (*Pearson*) statistics $\chi^2_{\alpha;\upsilon} = \chi^2_{0.01;14}$, found from the Montgomery table [32]:

$$W_{min} = \frac{\chi^2_{\alpha;\upsilon}}{n(m-1)}. \tag{4}$$

The concordance coefficient W can be calculated based on the ranks of the criteria assigned by experts. If their estimates were presented in any other form, they should be ranked. The ranking is a procedure when the most significant criterion is given the rank of one, the second is ranked by two, etc. The ranking of all criteria groups, according to their significance, carried out by 15 experts and respondents is demonstrated in Table 2.

Table 2. The values of the concordance coefficient W the calculated value χ^2 and the lowest value of the concordance coefficient W_{min} for the group of criteria and the sub-criteria obtained in the questionnaires elicited from respondents and experts.

Parameter	Group	Experts (15)	Respondents (15)	Group	Experts (15)	Respondents (15)
W	**Groups of criteria** (5 criteria)	0.9266	0.37	**C group - Technical Characteristics of Tire** (11 sub-criteria)	0.4808	0.2058
X^2 ($X^2_{\alpha\upsilon}$)		55.6 (13.28)	66.6 (13.28)		72.12 (23.21)	30.88 (23.21)
W_{min}		0.2213	0.2213		0.1547	0.1547
W	**A group - Tire Seller Characteristics** (6 sub-criteria)	0.76	0.3153	**D group - Commercial Factors in Tire Sales** (11 sub-criteria)	0.3757	0.4972
X^2		57.46 (15.09)	23.62 (15.09)		56.36 (23.21)	74.58 (23.21)
W_{min}		0.2012	0.2012		0.1547	0.1547
W	**B group - Tire Manufacturer's Brand** (7 sub-criteria)	0.6968	0.2615	**E group - Factors Increasing the Demand for Tires** (6 sub-criteria)	0.3839	0.4652
X^2 ($X^2_{\alpha\upsilon}$)		62.71 (16.81)	23.54 (16.81)		28.79 (15.09)	34.89 (15.09)
W_{min}		0.1867	0.1867		0.2012	0.2012

For that ARTIW method was used, and its estimated results are presented in Table 3.

Table 3. Ranking of criteria in expert questionnaires and criteria weights calculated by ARTIW method.

		Significance of criteria A, B, C, D, E, (j = 1, 2, …, m)					
	Expert number	A	B	C	D	E	Total
(i = 1,2, …, n)	E1	5	3	1	2	4	15
	E2	3	2	1	5	4	15
	E3	3	2	1	5	4	15
	E4	5	1	2	3	4	15
	E5	3	2	1	5	4	15
	E6	2	3	1	5	4	15
	E7	3	2	1	5	4	15
	E8	4	2	1	5	3	15
	E9	3	2	1	5	4	15
	E10	2	3	1	4	5	15
	E11	3	2	1	4	5	15
	E12	4	1	2	5	3	15
	E13	3	2	1	4	5	15
	E14	3	2	1	5	4	15
	E15	4	1	2	3	5	15
	Sum of ranks R_{ij}	50	30	18	65	62	225
	Mean $\overline{R_j}$	3.3333	2	1.2	4.3333	4.1333	15
	W	0.9266					
	X^2, ($X^2_{\alpha\upsilon}$)	55.6 (13.28)					
	W_{min}	0.2213					
	ω_j (by ARTIW)	0.1777	0.2666	0.32	0.1111	0.1244	1
	PRIORITY	**3**	**2**	**1**	**5**	**4**	-

The criteria subjective weights can be defined by ARTIW methodology (Sivilevičius (2011) and Sivilevičius, Maskeliūnaitė (2018) [33, 34] by which the importance (weight) of criteria A, B, C, D, E and it's sub-criteria is defined. For that, the following equation is applied:

$$\omega_j = \frac{(m+1) - \overline{R_j}}{\sum_{j=1}^{m} \overline{R_j}}, \tag{5}$$

where $\overline{R_j}$ – average rank of j criterion or sub-criterion.

Rank averages do not indicate the superiority of one criterion over another. Besides, in practice, when evaluating criteria priorities, it is more convenient to apply their importance indicators – their best value (the most important indicator) is the highest value.

3 Results and Discussion

The criteria weights determined by the ARTIW and AHP methods have the same trend, but their weights have different values. The weights of the most important criteria determined by the AHP method differ more from the weights of the least important criteria than the weights of these criteria determined by the ARTIW method. The AHP method is more sensitive to finding the very first most important criteria, so the difference between the results of the most important criteria of the AHP and ARTIW method is much larger.

Averages value of the sum of the results obtained by ARTIW and AHP methods were used to compare the selection indicators of drivers (respondents) and experts in order of importance. The average opinion of the group of respondents summarized in the research is accepted as a possible solution to the problem or the result of the solution only when the opinions of all respondents are agreed (impeccable). The averages of wights of the criteria and sub-criteria determined by the two methods (ARTIW and AHP) was calculated according to the following formula:

$$\omega_j = \frac{\omega_{jARTIW} + \omega_{jAHP}}{2}, \tag{6}$$

where ω_{jARTIW}– normalized weight of jth criterion or sub-criteria determined by the ARTIW method; ω_{jAHP} normalized weight of jth criterion or sub-criteria determined by the AHP method.

After calculation the significance of the criteria, it was found that the priority of respondents and experts was different (from the most important factor to the least important) (Fig. 3a). The priority of the experts was C≻B≻A≻E≻D and the priority of drivers was B≻C≻A≻D≻E. The experts identified the most important criteria as the group of technical parameters of the tire (weight average – 0.355), and respondents chose the criteria of the tire manufacturer as the most important (weight average – 0.29).

A bar diagram, showing the significance of the 6 sub-criteria in the group A to the experts and respondents is given in Fig. 3b. As shown in this figure, the sub-criteria A4 describing long-term positive experience when using the services of the same seller more than once are the most important for both groups, while the A6 sub-criteria (Smooth fulfillment of the seller's financial and tax obligations) are assessed as insignificant with weight average 0.071–0.051.

As shown by the diagram, the highest ranks of the 7 sub-criteria of group B (Tire Manufacturer's Brand) are assigned by experts to the sub-criteria B3 (High marks of the brand tire, confirmed by (ADAC) certified tests), B4, and B1 (Fig. 3c) and by respondents to the sub-criteria B3, B5 and B1, while the lowest ranks by both groups are given to the sub-criteria B6 and B7.

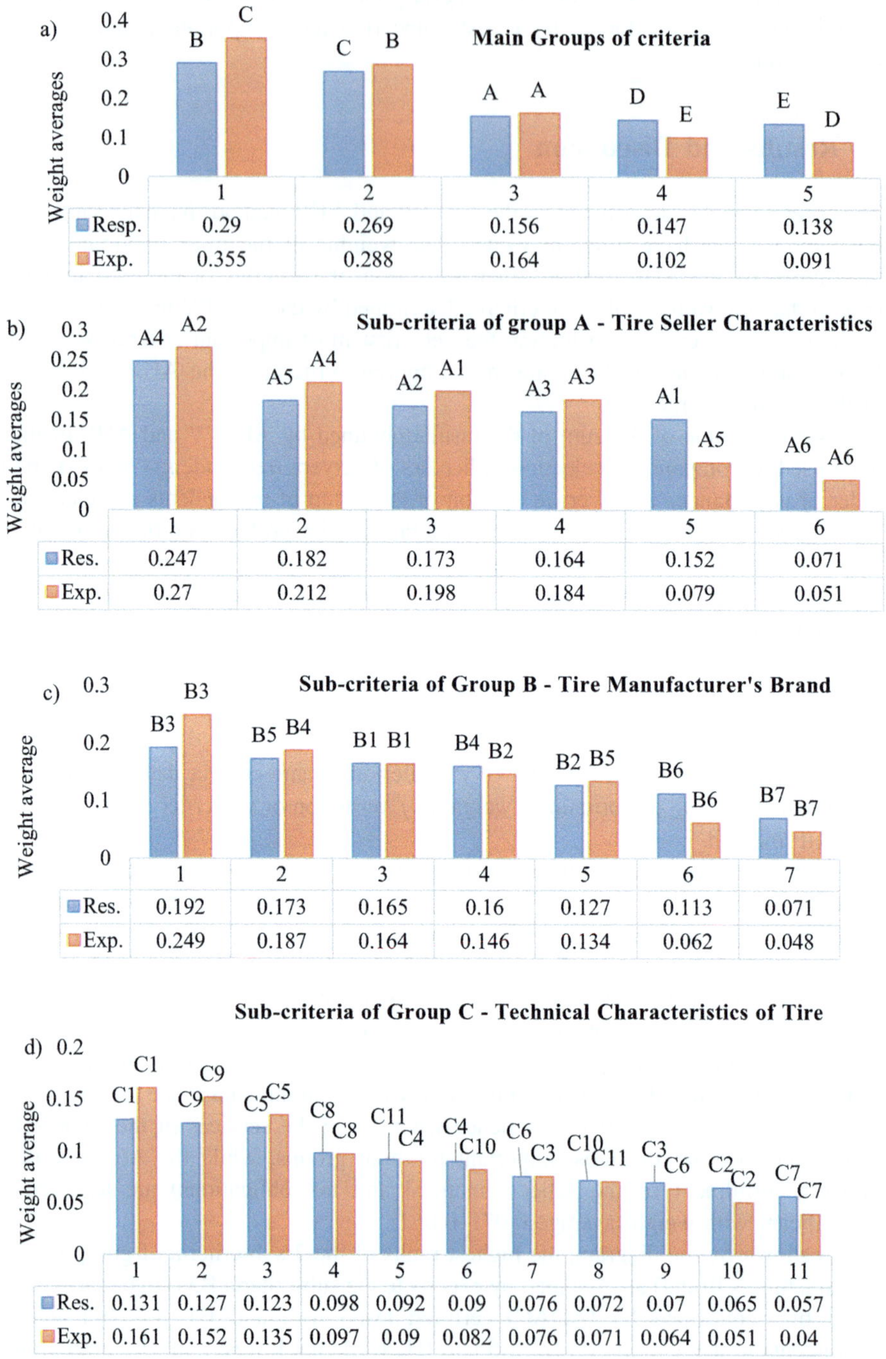

Fig. 3. Comparison of results.

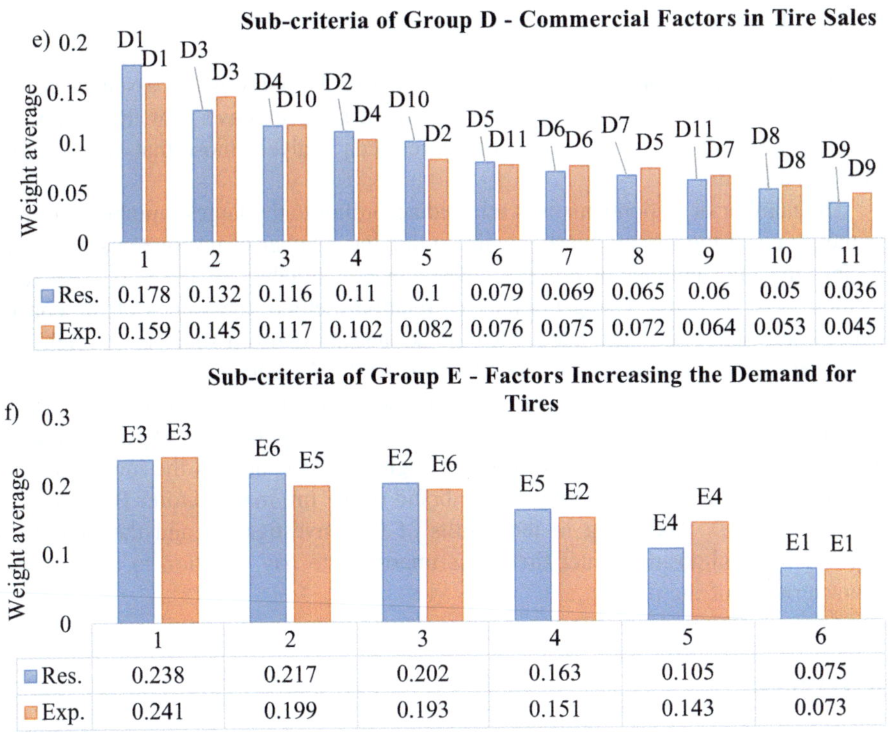

Fig. 3. (*continued*)

A bar diagram evaluating the significance of the 11 sub-criteria in group C is given in Fig. 3d. It can be observed that the criteria C1, C9, C5, and C8 are significant equally for experts and respondents because the criteria for the technical characteristics of the tires were the most important of all sub-criteria in all groups.

A results showing the significance of the 11 sub-criteria of the group D does not indicate a significant difference of opinion (Fig. 3e). The priority of the experts was D1 (weight average 0.159) ≻D3 (0.145) ≻D10≻D4≻D2≻D11≻D6≻D5≻D7≻D8≻D9 and the priority of respondents was D1 (0.178) ≻D3 (0.132) ≻D4≻D2≻D10≻D5 ≻D6≻D7≻D11≻D8≻D9.

Finally, as shown by the diagrams, the sub-criteria group E - Factors Increasing the Demand for Tires as group D - Commercial Factors in Tire Sales were least important to both experts and respondents. Here the highest ranks of participants were assigned to the sub-criteria E3 with a weighted average of 0.238 (experts) and 0.241 (respondents).

4 Conclusions

An analysis of the work of other authors related to the analysis of passenger car tire properties and tire interaction with the pavement surface shows that the criteria for selecting tires are various, but most of the research analyzes those that affect driving safety and comfort.

The number of respondents was selected according to the largest number of criteria in the group of the survey questionnaire so that the number of respondents would not be less than the number of criteria. Questionnaires suitable for the rank correlation method were prepared. Using the rank correlation method, the values of the concordance coefficient were calculated. Mean ranks and normalized subjective weights for all criteria were calculated using the AHP and ARTIW methods. The arithmetic mean of the sum of the results of the multi-criteria methods was calculated for comparison.

After examining the significance of the groups of factors, the experts chose the technical characteristics of C (average of weight – 0.355) tires as the most important criteria, and the respondents selected the brand of the tire manufacturer B (average of weight – 0.290). According to the results of this investigation and the opinion of experts, when choosing winter tires, it is important to pay attention to the technical characteristics of the tires.

References

1. OICA. World vehicles in use - all vehicles (2016). http://www.oica.net/wp-content/uploads//Total_in-use-All-Vehicles.pdf, Accessed 21 May 2021
2. Nowakowski, P., Król, A.: The influence of preliminary processing of end-of-life tires on transportation cost and vehicle exhausts emissions. Environ. Sci. Pollut. Res. **28**(19), 24256–24269 (2020). https://doi.org/10.1007/s11356-019-07421-y
3. Ateekh-Ur-Rehman, A.M.: Automobile tire assessment: a multi-criteria approach. Eng. Technol. Appl. Sci. Res. **7**(1), 1363–1368 (2017)
4. Ciuperca, R., Persu, C., Nedelcu, A., Popa, L., Zaica, A., Stefan, V.: Researches On the monitoring of air pressure in the tires at road transport means. Inmateh-Agric. Eng. **56**(3), 75–82 (2018)
5. Hjort, M., Eriksson, O., Bruzelius, F.: Comprehensive study of the performance of winter tires on ice, snow, and asphalt roads: the influence of tire type and wear. Tire Sci. Technol. **45**(3), 175–199 (2017). https://doi.org/10.2346/tire.17.450304
6. Chen, S.S., Lamanna, M.F., Tabler, R.D., Kaminski, D.F.: Computer-aided design of passive snow control measures. Transp. Res. Rec. **2107**, 111–120 (2009). https://doi.org/10.3141/2107-12
7. Malmivuo, M., Luoma, J., Porthin, M.: Studded and unstudded winter tires in fatal road accidents in Finland. Traffic Inj. Prev. **18**(5), 550–555 (2017). https://doi.org/10.1080/15389588.2016.1243239
8. Ružinskas, A., Sivilevičius, H.: Magic formula tyre model application for a tyre-ice interaction. Procedia Eng. **187**, 335–341 (2017). https://doi.org/10.1016/j.proeng.2017.04.383

9. Fichtinger, A., Edelmann, J., Plochl, M., Holl, M.: Aquaplaning detection using effect-based methods: an approach based on a minimal set of sensors, electronic stability control, and drive torques. IEEE Veh. Technol. Mag. **16**(3), 20–28 (2021). https://doi.org/10.1109/MVT.2021.3085536
10. Fwa, T.F., Kumar, S.S., Ong, G.P., Huang, C.J.H.: Analytical modeling of effects of rib tires on hydroplaning. Transp. Res. Rec. **2068**, 109–118 (2008). https://doi.org/10.3141/2068-12
11. Fwa, T.F., Kumar, S.S., Anupam, K., Ong, G.P.: Effectiveness of tire-tread patterns in reducing the risk of hydroplaning. Transp. Res. Rec. **2094**, 91–102 (2009). https://doi.org/10.3141/2094-10
12. Douglas, R.A., Woodward, W.D.H., Rogers, R.J.: Contact pressures and energies beneath soft tires: modeling effects of central tire inflation–equipped heavy-truck traffic on road surfaces. Transp. Res. Rec. **1819B**, 221–227 (2003). https://doi.org/10.3141/1819b-28
13. Luo, R., Prozzi, J.A.: Effect of measured three-dimensional tire–pavement contact stress on pavement response at asphalt surface. Transp. Res. Rec. **2037**, 115–127 (2007). https://doi.org/10.3141/2037-11
14. Park, D.-W., Martin, A.E., Jeong, J.-H., Lee, S.-T.: Effects of tire inflation pressure and load on predicted pavement strains. Baltic J. Road Bridge Eng. **3**(4), 181–186 (2008)
15. Soon, S.-C., Drescher, A., Stolarski, H.K.: Tire-induced surface stresses in flexible pavements. Transp. Res. Rec. **1896**, 170–176 (2004). https://doi.org/10.3141/1896-17
16. Wallaschek, J., Wies, B.: Tyre tread-block friction: modelling, simulation and experimental validation. Veh. Syst. Dyn. Int. J. Veh. Mech. Mob. **51**(7), 1017–1026 (2013)
17. Li, S., Zhu, K., Noureldin, S., Harris, D.: Identifying friction variations with the standard smooth tire for network pavement inventory friction testing. Transp. Res. Rec. **1905**, 157–165 (2005). https://doi.org/10.3141/1905-18
18. Xia, K.: A finite element model for tire/pavement interaction: application to predicting pavement damage. Int. J. Pavement Res. Technol. **3**(3), 135–141 (2010)
19. Chen, J.-S., Huang, C.-C., Chen, C.-H., Su, K.-Y.: Effect of rubber deposits on runway pavement friction characteristics. Transp. Res. Rec. **2068**, 119–125 (2008). https://doi.org/10.3141/2068-13
20. Simic, V., Dabic-Ostojic, S.: Interval-parameter chance-constrained programming model for uncertainty-based decision making in tire retreading industry. J. Clean. Prod. **167**, 1490–1498 (2017). https://doi.org/10.1016/j.jclepro.2016.10.122
21. Vieira, T., Sandberg, U., Erlingsson, S.: Acoustical performance of winter tyres on in-service road surfaces. Appl. Acoust. **153**, 30–47 (2019). https://doi.org/10.1016/j.apacoust.2019.03.025
22. Krezel, Z.A., McManus, K.: Environmentally friendly sound absorbing noise barrier made from concrete waste– further developments. Int. J. Pavement Res. Technol. **3**(4), 223–227 (2010)
23. Crocker, M.J., Hanson, D., Li, Z., Karjatkar, R., Vissamraju, K.S.: Measurement of acoustical and mechanical properties of porous road surfaces and tire and road noise. Transp. Res. Rec. **1891**, 16–22 (2004). https://doi.org/10.3141/1891-03
24. Shukla, A.K., Jain, S.S., Parida, M., Srivastava, J.B.: Performance of FHWA model for predicting traffic noise: a case study of metropolitan city, Lucknow (India). Transport **24**(3), 234–240 (2009)
25. Ratrout, N.T.: Evaluation of passenger car tire failure in Saudi Arabia. Arab. J. Sci. Eng. **36**(5), 749–760 (2011). https://doi.org/10.1007/s13369-011-0069-1
26. Saaty, T.L.: Fundamentals of Decision Making and Priority Theory with the Analytic Hierarchy Process, 1st ed, p. 477. RWS Publications (2000)

27. Saaty, T.L., Vargas, L.G., Dellmann, K.: The allocation of intangible resources: the analytic hierarchy process and linear programming. Socioecon. Plann. Sci. **37**(3), 169–184 (2003). https://doi.org/10.1016/S0038-0121(02)00039-3
28. Farhan, J., Fwa, T.F.: Pavement maintenance prioritization using analytic hierarchy process. Transp. Res. Rec. **2093**, 12–24 (2009). https://doi.org/10.3141/2093-02
29. Podvezko, V.: Application of AHP technique. J. Bus. Econ. Manag. **10**(2), 181–189 (2009). https://doi.org/10.3846/1611-1699.2009.10.181-189
30. Sivilevičius, H.: Modelling the interaction of transport system elements. Transport **26**(1), 20–34 (2011). https://doi.org/10.3846/16484142.2011.560366
31. Sivilevičius, H., Maskeliūnaitė, L.: The criteria for identifying the quality of passengers' transportation by railway and their ranking using AHP method. Transport **25**(4), 368–381 (2010). https://doi.org/10.3846/transport.2010.46
32. Montgomery, D.C.: Introduction to Statistical Quality Control, 6 ed., p. 734. Wiley (2008)
33. Sivilevičius, H.: Application of expert evaluation method to determine the importance of operating asphalt mixing plant quality criteria and rank correlation. Baltic J. Road Bridge Eng. **6**(1), 48–58 (2011). https://doi.org/10.3846/bjrbe.2011.07
34. Sivilevičius, H., Maskeliūnaitė, L.: Multiple criteria evaluation and the inverse hierarchy model for justifying the choice of rail transport mode. Promet Traffic Transp. **30**(1), 57–69 (2018). https://doi.org/10.7307/ptt.v30i1.2417

Author Index

O. Prentkovskis et al. (Eds.): TRANSBALTICA 2021, LNITI, pp. 839–841, 2022.
https://doi.org/10.1007/978-3-030-94774-3

9783030947736